ENCYCLOPEDIA OF APPLIED PHYSICS

SPONSORS

AMERICAN INSTITUTE OF PHYSICS
DEUTSCHE PHYSIKALISCHE GESELLSCHAFT
JAPAN SOCIETY OF APPLIED PHYSICS
PHYSICAL SOCIETY OF JAPAN

AMERICAN INSTITUTE OF PHYSICS

DEUTSCHE PHYSIKALISCHE GESELLSCHAFT

JAPAN SOCIETY OF APPLIED PHYSICS

PHYSICAL SOCIETY OF JAPAN

ENCYCLOPEDIA OF APPLIED PHYSICS

VOLUME 4
Combustion to Diamagnetism

Edited by
GEORGE L. TRIGG

Associate Editors
EDUARDO S. VERA

WALTER GREULICH

Managing Editor
EDMUND H. IMMERGUT

Assistant Managing Editor
GORDON A. DANA

VCH

George L. Trigg
275 Beaver Dam Road
Brookhaven, New York 11719

Edmund H. Immergut
2 Sidney Place
Brooklyn, New York 11201

Walter Greulich
Zügelweg 30
D-6943 Birkenau
Federal Republic of Germany

Gordon A. Dana
2 Sidney Place
Brooklyn, New York 11201

Eduardo S. Vera
NTT Interdisciplinary Research Laboratories
Nippon Telegraph and Telephone Corporation
Musashino-shi, Tokyo 180, Japan

Library of Congress Cataloging-in-Publication Data

Encyclopedia of applied physics.
(Revised for Vol. 4)

"Sponsors, American Institute of Physics...[et al]."
Includes bibliographical references and index.
1. Physics—Encyclopedias. 2. Engineering—Encyclopedias. I. Trigg, George L. II. Vera, Eduardo S. III. Greulich, Walter. IV. American Institute of Physics.
QC5.E543 1991 530′.05 91-8738
ISBN 1-56081-063-7 (VCH Publishers)

British Library Cataloguing in Publication Data

Encyclopedia of applied physics.
Vol. 4
I. Trigg, George L. (George Lockwood) II. Vera, Eduardo S. III. Greulich, Walter
621
ISBN 1-56081-058-0 (set)
ISBN 1-56081-063-7 (vol. 4)

Printed in the United States of America.

ISBN 1-56081-063-7 (volume 4) VCH Publishers
ISBN 1-56081-058-0 (set) VCH Publishers
ISBN 3-527-28126-6 (volume 4) VCH Verlagsgesellschaft
ISBN 3-527-26841-3 (set) VCH Verlagsgesellschaft

Printing History:
10 9 8 7 6 5 4 3 2 1

Published jointly by:

VCH Publishers, Inc.
220 East 23rd Street
Suite 909
New York, NY 10010

VCH Verlagsgesellschaft mbH
P.O. Box 10 11 61
D-6940 Weinheim
Federal Republic of Germany

VCH Publishers (UK) Ltd.
8 Wellington Court
Cambridge CB1 1HZ
United Kingdom

ADVISORY BOARD

EDITORIAL CONSULTANTS

MAIN ENTRIES

The subject matter in the *Encyclopedia of Applied Physics* is presented in approximately 500 individual articles, arranged alphabetically. The topics can be classified into 20 sections, similar to the AIP Physics and Astronomy Classification Scheme (PACS):

01	General Aspects: Mathematical, Computational, and Information Techniques
02	Measurement Science, General Devices and/or Methods
03	Nuclear and Elementary Particle Physics
04	Atomic and Molecular Physics
05	Electricity and Magnetism
06	Optics (classical and quantum)
07	Acoustics
08	Thermodynamics and Properties of Gases
09	Fluids and Plasma Physics
10	Condensed Matter A: Structure and Mechanical Properties
11	Condensed Matter B: Thermal, Acoustic, and Quantum Properties
12	Condensed Matter C: Electronic Properties
13	Condensed Matter D: Magnetic Properties
14	Condensed Matter E: Dielectrical and Optical Properties
15	Condensed Matter F: Surfaces and Interfaces
16	Materials Science
17	Physical Chemistry
18	Energy Research and Environmental Physics
19	Biophysics and Medical Physics
20	Geophysics, Meteorology, Space Physics, and Aeronautics

Each article has been assigned a code number consisting of two digits which denotes the section, and a letter which gives the type of article. There are six types: A = Devices, Equipment; B = Materials; C = Methods, Processes; D = Phenomena, Effects; E = Scientific or Technological Fields; F = Institutions, Companies, Societies and other organizations.

CONTRIBUTORS

Robert H. Barth, Chief, Hemodialysis, VA Medical Center, 800 Poly Place, Brooklyn, New York 11209, U.S.A.
Dialysis

Charles P. Beetz, Jr., Advanced Technology Materials, Inc., 7 Commerce Drive, Danbury, Connecticut 06810, U.S.A.
Composite Materials

H. Behrens, Fachinformationszentrum Karlsruhe, D-7514 Eggenstein-Leopoldshafen 2, Federal Republic of Germany
Computer Databases

Herbert J. Bernstein, Bernstein + Sons, P.O. Box 177, Bellport, New York, 11713, U.S.A.
Computer Programming Languages

Hans M. Bozler, Department of Physics, University of Southern California, Los Angeles, California 90089-0484, U.S.A.
Cryogenics

John W. Carr, III, Department of Computer and Information Science, 200 So. 33rd St., University of Pennsylvania, Philadelphia, Pennsylvania 19104-6389, U.S.A.
Computers

Chung Chieh, Department of Chemistry, University of Waterloo, Waterloo, Ontario N2L 3GI, Canada
Crystallography

Nathan L. Church, International Copper Association, 708 Third Avenue, New York, New York 10017, U.S.A.
Copper

Bernard R. Cooper, Department of Physics, West Virginia University, Morgantown, West Virginia 26506, U.S.A.
Copper

Mary W. Covington (deceased)
Copper

Cyril Domb, Department of Physics, Bar Ilan University, Ramat-Gan, 52001 Israel
Critical Phenomena

Antun Domic, Digital Equipment Corporation, 77 Reed Road, Hudson, Massachusetts 01749, U.S.A.
Computer-Aided Design in Electronics

William H. Dresher, International Copper Association, 708 Third Avenue New York, New York, 10017, U.S.A.
Copper

Christian W. Fabjan, CERN, CH-1211 Geneva 23, Switzerland
Detectors, Particle Calorimetric

L. M. Falicov, Department of Physics, University of California, Berkeley, California 94720, U.S.A.
Diamagnetism

Walter Gekelman, Department of Physics, University of California, Los Angeles, Califoria 90024-1547, U.S.A.
Data Aquisition

Alan I. Goldman, Ames Laboratory, U.S. Department of Energy and Department of Physics, Iowa State Univerisity, Ames, Iowa 50011, U.S.A.
Crystalline State

Christopher M. Gould, Department of Physics, University of Southern California, Los Angeles, Califoria 90089-0484, U.S.A.
Cryogenics

Kazuaki Iwata, Department of Mechanical Engineering for Computer-Controlled Machinery, Osaka University, 2-1 Yamada-oka Suita-shi, Osaka 565, Japan
Computer-Aided Manufacturing

George H. Mackenzie, TRIUMF, 4004 Wesbrook Mall, Vancouver, British Columbia V6T 2A3, Canada
Cyclotrons

Bernard S. Meyerson, IBM, T. J. Watson Research Center, Yorktown Heights, New York 10598, U.S.A.
CVD (Chemical Vapor Deposition)

Bruce F. Naylor, AT&T Bells Labs, 600 Mountain Ave., Murray Hill, New Jersey 07974, U.S.A.
Computer Graphics

Arun N. Netravali, AT&T Bells Labs, 600 Mountain Ave., Murray Hill, New Jersey 07974, U.S.A.
Computer Graphics

Eric B. Norman, Nuclear Science Division, Building 88, Lawrence Berkeley Laboratory, Berkeley California 94720, U.S.A.
Detectors, Scintillation

William B. Person, Department of Physics, University of Waterloo, Waterloo, Ontario N2L 3GI, Canada
Crystallography

Paul W. Schmor, TRIUMF, 4004 Wesbrook Mall, Vancouver, British Columbia V6T 2A3, Canada
Cyclotrons

Harvey R. Schneider, TRIUMF, 4004 Wesbrook Mall, Vancouver, British Columbia V6T 2A3, Canada
Cyclotrons

Erich Schönherr, Max–Planck-Institute für Festkorperförschung, Heisenbergstrass 1, Postfach 80 06 65, D-7000 Stuttgart 80, Federal Republic of Germany
Crystal Growth

Joseph E. Shepherd, Department of Mechanical Engineering, Rensselaer Polytechnic Institute, Troy, New York 12180-3590, U.S.A.
Compressible Flows

Kenneth H. Slater, Digital Equipment Corporation, 77 Reed Rd., Hudson Massachusetts 01749 U.S.A.
Computer Hardware

Barry N. Taylor, National Institute of Standards and Technology, Gaithersburg, Maryland 20899, U.S.A.
Constants, Fundamental

J.W. Tepel, Fachinformationszenetrum Karlsruhe, D-7514 Eggenstein-Leopoldshafen 2, Federal Republic of Germany.
Computer databases

F. John Walter, IntraSpec, Inc., 4579, Oak Ridge, Tennessee 37831, U.S.A.
Detectors, Semiconductor

Charles K. Westbrook, Physics Department, Lawrence Livermore National Laboratory, Livermore, California 94550, U.S.A.
Combustion

S. Scott Zimmerman, Department of Chemistry and Graduate Section of Biochemistry, Brigham Young University, Provo, Utah 84602, U.S.A.
Conformational Analysis

COLLISION PROCESSES OF ATOMS AND MOLECULES

See MOLECULAR AND ATOMIC COLLISION PROCESSES

COMBUSTION

CHARLES K. WESTBROOK, *Physics Department, Lawrence Livermore National Laboratory, Livermore, California, U.S.A.*

INTRODUCTION

Over millennia, combustion has provided society with much of its practical energy production. From the fire that cooked food for cavemen, to the forges that enabled prehistoric man to emerge from the Stone Age, to the gunpowder that transformed relations between societies in the Middle Ages, and to the boilers and furnaces that drove the Industrial Revolution, combustion has been intimately interwoven with the technological factors that have moved civilization. Even today, although alternative sources of energy production are being intensively pursued, a large fraction of practical energy production is the result of combustion of hydrocarbon fuels.

Combustion has also been responsible for many of the environmental problems associated with modern civilization. The killer fogs and smogs of the past century and the problems of modern air pollution and acid rain are products of the use of combustion for practical energy production. Even more recently, the production of carbon dioxide, an apparently benign product of complete combustion, is now being associated with global warming and the so-called greenhouse effect. (See AIR POLLUTION: COMPONENTS, CAUSES, AND CURES.)

Combustion interacts with society in many ways, influencing economics, foreign policy, and other political and sociological problems. The import and export of hydrocarbon combustion fuels has an enormous impact on the balance of payments for many countries, and access to fuels has played a large role in many foreign policy decisions over the past hundred years.

1. COMBUSTION FUNDAMENTALS

Combustion consists of the reaction of one or more chemical species to produce other species, together with the release of heat. As stated by Strehlow (1984), combustion occurs when "...weak molecular bonds are broken and replaced by stronger bonds and the excess

1-56081-063-7/92/$5.00+.50

bond energy is liberated to the system, usually causing a very large increase in temperature." This heat can then be used to do work in a thermodynamic sense, driving a practical system such as an internal combustion engine or a furnace.

1.1 Fuels

In most familiar systems, combustion takes place between a fuel, often a hydrocarbon species, and an oxidizer, usually air. The physical properties of the hydrocarbon fuel can vary considerably. Many fuels exist as gases under normal conditions. For example, natural gas consists primarily of methane (CH_4) with small amounts of ethane (C_2H_6), propane (C_3H_8), and other trace species, all of which are gases at room temperature and pressure. Automobile gasoline and diesel fuel are complex mixtures of larger hydrocarbons, with components as large as toluene (C_7H_8), octane (C_8H_{18}), and dodecane ($C_{12}H_{26}$), all of which are liquids at room temperature. Familiar fuels such as firewood, paper, gunpowder, and many others are solids under normal conditions. Fuel storage and distribution, as well as the configuration of the combustion system that consumes them, will then depend on the nature of the fuel.

The fuel need not consist of a hydrocarbon species; one simple fuel is a mixture of molecular hydrogen and carbon monoxide. Other fuels can include N-H species such as ammonia (NH_3) or hydrazine (N_2H_4), sulfur-bearing species such as hydrogen sulfide (H_2S) and carbon disulfide (CS_2), and mixed species such as hexyl nitrate ($C_6H_{13}NO_2$) and hydrogen cyanide (HCN).

In the familiar case of toxic waste incineration, the "fuel" can consist of any of a wide variety of chlorinated hydrocarbon species. Some of the most common of these are methyl chloride (CH_3Cl), vinyl chloride (C_2H_3Cl), and trichloroethylene, TCE (C_2HCl_3). Combustion of such species leads to carbon dioxide, water, and hydrogen chloride (or hydrochloric acid), HCl.

1.2 Oxidizers

Similarly, in the most common combustion situations, the fuel reacts with oxygen, either that present in air (which includes 21% O_2) or in the form of pure oxygen. From a historical point of view, combustion has been nearly synonymous with oxidation of these fuels in air, and this is still the basis of most practical combustion. However, when higher combustion temperatures are desirable, it is possible to use pure oxygen as an oxidizer, and other species such as N_2O_4 and ozone (O_3) are sometimes used as oxidizers for specialized applications. The key to all of these systems is that the reaction of the fuel with the oxidizer releases a significant amount of thermal energy.

1.3 Concept of Stoichiometry

The final product temperature and composition of a combustible mixture, and often the overall rate of combustion, depend on the ratio of fuel to oxidizer. A *stoichiometric* mixture is defined as one in which there is exactly the correct amount of oxidizer present to convert all of the fuel to final products. For example, if propane (C_3H_8) is the fuel, five oxygen (O_2) molecules are required to convert each propane molecule to CO_2 and H_2O:

$$C_3H_8+5O_2=3CO_2+4H_2O. \quad (1)$$

Therefore, a stoichiometric mixture of propane and oxygen would consist of $\frac{1}{6}$ propane and $\frac{5}{6}$ oxygen. If the oxidizer in the same example is air instead of oxygen, then the overall reaction is given by

$$C_3H_8+5O_2+18.8N_2=3CO_2+4H_2O+18.8N_2, \quad (2)$$

since the ratio of N_2 to O_2 in air is approximately 3.76, and a stoichiometric mixture of propane and air would include approximately 4% propane [i.e., one molecule of propane in the 24.8 molecules on the reactant side of reaction (2) above]. The same number of chemical bonds are broken and made in reactions (1) and (2), and so the total heat release is equal in the two reactions. Since that heat must be shared by more product molecules in reaction (2), the product temperature in reaction (1) will be considerably higher than that of reaction (2).

A given fuel-oxidizer mixture can include more or less fuel than a stoichiometric amount. The ratio of the fuel concentration actually present to that which would be the stoichiometric amount is termed the *equivalence ratio*, usually indicated by the symbol Φ, and a mixture with more than a stoichiometric amount of fuel (fuel-rich) has $\Phi > 1$, whereas a mixture with less than a stoichiometric amount of fuel (fuel-lean) has an equivalence ratio $\Phi < 1$.

The importance of the equivalence ratio can be illustrated by showing the variation in product temperature for a methane-air mixture in Fig. 1. The product temperature is defined as the final temperature achieved by a mixture of fuel and oxidizer at a particular equivalence ratio when all of the heat released by the reaction is used to heat the reaction products. This condition is also known as the *adiabatic* condition, and the product temperature is known as the *adiabatic flame temperature, T_{ad}*. The maximum value of T_{ad} is achieved for values of the equivalence ratio close to stoichiometric, falling quite rapidly as Φ moves towards lean or rich. Values of T_{ad} for many fuel-oxidizer mixtures have been collected and tabulated by Lewis and von Elbe (1987) and by Gaydon and Wolfhard (1979). Generally speaking, fuel-oxygen mixtures have much higher values for T_{ad} than fuel-air mixtures, but for both fuel-air and fuel-oxidizer mixtures maximum product temperatures are achieved at nearly stoichiometric values of the equivalence ratio. One result of this trend is that the greatest rate of combustion and the highest power production are achieved for stoichiometric conditions. Unfortunately, for combustion in air (which contains nearly 80% molecular nitrogen), production of nitric oxide (NO), a major chemical pollutant caused by combustion, grows very rapidly with increasing temperature, so that the rate of NO production is also greatest for stoichiometric conditions. This results is one of the great conflicts in much of practical (i.e., air-breathing) combustion: the greatest efficiency and power production occurs at $\Phi = 1$, but the greatest rate of pollutant production also occurs at $\Phi = 1$. One must often sacrifice some degree of performance in order to reduce the unwanted pollutant emissions.

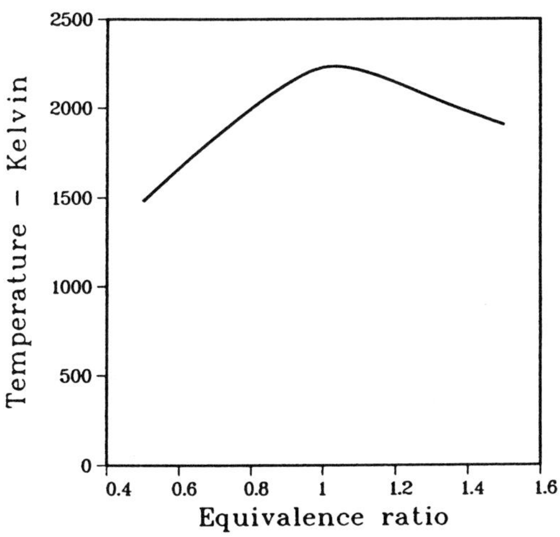

FIG. 1. Variation in product temperature for methane-air mixtures with equivalence ratio.

In addition to the final temperature, the product chemical composition depends strongly on Φ. In particular, on the fuel-rich side of stoichiometric, insufficient O_2 results in incomplete oxidation of the fuel, and products such as carbon monoxide and soot will result.

2. MIXING OF FUEL AND OXIDIZER

In order for combustion to occur, the fuel and oxidizer must interact on a molecular level. There are two general situations that can be used to characterize most combustion environments, based on the mechanism by which the fuel and oxidizer are combined. In the first limit, the fuel and oxidizer are already mixed prior to combustion, whereas in the opposite limit the fuel and oxidizer are initially separated and must be mixed during the combustion process.

2.1 Premixed Systems

In so-called premixed systems, molecular-scale mixing is accomplished prior to combustion. For example, in the familiar case of a laboratory Bunsen burner, the fuel and air are first mixed and stored in a single container, from which the combustible mixture flows into the burner itself. Another perhaps less obvious premixed system is found in the conventional internal combustion engine, where the initially liquid fuel is vaporized and mixed with air in the carburetor immediately prior to being introduced into the engine combustion chamber.

2.2 Laminar Burning Velocity

Premixed combustion represents an instructive limiting condition, and many fundamental combustion properties are defined in terms of premixed processes. For example, if a stoichiometric premixed fuel-air mixture is confined in a linear tube, and the mixture is ignited at one end of the tube, a flame will propagate down the tube. If the gas mixture is initially quiescent, and heat and mass transport processes take place on a molecular level (the laminar limit), then the rate of propagation of this flame is a fundamental combustion characteristic of the fuel, termed the laminar burning velocity, S_u. The laminar burning velocity is found experimentally to depend on the equivalence ratio, on the gas pressure, and on the unburned gas temperature. It also depends on the chemical properties of the fuel, so that S_u varies from one fuel to another. In Fig. 2, taken from Westbrook *et al.* (1983), the laminar burning velocity of an ethylene-air mixture is plotted as a function of equivalence ratio at two different pressures. The shapes of these curves are seen often in combustion problems, having maximum values for mixtures close to stoichiometric and falling for both rich and lean conditions.

Different fuels and oxidizers can be compared by noting their respective laminar burning velocities. Because heat release rates and adiabatic flame temperatures are higher for fuel-oxygen mixtures than for fuel-air mixtures, fuel-oxygen mixtures are found to have larger laminar burning velocities. The laminar burning velocities for a variety of stoichiometric fuel-oxidizer mixtures are shown in Table 1.

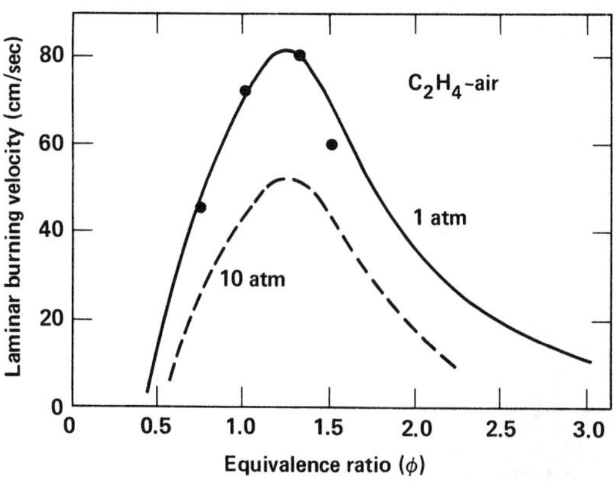

FIG. 2. Variation in laminar burning velocity of ethylene-air mixtures with equivalence ratio at atmospheric pressure and at 10-atm pressure.

Table 1. Laminar burning velocities for different stoichiometric mixtures of fuel and oxidizer.

Fuel-oxidizer mixture	S_u (cm/s)	Fuel-oxidizer mixture	S_u (cm/s)
H_2-air	310	H_2-oxygen	1100
Methane-air	38	Methane-oxygen	450
Propane-air	40	Acetylene-oxygen	1140
Ethylene-air	70	Ammonia-oxygen	115
Methanol-air	44		
Octane-air	40		

2.3 Flammability Limits

In Fig. 2, as the equivalence ratio differs from nearly stoichiometric for either lean or rich conditions, the laminar burning velocity decreases steadily from its maximum value. Extrapolating this curve in either direction suggests that S_u will fall to zero for a critical lean value Φ_L and a critical rich value Φ_R of the equivalence ratio. Experimentally, it is observed that for fuel-oxidizer mixtures outside these values of equivalence ratio, flame propagation is impossible to achieve. These limiting values Φ_L and Φ_R represent the lean and rich flammability limits for premixed flame propagation and are often used as fundamental parameters for combustion of different fuels. Lean and rich flammability limits for a variety of fuel-oxidizer mixtures are given in Table 2. Extensive studies of lean and rich flammability limits have been made by Simon (1951) and by NACA (1959).

2.4 Detonations

In a manner similar to laminar flame propagation, it is found that a detonation wave can propagate down a linear tube filled with a premixed fuel-oxidizer mixture. The detailed properties of detonations are quite complex (e.g., Fickett and Davis, 1979; Nettleton, 1987), but the fundamental feature of a detonation is that combustion is caused by the propagation of a shock wave into the reactive mixture (see EXPLOSIVES).The shock wave compresses and heats the premixed gases to a point where they react rapidly, releasing heat and reinforcing the propagating shock wave. The structure of such a detonation wave is characterized by the existence of cell-like structures that can be recorded as the detonation proceeds down the tube. A typical record of detonation cells is shown in Fig. 3. The sizes of these cells depend

Table 2. Lean and rich flammability limits; values shown are percentages of fuel in the total mixture.

Fuel-oxidizer mixture	Lean limit	Rich limit
Hydrogen-air	4.0	75.0
Methane-air	5.0	15.0
Propane-air	2.1	9.5
Ethylene-air	2.7	36.0
Hydrogen-O_2	4.0	95.0
Methane-O_2	5.1	61.0
Ammonia-air	15.0	28.0
Methanol-air	6.7	36.0
Acetylene-air	2.5	100.0

on the fuel-oxidizer equivalence ratio and on the unreacted mixture temperature and pressure. As in the case of laminar flame propagation, there are lean and rich limits to detonability that represent fundamental combustion properties of a given fuel.

2.5 Turbulent versus Laminar Combustion

The fluid mechanical features of combustion systems depend very strongly on the magnitude of gradients of various quantities, including the temperature, pressure, or fluid velocity field. When these quantities vary slowly throughout the physical domain of interest, then the fluid flow is laminar, and momentum, mass, and heat transport are controlled by molecular properties of the medium. However, when there are significant gradients, particularly in fluid velocity, then fluid mechanical instabilities are observed in which the flow system is found to break up spontaneously into chaotic, statistically random motions. Under such conditions, the fluid system is made up of three-dimensional structures often called turbulent eddies that move and interact among themselves, exchanging heat, mass, and momentum while retaining their identity as eddies for relatively long periods of time. These eddies are much larger than characteristic length scales encountered in laminar systems, and net transport of heat, mass, and momentum is found to be much more efficient and rapid than in laminar environments.

The most reliable quantitative measure that divides laminar from turbulent regimes is the *Reynolds number*, denoted as *Re*. The Reynolds number of the fluid is defined (Williams, 1965) in terms of the fluid velocity V, a characteristic length scale L of the system being studied, the gas density ρ, and the viscosity μ, as

$$Re = LV\rho/\mu = LV/\nu, \tag{3}$$

where ν is the kinematic viscosity, $\nu = \mu/\rho$. The Reynolds number can be interpreted as the

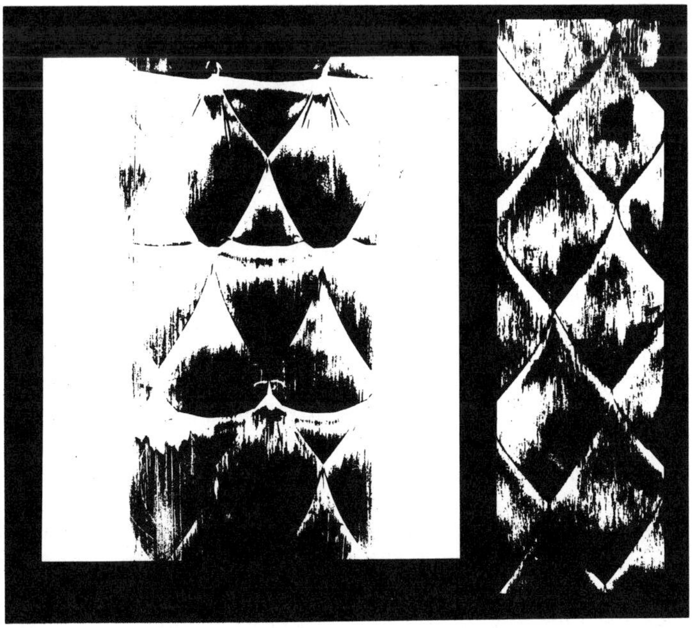

FIG. 3. Detonation cells observed using smoked foils in a detonation tube. Picture provided by Prof. R. A. Strehlow, University of Illinois.

ratio of inertial fluid forces to viscous forces in the fluid, with the inertial forces driving the system unstable and viscous forces trying to restore stability to the flow. When *Re* is small, laminar conditions are observed, and when *Re* is large, turbulent regimes are observed. Between these limits, the flow is found to be intermittently unstable. Values of *Re* in typical laminar flows are generally less than 100, values in fully turbulent flows are often 10^3 or greater, with the intermittent regime characterized by values of *Re* between these values.

In turbulent premixed combustion, the turbulent unburned fuel-oxidizer mixture produces a highly wrinkled flame surface. These wrinkles vary randomly in time and magnify the total amount of surface area between burned and unburned gases, thereby increasing dramatically the overall rate of heat release that can be achieved.

2.6 Diffusion Flames (Nonpremixed)

In contrast with premixed combustion systems, often the fuel and oxidizer are initially separate and then mixed together as they burn. This is actually the case in the great majority of practical combustion systems. For example, in a common gas-burning kitchen stove, jets of pure fuel gas emerge from the burner, where the fuel mixes with the air in the room. If an ignition source is present (e.g., a match or pilot flame), an essentially stationary flame is established, and the location of this flame coincides very closely with the position where fuel and oxidizer exist in stoichiometric proportions.

This type of flame is thus characterized by a high-temperature flame zone in which fuel and oxidizer react rapidly. Fuel approaches the flame from one side and oxidizer from the opposite direction. The physical process by which the reactants approach the flame and mix is diffusion, resulting in the term *diffusion flame* for this type of flame. Diffusion flames can be either laminar or turbulent. In *laminar* diffusion flames, molecular diffusion of fuel and oxidizer is the dominant transport process, and the features of these flames are controlled by molecular quantities. In *turbulent* diffusion flames, the mixing is dominated by the fluid mechanics of the combustion environment.

In addition to gas stoves, some other familiar diffusion flames include furnaces, gas turbines, and large-scale industrial burners. A forest fire and the combustion of a fireplace log are diffusion flames, as are a candle flame and a simple wooden match.

In spite of the ubiquity and enormous economic importance of turbulent diffusion flames, and the fact that they have been studied in one form or another for thousands of years, there is still a great deal of theoretical activity being devoted to these systems, and there is considerable controversy as to their controlling physical and chemical factors. The mathematics of turbulence phenomena is very complex, and these systems are very "stiff" (Gear, 1971)—that is, the characteristic time and space scales vary over many orders of magnitude, from a molecular level to the size of a gas furnace, and from a chemical reaction time scale of a fraction of a microsecond to the time required for a turbulent flame to cross an engine chamber. These problems cannot be resolved with current computational techniques, and a thorough theoretical understanding of turbulent combustion is still not available.

2.7 Spray Combustion and Diffusion Flames

Many hydrocarbon fuels are liquids at room temperature and pressure. Since rapid combustion requires the fuel to mix with oxidizer at a molecular level, it is necessary for such fuels to evaporate at the time of combustion, and the most common means of vaporizing the fuel is to disperse it into the combustion environment in the form of a spray. There are a considerable number of practical combustion systems in which the fuel is initially present in the form of a spray of liquid droplets. For example, in a diesel engine (Heywood, 1988), a liquid fuel spray is injected into the combustion chamber where the fuel spray droplets begin to evaporate. The vaporized gaseous fuel then diffuses away from the droplet surface into the surrounding air. At some distance away from the droplet, the ratio of fuel to oxygen achieves a stoichiometric value, and a flame zone will be established there (see DIESEL ENGINES). Subsequently, this diffusion flame is supplied by fuel evaporating from the surface of the droplet and diffusing out to the flame zone. Of course, in situations where the spray has enough time to vaporize completely and mix with the air prior to combustion, premixed conditions will prevail, but in most practical systems, the sequential processes of evaporation, diffusion, and oxidation all occur during the combustion process.

3. COMBUSTION CHEMISTRY

It is often convenient to write a simple expression for the overall reaction in a combustion system. This was done above as reaction (2) for the combustion of propane in air. However, on a molecular level the actual chemistry of combustion is very much more complex. For example, in the case of methane (CH_4), oxidation begins by removing a single hydrogen atom from the methane molecule. The resulting CH_3 species then can be consumed by any of several possible reactions, leading ultimately to CH_2O, which then produces HCO, CO, and finally CO_2. A complete chemical kinetic description of methane oxidation requires a detailed reaction mechanism with approximately 30 distinct chemical species, including both stable and radical species, and approximately 200 elementary chemical reactions between these species (Warnatz, 1984). For fuels with more carbon and hydrogen atoms, reaction mechanisms can be much larger; for heptane C_7H_{16}, a reasonably complete reaction mechanism must consider 350 chemical species and 2000 elementary chemical reactions (Westbrook *et al.*, 1991).

Combustion chemistry is best understood as a classical chain reaction (Semenov, 1959), in which the chain carriers are radical species. In hydrocarbon systems these radical species include, roughly in order of importance, OH, H, O, HO_2, CH_3, CH_3O_2, and other less significant radicals. In other systems radicals of importance can include NH_2, NH, N, HS, CN, NCO, and many others. Chain-propagation reactions maintain the total number of radical species, such as in the reaction of H_2 with OH:

$$H_2+OH=H_2O+H, \qquad (4)$$

which consumes OH and produces H radicals. Termination reactions reduce the size of the so-called radical pool, as in the reaction

$$H+OH=H_2O. \qquad (5)$$

In reaction (5) two radical species are consumed and none produced, resulting in a net loss of radicals to the combustion system. Chain termination can also result from nonkinetic processes such as molecular diffusion of radicals away from a flame zone.

Perhaps most important in understanding the chemistry of combustion are the chain-branching reactions, including

$$H+O_2=O+OH, \qquad (6)$$

$$CH_4+O=CH_3+OH, \qquad (7)$$

and numerous others in which the number of radical species increases. Reaction (6) is of particular significance for hydrocarbon combustion, and many practical combustion phenomena can be understood by the ways they interact with it. Reactions and processes that produce H atoms lead to accelerated rates of combustion because they promote reaction (6), whereas reactions that consume H atoms tend to retard combustion.

For illustration, it is well known that addition of small amounts of halogenated species to flammable mixtures dramatically reduces the rate of combustion and often makes them nonflammable (Westbrook, 1980, 1982). For example, the active chemical in commercial flame extinguishers is often CF_3Br. Chemical kinetic analysis of the influence of such additives show that they inhibit flame propagation through a sequence of reactions such as

$$HBr+H=H_2+Br, \qquad (8a)$$

$$H+Br_2=HBr+Br, \qquad (8b)$$

$$Br+Br=Br_2, \qquad (8c)$$

where bromine is the inhibitor. Most often, HBr is the form in which Br is present as an inhibitor, but even when molecular bromine (Br_2) is used, the above reactions rapidly establish an equilibrium among the Br-containing species so that these reactions proceed in the directions shown here. The net result of these reactions is the conversion of two H atoms into one H_2 molecule, and each H atom that is consumed in this way is therefore not able to react with molecular oxygen in reaction (6) and provide chain branching to the reaction mechanism.

Another illustration is provided by the observation that the laminar burning velocity S_u decreases with increasing pressure. An important part of the analysis of this effect is that a reaction that competes with reaction (6) for H atoms is

$$H+O_2+M=HO_2+M, \qquad (8d)$$

where M refers to any so-called third-body species required to conserve energy and momentum in the reaction. Since this reaction depends on the cube of the density and converts three reactants into two products, its rate increases more rapidly with density than the rate of reaction (6). Reaction (8d) is a chain-propagation step, producing one HO_2 radical for each H atom radical consumed. In addition, HO_2 radicals are considerably less reactive than the O and OH products of reaction (6). For these reasons, increasing the rate of reaction (8d) relative to the rate of reaction (6), which happens as the pressure and density increase, results in overall inhibition of the overall rate of combustion.

In some environments, a sequence of several reactions is responsible for chain branching, as in the intermediate-temperature pair of reactions

$$CH_4+HO_2=CH_3+H_2O_2, \qquad (9)$$

$$H_2O_2+M=OH+OH+M, \qquad (10)$$

which consumes one HO_2 radical and produces three radicals, including one CH_3 and two OH radicals.

In combustion chain reactions, as in nuclear and other chain reactions, a constant level of chain carriers (radical species) results in a steady rate of reaction, whereas a net loss of radicals due to termination will extinguish the reactive system, and a net increase in radical levels due to chain branching can lead to very rapid reaction or explosion.

The reactions in a detailed mechanism fall into several general classifications, consisting of initiation, abstraction, radical decomposition, and radical recombination (Westbrook and Dryer, 1984).

3.1 Initiation Reactions

A fuel molecule can decompose thermally to produce two radical species. Examples of such reactions are

$$C_3H_8=C_2H_5+CH_3, \qquad (11)$$

$$C_5H_{12}=C_2H_5+C_3H_7, \qquad (12)$$

for propane and pentane. These reactions require strong bonds to be broken and are therefore very slow except under high-temperature conditions. Their primary task in combustion is to provide a small number of radical species, following which other reactions carry out the process of fuel consumption.

3.2 Abstraction Reactions

Most fuel consumption is accomplished in combustion systems through reactions with the radical species listed previously. Radicals remove (abstract) a single H atom from a hydrocarbon fuel species, as in the case of ethane reacting with OH or O:

$$C_2H_6+OH=C_2H_5+H_2O, \qquad (13)$$

$$C_2H_6+O=C_2H_5+OH. \qquad (14)$$

Abstraction reactions can function as either chain-propagating, as in reaction (13), or chain-branching reactions, as in reaction (14).

There are three types of C–H bonds in hydrocarbon fuels, varying according to the structure of the fuel molecule and termed primary, secondary, and tertiary bonds. In primary C–H bonds, the carbon atom is connected to only one other carbon atom; in secondary C–H bonds, the carbon atom is bonded to two other carbon atoms; and in tertiary C–H bonds, the carbon atom is connected to three other carbon atoms. Primary C–H bonds are strongest and tertiary bonds are the weakest. These can be shown as

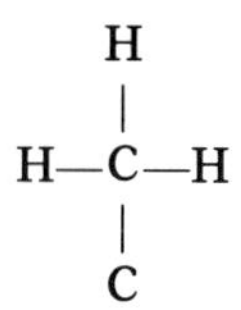

primary
C atom is connected
to only one
other C atom

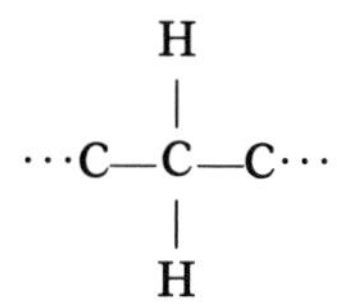

secondary
C atom is connected
to two other
C atoms

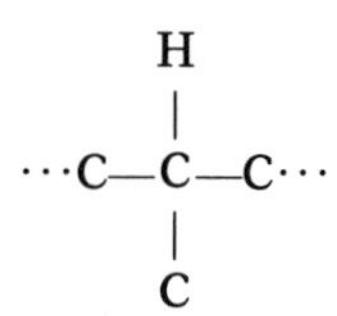

tertiary
C atom is connected
to three other
C atoms

Abstraction of H atoms from primary C–H bonds is slower than abstraction from secondary C–H sites, and both are slower than abstraction from tertiary sites. These distinctions are responsible for variations in ignition and combustion rates among fuels that differ only in their molecular structure. For example, two isomeric forms of butane each consisting of 4 carbon atoms and 10 hydrogen atoms, differing only in the way they are put together, are *n*-butane and iso-butane, shown as

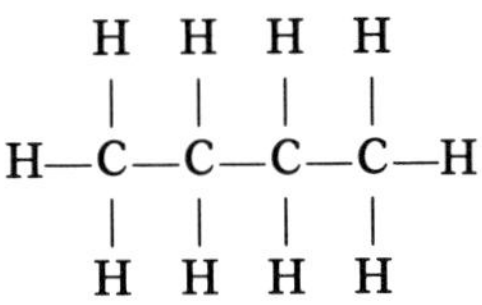

n-butane

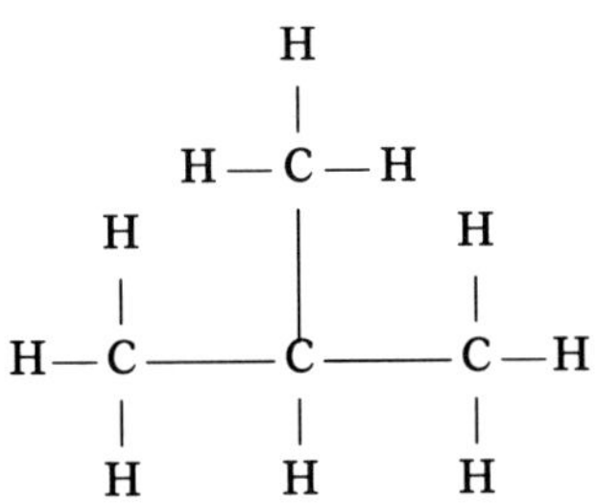

iso-butane

Iso-butane has nine H atoms at primary sites and is therefore more resistant to H-atom abstraction than *n*-butane with only six such primary H-atom sites. As a result, iso-butane ignites more slowly than *n*-butane in combustion environments.

3.3 Radical Decomposition

After an H atom has been abstracted from the fuel, the resulting large radical species will in most cases decompose to produce smaller radical and stable species. The product radical species then decompose further until only small radical and stable species remain. Thus even for large hydrocarbon fuels, this process quickly produces small species with one to three carbon atoms. In this sense, a large fraction of the combustion chemistry of virtually all hydrocarbon fuels eventually is reduced to the reactions of small-fragment hydrocarbon species. Many of these decomposition reactions produce H atoms, including the following in the cases of the ethyl, iso-propyl, and formyl radicals:

$$C_2H_5=C_2H_4+H, \quad (15)$$

$$iC_3H_7=C_3H_6+H, \quad (16)$$

$$HCO=H+CO. \quad (17)$$

The H atoms that are produced in these decomposition reactions can then react with O_2 via reaction (6) to provide chain branching to the overall reaction system.

3.4 Radical Recombination

Rather than decomposing, radical species can recombine to produce larger, more stable hydrocarbon species. Perhaps the most important of these recombination reactions involves the recombination of two methyl radicals to produce ethane:

$$CH_3+CH_3=C_2H_6 \quad (18)$$

This process reduces the number of radical species by two and provides chain termination, slowing the overall rate of reaction.

3.5 Pollutant Chemistry and Emissions

In addition to desirable products such as heat and radiation, combustion processes also generate unwanted and harmful products. These emissions can include products of incomplete combustion such as carbon monoxide, soot, and hydrocarbon species, as well as combustion by-products such as oxides of nitrogen or sulfur. Finally, there is some concern that an otherwise benign product of complete combustion, carbon dioxide, may be leading to global warming through the greenhouse effect and may ultimately be the most serious combustion problem to be considered.

Incomplete combustion results either when insufficient oxidizer was supplied to the fuel or when the fuel-oxidizer mixture was given inadequate time to complete its reaction. Fuel-rich hydrocarbon combustion produces carbon monoxide, even with ample time to achieve chemical equilibrium, and some CO is produced even under stoichiometric and lean conditions. In nonpremixed systems, if the time for mixing of fuel and oxidizer is not considerably shorter than the residence time in the combustion zone, then a variety of partially oxidized species will result. Coal combustion produces flyash and larger particulates, and many combustion systems produce particulates loosely referred to as soot.

Production and destruction of soot particles in combustion are controlled by both chemical kinetics and transport processes (Harris and Weiner, 1983; Frenklach *et al.*, 1986). Gas-phase kinetic processes lead to production of large aromatic hydrocarbon species that are often termed soot precursors. The chemistry involved in production of soot precursors is a current research topic and apparently involves reactions of both neutral and ionized chemical species, leading to species with molecular weights of nearly 10 000. At this size, these molecules are almost large enough to be considered particles, and they form nuclei upon which small hydrocarbon species begin to collect, eventually growing by many orders of magnitude in mass over time scales of milliseconds.

Soot can be both harmful and necessary. In furnaces and boilers, soot is an essential element in transferring heat of combustion to the metal surfaces that in turn heat the working fluid, usually air or water. Much of the energy released by combustion is absorbed by soot particles in the flame and then radiated from the luminous soot particles to the walls of the combustor. Without soot, the hot product gases are much less efficient in providing this transfer of heat. The soot is intended to be oxidized to CO_2 and water later in the combustion process, downstream from the luminous flame zone. However, soot that is not consumed or is produced unintentionally and then emitted along with the other exhaust gases can be a source of visible smoke, odor, and toxic and carcinogenic chemical species.

Chemical species that are present in the fuel in small amounts often lead to pollutant emissions. The best example of this is provided when liquid hydrocarbon fuels and coal and coal-derived fuels contain quantities of sulfur. Combustion temperatures and the presence of oxygen then convert whatever sulfur is present into oxides of sulfur, particularly SO_2, with small fractions of SO_3, denoted collectively as SO_x. If these species are allowed to escape into the air, they combine with water in the atmosphere and lead to acid rains. To remove SO_x from combustor exhausts, the most common strategy is to provide calcium carbonate or lime with the exhaust and produce $CaSO_4$ that can be collected in the solid phase.

The other common constituent of acid rain consists of oxides of nitrogen or NO_x. Some NO_x is produced from nitrogen that is present in the fuel, as in the case of SO_x above. However, the majority of NO_x production from combustion results from the oxidation of N_2 in the air via the catalytic series of reactions called the Zel'dovich mechanism,

$$N_2 + O = NO + N, \tag{19}$$

$$N + O_2 = NO + O. \tag{20}$$

Since both reactions require breaking strong chemical bonds, especially the triple bond between the two N atoms, these reactions proceed rapidly only at elevated temperatures, and their rates are strongly temperature dependent. Thus most NO_x production in a typical combustion system takes place only in the very highest temperature parts of the system.

Combustors can be modified to reduce the production rate of NO_x by removing or limiting the regions of high temperature. Based on the variability of product temperature with equivalence ratio as shown in Fig. 1, the most common strategy is to carry out the combustion in two or more stages, none of which is stoichiometric with its associated high temperature. In industrial burners and furnaces, often the first combustion zone is fuel-rich ($\Phi > 1.3$), resulting in temperatures considerably lower than the peak value at $\Phi = 1$, together with large fractions of CO and other products of incomplete combustion. Then, downstream from this rich combustion zone, additional air is injected into these gases in what is effectively a fuel-lean combustion zone, completing the oxidation of the fuel and the overall heat release process, but again avoiding high temperature that would promote production of NO_x.

Another strategy used to reduce NO_x emissions is to dilute the reactant gases with inert gases, a process called exhaust-gas recirculation (EGR). The principle involved was noted earlier, where it was observed that product temperatures in fuel-O_2 mixtures are considerably higher than in fuel-air mixtures because the additional N_2 in air dilutes the mixture and must be heated by the reaction along with the chemical products. In exactly the same way, other inert species such as H_2O and CO_2 can also be added to combustible reactants and will result in lower combustion temperatures. These lower product temperatures will then result in lower production rates of NO_x. This process is often very simple to implement, since the combustion exhaust consists of mixtures of N_2, H_2O, and CO_2, all of which are suitable as noncombustible diluents. This strategy is used in virtually all modern automobile engines to lower NO_x production.

Most of these techniques for NO_x control focus on temperature reduction to limit the rate of NO production during combustion. However, it is also possible, under tightly controlled conditions, to consume some of the NO that has already been produced. Several chemical kinetic schemes consume NO and produce N_2 by addition of nitrogenous species such as ammonia (NH_3) (Lyon, 1976), urea, and cyanuric acid (Perry and Siebers, 1986) to the product gases of a combustor. The details of these processes are chemically quite complex and have only recently become reasonably well understood (Miller and Bowman, 1989).

Finally, even the products of chemically complete combustion are now suspected of having serious harmful effects as leading to the greenhouse effect and causing long-term warming of the earth's climate. Worldwide rates of use of hydrocarbon fuels and their resulting production of CO_2 have risen steadily since the Industrial Revolution and are still increasing at present. Some computer simulations of the results of ever-increasing CO_2 emissions suggest a direct connection between such emissions and global warming, although many of the details of CO_2 circulation throughout the earth's land, ocean, and atmospheric systems are still extremely uncertain.

4. PRACTICAL COMBUSTION SYSTEMS

Although at first glance very different from each other, essential elements of most practical combustion systems are often quite similar.

4.1 Internal Combustion Engines

The fuel generally consists of a mixture of liquid hydrocarbons (Heywood, 1988). Both premixed and diffusion combustion operation can be utilized. In carbureted engines, the fuel is vaporized and mixed with air in the carburetor and then drawn into the combustion chamber. This mixture is ignited by means of localized deposition of energy, usually a spark plug. However, other ignition systems can and are used, including plasma and laser discharges. The mixture must be within the flammability limits or ignition and flame propagation are not possible. A premixed turbulent flame then propagates across the combustion chamber. The turbulence is the result of fluid-dynamic instabilities caused by the large Reynolds number of the inlet flow into the combustion chamber. This situation is illustrated in Fig. 4, showing the combustion flame front in a typical engine chamber (Cowart *et al.*, 1990), both from above and in a side view. In the top view, the circle at the left indicates the inlet and exhaust valve, while the circle at the right indicates the location of the spark plug. Heat release from the rapid chemical reactions causes a large pressure increase within

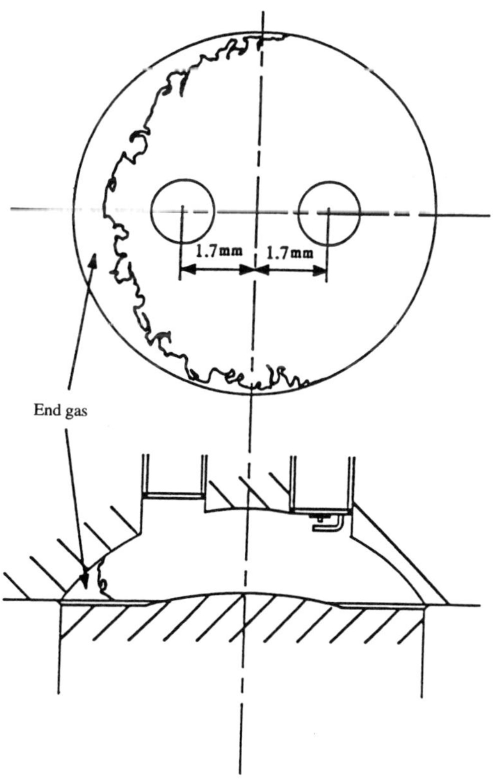

FIG. 4. Side view and top view of an automotive engine chamber, showing the flame propagating away from the spark plug where the spark ignition started the flame. The end gases are also indicated.

the combustion chamber that results in a strong force exerted on the piston, creating torque and driving the motor.

As the flame propagates across the combustion chamber, the Zel'dovich reactions produce NO from the nitrogen in the combustion air. In modern automobile engines, cold exhaust gases, consisting almost entirely of water, carbon dioxide, and nitrogen from previous combustion cycles, are recirculated into the combustion chamber together with the reactant gases, and this dilution, by exhaust gas recirculation (EGR), lowers the combustion temperatures in order to reduce the production rate of NO_x. Carbon monoxide and unburned hydrocarbon species that are present in the exhaust gases pass through a catalytic treatment that completes their oxidation at low temperature, reducing the hydrocarbon emissions without producing further NO_x.

In other cases the fuel is injected directly into the air already present in the combustion chamber. Most often this fuel is normally a liquid and is injected as a spray that then vaporized and mixes with air, creating a mixture in which the local equivalence ratio varies as a function of position in the combustion chamber. If there were sufficient time, eventually all of the liquid fuel would vaporize and the entire combustion chamber would consist of a homogeneous mixture of fuel and air very much like the carbureted engine. However, in the fuel-injected engine, ignition occurs quite early in the cycle, again generally by means of a spark plug. A difficulty in this type of operation is that the spark plug must "see" a flammable mixture when it discharges; otherwise no stable flame will result and an engine misfire will result.

A motivation for this type of engine configuration is its potential for fuel economy and lower NO_x emissions. The quantity of fuel to be injected can be matched accurately to the power needs of the engine, and by confining the fuel to a subvolume of the combustion chamber in which some combustion will take place under rich, and at other places, under lean conditions, the higher temperatures of stoichiometric combustion can be avoided, lowering NO_x production.

Under these conditions all of the flammable mixture burns very quickly, long before all of the remaining fuel spray has evaporated and mixed with air. There is then an extended period during which this remaining fuel burns as a diffusion flame, with fuel and air approaching the relatively stationary flame zone from opposite sides. Otherwise, combustion in this system is very much the same as in the carbureted engine. The heat release drives the piston, NO_x is produced via the Zel'dovich mechanism and controlled with EGR, and unburned hydrocarbons and CO are consumed in a catalyst.

4.1.1 Engine Knock. During the combustion cycle illustrated in Fig. 4, a premixed turbulent flame crosses the combustion chamber. The unreacted gases farthest from the spark plug, called the "end gases," are therefore the last gases to be consumed by the flame. These end gases, shown in Fig. 4, are steadily compressed and heated both by the motion of the piston and by the heat release of the propagating flame. As a result, although they are not in contact with the flame front, these end gases begin to react and, given sufficient time, would spontaneously ignite. Under normal operating conditions, the flame front arrives at the end gases and consumes them, resulting in smooth combustion. However, if these end gases are sufficiently reactive, or if the flame propagation is slow, then the end gases will explode. This explosion, and the pressure wave that results, produce the characteristic ringing phenomenon known as engine knock.

Engine knock can be very destructive if permitted to persist. An example of an engine piston damaged by prolonged knocking is shown in Fig. 5. Therefore, it is desirable to avoid operating conditions that encourage knocking performance. Two of the engine variables that influence knock are the compression ratio and the chemical nature of the automotive fuel. The compression ratio CR is the ratio of the combustion chamber volume at its greatest value [when the piston is at the bottom of its cycle, called *b*ottom *d*ead *c*enter (BDC)] to its smallest value at *t*op *d*ead *c*enter (TDC). The overall efficiency of engine combustion is known to increase steadily with increasing compression ratio, and so there is motivation to operate at as high as possible a value of CR. However, the rate of end-gas ignition also increases rapidly with CR, even-

Fig. 5. Photograph of an automobile engine piston which has been severely damaged by engine knock. (Courtesy of P. Jessup, UNOCAL.)

tually providing a limit to the increase in efficiency that could otherwise be achieved. The use of antiknock additives slows the rate of end-gas ignition, permitting the engine to increase its compression ratio further and to achieve a higher practical efficiency.

The chemical nature of the fuel also has a great influence on its tendency to knock (Lovell, 1948). That tendency is measured by the octane number of the fuel, with high values of octane number signifying resistance to knock and low values indicating significant tendency to knock. Straight-chain hydrocarbon species, with large percentages of secondary C–H bonds, have very low octane ratings. For example, the fuel that has, by definition, an octane rating of zero, is the straight-chain species *n*-heptane, indicated schematically as

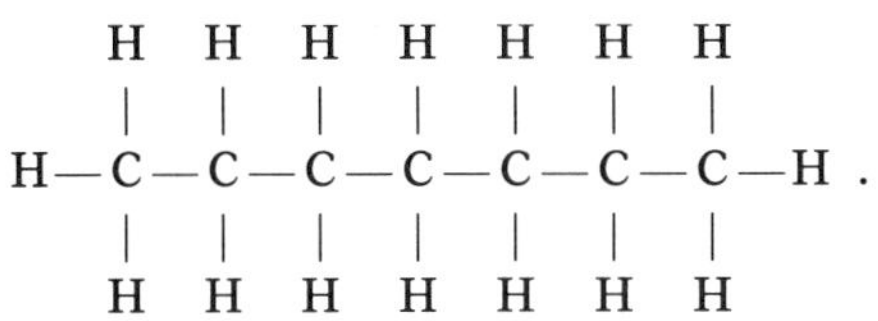

In contrast, highly branched fuels with large numbers of primary C–H bonds are extremely knock resistant and have large octane numbers. The fuel that has, again by definition, an octane number of 100, is iso-octane:

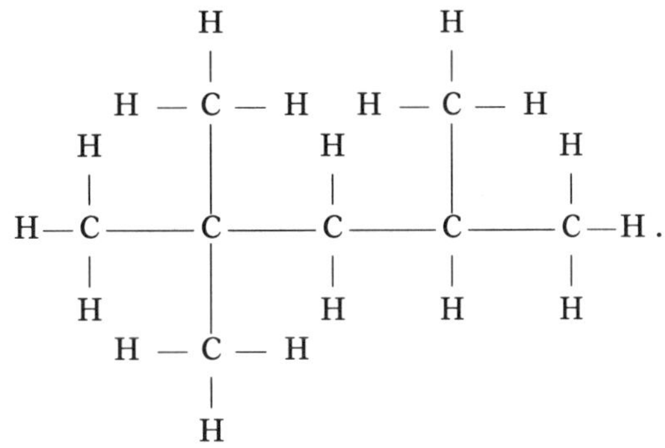

Although the octane numbers and knock tendencies of pure-component fuels have been known for many years and some structural trends related to knocking are recognized, only very recently have these properties been placed on a quantitative and predictive basis (Westbrook *et al.*, 1991). Other factors can influence engine knock, including carbon deposits on engine chamber walls and fluid mechanical details in the flow patterns in the engine, but the above mechanism is the most important factor leading to engine knock in modern spark-ignition engines.

4.1.2 Diesel Engines. From a point of view of combustion principles, the diesel engine is very similar to the spray-injection spark-ignition engine above. One obvious difference is that ignition is accomplished by compression, much like in engine knock, rather than by means of a spark plug. Combustion is believed to take place in a combination of partially premixed, partially diffusion regime. It has been suggested that a diffusion flame exists around each fuel droplet, but it has not been possible to observe the phenomenon in sufficient detail to determine for certain the structure of the combustion zone in a diesel engine.

Diesel engines operate at much higher compression ratios than spark-ignition engines. This is possible because compression ignition or knock is effectively encouraged in the diesel engine. Desirable diesel fuels are denoted with high values of the *cetane* number, indicating good compression ignition tendency. As a result, often fuels with poor or low octane numbers have good or high cetane numbers.

4.2 Burners, Boilers, and Furnaces

A great number of practical combustors consist of stationary systems (as opposed to mobile systems such as automobiles), each of which can produce an enormous amount of thermal energy. These systems heat product gases that then heat a working thermal fluid, usually water, which in turn does the work desired.

All of these systems consist of diffusion flames, although the fuel can be gaseous natural gas, liquid fuel oil, or solid coal particles. Some systems in fact burn mixed fuels such as coal slurries, solid coal particles suspended in a liquid. Most such systems are turbulent, including virtually all of the industrial-scale combustion systems. This is because turbulent systems are able to produce much greater overall rates of heat release per unit volume than any laminar system of comparable combustor volume.

Because the combustion system is large and stationary, it is possible to apply a wider variety of pollution control equipment and technologies to these larger systems than to smaller units such as automobile engines. This includes scrubbing units, precipitators, and other devices to remove SO_x, NO_x, and burned hydrocarbons from exhaust gases.

4.3 Novel Combustion Systems

In addition to the familiar combustors already described, there are many other types of systems that employ the same combustion principles but have different boundary and other operating conditions. In supercritical water oxidation (SCWO), the fuel and oxidizer are dissolved in water at pressures well above the critical point of the water. Under such conditions of approximately 250 atm, water is strictly neither a liquid nor a gas, and such a closed environment has been used to consume toxic waste materials that must not be released into the atmosphere.

Pulse combustion systems have been used for many years, but only recently have the principles of pulse combustors become understood. In such systems, combustion in a chamber forces product gases out through a tailpipe assembly, and a rarefaction wave in the combustion chamber then sucks fresh fuel and air in through flapper valves. These fresh reactants then mix with hot combustion products from the previous combustion event, leading to a spontaneous ignition of the fresh mixture. This then drives a new cycle, and the process repeats itself continuously. Pulse combustors are now used in propulsion systems, in industrial processes, and in home heaters.

4.4 Incineration and Toxic Wastes

There are a wide variety of chemical species that are closely related to hydrocarbon fuels, but in which one or more of the H atoms have been replaced by chlorine atoms. These species, many of which are poisonous or carcinogenic, are used widely in industrial processes. Examples of such species are vinyl chloride, trichloroethylene, and chlorobenzene. When compounds containing these species, such as excess plastic pipe, used transformer fluids, and many others, are no longer needed, there is a need to consume them in an environmentally safe way. A simple method of consumption is combustion, in which the hydrogen and carbon constituents are oxidized to water and carbon dioxide, and the chlorine atoms are converted to HCl. The HCl products can then be dissolved in water and the acid water then concentrated and removed to be treated in a sensible and safe way.

The difficulty with this process of combustion, or incineration, is that all combustion processes are subject to the same problems noted above for simple hydrocarbon combustion. Specifically, incomplete combustion, which in the case of hydrocarbon fuels leads to soot production, in the case of toxic species can produce very harmful toxic wastes such as dioxin and phosgene (mustard gas). Thus the combustion devices intended to consume toxic species are much less tolerant to incomplete combustion than simple hydrocarbon fuels, and the environmental penalties for combustion problems are very severe. The thermochemical product of chlorine atom combustion is definitely HCl, but the necessity of producing only HCl and the opportunities for producing much-less-desirable other products make incineration a problematical science.

5. SUMMARY

Combustion is a part of everyday life, from the production of home heating and home cooking, through the energy required to propel the automobile, the bus, the truck, the train, and the airplane, to the firing of electricity generating plants, industrial boilers, and furnaces. In addition, combustion produces much of our air pollution and other environmental problems.

It is likely that combustion will continue to be an important source of energy and power production for many years, and it is very important to understand the fundamental processes, chemical and physical, that control combustion and its associated processes.

GLOSSARY

Cetane Number: Comparative rating of ignition for fuels under diesel conditions. Higher cetane numbers indicate greater ease of ignition; small cetane numbers indicate a fuel difficult to ignite.

Compression Ratio (CR): Ratio of the combustion chamber volume at its greatest value, at bottom dead center, to its volume at its smallest value, at top dead center.

Detonation: Reactive shock wave.

End Gases: The last fuel-air charge to be consumed in the combustion chamber; the reactants that can explode to produce engine knock.

Equivalence Ratio (Φ): Ratio of fuel concentration actually present to that required to consume exactly the available oxidizer.

Exhaust Gas Recirculation (EGR): Practice of dilution of reactants with reaction products, generally consisting of CO_2, H_2O, and molecular nitrogen, usually employed to reduce product temperatures and slow production of NO_x.

Laminar: Combustion regime in which transport of heat and mass occurs via molecular diffusion and conduction.

NO_x: Nitrogen oxides ($NO+NO_2$).

Octane Number: Relative measure of susceptibility to knocking behavior. Fuels with large octane numbers resist knock; fuels with small octane numbers are likely to knock.

Stoichiometric: Conditions in which the ratio of fuel to oxidizer is exactly correct to produce complete oxidation.

Turbulent: Combustion regime in which transport of heat and mass occurs via unstable large-scale motions of the fluid.

List of Works Cited

Cowart, J. S., Keck, J. C., Heywood, J. B. Westbrook, C. K., Pitz, W. J. (1990), "Engine Knock Predictions Using a Fully-Detailed and a Reduced Chemical Kinetic Mechanism," in: *Proceedings of the 23rd Symposium (International) on Combustion,* Pittsburgh; The Combustion Institute, pp. 1055–1062.

Fickett, W., Davis, W. C. (1979), *Detonation,* Berkeley: University of California Press.

Frenklach, M., Clary, D. W., Gardiner, W. C., Jr., Stein, S. (1986), "Effect of Fuel Structure on Pathways to Soot," in: *Proceedings of the 23rd Symposium (International) on Combustion,* Pittsburgh: The Combustion Institute, pp. 1067–1076.

Gaydon, A. G., Wolfhard, H. G. (1979), *Flames, Their Structure, Radiation, and Temperature,* London: Chapman and Hall.

Gear, C. W. (1971), *Numerical Initial Value Problems in Ordinary Differential Equations,* New York: Prentice-Hall.

Harris, S. J., Weiner, A. M. (1983), "Surface Growth of Soot Particles in Premixed Ethylene/Air Flames," *Combust. Sci. Tech.* **31**, 155–167.

Heywood, J. R., 1988, *Internal Combustion Engine Fundamentals,* New York: McGraw-Hill.

Lewis, B., von Elbe, G. (1987), *Combustion, Flames, and Explosions of Gases,* Orlando: Academic.

Lovell, W. G. (1948), "Knocking Characteristics of Hydrocarbons," *Indus. Eng. Chem.* **40**, 2388–2438.

Lyon, R. K. (1976), *Int. J. Chem. Kinet.* **8**, 315–318.

Miller, J. A., Bowman, C. T. (1989), "Mechanism and Modeling of Nitrogen Chemistry in Combustion," *Prog. Energy Combus. Sci.* **15**, 287–338.

National Advisory Committee for Aeronautics (1959), Report No. 1300, "Basic Considerations in the Combustion of Hydrocarbon Fuels with Air," Cleveland: Lewis Flight Propulsion Laboratory.

Nettleton, M. A. (1987), *Gaseous Detonations, Their Nature, Effects, and Control,* London: Chapman and Hall.

Perry, R. A., Siebers, D. L. (1986), *Nature,* **324**, 657–658.

Semenov, N. N. (1959), *Some Problems in Chemical Kinetics and Reactivity* (trans.), Princeton: Princeton University Press.

Simon, D. M. (1951), "Flame Propagation," *Indus. Eng. Chem.* **43**, 2718–2725.

Strehlow, R. A. (1984), *Combustion Fundamen-*

tals, New York: McGraw-Hill.

Warnatz, J. (1984), "Rate Coefficients in the C/H/O System," in *Combustion Chemistry,* edited by W. C. Gardiner, Jr., New York: Springer-Verlag.

Westbrook, C. K. (1980), "Inhibition of Laminar Methane-Air and Methanol-Air Flames by Hydrogen Bromide," *Combus. Sci. Tech.* **23**, 191–202.

Westbrook, C. K. (1982), "Inhibition of Hydrocarbon Oxidation in Laminar Flames and Detonations by Halogenated Compounds," in: *Proceedings of the 19th Symposium (International) on Combustion,* Pittsburgh: The Combustion Institute, pp. 127–141.

Westbrook, C. K., Dryer, F. L. (1984), "Chemical Kinetic Modeling of Hydrocarbon Combustion," *Prog. Energy Combus. Sci.* **10**, 1–57.

Westbrook, C. K., Dryer, F. L., Schug, K. P. (1983), "Numerical Modeling of Ethylene Oxidation in Laminar Flames," *Combus. Flame* **52**, 299–313.

Westbrook, C. K., Pitz, W. J., Leppard, W. M. (1991), "The Autoignition Chemistry of Paraffinic Fuels and Pro-Knock and Anti-Knock Additives: A Detailed Chemical Kinetic Study," Society of Automotive Engineers Paper No. SAE-912314.

Williams, F. A. (1965), *Combustion Theory,* Reading, Mass.: Addison-Wesley.

Further Reading

Glassman, I. (1987), *Combustion,* Orlando: Academic.

Heywood, J. B. (1988), *Internal Combustion Engine Fundamentals,* New York: McGraw-Hill.

Lewis, B., von Elbe, G. (1987), *Combustion, Flames, and Explosions of Gases,* Orlando: Academic.

Sawyer, R. F. (1981), "The Formation and Destruction of Pollutants in Combustion Processes: Clearing the Air on the Role of Combustion Research," in: *Proceedings of the 18th Symposium (International) on Combustion,* Pittsburgh: The Combustion Institute, pp. 1–22.

Spalding, D. B. (1979), *Combustion and Mass Transfer,* New York: Pergamon.

Strehlow, R. A. (1984), *Combustion Fundamentals,* New York: McGraw-Hill.
Williams, F. A. (1965), *Combustion Theory,* Reading, Mass.: Addison-Wesley.

COMMUNICATION ENGINEERING

See TELECOMMUNICATION

COMPOSITE MATERIALS

CHARLES P. BEETZ, JR., *Advanced Technology Materials, Inc., Danbury, Connecticut, U.S.A.*

INTRODUCTION

History of Composite Materials

Since the earliest of times it has been recognized that certain combinations of different materials produce a composition whose properties are superior to those of the constituents alone. An ancient example is the use of straw to reinforce mud to produce reinforced bricks for construction purposes. Another important structural composite material was concrete developed and used by the Romans in the 2nd or 3rd century B.C. The Romans developed an extensive knowledge of the chemistry of cementitious limes as recorded in the writings of Vitruvius (Murdok and Brook, 1979). The Roman concrete was a combination of volcanic ash, lime, sand, gravel or crushed rock, and water. This composition, when properly mixed and dried, produced an extremely strong composite material that made possible the construction of truly massive structures. Another advantage of concrete was the ability to cast it into any desired shape to

1-56081-063-7/92/$5.00+.50

produce the great arches, domes, and vaults characteristic of much of Roman architecture. The great aqueducts, the dome of the Pantheon, and the Colosseum are examples of the large-scale application of man-made composite materials.

The use of concrete declined with the fall of the Roman Empire and did not reappear on a similar scale until the late 19th century, when the addition of steel reinforcing bars to concrete provided a major advancement. The steel reinforcement extended the use of concrete to areas such as beams, columns, and floor slabs subject to bending loads that place the structural element in a state of tension on one side and compression on the other. The tensile load is transferred to the steel reinforcement, which has a much greater tensile strength than the unreinforced concrete.

Another example of an early man-made composite material is paper, invented by the Chinese over 1800 years ago. The early Chinese papers were composed of pulp from the bark of mulberry trees, hemp, and cloth fibers. The pulp and fibers were mixed together in water, precipitated on a cloth screen, and dried to form the paper sheet. More recent examples since the middle of the 20th century include wood laminates such as plywood, directionally solidified eutectics, fiber-reinforced polymers such as fiberglass, and fiber-reinforced concrete, ceramics, and metals.

Constituent Materials and Properties of Composites

Composites encompass an extremely wide variety of constituent materials, as illustrated in the brief historical introduction. This extreme diversity makes it difficult to develop an acceptable definition of a composite material. In this article we will limit the scope of our definition of a composite material to those materials that are man-made by combination of two or more constituent materials to form a third material that has specific properties not possessed by the constituent materials individually. Thus we will not consider, for example, naturally occurring composite materials such as wood. This still leaves an almost unlimited range of combinations of constituent materials that could be considered as composite materials. In spite of this great diversity, the commonly used constituents in composites fall into three basic categories: reinforcements, fillers, and matrix materials. The reinforcing constituents are usually fibers, particulates, or flakes, which provide the principal structural properties of the composite. The matrix material provides the means of encapsulating and binding the reinforcing constituents, giving the composite material its bulk shape and form. Filler constituents are usually added to the matrix material to provide specific engineering, functional, or processing properties.

The properties of the composite material are derived from

1. the basic properties of the constituents,
2. their geometrical arrangement, and
3. the resulting complex interactions between the constituents.

The central idea encompassing all composite materials development is that different constituent materials can be combined to produce a material having specific properties or combinations of properties that are different from those of the constituents. As a simple example, consider a fiber-reinforced composite in which the fiber is unidirectionally oriented in an isotropic matrix material. The tensile strength of the composite measured along the fiber direction will in general be greater than that of the matrix and representative of the fiber, assuming that we use a fiber whose tensile properties are greater than those of the matrix. The matrix transfers the load to the fibers that support the applied load. In structural composites, the fibers form the majority phase, uniformly distributed and separated from one another by a thin layer of the matrix that serves to maintain the fibers in a fixed orientational relationship and to distribute the load uniformly over the fibers. Continuous-fiber–reinforced composite materials are in general anisotropic; properties measured along the fiber direction are different from those measured perpendicular to the fiber direction. An interesting and important aspect of fiber-reinforced composites is the ability to design the composite to make use of the anisotropy and tailor the properties of the composite to carry increased loads along specific directions. The properties of the composite ultimately depend on the constituents and the orientational relationship of the reinforc-

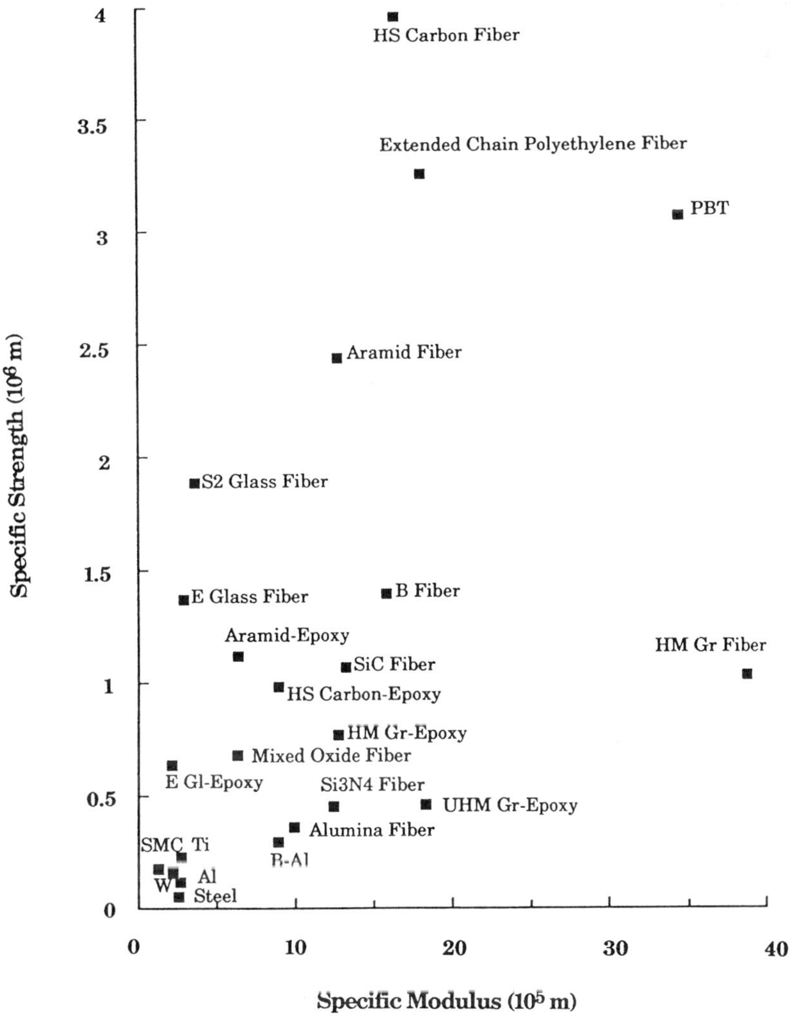

FIG. 1. A comparison of the specific ultimate strength and specific tensile modulus of a number of commonly used structural materials with various uniaxial fiber-reinforced composite materials (abbreviations: HS, high strength; HM, high modulus; UHM, ultrahigh modulus; SMC, sheet molding compound; see text for others.

ing fiber relative to the direction of the applied load. Figure 1 compares the specific strength (ultimate strength divided by density) and specific modulus of elasticity (Young's modulus divided by density) for a variety of commonly used structural and composite materials. Composite materials constructed from light-weight, high-performance fibers and polymeric matrix materials can exhibit specific strength and stiffness 3 to 4 times higher than those of commonly used structural materials such as steel and aluminum.

1. TYPES OF COMPOSITE MATERIALS

Composite materials can be classified (Schwartz, 1984) in several ways, such as by the type of matrix material, by the distribution of the reinforcement, e.g., continuous or discontinuous, or according to the functional purpose such as high-temperature, structural, or conductive composite materials. Composites can be most broadly classified according to the nature of the structural constituent. There are three basic categories of composite materials, as illustrated in Fig. 2. The three

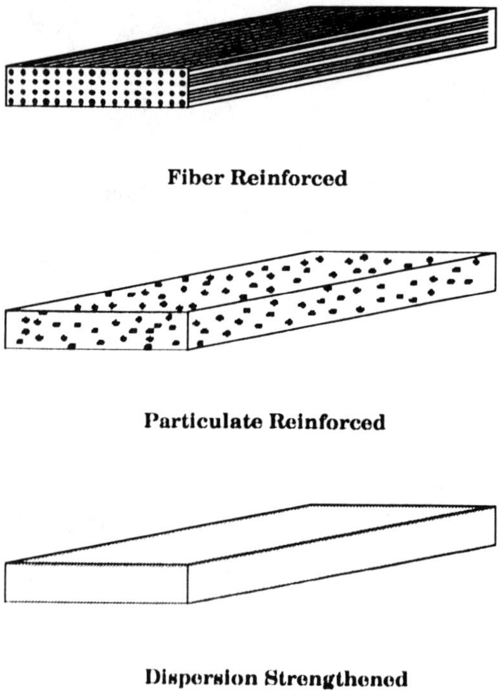

FIG. 2. An illustration of the three basic categories of composite materials: fiber-reinforced, particulate-reinforced, and dispersion-strengthened matrices.

categories are (1) fiber-reinforced composites, (2) particulate-reinforced composites, and (3) dispersion-strengthened composites. The effect of the size of the reinforcing agent on the composite strength is illustrated in Fig. 3 for particulate- and fiber-reinforced composites (Broutman and Krock, 1967). In each case very different mechanisms are responsible for strengthening and stiffening of the matrix material. In the case of fiber-reinforced composites, the fiber must have an aspect ratio (length divided by diameter) that is at least ~10 and preferably >100 to transfer the load from the matrix to the fiber completely. The three categories of composite materials are discussed in the following section and are readily distinguished by the range in size and volume concentration of the reinforcing phase.

1.1 Fiber-Reinforced Composites

In the area of structural composites, fibers are the most important class of reinforcement. This is driven by the necessity to produce ever stronger, stiffer, and more lightweight materials. Fibers provide a unique opportunity to obtain the ultimate mechanical properties of a given type of material. In general, materials when made in fiber form tend to exhibit more nearly intrinsic tensile properties than do bulk forms. The reasons for this will be discussed in Sec. 2. Glass fibers, for example, have much higher tensile strength than bulk glass plates. Similar statements can be made for metal, carbon, ceramic, and polymeric fiber forms. The most weight-efficient method for using the large Young's modulus and tensile strength of fibers is to combine them in a matrix of a lightweight material usually, but not necessarily, having lower tensile strength and modulus. Glass-fiber–reinforced resin is one of the first modern structural composites and has been devel-

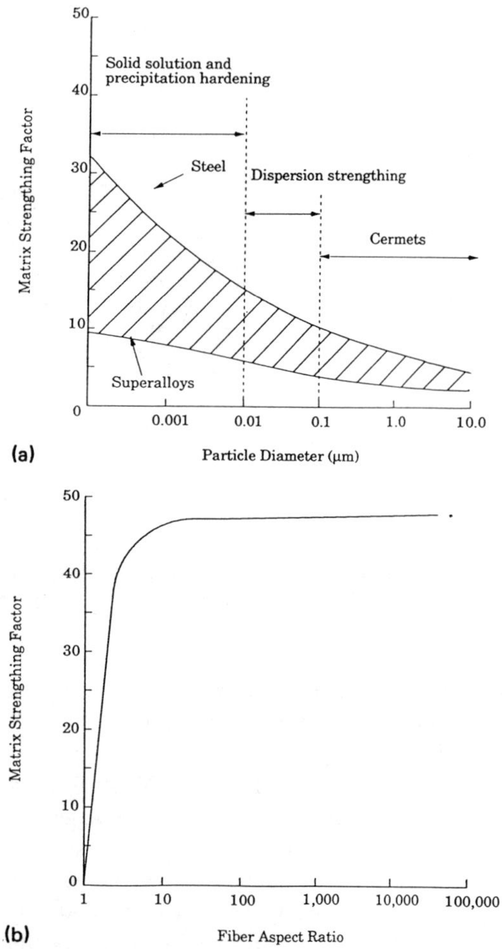

FIG. 3. Effect of the size of (a) particulates and (b) fiber aspect ratio (aspect ratio is defined as the ratio of fiber length to diameter) on the composite strength; after Broutman and Krock (1967).

oping an ever-increasing scope of utilization since the end of World War II. There have been significant developments since the 1960s in the area of carbon fibers and carbon-fiber–reinforced composites, resulting in some of the highest-strength lightweight materials. Carbon fibers have been used to reinforce a variety of polymeric resins and lightweight metals such as aluminum and magnesium. Besides reinforcing lower-modulus matrix materials, there has also been significant progress in using fibers to provide toughness to brittle matrix materials such as ceramics and glasses.

The performance of fiber-reinforced composites is determined by many variables such as

1. fiber orientation, length, shape, and composition,
2. composition of the matrix material,
3. the volume fraction of the reinforcing phase,
4. the interaction between the fiber and the matrix, and
5. the presence of any additional filler materials (Weeton *et al.*, 1987).

The mechanical properties of the composite are highly anisotropic and depend on the orientation of the fiber in the matrix. The strength and modulus of a uniaxially oriented fiber-reinforced composite will be largest along the direction of the fiber. If the applied load is at an angle that is not along the fiber direction,

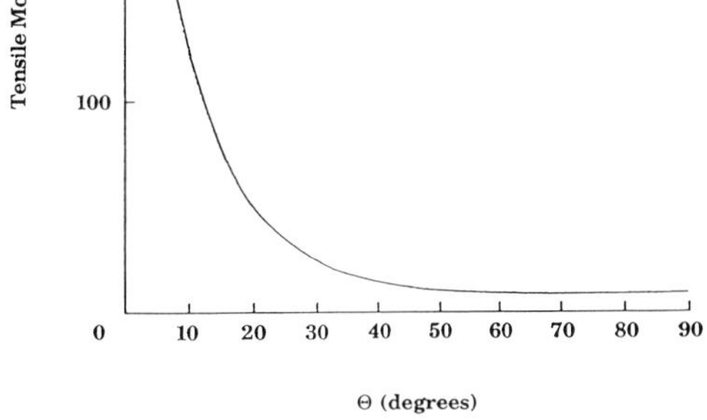

FIG. 4. Variation of the tensile modulus of a uniaxial composite as a function of the angle between the fiber direction and the applied load. After Hull (1981).

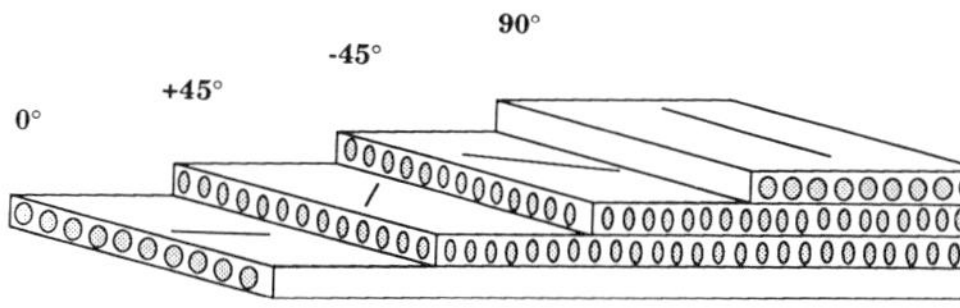

FIG. 5. Example of an angle-plied laminate; the angles indicate the amount of rotation of the fiber orientation in each lamina from the uniaxial direction.

the strength and modulus of the composite will be dramatically reduced. Figure 4 illustrates the directional dependence of the tensile modulus as a function of angle between the applied load and the fiber direction for a carbon-fiber–reinforced epoxy composite. Since in actual applications structural elements are rarely subject to pure uniaxial loads, it is necessary to combine different fiber orientations in a single composite. This is typically done by forming a laminate made up of uniaxial plies oriented at various angles to one another. A typical laminate construction is illustrated in Fig. 5. The ability to tailor the mechanical properties of the composite to accommodate complex loads is a major advantage of composite materials over isotropic materials. The directional selectivity available with composite material design allows for more mass-efficient structures. The stiffness and strength of a uniaxial fiber-reinforced composite are directly proportional to the volume fraction of the fiber in the composite and can be described in a first-order approximation by a simple rule-of-mixtures expression such as

$$E_c = E_f V_f + E_m V_m,$$

where E_c, E_f, and E_m are the moduli of elasticity of the composite, fiber, and matrix, respectively, and V_f and V_m are the volume fractions of the fiber and matrix in the composite. Similar expressions also can be written for the strength, density, Poisson's ratio, etc., of the composite.

1.2 Particulate and Flake-Reinforced Composites

1.2.1 Particulate Composites. Particulate composites include cermets, metal and ceramic particles in metal, and plastic- and ceramic-matrix materials. Particulate composites differ from fiber-reinforced composites in that the reinforcing constituent does not have a shape anisotropy. The particulate can be viewed in a first approximation as spherical in shape, although in actuality the particulate may be spherical, disk shaped, cubic, or polyhedral. In cases where the particulate is randomly distributed in the matrix and lacks shape anisotropy, the properties of the resulting composite material are isotropic. Since the particulate composite does not have a homogeneous composition, the properties depend on the size, shape, concentration, and composition of the particulate as well as the strength of the interfacial bond with the matrix.

An important class of particulate-reinforced composites intended for use at high temperatures is comprised of cermets, a combination of the words ceramic and metal. Cermets are usually composed of ceramic particles 10%–97% by volume, supported in a metal matrix. Examples are cobalt-cemented tungsten carbide, nickel-cemented titanium carbide, alumina-reinforced chromium, and uranium-plutonium cermets. Cermets are usually produced by powder metallurgical techniques. The bonding between the particulate and the metal matrix is usually due to partial solubility of the two phases in one another. The properties of the cermets are variable over a wide range through variation of the ceramic-to-metal ratio in the starting powder mixture and the processing conditions. There are two commercially important classes of cermets, based on

1. metal carbide and
2. metal oxide particulates.

Many other cermet materials have been produced on a laboratory scale for applications as diverse as selective optical absorbers for solar collector panels.

1.2.1.1 Metal Carbide Cermets. A commonly used metal carbide cermet is cemented tungsten carbide, which finds broad application as a cutting-tool material. The most commonly used binder metal is cobalt. Cobalt-cemented tungsten carbides in general are abrasive and have high compressive strength, hardness, and rigidity. These properties can be varied through changes in the amount of cobalt binder used. Typically the amount of cobalt ranges from 1% to 35% by volume. Increased cobalt content results in increased toughness and ductility and decreased hardness, rigidity, and abrasiveness. Most commercial tungsten carbide cutting tools have small amounts of TiC, NbC, and TaC powders (0%–1%) to improve the high-temperature mechanical properties and the chemical wear resistance of the tool (North, 1987). Typical particle sizes are on the order of 1–10 μm, although new fine-grained tungsten carbide cobalt cermets are being developed that have superior properties. Titanium-carbide-particulate–reinforced nickel or nickel-alloy–matrix cermets have high stiffness and strength at temperatures in excess of 2200 °C.

1.2.1.2 Metal Oxide Cermets. Besides the carbide cermets, metal oxide cermets also find wide commercial application. Oxide cermets find use as high-speed cutting tools, low-friction mechanical seals, nuclear reactor fuel elements, and selective solar absorber coatings. The oxide cermets have the unique property that the metal and the ceramic may be interchanged as either the particulate or the matrix material. A typical metal oxide cermet composition is 72% Cr, 28% Al_2O_3; the properties of the cermet can be controlled by altering the ratio of metal to ceramic. The oxide cermets are typically formed by powder metallurgical manufacturing methods.

1.2.1.3 Dispersion-Strengthened Systems. Dispersion-strengthened composite materials consist of a matrix material that has fine particles of a second phase, typically 0.01–0.1 μm in diameter, dispersed uniformly throughout the matrix. In contrast to the particulate or fiber-reinforced composites, the dispersed particles constitute only a small fraction of the total volume of the composite, usually less than 3%. The strengthening mechanism in dispersion-strengthened systems is very different from that in either fiber- or particulate-reinforced materials. The fine particles reinforce the matrix by restricting the motion of dislocations. An excellent example of a dispersion-strengthened (DS) metal is commercial DS copper, which consists of pure copper that contains fine Al_2O_3 particles (Boyer and Gall, 1985). The effects of varying the Al_2O_3 content on the mechanical and electrical properties of the copper are shown in Fig. 6. The DS copper alloys provide excellent high-temperature stability over pure copper and copper alloys. Other metals in which dispersion strengthening has been successfully employed include

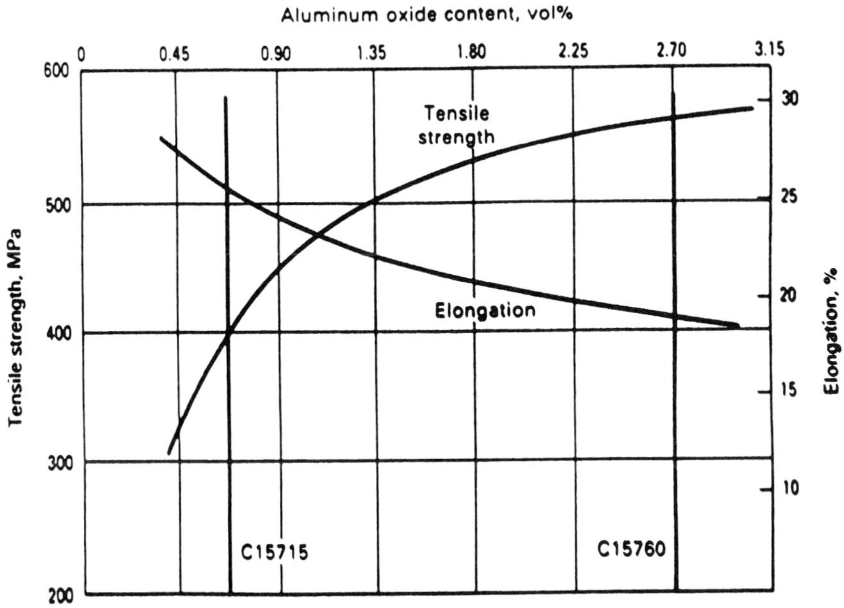

FIG. 6. The effect of aluminum oxide content on the mechanical properties of dispersion-strengthened copper. Reproduced from Boyer and Gall (1985) with permission from ASM International.

sintered aluminum powder alloys, thoria dispersed in tungsten, and thoria dispersed in nickel.

1.2.2 Flake-Reinforced Composites. A special case of a particulate-reinforced composite occurs when the particulate is in the form of a flake. A flake is a particulate that exhibits two-dimensional shape anisotropy, most easily visualized in the form of a plate. In the plane of the flake the shape is isotropic, and perpendicular to the plane of the flake the shape is highly anisotropic, resembling a fiber. The aspect ratio of the flake (diameter divided by thickness), just as in the case of fiber-reinforced materials (Fig. 3), is an important variable that determines how effective the flake is at supporting the load applied to the composite. The two-dimensional shape anisotropy of the flake leads to interesting composite properties provided

1. the aspect ratio is large ($\sim$100) and
2. the flakes can be incorporated into the composite such that they orient parallel to one another.

An oriented flake-reinforced composite exhibits isotropic mechanical properties in the plane of the flakes, in contrast to fiber-reinforced materials where special laminating or filament-winding procedures must be used to realize in-plane isotropic properties. The two-dimensional shape anisotropy allows the flakes to pack more closely together in the composite, yielding a higher volume fraction of reinforcement and increasing the modulus and strength. The most commonly used flakes are metal, graphite, glass, and mica. Metal flakes are typically used to increase the electrical and thermal conductivity of a polymer matrix material. Mica and glass flakes are used to provide high-temperature insulation properties. Flake-reinforced materials also find use as impermeable barrier materials because of the high packing densities and overlapping of the flakes, preventing fluids and gases from penetrating the composite.

2. FIBERS AND MATRIX MATERIALS

2.1 Fiber Types

A variety of materials can be produced in fiber form and used as reinforcing agents in a composite material. Consequently, fibers come in many sizes, shapes, and structures that usually reflect the manufacturing method. The range of structures commonly found in fibers is illustrated in Fig. 7, ranging from amorphous to single-crystalline to multiphase. Fibers are commonly classified according to material type such as glass, polymer, inorganic, metallic, or carbonaceous. Since many materials can be produced as continuous fibers, this section will describe some commonly used fibers for structural composite materials and recent trends in fibers for high-temperature composite applications.

2.1.1 Glass Fibers. The most frequently used material for manufacturing fibers for structural composite applications is glass. Glass fibers consist primarily of silica (SiO_2) with varying amounts of other metal oxides such as calcium, boron, sodium, aluminum, and iron. There are four major types of glass fiber:

1. A-glass (high alkali content) for good chemical resistance,
2. C-glass (chemical) for extremely good resistance to chemical attack,
3. E-glass (electrical) for high electrical resistivity for insulation purposes (also, because of its strength and resistance to attack by water, it is the most frequently used

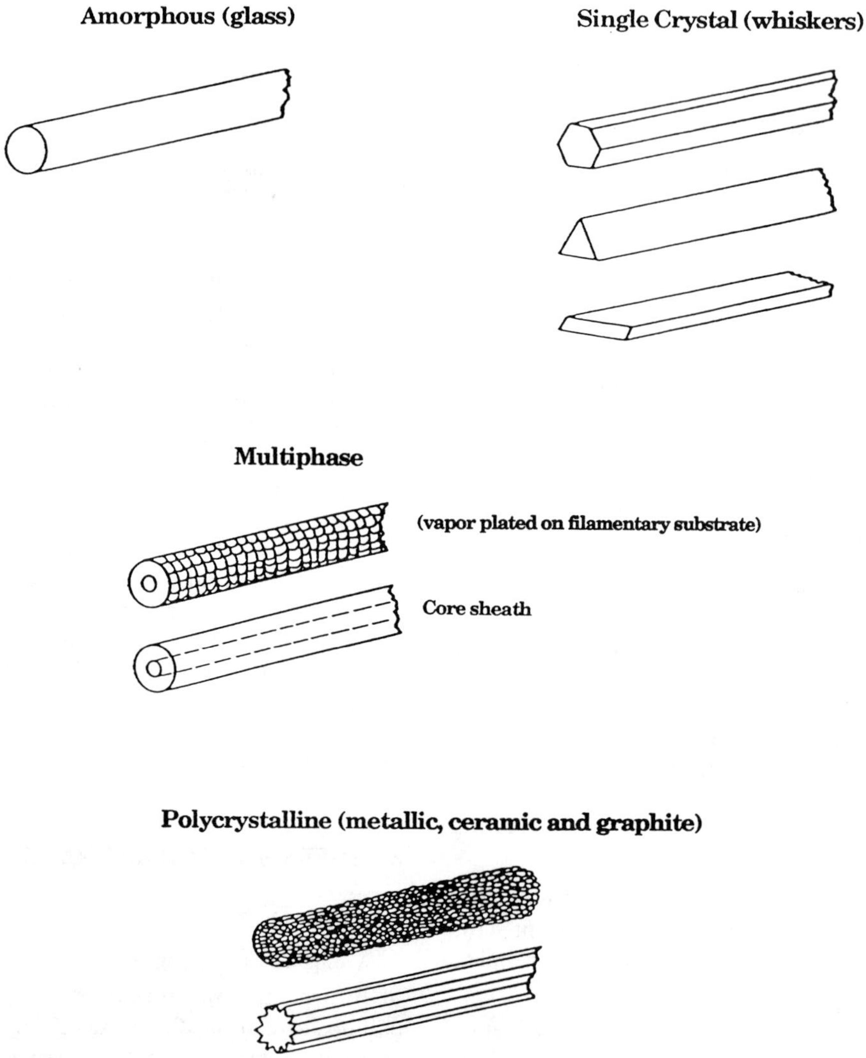

FIG. 7. Typical structures exhibited by fibrous reinforcements. Reproduced from Weeton *et al.* (1987), with permission from ASM International.

Table 1. Composition of the major types of glass fibers (in percent).

	Glass fiber type			
Constituent	A	C	E	S
Silicon oxide	72.0	64.6	54.3	64.2
Aluminum oxide	0.6	4.1	15.2	24.8
Ferrous oxide	...	...	...	0.21
Calcium oxide	10.0	13.2	17.2	0.01
Magnesium oxide	2.5	3.3	4.7	10.27
Sodium oxide	14.2	7.7	0.6	0.27
Potassium oxide	...	1.7	...	...
Boron oxide	...	4.7	8.0	0.01
Barium oxide	...	0.9	...	0.2
Miscellaneous	0.7	...	...	...

glass fiber for structural purposes), and
4. S-glass for structural applications where higher strength is needed.

S-glass also has higher strength at elevated temperatures. The compositions of the major types of glass fibers are shown in Table 1.

Glass fibers are produced by melting mixtures of the appropriate oxides. The molten glass is extruded under the action of gravity through a heated die containing many small-diameter holes; see Fig. 8. The purpose of the die is to ensure that the fibers have a uniform and well-defined diameter. The extruded glass is drawn away from the die at high speed, resulting in a large number of small-diameter fibers. The fibers are then given a coating of a sizing agent and taken up on a winder. This produces a continuous roving that is used to make yarns, tapes, and various forms of fabrics used for composite material fabrication. Other methods disrupt the flow of molten glass through the heated die using a gas jet to produce short discontinuous staple fibers. Such fibers are used in making fiber mats and nonwoven materials.

Glass fibers produced by spinning processes have an amorphous structure that exhibits no long-range order typically characteristic of crystalline solids. The glass structure consists of a random network of silicon atoms covalently bonded to oxygen atoms. The randomness of this arrangement results in a completely isotropic structure whose properties are independent of direction, i.e., parallel or perpendicular to the fiber axis. The modulus of the glass fiber is primarily determined by the silicate network. Addition of other metal oxides can disrupt or strengthen the silicate network as illustrated in Table 2. Glasses having modulus and strength values comparable to or greater than S-glass have been produced using beryllium oxide additions (Diefendorf, 1985); however, because of toxicity concerns, these glasses have not received wide utilization.

2.1.2 Carbon Fibers. Over the course of the past 30 years, carbon fibers have gone from being a laboratory curiosity to a major industry. Carbon-fiber–reinforced composites are used in a variety of applications ranging from advanced military fighter aircraft to sporting equipment such as fishing rods, tennis racquets, and golf clubs. Carbon fibers have unsurpassed specific strength and stiffness that make them attractive for applications where mass efficiency is important. Commercially available carbon fibers are produced by controlled pyrolysis, carbonization, and graphitization of one of several precursor fibers:

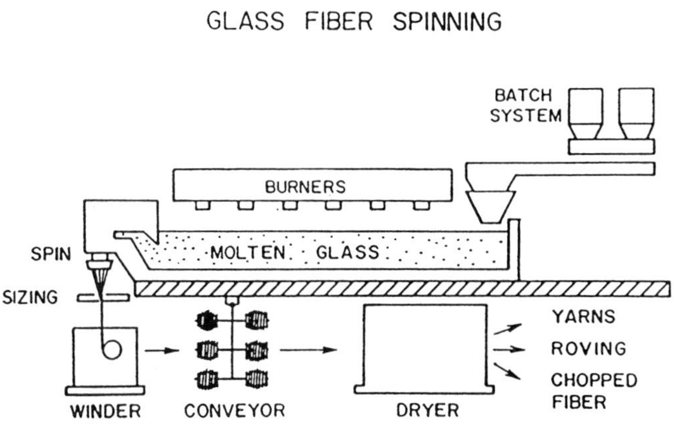

FIG. 8. Schematic diagram of the process used for the production of continuous glass fiber. Reproduced from Diefendorf (1983), with permission.

Table 2. Variation of glass fiber properties with composition (Rauch *et al.*, 1966).

	E-glass	I	II	III	IV	V
Constituent	Composition (%)					
SiO_2	54.3	45.5	45.5	50.7	47.7	42.0
B_2O_3	8.0	19.0	19.0	17.2	14.2	18.2
LiO_2	...	0.5	0.5	0.45	0.3	0.5
MgO	4.7	17.5	10.6	6.30	15.6	12.7
Al_2O_3	15.2	17.5	24.4	25.3	22.2	19.5
BeO	...	...	...	...	...	6.6
Fe_2O_3	...	...	...	...	...	0.5
Tensile Strength (GPa)	3.45	4.05	3.97	4.06	4.36	5.03
Young's Modulus (GPa)	74.5	75.2	62.7	62.1	73.1	82.0
Density (g/cm^3)	2.54	2.33	2.05	2.00	2.20	2.20

1. polyacrylonitrile (PAN),
2. petroleum or coal-tar-pitch–based precursors, or
3. rayon (cellulose).

Of these precursors, PAN is the most widely used for commercial carbon-fiber production. A schematic diagram of the manufacturing process is shown in Fig. 9 for both the PAN- and pitch-based fibers. In the PAN process, a starting fiber of polyacrylonitrile is subjected to a stabilization heat treatment during which the fiber is stretched, orienting the polymer chains parallel to the fiber axis. During the oxidative stabilization process, the linear carbon-nitrogen backbone of the polymer is converted to a ladder polymer of aromatic hexagonal rings. The stabilization process produces a fiber whose glass transition tem-

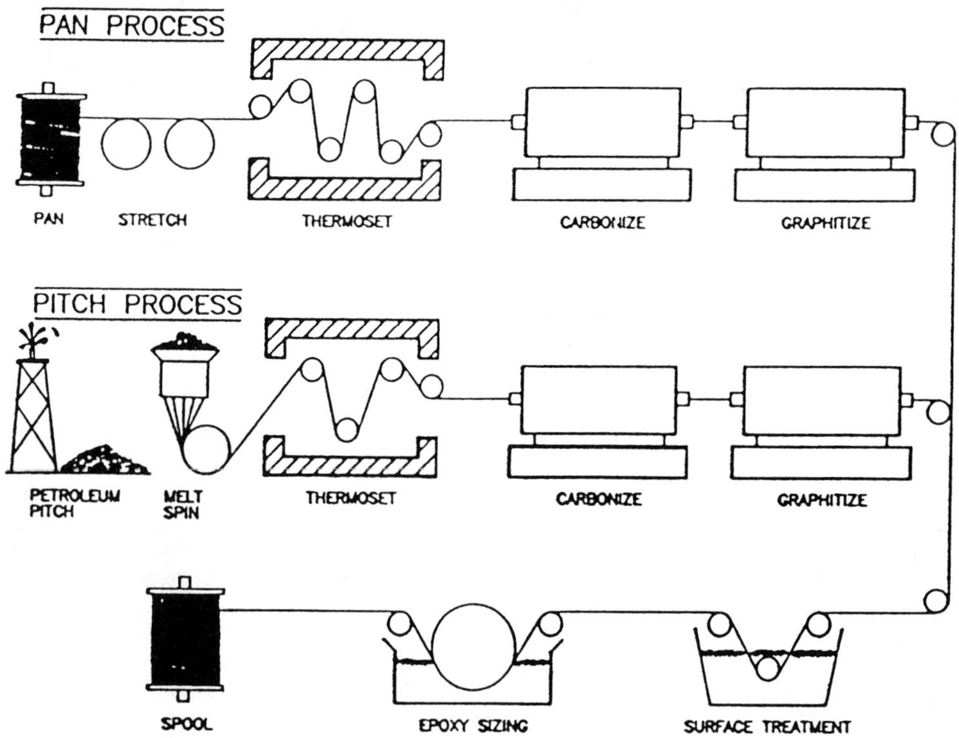

FIG. 9. Schematic diagram of the processes used for the production of PAN- and pitch-based carbon fibers. Reproduced from Diefendorf (1983), with permission.

perature is well above its decomposition temperature. The stabilized fiber can then maintain the orientation of the ladder polymer structure during the subsequent carbonization process. The carbonization heat treatment drives off hydrogen, oxygen, and nitrogen, leaving behind a partially oriented graphitelike network of carbon atoms. The alignment and crystalline perfection of the carbon network are substantially improved by giving the fibers a high-temperature heat treatment, frequently termed the graphitization heat treatment. Fibers manufactured from PAN precursors in a strict sense do not actually develop a fully three-dimensional graphite lattice until heated to temperatures on the order of 3000 °C.

Pitch-based carbon fibers are produced by conversion of a petroleum or coal-tar-pitch–based precursor fiber into a carbon fiber through a series of heat-treatment processes similar to the PAN process. The pitch-based precursor fiber is produced by first extracting a liquid-crystalline mesophase that is melt-spun into a precursor fiber. The precursor fibers have a high degree of orientation of large platelike aromatic hydrocarbon molecules. This precursor fiber is then subjected to a series of heat treatments analogous to the PAN precursor fiber. The large platelike hydrocarbons become aligned parallel to the fiber axis and develop three-dimensional ordering into a true graphite structure at much lower temperatures than PAN-based fibers.

The Young's modulus of the fiber is highly dependent on the temperature of the graphitization heat treatment. The Young's modulus increases with increasing orientation of the carbon planes and therefore with increasing heat-treatment temperature. Carbon fibers can be produced with a wide range of properties, depending in detail on the process conditions. The higher degree of preferred orientation of the graphite planes parallel to the fiber axis in pitch-based fibers permits pitch-based fibers to develop higher values of Young's modulus at lower values of graphitization heat-treatment temperature.

The tensile strength of the fiber is dependent not only on the degree of orientation of the carbon planes parallel to the fiber axis but also on the number of critical defects present in the fiber volume. Many types of structural defects such as voids, particulate inclusions, misoriented crystallites, and microcracks are present in the fibers as a result of the fiber-manufacturing process. These defects limit the strength of the fiber in much the same way as the weakest link limits the strength of a chain. The tensile strength of the fiber increases, for example, with decreasing fiber diameter as a result, in part, of the reduced number of critical defects present in the smaller fiber volume.

The degree of orientation of the carbon planes parallel to the fiber axis and the concentration of defects present per unit volume can be controlled through variation of the process parameters so that carbon fibers having a variety of properties can be produced. Most carbon fibers fall into one of three different grades:

1. aerospace grade,
2. intermediate modulus, and
3. high modulus.

The Young's modulus and tensile strength of these three grades are given in Table 3.

2.1.3 Polymer Fibers. In carbon fibers, the high degree of preferred orientation of the sp^2-bonded carbon planes parallel to the fiber axis accounts for the extremely large modulus values achieved in these fibers. An analogous situation can be realized using organic polymers, such as paraphenylene terephthalamide (PPT), that can be spun from a nematic liquid-crystalline solution into fibers having a high degree of preferred molecular orientation. In the nematic liquid-crystalline phase, the polymer chains are oriented parallel to one another in local domains. The liquid-crystalline domains are randomly oriented with respect to one another. During spinning, the molecular orientations of the liquid-crystal domains are redirected so that they are parallel to one another, aligning the polymer chains parallel to the fiber axis. The perfection

Table 3. Properties of the three major grades of commercial carbon fibers.

Fiber grade	Modulus range(GPa)[a]	Strain to failure (%)
Aerospace	220–241	1.5–2.0
Intermediate modulus (IM)	275–345	~2.0
High modulus (HM)	379–586	~1.0

[a]Young's modulus.

of the molecular orientations in the fiber can be further improved by a post-spinning heat treatment. The first commercially available fiber based on this process was Kevlar™. The specific strength and modulus of Kevlar 49 fibers are ~5× and ~2× those of steel, respectively. More recently a new grade of aramid fibers has been introduced that has a Young's modulus of 179 GPa and tensile strength of 3.45 GPa at a density of 1.44 g/cm^3 (see Table 4). These fibers have specific properties comparable to those of aerospace-grade carbon fibers. The compressive properties of aramid fibers are somewhat low because of weak intermolecular bonding in the structure that allows it to buckle at low compressive stresses (Diefendorf, 1985). Kevlar-fiber–reinforced composites find use in tensile-dominated loading applications such as rocket motor casings, belts in radial car tires, ropes, and cables. Kevlar-reinforced composites have excellent impact damage resistance and are also used in ballistic protection applications.

Since the discovery of the aramid class of solution-spinnable polymers in the early 1970s, there have been other rigid-rod polymers discovered that also can be solution-spun into highly oriented fibers. These include the aromatic heterocycles such as polyphenylene bisbenzthiazole (PBT) (Helminiak *et al.*, 1975). PBT is a rigid-rod polymer that does not have the kink structure of PPT and therefore produces a fiber that has a substantially larger Young's modulus; see Table 4. Besides solution spinning, other processes have been developed for producing fibers composed of highly oriented rigid-rod molecules such as melt spinning of liquid-crystalline polymers (LCP). This technique has recently been successfully demonstrated on a commercial scale (Vectran™) using thermotropic copolyesters whose properties are comparable to those of the aramid-based fibers (Table 4). The organic fibers offer strength and stiffness intermediate between glass and carbon fibers.

The molecular bonds in linear chain polymers are very strong and when oriented appropriately can produce mechanical properties comparable to those of steel. Calculations of the modulus of fully extended polyethylene, for example, have predicted values ranging from 180 to 285 GPa (Lindenmeyer, 1976). Recently very-high-modulus extended-chain polyethylene fibers have been manufactured using a solution-spinning process. These fibers have a Young's modulus of 172 GPa, tensile strength of 3 GPa, and a density of 0.97 g/cm^3, in good agreement with the early predictions. The key step in the production of this type of fiber was developing a spinning solution that kept the long-chain polyethylene molecule extended, preventing it from folding back on itself. These new fibers are being used in applications such as ropes, cables, ballistic fabrics, and sporting equipment.

2.1.4 Inorganic Fibers. This section describes the major types of ceramic fibers that can be used to reinforce metal, glass, and ceramic composite materials. These fibers are usually intended for use in applications requiring high temperature stability (mechanical and chemical) that cannot be realized using polymeric composites. Many of the ceramic fibers discussed have also found application in polymer-based composites. Although

Table 4. Properties of several high-performance organic fibers.

Fiber type	Modulus[a] (GPa)	Tensile strength (GPa)	Density (g/cm^3)
Kevlar™ 29[b]	58	3.62	1.44
Kevlar 49	124	3.62	1.45
Kevlar 149	179	3.45	1.44
PBT	304	2.7	...
LCP (Vectran™)[c]	62–86	2.5–3.1	1.40
Extended-chain polyethylene (Spectra 1000™)[d]	172	3.09	0.97

[a]Young's modulus.
[b]Kevlar is a registered trademark of E. I. duPont de Nemours & Co., Inc.
[c]Vectran is a registered trademark of Hoechst Celanese Corp.
[d]Spectra 1000 is a registered trademark of Allied Signal, Inc.

many different types of ceramic fibers have been produced over the years, only the three most commonly used fibers will be discussed. The most frequently used ceramic fibers are boron, silicon carbide, and alumina. The reader is directed to Bracke *et al.* (1984) and Sowman (1988) for a more complete discussion of past work and future trends in ceramic fibers and composites.

2.1.4.1 Boron Fibers. Boron is an attractive candidate material for producing in fiber form as it has an atomic weight less than carbon, a Young's modulus about twice that of steel, and a density of 2.34 g/cm^3. The preparation of fibers from refractory materials such as boron via direct melt spinning as in the case of glass fibers or by pyrolysis of a precursor fiber as in the case of carbon fibers is usually very difficult. Consequently, commercial boron fibers are produced by chemical vapor deposition (CVD) of a thick boron coating onto a continuous core of tungsten wire. The boron fiber is itself a composite consisting of a small-diameter tungsten wire core coated with a thick layer of polycrystalline boron. Boron fibers are produced as a monofilament instead of a tow of many small fibers. The diameter of the tungsten wire core is typically ~20 μm and the final diameter of the boron filament is variable but is usually ~100–140 μm. A schematic diagram of the CVD process for making boron fibers is shown in Fig. 10. Boron is deposited by thermal decomposition of a boron-containing gas such as boron trichloride on the tungsten wire that is resistively heated by passing a current through the wire. The thickness of the boron coating increases as the wire is drawn through the reactor tube. The boron coating is built up to the desired thickness by passing the wire through many reactors placed in series. The boron fibers produced by this method have a Young's modulus of 400 GPa and tensile strength of 3.6 GPa with a density of 2.57

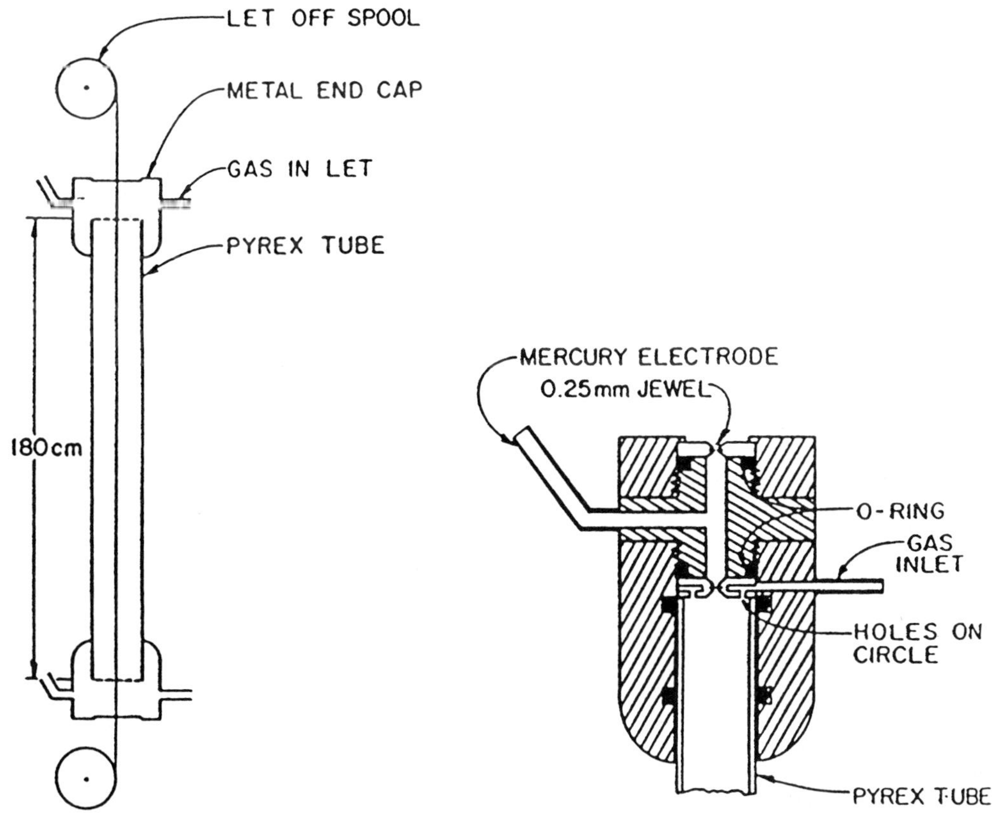

FIG. 10. Apparatus for chemical vapor deposition of a boron fiber onto a tungsten wire. The wire is resistively heated to temperatures high enough to decompose a boron-containing gas such as BCl_3 in hydrogen to deposit elemental boron on the tungsten wire. Reproduced from Diefendorf (1983), with permission.

g/cm^3. One of the largest applications for boron fibers is in reinforcing epoxy matrices for sporting goods such as lightweight racquets and shafts for golf clubs. Boron is also used as a reinforcement in aluminum, magnesium, and titanium metal matrix composites.

2.1.4.2 Silicon Carbide Fibers. Silicon carbide fibers can be produced by CVD on a core carbon monofilament or by pyrolysis of a precursor fiber. Silicon carbide fibers prepared by CVD are most frequently deposited on a carbon monofilament using a process very similar to that for CVD boron fibers. The SiC coating proceeds by reaction of a mixture of chlorinated alkyl silane compounds with hydrogen on the surface of the resistively heated carbon core. The carbon monofilament is ~50 μm in diameter, much larger than the tungsten wire used for boron fibers. The CVD SiC fibers have a Young's modulus of 400 GPa, tensile strength of 3.45 GPa, and a density of 3.3 g/cm^3.

Silicon carbide fibers also can be produced by pyrolysis of Si-containing organometallic polymers such as polycarbosilane. The polymer is first melt-spun into a fiber that undergoes a curing process, which produces a high degree of cross-linking; the cured fiber then is pyrolyzed forming the product SiC fiber. There are many difficulties in the curing process such that easily spinnable high-molecular-weight Si polymers cannot be used because during the curing step they produce structures that result in breakup of the fiber structure. Thus the polymers used contain elements other than Si and C such as nitrogen and oxygen that introduce stable compounds, resulting in a nonstoichiometric silicon carbide fiber. The properties of these fibers are significantly lower than CVD-produced fibers, i.e., Young's modulus ~196 GPa, tensile strength ~2.9 GPa, and density ~2.55 g/cm^3.

2.1.4.3 Alumina Fibers. Fibers of polycrystalline Al_2O_3, Al_2O_3-SiO_2, and Al_2O_3-SiO_2-B_2O_3 mixtures are commercially produced using slurries or sols containing the inorganic phases, various salts, and polymers to control spinnability of the fiber. The solution-spun fibers are given a pyrolysis heat treatment that nucleates the crystallization of the oxide ceramic phase and concurrently removes the polymers or other binders. On further heat treatment, the ceramic particles sinter together, densifying the structure; increased grain growth then follows. The fibers produced in this manner have a very fine grain structure, with fibers produced from sols having the finest microstructures. The properties of fiber FP (Al_2O_3) produced by DuPont, using a slurry-modified sol process, are Young's modulus ~380 GPa, tensile strength ~1.4 GPa, and density 3.9 g/cm^3. Although the tensile strength of these fibers is low, they have excellent compressive properties. The fiber FP also has good stability up to the temperature of 1000 °C and can be processed in aluminum- or magnesium-matrix materials.

2.1.5 Whiskers. It has been known since the early 1950s that materials in whisker form can exhibit superior mechanical properties over bulk forms. Whiskers are small, elongated, single crystals that are almost free of macroscopic defects common to bulk forms of the material. The strength of whiskers increases with decreasing whisker diameter: at diameters less than ~1 μm, whiskers exhibit strengths approaching theoretical estimates; above ~100 μm, whiskers exhibit strengths comparable to bulk forms. The nearly ideal mechanical properties of small whiskers have made them attractive for composite material applications, especially since whiskers of many interesting materials can be grown that have reasonable aspect ratios. Many materials have been grown in whisker form and evaluated in one way or another for composite-material applicability; in this section we will focus on the most promising whisker materials. The reader is directed to Evans (1972) for a review of much of the early work on whiskers.

Primarily for economic reasons, the only commercially available whisker reinforcements are composed of silicon carbide. The SiC whiskers are produced by the controlled pyrolysis of rice hulls that inherently have a high silicon content. SiC whiskers have also been produced by the vapor–liquid–solid (VLS) process in which a solid catalyst particle is placed in a growth reactor and exposed to the reactive gases H_2, CH_4, and SiO. The catalyst particle becomes saturated with Si and C and forms a molten ball that mediates the growth of the whisker through the liquid–solid interface. The crystal grows by precipitation from the supersaturated liquid. The growing crystal is fed from the vapor by the reaction of the constituent molecules at the vapor–liquid interface. β-SiC whiskers having tensile strengths over 8 GPa and a Young's modulus of 581 GPa have been reported (Petrovic *et al.*, 1985). The growth mechanism of the rice-hull–produced SiC whiskers is a variant of the VLS process.

Another whisker material that has recently shown promise as a potentially low-cost reinforcement for polymer-based composites is carbon (Tibbetts *et al.*, 1986). Carbon whiskers can be produced by a vapor growth process that uses a small-diameter iron catalyst particle to grow the whisker. The iron particle is exposed to a mixture of CH_4 in H_2 at elevated temperatures, saturating the iron particle with carbon. The carbon then precipitates from the particle as a hollow graphite tube approximately the diameter of the catalyst particle. The graphite tube produced by this process has the structure of high-quality graphite and is extremely small, ~30 nm in diameter, and difficult to handle. The length of the hollow graphite tube is dependent on the processing conditions, but tubes tens of centimeters in length have been produced. To produce larger-diameter whiskers, a chemical vapor deposition step is carried out that increases the diameter to the desired value, typically ~10 μm. This step represents a departure from the traditional whisker process, and the structures produced are more appropriately called fibers. The properties of the resulting CVD-coated graphite tubes are highly dependent on the final diameter but typically fall in the range of aerospace-grade carbon fibers. The CVD carbon coating is highly imperfect as deposited and accounts for the reduction of the fiber properties from those expected for single-crystal graphite. However, high-temperature heat treatment of the fibers produces properties approaching those of single-crystal graphite. Fibers produced by this method have been shown to have thermal conductivities comparable to single-crystal graphite and should be useful for composite applications requiring this property.

2.1.6 Single-Crystal Fibers. An outstanding deficiency of currently available ceramic fibers is creep at high temperatures. Above ~1000 °C, the structure of polycrystalline Al_2O_3 and SiC fiber changes, causing slippage along grain boundaries, greatly reducing the physical properties of the fiber. Many of the mechanisms responsible for high-temperature creep, such as grain growth and slippage at grain boundaries, can be eliminated through the use of single-crystal fibers. Single-crystal fibers can be produced by many different techniques, such as growth from solutions, melts, and the vapor phases. Production of single-crystal fibers is difficult, and it is likely that only three techniques will be exploited in the future: controlled drawing from the melt, vapor growth, and zone refining.

The most versatile method for producing single-crystal fibers is the recently developed laser-heated pedestal-growth (LHPG) technique (Feigelson, 1988). This is a variation on the float-zone pedestal-growth process. In the LHPG technique, a source rod having the desired composition is heated by a laser focused onto one end of the rod, creating a molten zone from which the fiber is pulled. The process does not employ a crucible. The position of the laser-heated zone is fixed in space while the source rod is translated into the laser beam at a rate necessary to conserve the melt volume. The principal advantage of this process is the ability of the laser to produce extremely sharp temperature gradients (~1000 °C/cm) that account for the high growth rates, from millimeters to tens of millimeters per hour. In addition, the steep temperature gradients, fast growth rates, and high surface-to-volume ratio of the fibers imply that the fibers cool very rapidly. This helps to suppress high-temperature solid-state phase transformations and thus can be important for stabilizing metastable forms of the material for nonstructural applications. Another of the major advantages of this technique is the complete absence of hot furnace elements that can cause contamination of the fiber. The field of single-crystal fiber growth is relatively new and has been primarily focused on fibers for optical applications. There is still need for much more work to be done relating processing conditions to defects in the crystal structure and how these defects influence fiber properties. However, the LHPG technique has demonstrated potential for producing high-quality single-crystal fibers. Czochralski fibers, by comparison, suffer from low pull rates, and therefore, without significant process improvements, it is highly likely that fibers produced by this method will be very expensive.

2.1.7 Metal Wires. Metal wires have also been used to reinforce polymer, metal, and ceramic matrices. Wires are typically made by drawing a rod of the metal through a series of dies of decreasing diameter; typical diameters range from several millimeters to $\sim 10\ \mu m$. Wires of steel, tungsten, molybdenum, and niobium and alloy wires such as tungsten-rhenium-hafnium-carbon have been evaluated for the reinforcement of superalloy materials such as iron-, nickel-, and cobalt-based alloys. The reader is guided to Jones (1989) for a review, a compilation of properties, and a comment on the status of wire-reinforced composites.

2.2 Matrix Materials

The purpose of the matrix phase is varied, depending on the intended use of the composite structure. The role of the matrix material in structural composites is to distribute and transfer the applied load to the reinforcement. Even this statement needs to be clarified, as it applies only to those composite systems in which properties of the reinforcement such as tensile strength and modulus are substantially greater than those of the matrix. Examples of such systems are carbon- or glass-fiber–reinforced polymers. Here the strength and modulus of the composite are dominated by the properties of the reinforcement. At the other extreme are systems such as ceramic matrix composites that seek to use the low-density, high-temperature, and environmental stability of bulk ceramic materials while employing a second-phase fiber reinforcement to thwart crack propagation and prevent catastrophic failure modes common to bulk ceramic materials. In both cases, the interfacial reactions between the matrix and the reinforcement are critical to developing the expected properties. Most fibers, for example, are given a special surface treatment to facilitate the interaction of the fiber with the matrix material. In some metal- or ceramic-matrix composites, a thin coating is applied to the fiber to prevent the matrix from aggressively attacking the fiber and degrading its properties. The physical and chemical nature of the fiber surface and how it interacts with the matrix both during the fabrication of the composite and during the actual in-service use determine how effectively the reinforcement is utilized and governs the primary performance of the composite; see Jones (1989), Metcalf (1974), and Kerans *et al.* (1989).

Besides the reinforcement aspects, matrix materials also must provide a whole set of process-oriented properties that permit them to be readily fabricated into a composite. These are properties such as matrix viscosity that permit infiltration of the fiber tows, processing temperatures that are compatible with the thermodynamic stability of the reinforcement, and interaction with and bonding to the reinforcing phase. These comments apply to all four basic classes of matrix materials—polymers, metals, and ceramics including glass and carbon.

2.2.1 Polymers. Polymers represent the most commonly used matrix materials. This is due to their low cost, low density, and acceptable mechanical properties, and the relative ease with which they can be processed. Polymer matrices are readily separated into two types, thermoplastic polymers and thermosetting polymers, depending on their thermorheological behavior.

Thermoplastic polymers have a relatively well-defined melting temperature and exhibit reversible softening or melting behavior. Thermoplastic matrices are processed by traditional polymer processing techniques such as compression molding and extrusion. Examples of traditional thermoplastic polymers are polyethylene, polypropylene, nylon, and polystyrene. These polymers have not been extensively used in structural composite applications, primarily because of creep at elevated temperatures. Thermoplastic matrices are commonly used in noncritical structural applications such as equipment housings and instrument panels in automobiles. Recently, polymers such as polyetherimide, polyphenylene sulfide, and polyetheretherketone have been explored as high-temperature thermoplastic matrices. The excellent temperature resistance and superior impact strength exhibited by these polymers should allow them to be used in higher-temperature structural applications (National Materials Advisory Board, 1987).

Thermosetting polymers do not exhibit the reversible melting behavior of thermoplastics. On initial heating or curing, thermosets develop extensive cross-linking of the polymer chains, and on subsequent heating, the polymer will decompose before melting. The rheological properties of the uncured polymer and the conditions necessary for initiation of the cross-linking reactions can be varied by changing the basic formulation of the thermoset. This is especially important for impregnation of fiber bundles and tows before curing. An interesting property of thermosets that makes them very useful for composite-material fabrication is their ability to be placed in a state of partial cure and held in that condition for long periods of time before being fully cured. This ability to be carried to a stage of partial cure gives the impregnated fiber sheet mechanical properties similar to those of a fabric, facilitating fabrication of complex laminated structures. Typical thermosetting polymers are epoxy, polyester, and phenolic polyimide resins. Thermosetting polymers are widely used in structural composite applications such as military and civilian aircraft, boat hulls, and automotive parts.

2.2.2 Metals. In contrast to polymers, metals offer greater strength and stiffness and higher use-temperature capability. The major purpose for fiber and particulate reinforcement in metals is to improve their high temperature mechanical properties. The most investigated metals are aluminum, magnesium, titanium, and copper, although there has also been a considerable amount of research on superalloy matrices such as nickel-based, cobalt-based and iron-nickel–based alloys (Islam *et al.*, 1988). These metals, with the exception of copper and the superalloys, are primarily chosen for their low densities. The most commonly used reinforcements are boron, silicon carbide, alumina, and carbon fibers. The fabrication of fiber-reinforced metal requires high temperatures for melting or diffusion bonding of fiber-interleaved metal foils. The interaction of the molten or hot metals with the fiber can significantly degrade fiber properties. Special diffusion-barrier coatings may be required, such as SiC coatings on boron fibers for use in titanium matrices. Similarly, carbon fibers require a protective coating before they can be used in the fabrication of aluminum-matrix composites.

The introduction of a continuous fiber into a metal matrix results in an anisotropic composite whose strength and modulus are greater along the fiber axis than in the transverse direction. Thus continuous-fiber–reinforced metals cannot be readily fabricated using traditional metal-forming machinery. The use of SiC whiskers or particulate reinforcements in aluminum has resulted in metal-matrix composites that are more nearly isotropic. A commercial application of metal-matrix composites is the use of alumina fiber to reinforce and provide superior wear resistance to the upper piston-ring groove region of an aluminum diesel engine piston (U.S. Congress, Office of Technology Assessment, 1988).

2.2.3 Ceramics. Bulk ceramic materials have excellent performance characteristics, such as superior retention of strength and stiffness at high temperature, high hardness, low density, and excellent oxidation resistance. The problems with bulk ceramic materials are their poor tensile strength, poor thermal shock resistance, and poor impact resistance, significantly limiting their scope of applicability. The development of ceramic-matrix composite materials is an attempt to impart toughness to these brittle materials through the use of fibrous and particulate reinforcements to alter the failure mode of the ceramic. The inclusion of fibers and particulates into ceramic matrices can alter the failure mode and toughen the ceramic in one of several ways (Diefendorf, 1985):

1. by increasing the local driving forces necessary to propagate a crack to failure,
2. by increasing locally the amount of mechanical energy per unit area necessary to propagate a crack, and
3. by decreasing the local strain in the matrix by cracking along fiber-matrix interfaces that reduces the stress concentration, preventing catastrophic failure.

In most structural composites, the fibers are needed to provide improved stiffness and strength. In many ceramic matrices, however, the modulus of the matrix is comparable to that of the fiber. This creates a situation in which the strain necessary for transferring the load from the matrix to the fiber is larger than the strain to failure of the matrix; thus addition of reinforcing fibers does not always result in improved properties. The addition of fibers to improve the brittle fracture characteristics of the ceramic may be viewed as the introduction of a homogeneous population of defects that impede crack formation and propagation and may require compromising other properties of the ceramic such as tensile strength. Many ceramic materials have been explored as matrices for composite structures (Schioler and Stiglich, 1986), including glass, silicon carbide, silicon nitride, alumina, mullite, and cements. One system that has demonstrated encouraging results is the use of carbon and silicon carbide fibers to reinforce glass and glass ceramic matrices such as borosilicates and aluminosilicates (Prewo, 1982). The glass and glass ceramic matrices have an important advantage over many other ceramic materials in that their viscosity can be substantially reduced by heating above the softening point so that the matrix can flow and fully impregnate the fiber bundles, producing fully dense composite structures.

2.2.4 Carbon. Almost since the introduction of carbon fibers as reinforcements for composite materials, carbon-fiber–reinforced carbon matrix (carbon/carbon) composites were envisioned (Stoller *et al.*, 1974). Carbon as graphite has excellent high-temperature properties. The high strength and stiffness of graphite are maintained to temperatures well in excess of 2700 °C. Applications such as rocket nozzles and space reentry vehicles, which experience very high service temperatures, have made great use of carbon/carbon composites. A major area of research activity in carbon/carbon composites is environmental protection. Carbon/carbon composites are susceptible to oxidative erosion when heated in oxygen-containing environments and require a protective coating such as silicon carbide.

Carbon/carbon composites are usually fabricated by liquid infiltration of a fiber preform with coal-tar or petroleum pitches. The infiltrated preform is heated to pyrolyze the pitch, leaving behind the carbon matrix. This process must be repeated several times to reduce the porosity in the carbon matrix caused by the release of volatile degradation products during the pyrolysis of the pitch. Other fabrication methods include vapor infiltration in which a carbonaceous gas infiltrates and decomposes on the heated fiber preform. This approach also has problems with porosity due to incomplete filling of the interior of the preform and sealing off the porous interior regions, as the gas phase is most rapidly depleted of the carbon species on the outermost regions of the preform.

Carbon/carbon composites have also received attention for use in biomedical applications such as artificial implant structures for replacing bone. Carbon/carbon composites are attractive for this purpose because of their light weight, excellent mechanical properties, and chemical inertness.

3. COMPOSITE FABRICATION

3.1 Fabrication Processes

There are many methods for fabrication of composite materials. This section describes the more commonly used processes for fiber-reinforced composites in a generic sense and directs the reader who desires a more in-depth description to specific references. The fabrication of a fiber-reinforced composite structure can be broken down into two basic operations:

1. placement and orientation (lay-up) of the various layers of fibers impregnated with resin and
2. consolidation of the laminate to induce bonding between the fibers and the matrix.

The fabrication of laminated structural composites is very hand-labor intensive, although significant progress has been made in automating the hand lay-up procedures.

3.1.1 Molding Processes

3.1.1.1 Contact Molding. The simplest and one of the most widely practiced forms of composite fabrication is contact molding, more commonly referred to as wet lay-up. This molding technique is more suited for fabrication using fibers in the form of fabrics and is not generally used for unidirectional composites because of limitations with selecting fiber orientation. The basic raw materials for this process are a two-part thermosetting resin system and fibers usually in the form of mats or woven fabrics. In this process, the resin is coated onto or into a mold. Fibers in the form of mats or woven fabric sheets are hand placed over the resin-coated mold, and more resin is applied. The resin is worked into the fiber mats, and air bubbles are removed through the use of squeegees or rollers. This process is repeated until the required thickness is achieved. The fiber-and-resin–coated mold is then covered with a release film to provide a smooth surface to the exposed surface. The lay-up is then allowed to cure either at room temperature or by heating.

3.1.1.2 Vacuum-Bag Molding. Vacuum-bag molding is widely used for fabrication of laminates of simple to complex geometries. The technique is an extension of the wet lay-up process, using a simple vacuum technique to remove air bubbles and to apply a modest hydrostatic pressure to the laminate to aid in consolidation and removal of excess resin. The vacuum-bag process is shown schematically in Fig. 11. Fibers in the form of resin-preimpregnated mats, tapes, fabrics, or tows are placed on a mold that has the desired form or shape. The fibers are placed on the mold in the desired orientation, and a laminate having the desired number of layers is built up; a porous release film is then placed over the laminate and covered with a bleeder fabric that takes up excess resin as the laminate structure cures. A nonpermeable bag material such as a polyvinyl alcohol film is placed over the entire structure and sealed around the edges, and a system of vents or passages is provided so that a vacuum can be drawn in the space between the mold and the vacuum bag. The bagged mold is then placed in an oven while the vacuum is maintained to consolidate and cure the laminate. Curing also can be effected by heating the mold. After curing, the vacuum bag, bleeder cloth, and release film are removed, and the laminate can be removed for further processing such as edge trimming, cutting, or further machining. The reader is guided to Schwartz (1984) for a more detailed discussion of the vacuum-bag process.

3.1.1.3 Expansion- and Pressure-Bag Molding. Pressure-bag molding is a variation of the vacuum-bag molding technique in which the fiber lay-up is constructed on a mold in a manner similar to the vacuum-bag technique. The lay-up is covered with a rubber bag that is pressurized to remove voids and excess resin during the curing step. The process is shown schematically in Fig. 12(a). An interesting modification of this process, shown in Fig. 12(b), is the use of silicone rubber inserts in the mold that expand when heated and apply pressure to the lay-up during the curing cycle. The vacuum fitting removes entrapped air and volatiles evolved during curing of the resin. This technique is frequently used in the fabrication of complex structures.

3.1.1.4 Autoclave Molding. The majority of high-performance structural composites for aerospace applications are produced by autoclave molding. Autoclave molding is also very similar to the vacuum-bag process, the difference being that the resin is cured in an autoclave held at elevated temperature and pressure. The vacuum bag allows the removal of air early in the curing cycle while the high pressure permits a higher degree of consolidation than either vacuum- or pressure-bag molding.

3.1.1.5 Injection Molding. Injection molding is widely used for volume production of short-fiber–reinforced articles. The process involves heating a thermoplastic polymer that contains reinforcing materials such as short fibers, flakes, or particulates and injecting the heated charge into a mold. The mold is cooled so that the polymer material rapidly solidifies. The mold is opened and the part is removed. This process is very flexible and has been

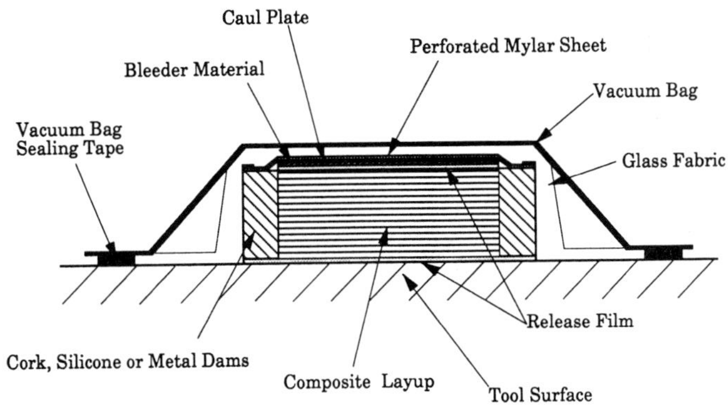

FIG. 11. Schematic diagram of the vacuum-bag molding process. After Schwartz (1984).

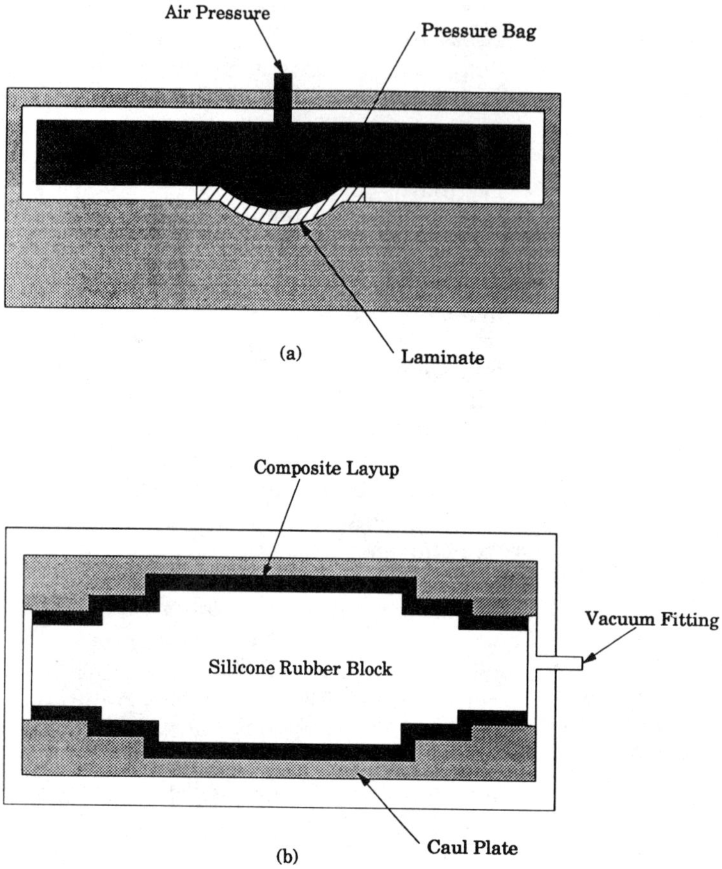

FIG. 12. Schematic diagrams of the (a) pressure-bag and (b) expansion-molding processes. After Schwartz (1984).

adapted to a variety of shapes and sizes. The process cycle time can be very short and therefore attractive for high-volume applications such as glass-fiber–reinforced automotive parts. An interesting modification of the injection-molding process is reaction injection molding, in which a two-component thermosetting polymer such as an epoxy resin is injected into a mold that is prefilled with reinforcing fibers. The two-component resin system is mixed together just before the injection point into the mold. The two components begin to react as they are forced into the mold. On completion of the injection process, the mold is then heated to complete the curing of the resin. This process is very adaptable for complex composite forms that involve the use of several different reinforcements used at selected places in the composite article. The process has been used for limited production of automotive body components and large structures such as boat hulls. The main limitation of the process is the long period necessary to cure the polymer matrix.

3.1.1.6 Compression Molding. This process is primarily used for volume production of fiber-reinforced components such as automotive fascias, fenders, hoods, and trunk lids. The process consists of loading a mold with a charge of a fiber-filled thermosetting resin commonly called sheet molding compound, closing the mold, and forcing the charge to conform to the shape of the mold, followed by heating to effect the cure of the thermoset resin. The cured part is then ejected from the mold and the process is repeated. Typical cycle times are about 30 s to several minutes.

3.1.2 Filament Winding. The filament-winding fabrication process has received a great amount of attention initially as a method for high-rate production of composite tube structures. The process consists of winding fiber tows or tapes onto a rotating mandrel. Either the winding can be carried out using preimpregnated fibers or the resin can be applied to the fiber by passing it through a resin bath as it is being wound onto the mandrel. In either case, the mandrel is covered with the desired number of plies of fiber and then placed in an oven for curing. The filament winding has been computer-automated to a large extent so that very complex shapes can be produced using multiple-axis filament-winding machines (Schwartz, 1984). The concept of automated filament winding has been generalized to the extent that machines have been produced that can lay down fiber tapes or tow in very complex patterns, removing much of the hand-labor–intensive work and producing composites that have a high degree of accuracy and precision in fiber placement.

3.1.3 Pultrusion. Pultrusion is a technique that was initially developed to produce solid shapes having a constant cross section such as round rod (Schwartz, 1984). In this process, continuous-fiber tows are pulled through a heated die that is the shape of the desired cross section. The resin cures during this heating step and a solid, uniaxially rein forced rod is produced. The major advantage of the pultrusion process is speed.

3.1.4 Infiltration and Diffusion Bonding. These techniques are commonly used for the fabrication of metal-matrix composite structures. In the vacuum infiltration process, molten metal is drawn, through the use of a vacuum, into a mold containing a fiber preform. The fibers are placed in the mold in the desired locations and orientations and the molten metal is allowed to fill the mold. The mold is cooled to solidify the metal after which the composite is removed from the mold. This process is difficult to carry out if selective positioning of the fibers is required so that they are not uniformly distributed in the mold. The process is very dependent on the viscosity of the molten metal and its ability to wet the surface of the fiber to facilitate full impregnation of the fibers to make a fully dense composite.

In the liquid-metal infiltration process the fiber tow is pulled through a crucible containing the molten metal. The infiltrated tow passes through an orifice having the desired diameter, and the metal is cooled quickly on passing through the orifice. The resulting uniaxially reinforced wire is then wound up on a large-diameter spool. The wire can be cut and used at a later time as the raw material in a hot-pressing consolidation process to fabricate a composite structure.

In the diffusion-bonding process, fibers are attached to metal foils by use of an organic binder. The foils are stacked up to form a preform of the laminate having the desired fiber orientations. The preform is placed in a collapsible container and subjected to high temperature and pressure to cause the metal matrix to flow around the fibers and bond the laminae together. The organic binder is pyrolyzed to burn off volatile components during the beginning of the heating cycle. Diffusion-bonding processes have been used to make boron-reinforced aluminum and SiC-reinforced titanium composites.

4. COMPOSITE APPLICATIONS

4.1 Military Aircraft

Historically, military aircraft have provided the driving force behind the research, development, and utilization of composite structures. Composite construction on aircraft is continually increasing from 1% of the structural weight on the F-14 to 26% on the AV-8B (Schwartz, 1984). This trend does not show any sign of changing in the near future, as many of the new military aircraft programs on composite structures are aimed at further mass reduction. For example, the goal of the Air Force Ultra-Lightweight Structures program is to reduce the structural weight of several baseline aircraft by 50%. The extended and successful use of composites in military aircraft is now spilling over to other military applications—for example, the use of ceramic-matrix composites for vehicle armor and of polymer-matrix composites for structural components on ships, submarines, and other weapon systems.

4.2 Aerospace

Carbon-fiber–reinforced polymer and metal-matrix composites have played an important role in providing lightweight, dimensionally stable platforms in space for communications satellites. These applications are primarily performance-driven by the need to reduce the mass of structural components and increase the amount of active payload. The space shuttle orbiter uses many composite components to provide the performance necessary for a reusable reentry vehicle. Programs such as the National Aerospace Plane, a passenger version of the space shuttle for very fast transcontinental flight, are relying on composite materials to provide enabling technologies. The NASA-proposed space station project will require major advances in composite-material fabrication and processing to enable a massive structure to be assembled in orbit.

4.3 Commercial Aircraft

The use of composite materials on commercial aircraft has increased as experience with composites on military aircraft has grown, with both cost and safety mediating the rate of acceptance. Fiberglass-epoxy composite fairings and control-structure applications have increased dramatically and are now being supplanted by more advanced Kevlar-epoxy composites. Graphite-reinforced epoxy composites are also being more extensively used in secondary and primary structural applications such as elevator structures, rudders, and vertical fin structures. The use of carbon/carbon composite brake disks on commercial aircraft has resulted in significant weight savings (~590 kg) and lower wear rates. The use of advanced composites on commercial aircraft is expected to increase, especially as lower-cost materials and processing are developed. The increased penetration of carbon-fiber–reinforced polymer composites into primary load-bearing structures on commercial aircraft depends on the ability to improve the properties of the present composites such as

1. toughness,
2. higher temperature capability,
3. strength and modulus,
4. environmental resistance, and
5. compressive strength.

If these problem areas are solved, the use of carbon-fiber–reinforced polymers will also increase in other applications as well.

4.4 Ground Transportation

One area in which composites potentially could have large impact is automotive applications. At present there is significant use of nonstructural composite materials on automobiles, such as sheet molding compounds reinforced with chopped glass fiber for the front and rear fascias, hoods, trunk lids, and body panels. There have been a few small-volume applications of carbon-fiber–reinforced drive shafts in light-duty vans and trucks. Continuous-filament–wound, glass-fiber–reinforced polyester leaf springs are now used in a variety of passenger cars and light trucks. The composite leaf spring represents a major structural application of composites, having a superior fatigue lifetime over the metal component. Composites have been used in a number of prototype engine parts such as pistons and connecting rods made from both graphite-fiber and silicon-carbide–particulate reinforced aluminum. Prototype engines have been made that are constructed almost entirely from polymer-matrix composites including the engine block. Carbon/carbon composite disk brakes are presently used in racing cars because of their superior performance characteristics and lighter weight. Although many prototype parts have been evaluated, carbon-fiber–reinforced composites have not penetrated the automotive marketplace primarily because of the high costs associated with the raw materials and composite fabrication. The real potential of composites to lower manufacturing and assembly costs through integration of many individual components into a single composite structure has not yet been realized because of the absence of high-rate fabrication processes, and remains a challenge to future scientists and engineers.

4.5 Recreation and Sporting Equipment

In direct contrast to the cost-sensitive automotive applications, sporting and recreational applications of composites are performance-driven. Glass fibers were first used in composite fishing rods to improve performance over bamboo and metal poles. The fiberglass was supplanted by graphite and Kevlar fibers. Tennis racquets that are constructed from graphite-epoxy composite are stiffer than wooden or aluminum racquets and do not dissipate as much energy in deformation of the racquet, allowing more energy to be given to the ball, giving the player a definite performance advantage. Boron-fiber–reinforced epoxy is also experiencing increased interest for tennis racquet applications. Advanced composites are being used in nearly every possible sporting-goods application including fishing rods, snowmobiles, motorcycles, bicycles, skis, track and field equipment, archery bows and arrows, protective helmets, and golf clubs, to mention a few. The application of advanced composites to sports and recreational equipment will continue to grow as new materials are developed, such as the extended-chain polyethylene fibers.

5. FUTURE OUTLOOK FOR COMPOSITE MATERIALS

The use of composite materials is cost- and performance-benefit–driven. Composite materials usually cost more than metals so that direct replacement of metal parts with their composite counterparts is not an effective strategy unless the benefit such as mass reduction, increased lifetime, additional performance, etc., can offset the additional cost of the composite. Substitutional strategies work for performance-driven applications such as aircraft where mass savings have a direct economic impact, such as improved fuel economy or increased payload capability. In cost-driven applications such as passenger automobiles, mass reduction does not result in a sufficient savings to the user in terms of either decreased operating costs or increased payload capability that can offset the increased cost of the composite structure. Thus, the composite must provide an economic benefit in terms of lower overall vehicle construction cost through parts consolidation and lower assembly costs; simple part-substitution strategies do not work, because of the higher cost of the composite part. In order to address high-volume, cost-driven applications, radically new engineering approaches are necessary that depart from traditional engineering practices used for metallic structures. Such new approaches are being explored as high-speed resin-transfer molding of highly integrated structures, high-speed reaction injection molding of large integrated structures, and filament winding of thermoplastic structural components (U.S. Congress, Office of Technology Assessment, 1988).

The development of new low-cost composite fabrication techniques to supplant traditional fabrication practices in cost-driven high-volume applications represents an important goal for domestic industries to achieve in order to remain competitive in the world marketplace.

In addition to the development of new composite-fabrication technology, the development of lower-cost, high-performance constituent materials will also lead to the more widespread utilization of composites. As the use of composite materials has increased, the need for a more basic understanding of the science of composite materials has also increased. Since the early 1980s, our understanding of the complex interactions between the constituents in composites has dramatically improved through the application of more sophisticated analytical techniques. The improved understanding in turn challenges researchers to develop new materials and processes.

GLOSSARY

Anisotropic: Having properties that are different when tested along different directions relative to a coordinate system fixed in the material.

Aspect Ratio: The ratio of the length of a fiber to its diameter.

Autoclave: A pressure vessel that can maintain a prescribed pressure and temperature of a gas for curing polymeric matrix composites.

Carbonaceous: Relating to or composed of carbon.

Cermet: A composite material made by mixing, pressing, and sintering a mixture of metal and ceramic powders.

Die: A tool or mold used to impart shapes to or form impressions on materials such as metals, glasses, polymers, and ceramics. An example is a die used in spinning glass fibers; a molten glass charge is forced through small orifices in the die, which shapes the glass into small continuous cylinders or fibers.

Fabric: Planar sheets of woven or non-woven materials consisting of intertwined yarns, fibers, or filaments.

Fiber: A term usually used to describe short lengths of a material having very small cross-sectional areas and aspect ratios > 1.

Filament: A fiber having an infinite aspect ratio, such as a continuous fiber.

Laminate: A composite material consisting of multiple layers of a single material or different materials bonded together.

Matrix: The principal phase in which the reinforcing phase is embedded.

Petroleum Pitch: A distillate product from petroleum used in the production of carbon fibers.

Poisson's Ratio: A constant relating the change in sample cross-sectional area to the change in length when a material is elastically deformed.

Polyacrylonitrile (PAN): A polymer of acrylonitrile used in the manufacture of carbon fibers.

Stiffness: A general term used to describe the ratio of the stress applied to a material and the deformation produced. The stiffness of an elastic material subjected to a uniaxial stress is referred to as Young's modulus.

Superalloys: Alloys based on iron, nickel, and cobalt that are an extension of stainless-steel alloys. Superalloys have higher tensile strength than stainless-steel alloys at elevated temperatures and are primarily for use at temperatures above 540 °C.

Tensile Strength: The maximum tensile force per unit area that a sample can sustain.

Thermoplastic: A material that can be repeatedly softened by increasing or hardened by lowering its temperature.

Thermosetting Polymer: A polymeric material that irreversibly converts into an infusible and insoluble solid on curing by heating or chemical means.

Tow: An untwisted loose collection of continuous filaments.

Yarn: Strands of fibers or filaments suitable for weaving.

Young's Modulus of Elasticity: The ratio of the tensile stress to the elastic strain or deformation produced.

List of Works Cited

Boyer, H. E., Gall, T. L. (1985), *Metals Handbook*, Metals Park, OH: American Society for Metals.

Bracke, P., Schurmans, H., Verhoest, J. (1984), *Inorganic Fibers and Composite Materials*, New York: Pergamon.

Broutman, L. J., Krock, R. H. (1967), *Modern Composite Materials*, Reading, MA: Addison-Wesley.

Diefendorf, J. R. (1983), *Fiber Producer*, August, pp. 56–68.

Diefendorf, J. R. (1985), in: W. Watt, B. V. Perov (Eds.), *Strong Fibers*, Vol. 1 of *Handbook of Composites*, A. Kelly, Yu. N. Rabotnov, series eds. Amsterdam: North Holland, p. 46.

Evans, C. C. (1972), *Whiskers*, M & B Monograph ME/8 series, J. G. Cook (Ed.), London: Mills & Boon.

Feigelson, R. S. (1988), *Mater. Sci. Eng.* **B1**, 67–75.

Helminiak, T. E., Arnold, F. E., Brenner, C. L. (1975), *Polym. Prepr. Am. Chem. Soc. Div. Polym. Chem.* **16**, 659–671.

Hull, D. (1981), *An Introduction to Composite Materials*, Cambridge: Cambridge University Press, p. 111.

Islam, M. U., Wallace, W., Kandell, A. Y. (1988), *Artificial Composites for High Temperature Applications*, New Jersey: Noyes Data Corporation.

Jones, F. R. (Ed.) (1989), *Interfacial Phenomena in Composite Materials 89*, London: Butterworth.

Kerans, R. J., Hay, R. S., Pagano, N. J., Parthasarathy (1989), *Ceram. Bull.* **68**, 429–442.

Lindenmeyer, P. H. (1976), in: L. E. Murr, C. Stein (Eds.), *Frontiers of Material Science Distinguished Lectures*, New York: Dekker, Chap. 6, p. 219.

Metcalfe, A. G. (Ed.) (1974), *Interfaces in Metal Matrix Composites*, New York: Academic.

Murdok, L. J., Brook, K. M. (1979), *Concrete Materials and Practice*, New York: Wiley.

National Materials Advisory Board (1987), Report of the Committee on Thermoplastic Composites as Structural Components, *The Place for Thermoplastic Composites in Structural Components*, Washington, D.C.: National Academy (NMAB-434).

North, B. (1987), *Int. J. High Technol. Ceram.* **3**, 113–127.

Petrovic, J. J., Milewski, J. V., Rohr, P. L., Gac, F. D. (1985), *J. Mater. Sci.* **20**, 1167–1173.

Prewo, K. M. (1982), *J. Mater. Sci.* **17**, 3549–3557.

Rauch, H. W., Sutton, W. H., McCreight (1966), *Survey of Ceramic Fibers and Fiberous Composite Materials*, Springfield, VA: National Technical Information Service. AFML-TR-66-365.

Schioler, L. J., Stiglich, Jr., J. J. (1986), *Am. Ceram. Soc. Bull.* **65**, 289–292.

Schwartz, M. M. (1984), *Composite Materials Handbook,* New York: McGraw-Hill.

Sowman, H. G. (1988), *Am. Ceram. Soc. Bull.* **67**, 1911–1916.

Stoller, H. M., Butler, B. L., Theis, J. D., Phillips, D. C. (1974), in: J. W. Weeton, E. Scala (Eds.), *TMS-AIME Proceedings of Symposium on Composites: The State of the Art,* Warrendale, PA: The Minerals, Metals & Materials Society, p. 69.

Tibbetts, G. G., Beetz, C. P., Endo, M. (1986), *SAMPE J.* **22**, 30–36.

U.S. Congress, Office of Technology Assessment (1988), *Advanced Materials by Design,* Washington, D.C.: U.S. GPO (OTA-E-351), p. 106.

Weeton, J. W., Peters, D. M., Thomas, K. L. (1987), *Engineer's Guide to Composite Materials,* Metals Park, OH: American Society for Metals.

Further Reading

Jones, F. R. (Ed.) (1989), *Interfacial Phenomena in Composite Materials 89,* London: Butterworth.

Kelly, A. (Ed.) (1990), *The Concise Encyclopedia of Composite Materials,* Cambridge, MA: MIT Press.

Lee, S. M. (Ed.) (1990), *The International Encyclopedia of Composites,* New York: VCH.

Schioler, L. J., Stiglich, J. J., Jr. (1986), *Am. Ceram. Soc. Bull.* **65**, 289–292.

Schwartz, M. M. (1984), *Composite Materials Handbook,* New York: McGraw-Hill.

Weeton, J. W., Peters, D. M., Thomas, K. L. (1987), *Engineer's Guide to Composite Materials,* Metals Park, OH: American Society for Metals.

COMPRESSIBLE FLOWS

J. E. SHEPHERD, *Department of Mechanical Engineering, Rensselaer Polytechnic Institute, Troy, New York, U.S.A.*

INTRODUCTION

Compressible flow is the study of fluid motion with significant induced density changes. Sufficiently strong forces will cause any real material to flow as a compressible fluid. This covers the range from the nonlinear elastic deformation of solids under high explosive loading to the flow of compressed air from an automobile tire. Compressible flows are significant in a wide range of technical devices and natural phenomena. These include high-speed flight, detonations and explosions, gas turbines, rocket engines, nuclear weapons, blast waves and sonic booms, high-velocity impact, power generation, chemical synthesis, laser fusion and high-pressure research, volcanic eruptions, and shock waves within supernovae.

Compressible flows are essentially nonlinear phenomena that arise from the interaction of thermodynamics and fluid dynamics. This article describes the general theory of compressible flow starting from the basic thermodynamic properties of sound speed, the fundamental derivative, the Grüneisen coefficient, and the conservation relations for a compressible fluid. The special case of adiabatic, inviscid flow governed by the Euler equations is given particular attention. Many practical flows can be analyzed in this approximation, since the effects of viscosity and thermal conduction are often confined to thin layers near walls or at slipstreams. Solutions to the Euler equations are observed to consist of smooth regions of isentropic flow joined by discontinuous jumps in properties through shock waves or contact surfaces.

This article begins with basic considerations about thermodynamics and fluid dynamics. The two basic elements of compressible flow, isentropic motion and shock waves, are then introduced. Steady flow within a converging–diverging or de Laval nozzle is discussed with one-dimensional theory. The complexity of real nozzle flows is illustrated with selected experimental results. Shock waves are treated from both the classical Rankine–Hugoniot discontinuity approach and the molecular viewpoint. Detonations are treated as shock waves followed by chemical reaction zones.

One-dimensional unsteady flow is illustrated with a discussion of the problem of shock-wave formation, the theory of shock tubes, and the propagation of blast waves. Steady supersonic flows in two dimensions are characterized by regions of isentropic flow

1-56081-063-7/92/$5.00+.50

(uniform motion or expansion waves) interrupted by contact surfaces or shock waves. If the shock waves are curved, a continuous variation of entropy and vortical flow will be produced. Simple flows without vorticity can be analyzed by the method of characteristics in the isentropic regions, supplemented by the shock jump conditions and matching conditions at contact surfaces.

Aerodynamic applications are considered in the context of two-dimensional steady flows over streamlined bodies. Corrections to the lift and drag due to compressibility are derived from linearized equations. Optimal airfoil design for modern aircraft illustrates the issues associated with flow near the sound speed and the interaction between shock waves and boundary layers. Flows at very high velocities are characterized by strong shock waves. Lift and drag on reentry vehicles are determined by the shock configuration and viscous heating at the body surface.

Unsteady compressible flows in two dimensions are illustrated by the problem of Mach reflection. Experimental and numerical results on the interaction of a shock wave with a ramp are discussed. The article ends with some selected results on compressible flow with phase change.

A few comments on the history and origins of compressible flow are appropriate here. Early efforts, starting with Newton, originated in the study of sound generation and propagation. Later the technological problems of steam engines and steam turbines provided motivation for the engineers and scientists of the 1800s. A genuine theory of compressible flow was possible only after the foundations of thermodynamics had been completed in 1850. Ideas from classical fields of mechanics, hydrodynamics, elasticity, and acoustics were ultimately combined with thermodynamic notions to obtain the understanding of today.

The basic foundations of compressible flow were developed in the period 1850–1910. A significant step was the experimental observation of "choking" by Napier (1867) and others and the theoretical explanation using isentropic flow ideas by Reynolds (1885). The problem of nonlinearity and wave steepening was formulated by Riemann in 1860, and the theory of the shock wave as a discontinuity was completed by Rankine (1870) and Hugoniot (1889). Supersonic flow and shock waves were revealed by Mach's experimental observations (1870–1900) and analysis of shock waves produced by high-speed bullets, spark discharges, and gas jets. Berthelot and Vielle discovered (1878–1883) the nature of the detonation process in gases and measured detonation velocities. Chapman (1899) and Jouguet (1904) analyzed the solutions to the shock jump conditions for explosive gases and determined that the minimum–wave-speed solution agreed with the experimental measurements. Vielle (1900) experimentally confirmed Hugoniot's theory for the relation between pressure rise and wave speed for moving shock waves.

Stodola's experiments and analysis of the de Laval nozzle flow (1903) clarified the role of shock waves in steady flows. This was followed by Prandtl's comprehensive experiments and analysis of the nozzle flow problem (1906). Prandtl's analysis (1908) of oblique shock waves and steady expansion waves in two-dimensional flow allowed a complete explanation of the jet phenomena first observed by Mach. These basic ideas were then extensively applied to gas turbines and supersonic flight from 1910 to 1940. Since that time, the technological problems of high-speed aircraft, gas turbines and rockets, and explosives have continued to motivate the further developments of experimental techniques and more sophisticated theory. In the last 20 years, numerical computational methods have played an increasingly important role in both engineering and research.

1. THERMODYNAMICS OF COMPRESSIBILITY

Most relevant to compressible flows is the relation between pressure and volume of a fluid, which can be determined from classical thermodynamics. The relative change in specific volume v for small changes in pressure P and entropy s about an undisturbed state $(\ )_0$ is given by

$$\frac{v-v_0}{v_0}=-K_s(P-P_0)+\Gamma K_s^2(P-P_0)^2 + G\frac{K_sT_0}{v_0}(s-s_0)+\cdots. \tag{1}$$

The quantity K_s is the *adiabatic compressibility*,

$$K_s = -\frac{1}{v}\left(\frac{\partial v}{\partial P}\right)_s, \tag{2}$$

which is inversely related to the velocity of small-amplitude adiabatic (acoustic) disturbances, the *sound speed* $c=\sqrt{v/K_s}$. Compressibility of low-pressure gases such as ordinary air is on the order of the inverse pressure or about 10^{-5} Pa^{-1} at normal atmospheric pressure. Liquid and solid compressibilities are much smaller, typically 10^{-10} Pa^{-1}.

The *fundamental gasdynamic derivative* Γ is defined (Thompson, 1988) to be

$$\Gamma = \frac{c^4}{2v^3}\left(\frac{\partial^2 v}{\partial P^2}\right)_s, \tag{3}$$

a nondimensional measure of the nonlinear response of the fluid. Values of Γ range from slightly above unity for ordinary gases up to 4–7 for typical organic liquids. Normal fluids have values of $\Gamma > 1$. In some circumstances Γ can be less than 1 and in exceptional cases Γ can be negative. The *Grüneisen coefficient* G is a nondimensional measure of thermal expansion,

$$G = \frac{1}{c_p K_s}\left(\frac{\partial v}{\partial T}\right)_P, \tag{4}$$

which has a value between 0 and 0.67 for gases and around 2 in most solids.

K_s, Γ, and G are thermodynamic properties that are determined by the molecular composition of the fluid and the extent of intermolecular interactions. Representative values of these and related properties are given in Table 1 for selected substances at conditions of $P=1$ bar (10^5 Pa) and $T=20$ °C.

Specific volume may change as a result of either applied pressure or entropy increases that are the consequence of irreversible processes such as external heating or shock waves. Many compressible flow problems of significance to applied physics are adiabatic and consist of smooth regions of *isentropic flow* with zero entropy change interrupted by thin layers or *shocks* in which the entropy irreversibly increases.

It is the variation of the sound speed with thermodynamic state that provides compressible flow with its true nonlinear character. There are two aspects of nonlinearity: the strength of the disturbances, and the degree of nonlinear response of the fluid to these disturbances. Disturbance strength is quantified by the ratio of the amplitude of the pressure variation ΔP to the quantity $1/K_s = \rho c^2$. Material response will be nonlinear when $\Delta P/\rho c^2$ becomes nonnegligible. This will occur for disturbance pressures of >0.2 bar for atmospheric air and pressures of thousands of atmospheres for substances such as water.

Nonlinearity is intimately related to Γ, both as the coefficient of the second term in Eq. (1) and by the relation to the variation (Thompson, 1971) of sound speed with pressure,

$$\left(\frac{\partial c}{\partial P}\right)_s = \frac{v}{c}(\Gamma - 1). \tag{5}$$

Wave propagation in compressible flows is profoundly influenced by nonlinearity. The dependence of propagation velocity on amplitude results in the creation of propagating sharp jumps, *shock waves*, or continually spreading *expansion waves* in the fluid state. The character and degree of nonlinearity are determined by the numerical value of Γ. The sign of Γ determines the curvature of the isentropes in (P,v) coordinates and the type of shock waves that are formed.

Table 1. Properties of some compressible materials.

	ρ (kg/m^3)	c (m/s)	K_s (Pa^{-1})	Γ	G
Air	1.205	344	7.01×10^{-6}	1.20	0.40
Water	998.2	1484	4.55×10^{-10}	4.4	0.11
Ethyl Alcohol	790	1159	9.42×10^{-10}	6.4	0.64
Steel (304S)	7896	4569	6.07×10^{-12}	2.98	2.47

2. FLUID DYNAMICS—THE EULER EQUATIONS

Changes in specific volume or mass density $\rho=1/v$ will produce fluid motion. Conservation of mass applied to a fluid element requires that a local motion inward or outward, indicated by a nonzero divergence in the velocity field $\mathbf{u}$, be produced by any change in mass within that volume:

$$\mathrm{div}\mathbf{u}=-\frac{1}{\rho}\frac{D\rho}{Dt}. \tag{6}$$

The *substantial* or *convective* derivative used in this expression is defined as $D(\cdots)/Dt=\partial(\cdots)/\partial t+\mathbf{u}\cdot\mathrm{grad}(\cdots)$ and represents the time derivative taken in a frame of reference moving with the fluid.

The complete description of the motion is determined by the thermodynamic properties discussed above together with the balance equations for mass, momentum, and energy including the appropriate boundary and initial conditions, and thermodynamic and transport properties (see FLUID DYNAMICS, EQUILIBRIUM). An approximation of wide applicability is that of inviscid, adiabatic flow. Such flows are determined by solutions to the *Euler equations*:

$$\frac{\partial}{\partial t}\rho+\mathrm{div}(\rho\mathbf{u})=0, \tag{7}$$

$$\frac{\partial}{\partial t}\rho\mathbf{u}+\mathrm{div}(\rho\mathbf{uu})=-\mathrm{grad}P, \tag{8}$$

$$\frac{\partial}{\partial t}\rho\left(e+\frac{|\mathbf{u}|^2}{2}\right)+\mathrm{div}\left[\rho\mathbf{u}\left(e+\frac{P}{\rho}+\frac{|\mathbf{u}|^2}{2}\right)\right]=0. \tag{9}$$

Pressure plays a special role. As in incompressible flow, the pressure field plays a mechanical role in the balance of momentum as a force field that accompanies material acceleration. An equally important role for pressure in compressible flows is as a thermodynamic variable, which is implicitly coupled to both internal energy $e=e(P,\rho)$ and the mass density through the Euler equations. This dual role of pressure is responsible for many of the special features of compressible flow, including the phenomena of wave propagation and the interconversion of thermal and mechanical energy.

An equation of state for the internal (thermodynamic) energy $e=e(P,\rho)$ must be supplied to complete the Euler-equation model of compressible flow. High-pressure experiments or molecular-dynamics simulations are required for solids and liquids to obtain the necessary thermodynamic information. Ideal models of thermodynamic behavior can often be applied in the case of gases. Most often studied has been the compressible flow of a *perfect gas*, an ideal gas with constant specific heats. Ordinary air and most other gases can be described by this model up to moderate pressures (50 bar) and temperatures (1000 K). Only one parameter is required for this model, the ratio of specific heats $\gamma=c_p/c_v$, which takes on values between 1 and 1.667. The equation of state is

$$e=Pv/(\gamma-1). \tag{10}$$

Temperature can be determined by the usual ideal-gas law $P=\rho RT$. The various derivatives significant to compressible flow behavior are

$$K_s=1/\gamma P,$$

$$c=\sqrt{\gamma Pv},$$

$$\Gamma=\tfrac{1}{2}(\gamma+1),$$

$$G=\gamma-1. \tag{11}$$

Solution of the Euler equations with the perfect-gas equation of state is the subject of classical *gasdynamics*. These equations must be solved numerically except for special cases in which analytic solutions have been obtained for steady flows and self-similar transient flows. Solutions to compressible flow problems have the special feature of the spontaneous generation of discontinuities such as shock waves and contact surfaces.

Fluid composition must be specified in order to analyze chemically reacting flows such as detonations. This requires an equation of state appropriate to a mixture of species and an evolution equation for the composition. Changes in the species concentrations are due to simultaneous material transport and chemical reaction. A reaction-rate model must be developed from molecular considerations of the processes of reactive collisions (combustion chemistry), vibrational–translational energy exchange, and electronic excitation or ionization dynamics. Supersonic and hypersonic vehicle aerodynamics, spacecraft heating during reentry, high-speed flames in gases, and detonation of explosives (gaseous and condensed) (see EXPLOSIVES) offer a rich variety of applications that motivate the study of *compressible reacting flows*.

3. CRITERIA FOR COMPRESSIBILITY

Under what circumstances is a flow considered compressible? Combining our thermodynamic analysis with the Euler equations shows that the nondimensional pressure $\Delta P/\rho c^2$ is of the same magnitude as the *Mach number* $M=u/c$ for steady flows. Therefore, the Mach number must exceed some minimum value for compressibility to be considered an essential feature of a steady flow. A value of $M>0.3$ is often taken as a criterion for ordinary gases. A more important distinction is between *subsonic* flow, $M<1$, and *supersonic* flow, $M>1$, with the case of *sonic* flow, $M=1$, playing a special role.

In the case of unsteady flows, compressibility effects must be considered whenever the acoustic wave transit time L/c across a characteristic distance L is comparable to or greater than other significant time scales present in the flow. Small-amplitude motion with $M\ll 1$, $\Delta\rho/\rho \ll 1$, and $\Delta P/\rho c^2 \ll 1$ can be described by the theory of *acoustics* (see ACOUSTICS, LINEAR). The approximation of acoustics is to neglect the convective derivative terms compared to the unsteady derivatives in the Euler equations and to treat the motion as isentropic. Acoustic flows are described by the linear wave equation, for which a large number of exact solutions can be obtained. Large-amplitude motions are intrinsically nonlinear and require experimental or numerical investigation (see ACOUSTICS, NONLINEAR).

4. SIMPLE FLOWS

Many technical applications of compressible flow require only simple notions about the dynamics of motion in one dimension. Useful information can be obtained by considering only variations in one spatial direction even in the case of apparently multidimensional flow within a nozzle or a tube of variable cross section. Transient flows in ducts of a constant cross section play an important role in the operation of devices such as shock tubes, mufflers, and the phenomenon of water hammer. Shock waves can be analyzed by considering only the one-dimensional structure normal to the wave front, even in cases of curved waves. Detonation waves and transient shock waves (blast waves) resulting from explosions can also be productively examined from a one-dimensional viewpoint.

4.1 Quasi One-Dimensional Flows

A typical application of compressible flow is accelerating fluids from rest to high velocities. This can be accomplished through a de Laval nozzle, first used in steam turbines. It has since been widely applied in the generation of supersonic flow in devices such as wind tunnels and rocket motors. Fluid at high pressure is accelerated through a duct that converges and then diverges as shown in Fig. 1.

Simple explanations of the principal features of these flows are possible by assuming a slowly varying duct cross section and steady-state ($\partial/\partial t=0$), adiabatic flow. This ensures that the flow is almost parallel and that the

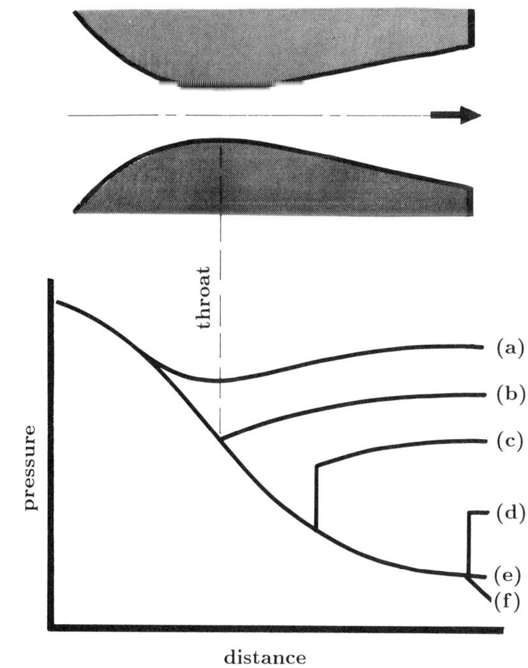

FIG. 1. Flow through a de Laval nozzle and the associated changes in the fluid pressure. Each curve represents a different downstream pressure as discussed in the text.

accelerations perpendicular to the main flow direction are also small. In this *quasi one-dimensional* approximation, the Euler equations can be compactly written as

$$\frac{d}{dz}(\rho u A)=0,$$

$$\frac{dP}{dz}=-\rho u\frac{du}{dz},$$

$$\frac{d}{dz}\left(h+\frac{u^2}{2}\right)=0, \tag{12}$$

where $A(z)$ is the duct cross-sectional area perpendicular to the main flow direction z. In this situation, the total mass flow rate is constant and equal to ρuA. The energy equation is written in terms of the *enthalpy* $h=e+P/\rho$. The combination $h_t=h+u^2/2$ is known as the *total* or *stagnation* enthalpy, which is constant in adiabatic, steady flows.

If the motion is inviscid and no shock waves are present, adiabatic flow can be further simplified as isentropic, and variations in pressure and density can be related by Eq. (1) to obtain the isentropic flow equations

$$\frac{d\rho}{dz}=\frac{1}{c^2}\frac{dP}{dz}, \quad \frac{1}{u}\frac{du}{dz}=\frac{1}{M^2-1}\frac{1}{A}\frac{dA}{dz}. \tag{13}$$

For initially subsonic flow, $M<1$, as in the case of accelerating a fluid from rest, a converging duct $dA/dz<0$ will increase the fluid velocity as the pressure and density decrease. This acceleration is equivalent to the conversion of thermal energy into mechanical energy, as shown by the integrated form of the energy equation $u=\sqrt{2(h_t-h)}$, where the total enthalpy h_t is just the enthalpy of the stationary fluid in the reservoir supplying the duct.

The *stagnation state* is a hypothetical state produced by isentropically bringing a flow to rest (stagnation). If the fluid reservoir supplying the nozzle is large enough, then the pressure P within the reservoir will be equal to the *stagnation pressure* P_t. Stagnation pressure in a steady, isentropic flow is constant. The stagnation pressure in a steady flow can be changed only by external work, energy addition, or dissipative processes such as friction or shock waves.

The Mach number will increase monotonically with the velocity for usual fluids, which have $\Gamma>1$, since

$$\frac{du}{u}=\frac{1}{1+(\Gamma-1)M^2}\frac{dM}{M}. \tag{14}$$

Given sufficient stagnation pressure, continuous contraction in the duct cross section will result in the Mach number approaching 1, a phenomenon known as *choking*. In a constant-area duct, choking can also occur as a result of frictional forces or heat addition.

It is apparent from the differential relation equation (13) for velocity that transition through the *sonic point*, $M=1$, is possible only at an area extremum. This takes the shape of a minimum forming a *throat* in the nozzle for the case of normal fluids, $\Gamma>0$. If the downstream pressure is sufficiently low, as in cases (c)–(f) of Fig. 1, sonic flow will occur at the throat and the flow downstream will be supersonic for some distance. In the supersonic flow downstream of the throat, increasing the duct area results in further velocity and Mach-number increases. Pressure, temperature, and density will decrease with increasing Mach number in a supersonic flow.

For a given nozzle and stagnation pressure, the flow within the nozzle depends crucially on the pressure at the exit of the nozzle. If the flow is subsonic at the nozzle exit, the exit pressure will be approximately equal to the ambient pressure in fluid outside the nozzle. The situation is more complex for sonic or supersonic exit flows and the exit pressure is, in general, not equal to the ambient pressure.

Several possible situations are shown in Fig. 1. First, in case (a), the ambient pressure must be less than the stagnation pressure in order for flow to take place. Second, if the ambient pressure is not too low, then the flow remains subsonic throughout. This is the case for exit pressures above case (b) of Fig. 1. Third, when the ambient pressure is sufficiently low, cases (c)–(f), the flow will be *sonic* at the nozzle throat ($M=1$) and supersonic ($M>1$) for some distance downstream. Sonic flow at the throat will occur when the ratio of stagnation pressure to throat pressure exceeds a critical value. This value is determined by the thermodynamic properties of the fluid and can be estimated to be about 1.9 for air, 5300 for water, 2000 for ethyl alcohol, and 5.21×10^5 for steel. The pressure required at the nozzle exit to first produce this state, case (b), must be determined from the specific nozzle geometry and operating conditions.

Supersonic flow will exist for some distance downstream of the throat for all cases with an exit pressure lower than case (b). For exit pressures between cases (b) and (e), the nozzle flow is termed *overexpanded.* For exit pressures between those of cases (b) and (d), a special situation within the nozzle must exist. A sudden increase in pressure occurs through a nonisentropic *shock wave,* discussed in detail in the next section. The shock wave compresses and decelerates the fluid, resulting in subsonic flow downstream with the pressure increasing with distance up to the exit value. There is a loss of total pressure across the shock wave and an increase of entropy. The shocks are idealized as sharp discontinuities normal to the flow in Fig. 1, but in reality these waves are usually obliquely oriented to the flow (described in Sec. 7.1). An additional feature of real nozzle flow is the interaction between the shock and the boundary layer at the nozzle surface. This interaction can result in separation of the flow from the nozzle surface and the formation of a recirculation or reverse-flow region downstream of the interaction point.

For pressures between cases (d) and (e), shock waves will form in the jet outside the nozzle. The *pressure-matched* situation of case (e) has a smooth acceleration of flow up to nozzle exit where the pressure is exactly equal to ambient. Case (f) is an *underexpanded* flow and the nozzle exit pressure is higher than ambient. Further expansion must take place outside the nozzle. Conservation of energy limits the maximum fluid velocity that can be achieved by expansion to zero pressure to be $u_{\max} = \sqrt{2h_t}$. However, for ordinary fluids, the Mach number will be infinite in this state.

Analytic and tabulated solutions (Brower, 1990) are available for the thermodynamic states and flow variables of quasi one-dimensional isentropic flow of a perfect gas. For other fluids, each case can be treated numerically given sufficient thermodynamic information to compute enthalpies and sound speeds on isentropes. Approximate similarity principles based on Γ are available (Thompson and Sullivan, 1977) as guidelines for liquids and dense gases.

Real fluid flows in nozzles are multidimensional and have complicating factors such as condensation in moist air and a viscous boundary layer on the duct walls. Accurate computations of the flow should include the displacing effect of this layer and the motion transverse to the main (axial) flow direction. Realistic two-dimensional steady-state computations including these effects have been carried out and are in quantitative agreement with experiments. Figure 2 shows a schlieren photograph of the flow (Schnerr, 1989) of moist air through a two-dimensional circular arc (radius 100 mm) nozzle with a throat width of 120 mm. The flow is from left to right. Dark regions in the photograph are expansions; light regions represent compressions.

The fluid in the reservoir was at a pressure of 1 bar and temperature of 295 K, and had a relative humidity of 37.2%. The rapid cooling rate ($>10^6$ K/s) due to the expansion in the nozzle causes supersaturation and homogeneous nucleation of the water vapor. This occurs downstream of the throat in the supersonic flow region near a Mach number of 1.2 to 1.3, resulting in energy addition to the flow from the heat of condensation. Addition of thermal energy to the supersonic flow causes

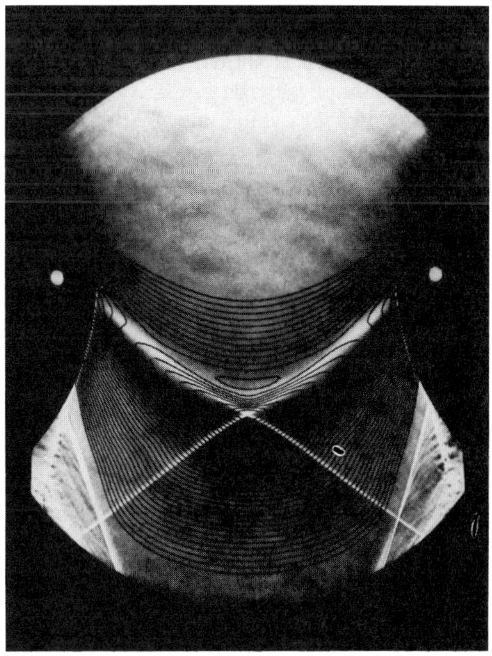

FIG. 2. Experimental visualization of the flow of moist air through a circular arc nozzle with superimposed calculated iso-Mach contours.

the density to increase and oblique shocks to form. These form the bright X-shaped pattern visible in the photograph. Directly downstream of the shocks, a rapid expansion takes place. Unrelated to condensation, but equally important in real nozzles, is the wedge-shaped separated-flow region visible at the downstream edges of the nozzle arcs. Lines of constant Mach number are superimposed on the flow image. These were computed from the numerical solution (Schnerr and Dohrmann, 1990) of the Euler equations supplemented with a condensation model. These contours start with the sonic line, $M=1$, at the left and increase in increments of 0.02 to the right.

The supersonic acceleration process in a nozzle is limited only by the decoupling of translational and internal energy modes that occurs when the density becomes too low for molecular collisions to be effective. When the supersonic expansion occurs in a very low-pressure region (vacuum), the resulting flow is known as an *underexpanded supersonic jet*, illustrated in Fig. 3. This is the multidimensional flow resulting from a converging nozzle with choked flow ($M=1$) at the exit. The sonic exit flow is further accelerated through an expansion fan (discussed subsequently in Sec. 7) after exiting from the nozzle.

The effect of a small but not negligible ambient pressure is shown in Fig. 3. Within the expansion flow, the pressure decreases below the ambient value and is brought back up by a complex system of shock waves. This shock "cell" consists of a barrel-shaped shock surrounding the jet and a nearly planar normal shock, or Mach disk, downstream. The boundary between the jet and ambient fluid is outside of the barrel shock and actually is a region of mixing or a *shear layer*, as indicated by the shaded region. There are large changes of density across this boundary, resulting in the pressure-wave reflections shown in Fig. 3. There are a series of shock cells that can be observed within the jet flow at lower pressure ratios in sonic and supersonic jets, corresponding to cases between (d) and (e) of Fig. 1.

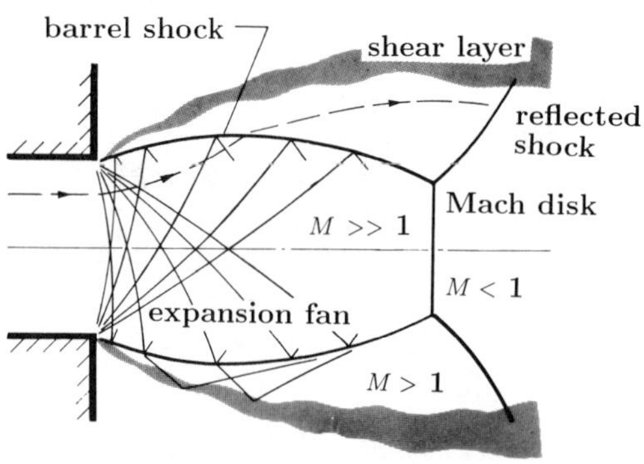

FIG. 3. Features of a free jet expansion from a sonic flow produced by a converging nozzle. The pressure at nozzle exit is much higher than the pressure of the stationary ambient fluid.

Free jet expansions are used to create very low-temperature, supersonic molecular beams essential to modern chemical physics (Levine and Bernstein, 1987). A population inversion can be produced in the molecular vibration–rotation states if the expansion process is sufficiently rapid. This can create the conditions necessary for light amplification, the principle behind the *gasdynamic laser* (Anderson, 1976) (see LASERS, GAS).

4.2 Shock Waves

Steady supersonic flow and transient compressible flows are distinguished by the potential for spontaneous generation of shock waves. Shock waves occur in steady supersonic flows whenever flow deflections are associated with pressure increases. In transient flows, rapid energy addition or boundary motions will produce traveling compressive disturbances that may form shock waves if sufficiently strong. Steady and unsteady shocks are a consequence of the nonlinearity of compressible flow and the underlying *hyperbolic* nature of the Euler equations. In the context of the Euler equations, shock waves appear as jumps or discontinuities in the solutions. Physically, shock waves are found to have a finite thickness, on the order of microns for gases at atmospheric pressure and nanometers for liquids or solids.

All shock waves can be locally analyzed by focusing on a small spatial region (control volume) surrounding the wave and by transforming to a coordinate system moving at the local shock velocity. Conservation of momentum implies that only the component of velocity u normal to the wave experiences change. The tangential component of velocity must be continuous. For simplicity, the tangential component is set equal to zero in this analysis. Including a nonzero tangential velocity leads to the concept of oblique shock waves, discussed in subsequent sections. The analysis carried out here is valid for the normal component of velocity in both cases. Irrespective of the shock orientation or motion, only a single set of equations is required.

Integration of the Euler equations through the thin control volume around the shock yields the *jump conditions*

$$\rho_1 u_1 = \rho_2 u_2, \quad (15)$$

$$P_1 + \rho_1 u_1^2 = P_2 + \rho_2 u_2^2, \quad (16)$$

$$h_1 + \frac{u_1^2}{2} = h_2 + \frac{u_2^2}{2}, \quad (17)$$

in which the subscripts denote states upstream (1) and downstream (2) of the wave and u is the fluid velocity normal to the wave surface in a wave-fixed coordinate system. These equations can be combined into a single expression, the *Rankine–Hugoniot equation*:

$$h_2 - h_1 = (P_2 - P_1)(v_2 + v_1)/2. \quad (18)$$

In the wave-fixed frame, the velocity component normal to the wave is always supersonic, $M_1 = u_1/c_1 > 1$, upstream and subsonic, $M_2 = u_2/c_2 < 1$, downstream. The flow can be supersonic downstream of shock waves oriented obliquely (see Sec. 7.1) to a supersonic flow, since only the normal component of velocity changes across a shock. Flow behind a moving shock wave may also be supersonic relative to a stationary observer. Pressure, density, and temperature always increase across shocks in normal fluids.

Solutions to this equation are known as *shock adiabats* or *Hugoniots*. Shock adiabats can be determined from thermodynamic data $h(P,v)$ by solution of the Rankine–Hugoniot equation. In the case of the perfect-gas model, analytic solutions can be obtained and tabulations are available for many values of the specific-heat ratio γ. The results for $\gamma = \frac{7}{5}$, used for low-temperature air, are shown in Fig. 4. For perfect gases, pressure and temperature increase indefinitely as the shock strength M_1 is increased. However, the density and Mach number reach finite, limiting values ($\rho_2/\rho_1 \to 6.0$ and $M_2 \to 0.378$) for large shock strengths. For a general perfect gas,

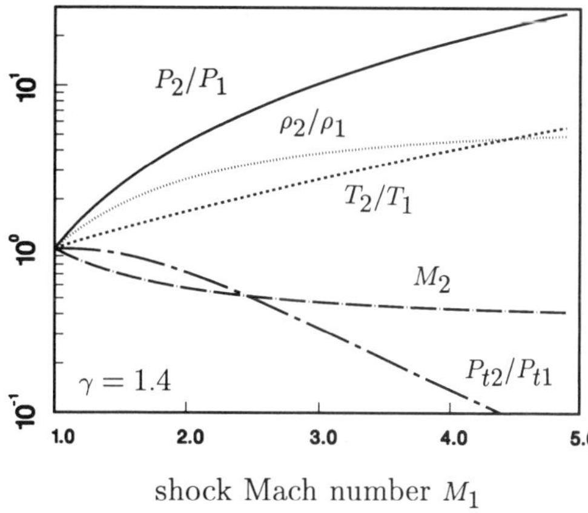

FIG. 4. Solutions to shock jump conditions, perfect-gas model, $\gamma = 1.4$. The Mach numbers M_1 and M_2 are for the velocity component normal to the shock front as measured in a coordinate system attached to the wave.

$$\frac{\rho_2}{\rho_1} \to \frac{\gamma+1}{\gamma-1} \text{ and } M_2 \to \sqrt{\frac{\gamma-1}{2\gamma}} \text{ as } M_1 \to \infty. \quad (19)$$

These limits are consequences of the equipartition of the energy in the incoming flow into internal and kinetic energies in the shocked fluid when the shock strength is large. Stagnation pressure, P_{t2}/P_{t1} in Fig. 4, always decreases across shock waves. This is due to the entropy increase across the shock and represents a source of losses in supersonic flows containing shocks.

For dense fluids and liquids, numerical solutions can be readily obtained using tabulated or empirical equations of state. In the case of solids, the experimental measurements needed for Rankine–Hugoniot computations are often obtained from actual shock-wave propagation measurements. Consider a shock wave propagating with velocity U_s through a material initially at rest. Behind the shock wave, the fluid particles are set in motion with velocity u_p. This moving shock can be considered as equivalent to a stationary wave through a Galilean transformation, yielding the velocities $u_1 = U_s$, $u_2 = U_s - u_p$. Shock data for solids are usually given in form of tabulated (U_s, u_p) pairs. These data (March, 1980) are often found to obey the empirical expression $U_s = c_0 + s u_p$, in which s is an empirical parameter, approximately equal to $\Gamma/2$. Shock data are frequently combined with information about G and K_s to produce an approximate *thermoelastic* or *Mie–Grüneisen* (Zel'dovich and Raizer, 1966) equation of state used for numerical solutions of the Euler equations.

A shock wave is an example of an irreversible natural process and, in addition to the jump conditions, must satisfy the second law of thermodynamics,

$$s_2 > s_1. \tag{20}$$

The amount of entropy generated by the shock depends on the strength of the shock wave. This can be measured by either the shock Mach number U_s/c_1 or any of the nondimensional jumps such as $[P]/\rho_1 c_1^2$ or $[v]/v_1$, where $[P]=P_2-P_1$, $[v]=v_2-v_1$. Shock adiabats are initially close to isentropes, so that weak shock waves can be considered as acoustic disturbances. Shock adiabats progressively deviate from the isentropes as the shock strength is increased. The entropy jump $[s]=s_2-s_1$ is third order in the shock strength,

$$\frac{T_1[s]}{c_1^2}=\frac{\Gamma_1}{6}\left(\frac{[P]}{\rho_1 c_1^2}\right)^3+\cdots , \tag{21}$$

so that the effect of nonlinearity [the second term in Eq. (1)] is more significant than entropy generation in weak shock waves. It is this condition that dictates that shock waves in normal fluids ($\Gamma > 0$) are compressive waves and in unusual fluids ($\Gamma < 0$) are expansive (rarefaction) waves.

A mechanical interpretation of the entropy jump across the shock is possible if we consider the molecular processes involved. Large gradients of temperature and velocity occur within the shock. Molecular transport of energy and momentum across these gradients results in dissipation throughout the shock front region. The dependence of $[s]$ on shock strength given above suggests that the shock layer becomes very thick for weak shocks, $M_1 \to 1$. Experimental studies of shock structure in gases confirm this, and simple computations demonstrate (Thompson, 1983) that the thickness of strong shocks in gases can even be predicted by continuum estimates of the dissipation rate. The detailed structure of such waves does require the solution of kinetic equations (Bird, 1976) describing the evolution of the distribution function of the molecular velocities. In the case of shocks in complex molecular systems, coupling between vibration–rotational and translation states can result (Vincenti and Kruger, 1986) in unusual wave structures. Low-amplitude waves are *fully dispersed* structures that contain continuous compression waves. Intermediate-amplitude waves are *partially dispersed* structures with both continuous compression and shock jumps. Large-amplitude waves consist of a shock followed by a zone of equilibration ("relaxation") between energy excitation modes.

Compression and entropy generation in strong shock waves will result in a substantial temperature increase across the shock wave. Strong shock waves ($M > 5$) can be produced in gases by use of laboratory shock tubes. The temperature jump across such waves is sufficient (> 1200 K) to initiate molecular dissociation and chemical reactions rapidly (Zel'dovich and Raizer, 1966). Explosion (see EXPLOSIVES) or detonation can occur when strong shock waves are created in combustible mixtures. Ionization and radiative transport of energy are important features of very strong shocks (Zel'dovich and Raizer, 1966), particularly in determining the structure of the shock front.

High explosives or high-velocity impact are used in solids to generate strong shock waves. A variety (Davison and Graham, 1979) of unique electrical, mechanical, and chemical effects can be produced in this fashion. It is possible to simulate conditions (pressures of up to 3.5 Mbar) encountered in geological materials (Jeanloz, 1990) deep within planetary cores. Unique phenomena (Liepmann and Laguna, 1984) occur when superfluid liquid helium II is shocked. The two-fluid nature of helium II gives rise to two shock waves that travel at distinct velocities. One shock wave is a pressure wave with very little temperature rise and the other is a temperature wave at nearly constant pressure.

5. DETONATION WAVES

Detonation waves can be envisioned as shock waves induced and sustained by exothermic chemical reaction, i.e., combustion. The high temperature behind the shock wave initiates the combustion reactions, and the volume expansion resulting from the reaction acts as a piston to drive ahead the shock wave. Gas, liquid, and solid explosives are all capable of detonation and have been extensively studied (Fickett and Davis, 1979; Gibbs and Popalato, 1980). Detonation waves are observed under many conditions to travel with a constant velocity characteristic of the material and the initial conditions. Detonation velocities are 2–3 km/s for gases and up to 8 km/s in solids or liquids. Pressures behind gaseous detonations are 15–30 times the initial pressure. Pressures behind solid or liquid explosives can be as large as 400 kbar.

Detonation velocities can be predicted by supposing that the fluid changes composition within the shock layer and is in chemical equilibrium downstream. Detonation adiabats can be obtained by combining the shock jump conditions with an appropriate multicomponent equation of state and chemical equilibrium relations for the downstream state. The solution for shock velocity is nonunique with the exception of a single distinguished case, the minimum possible wave speed, known as the *Chapman–Jouguet velocity*. The Chapman–Jouguet solution has the special feature that fluid velocity downstream is sonic with respect to the wave and the entropy jump is an extremum. Physically, this can be interpreted as providing a sonic barrier to prevent acoustic disturbances from disrupting the wave from the rear.

Computations of the Chapman–Jouguet (CJ) velocity are readily carried out for gases by numerical methods. Experimental velocities are found to be close (0–5 % below) to the theoretical values computed in this fashion. Detonation products for solid and liquid explosives are in a highly compressed dense-gas state and the thermodynamic computations are less reliable than in gases. Modern statistical mechanical methods based on thermodynamic perturbation theory (Ree, 1984) are moderately successful for detonation velocity prediction in these cases. Of great importance to high explosives is the yielding of the confining container, which can produce a significant influence on the propagation velocity and even result in detonation failure. As the diameter of a cylindrical charge of high explosive is decreased, the detonation velocity decreases until a *critical* or *failure diameter* is reached. Steady detonation is found to be impossible for charges smaller than the failure diameter (Gibbs and Popalato, 1980).

Zel'dovich, von Neumann, and Döring (ZND) independently formulated the theory (Fickett and Davis, 1979) of steady, laminar detonation waves. In this ZND model, the detonation consists of a nonreactive shock wave traveling at the CJ velocity, followed by a reactive boundary layer or reaction zone in which reactants are converted to products. However, both the CJ and ZND models conceal the true unsteady character of detonations. It is observed experimentally and predicted from stability analyses that transverse shock-wave disturbances spontaneously appear on a planar detonation front and that propagating gaseous detonations almost invariably display this instability. The quasiperiodic system of transverse waves forms a cellular pattern with a characteristic wavelength that is indicative of the chemical sensitivity of the mixture. Computations using detailed chemical kinetic models of the combustion process demonstrate (Westbrook and Urtiew, 1982) that the instability wavelength is proportional to the length of the chemical reaction zone computed on the basis of the ZND model. Other types of transient behavior such as initiation and failure are intimately linked (Lee, 1984) to the length scale and dynamics of the reaction zone. The simplest notion is that explosives with larger reaction zones are the least sensitive, have the greatest instability wavelength, and have the largest failure diameters. These ideas are relevant to both gaseous and condensed explosives.

6. ONE-DIMENSIONAL UNSTEADY FLOWS

Analytic solutions to the unsteady Euler equations can be obtained for flows with spatial variation in only one direction. This is particularly true for planar flow since the equations can be written in *characteristic form* and solved exactly for several cases of technical importance. The characteristic form of the Euler equations can be compactly written in the case of nonreactive flow as

$$\frac{du}{dt} \pm \frac{1}{\rho c}\frac{dP}{dt} = \mp \frac{n}{x} cu \quad \text{on} \quad \frac{dx}{dt} = u \pm c, \tag{22}$$

$$\frac{ds}{dt} = 0 \quad \text{on} \quad \frac{dx}{dt} = u. \tag{23}$$

The dimension index n takes on a value of 0 for planar flow, 1 for cylindrical flow, and 2 for spherical flows. The time derivatives in Eqs. (22) and (23) are to be interpreted as derivatives obtained by an observer traveling on the lines in space-time defined by the velocities $u \pm c$ and u, known as the *characteristics* $C^{\pm}$ and C^0. Physically, the characteristics associated with velocities $u \pm c$ correspond to the speed of adiabatic disturbances in pressure and velocity *relative to the flow*. Entropy disturbances are associated with fluid motion itself and travel with the particle velocity u. In the present case of adiabatic flow, the entropy of each element remains unchanged with the exception of passage through shock waves. Equation (23) does not apply at the location of the shock, and the jump conditions, Eqs. (15)–(17), must be applied to determine the changes across the shock.

A set of characteristics passes through each point in the flow. The wedge-shaped region determined by characteristics C^+ and C^- coming from the past defines the region of fluid that can influence motion at that point. The projection of the characteristics into the future determines the region that will be influenced by disturbances at that point. For adiabatic, nonreactive flows it is possible to define a quantity F

$$F(P,s_0) = \int_{P_0}^{P} \frac{dP'}{\rho c}, \tag{24}$$

and to simplify further the equations for pressure and velocity to read

$$\frac{d}{dt}(u \pm F) = -\frac{n}{x}\, cu \quad \text{on} \quad \frac{dx}{dt} = u \pm c. \tag{25}$$

The right-hand side vanishes in planar flows, $n=0$, and the quantities $u \pm F$ are known as the *Riemann invariants*. The Riemann invariants are constant on the corresponding characteristics, except when the characteristics cross shock waves. The value of the invariant must change across the shock in order to be consistent with Eqs. (15)–(17). For the perfect-gas model, the quantity F can be computed in closed form and is equal to $2c/(\gamma - 1)$.

As an application of the theory, consider the initial-value problem associated with a discontinuity in pressure separating uniform regions of gas, the so-called *Riemann problem* of compressible flow. This situation occurs in *shock tubes* (Oertel, 1966), used in the laboratory study of shock waves. A closed, gas-filled tube is divided into two sections by a thin diaphragm of metal or plastic. The *driver* section is filled with high-pressure gas. On the other side of the diaphragm is the low-pressure *driven* section. A space-time diagram and the characteristic network associated with this process are shown in Fig. 5.

The diaphragm is ruptured by either overstress or mechanical puncturing. Referring to Fig. 5, a compression wave (shock) propagates into the low-pressure region (1) and an expansion wave, denoted by the "fan" of characteristics C^-, propagates into the high-pressure region (4). A contact surface (dashed line) separates shocked gas (2) from the accelerated driver gas (3). The shock reflects from the end wall to produce a stagnant region (5) in which testing is usually carried out. The reflected-wave interaction (a) with the contact surface produces a secondary shock between regions (5) and (6) that limits the test time to a value τ as shown. The actual test time is significantly influenced by real fluid effects such as the boundary layer behind the shock wave. By varying the pressure and gas composition in driver and driven sections, a wide range of shock strengths can be produced.

For an ideal shock tube, a shock wave is formed immediately on the low-pressure side of the diaphragm. However, if the diaphragm in our example shock tube were to rupture slowly, a shock-free compressive wave could be produced that ultimately would steepen and form a shock wave at some distance from the diaphragm. This process is illustrated in Fig. 6 for a piston (P), analogous to the contact surface in the shock tube, being accelerated into an initially undisturbed region. The particles are accelerated when the first signal arrives along the C^+ characteristic as shown by the trajectory (dashed line) or particle path PP. The slope of the C^+ characteristics decreases as time increases and the piston accelerates. Adjacent characteristics first cross at point (a), producing a shock wave. The point of shock formation can be computed from the piston path and the Riemann invariant on both C^+ and C^-.

The nonlinearity of convection and the thermodynamic dependence of sound speed [Eq. (5)] are both significant to shock formation. Solution of the initial-value problem for

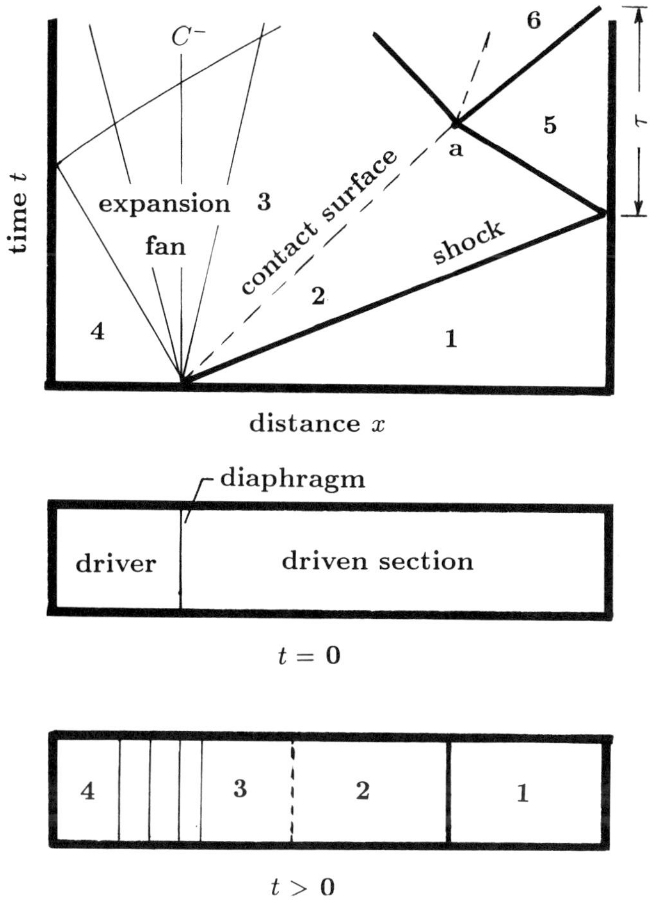

FIG. 5. Space-time (x-t) diagram for shock-tube operation.

a simple compressive wave, as shown in Fig. 6, yields the result that the characteristic velocity changes with pressure,

$$\frac{d}{dP}(u+c)=\frac{\Gamma}{\rho c}. \tag{26}$$

This leads to the following conclusions about the evolution of pressure disturbances. Compression shocks form when $\Gamma > 0$, shocks never form in the exceptional case $\Gamma=0$, and expansion shocks form if $\Gamma<0$.

Another feature of the Riemann problem is the spreading wave or *expansion fan* that propagates into the high-pressure region. This flow is characterized by a self-similar motion dependent only on the coordinate x/c_0t, which results in acceleration of fluid towards the shock wave. The fluid isentropically expands as it accelerates. This unsteady expansion process will result in a maximum particle velocity or *escape speed* of $u_e=2c_0/(\gamma-1)$ for perfect gases. This is achieved when the low-pressure region is a vacuum. An interface or *contact surface* is created between the fluid originally in the high-pressure region and the shocked fluid. Pressure and velocity are continuous across this interface, but the entropy and composition may change arbitrarily.

In the Riemann (shock-tube) problem, the key elements of many compressible flow problems appear: spontaneous formation of contact surfaces, expansion waves, and shock waves. These features are universal and appear in both one- and higher-dimensional flows. The shock-tube flow is the prototype for many similar technical problems such as *water hammer*, which is simply wave motion in fluid-filled pipes induced by sudden valve operation or other transients. These problems can be analyzed (Moody, 1990) by characteristic methods or other numerical techniques. For small-amplitude waves, a linear analysis suffices and the characteristic relations reduce to $dP= \pm\rho_0c_0du$ for waves moving to the right

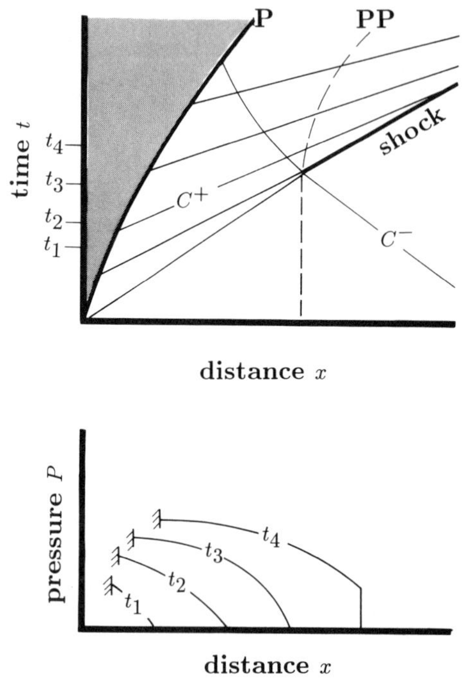

FIG. 6. Compression wave steepening to form a shock, illustrating crossing of characteristics.

($dx/dt = +c_0$) and the left ($dx/dt = -c_0$), respectively. The quantity ρc is referred to as the *acoustic impedance* of the fluid. The ratio of the fluid and solid acoustic impedances determines the pressure produced when a wave interacts with a fluid–solid boundary. At hard boundaries, a large increase in impedance occurs and total reflection results in doubling the pressure. At soft boundaries, a large decrease in impedance occurs and the pressure changes sign. The reflected wave has equal magnitude to the original, i.e., a compressive wave reflects as an expansion and vice versa.

Similarity solutions have played an important role in the development of compressible-flow theory. Perhaps the best-known example is the solution (Taylor, 1950) for the propagation of a decaying spherical shock wave produced by a rapid release of energy within a uniform atmosphere. This solution utilizes the strong shock approximation to the jump conditions, in which the initial pressure in the atmosphere is neglected. If an amount of energy E is suddenly released into the atmosphere of density ρ_0, the radius R_s of the resulting shock wave is proportional to $(E/\rho_0)^{1/5}t^{2/5}$. The pressure P_s immediately behind the shock can be shown to decay inversely with the cube of the distance, $P_s/P_0 \sim EP_0^{-1}R_s^{-3}$. Many variations on this problem can be solved analytically through similarity transformations, including the inverse problem of collapsing or imploding shock waves (Whitham, 1974).

The Taylor solution is an example of the decay of a transient shock or *blast wave* created by an explosion. Blasts from accidental and deliberate explosions have been extensively studied (Kinney and Graham, 1985). Shock pressures from intense explosions follow scaling suggested by the Taylor solution $P_s/P_0 \sim (R_e/R_s)^n$ in which the scaling distance $R_e = (E/P_0)^{1/3}$. In regions close to the energy source, the exponent $n \sim 3$ as given by the Taylor solution. Far from the source, the exponent $n \sim 1$ as given by the acoustic solution. Solutions uniformly valid in the near and far fields can be constructed by the technique of *nonlinearization* (Whitham, 1974). Linear (acoustic) solutions are used for amplitudes but are evaluated on the nonlinear characteristics described by Eq. (25). Shock waves are inserted at the points of characteristic crossing. The shock amplitude is chosen so that the conservation laws, Eqs. (15)–(17), are satisfied.

The dependence of the scaling distance on energy released leads to the concept of *cube-root* scaling for blasts produced by high explosives. Blast pressures from two different charges (mass W) will be equal if the scaled distance $R_s W^{-1/3}$ are the same. Such scaling relations are valuable engineering tools (Kinney and Graham, 1985) for estimating blast waves in air and water. In addition to similarity solution methods, the very useful approximate technique of *shock dynamics* (Whitham, 1974) can be used to solve for shock-wave motion in complex geometries (Schwendeman, 1988).

Because of the importance of one-dimensional compressible flow in technical problems as diverse as plumbing systems and nuclear weapons, numerous methods have been developed for numerical solutions of these flows. A reliable, but cumbersome, method is based directly on the method of characteristics (Ritchmeyer and Morton, 1967) with insertion of shock waves when dictated by characteristic crossing. More flexible but potentially less accurate methods are based on finite-difference approximations to the Euler equations. The accuracy of the technique is determined by the ability to resolve shock waves and contact surfaces. A crude but useful version is to employ *artificial viscosity* to smooth out shock waves. Many more sophisticated techniques such as the methods of random choice (Sod, 1985) and of flux-corrected transport (Oran and Boris, 1987) and the piecewise parabolic method (Collela and Woodward, 1984) have been developed in recent years. Numerical techniques developed for one-dimensional nonreactive flows can be extended to solve reactive, multidimensional flows by the techniques of operator splitting. A solution is built up through a sequence of one-dimensional solutions for each physical process and dimension in space. Very complex problems such as transient supersonic jets, oblique shock reflection, and detonation instability can be treated in this fashion.

7. MULTIDIMENSIONAL FLOWS

Compressible flow in two and three dimensions is of great significance in designing aircraft, space vehicles, propulsion systems, and experimental facilities such as wind tunnels. Realistic design requires including transients, real gas effects, and complex geometrical configurations. Solution of such problems requires a concerted effort with both experimental investigation using scale models and numerical solutions to some approximation such as the Euler equations. However, it is possible to indicate the features of such flows by examining the analytic solutions possible for steady flows of a perfect gas.

7.1 Steady Supersonic Flows

An essential distinction for steady multidimensional flows is the presence or absence of *vorticity*, $\mathbf{\Omega}=\text{curl}\,\mathbf{u}$, in the flow. As in low-speed, incompressible flow, vorticity is a measure of rotational relative motion (rate of strain) within the fluid. Rotational motion can only be created in an initially irrotational fluid by the action of viscous forces at boundaries or by the existence of entropy gradients within the flow. The Euler equations can be used to deduce a transport equation for vorticity, the *Crocco–Vazsonyi equation*:

$$\frac{\partial \mathbf{\Omega}}{\partial t}+\mathbf{\Omega}\times\mathbf{u}=T\,\text{grad}\,s-\text{grad}\left(h+\frac{u^2}{2}\right). \tag{27}$$

For steady flow, this implies that vorticity exists solely due to gradients in total enthalpy or entropy in the direction transverse to the local flow direction $\mathbf{\Omega}\times\mathbf{u}=T\,\text{grad}\,s-\text{grad}(h+u^2/2)$. Total enthalpy is conserved in steady adiabatic flows (with or without shocks) since the Euler equations directly yield $\text{div}(h+u^2/2)=0$. Therefore any variation in h_t must originate upstream and be a consequence of nonuniform boundary conditions. Transverse entropy variations can spontaneously arise from the presence of shock waves of varying strength. A single curved shock wave or a system of transversely nonuniform waves are the most common origins of such entropy variation. We conclude that general multidimensional flows containing curved shocks must also be rotational flows and cannot be isentropic even if the upstream flow is uniform.

We conclude that steady, adiabatic, supersonic flow can be treated as isentropic in the absence of curved shock waves or viscous effects. Therefore the Euler equations can be written in characteristic form. For simplicity, only two-dimensional flows will be considered here. This is most readily done in the *natural coordinate system* (s,n) in which s is distance measured along the flow direction or *streamline* and n is the distance measured normal to the streamlines; see Fig. 7. Characteristic directions are oriented at angles $\pm\mu$ to the main flow direction (Fig. 7), where μ is the angle that a weak disturbance or *Mach line* will make with a uniform flow,

$$\mu=\sin^{-1}(1/M). \tag{28}$$

These characteristic directions in space are analogous to the characteristic directions in space-time introduced in the previous discussion on unsteady one-dimensional flow. The similarity of the governing equations results in a very close analogy between one-dimensional unsteady flow and two-dimensional steady flow. Primitive variables for the characteristic equations in natural coordinates are the magnitude of the flow velocity u and its orientation θ with respect to a fixed axis. The equations can be compactly written as

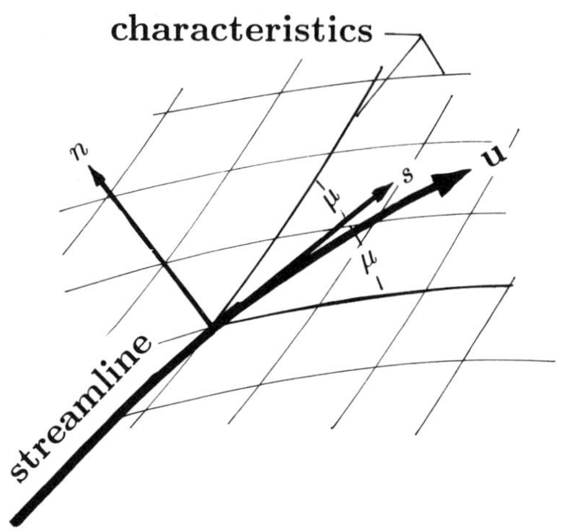

FIG. 7. Natural coordinate system (*s*,*n*) for two-dimensional flow showing Mach lines for supersonic flow.

$$\frac{\partial}{\partial s}(\omega \pm \theta) \mp \tan\mu \frac{\partial}{\partial n}(\omega \pm \theta) = 0, \tag{29}$$

using the *Prandtl–Meyer function* ω defined by

$$d\omega = \frac{dM}{M} \frac{\sqrt{M^2-1}}{1+(\Gamma-1)M^2}. \tag{30}$$

The quantities $\omega \pm \theta$ are the Riemann invariants for this theory. The results of quasi one-dimensional flow theory find direct application in steady multidimensional problems. The differential relationships between thermodynamic and flow variables defined in Sec. 4.1 are valid along the *stream tubes* defined by surfaces of constant n. The energy equation (12) can be used to relate the Mach number M to the thermodynamic state in order to evaluate Γ and complete the integration for the Prandtl-Meyer function.

The Mach number appears as a parameter just as in one-dimensional flow. The distinction between subsonic and supersonic flows is of great importance. Characteristic directions and shock waves can exist only when the flow is locally supersonic. Mathematically, this is a result of the *hyperbolic* character of equations in which all characteristics are real. For steady subsonic flows, the equations are *elliptic* and no real characteristics exist.

The characteristic directions, $dn/ds = \pm\tan\mu$, define the cone of influence for a disturbance generated at a point. Only locations inside of this *Mach cone* can be influenced by events at the apex of the cone. These directions also define the characteristics along which the Riemann invariants $\omega \pm \theta$ are constant. The shock jump conditions discussed in Sec. 4.2 imply that the Riemann invariants discontinuously change when the characteristics cross shock waves.

Solutions for simple changes in flow direction can be constructed for initially uniform flows by use of the Riemann invariants originating in the undisturbed flow. Consider a flow moving from left to right in the x direction. Flow deflections $d\theta$ are related to changes in the Prandtl–Meyer function and velocity by

$$d\omega = \frac{du}{u} \sqrt{M^2-1} = \pm d\theta \tag{31}$$

along the characteristic directions $dy/dx = \pm\tan\mu$. Compressive turns $du<0$ result in velocity decreases associated with stream-tube contraction in supersonic flow. Expansive turns $du>0$ result in velocity increases associated with stream-tube expansion in supersonic flow. For the perfect-gas model, $\omega(M)$ can be computed in closed form. Tabulations are available for many values of γ. A maximum change in ω is associated with the maximum velocity that can be obtained in quasi one-dimensional flow expansion. For a value of $\gamma=1.4$, $\omega_{max}=130.45°$ for acceleration from an initially sonic flow.

A strong compressive turn will result in the crossing of adjacent characteristics of the same family, generating a shock wave just as in the one-dimensional unsteady case. This wave will, in general, be an *oblique shock*. That is, the wave is not normal to the flow direction. As discussed in Sec. 4.2, only the component of the flow velocity in the direction normal to the shock is changed. An oblique shock at an angle β (Fig. 8) with respect to the incident flow (velocity U_1) is therefore equivalent to a shock at normal incidence to a flow with velocity $u_1 = U_1\sin\beta$. The flow is deflected through an angle θ which depends on β and u_1.

With this simple geometric transformation, all oblique shocks can be analyzed as equivalent normal waves using the methods described previously. A consequence of the transformation is that the flow deflection is a multivalued function of wave angle and a maximum flow deflection exists at a critical

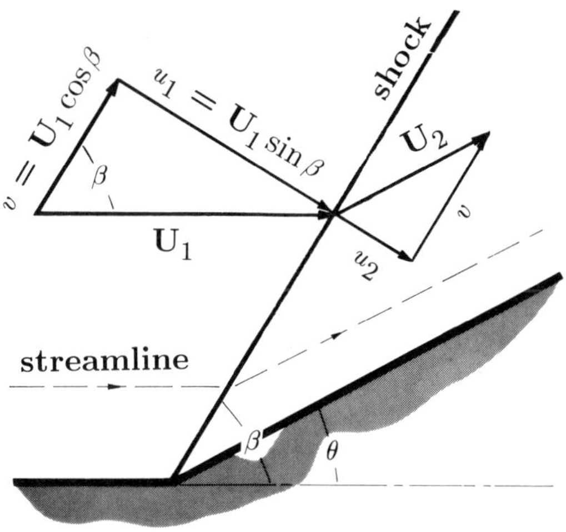

FIG. 8. Oblique-shock velocity diagram, showing wave angle β and resolution into normal and tangential components.

value of the wave angle, which is dependent on the shock strength. For strong oblique shocks in a perfect gas, $\gamma = 1.4$, the maximum flow deflection has the largest value of 45.5° and occurs at $\beta = 68°$. The transition between downstream subsonic and supersonic flow occurs at a critical wave angle β^* that is close to occurrence of maximum deflection. Oblique shock waves with $\pi/2 > \beta > \beta^*$ result in locally subsonic flow downstream of the wave and are known as *strong* waves. Oblique shocks with $\beta^* > \beta > 0$ are locally supersonic downstream and are known as *weak* waves. These properties of oblique shocks play an important role in determining the flow over supersonic projectiles.

7.2 Aerodynamics

Aerodynamics (*q.v.*) is the study of fluid motion over bodies such as wings, fuselages, missiles, and entire airplanes. Flow over streamlined surfaces is characterized by the absence of flow *separation* from the body, and the effects of viscosity are confined to thin *boundary-layer* and wake regions. The flow is usually assumed to consist of two regions. Far from the surfaces and the wake is the *outer* region which is usually treated as inviscid and adiabatic, approximately governed by Euler's equations. Within the wake and near the surfaces is an *inner* region in which the effects of viscosity and heat conduction dominate. Some form of the Navier–Stokes equations must be applied within these regions, and the problem of turbulent flow must usually be addressed.

The outer region is initially computed by supposing that the viscous layer is infinitely thin and that the flow next to the surface must follow the surface contour but *slips* on the surface, i.e., has a nonzero velocity parallel to the surface. The inviscid flow is then used to compute the viscous boundary layer. The displacing effect of the inner boundary layer can then be used to correct the outer flow for viscous effects. In supersonic flow, this procedure is complicated by the presence of shock waves in the outer flow and the interaction of shocks with the boundary layers, which in some cases, can lead to flow separation.

7.2.1 Linearized Flow Theory. Aerodynamic design is a mature subject that uses sophisticated numerical simulations and wind-tunnel testing to develop optimal shapes for lifting surfaces (wings) and entire vehicles. There is also a substantial body of theory on approximate analytical solutions for *thin* or *slender* shapes such as wings and fuselages. One class of theories involves the actual mechanism of lift in subsonic flight, intimately related to the production and maintenance of vorticity in the fluid. Another class of theories attempts to describe the influence of compressibility on the flow, without explicitly considering vorticity.

We are concerned with the second category, which analyzes *irrotational* and isentropic motion. Shock waves can be accommodated by this theory as long as they are weak. Solutions based on potential flow notions are very useful in aeronautics. For wing sections, these theories give very useful estimates of lift, and the drag induced by lift. They also provide the outer boundary conditions necessary for the viscous boundary-layer computations. Linearized analyses usually come to the happy conclusion of an analytical solution that is invaluable in correlating and understanding the experimental data. The irrotational and compressible analyses have particular relevance to near-sonic and supersonic flight. For supersonic and hypersonic flight, the actual mechanism of lift (and drag) is almost exclusively determined by compressibility effects rather than the vorticity mechanism of subsonic flight.

Consider a streamlined body at rest being approached by a flow that is uniform with velocity U_∞ at large distances upstream from the body. If the flow is irrotational, curl$\mathbf{u}=\mathbf{0}$, then we can use a velocity potential Φ to define $\mathbf{u}=\text{grad}\,\Phi$. Linearized flow is the case in which the body has a small aspect ratio (transverse thickness divided by length) and therefore produces only a small perturbation to the incoming flow. Using a velocity potential $\Phi=U_\infty x+\phi$ and linearizing the momentum equation for isentropic flow in a perfect gas, we find that in Cartesian coordinates (x,y,z) the approximate equation for the perturbation potential ϕ can be written

$$(1-M_\infty^2)\frac{\partial^2\phi}{\partial x^2}+\frac{\partial^2\phi}{\partial y^2}+\frac{\partial^2\phi}{\partial z^2}=(\gamma+1)M_\infty^2\,\frac{1}{U_\infty}\frac{\partial\phi}{\partial x}\frac{\partial^2\phi}{\partial x^2} \tag{32}$$

in which $M_\infty=U_\infty/c_\infty$ is the freestream Mach number. In the case of two-dimensional flows, used in idealized wing theory, the z derivative can be omitted or else the z and y terms can be combined to produce a radial derivative used in the axially symmetric theory for fuselages and missiles. The term on the right-hand side is nonlinear and required only in the *transonic flow* regime, $M_\infty\sim 1$. Solutions for subsonic and supersonic flow are greatly simplified by omitting this term. The main results of the theory, based on solutions to the linearized potential equation, are outlined in the subsequent sections.

7.2.2 Subsonic Flow. Subsonic, compressible flow, $M<1$, is of great technical significance in the design of the modern passenger aircraft, in which $M_\infty\sim 0.78$–0.82 during high-altitude cruise. Acceleration of the flow over the lifting surfaces of the aircraft results in local regions of sonic and supersonic flow, which must be treated as a transonic flow problem, discussed in Sec. 7.2.4. A critical flight Mach number is reached when sonic flow first occurs on the lifting surface. Compressibility effects are manifested before the critical Mach number as an increase or apparent *divergence* in the drag coefficient. This important effect can be understood by simple dimensional analysis of the subsonic potential equation.

The linearized potential equation for subsonic flow can be written as a *Laplace equation*

$$\frac{\partial^2\phi}{\partial x^2}+\frac{\partial^2\phi}{\partial y'^2}+\frac{\partial^2\phi}{\partial z'^2}=0, \tag{33}$$

in which the scaled coordinates are $y'=y\sqrt{1-M_\infty^2}$ and $z'=z\sqrt{1-M_\infty^2}$. The pressure P is conventionally presented in aerodynamic analyses as a nondimensional *pressure coefficient*

$$C_P=\frac{P-P_\infty}{\frac{1}{2}\rho_\infty U_\infty^2}\approx-\frac{2}{U_\infty}\frac{\partial\phi}{\partial x}. \tag{34}$$

Consider a low-Mach-number ($M_\infty\to 0$) subsonic flow which results in a measured pressure coefficient $C_{P,0}$. If this flow is governed by the subsonic potential equation, the scaling implicit in the spatial transformation used to obtain the Laplace equation implies that for any finite Mach number the pressure coefficient will be given by

$$C_P=C_{P,0}/\sqrt{1-M_\infty^2}\,. \tag{35}$$

This is the *Prandtl–Glauert rule* of subsonic flow, clearly indicating the divergence effect as $M_\infty\to 1$. Since lift and drag are simply the net forces, pressure and viscous shear, integrated over the body surface, the drag divergence phenomenon follows. The viscous contribution will also change with Mach number but not directly in a divergent fashion; rather, indirectly through the phenomenon of boundary layer separation. The drag in question is induced drag due to lift since the form or shape drag is known to be zero (d'Alembert's paradox) for potential flows. More sophisticated drag correction rules can be obtained by integrating the correction factor over the surface or using other approximations to the linearized potential equation.

The drag divergence effect described by the linearized potential equation arises purely because of the rise in surface pressures as the local Mach number approaches unity. This is a reliable guide as long as the flow everywhere remains subsonic. For supercritical flows, realistic computations of the compressibility effect must incorporate other significant effects. There is a loss in total pressure if shock waves are formed. The pressure rise on the surface will result in thickening or even separation of the boundary layer due to the adverse pressure gradient. Separation of the boundary layer will result in a large *form* drag associated with the change in the pressure distribution due to the presence of reverse flow region downstream of the flow separation point. It is this effect which is primarily responsible for the severe degradation of performance when airfoils designed for subsonic flight are used near $M=1$.

7.2.3 Supersonic Flow. Supersonic flow, $M>1$, results in a wave equation for the potential function

$$\frac{\partial^2\phi}{\partial y'^2}+\frac{\partial^2\phi}{\partial z'^2}-\frac{\partial^2\phi}{\partial x^2}=0 \qquad (36)$$

in which $y'=y\sqrt{M_\infty^2-1}$ and $z'=z\sqrt{M_\infty^2-1}$ are now the scaled coordinates. Solutions to the wave equation are fundamentally different from those of the Laplace equation. However, it is clear from the coordinate transformations in the supersonic case that a divergence in pressure coefficients will also occur as $M_\infty \to 1$. Consideration of the transonic version of the potential equation is required to resolve this singularity.

Thin wings in supersonic flight can be analyzed via the wave equation, and the results are equivalent to a linearized solution using characteristics. The pressure coefficient of a wing element at a small angle θ to the flow can be computed from Eq. (31),

$$C_P=2\theta/\sqrt{M_\infty^2-1}\,. \qquad (37)$$

Airfoils designed for supersonic fight have sharp leading and trailing edges in contrast to the blunt leading edge of subsonic airfoils. A blunt leading edge in supersonic flight is undesirable since it will be proceeded by a detached shock wave with a substantial loss in total pressure. Shock waves emanate from the leading and trailing edge of a thin, sharp-edged airfoil in a supersonic flow. Expansion waves exist between the two shocks, resulting in the *shock-expansion* configuration characteristic of flow field around supersonic bodies. Far from the wing or aircraft, a pressure wave is created that starts and ends with shock waves. The pressure decreases between the shocks, leading to the characteristic *N-wave* pressure profile of a *sonic boom* produced by a supersonic aircraft. The dissipative nature of the shock waves results in an inviscid contribution to the drag.

Flow near the front of a supersonic wedge or cone is of great interest since this is a common configuration, shown in Fig. 9, in vehicles and projectiles. Deflection of the flow by the body results in a shock wave ahead of the body surface. Depending on the angle subtended by the wedge or cone, this shock wave may either be *attached* to the tip of body or *stand off* from it. This behavior is a simple consequence of the maximum turning angle that is possible through an oblique shock, mentioned in the previous discussion. It is possible to obtain similarity solutions for flows over simple wedges and cones when the shock is attached.

The supersonic wedge flow consists of an oblique shock, at a constant angle to the incoming flow, of the appropriate strength to turn the flow parallel to the wedge, and a uniform flow between the shock and the wedge. Supersonic cone flow is similar in that an oblique shock at a constant angle to the incoming flow emanates from the cone tip. However, the flow between the shock and the cone surface is not uniform. The flow does have a similarity character, and the flow field can be obtained from a single differential equation with the similarity variable r/x. This solution reveals that a smooth transition between subsonic and supersonic flow can exist in these conical flows. Such transitions and the associated sonic lines play an important role in the case of a wedge or cone that is blunted or has an included angle that exceeds the maximum turning angle for an oblique shock. In these situations, a subsonic region exists between the shock and body near the tip of the body since the flow behind a normal shock wave is subsonic. Acceleration of the flow over the body can result in a transition to supersonic flow, as shown in Fig. 9. Analysis of such flows is complicated by both the transonic character and the nonlinearity originating in the substantial magnitude of the flow deflections.

7.2.4 Transonic Flow. Transition between subsonic and supersonic flow, the *transonic* flow regime, is of particular significance to modern aerofoil technology. Transonic flow problems are also very challenging to analyze (Cole and Cook, 1986). They are intrinsically nonlinear, i.e., the right-hand side of the po-

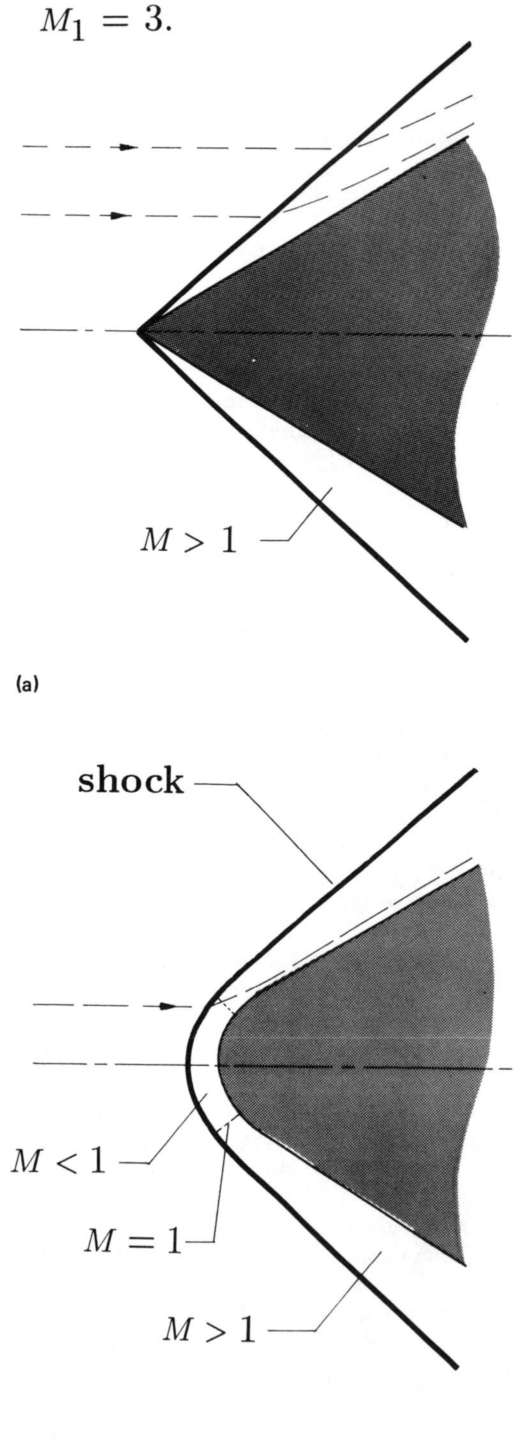

FIG. 9. Flow near the tip of (a) slender and (b) blunt supersonic bodies, illustrating the transition between subsonic and supersonic flow, the distinction between attached and detached shocks, and the streamlines in self-similar flows.

tential equation, Eq. (32), cannot be neglected. The character of the flow and solution methodology changes type, elliptic to hyperbolic, when the sonic line is crossed. Experiments on transonic flows require special care since the cone of influence is quite broad ($\mu \sim 90°$) for near-sonic flow. As a consequence, the configuration of the wind tunnel or testing range is quite significant and substantial corrections for *tunnel interference* are often required.

Supersonic flow will occur over portions of the wings on modern subsonic passenger aircraft. This can result in a shock wave associated with the ultimate deceleration of the supersonic flow. A bubble-shaped region of supersonic flow occurs between the upper surface of the wing, the sonic line, and the shock wave as shown in Fig. 10. By careful choice of the airfoil shape, the critical freestream Mach number at which the shock wave occurs can be increased substantially over conventional designs. With highly optimized surfaces, smooth deceleration from supersonic to subsonic flow, *without a shock wave*, can be obtained. These airfoils, known as *supercritical* designs, can result in substantial drag reduction over more conventional designs and are extensively used on modern aircraft. Design procedures for supercritical airfoils have been developed in some detail (Sobieczky and Seebass, 1984). The design process requires determining an inviscid flow through a solution of the Euler equations, then computing an actual wing profile by correcting for the computed viscous boundary-layer thickness.

7.3 Hypersonic and Hypervelocity Flows

When the Mach number of a vehicle or projectile exceeds a value of $M \sim 5$, the flow is considered to be *hypersonic* (Anderson, 1989), a flow regime encountered in spacecraft launch and reentry. In such flows, the characteristic features are very strong shock waves that dominate the flow field. Lift and drag in hypersonic flows are almost exclusively produced by the high-pressure region generated by the shock on the lower surface of a lifting body. The upper surface is at a very low pressure due to the strong Prandtl–Meyer expansion flow created at the leading edge. Hypersonic flows are relatively simple for vehicles with a sharp nose or leading edge and

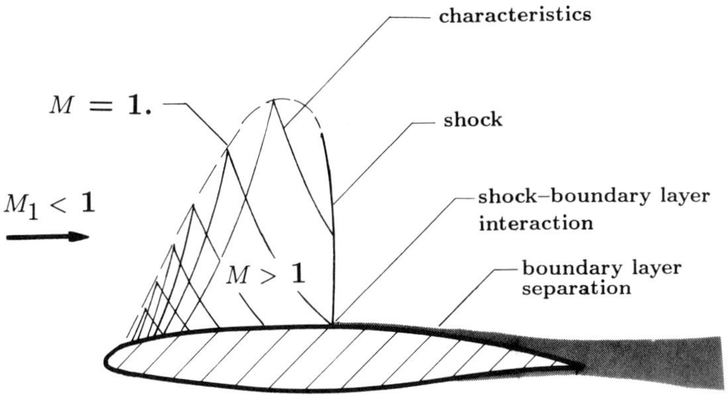

FIG. 10. Principal features of transonic flow over an airfoil.

a small inclination (angle of attack) to the incoming flow. In that case, the region between the shock and vehicle is very narrow since the Mach lines and shock waves are inclined only slightly relative to the flow direction. The region between the shock and the lower surface of the body contains high-enthalpy fluid that has been processed by the oblique shock. Closer to the vehicle can be an *entropy* layer, consisting of fluid processed by the bow shock; and finally, adjacent to the vehicle will be a viscous boundary layer in which friction heating further increases the temperature of fluid. Vehicles with blunt noses (reentry vehicles) will have detached shock and a rather thick associated entropy layer.

Hypersonic flows are distinguished by a large Mach number, although the actual flow velocity may be modest, $u<4$ km/s. This is typical of flows in laboratory shock tunnels with stagnation enthalpies less than 8 MJ/kg. Molecular dissociation and ionization effects are often negligible under these conditions and the gas can be treated as ideal. Hypervelocity flows are distinguished by large velocities, $u \geqslant 8$ km/s, but could have modest Mach numbers. Recall that it is the kinetic energy, $u^2/2$, that is significant in the energy equation, so that these two regimes differ by a factor of 4 in energy. More significant is the strong dependence of chemical reaction rates on temperature. Chemical reaction is much more important in hypervelocity flows, and effects such as dissociation can dominate the observed flow phenomena (Stalker, 1989). In hypersonic flows, the shocks can all be described as strong and the downstream fluid characterized by the limiting density ratio, but hypervelocity flows are better characterized (Stalker, 1989) by a constant-temperature flow downstream of the shocks. It is possible for the two flow regimes to occur simultaneously in the flow over a single body such as a blunt cone.

High-speed flight and atmospheric reentry of space vehicles are the primary applications of hypersonic and hypervelocity research. Atmospheric reentry occurs under both conditions simultaneously, $M=25$ and $u=8$ km/s for a typical near-earth orbit. Behind the shock waves, the temperatures are high enough that dissociation produces atomic nitrogen N and oxygen O, and chemical reaction produces NO. Gas-surface reactions can also be important (Park, 1990) to the survival of reentry vehicles. Scaling methods have been developed so that laboratory hypervelocity experiments (Stalker, 1989) can be used, particularly in problems where near-normal shocks and blunt bodies are important.

A simple means of estimating lift and drag in hypersonic flows is to assume that the fluid molecules inelastically collide with the lower surface of the vehicle and the component of momentum normal to the surface is completely transferred to the vehicle. This is known as the *Newtonian theory* after Newton's early efforts to model low-speed flows. The shock wave and resulting flow next to the body constitute the physical mechanism that produces the inelastic collision effect. If we apply these notions to a surface inclined at an angle θ to the freestream, the local pressure coefficient is found to be

$$C_p=2 \sin^2 \theta. \tag{38}$$

When we apply this concept to a flat plate with velocity U and an angle of attack α in an atmosphere of density ρ_0, the lift L and drag D forces are found to be

$$L=\rho_0 U^2 \sin^2 \alpha \cos \alpha, \tag{39}$$

$$D=\rho_0 U^2 \sin^3 \alpha. \tag{40}$$

These results can be extended to more complex bodies by taking account of the pressure field induced by curved stream lines (centripetal force) and the nonuniformity of the flow between the shock and the body. Another solution technique is indirect and specifies a *waverider* body shape that is a stream surface of a known flow behind a shock wave of specified shape (Anderson, 1989). Yet another solution method is possible for slender bodies, termed *hypersonic small-disturbance theory* (Hayes and Probstein, 1966). Linearizing the equations in the cross-stream direction effectively reduces two-dimensional hypersonic flow problems to an equivalent one-dimensional unsteady flow problem. Numerical and similarity solution methods developed for one-dimensional unsteady flow can then be directly applied to solutions of hypersonic flow.

Hypersonic flows can be created in the laboratory by a variety of steady-flow wind tunnels, shock tubes, and blowdown wind tunnels. All of these systems create in some fashion, either by combustion, shock heating, or electrical heating, a high-enthalpy, high-pressure reservoir of gas. This gas is then expanded through a de Laval nozzle to produce the desired hypersonic flow. Hypervelocity facilities have the highest stagnation enthalpies, up to 30 MJ/kg. These systems are usually transient in operation since the high-enthalpy fluid must be created by a dynamic process such as shock compression. An example of this type of facility is the *free piston compressor* (Stalker, 1990), in which a massive piston is driven by compressed gas into the driver fluid. The compressed driver gas ($P \sim 1$ kbar) then ruptures a diaphragm and drives a shock through the test gas and out the nozzle at the end of the test section, which initiates a quasisteady flow through the nozzle. The model or experiment is conducted in a vacuum tank surrounding the nozzle. Another technique for generating the driver gas is to use a conventional shock-tube arrangement with high-pressure helium or to burn a combustible mixture initially at high pressure. Finally, an alternative method for producing hypersonic flows is the ballistic range. A projectile is accelerated by a gas or powder gun to hypersonic velocities and injected into a test chamber filled with the atmosphere of interest.

7.4 Unsteady Flows

Many flows of interest in applied physics are unsteady flows with significant multidimensional motion. Great advances have been made in understanding these flows through careful experimentation and numerical simulation of the Euler equations. These simulations are based on numerical solution techniques that can either *capture* shock waves as part of the general solution algorithm or attempt to isolate and treat shock waves and contact surfaces separately through special *front tracking* techniques. Common to all methods is the use of the *conservation form* of the equations, generically written as

$$\frac{\partial f}{\partial t}+\text{div}\mathbf{F}=0. \tag{41}$$

The quantities f and $\mathbf{F}$ are the generalized densities and fluxes. For the Euler equations in two-dimensional Cartesian coordinates with velocity components $\mathbf{u}=(u,v)$,

$$f=\begin{pmatrix} \rho \\ \rho u \\ \rho v \\ \rho u\left(e+\dfrac{u^2}{2}\right) \end{pmatrix},$$

$$\mathbf{F}=\begin{pmatrix} \rho u & \rho v \\ \rho u^2+p & \rho uv \\ \rho vu & \rho v^2+p \\ \rho u\left(h+\dfrac{u^2+v^2}{2}\right) & \rho v\left(h+\dfrac{u^2+v^2}{2}\right) \end{pmatrix}. \tag{42}$$

Typical applications are to the problems of unsteady transient jets, blast propagation, and shock diffraction and refraction. As an example, consider the problem of unsteady shock reflection from a wedge. This is a configuration that often occurs when blast waves interact with structures.

Experimental studies (Hornung, 1986) have shown that there are at least two distinct types of reflection, *regular* [Fig. 11(a)] and *Mach* [Fig. 11(b)]. Regular reflection produces a

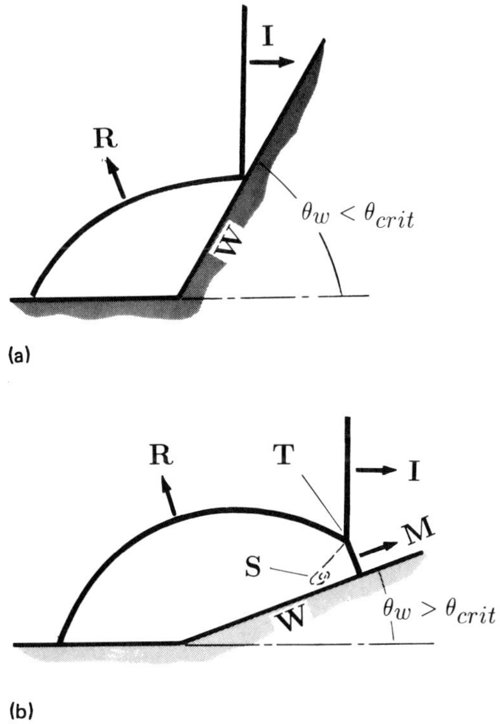

FIG. 11. (a) Regular and (b) Mach reflection of an unsteady shock wave interacting with a wedge.

reflected wave (R) at the point where the incident wave (I) impacts the wedge (W). Mach reflection produces a *Mach stem* (M), a shock wave that is oriented nearly normal to the ramp and intersects the incident (I) and reflected waves (R) in a *triple point* (T), from which a contact surface and associated vorticity [manifested as the shear layer (S)] emerge into the shocked fluid. Mach reflection is observed for wall angle θ_W smaller than a critical value; above that, transition to the regular configuration occurs. The critical wedge angle depends on the shock Mach number. Its determination is apparently complicated (Hornung, 1986) by the displacement effect due to the viscous boundary layer on the ramp surface and the shear layer at the contact surface.

If the viscous effects are ignored, the problem can be viewed as pseudosteady by transforming to a frame attached to the triple point. Oblique shock adiabats (known as *shock polars*) can then be used to construct a simple approximation to the flow field. The critical angle for transition can be determined by applying a causality condition on disturbance originating from the tip of the wedge. Despite the success of the pseudosteady theory and the appealing simplicity of a classification into either Mach or regular reflection, a rich variety of flows are actually observed experimentally. Notable features include secondary and tertiary triple points, vortex structures, and jetlike flows behind the Mach stem. Numerical simulation based on the Euler equations offers insight into the rich variety of these flows. Shown in Fig. 12 is a comparison between observed density contours (isopycnics) in an interferogram image [Fig. 12(a)] and computed isopycnics [Fig. 12(b)] for a Mach 3.51 shock wave in SF_6 incident on a 20° wedge. This is an example of double Mach reflection that clearly shows the "kink" in the reflected wave where the second Mach stem occurs. Other features on the figure include the incidence shock separating undisturbed fluid (0) from shocked fluid (1), a Mach stem between (0) and (3), a reflected wave between regions (2) and (1), and a shear layer between (3) and (2). A vortex is clearly visible behind the curved Mach stem. Letters label the corresponding isopycnics in the experimental and calculated results.

8. FLOWS WITH PHASE CHANGE

A recent development in compressible flows is the exploration of adiabatic phase change induced by rapid compression. While the phenomenon of water condensing from moist air upon isentropic expansion (shown in Fig. 2) is well known, more startling and less widely appreciated is the possibility of liquefying a pure substance by adiabatic compression. Shock-induced liquefaction can occur (Detleff *et al.*, 1979) for fluids with sufficiently large molar heat capacity. Partial and complete liquefaction shocks have been observed in a shock tube when the test (driven) fluid is composed of molecules similar to, or more complex than, perfluoroheptane or octane.

A unique feature of the complete liquefaction shocks is the presence of an extraordinary number of vortex rings in the liquid behind the shock wave. Photographic observations (Thompson *et al.*, 1986) of partial liquefaction shocks demonstrate that the shock wave may be *split* into a precursor shock and a separate condensation wave. These waves

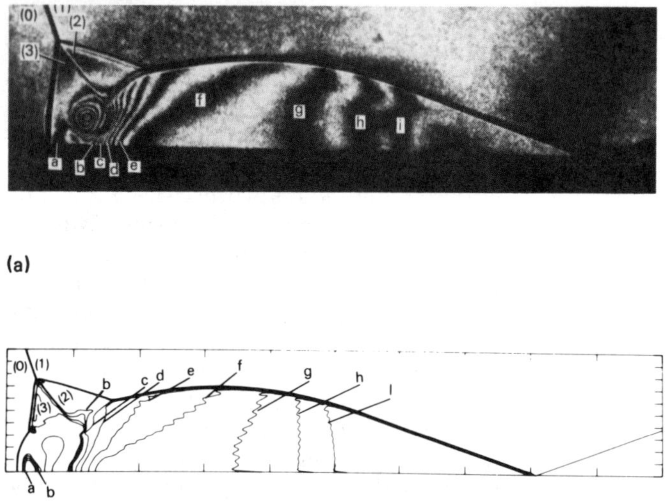

(a)

(b)

FIG. 12. (a) Experimental and (b) computational results for double Mach reflection of a shock wave in SF_6 (Glaz *et al.*, 1986).

can also be unstable, as manifested by a highly wrinkled wave front. A mild version of this instability is exhibited by the wave shown in Fig. 13. This photograph is an oblique image of a partial liquefaction wave front produced behind a $M=2.63$ shock in isooctane initially at 383 K and 1.1 bar.

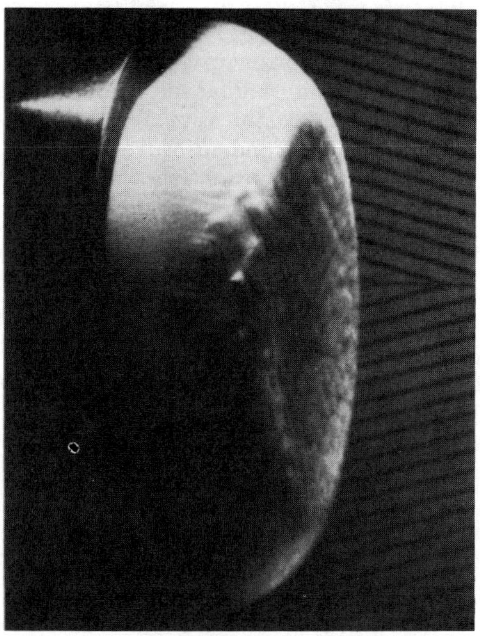

FIG. 13. Diffraction of an unstable liquefaction shock front from a tube (Thompson *et al.*, 1986).

Wave splitting and spontaneous wave-front distortions are two features that may be related to the more general problem (Menikoff and Plohr, 1989) of shock stability. It is known that under certain circumstances, often associated with phase transformations (Duvall and Graham, 1977), shock waves can be unstable to both longitudinal and transverse disturbances. Such instabilities are associated with unstable feedback mechanisms between the mechanics of fluid flow and the thermodynamics of adiabatic compression. Additional instabilities are possible due to the wave structure associated with the finite rate of phase transformation. These instabilities may be similar to those observed in detonations; however, the origin is presently unknown.

Another type of compressible flow with phase change is a two-phase mixture of a liquid with its vapor, typically characterized as a bubbly liquid. Experimental observations (Nakaryakov *et al.*, 1979) of shock waves propagating in such mixtures indicate that many different types of wave structures can exist. The large difference between the vapor and liquid properties combined with the dynamic nature of the phase change results in dispersive waves with complex pressure profiles, including smooth compressions and oscillatory tails, rather than just simple shocks. If the bubbles also contain a combustible mixture, then explosions within the bubbles can result in a self-sustaining wave (Hasegawa and Fujiwara, 1982). Unique wave phenomena (Kutateladze *et al.*, 1987) have also been observed in a pure substance near its thermodynamic critical point. It is possible to obtain very low or even negative values of the fundamental gasdynamic derivative, making possible the demonstration of *rarefaction shock waves*, a possibility hinted at by Eq. (21).

GLOSSARY

Acoustic Impedance: Product of mass density and sound speed ρc, proportionality constant between pressure and velocity changes for acoustic waves.

Adiabatic Compressibility: Proportionality between fractional specific volume changes and pressure changes, $K_s = -(1/v)(\partial v/\partial P)_s$.

Blast Wave: Transient shock wave created by explosion.

Chapman–Jouguet Velocity: Theoretical velocity at which a self-sustaining detonation propagates.

Characteristics: Directions in space-time along which information travels.

Choking: Limitation of mass flow rate due to fluid velocity approaching the sound speed.

Contact Surface: Physical surface in flow separating fluids of different composition, density, or entropy. Pressure and velocity are continuous across contact surfaces.

de Laval Nozzle: Duct containing a converging–diverging section that is used to accelerate the flow to supersonic velocities.

Expansion Wave: Isentropic steady or unsteady flow resulting in a decrease in pressure.

Fundamental Gasdynamic Derivative: Basic measure of the nonlinearity of the fluid, $\Gamma = (c^4/2v^3)(\partial^2 v/\partial P^2)_s$.

Grüneisen Coefficient: Measure of the thermal expansion of a fluid, $G = (1/c_p)K_s(\partial v/\partial T)_P$.

Hugoniot: Relationship between state variables for a shocked fluid, also known as shock adiabat.

Hypersonic Flow: Fluid motion where the velocity greatly exceeds the sound speed, i.e., $M > 5$.

Mach Number: Ratio of fluid velocity to sound speed, $M = u/c$.

Nonlinearization: Process of using nonlinear characteristics together with linear wave equation solutions to obtain approximate solutions to blast wave problems.

Prandtl–Meyer Function: Function used to relate flow deflection angle to velocity changes in steady two-dimensional isentropic flow.

Quasi One-Dimensional: Approximation used to analyze flow in variable-area ducts.

Shock Wave: Adiabatic jump in fluid properties that occurs in a propagating thin layer.

Sonic Flow: Condition in which the Mach number is 1 i.e., the fluid velocity equals the sound speed.

Stagnation State: Thermodynamic state reached when a fluid is isentropically brought to rest.

Substantial Derivative: Derivative following the path of the fluid particles.

Supercritical Airfoil: Airfoil capable of shock-free flight even though the Mach number locally exceeds 1.

Supersonic Flow: Fluid motion with the velocity greater than the sound speed, i.e., $M > 1$.

Thermoelastic Equation: Also known as the Mie–Grüneisen equation, an equation of state used for shock-wave calculations in solid materials.

Transonic Flow: Fluid motion near the speed of sound.

Water Hammer: Shock waves generated by transients in plumbing systems.

Waverider: Shape of vehicle designed by examining streamline solutions for hypersonic flow problems with specified shock profiles.

List of Works Cited

Anderson, J. D., Jr. (1976), *Gasdynamic Lasers: An Introduction*, New York: Academic.

Anderson, J. D., Jr. (1989), *Hypersonic and High-Temperature Gas Dynamics*, New York: McGraw-Hill.

Bird, G. A. (1976), *Molecular Gasdynamics*, Oxford: Clarendon.

Brower, W. B., Jr. (1990), *Theory, Tables and Data for Compressible Flow*, New York: Hemisphere.

Cole, J. D., Cook, L. P. (1986), *Transonic Aerodynamics*, Amsterdam: Elsevier.

Collela, P., Woodward, P. D. (1984), *J. Comp. Phys.* **54**, 174–201.

Davison, L., Graham, R. A. (1979), *Phys. Rep.* **55**, 255–379.

Detleff, G., Thompson, P. A., Meier, G. E. A., Speckmann, H.-D. (1979), *J. Fluid Mech.* **95**, 279–304.

Duvall, G. E., Graham, R. A. (1977), *Rev. Mod. Phys.* **49**, 523–579.

Fickett, W., Davis, W. C. (1979), *Detonation*, Berkeley: University of California Press.

Gibbs, T. R., Popalato, A. (1980), *LASL Explosive Property Data*, Berkeley: University of California Press.

Glaz, H. M., Walter, P. A., Glass, I. I., Hu, T. C. J. (1986), in: J. R. Bowen, J.-C. Leyer, R. I. Solouhkin (Eds.), *Dynamics of Explosions*, Progress in Astronautics and Aeronautics, Vol. 106. New York: AIAA, pp. 359–387.

Hasegawa, T., Fujiwara, T. (1982), *19th Symposium (International) on Combustion*, Pittsburgh: The Combustion Institute, pp. 675–683.

Hayes, W. D., Probstein, R. F. (1966), *Hypersonic Flow Theory*, New York: Academic.

Hornung, H. A. (1986), *Annu. Rev. Fluid Mech.* **28**, 33–58.

Jeanloz, R. (1990), *Annu. Rev. Earth Planetary Sci.* **18**, 357–386.

Kinney, G. F., Graham, K. J. (1985), *Explosive Shocks in Air*, 2nd ed., Berlin: Springer.

Kutateladze, S. S., Nakoryakov, V. E., Borisov, A. A. (1987), *Annu. Rev. Fluid Mech.* **19**, 577–600.

Lee, J. H. S. (1984), *Annu. Rev. Fluid Mech.* **16**, 311–336.

Liepmann, H. W., Laguna, G. A. (1984), *Annu. Rev. Fluid Mech.* **16**, 139–177.

Levine, R. D. Bernstein, R. B. (1987), *Molecular Reaction Dynamics*, Oxford: Oxford University Press, pp. 243–247.

March, S. P. (Ed.) (1980), *LASL Shock Hugoniot Data*, Berkeley: University of California Press.

Menikoff, R., Plohr, B. J. (1989), *Rev. Mod. Phys.* **61**, 75–130.

Moody, F. J. (1990), *Introduction to Unsteady Thermofluid Dynamics*, New York: Wiley.

Nakoryakov, V. E., Pokusaev, B. G., Shreiber, I. R., Pribaturin, N. A., (1988), *Int. J. Multiphase Flow* **14**, 655–677.

Oertel, H. (1966), *Stossrohre*, Berlin: Springer.

Oran, E. S., Boris, J. P. (1987), *Numerical Simulation of Reactive Flow*, Amsterdam: Elsevier.

Park, C. (1990), *Nonequilibrium Hypersonic Aerothermodynamics*, New York: Wiley.

Ritchmeyer, R. D., Morton, K. W. (1967), *Difference Methods for Initial Value Problems* , 2nd ed., New York: Wiley.

Ree, F. H. (1984), *J. Chem. Phys.* **81**, 1251–1263.

Schnerr, G. (1989), *Exp. Fluids* **7**, 145–146.

Schnerr, G., Dohrmann, U. (1990), *AIAA J.* **28**, 1187–1193.

Schwendeman, D. W. (1988), *Proc. R. Soc. London Ser. A* **416**, 179–198.

Sobieczky, H., Seebass, A. R. (1984), *Annu. Rev. Fluid Mech.* **16**, 337–364.

Sod, G. A. (1985), *Numerical Methods in Fluid Dynamics*, London: Cambridge University Press.

Stalker, R. J. (1989), *Annu. Rev. Fluid Mech.* **21**, 37–60.

Stalker, R. J. (1990), in Y. W. Kim (Ed.), *Current Topics in Shock Waves*, New York: AIP.

Taylor, G. I. (1950), *Proc. R. Soc. London Ser. A* **201**, 159–174.

Thompson, P. A. (1971), *Phys. Fluids* **14**, 1843–1849.

Thompson, P. A. (1983), *Phys. Fluids* **26**, 48–49.

Thompson, P. A. (1988), *Compressible-Fluid Dynamics*, Troy, NY: RPI Bookstore.

Thompson, P. A., Sullivan, D. A. (1977), *Phys. Fluids* **20**, 1064–1071.

Thompson, P. A., Carofano, G. C., Kim, Y.-G. (1986), *J. Fluid Mech.* **166**, 57–92.

Vincenti, W. G., Kruger, C. H. (1986), *Introduction to Physical Gasdynamics*, Malabar, FL: Krieger, pp. 281–293.

Westbrook, C. K., Urtiew, P. A. (1982), *19th Symposium (International) on Combustion*, Pittsburgh: The Combustion Institute, pp. 615–623.

Whitham, G. B. (1974), *Linear and Nonlinear Waves*, New York: Wiley.

Zel'dovich, Ya. B., Raizer, Yu. P. (1966), *Physics of Shock Waves and High-Temperature Hydrodynamic Phenomena*, Vols. I and II, New York: Academic.

Further Reading

Anderson, J. D., Jr. (1976), *Gasdynamic Lasers: An Introduction*, New York: Academic.

Anderson, J. D., Jr. (1989), *Hypersonic and High-Temperature Gas Dynamics*, New York: McGraw-Hill.

Anderson, J. D., Jr. (1990), *Modern Compressible Flow*, New York: McGraw-Hill.

Cole, J. D., Cook, L. P. (1986), *Transonic Aerodynamics*, Amsterdam: Elsevier.

Emanuel, G. (1986), *Gasdynamics: Theory and Application*, New York: AIAA.

Fickett, W., Davis, W. C. (1979) *Detonation*, Berkeley: University of California Press.

Liepmann, H. W., Roshko, A. (1957), *Elements of Gasdynamics*, New York: Wiley.

Moody, F. J. (1990), *Introduction to Unsteady Thermofluid Dynamics*, New York: Wiley.

Oertel, H. (1966), *Stossrohre*, Berlin: Springer.

Park, C. (1990), *Nonequilibrium Hypersonic Aerothermodynamics*, New York: Wiley.

Sedov, L. I. (1959), *Similarity and Dimensional Methods in Mechanics*, New York: Academic.

Stanyucovich, K. P. (1960), *Unsteady Motion of Continuous Media*, London: Pergamon.

Sod, G. A. (1985), *Numerical Methods in Fluid Dynamics*, London: Cambridge University Press.

Thompson, P. A. (1988), *Compressible-Fluid Dynamics*, Troy, NY: RPI Bookstore.

Vincenti, W. G., Kruger, C. H. (1986), *Introduction to Physical Gasdynamics*, Malabar, FL: Krieger.

Whitham, G. B. (1974), *Linear and Nonlinear Waves*, New York: Wiley.

Zel'dovich, Ya. B., Raizer, Yu. P. (1966), *Physics of Shock Waves and High-Temperature Hydrodynamic Phenomena*, Vols. I and II, New York: Academic.

COMPUTER-AIDED DESIGN IN ELECTRONICS

ANTUN DOMIC, *Digital Equipment Corporation, Hudson, Massachusetts, U.S.A.*

INTRODUCTION

The application of computers to the design of electronic systems has steadily increased since its introduction in the late 1950s. During the last two decades, the prices of minicomputers and desktop machines (workstations and personal computers) have been steadily decreasing while their overall performance has been continuously improving. This upward trend in computer performance has been accelerated by rapid advances in microprocessors, which have doubled their execution speeds approximately every 18 months, and faster and larger-capacity semiconductor memories. Increased performance has allowed the development of sophisticated new applications whose functionality, in turn, makes possible the efficient design of more powerful machines. This cycle will continue.

To illustrate the increase in workstation computational power, we cite two sets of numbers: The Motorola 68020 family of CISC microprocessors (complex instruction set computer), widely used in these machines a few years ago, had in 1986 an execution speed of approximately 4 MIPS (million instructions per second). Two machine generations later, by mid-1990, Motorola's 68040 microprocessor had achieved a speed of 20 MIPS. In four years, processing power had grown five times. The progress in chips implementing RISC (reduced instruction set computer) architectures has been even more rapid. The first workstation that implemented SUN Microsystems' SPARC architecture appeared in the second half of 1987 and offered 10 MIPS, while in the first half of 1991 Hewlett-Packard's Precision Architecture machines achieved 70 MIPS. That is a sevenfold increase in approximately four years. Hardware prices have decreased to the point where the difference between the low end of the workstation market and the high-end PCs (personal computers) has disappeared.

The progress in the application of computers to the design of computer systems has been equally impressive. New and better computer-aided design (CAD) tools have been one of the key factors making possible the successful design of extremely complex integrated circuits (ICs or chips). To point out the increase in IC complexity, we offer the following illustration. In 1985, the microprocessor designed by Digital Equipment Corporation for use in its MicroVAX II series machines contained about 125 000 transistors. By 1990, Intel's 80486 microprocessor, used in many personal computers, contained around 1 200 000 transistors. Without important ad-

1-56081-063-7/92/$5.00+.50

vances in computer-aided design, systems of such complexity would be impossible to design.

The field of computer-aided design includes a wide span of applications. We will cover only aspects related to the design of electronic systems, with emphasis on the design of integrated circuits. Given the many areas of interest, we are forced to mention some of them only briefly. In general, CAD refers to specialized software that allows data entry and verification, simulation of the behavior of pieces of a system, automatic synthesis of increasingly detailed descriptions, and a number of data management facilities. Examples include graphical editors to enter logic schematics and chip layouts, programs to verify that layout design guidelines have been respected, systems to simulate the electrical behavior of a set of interconnected transistors, or programs that automatically generate geometries in order to connect or "route" the desired pins among components placed on a printed-circuit board.

As its name indicates, CAD is an aid to design. So it is not a discipline with its own set of problems, even though many CAD problems have models with great theoretical importance. CAD's effectiveness is measured by what it makes possible. Two areas are emphasized: the ability to design very-large-scale integration (VLSI) chips and systems using them, and the contribution to productivity, such as savings in the size of design teams and overall time-to-market schedules. In consequence, developments and topics of interest have changed through time, tracking the leading-edge technologies used in the design of computers and integrated circuits. Nevertheless, one tendency has been consistent. As systems become more complex, the design process moves toward more and more abstract levels of description. This phenomenon has required a corresponding increase in the development and use of automatic synthesis and simulation tools. While advances will continue in most CAD areas, the most visible increases in design productivity will be achieved through synthesis (at logic and layout levels) and innovative uses of simulation and other verification techniques.

1. HISTORY

CAD, as part of the broader field of design automation, started in the mid-1950s. The use of transistors in computer circuits created new problems, and to solve them, designers took advantage of the wider availability of computers. The first CAD paper probably was by Cray and Kisch in 1956 (see Breuer, 1972), describing a system to check and simulate Boolean equations. Another area of early interest was the entry of data to machines that automatically placed wires to connect components on a board, and programs to verify these data. In these early beginnings we immediately recognize topics that have persisted until today: logic simulation, interconnect estimation, and entry and verification of layout-related data.

In the 1960s, the problems of interest were related to board design. The placement and routing of components on a board were important issues, and we can see early work that is still relevant today (Lee, 1961; Hightower, 1969). Simulation of Boolean descriptions (gate-level simulation in today's terminology) and generation of appropriate input values to exercise the logic (test vectors) became an alternative to the building of increasingly complicated hardware prototypes, often called "bread boards," for system verification. The desire to automate steps in the design process—for example, the automatic generation of logic—also received attention quite early (Friedman and Yang, 1968).

Traditionally, the development of CAD tools was confined to internal efforts in computer and semiconductor companies and research carried out at universities. The CAD development groups were closely tied to the design groups that used their tools, and they were considered truly part of the overall system- or chip-development effort. In the 1970s, the appearance of minicomputers that could be adapted to dedicated tasks in a cost-effective way created new opportunities. Independent commercial companies appeared that offered systems, combining hardware and software, for design entry and for automating some design tasks. The first systems were devoted to the editing and routing of boards and, with the growing importance of the IC, to IC layout entry and geometrical design checking. Schematic entry was the next category of systems that attracted these companies.

Important advances in all areas were achieved in the 1970s. A circuit simulator developed at the University of California, SPICE (Nagel, 1975), while having gone through many enhancements, is still in use today. New ways to do IC layout were proposed (Williams, 1978), and significant attempts to shift the design process toward more abstract levels were reported (Johanssen, 1979). The increased transistor density in ICs allowed by the newest metal oxide semiconductor (MOS) technologies brought the introduction of the term VLSI to describe these new chips that could really be considered a fairly complex computer system. The appearance of the book *Introduction to VLSI Systems* by Mead and Conway (Mead and Conway, 1980) significantly turned the perspective in IC design from a pure "device" to a "system" orientation. Clearly, this approach emphasized the importance of automation. In the United States, the funding provided by the Defense Advanced Research Projects Agency (DARPA) to support research through its VLSI projects had a major impact on CAD development through the mid-1980s. Technology advances, such as the emergence of denser MOS gate arrays, shifted the focus even more towards CAD for ICs.

In the early 1980s, the success of the microprocessor provoked a shift from minicomputers to cheaper workstations. In the commercial CAD world, companies that were not able to port software from their proprietary, minicomputer-based hardware to these new machines ended up disappearing from the market, replaced by new players that competed on the basis only of their software, with UNIX as the dominant operating system. As the entry barriers for developing software are much lower than those for hardware, a large number of companies were formed, most of them offering narrowly focused programs (layout verification, simulation, schematic editors, etc.). But as the decade progressed, two phenomena occurred. Internal CAD groups at computer and semiconductor design houses started to rely increasingly on CAD companies for important pieces of their software, and the CAD companies went through a process of consolidation that continues until today (1992). The survivors of this consolidation process, such as Mentor Graphics (Beaverton, OR) and Cadence (San Jose, CA), are large, publicly traded companies with yearly sales on the order of hundreds of millions of dollars. These companies have concentrated their offerings on fairly wide, integrated systems for design instead of isolated point tools.

Today, CAD is advancing along three different lines. Universities continue to play an important role in new research. The University of California at Berkeley, Stanford University in Palo Alto (both in CA), and Carnegie Mellon University in Pittsburgh, PA, have been leading centers in the United States, while the Universities of Edinburgh in Scotland, Grenoble in France, Louvain in Belgium, and Eindhoven in The Netherlands have been important European contributors. In addition, research centers associated with companies, such as IBM's Thomas J. Watson Research Center at Yorktown Heights, NY, and AT&T's Bell Laboratories in Holmdel, NJ, have played a key role. Commercial CAD companies and internal CAD groups have traditionally functioned using different models. Internal CAD groups have focused more on specialized applications that serve their companies's technologies, particularly in areas that today are leading edge but soon will be more generic, while CAD companies have been generally interested in providing systems with the widest possible appeal. Currently, however, most computer and semiconductor companies use a combination of commercial and in-house tools. The evolution of this mix is not clear at this time, but it will determine much of the CAD development in the coming years.

2. COMPUTER-AIDED DESIGN IN ELECTRONIC SYSTEMS

We will consider only certain aspects in the design of electronic systems, in particular, digital computer systems. We will include CAD tools for the design of a system as a logical unit, including component ICs, for the simulation of the system, and for the design of boards. We will exclude the semiconductor process design and manufacturing, thermal analysis, testing and reliability, cabinet design, specialized software, and other topics that, while significant in the design of a computer system, fall outside our description of a "logical unit."

As already indicated, the focus of CAD has changed through time. While most of the efforts in the 1960s were dedicated to board-level design, the phenomenal increase in the complexity of integrated circuits in the 1980s has shifted its focus. In 1990, a state-of-the-art microprocessor contained between one and two million transistors. It must be noted that, while much more complex than other parts, the microprocessor is only one component in a system. Many CAD tools are equally applicable to boards and IC design, even though the emphasis can be different. Examples are simulation tools, or placement and routing systems that use similar algorithms. Our descriptions will mainly address the IC case, although we will mention differences and applicability to systems.

2.1 Electronic Design of Integrated Circuits

There is a large variety of integrated circuits, using different semiconductor technologies and having equally different kinds of circuit functionality. The first important distinction among the types of ICs is between digital and analog chips. As these names indicate, the main difference is whether the signals can assume a continuous range or are restricted to two discrete values (high and low, or 0 and 1) for basic design purposes. A microprocessor is a mostly digital chip, while a filter can be purely analog. A second distinction is made according to the semiconductor technology used. The most common technologies today are CMOS (for complementary metal oxide semiconductor) and bipolar technologies such as ECL (emitter coupled logic). Lastly, a third distinguishing characteristic is the IC's design style. The most important categories of design styles are custom, semicustom, and gate arrays. Special chips such as memories do not fit any of these types, and traditionally their advances have been made possible through progress in semiconductor-processing technology rather than through CAD. Many developments in IC CAD tools, particularly related to synthesis versus entry and simulation, have been motivated by applications to semicustom and gate-array MOS chips.

The distinctions among these design styles are based on the degrees of freedom we assume. In the case of custom design, maximum flexibility is assumed: Essentially each transistor can be individually tailored. While it achieves the best use of a technology, it levies a large design-effort cost. So it is prevalent in high-speed microprocessors and analog parts. In the case of semicustom design, there is extensive use of a predesigned library of functions, referred to as standard cells, with each usually consisting of a few logic gates. Larger units, called macros, are also employed as single entities. This approach causes a number of inefficiencies, but it allows much more automation. Frequently, custom and semicustom approaches are combined. Finally, in the case of gate arrays, the transistors are laid out in a fixed, regular fashion, and a library of basic gates is constructed, very much as in the case of a semicustom chip but usually with lower performance. A key difference is that in a gate array, we "design" only the interconnection layers, bringing an important economic advantage as only a few layers are customized. Technology advances have increased significantly the capacity of the CMOS sea-of-gates types of arrays, due to the use of more interconnect layers (usually three metal layers), resulting in a higher percentage of utilization of the available gates, and reduced transistor sizes, increasing the density in the chip. Today, the largest gate-array chips contain hundreds of thousands of logic gates and are offered by many semiconductor vendors. Gate arrays constitute the largest component of the market for special-purpose chips (application-specific ICs or ASICs).

2.2 The Design Process

We give now a brief overview of the design process, with the intent of relating design steps to the different CAD tools. The steps involved vary in importance among the types of systems and chips, but they apply quite generically to most computer systems. The design process is a sequence of transformations, each of which produces a description that progressively refines the previous one by adding further detail. This process is usually referred to as "top-down" design.

First, certain technology choices are made and translated into design parameters. For example, a system intended for a certain market will have specific requirements for speed, power consumption, and overall capabilities. A suitable technology is selected to meet these requirements, and the system's basic characteristics are established. In the case of a microprocessor, basic problems include establishing an acceptable clock rate, determining the amount of memory that can go inside the chip, calculating the total power dissipation, and other global factors.

The design proceeds in stages, with each adding more detail in terms of the actual system implementation. Usually, a high-level or "behavior" model is constructed and extensively simulated. From here, logic and circuit design is done to transform the high-level model into Boolean functions and their circuit implementations. In the case of chips, the circuits are laid out, that is, the different mask layers required by manufacturing are drawn. On the other hand, for boards, the positions of the components and the wires connecting them are determined. At each step, extensive verification tasks are performed to establish compliance to technology rules and to establish the equivalence of the different descriptions. Now, we will proceed to describe in greater detail these major steps and the CAD tools employed in the process.

3. HIGH-LEVEL DESIGN

The logic design process can be divided into two major phases. First, a high-level description of the system or IC is usually created using a special language referred to as hardware description language (HDL). Depending on its level of detail, this model is called a behavior or an RTL (register transfer level) model. Second, this high-level or functional model is converted, through manual or automated steps, into a logic description. This description includes the precise components being used at a level of logical units such as Boolean gates, ALUs (arithmetic logic units), etc., and, in the case of boards, components such as ICs that are treated only from the point of view of their input/output behavior.

The functional model indicates what the system does, but does not indicate exactly how it does it. The difference between a pure "behavior" and an "RTL" model is the level of detail, and it is not clear cut. A way to indicate the distinction is through an example: We may want to add two numbers, and we write such a statement in a behavior model, while in the RTL version we indicate exactly which arithmetic unit and registers are actually involved in performing the operation. A functional model resembles a software program. It is written in an HDL that includes a simulation environment so that the system's behavior under specific input patterns can be observed. Architectural decisions and their impact on system performance are usually evaluated by use of a behavior model and extensive simulation.

Many specialized languages, such as DPL or DECSIM BDS (see De Micheli *et al.*, 1987) have been used as HDL. Until the early 1980s, most of the HDLs used by industry were developed by computer or semiconductor companies. In some cases, general software programming languages (such as C) have been used directly, even though only a restricted subset of the language's constructs is allowed. The key tradeoff in the design of an HDL and its simulation environment has been between flexibility and execution speed. Implementations based on interpreters allow excellent observability and user interaction but at a serious cost in simulation speed.

An effort to create an HDL that could be widely used was initiated by the U. S. Department of Defense as part of their VHSIC program (very-high-speed integrated circuits). This resulted in the definition of VHDL (VHSIC HDL), now an IEEE standard (Dewey and Gadient, 1986). VHDL has gained increased acceptance as a natural candidate for standardization. Commercial CAD companies had been offering their own languages and simulators for years, contributing to the creation of a large installed base of models that are exchanged among the system or chip users. Cadence's Verilog HDL, for instance, has played a significant role due to its large market share (Cadence Design Systems, San Jose, CA).

The functional description has taken another increasingly important role as input to the logic synthesis tools now being used in IC design. The logic synthesis process consists of the transformation of a functional description into a structural representation, usually at the gate level, including steps to minimize the logic and to meet timing constraints. This has caused reexamination of the HDL, as the exact interpretation of its statements in terms of Boolean equations has become very important.

4. LOGIC DESIGN

The creation of a representation of the design using only units whose functionality corresponds essentially to Boolean equations and storage elements is called logic design. The most usual description is a set of schematic diagrams involving simple logic gates (such as multiple-input AND, OR, NOT, etc.) and registers. In the case of systems, these functions are divided into components. The representation of a design in these terms is also called a structural description. Many systems, particularly before the advent of VLSI, were designed with schematics, not functional descriptions, as their starting point. Only as complexity increased has the use of functional models become the norm.

Specialized editors to enter gate- and transistor-level descriptions are called schematic editors. Output from the schematic editors are lists of components or cells and their interconnections (netlists), including several levels of detail ranging from large library items (such as an ALU) to transistor-level descriptions, and plots. The netlists are used by simulators to investigate the behavior of the circuits. Special software and computers for schematic editing followed a development path similar to layout editors. This will be described in a later section. Today, software systems for editing on workstations and PCs are offered by many companies, including component libraries for the major gate-array vendors.

4.1 Circuit Design

Circuit design consists of determining the exact implementations of the logic functions and storage elements in terms of the primitive elements of the specific technologies. In the case of ICs, these primitive elements are transistors and other basic elements, including their electrical parameters. Each technology has its own specific circuit techniques. For the case of CMOS the reader can consult Glasser and Dobberpuhl (1986). A design comprises a library of items, each implementing a logic function and parameters such as delays, drive, and power consumption; a method for distributing special signals such as clocks; and general usage rules. In the case of a custom design, there is much more flexibility in the creation of circuits. For example, some tools have been developed to assist the circuit design process, such as CMOS transistor sizers (Hill *et al.* 1989), but it still requires a large amount of trial and detailed simulation.

4.2 Logic and High-Level Synthesis

As already mentioned, the use of logic synthesis has become a part of the design process. Even though work in aspects of logic synthesis started decades ago, it was during the 1980s that the extraordinary progress achieved in this area changed the logic design process. Most of the work has been centered on the generation of the combinational logic (excluding the storage elements that convert the design into a sequential type). The process is divided into two steps: a technology-independent step, where Boolean logic is minimized, followed by a technology-mapping phase where the logic is implemented using a set of predefined gates dependent on specific technology choices.

A Boolean function of N variables can be represented by three subsets of the N-dimensional Boolean cube (the set product of $\{0,1\}$ with itself N times): the points where its value is 1, or ON set, the points where its value is 0, or OFF set, and optionally a "don't care" set, that is, those input value combinations for which the function's output value is unimportant. This case occurs typically when we know that a specific combination of input values will never happen. Minimizing the function is closely related to the problem of "covering" the ON set optimally by subsets of the N-dimensional cube that are themselves lower-dimensional cubes (or "faces") with particular properties. Techniques to obtain a minimal realization of a Boolean function using two levels of logic date back to the 1950s (McCluskey, 1956). One subproblem encountered in this process (finding the collection of optimal special subsets) is "*NP* complete." In loose terms, *NP*-complete problems cannot be solved by any algorithm the number of whose steps is bounded by a polynomial function of the problem variables (Garey and Johnson, 1979), making it necessary from a computational point of view to develop reasonable heuristics. When MOS technology started to predominate in VLSI chips, interest in two-level representations was renewed, as it was possible to produce efficient layouts for a structure called PLA (programmable logic array) to implement two-level logic functions. A detailed description of a widely used program, ESPRESSO-II, and problems found in the logic minimization of PLAs is given in Brayton *et al.* (1984).

Logic synthesis for the general case of a multilevel network, including sequential elements, has been done with a variety of techniques. Systems such as IBM's LSS (Darringer *et al.*, 1984) have combined "local" and "global" transformations to modify the logic until the desired constraints are met. Local transformations examine only a small portion of the circuit, an example being the application of equivalences between AND-NOT and NOT-OR combinations (DeMorgan laws; see Dietmeyer, 1978). Global ones take a larger view, such as following input–output signal flow to identify particular relations among variable values, or collapsing a selected portion of the circuit into two levels to apply special minimization. Several objectives are considered, such as area minimization, timing optimization, and testability. General Electric's SOCRATES (de Geus and Cohen, 1985) employed a rule-based approach to the technology-dependent phase following Boolean minimization. A completely algorithmic approach is followed in MISII (Brayton *et al.*, 1987), applied to strictly combinational circuits. The program executes a user-controlled sequence of optimizations that include tradeoffs between area and time. The technology-mapping phase is based on a technique introduced in DAGON (Keutzer, 1987), where the predefined logic gates and the network are decomposed into tree graphs and matched using local analysis.

A number of new areas closely related to logic synthesis are being actively researched. Because of space constraints, we can only mention them briefly: the relation between the testability of a circuit and its minimal Boolean implementation, the problem of encoding a finite-state machine (see, for example, Dietmeyer, 1978), improved timing optimization algorithms, and the possibility of looking at the full sequential circuit and taking advantage of operations that cross the register boundaries. The interested reader is referred to the CAD conferences cited for further reading.

A different type of logic synthesis is usually referred to as high-level synthesis. The intention is to find a structural implementation for a given behavior and set of constraints, but the basic elements from which we are selecting have a higher level of complexity than logic gates—typically registers, ALUs, multiplexers, and buses. The intended output is a data-path structure and the specification of its control.

Two important problems in high-level synthesis are scheduling and allocation. In scheduling, we determine which operations take place at each control step under a set of restrictions. The restrictions can include, for instance, the number of hardware units available, the total delay that operations can take, or the estimated power consumption. Many algorithms have been investigated, but most fall into two categories. These are exhaustive search combined with some heuristics or standard mathematical techniques such as linear programming (Brayton *et al.*, 1985; Hafer and Parker, 1983), and constructive approaches that proceed by building the operations incrementally (Marwedel, 1986). During allocation, we choose the units in the data path and explicitly assign the operations. Constraints are total area, number of units, and bus interconnects, among others. Techniques used for allocation can be separated into constructive and global ones. In the latter case, graph models have been employed (Hafer and Parker, 1983). Given the generality of the problems, rule-based expert systems have also been applied to high-level synthesis (Thomas *et al.*, 1988).

5. SIMULATION AND VERIFICATION

The need to verify that a system description actually performs as intended under sequences of input values is obvious. Extensive simulations are performed at all levels of system description to ensure two key objectives: that the descriptions match the design intention, and that the different descriptions are actually equivalent. The first of these objectives is usually called the validation problem, even though, unless we are able to do exhaustive analysis, we cannot totally validate a design. The second, or verification, is used in a looser sense to describe approaches that are used to compare different descriptions and obtain a reasonable degree of confidence that the design performs as intended. While detailed electrical simulations—for instance, at the transistor level—would give us a larger degree of assurance that a complex chip will operate properly, the time it would take a computer to perform such a simulation makes this approach completely unfeasible. Thus simulations are performed with models that have very different levels of accuracy in order to cover a large chip or system "adequately."

The most general description available is an HDL program. This representation is usually simulated at the functional or RTL level. The simulation parallels closely the execution of a general software program, and the data that we are observing can be quite general (such as an integer, for instance). The advantages of functional simulation include high execution speed and the ability to perform extensive simulations before the design of the actual circuits has progressed sufficiently to be useful.

A more detailed description is obtained at the logic level. In this case, the description consists of logic gates (ANDs, ORs, etc.) and the values we observe are 0 and 1, plus some "undefined" states, producing what is referred to as gate-level simulation. The concept of time can be now introduced in a much more precise way. Relations are established in discrete units among the different clocks, the behavior of the different register elements is established, and delays through each gate type are indicated. In addition, delays due to interconnects can be modeled. These are usually based on extracted wire lengths or total capacitance including parasitic effects such as couplings. Gate-level simulation has been the most prevalent way to establish a design's correctness, at least from a functional perspective that includes simple timing models, even though in most cases no proof of correctness is achieved. Gate-level descriptions are often combined with RTL sections, creating a "mixed-mode" simulation. An example of such a system, used widely by a large computer company, is found in Insinga (1984). For topics closely related to gate-level simulation, such as test-pattern generation and fault simulation, the interested reader is referred to Abramovici *et al.*(1990).

Many techniques have been used to execute gate-level simulations. We mention the two most basic: compiler and event driven (see Saleh and Newton, 1990). In a compiler-driven simulator, the elements in the circuit are ordered according to the time at which they may execute a transition, usually according to their logic level. Code is then generated to evaluate all gates sequentially. In an event-driven simulation, we try to take advantage of the fact that the ratio between the total number of nodes and the nodes that actually change values is normally below 10% so that it is not necessary to simulate all nodes at each time step. On the basis of the inputs that have changed values, we schedule evaluations only for those nodes that may be affected. As a way to accelerate gate-level simulations, special-purpose hardware accelerators have been built by computer companies for their internal use, and some small companies have offered these machines commercially. Two examples can be found in Denneu (1982) and Agrawal *et al.* (1987). Many attempts to use parallel processing (see PARALLEL COMPUTING) have also been described. The possibility of building systems based on field-programmable gate arrays (FPGA) to do simulation (versus the actual implementation of the chip or system) is being actively researched.

In the case of MOS circuits, the use of switch-level simulators has been quite successful as a way to combine a simulation in terms of obtaining digital information (high and low values at each node) with the inclusion of electrical and timing information. The idea is to model the transistors as fixed-value resistors and capacitors (*RC* model), controlled by a switch (the value on the transistor's gate), and wires as ideal conductors. This simplification allows the use of an efficient approximate model to calculate delays (White and Sangiovanni-Vincentelli, 1987). In addition to high and low values, some electrical notions can be introduced. For instance, based on the transistor's width and length, a concept of "strength" can be introduced to model behavior in a feedback loop. A problem has been the fact that in many common MOS circuits, a transistor can apparently act as a bidirectional device. While this actually may not happen during the circuit operation, to determine automatically the direction of signal propagation is difficult. Efforts to improve the actual processing of the transistor interconnection list and to transform it into a set of Boolean equations has produced important speed improvements in simulation times (Bryant *et al.*, 1987).

In addition to the use of the switch-level model for simulation, it has become common to use it for chip-level timing verification of MOS circuits. The problem here is different: to determine paths that do not meet a timing constraint—for example, whether a signal is arriving at the input of a storage element in time with respect to the clock controlling the element. The analysis now is not based on actual input patterns but on looking at worst-case delays using the *RC* model. In addition to the directionality problem, another concern is the appearance of "false paths," that is, paths that will never actually be logically exercised. An early example of such a system was RSIM (Terman, 1983). The applicability of these techniques to the processing of very large transistor networks has continued to make them extremely attractive (Pan *et al.*, 1991). Many attempts have been made to speed up the analysis of these *RC* networks. Very large networks can be obtained by detailed layout extraction to include the effects of actual interconnect delays. A recent technique called asymptotic wave-form estimation (Pillage and Rohrer, 1990) has been successfully applied to networks with hundreds of thousands of elements.

The most detailed simulations are done at the device level. In this case, very accurate models of transistors, transmission lines, resistors, and other basic circuit elements are used. Analysis of such aspects as transients, steady state, or sensitivity can be performed with great accuracy but involve significantly increased simulation times. In the case of ICs, the number of transistors that can be simulated at this level in a reasonable amount of time rarely exceeds a few thousand. Probably the most widely used circuit simulator is SPICE and its many successive versions from the University of California and commercial CAD companies (Nagel, 1975).

The main components in a circuit simulator are as follows: The complex elements, such as transistors, are broken into simpler components, usually two-terminal linear and nonlinear resistors, capacitors, inductances, and controlled sources, using predefined models (for the case of SPICE MOS devices, see Vladimirescu and Liu, 1980). By use of Kirchhoff's current and voltage laws, systems of differential equations can be written and solved by numerical integration methods. The numerical techniques used account for the large amount of computer resources needed by circuit simulators, the difficulties in solving very large circuits, and problems related to convergence and stability of the solutions. For a more detailed discussion we refer the reader to White and Sangiovanni-Vincentelli (1987) and Saleh and Newton (1990). Also, for a deeper level of analysis of the detailed simulation of devices such as transistors as two- and three-dimensional structures, the interested reader can consult Sze (1984).

A large amount of simulation is done just to ensure that small logic changes have not altered the circuit's behavior. Recently, new techniques have been applied to prove formally the equivalence of two Boolean networks, avoiding multiple iterations through a simulation loop. The general problem of deciding if a Boolean expression is a tautology, that is, is true for all possible input values, is *NP* complete (Garey and Johnson, 1979). But special representations of a function that allow fast comparisons using reasonable amounts of memory for important subsets of functions, particularly those generated by logic synthesis tools, have been introduced (Bryant, 1986; Brace *et al.*, 1990). These functions are given canonical representations as special acyclic graphs called binary decision diagrams. This representation has allowed the verification of equivalence inside logic synthesis programs such as MISII. Combined with the ability to abstract a transistor-level netlist into a Boolean representation, the technique offers new possibilities for the functional verification of MOS circuits.

6. LAYOUT EDITING AND VERIFICATION

An important piece of the design process is the representation of the detailed data describing the exact geometries for the actual masks used to manufacture ICs, or the metal areas and holes for a board. Before the appearance of electronic editors, these large amounts of data were manually transferred (or "digitized") from drawings into computers, so that they could be further processed and used in manufacturing. The basic problem in the design of the first electronic editors was the actual hardware, particularly the graphic display systems. These displays were connected to general-purpose computers or were sold as part of a complete system that included the editing software and a modified minicomputer. Plotting hardware followed a similar path. In the 1970s and early 1980s, several companies offered these very successful systems, running on proprietary hardware closely tied to their specific programs. Then in the 1980s, the appearance of the workstation changed this picture radically. Layout editors could operate on these smaller computers that, in addition, were used for a multiplicity of other tasks. A number of companies, unable to port their software quickly to the new computers, simply disappeared from the marketplace and were replaced by new, software-oriented players.

6.1 Layout Editing

The representation of layout data has two aspects. First, we need standard exchange formats in order to transfer data. Second, each editor needs efficient internal data representations in order to manipulate data in an interactive fashion. Some standard exchange formats, such as CIF (Caltech intermediate form) and GDSII (originally from Calma Company), widely used for IC layout data, Gerber (property of Gerber Scientific Instrument), and the more general EDIF (electronic design interchange format) are described in Rubin's book (Rubin, 1987). As the amount of data increased significantly, dealing with flat polygon representations became difficult. The use of replication and hierarchy became common in IC layout, reducing complexity and memory usage. New data structures, able to include relations among polygons, began to be used. An editor widely used in academia, at least in the United States, is MAGIC (Osterhout *et al.*, 1984), where the concept of links among some neighboring polygons (called "corner stitching") was exploited to perform operations beyond the usual polygon editing such as geometrical design rule checking. Other two-dimensional data organizations, such as dividing the available area into quadrants and storing the data in related trees ("quad-trees"; see Rosenberg, 1985), have been used to allow quick searches.

Entering polygons to represent an integrated circuit is an extremely time-consuming and error-prone task. While reliable data are difficult to obtain, the problem being similar to the question "How many lines of code does a programmer write per day?," the figure is generally a few dozen unique transistors per day. Representing the masks at an abstract level could simplify the process and increase productivity significantly. The concept of symbolic layout (Williams, 1978) was introduced in the context of MOS IC design, even though the idea of representing basic objects as symbols is older. Connections are represented by dimensionless wires along centerlines on the correct process layers, while primitive elements are treated as objects with design parameters. For instance, a MOS transistor would be described in terms of its type, channel width, and length. The editing functions are simplified greatly as the designer does not need to enter the actual polygons and is not concerned with the exact spacings but only with the topology of the layout. A program acts on the symbolic description and expands it into the actual polygons with required spacings based on the technology rules. Unfortunately, the density of the layout obtained after this simple expansion is generally very poor. It has been necessary to alter the relative position of transistors and interconnect in order to reduce the total layout area. This step is called compaction. A compactor program creates a mathematical model to represent the many geometrical constraints (such as spacings, avoiding the intersection of some layers, or keeping some elements fixed) among the large number of basic elements (transistors and connection polygons) and uses an optimization technique to minimize the total layout area subject to the derived constraints. Many algorithms for performing this task continue to be proposed, starting from the first, one-dimensional graph models, to flexible probabilistic ones (Boyer, 1988). In spite of the progress achieved, the use of this technique to design large circuits has not been widespread. The best successes reported have been in the use of symbolic layout to design at the cell level (circuits up to a few dozen transistors), particularly because of the simplicity in transporting the cell layouts across similar semiconductor processes.

Another way of describing layout data is by creating a procedural description that, when executed as a program, generates the actual polygons (Johanssen, 1979; Batali and Hartheimer, 1980). The layout is then represented by text. Typical basic elements, such as a transistor or a contact, could be instantiated by calls of the parametrized procedures. While there is some loss of design freedom as we cannot operate directly on the polygons, the layout representation could take much less memory and storage than the full polygon expansion. Another advantage is the possibility of using more abstract geometrical concepts to express relative positions, such as "to_the_right_of." A useful abstraction introduced in the late 1970s was the simplification of the geometrical design rules so that they could all be described as multiples of a basic unit, resulting in the so-called "lambda style rules" (Mead and Conway, 1980). The use of procedural layout languages and simplified design rules has been especially fruitful in the design of special array-type compilers, such as memory or data-path generators. They form a very important part of today's layout systems, including those offered by commercial CAD companies.

6.2 Layout Verification

The drawings describing the mask layers which form the integrated circuit contain millions of polygons. Whether these are generated by hand or by software programs, errors can be introduced making it necessary to perform exhaustive analysis on the data. One looks for three types of problems: compliance to the geometrical design rules, equivalence with the schematic data, and appearance of electrical effects detrimental to the chip performance.

Each semiconductor process has a set of geometrical design rules describing its layers, the minimum dimensions acceptable for each layer, the minimum separation among polygons, and the interactions between the different layers. These interactions allow, for example, the overlapping of a number of layers to form transistors or the prohibition of intersection of certain layers to prevent connections. A geometrical design rule checking (DRC) program verifies the polygon database measuring polygon sizes, distances, and intersections and indicates the places where a rule has been violated. Usually, the violations are superimposed on the layout in such a way that they can be displayed in the layout editor. Some editors, such as MAGIC (Osterhout *et al.*, 1984), allow running a DRC program interactively as the layout is being done.

The second problem that needs to be resolved is to ensure the equivalence of the devices formed by the masks with the database generated from the schematics (or some other data). Again, the process has rules that describe how electrical elements are formed and how the geometrical dimensions are translated into electrical parameters. A typical case in MOS technologies is how a transistor is formed by intersecting some rectangles in specific layers, and how the resulting rectangle's dimensions relate to the transistor's basic parameters, its channel width and length. So, we are considering two related problems: conversion of the geometrical database into devices and interconnections, and proof that the extracted device database is actually equivalent to another of the same basic type. A CAD program performing the first step is called a layout or node extractor, while the checking one is a wirelist comparator. Unfortunately, another set of circuit elements is created by the layout. For instance, rectangles overlapping in different layers, such as the metal layers used for interconnect, will create a capacitive coupling between the corresponding nets. Each technology has a number of effects introducing unintended circuit elements called parasitics. The parasitics can alter the circuit performance or, worse, the actual function that the circuit was designed to implement. The layout extractors usually extract the parasitics of importance to the design, and these are examined according to design restrictions and passed to simulators.

The approaches to rule checking and extraction have been similar. We can operate on the polygon database through a set of basic operations, such as intersections, unions, differences, and expansion, and express the extraction and spacing rules in terms of these operations. Another approach has been to rasterize the polygon data and "slide" a window recognizing specific patterns, taking advantage of the fact that the rules can be localized. In all cases, use of sorting techniques and special representations has been common. Since many rules are independent and apply on a layer-by-layer basis, the checking process lends itself well to the use of multiprocessors. A much more difficult issue has been the checking of three-dimensional effects, such as conductor couplings, where recognition of particular geometry patterns and scaling are used to avoid solving complex differential equations.

7. AUTOMATIC LAYOUT

Efforts to automate the layout tasks (as opposed to simplifying the entering and checking of data) by use of special algorithms started in the early days of CAD. Traditionally, layout tasks have been divided into placement and routing. Placement consists of finding the exact locations that the basic components or cells (small collections of gates that are treated as a single unit) will occupy on a chip or a board. Routing consists of determining the precise paths, including layers, that nets will follow to connect the basic elements' pins. Since many placement and routing problems fall into the class of *NP*-complete problems, it became necessary to develop heuristic algorithms to obtain solutions in reasonable computer times.

7.1 Placement

A key difficulty in placement problems is the huge search space. If we try to place N elements into N slots, there are $N!$ possible configurations. Today, a block in a chip can easily contain 10 000 cells, making it impossible to do anything remotely close to an exhaustive search. Algorithms to place components on a board started to appear in the 1960s. The basic ideas in the early approaches were first to cluster elements according to some specific criteria, such as having a large number of nets in common, and then to perform element exchanges to improve the configuration on the basis of some measure of its "goodness" or cost. The selection of which components to exchange was based on connectivity considerations. The cost function tried to reflect the congestion of the nets in the areas available for routing and some easily computable approximation to the actual net lengths that would result after routing. The simplest approximation to net length used is the half perimeter of a rectangle circumscribing all pins, followed in complexity by some special trees. The expectations were that routes would be short, and so the same tracks could be used for several nets achieving full routing within the available space. Examples of these approaches are described in Breuer's book (Breuer, 1972). One key deficiency in these initial approaches is that they could only drive the placement into a local minimum of the cost function.

A more elaborate model involved considering a placement as a graph, with the components as vertices and the nets as edges, and application of graph-partitioning techniques. In the early 1970s, the "min-cut" approach was introduced (Kernighan and Lin, 1970). As its name indicates, we try to exchange components so that if we "cut" the graph into two, the number of nets crossing between the two sides is minimized. In the process, many iterations are performed, and we are allowed to go through configurations that are not always better than the previous ones, allowing an escape from local minima. There have been multiple refinements of this basic idea. These include the use of elaborate net-weighting schemes to give preferences to some nets over others, the design of special software data structures to simplify the algorithmic complexity of the procedure (Fidduccia and Mattheyses, 1982), and the introduction of fictitious pins in a recursive approach to divide the available area (Dunlop and Kernighan, 1985).

In the early 1980s, a probabilistic approach to solving combinatorial problems was introduced (Kirkpatrick *et al.*, 1983). The overall process was analogous to the melting of a solid at high temperature and then slowly cooling it to form a crystal (annealing to a state of minimum energy), thus the name simulated annealing. The method had a number of important advantages: the cost function could be quite general, the method allowed escaping from local minima, and the process could be stopped at any desired point. More general schemes following the annealing basic principles were called probabilistic hill-climbing algorithms. The algorithms required determining three basic operations: the generation of new configurations, the modification of the algorithm's control parameters, and the configuration acceptance or rejection criteria. Formal proofs of convergence with probability equal to 1 to a global minimum were given under fairly general assumptions on the specific problem, based on Markov chain theory (Romeo and Sangiovanni-Vincentelli, 1985). In the case of simulated annealing, the control parameter is called temperature. Initially, the temperature was decreased by a constant factor in a geometrical fashion (new_temperature = old_temperature×ratio, where $0 < \text{ratio} < 1$), but a number of adaptive schemes have been introduced. The acceptance criterion is based on the comparison of an exponential function of the cost difference between the current and newly generated configuration, exp(−cost_difference/temperature), with a randomly generated value in the range of 0 to 1. Many problem-specific stopping criteria have been described.

Simulated annealing has been applied with great success to the placement of row-style, standard cell blocks in integrated circuits. A large number of refinements, particularly related to techniques to cut down the computer run times of the algorithm and to more general placement problems, have been introduced. Timberwolf, a system originally developed in the University of California at Berkeley and subsequently at Yale University, deserves particular mention due to its industrial acceptance (Sechen, 1988).

In general, the basic measure used in placement algorithms has been a combination of total area after routing and average interconnect net length. Recently, interest has shifted toward "performance-driven" placement, meaning taking into account the effects of the placement on the overall performance of the circuit in a more direct fashion. For instance, nets can have very different net lengths (roughly proportional to delay) and still meet the timing requirements because of their specific path constraints (Benkoski and Strojwas, 1991). While minimal area and short connections are certainly related to these objectives, the emphasis and resolution have changed. Closer ties to the timing verification systems, and possibly changes in the circuit parameters, may be required to meet all constraints automatically.

7.2 Routing

Routing is the process of determining the actual paths assigned to the interconnections among the pins of the placed components. Each technology allows a certain number of conducting layers to carry nets and of "vias" or "contacts" to switch among the layers. Each layer has its own electrical and geometrical characteristics, such as resistivity per square, fixed or variable widths, and many others. The vias also vary widely; in some cases we can jump directly across several layers, in others only between neighboring ones. To illustrate the ranges, we give two examples. First, until recently, CMOS technologies only allowed the use of up to two metal layers for routing, while boards had dozens of layers available. Second, other typical constraints are that all paths must fit inside a fixed area, that paths must never get closer than a given minimum spacing, or that certain layers are assigned a fixed direction (north–south or east–west). A router is a program that assigns paths to the nets so that paths corresponding to different nets never overlap on the same layers.

A common situation is that only two directions are allowed, so that "wires" can turn 90°, resulting in what is referred to as "Manhattan geometry." If we add to this assumption a fixed width and minimum spacing per layer, then each layer can be modeled as a grid. The first algorithm reported to find paths, still of interest today, is the Moore–Lee algorithm (Lee, 1961), based on graph-theory shortest-path algorithms. The basic idea is to expand a net, starting at its pins, to the neighboring grid points successively until all required connections are achieved. Routers based on this technique are called maze routers. While very flexible, their results depend on the order in which the nets are routed, and their implementation can consume large amounts of memory since we need to store a matrix representing the grid points. A new algorithm was proposed on the basis of a technique called line search (Hightower, 1969). Instead of advancing by a minimum unit, we try to determine in one step a whole line segment containing the "current" point, extending the segment until an obstacle or a target is found. Once a segment is constructed, we explore new segments in the direction orthogonal to the previous one, starting from each of its end points. The search is conducted from the two points we wish to connect, and when the two set of segments intersect, the connection has been achieved. While the algorithm does not guarantee that a path is found, or even that a path exists, it is much faster than a Moore–Lee type algorithm and it takes less memory. Many refinements and combinations of these basic approaches have been used through the years, particularly for board routing. Another problem frequently studied was how to "rip out" a connection already made to allow the completion of others and to "reroute" the previous one.

As integrated circuits moved into the VLSI scale in the late 1970s, routing among their basic cells or blocks required dealing with areas that, properly scaled for the interconnect layer dimensions, were much larger than those found in boards. The routing problem was then split into subproblems, each with a different objective. The available area was divided into routing regions, usually rectangular, with special characteristics so that different algorithms could be applied in each type of area. The overall routing process was divided into phases, each adding increased levels of detail. In the case of ICs, the division of the chip into major functional blocks and routing regions is called floor planning.

The most important division introduced was between global and detailed routing, a concept applicable to IC and board problems. Global or "loose" routing is the process of simply distributing nets among the identified routing areas. Once this assignment is complete, the detailed routing phase determines the exact paths inside each area using any appropriate router. For global routing, there are two key considerations: to minimize an estimated net length and to avoid excessive congestion in any area (the allocation of so many nets into the area that detailed routing becomes physically impossible). One common approach has been to model the routing areas as vertices in a graph whose edges connect only neighboring areas. The model is refined by adding capacity to the edges to represent the number of nets that can be allowed in the region, and a length to indicate an expected average distance that a path will traverse inside the region. Once a model like this is set up, many algorithms can be called to assign the nets to the graph arcs.

Some special areas have become typical in IC layout, generating considerable research to solve their specific problems. The most important has been channel routing. In this case, we have a rectangular region with pins located only at its top and bottom edges; the vertical separation can be changed, and two conducting layers are available, one for each orthogonal direction. The problem is to connect all pins using the minimum possible number of horizontal tracks or, equivalently, vertical spacing (Hashimoto and Stevens, 1971). Many algorithms have been developed to solve this problem (see Lengauer, 1990). Another important region is the switch box, differing from a channel essentially by having pins on all sides and nonrectangular shape.

A related problem is the interaction among the different regions through the routing process. We may be forced to enlarge a channel region, altering the overall structure of the initial division: New regions can easily appear. More integrated approaches to resolve routing without resorting to complex area divisions are starting to receive attention. A key issue is the ability to route these large regions in reasonable computer times. The most recent CMOS gate arrays, offering hundreds of thousands of gates and three metal layers for interconnect, have made the problem particularly interesting for industry.

8. HIGHER-LEVEL DESIGN TOOLS

A number of approaches have been developed to move the design process as much as possible toward higher levels of abstraction. To reach this objective successfully, we need to increase the automation of the "lower-level" design tasks. Systems that automatically produced layouts starting from high-level descriptions were referred to as silicon compilers (Johanssen, 1979, Siskind *et al.*, 1982). The name was based on an analogy with a compiler for a software language, the behavior-level or RTL description playing the role of the language and the mask-level layout the equivalent of the object code. A large amount of research was done on the topic, and the approach generated considerable controversy. Ultimately, these early tools failed in actual applications because of their too broad scope. For a general-purpose microprocessor, many problems needed to be resolved simultaneously at a time when none of them, considered individually, had a fully satisfactory solution. For more recent developments, the reader is referred to De Micheli *et al.* (1987), and Gajski (1988).

Some silicon compilers attempted to map a description into a specialized chip architecture, restricting their applicability but taking advantage of optimizations valid only in their more restrictive domains. The most successful area has been in compilers targeted to digital signal processing algorithms. An early effort, a system developed at Edinburgh called FIRST, targeted filter structures using a bit serial architecture implemented in MOS technology (Denyer and Renshaw, 1985). Other ongoing efforts in this area that can be mentioned are the CATHEDRAL system at the Interuniversity Micro-Electronics Center in Belgium (Note *et al.*, 1991) and LAGER at the University of California at Berkeley (Pope *et al.*, 1987).

Other types of specialized systems in existence today are also called "silicon compilers." In general, these tools generate layouts for special kinds of circuits starting from a schematic like description and using libraries of cell layouts. So most are really special module generators that perform, beyond assembling a layout, specific optimization steps and typically generate appropriate simulation models. Standard examples include PLA and memory structure generators and data-path assemblers. These tools are offered by commercial CAD vendors and have been widely used in industry.

GLOSSARY

Bus: A set of bit signals of related type—for instance, the 32-bit signals commonly used as a word length in computer architectures.

Cell: A set of gates that is treated as a single entity. Cells are usually placed in libraries, with multiple descriptions available such as layout, schematic symbols, and simulation models.

Circuit: A collection of interconnected electrical devices. It consists of electrical devices and connected pins (nets or nodes). See Network.

CMOS: Complementary metal oxide semiconductor, an integrated-circuit technology. See MOS and NMOS.

Contact: Vertical connection from one layout layer to another; also referred to as via.

Data Path: A regular structure used in layout, consisting of functional units and busses connecting them.

Electrical Device: Primitive circuit element, such as resistor, capacitor, inductance, or transistor.

Floor Plan: The physical layout of the chip in terms of its main units, such as data paths, memories, and power and ground busses.

Gate Array: An integrated circuit where transistors are arranged in a two-dimensional array, and the design is done by specifying only the metal interconnect layers.

Grid: Uniformly spaced array of points in a coordinate space. Usually displayed in layout editors to guide the editing process and indicate the minimum geometrical resolution available.

Interconnect: A conductor electrically joining two or more pins.

Layout: The physical description of a design, in terms of polygons attached to technology layers, forming interconnect and electrical devices such as transistors.

Logic Gate: A functional unit of combination Boolean logic, such as an inverter (NOT function) or a NAND-2 (logical AND followed by NOT of two input signals).

Mask: Glass usually covered with chrome etched away in particular areas where specific material will be placed on the wafer during the semiconductor manufacturing process.

MOS: Metal oxide semiconductor, an integrated-circuit technology. NMOS and CMOS are the two most common MOS technologies.

Net: A set of electrically connected pins.

Netlist: A list of all nets in a circuit.

Network: A collection of interconnected electrical devices. It consists of electrical devices and connected pins (nets or nodes). See Circuit.

NMOS: *n*-type metal oxide semiconductor, an integrated-circuit technology.

Pin: A connection point in a device.

Router: A software program used to create the interconnections in a network where the devices have physical locations.

Standard Cell: A cell whose layout conforms to some uniform characteristics (such as given height). Usually, they form libraries, and designs are done with choice restricted to only those elements.

Transistor: An active semiconductor device with three (or more) terminals. It is a basic element in integrated-circuit design.

Via: Vertical connection from one layout layer to another; also referred to as a contact.

VLSI: Very-large-scale integration, a term used to describe ICs containing over 100 000 transistors.

Wire: The physical realization of a connection, indicating a detailed conducting path among pins.

List of Works Cited

Abramovici, M., Breuer, M. A., Friedmann, A. D. (1990), *Digital System Testing and Testable Design,* Rockville, MD: Computer Science Press.

Agrawal, P., Dally, W. T., Fisher, W. C., Jagadish, H. V., Krishnakumar, A. S., Tutundjian, R. (1987), "MARS: A Multiprocessor Based Programmable Accelerator," *IEEE Design Test Comput.,* **4** (5), 28–36.

Batali, J., Hartheimer, A. (1980), *The Design Procedure Language Manual,* VLSI Memo No. 80-31, Artificial Intelligence Laboratory, MIT, Cambridge, MA: MIT Press.

Benkowski, J., Strojwas, A. J. (1991), "The Role of Timing Verification in Layout Synthesis," *Proc. 28th Design Automation Conference, 1991,* Los Alamitos, CA: IEEE Computer Society Press, pp. 612–619.

Boyer, D. (1988), "Symbolic Layout Compaction Review," *Proc. 25th Design Automation Conference, 1988,* Los Alamitos, CA: IEEE Computer Society Press, pp. 383–388.

Brace, K. S., Rudell, R. L., Bryant, R. E. (1990), "Efficient Implementation of a BDD Package," *Proc. 27th Design Automation Conference, 1990,* Los Alamitos, CA: IEEE Computer Society Press, pp. 40–45.

Brayton, R. K., Hachtel, G. D., McMullen, C. T., Sangiovanni-Vincentelli, A. L. (1984), *Logic Minimization Algorithms for VLSI Synthesis,* Boston: Kluwer Academic.

Brayton, R. K., Brenner, N., Chen, C., De Micheli, G., McMullen, C. T., Otten, R. (1985), "The Yorktown Silicon Compiler," *Proc. Int. Symposium Circuits and Systems, 1985,* New York: IEEE Publishing Services, pp. 391–394.

Brayton, R. K., Rudell, R., Sangiovanni-Vincentelli, A. L., Wang, A. (1987), "MIS: A Multiple-Level Logic Optimization System," *IEEE Trans. Comput.-Aided Design,* **CAD-6** (6), 1062–1081.

Breuer, M. A. (Ed.) (1972), *Design Automation of Digital Systems,* Vol. 1, Englewood Cliffs, NJ: Prentice-Hall.

Bryant, R. E. (1986), "Graph Based Algorithms for Boolean Function Manipulation," *IEEE Trans. Comput.* **C-35**, (8), pp. 677–691.

Bryant, R., Beatty, D., Brace, K., Cho, K., Sheffler, T. (1987), "COSMOS: A Compiled Simulator for MOS Circuits," *Proc. 24th Design Automation Conference, 1987,* Los Alamitos, CA: IEEE Computer Society Press, pp. 9–16.

Darringer, J. A., Brand, D., Gerbi, J. V., Joyner, W. H., Trevillyan, L. (1984), "LSS: A System for Production Logic Synthesis," *IBM J. Res. Devel.* **28**, (5), 537–545.

de Geus, A., Cohen, W. (1985), "A Rule-Based Expert System for Optimizing Combinational Logic," *IEEE Design Test* **2** (4), 22–32.

De Micheli, G., Sangiovanni-Vincentelli, A., Antognetti, P. (1987), *Design Systems for VLSI Circuits; Logic Synthesis and Silicon Compilation,* Dordrecht, Netherlands: Martinus Nijhoff.

Denneu, M. (1982), "The Yorktown Simulation Engine," *Proc. 19th Design Automation Conference, 1982,* Los Alamitos, CA: IEEE Computer Society Press, pp. 55–59.

Denyer, P., Renshaw, D. (1985), *VLSI Signal Processing: A Bit Serial Approach,* Reading, MA: Addison-Wesley.

Dewey, A., Gadient, R. (1986), "VHDL Motivation," *IEEE Design Test Mag.* **3** (1), 12–16.

Dietmeyer, D. L. (1978), *Logic Design of Switching Functions,* 2nd ed., Boston: Allyn and Bacon.

Dunlop, A. E., Kernighan, B. W. (1985), "A Procedure for Placement of Standard-Cell VLSI Circuits," *IEEE Trans. Comput.-Aided Design,* **CAD-4** (1), 92–98.

Fiduccia, C. M., Mattheyses, R. M. (1982), "A Linear-Time Heuristic for Improving Network Partitions," *Proc. 19th Design Automation Conference, 1982,* Los Alamitos, CA: IEEE Computer Society Press, pp. 175–181.

Friedman, T. D., Yang, S. C. (1968), "Methods Used in an Automated Logic Design Generator (ALERT)," *IEEE Trans. Comput.* **C-18**, 593–613.

Gajski, D. (Ed.) (1988), *Silicon Compilation,* Reading, MA: Addison-Wesley.

Garey, M., Johnson, D. (1979), *Computers and Intractability: A Guide to the Theory of NP-Completeness,* San Francisco: Freeman.

Glasser, L. A., Dobberpuhl, D. W. (1985), *The Design and Analysis of VLSI Circuits,* Reading, MA: Addison-Wesley.

Hafer, L. J., Parker, A. C. (1983), "A Formal Method for the Specification, Analysis, and Design of Register-Transfer Level Digital Logic," *IEEE Trans. Comput.-Aided Design* **CAD-2** (1), 4–18.

Hashimoto, A., Stevens, J. (1971), "Wire Routing by Optimizing Channel Assignment within Large Apertures," *Proc. 8th Design Automation Conference, 1971,* Los Alamitos, CA: IEEE Computer Society Press, pp. 155–169.

Hightower, D. W. (1969), "A Solution to Line Routing Problems on the Continuous Plane," *Proc. 6th Design Automation Conference, 1969,* Los Alamitos, CA: IEEE Computer Society Press, pp. 1–24.

Hill, D., Shugard, D., Fishburn, J., Keutzer, K. (1989), *Algorithms and Techniques for VLSI Layout Synthesis,* Boston: Kluwer Academic.

Insinga, A. (1984), "Behavioral Modeling inside a Structural Level Simulator," *Proc. International Conference on CAD, 1984,* Los Alamitos, CA: IEEE Computer Society Press, pp. 42–44.

Johanssen, D. (1979), "Bristle Blocks: A Silicon Compiler," *Proc. 16th Design Automation Conference, 1979,* Los Alamitos, CA: IEEE Computer Society Press, pp. 310–313.

Kernighan, B. W., Lin, S. (1970), "An Efficient Heuristic Procedure for Partitioning Graphs," *Bell Syst. Tech. J.* **49**, 291–308.

Keutzer, K. (1987), "DAGON: Technology Binding and Local Optimization," *Proc. 24th Design Automation Conference, 1987,* Los Alamitos, CA: IEEE Computer Society Press, pp. 341–347.

Kirkpatrick, S., Gelatt, C., Vecchi, M. (1983), "Optimization by Simulated Annealing," *Science* **220**, 671–680.

Lee, C. Y. (1961), "An Algorithm for Path Connections and its Applications," *IRE Trans. Electronic Comput.* **EC-10**, (3), 346–365.

Lengauer, T. (1990), *Combinatorial Algorithms for Integrated Circuit Layout,* New York: Wiley.

Marwedel, P. (1986), "A New Synthesis Algorithm for the MIMOLA System," *Proc. 23th Design Automation Conference, 1986,* Los Alamitos, CA: IEEE Computer Society Press, pp. 271–277.

McCluskey, E. J. (1956), "Minimization of Boolean Functions," *Bell Syst. Tech. J.* **35**, 1417–1444.

Mead, C. A., Conway, L. (1980), *Introduction to VLSI Systems,* Reading, MA: Addison-Wesley.

Nagel, L. W. (1975), *SPICE2: A Computer Program to Simulate Semiconductor Circuits,* ERL Memo 520, 1975, University of California at Berkeley.

Note, S., Geurts, W., Catthoor, F., De Man, H. (1991), "CATHEDRAL-III; Architecture Driven High-Level Synthesis for High-Throughput DSP Applications," *Proc. 28th Design Automation Conference, 1991,* Los Alamitos, CA: IEEE Computer Society Press, pp. 597–602.

Osterhout, J., Hamachi, R., Mayo, R., Scott, W., Taylor, G. (1984), "Magic: A VLSI Layout System," *Proc. 21th Design Automation Conference, 1984,* Los Alamitos, CA: IEEE Computer Society Press, p. 152.

Pan, J-W. Biro, L., Grodstein, J., Grundmann, W., Yen, Y-T. (1991), "Timing Verification of a 1.2 M Device Full-Custom CMOS Design," *Proc. 28th Design Automation Conference, 1991,* Los Alamitos, CA: IEEE Computer Society Press, pp. 551–554.

Pillage, L. T., Rohrer, R. A. (1990), "Asymptotic Waveform Estimation for Timing Analysis," *IEEE Trans. Comput.-Aided Design* **CAD-9**, (4), 352–366.

Pope, S. P., Rabaey, J., Broderson, R. W. (1987), "Automatic Generation of Digital Signal Processing Circuits," in: G. De Micheli, *et al.* (Eds.), *Design Systems for VLSI Circuits; Logic Synthesis and Silicon Compilation,* Dordrecht, Netherlands: Martinus Nijhoff.

Romeo, F., Sangiovanni-Vincentelli, A. (1985), "Probabilistic Hill-Climbing Algorithms: Properties and Applications," in: H. Fuchs (Ed.), *1985 Chapel Hill Conference on VLSI,* Rockville, MD: Computer Science Press, pp. 393–418.

Rosenberg, J. B. (1985), "Geographical Data Structures Compared: A Study of Data Structures Supporting Region Queries," *IEEE Trans. Comput.-Aided Design,* **CAD-4** (1), 53–67.

Rubin, S. M. (1987), *Computer Aids for VLSI Design,* Reading, MA: Addison-Wesley.

Saleh, R. A., Newton, R. A. (1990), *Mixed-Mode Simulation,* Boston: Kluwer Academic.

Sechen, C. (1988), *VLSI Placement and Global Routing using Simulated Annealing,* Boston: Kluwer Academic.

Siskind, J., Southard, J., Crouch, K. (1982), "Macpitts: Generating Custom VLSI Design from Succinct Algorithmic Description," in: P. Penfield (Ed.), *Proc. MIT Conference on Advanced Research on VLSI, 1982,* Dedham, MA: Artech House, pp. 28–39.

Sze, S. M. (Ed.) (1984), *VLSI Technology,* New York: McGraw-Hill.

Terman, C. J. (1983), "RSIM: A Logic-Level Timing Simulator," *Proc. International Conference on Computer Design, 1983,* Los Alamitos, CA: IEEE Computer Society Press, pp. 437–440.

Thomas, D., Dirkes, E. M., Walker, R. A., Rajan, J. V., Nestor, J. A., Blackburn, R. L. (1988), "The System's Architect Workbench," *Proc. 25th Design Automation Conference, 1988,* Los Alamitos, CA: IEEE Computer Society Press, pp. 337–343.

Vladimirescu, A., Liu, S. (1980), "The Simulation of MOS Circuits Using SPICE2," ERL Memo 80/7, 1980, University of California at Berkeley.

White, J. K., Sangiovanni-Vincentelli, A. (1987), *Relaxation Techniques for the Simulation of VLSI Circuits,* Boston: Kluwer Academic.

Williams, J. D. (1978), "STICKS, a Graphical Compiler for High Level LSI Design," *Proc. AFIPS Conference* **47**, Montvale, NJ: AFIPS Press, pp. 289–295.

Further Reading

De Micheli, G., Sangiovanni-Vincentelli, A., Antognetti, P. (1987), *Design Systems for VLSI Circuits; Logic Synthesis and Silicon Compilation,* Dordrecht, Netherlands: Martinus Nijhoff.

Dietmeyer, D. L. (1978), *Logic Design of Switching Functions,* 2nd ed., Boston: Allyn and Bacon.

Gajski, D. (Ed.) (1988), *Silicon Compilation,* Reading, MA: Addison-Wesley.

Glasser, L. A., Dobberpuhl, D. W. (1985), *The Design and Analysis of VLSI Circuits,* Reading, MA: Addison-Wesley.

IEEE Proceedings (1990), Special Issue on the Future of Computer-Aided Design, *IEEE Proc,* **78**, (2), February 1990.

Lengauer, T. (1990), *Combinational Algorithms for Integrated Circuit Layout,* New York: Wiley.

Lund, P. (1986), *Printed Circuit Board Precision Artwork Generation and Manufacturing Methods,* Westlake Village, CA: Bishop Graphic.

Noble, P. J. W. (1989), *Printed Circuit Board Assembly,* Milton Keynes: Open University Press.

Tredennick, Nick (1987), *Microprocessor Logic Design,* Bedford, MA: Digital.

Saleh, R. A., Newton, R. A. (1990), *Mixed-Mode Simulation,* Boston: Kluwer Academic.

Weste, N., Eshraghian, K. (1985), *Principles of CMOS VLSI Design: A Systems Perspective,* Reading, MA: Addison-Wesley.

COMPUTER-AIDED MANUFACTURING (CAM)

KAZUAKI IWATA, *Department of Mechanical Engineering for Computer-Controlled Machinery, Osaka University, Osaka, Japan*

INTRODUCTION

Manufacturing is defined as a function of transforming raw materials into valuable products that meet the public's needs, by means of facilities, information, and labor power. It is also expressed more concretely as a series of interrelated activities and operations involving the design, material selection, planning, production, quality assurance, management, and marketing of the product by the manufacturing industries (CAM-I, Standard Committee, 1978). Automation in manufacturing has made great strides since the invention of electronic computers. Some typical examples of manufacturing technologies developed so far in conjunction with computers are CAD (computer-aided design), CAM (computer-aided manufacturing), CAD/CAM, FA (factory automation), and CIM (computer integrated manufacturing). CAM, which is the subject of this article, is defined as execution of process and operation planning with the proper input data such as product design specification and production order, and control of production processes with the assistance of computers. The interrelationships between CAD, CAM, FA, and CIM are explained in Sec. 2 (see also Fig.1).

1-56081-063-7/92/$5.00+.50

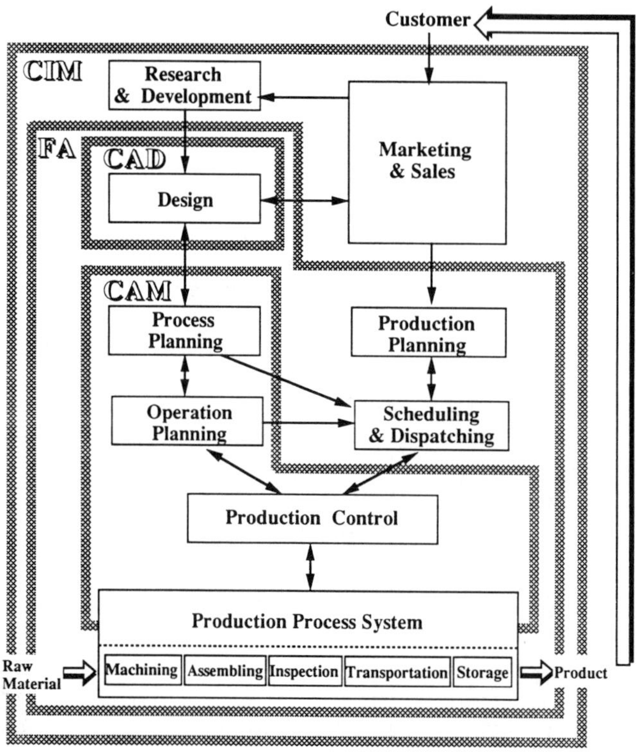

FIG. 1. Role of CAM within framework of CIM.

1. HISTORY

The history of CAM dates back to the early 1950s when research was initiated for the automation of cutting processes by means of NC (numerical control).

NC is a control method in which motions of machines are controlled according to a number of pulses that are generated by a digital circuit depending on input numeric data. Once a program that represents the numeric data is prepared, the operation of the machine can be controlled automatically. NC can be applied to various machines such as machine tools, industrial robots, plotters, etc. The control equipment for NC is generally called the NC controller.

The developments of numerical control have been made mainly in two areas. One is the development of the NC controller that automates production processes, and another is development in the NC program preparation that leads to automation in the area of information processing for manufacturing.

In the first generation of NC controllers, functions such as arithmetic operations, comparisons, and decisions were carried out by a hard-wired logic circuit. However, complicated circuits became necessary with the expansion of the functions. In the early 1970s, CNC (computer numerical control), which used a computer for performing these functions, was developed to improve functionality, expansibility, operability, and maintainability. Nowadays microprocessors are used. DNC (direct numerical control) was also developed in the late 1960s to control and coordinate a number of numerically controlled machines. The technical developments of NC controllers are summarized in Table 1.

Efforts have also been made to add new control features to the NC controller, such as adaptive control and learning control. An adaptive controller identifies the changes in the states of the manufacturing processes and system components, adjusts the control parameters so as to fulfill the control objectives, and then executes the control. Adaptive controllers have already been put into practical use in certain applications. In the next generation of NC controllers, the development of a learning controller is also desired. In addition to the adaptive control function, a learning

Table 1. Development of NC Controller.

Item \ Year	1976–1979	1980–1985		1986–1991	
		Microprocessor			
Processor	Bit slice	8 bit	16 bit	32 bit	CISC/RISC
Technical trend	Expansion of functions and improvement of human–machine interface through computerization of NC			Improvement speed and accuracy through the use of 32-bit processor Human friendly CNC	
	Expansion of NC functions, cutter compensation, fixed cycle, memory operation, etc.	Improvement of human–machine interface, color CRT, interactive symbolic programming, graphic simulation		High-speed, high-accuracy machining control, easy programming, intelligent control	
Servo system	DC analog servo		AC analog servo	AC digital servo	

controller has the function of accumulating and making use of the data obtained in the preceding operations to improve the control performance.

The automation of information processing for manufacturing originates in the development of a general-purpose NC programming language called APT (automatically programmed tools). It was developed at MIT for automatic tool path generation. Later on, IITRI (Illinois Institute of Technology Research Institute) took over the responsibility of the development program and improved the system. The project resulted in APTIII and APTIV, which have greatly contributed to the automation of NC data generation.

Tens of general purpose NC programming languages have been developed in the world since then. At the same time, a large number of special-purpose programming languages that can be applied to specific machine tools or products have been developed by many companies. The success of the APT system has stimulated the development of programming languages for many other manufacturing processes, such as those for plastic forming, assembling, and inspection.

The APT system can only generate tool paths of cutting processes. However, there has been a demand for the extension of the programming system functions to cover the automatic selection of tools and the determination of operating conditions. EXAPT (extended subset of APT), developed in the mid-1960s, is the first general-purpose system that was provided with such functions. Many special-purpose NC programming systems with these functions have been also developed.

The next step in the development of CAM information systems was to realize a system that could execute all necessary information processing from obtaining CAD data to generating data for production processes, such as NC data and process instructions. This kind of system is called the CAD/CAM system. The main problem of the system was that of developing CAPP (computer-aided process planning). Three approaches have been proposed for CAPP so far; they are a variant type, a generative type, and an expert type. The studies of these three types of approaches were started in the late 1960s, the mid-1970s, and the early 1980s, respectively.

In the early 1980s, people became interested in producing a variety of products; therefore, the integration of CAPP and production scheduling in addition to the integration of CAD and CAPP were proposed (e.g., Chryssolouris and Chan, 1985; Iwata and Fukuda, 1989).

In 1988, the concept of concurrent engineering or simultaneous engineering was proposed suggesting parallel processing of design and manufacturing information with constant interactions with each other rather than serial processing (Pennell and Winner, 1989).

FA came into practical use in the early 1980s, and from the mid-1980s, CIM (computer integrated manufacturing) has been put into practice. (The first book about CIM is Harrington, 1973.) With the advance of FA and CIM, CAM is regarded as the integrating component of the total manufacturing system.

2. BASIC CONCEPT AND MAJOR COMPONENTS OF CAM

CAM covers the functions of process planning, operation planning, production control, and production process systems as shown in Fig. 1. FA, on the other hand, includes design and production management (production planning, scheduling, and dispatching) in addition to the activities involved in CAM. Further expansion of the concept of FA leads to CIM, which includes R&D (research and development), marketing, and sales. Therefore, CAM can be understood as a subset of FA and CIM.

The information processing relevant to design, process planning, and operation planning is generally called technological information processing. Thus CAM plays a main role in technological information processing.

Process planning is the planning of production processes of parts or products based on the information and drawings generated in the design stage. Production processes include operations (machining, assembling, painting, packaging, etc.) and interprocess operations (transportation, storage, etc.). The results of the process planning are used as input information for operation planning and scheduling/dispatching.

In operation planning, operations comprising the process are identified and various items required for the executions of operations are determined, such as machining conditions and tool paths.

Production control is processing of control and management information for achieving efficient and reliable execution of various operations and interprocess operations. The production process is a series of operations to transform material into valuable products. CNC and DNC are the important elements of systems that execute the production processes.

It should be noted that in some countries the term CAM is used with a wider or narrower meaning than the definition given here. It may include the functions of production planning, scheduling, and dispatching in addition to the functions defined here, or on the contrary, it may exclude production process systems. The term process planning may also be used in a different way. It may involve functions of both the process planning and operation planning.

3. PROCESS PLANNING

3.1 Processing Flow of Process Planning in Machining for Machine Parts

As defined in the previous section, process planning is the process of generating the production process information such as information about production processes, production machines, jigs, fixtures, and so on, based on the data obtained from design information and drawings (geometry, dimension, tolerance, working conditions, etc.). The results of process planning are usually described in the form of a process chart. Figure 2 shows the information processing flow in the process planning of machine part production. Although the figure shows a straightforward process, there could be a number of feedback loops in the process. This processing flow can be applied to process planning for the production of other types of products such as semiconductors, printed circuit boards, apparel, food, etc., though there are specific features and particular considerations in the details of processing according to the type of product.

3.1.1 Understanding Design Information and Drawings, and Recognition of Production Features. The first step of process planning is to understand the information represented in assembly drawings, part drawings, and part lists, which are prepared in the design stage usually using CAD systems. The information is then checked to see whether it provides all the necessary and error-free information for the subsequent steps. At the same time, the features of products or parts are identified from the aspect of production; these are generally called production features. For example, typical forms or surfaces to be machined are recognized to prepare for the subsequent steps.

3.1.2 Determination of Production Processes and Their Sequence. In this process, possible production processes are selected on the basis of design information, drawings,

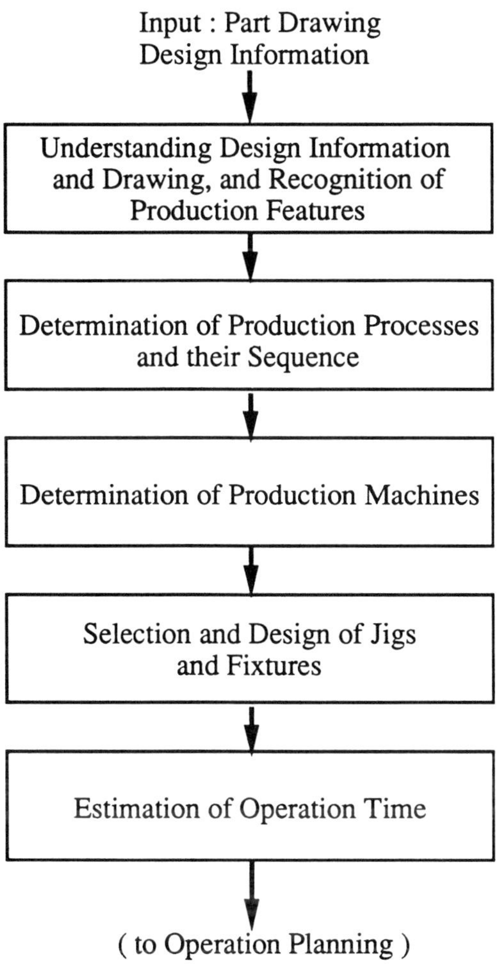

FIG. 2. Information processing flow of process planning of machine parts production.

and manufacturing features. There can be many possible combinations of various production processes such as deformation processing (casting, forging), removal processing (machining, grinding), joining (welding, adhesion), and treatment (surface treatment, heat treatment) for producing the same part with the same geometry. The optimum combination of production processes is determined on the basis of various evaluations in terms of quality, lot size, economic efficiency, and resource savings.

In the case of machining, for example, possible machining processes (such as turning, drilling, end milling, cylindrical grinding, etc.) are selected for each form or surface to be machined in the part. Another example is the selection of a welding method. There can be arc welding, resistance welding, electroslag welding, electron beam welding, laser welding, etc., depending on material and geometries of parts.

If there are multiple possibilities, the optimum processes and their sequences are selected in terms of available facilities, technological level and experience, economic efficiency, production time, etc.

3.1.3 Determination of Production Machines. The production machines (machine tools, forming machines, assembling machines, etc.) are determined corresponding to the production processes selected in the previous step. Since the forms and surfaces of the part are generated by the relative motion of the tool and the part, the possible tool geometry can be identified in this step.

It should be noted that one production machine may be able to perform multiple production processes. Therefore, there could be a number of possible combinations of machines. The optimum combination should be selected considering production time, the number of machines, load balance of the machines, economic efficiency, etc.

3.1.4 Selection and Design of Jigs and Fixtures. Fixtures for fixing workpieces on the pallets or tables of the production machines, and jigs for positioning and clamping the workpieces and for guiding the tools, are selected in this step. This is called auxiliary equipment and is used for improving accuracy, simplifying operations, and shortening the working time.

There are various technological and technical factors to be considered in designing and selecting the auxiliary equipment. These factors are classified into mainly three categories. They are the factors about the production machine such as the position and orientation of the spindle; the factors about the workpiece such as the gravity center and the reference planes for clamping, machining, and assembling; and the factors about the production operations such as vibration and thermal deformation.

3.1.5 Estimation of Operation Time. The operation time is estimated on the basis of the results of the steps described in Secs. 3.1.1–3.1.4. If there are multiple options in the process plan, the shortest operation time is adopted as the estimated time. The estimated time in this step is, however, a rough estimation based on standard times. The precise estimation is carried out after the operation planning.

3.2 Process Planning for Assembly of Printed Circuit Boards

Assembling methods, sequences, and machines are determined in assembly process planning. Few general-purpose process planning systems have been developed for assembling operations that differ from those for machining operations. Process planning systems developed so far are specific to the assembling operations by particular machines. In the following, the process planning for assembly of printed circuit boards, which are widely used in modern industries, is explained.

Figure 3 shows the processing flow of process planning for mounting electronic parts (resistors, condensers, ICs, etc.) on printed circuit boards. Obtaining the data from the CAD system, assembling processes are assigned first. Then collisions between insertion heads of inserters and parts are examined. This process, which is generally called collision check, is important, because the proper insertion sequence should be selected so as to avoid these collisions. Next the assembling processes are determined. Finally, the optimum operation sequence is determined, if there are multiple possibilities in determining insertion sequences. The total mounting time is usually adopted as a criterion for optimization (Japan Society for Precision Engineering, 1991).

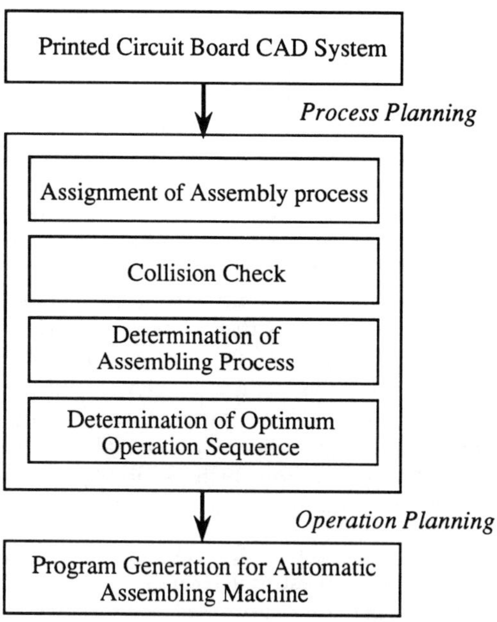

FIG. 3. Information processing flow of process planning for mounting electric parts on printed circuit boards.

3.3 Computer-Aided Process Planning System in Machining

Automation of process planning has been studied mainly for machining. The efforts in developing the CAPP system have contributed to the standardization of manufacturing processes and facilities, and for the rationalization of organizations. However, technologies have not matured to the point that current CAPP systems provide sufficient capabilities for CAD/CAM integration. Thus, the improvement of CAPP systems is an important subject for future factories.

The CAPP systems developed so far are categorized into the following three types:

1. variant type,
2. generative type, and
3. expert type.

3.3.1 Variant-Type CAPP System. The variant-type system uses a decision table for planning the production processes as shown in Fig. 4. The decision table represents the logic of the process planning that has been acquired from past experience. Many systems utilize the concept of GT (group technology) in which parts are grouped into families based on similarities in their production. The code representing the parts family is called the GT code. G2030 and H1819 in Fig. 4 are examples of GT codes. The decision table can be formed so as to represent the sequences of production processes corresponding to the specific GT codes. The systems of this type were the first of the three types to be developed. AUTOPROS (Nissen, 1969), MIPLAN (TNO, 1981), CAPP (CAM-I, 1976), and AUTAP (Eversheim and Fuchs, 1980) are some well-known systems of this type.

3.3.2 Generative-Type CAPP System. In the generative-type CAPP system, the process planning is performed on the basis of the logic that is obtained from the analysis of the part

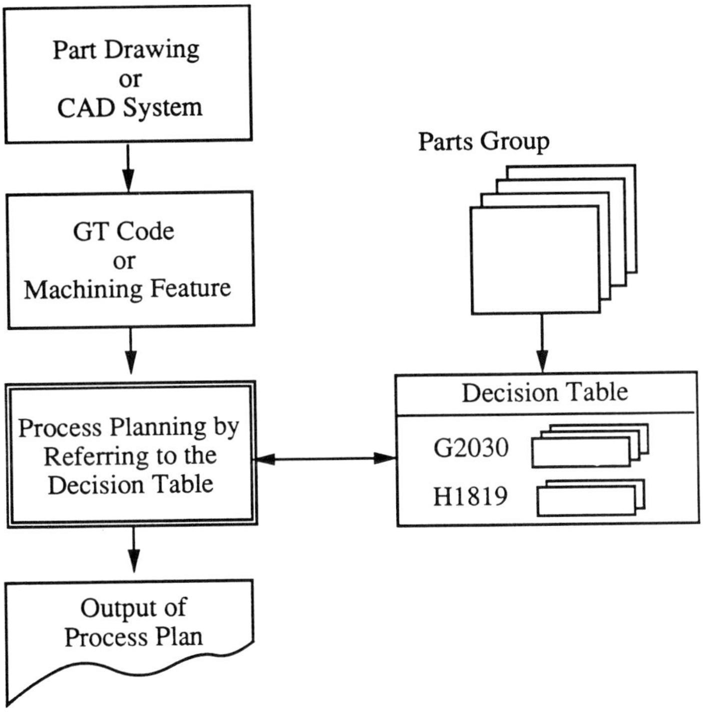

FIG. 4. Basic concept of variant-type process planning.

and the machine tool, in terms of part geometry, required accuracy, form-generating functions of the machine tool, and other related manufacturing technologies. Figure 5 illustrates the basic concept of the generative-type system. Although systems of this type have been studied for many years, further studies are required to develop practical systems, because the logical relations between the production features of the part and the production processes are not clarified.

Some typical research works about systems of this type are CAPSY (Spur *et al.*, 1978), CIMS/PRO (Iwata *et al.*, 1977), and CPPP (Mann, 1977).

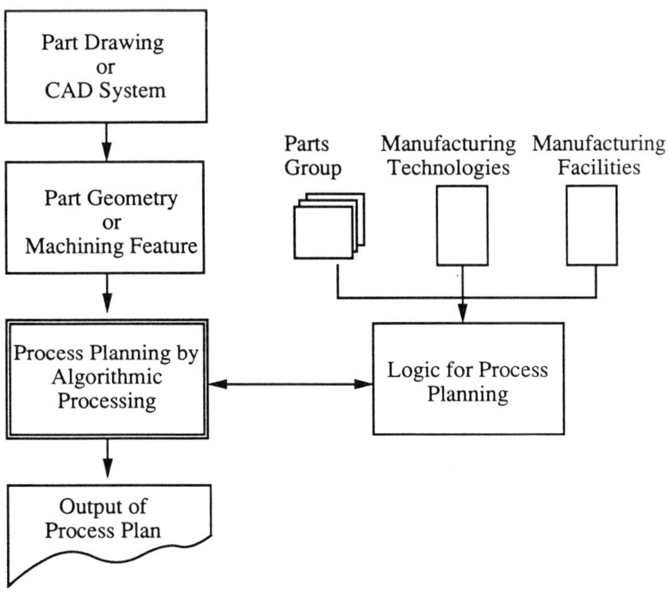

FIG. 5. Basic concept of generative-type process planning.

3.3.3 Expert-Type CAPP System. In recent years, as it became clear that the knowledge and know-how of the process planner played an important role in the practice of the process planning, many expert systems were developed. These systems consist of the inference engine, the knowledge base (containing the know-how and knowledge of the experts), the database (containing general data), the system that acquires the knowledge from the experts, and the interface between the CAD system and the users, as shown in Fig. 6. Many systems of this type have been developed, such as GARI (Descotte and Latombe, 1985), TOM (Matsushima *et al.*, 1982), CIMS/PRO (Iwata and Sugimura, 1984), XPLAN (Van't Erve and Kals, 1986), and KAPPS (Iwata and Fukuda, 1987). Current research work is mainly focusing on a feature-based approach. Process planning is associated with production features identified in the geometric models of parts that are generated in the design stage using CAD systems.

Until 1980, research and development work about CAPP was carried out in the field of removal processing, especially for machining. In the 1980s, however, research was extended to cover assembling and deformation processing (casting, forging, sheet-metal forming). Since 1990, it has been recognized that research on automation of process planning for all manufacturing processes is needed.

4. OPERATION PLANNING

4.1 Processing Flow in Operation Planning

The task of operation planning is to determine the detailed information necessary for the execution of production processes (such as machining and assembling), on the basis of the output of the process planning. The information processing flow for the operation planning of machine parts includes

1. determination of the operation sequence and operators within a process,
2. selection of tools,
3. determination of conditions of the operations,
4. determination of the tool path, and
5. postprocessing.

Operation planning of products or parts with simple geometries does not require much experience. However, for products with complicated geometries even the determination of

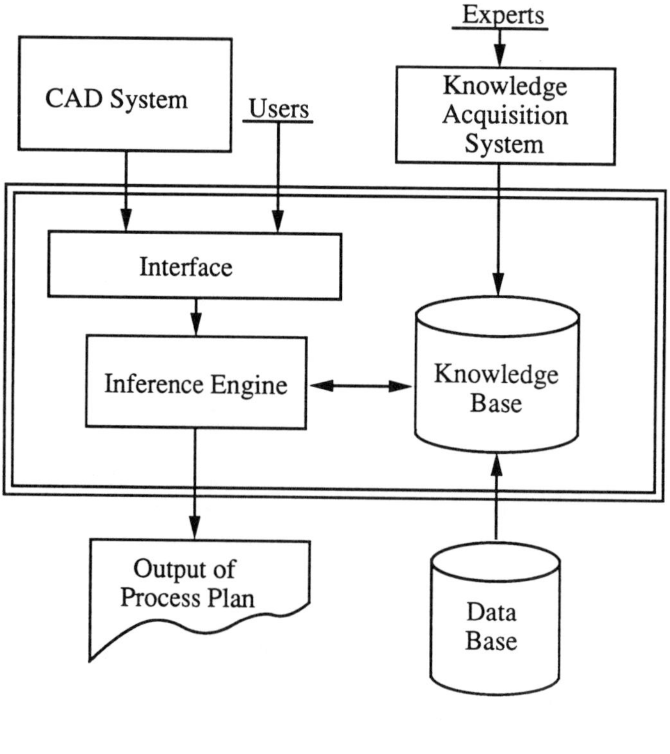

FIG. 6. Basic structure of expert-type process planning system

the tool path is a difficult task and requires a lot of work, time, and cost. A lot of effort has, therefore, been made on the automation of operation planning.

4.1.1 Determination of the Operation Sequence and Operators within a Process. Each production process determined by process planning is analyzed to identify what operations should be executed in which sequence and what actions of the operators are required within the process. Let us consider the operation of external cutting in turning, for example. The required operations are chucking the workpiece, clamping by the tail stock, air cutting, cutting (a number of times, if necessary), measuring, detaching the workpiece, cleaning the machine, disposing of chips, etc. In operation planning, such operations must be properly selected and their sequence must be determined. The series of operations for one production process is usually called the work cycle.

4.1.2 Selection of Tools. In this step, proper tools in terms of material, geometries, and dimensions are selected on the basis of the selected machines (type, size, rigidity, power, structural configuration, etc.), workpieces (material, hardness, rigidity dimension, and accuracy before and after machining, etc.), and types of operations.

4.1.3 Determination of Conditions of Operation. The proper conditions of the operations are determined in this step on the basis of the selected machine, tools, fixtures, and the role of the operators if operators are necessary. The conditions have to be determined so as to produce a product or part that meets the specifications and also to achieve efficient, economic, reliable, and safe operations.

Many studies have been carried out about the optimization of the condition of operations. Note that there are multiple operations to be considered in optimization, and different companies have different objective functions and constraints.

4.1.4 Tool Path Generation. In this step, proper tool path is generated according to the geometry of the part, the movement of the machine tool or assembling machine, and the geometry of the tools. The path must be determined to avoid collision between the tool and tool holder, on the one hand, and the workpiece and the jigs, on the other, and also to achieve minimum operation time.

In the past, the problem of tool path generation was dealt with independently whether the part geometry was $2\frac{1}{2}$ dimensional or 3 dimensional. However, efforts have been made to unify these problems in recent years. The term $2\frac{1}{2}$-dimensional is a special term in the field of manufacturing. It is defined as the dimensionality that is composed of a contouring motion in a 2-dimensional plane and a linear motion along the line perpendicular to the plane. Examples of $2\frac{1}{2}$-dimensional machining are drilling operations, face milling operations, and turning operations.

4.1.5 Postprocessing. In the previous steps, the tool path data were obtained in the form of cutter location data (CL data), which has a general format independent of the configuration of the machine tool. Postprocessing is a process to convert CL data into NC data adapted for the specific machine tool. The postprocessor should be designed so that NC data generated by it should not contain waste motions, should be adjusted to the timing of the mechanisms such as an automatic tool changer, and should be accompanied by proper documents that can be used for various checks by the operator.

The results of operation planning as described above are available through various media such as paper, tapes, and floppy disks. Nowadays they are often transmitted directly from the computer to the machine.

4.2 Computer-Aided Operation Planning Systems

Most computer-aided operation planning systems have been developed as special-purpose systems in companies; thus very few general-purpose systems have been developed.

The most famous general-purpose systems are APT (automatically programmed tools) and its subsets that were developed for the automation of tool path generation. EXAPT (extended subset of APT) is also well known as a general-purpose system that covers the determination of operation sequence, the selection of tools, and the determination of conditions of operations in addition to the tool path generation (e.g., Budde and Weisswellar, 1973).

The EXAPT processor reads a part program written in EXAPT language, executes technological calculations, and outputs the tool path data in the form of CL data, which are then processed by the post processor and converted to the NC data adapted to the specific machine. It can also issue work lists and work orders to the operator.

In the APT and EXAPT systems, NC language is used to input part programs into the system. Other types of input methods have also been developed such as a menu-driven interactive system and a direct interfacing to CAD system.

5. PRODUCTION CONTROL

As described previously, the objective of production control is to realize effective and safe operations of production processes such as machining, assembling, and inspection, and interprocess operations such as transportation and storage. As seen in Fig. 1, the data utilized for production control are classified into the following categories:

1. technological data generated by process planning and operation planning and management data prepared by production planning and scheduling, and
2. the data gathered in the production process (shop floor).

Figure 7 shows the software organization of the production control system which generates the control information using various relevant data. The system has five major functions: data management, operation control, operation monitoring, performance tracking, and status display.

The basic subfunctions and typical application-oriented subfunctions of each function are represented in the figure. The extent of the functions that are covered by a system is dependent on the level and organization of the system. Taking for example FMS (flexible manufacturing system: computer-controlled automated manufacturing system for mid-volume, mid-variety parts production, in which NC machine tools or NC manufacturing cells are combined organically by automatic transportation equipment), the operation control function is, so far, limited to the control of machining operations. There are very few systems that have the functions of assembling control and washing control. Recent systems, however, tend to be equipped with all these functions.

The operation monitoring function is an important function in order to realize automated operation of the production processes.

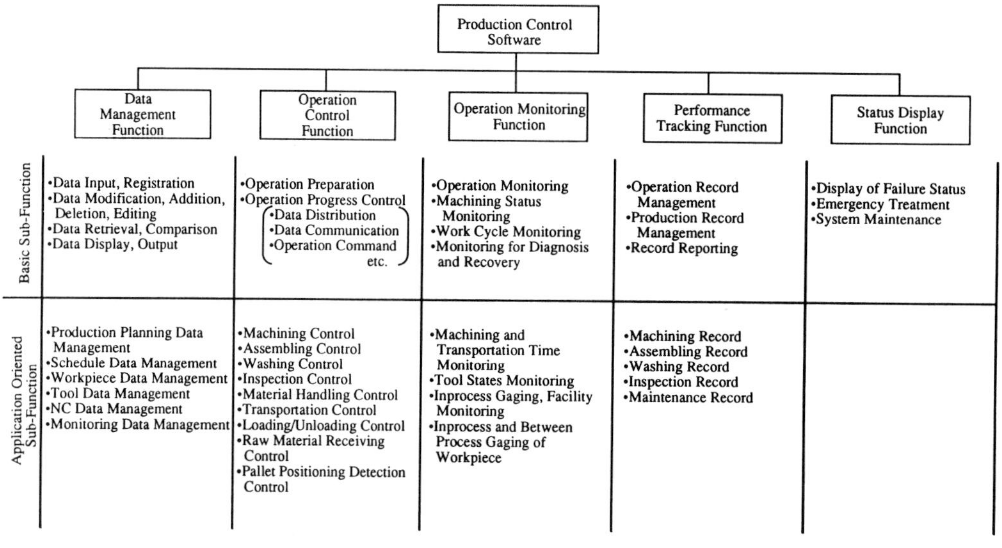

FIG. 7. Software organization for production control.

This function includes monitoring of machining and assembling operations, monitoring of work cycles and monitoring for failure diagnosis, and recovery actions.

It is also very important to pay attention to the human interface that provides the means of communication between the system component and personnel, especially when the system advances in automation.

6. COMPUTER NUMERICAL CONTROL AND DIRECT NUMERICAL CONTROL

Machines and equipment that comprise the production process system are commanded by production control. Computer numerical control (CNC) and direct numerical control (DNC) play a major role in controlling these machines and equipment.

6.1 CNC

The typical architecture of a CNC system is shown in Fig. 8. In a CNC system, most of the NC functions such as input/output, arithmetic and control, and servo control are performed by microprocessors instead of hard-wired circuits. As seen in Fig. 8, various processing units are contained, such as a communication control processor, a graphic processor, a peripheral equipment control processor for robot and sequence control, and a servo processor. It also has a PTR/FD interface and main spindle drive interface.

Input devices are dependent on the input method of information. Industrial robots, for example, often use a teaching pendant instead of a paper-tape reader by which an operator can instruct the controller in motions of robots. It was a common practice to use paper tapes or floppy disks containing NC programs for inputting the control information to the CNC controller, but recently, an interactive programming function has become available on CNC controllers. With this function, an operator inputs necessary data under the guidance of messages and menus displayed on a CRT in an operational panel. The CNC controller then generates NC programs automatically.

6.2 DNC

DNC systems have been developed for the purpose of controlling and coordinating a number of machine tools and robots on a shop floor by a central DNC computer. The DNC computer manages part programs and distributes them to the machines. The management functions of the part programs include the functions of input/output, storage, editing, copy, deletion, modification, renewal, retrieval, etc. While the DNC computer transmits the part programs to the machines, it monitors the operating status of the machines and coordinates their operations. At the same time, it can perform other functions such as production management and part programming. A DNC system is effective in improving the availability of machines because of the timely distribution of NC data. It is also effective in improving the reliability of distributing part programs, because it can transmit them directly to the machine without use of NC tapes that have been commonly used with NC machine tools.

To make the most of the DNC system advantage, it is important to use effective DNC computer software for production management, such as software for scheduling, controlling machine loading and material supply, and report generation (percentage of machine utilization, production performance, etc.).

7. DEVELOPMENT OF CIM

The application of computers in the manufacturing field has made rapid progress since the 1960s. Until the mid-1980s the systems were developed separately in each area of manufacturing such as design, production planning, production management, and shop floor control. In the late 1980s, however, there appeared a strong movement to the integration of the systems of different areas, which is called CIM (computer integrated manufacturing). The trend from CAD and CAM to CIM is illustrated in Fig. 9.

The several levels of CIM can be considered in the course of development as seen in Fig. 9. In the early 1990s, most CIM systems correspond to level 3, which integrates the activities from the acceptance of customer's

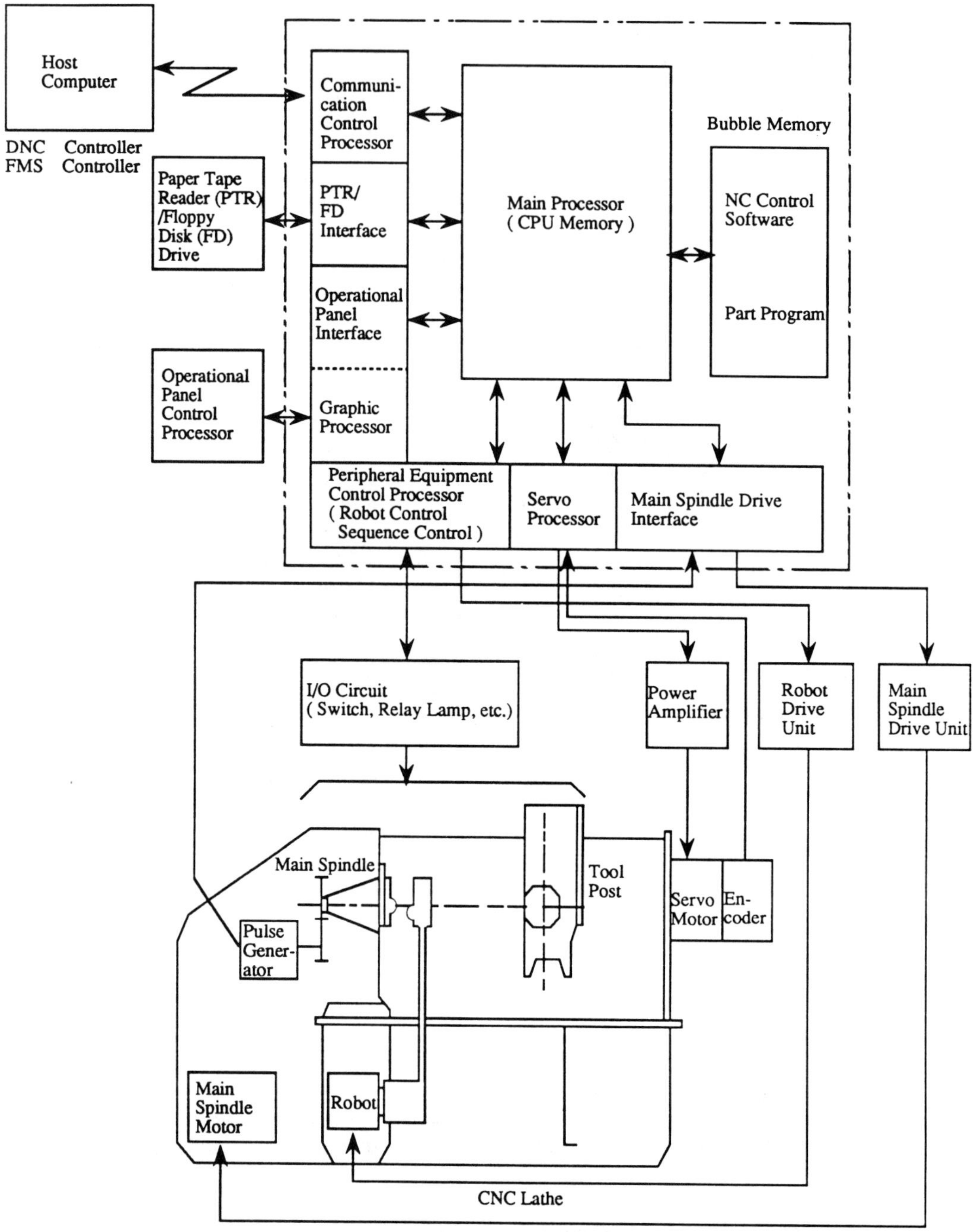

FIG. 8. Basic structure of CNC system (Courtesy of Okuma Iron Co., Japan).

needs and orders to shipping and distribution of products. The system aims to achieve the integrated processing of manufacturing tasks by realizing a coherent flow of information and material. As the number of companies and researchers engaged in CIM increases, however, the concept of CIM tends to diversify and to become ambiguous. The organization of level-3 CIM is represented in Fig. 1.

The base technologies that support CIM are

1. communication network,
2. database, and
3. interface.

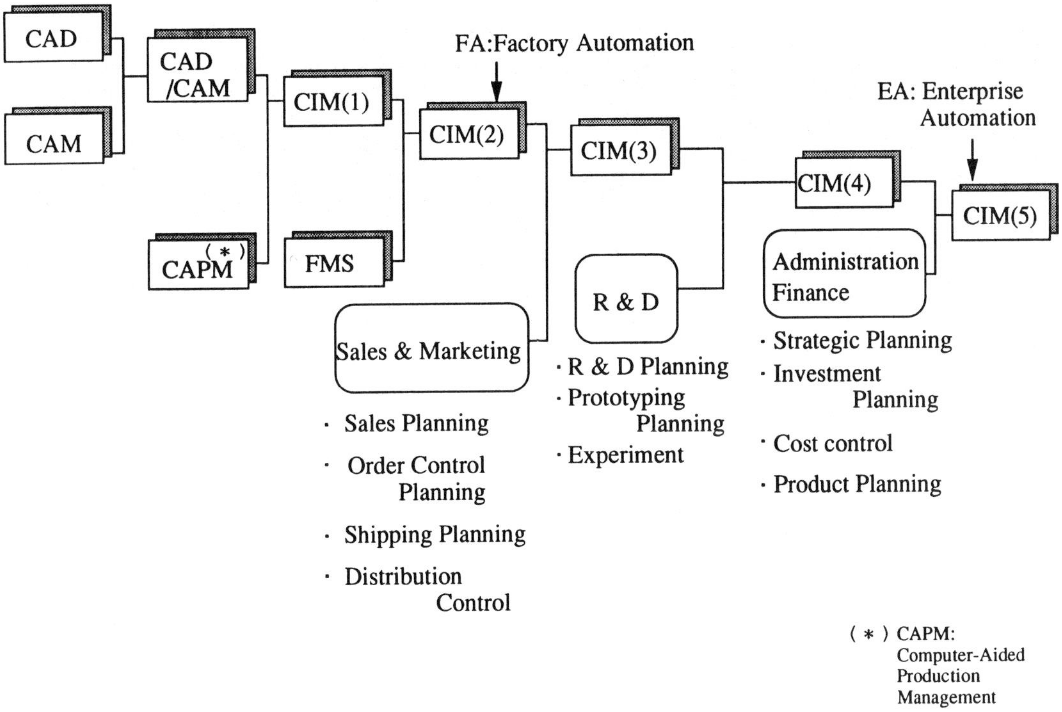

FIG. 9. Concept of CIM on five levels.

Besides CIM,

1. concurrent engineering or simultaneous engineering in which CAD and CAM are integrated and
2. integration of CAM and scheduling

are the noteworthy topics in relation to the CAM integration.

8. SUMMARY

CAM involves two aspects of computer application in manufacturing. One is the computer control of machines to automate the production processes, which originated in the development of NC machines. The other is the computer processing of production data, which involves computer-aided process planning, computer-aided operation planning, and production control.

Nowadays CAM is widely realized in most areas of manufacturing, and the level of automation is becoming more and more sophisticated. Along with this tendency, expert systems that can accumulate and utilize human knowledge and experiences relevant to CAM were recognized as a practical approach and have been rapidly developed in recent years. Automated operation of the factory has also made progress resulting in the automated operation of the 66 h from Friday evening to Monday morning. In addition to the development of each CAM technology, the integration of various manufacturing activities such as design, management, marketing, etc., has attracted strong attention. The trend is now moving from FA to CIM. CIM is, and will be, a key technology of manufacturing in the 1990s and will develop further, moving toward "intelligent CIM."

GLOSSARY

CAD: Computer-aided design.

CAM: Computer-aided manufacturing.

CAPM: Computer-aided production management.

CAPP: Computer-aided process planning.

CIM: Computer integrated manufacturing.

CNC: Computer numerical control.

DNC: Direct numerical control.

FA: Factory automation.

FMS: Flexible manufacturing system.

GT: Group technology.

IC: Integrated circuit.

Manufacturing: A series of interrelated activities and operations involving the design, materials selection, planning, production, quality assurance, management, and marketing of parts or products.

NC: Numerical control.

Operation: A defined action, i.e., the act of obtaining a result from one or more elements.

Operation Planning: Planning of operations based on the plan prepared by the process planning.

Process: A systematic sequence of operations to produce a specified result.

Process Planning: Planning of processes for the production of a part or product on the basis of the information and drawings generated in the design stage.

Production: Fabrication and assembly of a part or product.

Production Control: The function of directing or regulating the ordinary movement of parts or products through the entire production process from the requisitioning of raw materials to the delivery of the finished products.

List of Works Cited

Budde, W., Weissweiler, H. (1973), in: *Proceedings of the 14th International Machine Tool Design and Research Conference, Manchester, U. K.*, London: Macmillan, Paper 404, pp. 1–13.

CAM-I (1976), *Automated Process Planning System (CAPP), Systems Manual*, Version 2, Release 1. Arlington, TX: Cam-I, Inc.

CAM-I Standards Committee (1978), *Glossary of Computer Aided Manufacturing Terms*, R-78-SC-01. Arlington, TX: Cam-I, Inc.

Chryssolouris, G., Chan, S. (1985), *Ann. CIRP* **34**, 413–417.

Descotte, Y., Latombe, J. C. (1985), *Artif. Intelig.* **27**, 183–217.

Eversheim, W., Fuchs, H. (1980) *Advanced Manufacturing Technology*, Amsterdam: North-Holland, pp. 303–313.

Iwata, K., Kakino, Y., Ohba, F. (1977), in: *CIRP International Seminar on Manufacturing Systems, 6th, Cranfield, England*, Paris: Collége Internationale de Recherches pour la Production, pp. 25–40.

Iwata, K., Sugimura, N. (1984), in: *CIRP International Seminar on Manufacturing Systems, 16th Tokyo, Japan*, Paris: Collége Internationale de Recherches pour la Production, pp. 83–92.

Iwata, K., Fukuda, Y. (1987), in: *CIRP International Seminar on Manufacturing Systems, 19th, University Park, Pennsylvania*, Paris: Collége Internationale de Recherches pour la Produciton, pp. 287–294.

Iwata, K., Fukuda, Y. (1989), in: *CIRP International Workshop on Computer Aided Process Planning, Hannover, Germany*, Paris: Collége Internationale de Recherches pour la Production, pp. 73–83.

Japan Society for Precision Engineering (Ed.) (1991), *Manufacturing Software System*,Tokyo: Ohm, Inc. (in Japanese).

Mann, W. S. (1977), *A Computer Process Planning Technology*, R76-942625-14, East Hartford, CT: United Technology Co.

Matsushima, K., Okada, N., Sata, T. (1982), *Ann. CIRP* **31**, 329–332.

Nissen, K. (1969), in: *Proceedings of the CIRP International Conference on Applications of Computers in Manufacturing*, Paris: Collége Internationale de Recherches pour la Production, pp. 47–52.

Pennell, J. P., Winner, R. I. (1989), in: *Globecom '89. IEEE Global Telecommunications Conference & Exhibition. Conference Record*, New York: IEEE, Vol. 1, pp. 647–655.

Spur, G., Anger, H.-M., Kunzendorf, W., Stuckmann, G. (1978), in: Davies, B. J. (Ed.), *Proceedings of the Nineteenth International Machine Tool Design and Research Conference held in Manchester, 13th–15th September, 1978*, London: Macmillan.

TNO (1981), *Introduction to MIPLAN*, Appeldoorn, The Netherlands: Organization for Industrial Research, Inc.

Van't Erve, A. H., Kals, H. J. J. (1986), *Ann. CIRP* **35**, 325–329.

Further Reading

Ham, I., Lu, C.-Y. (1988), *Ann. CIRP* **37**, 591–601. (An overview of the current status and future trend in an area of CAPP.)

Japan Society for Precision Engineering (Ed.) (1991), *Manufacturing Software System*, Tokyo: Ohm Inc. (in Japanese). (A detailed text on computer aided manufacturing software.)

Rembold, U., Blume, C., Dillmann, R. (1985), *Computer-Integrated Manufacturing Technology and Systems*, New York and Basel; Marcel Dekker. (A detailed text on computer automation in manufacturing.)

Zozaya-Gorostiza, C., Hendrickson, C., Rehak, D. R. (1989), *Knowledge-Based Process Planning for Construction and Manufacturing*, San Diego; Academic. (An overview of expert-type CAPP systems and their research project.)

COMPUTER DATABASES

H. BEHRENS AND J. W. TEPEL *Fachinformationszentrum Karlsruhe, Eggenstein-Leopoldshafen, Germany*

INTRODUCTION

Data of every kind (citations, facts or numeric data) have been collected and handled since the beginning of the scientific era. Until a few decades ago, this handling was carried out by pencil, by ink, or in printed form, but all on paper. Thus we can speak of the paper age in science. In physics, data were published in journal articles, in monographs, in textbooks, and in data compilations. In addition, citations were already regularly compiled and published in abstract journals in order to facilitate the access of physicists to literature. Good examples were *Physikalische Berichte*, founded as *Fortschritte der Physik* in 1845, and *Physics Abstracts*, founded as *Science Abstracts* in 1898.

As far as numerical data are concerned, major efforts have been made to compile and evaluate these in handbooks. One of the best examples is Landolt–Börnstein, *Numerical Data and Functional Relationships*, of which the first volume was published in 1883. However, even in this paper age the physics community was confronted (as also holds true for other science communities) with an exponential growth of the literature and data, a fact that was, however, not realized by the community because the absolute numbers were not so large as they are today.

After the second World War, at the end of the 1940s and the beginning of the 1950s, the development of computers began (see COMPUTERS). In the middle of the 1950s we had the first mainframes, in the middle of the 1960s the first minicomputers, and in the middle of the 1970s the first microcomputers. In parallel, new supercomputers became available from the middle of the 1970s onwards. This computer development had two effects. On the one hand, it speeded up the increase in the amount of data and information generated, by some kind of feedback effect (e.g., because of the automatization of the measurement process), and, on the other hand, the capacity to handle and store numeric as well as alphanumeric information also increased. For the first time large amounts of data and information could be handed easily and inexpen-

1-56081-063-7/92/$5.00+.50

sively. Thus the electronic revolution offered the possibility to overcome at least partly the problems connected with the data and information explosion.

In the electronic age we now have data in electronic form or, as it is normally said, in machine-readable form. Usually a collection of electronically stored items is called a database. These electronically stored items can be citations, facts, numerical data, or even the full text of an article. Accordingly, one distinguishes between *bibliographic, factual, numerical*, and *full-text* databases. The bibliographic databases contain only references (titles, authors, abstracts, sources, etc.) together with other items for the corresponding subject analysis (classification, descriptors, etc.) The factual databases contain other structured but mainly non-numeric data. The numerical databases are defined by the fact that their content is dominated by structured numeric data. The full-text databases contain complete and unshortened texts from articles or other written papers. Next-generation databases will be bigger and more complex, containing a mixture of the above types of data, and will be used extensively for decision support purposes (Silberschatz *et al.*, 1990).

Let us now come back to the computers themselves: What are their advantages? They are ideal for handling data insofar as they are able to store, manipulate, display, and retrieve large amounts of data quickly. Even printed versions can be produced with very little human intervention, so that the transfer of data from the electronic form into the old-fashioned paper form is greatly facilitated. Another advantage of the computer technology development is that the costs of storage and access are constantly decreasing, that we can more easily replace old data by new data, i.e., it becomes easier to update our data collections, and that we can search more specifically for data in a database than was possible in the printed works via the indexes.

As was to be expected, the introduction of computers has also led to unwanted and less favorable developments. In the paper age each scientist and engineer interacted directly with each data item. He had to select the data and make all the necessary manipulations himself. Therefore the contact of each scientist and engineer with his data was very close. The introduction of computers had the consequence of a certain decoupling between humans and the data, which in some instances has led to inconsistencies, ambiguities and a loss of overall quality. On the other hand, whenever computational cross-checks can be made, typical human errors have been all but eliminated.

1. DATA MODELS AND FILE DESIGN

1.1 Data Storage

A computer is an excellent instrument for storing information. Even the simple $5\frac{1}{4}$-in. magnetic floppy disks, which have been around for many years, manage to store a respectable 360 kB (B=byte) of data. This represents about 140 typewritten pages. The $3\frac{1}{2}$-in. high-density disks popular with IBM-compatible personal computers store 1.46 MB, or as many as 580 pages. Hard disks with storage capacities of up to 300 MB are readily available for microcomputers. Optical disks or CD-ROMs (compact-disks read-only memory) have typical capacities of 550 MB.

With the advent of such cheap mass storage devices, storage of data is no longer a problem. Also, the user does not have to worry about the low-level physical organization of the data on disk, since this is taken care of by the operating systems. Care has to be taken, however, to preserve any relationships between the data items. A telephone directory is not created on a computer by storing at random names and telephone numbers. At the very least, one has to know which number "belongs" to which person. In addition, of what use is any data collection if the data needed cannot be retrieved? It is this aspect of computer-based data collections or databases that has become of increasing importance with the advent of the CD-ROM, needing special retrieval and indexing techniques in order to guarantee acceptable retrieval times.

Small data collections can easily be handled by modern text editors or word-processing systems. Most of these software packages are extremely flexible, they provide for rapid searching of text passages, they provide a nicely formatted output, and they have many other useful features. Managing, for example, a collection of 50 or so cooking recipes does not warrant the expense of a database management system. Any relationships among the data items can be recognized by just "looking" at the data, i.e., reading the recipe.

1.2 Data Models

More extensive data collections have to be structured in some way and cannot be retrieved by just "browsing" through them. The starting point in designing a database is an informal model of the real world, the *entity–relationship* (ER) model. This starts off by listing all relevant "objects" or *entities* and the *relationships* among them. An entity is a thing that exists and is distinguishable, such as a PERSON or an AUTOMOBILE. An entity has *attributes*, e.g., a PERSON has a name, age, address, etc., each of which in turn has a certain value from a *domain* of values for that attribute in the whole set of entities. The domain of the attribute AGE belonging to the entity PERSON is the set of integer numbers in the range 0 to 120—assuming that no person is older than 120 years. An attribute or a set of attributes that uniquely defines each entity in a set of entities is called a *key* for that entity set. A relationship among entity sets is simply an ordered list of entity sets. The list may consist of only two members, e.g., PERSONS and AUTOMOBILES with the relationship "has a" or "belongs to."

Entities and their relationships are best shown in the form of a diagram, the ER diagram (Fig. 1). Entity sets are represented by rectangles, their attributes by circles, and the relationships by diamond-shaped diagrams. Note that relationships may be of the kind one to one (1:1), many to one (n:1) or many to many (m:n). In our case we have a n:1 relationship between AUTOMOBILES and PERSONS when we assume that a person may own more than one car. The n:1 relationship is shown by the arrow in the diagram. More precisely, the arrow stands for "determines uniquely." A m:n relationship would exist if an AUTOMOBILE could in addition have more than one owner. More details on ER diagrams may be found in Ullman (1982).

Closely connected to the ER model are "data models," three of which, viz., the *relational*, *hierarchical*, and *network* models, have been used in practically all commercially available database systems (DBSs). These data models can be transcribed more directly into the physical structures and organization needed to make a DBS work on a computer.

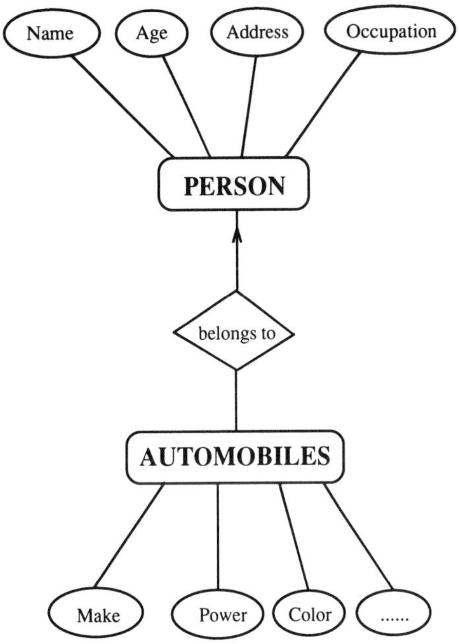

FIG. 1. Entity–relationship model of PERSONS and AUTOMOBILES. Objects (rectangles) together with their attributes (ellipses) are linked via undirected graphs (lines). The diamond-shaped relationship "belongs to" between AUTOMOBILES and PERSONS is of the "many-one" kind (n:1). This is expressed by the arrow into PERSON.

1.2.1 Relational Model. A relation is any subset of the Cartesian product $D_1 \times D_2 \times \cdots \times D_k$ of a list of domains $D_1, D_2, \cdots, D_k$. The members $(v_1, v_2, \cdots, v_k)$ of a relation, where v_i is a value from domain D_i, are called *tuples*. An example illustrates this more clearly. Let us assume that the entity PERSONS has the attributes NAME, AGE, and TELEPHONE. Let us assume that NAME consists of only two values, Smith and Jones, i.e., the domain D_1 = (Smith, Jones). Simplifying in a similar way, the domains for AGE and TELEPHONE might be D_2 = (10,24,60) and D_3 = (456 789, 234 567). The Cartesian product of the domains D_1 to D_3 or $D_1 \times D_2 \times D_3$ is given by the twelve 3-tuples (v_1, v_2, v_3):

(Smith, 10, 456 789), (Jones, 10, 456 789),
(Smith, 10, 234 567), (Jones, 10, 234 567),
(Smith, 24, 456 789), (Jones, 24, 456 789),
(Smith, 24, 234 567), (Jones, 24, 234 567),
(Smith, 60, 456 789), (Jones, 60, 456 789),
(Smith, 60, 234 567), (Jones, 60, 234 567).

Any subset of this set of 3-tuples including the set that might correspond to the real world,

(Smith, 24, 456 789),
(Jones, 60, 234 567)

is a relation of degree or arity 3. The relation scheme is written PERSONS (NAME, AGE, TELEPHONE). It helps to view relations simply as tables of values of given attributes, where the order of the attributes is unimportant.

The relational model may be used to construct a DBS for the entities PERSONS and AUTOMOBILES as follows. A relation scheme describing AUTOMOBILES might by written as AUTOMOBILES (AKEY, TYPE, COLOR, SERIAL__NO). Here AKEY serves to uniquely identify an automobile. It might consist of the letter *A* followed by a running number, e.g., A001, A002, ..., A376. Such a quantity plays the role of a "key" for accessing any particular automobile. To the PERSONS relation scheme we would add a unique person key PKEY consisting of, say, P001, P002, etc. Then the relationship between PERSONS and AUTOMOBILES is described by HAS__A__CAR (PKEY, AKEY), i.e., by tuples of the kind (P001, A002), (P001, A009), (P002, A001), etc., where we have assumed that person P001 owns two cars, A002 and A009. The three relations PERSONS(···), HAS__A__CAR(···), and AUTOMOBILES(···) suffice to describe the ER diagram in Fig. 1.

Note that by introducing a relation HAS__A__CAR(···) to describe the relationship between PERSONS and AUTOMOBILES, we can in effect handle *m*:*n* combinations: A person need not have a single car, nor is it necessary for a car to have one owner only. Restricting ourselves to *n*:1 relationships between automobiles and owners makes the relation HAS__A__CAR unnecessary: Augmenting the scheme AUTOMOBILES(···) with PKEY clearly defines the ownership of a car.

Relational database systems (RDBSs) have many advantages, and for this reason commercial systems for microcomputers are almost exclusively of this type. Data can be easily changed or updated. Deleting a car from the DBS would involve deleting a record from AUTOMOBILES without affecting the PERSONS file. Furthermore, the theory of RDBSs is well advanced and other data models can be reduced to the relational model at the cost of data storage (see below).

1.2.2 Network Data Model. This is the most general form in which data and their relationships can occur. Many real-life situations are of the network type. In order to avoid confusion with the relational model, it is best to talk here about logical records instead of tuples and logical record formats containing fields instead of relational schemes (Ullman, 1982). Instead of *n*:1 relationships one uses links connecting different record types. Figure 2 shows network data at the record level. The corresponding ER diagram (not shown) consists simply of the entities PARENTS, CHILDREN, and TOYS joined by undirected links. Parent records (P1,P2,...), children records (C1,C2,C3), and toy records are linked by pointers (not shown in the figure). Toy T4 is shared between children C2 and C3, whereas T5 is also owned by parent P4. Data are found by navigating along the links and not by performing relational algebra. In order to simplify implementation of networks on a computer, *m*:*n* relations are replaced by *m*:1 and 1:*n* relations by introducing intermediate record types containing the respective link information.

1.2.3 Hierarchical Model. The hierarchical model can be considered to be a special case of a network. Bidirectional links are replaced by parent-children 1:*n* links. The data structure looks like an inverted tree (Fig. 3), where the lower nodes and leaves can be reached only via the root. A linear physical record representing the hierarchical structure might look as shown in Fig. 4. The child and toy data are contained in a repeating group of fields called a periodic group (PG). In this case three such groups occur, i.e., the periodic

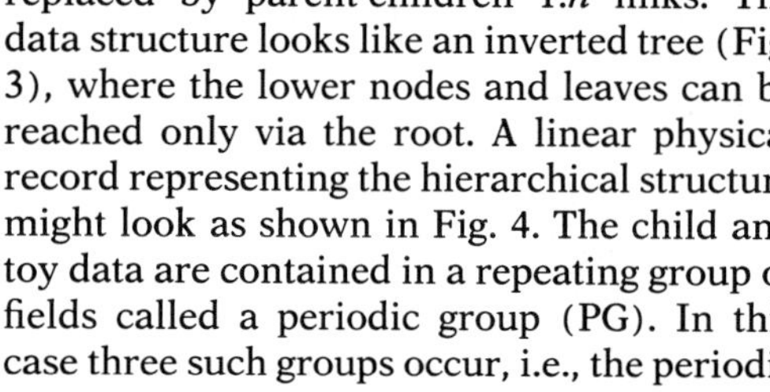

FIG. 2. Parent–child–toy diagram as an example of a network. The record types PARENTS, CHILDREN, and TOYS are linked by pointers. Note that toy T5 is owned by parent P4 and child C3. Complex relationships can be represented in a network diagram.

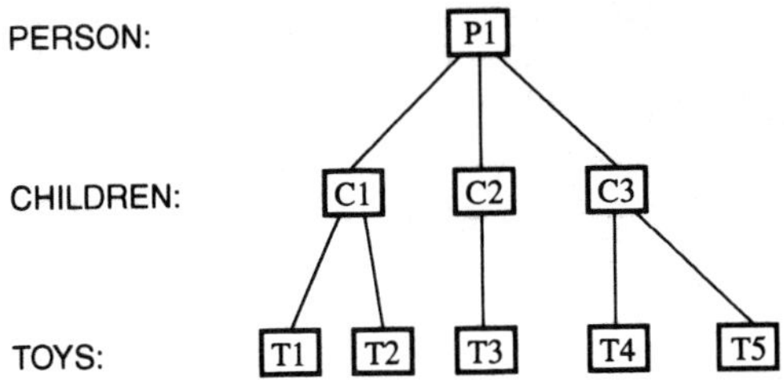

FIG. 3. Person–child–toy diagram as hierarchical structure.

group multiplier Pm is 3, since there is one instance of the group for each child. PG1 consists of data belonging to child C1 together with its toys T1 and T2. The toy multiplier Tm is needed to tell the computer the number of toy names that follow. Note that updating a record of this type is much more complex than in the relational case. Deletion of a node in Fig. 3 (say C1) eliminates the access paths (physical links) to the lower levels of the tree. Removing a child node from the tree also eliminates its toys. This can be considered to be an advantage or disadvantage depending on whether or not one argues that a toy is an entity even when not linked to its child owner.

Both the network and hierarchical data models can be transformed to the relational case. Linkages are replaced by relations. The hierarchical data in Fig. 4 can be represented by the relational schemes MOTHER(MKEY, ...), CHILDREN(CKEY, MKEY, ...), and TOYS(CKEY, ...). Here the three levels of hierarchy are linked by the unique keys CKEY and MKEY, identifying tuples in the CHILDREN and MOTHER relations, respectively.

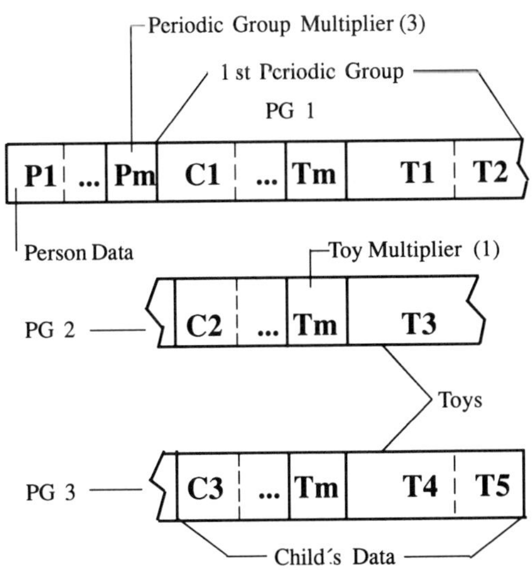

FIG. 4. Data of Fig. 3 arranged as a linear record (ADABAS).

1.3 File Design

Designing a database is an iterative process consisting of a number of steps shown in Fig. 5 (see Lockemann and Schmidt, 1987). The first step involves an analysis of information requirements in the broadest sense. What data are to be stored, who has access, and what information has to be provided by the DBS are some of the topics that have to be clarified. The *conceptual design* involves specification of objects and their relationships as in the entity–relationship model. At the *logical design* level, the data model most suited for the problem at hand is chosen, and the record and field formats for each of the files comprising the database are specified. The type and length of fields needed and the questions that have to be answered will determine to a large degree the database management system (DBMS) that can be used. The nature of a DBMS is explained in Sec. 2.2.

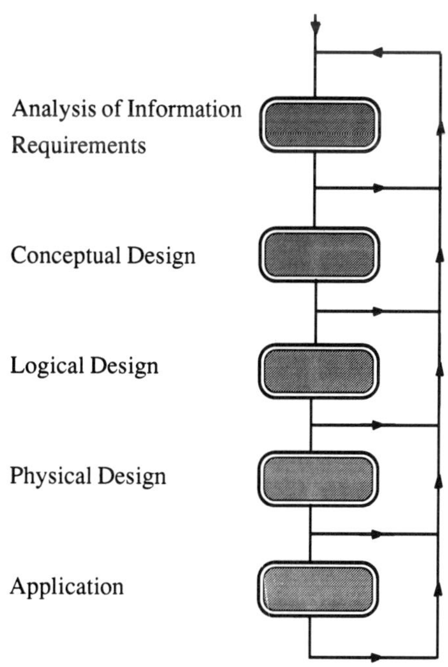

FIG. 5. Steps in designing a database.

The *physical design* is taken care of by the manufacturer of the DBMS—unless the whole system is to be developed from scratch, something that was only done for the older physics databases, where dedicated systems written in FORTRAN or PL/1 were used. Usually the data are organized sequentially in a direct access file with pointers to the organizational units of the data file (documents, fields) stored in one or more index files. Access to the data is usually by one or more keys constructed by concatenating the contents of some data fields. Lookup of these keys in the index files may be by use of a hashing algorithm, by an index sequential access method (ISAM) or by use of Bayer trees (BTREE/ISAM techniques) (see Bayer and McCreight, 1972); these terms are defined in the Glossary. Physical data organization is described in detail in the book by Ullman (1982).

The *applications* are the real test for the DBS. At this stage the gap between planning and reality becomes evident, and, depending on how well the database has been modeled to fulfill the overall requirements, some modifications will be necessary at the higher design stages of Fig. 5.

2. DATABASE MANAGEMENT SYSTEMS AND THEIR COMPONENTS

2.1 Dedicated Databases

Even today, most DBSs in use in physics are dedicated systems developed for some specific purpose and are usually written in a high-level language such as FORTRAN or PL/1 (see COMPUTER PROGRAMMING LANGUAGES). Many databases originated at a time when commercial DBMSs were either nonexistent or too expensive, and it is only natural that physicists used the programming language that they knew best and that they had been using for their numerical computations, FORTRAN. Systems written in PL/1 having more sophisticated input/output (I/O) routines and indexed access to the data emerged in the 1970s. Modern programming languages such as C are little used up to now in the physics community.

Dedicated DBSs can be very efficient as regards performance and the need for data storage. Their main drawback is their inability to adapt to changing user requirements. A new file design means major changes to the software. Updates are commonly done by generating the files anew. Developing a dedicated database or retrieval system is worthwhile only in very few isolated instances (Tepel and Müller, 1990).

2.2 Structure of Database Management Systems

A database management system (DBMS) is simply a database system (DBS) with additional management facilities. Some of these features are listed below (see Ross, 1978).

1. *Database security.* Provision has to be made to control user access to the data. This is achieved by introducing different external or user views of the data (in effect a filter allowing users to see only certain data fields), by disallowing certain commands that modify the data, by password protection, and by encoding of the data.
2. *Database integrity.* Routines have to be provided to allow backups to be made of the DBS. Larger DBMSs have transaction logging, checkpointing, and capabilities for restart after system failures.
3. *Data definition language (DDL).* The DDL provides a means of describing the data structures and maps them onto the storage media. As an example, Fig. 6 shows the ADABAS™ DDL module for Fig. 4. Low-

```
01,NA,20,A,DE             **Person's Name
01,AG,2,U,DE              **Age
01,AD,60,A                **Address
01,TL,12,U                **Telephone No
01,CH,PE                  **Periodic Group
      02,CN,20,A,DE       **Child's Name
      02,CA,2,U           **Age
      02,TY,30,A,MU,DE    **Toys
```

A = Alphanumeric Field	DE = Descriptor Field
U = Numeric Field	MU = Multiple Values

FIG. 6. ADABAS™ data definition module for the data of Figs. 3 and 4. Each entry consists of a level identifier (01 or 02), the field name, field length and type, and additional information such as whether the field is to be indexed (DE) or not.

level routines for disk access and file management are usually taken from the operating system.

4. *Data dictionary*. The data dictionary or database system catalog provides the user or database administrator with an overview of the DBS. It lists the files and data structures (tables and attributes) present in the system. Data definition modules appear in the data dictionary, where they are kept for future reference and modifications.
5. *Data acquisition*. Data input routines in batch mode and via masks depending on the data structures as described by the DDL are usually provided.
6. *Data manipulation language (DML)*. The DML is the command language of the DBS and allows operations on the database such as the creation or deletion of whole files or groups of files and the start and termination of user sessions as well as manipulations on the data itself, such as deletion, insertion, and updating of individual logical fields or records. The query language used for retrieving data is often considered to be a subset of the DML. Database producers often call their DMLs programming languages of the fourth generation, since they provide direct database support with commands such as FIND, ADD, and DELETE. In addition, most commercial DBMSs have interfaces to high-level programming languages such as PL/1 or C, and to Assembler. Well-known query languages are SQL, which is a de facto standard for relational database systems, and NATURAL™, which is used together with the DBMS ADABAS™ on mainframes.
7. *Report generator*. Most DBMSs include a report-generating facility that presents retrieved data as properly formatted reports. Personal computer (PC) systems often provide advanced features such as graphics capabilities and support for a variety of printers including systems using PostScript™.
8. *Support for multiple access*. An important feature of a DBS is the centralized storage of data. In addition to reducing data redundancy, this consolidates updating procedures, since only the master copy of the file has to be maintained. It is preferable to use a single database under a single control rather than having scattered collections of computer files maintained by individual groups or users. Multiple and simultaneous access to such a centralized system is important, and the DBMS software has to preserve the integrity of the data and prevent deadlock situations.

2.3 DBMSs for Personal Computers

The three types of DBMSs for use on PCs are the flat-file, relational, and network or hierarchical types. Of these, the relational database management systems (RDBMSs) are by far the most common.

Flat-file systems consist of linear arrays of data and permit no logical file couplings. Each database consists of a single file only. Records are usually subdivided into fields of fixed length, a certain number of which may be indexed.

Data acquisition and report generation are supported. These systems are inexpensive, are often available as public-domain or user-supported software, and are ideal for keeping simple lists of data. Through their menu-driven user interfaces they are very user-friendly.

A few professional network-type DBMSs exist for special applications on personal computers. An overview may be found in Seymour (1988).

Relational DBMSs for PCs are very common. Very widespread is dBASE™ by Ashton–Tate. Although more a DBS than a DBMS, it has a powerful programming language and is supported by a large number of software firms. Procedures can be rapidly developed and tested, since dBASE runs as an interpreter. These procedures can then be compiled with dBASE compilers, such as CLIPPER™, which produce executable files. dBASE relies heavily on the PC operating systems for file management.

An example of a full relational DBMS for PCs is ORACLE™. ORACLE was originally developed for minicomputers, but nowadays the whole spectrum of computer hardware is supported, from microcomputer and workstations to large mainframes. It uses SQL as query language (Date, 1987) and has interfaces to several high-level programming languages, in particular to C. Data management is reminiscent of mainframe systems, with password protection, clustering of files in definite partitions (areas on the hard disk), rollback and commit statements (*rollback* makes the changes to data at the end of a session undone, *commit* confirms the changes), and provision for a systems supervisor—the DBA or database administrator.

2.4 DBMSs for Mainframes

Most of the older types of DBMSs for mainframes are of the network or hierarchical type, e.g., Information Management System (IMS) of IBM. In recent years, relational DBSs have become more important. The advantages of such RDBMSs are easier maintenance of individual files and flexibility with respect to user queries. The data structure and linkages in a hierarchical system determine what type of questions can or cannot be answered, whereas the linear file design of relational systems allows future applications, which were not foreseen at the time of the file design. Possible drawbacks of RDBMSs, viz., the need to compute joins between many different relations with a corresponding loss in performance and the need to construct displays from relations stored on different physical regions of the hard storage devices, are partly offset in commercial systems such as ADABAS by allowing some types of repeating fields (multiple fields) and repeating groups (periodic groups) in each record of a file. In this way, retrieving a record for display purposes usually places all data at one's disposal, while still allowing indexing on individual values of multiple fields, such as author names in a bibliographic document.

The DBS used in the Messenger retrieval system of STN-International, FIDO, is a hierarchical system allowing linkage of terms on several different levels (word, sentence, paragraph, and linkage proximity). However, the requirements of a DBS for information retrieval systems used mainly for bibliographic and full-text databases (see below) are somewhat different from those of other DBSs.

2.5 Choosing a DBMS

Choosing a DBMS for a proposed application is not an easy task and needs a good understanding of the underlying principles of electronic data processing and data modeling. Unfortunately, often a choice has to be made before the scope of the application or the abilities of the DBMS software are known completely. Changing subsequently from one DBMS system to another can be very costly.

There is probably no single system that satisfies all the user requirements. A DBMS that works well with some type of data, e.g., records of fixed length subdivided into fields, performs very poorly or is unable to handle data of varying length. A number of criteria are helpful in determining which system is best suited for the particular purpose:

1. Structure of the data—free text, free text with some structure (e.g., a book), documents from on-line retrieval sessions, data in fixed format.
2. Data types—text, numeric, graphics, or a combination of these.
3. Lifetime of data—static or changing data (frequent updates).
4. Size of the database.
5. Searchability of the data—free text search, full text search, search by means of indexed fields or inverted lists.
6. Data relationships (1:1, *n*:1, *n*:*m*) and models.
7. User interfaces—what provision is made in the software package for user interaction with the data.

3. INFORMATION RETRIEVAL AND HYPERTEXT

3.1 Information Retrieval Systems (IRS)

Systems in use with bibliographic databases fall into the category IRS rather than DBMS. Of primary importance are the user interface and the retrieval possibilities for the documents. Less emphasis is placed on the DBM part of IRSs. The data are more static than with classical DBS, and much effort is spent in building index files for efficient data retrieval. The index parts of such files are normally larger than the actual data. Updates to older data are rarely done; files change because new documents are added at regular intervals. In addition, IRSs usually support structured thesauri containing hierarchical and associative relationships between terms (keywords), which makes document retrieval very effective. These terms are usually called descriptors and are classified into superordinate or broader terms, subordinate or narrower terms, and associative or related terms.

3.2 Hypertext

A comparatively recent development, hypertext combines the virtues of full-text DBSs with user-defined access paths to the data (Marchionini and Shneiderman, 1988; Franklin, 1989). In hypertext the underlying DB structure permits the user to traverse complex networks of information quickly. The user can select his own topics of interest, link these with related concepts, and navigate on-screen through the full text by choosing at each stage among a set of alternatives. The ability to browse through the text using a filter of selected topics is characteristic of hypertext systems.

3.3 Information Retrieval Models

In order to retrieve documents from an IRS the user formulates a query in the query language of the particular system. The goal of such a query is to maximize the *recall* of documents, defined as the ratio (number of relevant items retrieved)/(number of relevant items in the DB), while keeping the *precision*, defined as (number of relevant items retrieved)/(total number of retrieved items), as large as possible. Broad query formulations tend to have a high recall but a low precision, whereas narrow formulations have high precision but low recall figures.

IRSs used by large on-line hosts employ the so-called Boolean retrieval model. A search query is formulated and all documents are retrieved that exactly match this query. The query may contain terms connected by the Boolean operators AND, OR, and NOT, as in the following example from the database PHYS (comments have been added within the delimiters /*....*/):

```
/* user entries are expected at the system prompt ⇒  */

⇒ s database                              /* search for the term database */
L1      527  DATABASE                     /* found 527 documents */
     /* query results are labeled consecutively by L1,L2,...  */
⇒s information retrieval                  /* find information retrieval */
      37241  INFORMATION
       1039  RETRIEVAL
L2      387  INFORMATION RETRIEVAL        /* 387 documents found */
/* search for documents which contain both the terms DATABASE and
  INFORMATION RETRIEVAL */
⇒s         L1 AND L2
L3         31 L1 AND L2
⇒ s  L3 NOT optical                       /* of the 31 documents found
                                           previously show only
                                           those that do not contain
                                           the word OPTICAL  */
   125284 OPTICAL                         /* 125284 occurrences
                                           of OPTICAL in DB  */
L4       29 L3 NOT OPTICAL                /* final result of combined
                                           queries  */
```

Here documents are retrieved that contain the word "database" and "information retrieval," but not the term "optical." One of the documents found is displayed in Fig. 7.

Apart from the Boolean model, where an exact match between terms in the query and in the documents is required, other retrieval models such as the probabilistic model and the vector-space model relax this condition (see Salton, 1989). Here a measure of similarity is introduced between the terms in the query formulation and in the documents. Documents may be ranked according to the degree of overlap with the query. Recall is improved by these methods, and users can decide by themselves at which point in the

```
L4  ANSWER 2 OF 29
AN  10(14):66440  PHYS
TI  Data structures and organisation: Special problems
    in scientific applications.
AU  Read, B.J. (Rutherford Appleton Lab., Chilton
    (UK))
SO  Comput. Phys. Commun. (Dec 1989) v. 57(1-3) p.
    164-168
    International Conference on Computing in High Energy
    Physics, Oxford
    (UK), 10-14 Apr 1989
    ISSN 0010-4655; CODEN CPHCB
CY  NETHERLANDS
DT  Journal; Conference
TC  Computing
LA  English
AB  In this paper we discuss and offer answers to the
    following questions: What, really, are the benefits
    of databases in physics. Are scientific databases
    essentially different from conventional ones. What
    are the drawbacks of a commercial database manage-
    ment system for use with scientific data. Do they
    outweigh the advantages. Do database systems have
    adequate graphics facilities, or is a separate gra-
    phics package necessary. SQL as a standard language
    has deficiencies, but what are they for scientific
    data in particular. Indeed, is the relational model
    appropriate anyway. Or, should we turn to objecto-
    riented databases. (orig.)
CC  *0270
CT  *COMPUTER CALCULATIONS; COMPUTER CODES; COMPUTER
    GRAPHICS; PROGRAMMING LANGUAGES; *INFORMATION SYSTEMS;
    INFORMATION RETRIEVAL; DOCUMENTATION
```

FIG. 7. Example of a PHYS documentation unit.

ranked list they want to place their cutoff of useful documents. Little practical experience is available with these more advanced retrieval models.

4. BIBLIOGRAPHIC DATABASES

The databases, producers, and hosts mentioned in Secs. 4–7 are listed and described in the Cuadra directory of databases (Cuadra, 1990), which is updated regularly. For this reason, we have omitted the administrative details, such as addresses, from our overview.

As already mentioned in the Introduction, bibliographic databases are defined by the fact that they only contain references (titles, authors, sources, etc.) to published literature. Of course, there exist a very large number of small, mostly private files of this type. It is, however, not the intention to describe these in more detail, but to restrict ourselves to the large bibliographic databases covering whole fields of science, such as physics, chemistry, materials science, etc. They are produced professionally by large producers like the Institution of Electrical Engineers, Fachinformationszentrum Karlsruhe, Chemical Abstracts Service, etc., and are offered on line via so-called hosts.

Retrieving simply bibliographic citations is not enough for an on-line researcher. More is involved. Thus, in order to provide the prospective user with the details of the content of an original document and in order to make a specific and selective retrieval possible, each bibliographic item (documentation unit) also contains terms derived by what is known as subject analysis (see Fig. 8). This subject analysis, often also called indexing, is usually carried out manually by the document analysts. In some, very few, modern systems, parts of this analysis process (the indexing and adding other information) are nowadays also handled automatically (or semiautomatically) by application of suitable computer systems (automatic indexing and retrieval).

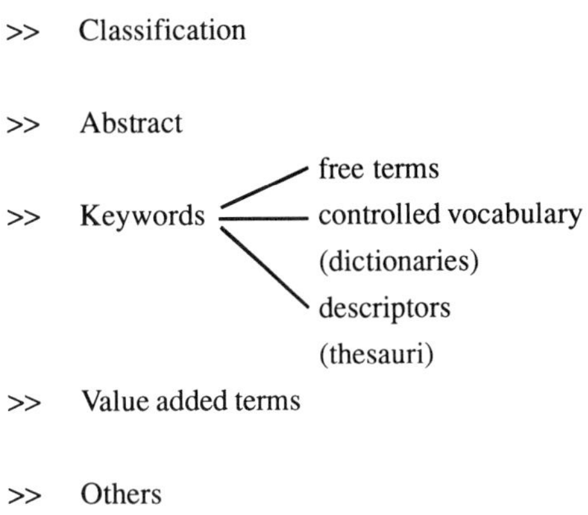

FIG. 8. The subject analysis of a document consists of adding the terms shown.

A normal documentation unit, which is the smallest unit of a bibliographic database and which contains all necessary information about an original document, consists of two parts, the technical bibliographical or subject cataloging part and the subject analysis part.

Documentation unit.

Technical bibliographical part

Field	Prefix
Accession number	AN
Authors, editors	AU
Title	TI
Source	SO
Journal title	JT
Corporate source or affiliation	CS
Publication year	PY
Publication date	PD
Meeting date	MD
Number of report	NR
Country of publication	CY
Language	LA
Document type	DT
Patent country	PC
Patent kind code	PK
Patent number	PN
Update code	UP

Documentation unit: Subject analysis part

Field	Prefix
Abstract	AB
Classification code	CC
Controlled terms (descriptors)	CT
Treatment code	TC
Value-added (e.g., element) terms	ET

A typical documentation unit making use of this scheme and taken from the database PHYS is shown in Fig. 7. Other information items, so-called value-added terms assigned in the process of subject analysis, are data flagging, chemical indexing with or without chemical structures, numerical indexing, astronomical objects, etc.

In physics, normally classifications are applied that are strongly related to the international ICSTI (International Council of Scientific and Technical Information) classification. It consists of a hierarchical scheme four levels deep of which the first two levels consist of 10 and 67 classes, respectively (see ICSTI, 1991).

For indexing free terms, controlled terms and descriptors from a thesaurus might be used. A thesaurus contains controlled terms only with additional hierarchical relationships in the form of so-called broader or narrower terms (see, e.g., Salton, 1983) and with so-called forbidden (synonyms) and other related terms. Typical examples in physics are the thesauri of the database INSPEC and of the database PHYS, which are described below.

Of the discipline-oriented bibliographic databases, those of major importance for the field of physics are the databases PHYS and INSPEC. Both have a long tradition because they already existed before the computer age in printed form as abstract journals.

PHYS originates from the *Physikalische Berichte*, which was founded in 1845 in Berlin under the title *Fortschritte der Physik*. In 1920 this abstract journal was renamed *Physikalische Berichte*; today its title is *Physics Briefs/Physikalische Berichte*.

In 1979 the abstract journal was computerized so that a database called PHYS today exists in parallel to the printed version with citations back to the publication year 1979. The producer since that time is the Fachinformationszentrum Karlsruhe or, briefly, FIZ Karlsruhe in Germany. Actualization takes place biweekly. In addition to the normal bibliographic terms such as title, authors with affiliations, sources, etc., this database contains document analysis terms such as abstracts, international physics classification codes, and descriptors from a thesaurus. For the field of astronomy the names of the astronomical objects are added as searchable key words. Searching for special substance designations is supported by a mixture of manual and automatic adding of substance names. The database PHYS covers the whole of physics, which includes elementary particles and fields, nuclear physics, atomic and molecular physics, classical areas of phenomenology, fluids, plasmas, discharges, condensed matter, cross-disciplinary physics, geophysics, astronomy, and astrophysics. As of September 1990, it contained about 1.4 million citations with an increase per year of about 125 000 entries.

INSPEC goes back to the abstract journal *Science Abstracts*, which was founded in 1898 in England. Very soon this abstract journal was divided into two parts, one for physics and one for electrical engineering, the physics part being named *Physics Abstracts*. At the end of the 1960s the production process for this abstract journal and other abstract journals was mechanized by the producer, the Institution of Electrical Engineers, with the consequence that a database was created in parallel to the printed version.

In the case of INSPEC, subject analysis is done by assigning classification codes (the classification used is similar to the ICSTI international classification) by indexing with a free and controlled vocabulary, and by chemical (for the designation of substances) and numerical indexing. The database covers the fields of electronics, computers, computing, electrical engineering, physics, and information technology. The physics part corresponds to the printed version *Physics Abstracts*. The total number of citations as of September 1990 was about 3.7 million (50%–60% physics). The updating is carried out biweekly, which results in an increase per year of about 270 000 citations.

In addition to those physics-specific databases that cover the field of physics completely, there are others that either are of a multidisciplinary character or focus on other fields like energy, chemistry, materials science, etc. A database of the first type is PASCAL, which covers fields such as chemistry, biology, engineering sciences, geoscience, medicine, and also physics (about 13.5%, without nuclear and elementary physics). The citations stored go back to the publication year 1973. PASCAL contains presently about 8 million citations with an increase per year of about 450 000. It is produced by l'Institut de l'Information Scientifique et Technique du Centre National de la Recherche Scientifique in Nancy, France. Titles are bilingual, French and original language; the abstracts are in French; and the descriptors are trilingual, French, English, and Spanish.

Other multidisciplinary databases are ENERGY [produced by an International Energy Agency (IEA) international cooperation], which covers energy research and technology, but contains also a large part of physics-relevant citations, and NTIS, which is devoted to reports on research and development sponsored by U.S. and non-U.S. governments. In this same context, the Science Citation Index (SCI or SCISEARCH), which contains references from approximately 3800 journals to worldwide literature across many disciplines, should also be mentioned. Discipline-oriented databases focusing on fields other than physics, but containing partly also physics-relevant citations, are CA (chemistry) produced by Chemical Abstracts Service in the U.S., INIS (nuclear research and technology) produced by an International Atomic Energy Agency (IAEA) international cooperation, METADEX (metallurgy and material sciences) produced by the Institute of Metals in London and by the American Society for Metals (ASM) in the U.S., MATH (mathematics) produced by FIZ Karlsruhe in Germany, and MathSci (mathematics) produced by the American Mathematical Society (for corresponding addresses see Cuadra, 1990).

What are significant quality criteria for bibliographic databases? First, the degree of in-depth subject cataloging and subject analysis, which varies from bibliographic database to bibliographic database, should be mentioned. Then, currency, completeness, and consistency are very important. The last points are self-evident since for researchers, for instance, timeliness is of major relevance because otherwise they will lose citations to the most recent publications. The same is true for completeness, because researchers must be sure not to miss citations to important papers in a well-defined area.

5. FACTUAL DATABASES

By definition, factual databases contain other structured but mainly non-numeric data. The factual databases are sometimes also alluded to as referral and textual-numeric databases. Conversely, factual and numeric databases together are sometimes called factual databases. The majority of their data consists of text data. This type of database refers to organizations, projects, experts, patents, products, conferences, etc. Typical examples are research databases, product databases, and conference databases.

Research databases contain information on research institutes and research projects (research in progress). The usual content of the file segment INSTITUTE is

- Institute (company) name and address
- Telecommunication numbers
- Classification code
- Director of the institution
- Staff (names)
- Description of the main research activities
- Facilities of the institute
- Accession numbers of projects carried out at the institute
- Update code

and of the file segment PROJECTS is

- Title of project
- Classification code
- Description of the project
- Performing organization
- Cooperating organizations
- Name of project leader
- Staff (names)
- Supporting organization
- Duration begin, date
- Duration end, date
- Number of contract
- Funding
- Update code

Typical examples relevant for physics are in Germany the databases VADEMECUM (information on institutes), INFOR (information on institutes and projects), and FORKAT (projects funded by the German Federal Ministry for Research and Technology); in Britain the database Current Research in Britain (descriptions of projects conducted at British universities); and in France the database Telelab (information on governmental and university laboratories).

Another prototypical example consists of the conference databases that refer to future meetings and conferences. The components of the record are the following:

- Accession number
- Meeting title
- Meeting organizer
- Meeting location
- Meeting date
- Meeting secretariat
- Classification code
- Status date

A representative example is the database CONF produced by the FIZ Karlsruhe in Germany. A record of this database is shown in Fig. 9. Perhaps it should be mentioned that many of the factual databases are based on questionnaire actions for the acquisition of the necessary information needed for the input. The examples shown before show this very clearly. It is here our intention not to give a complete collection of the existing factual databases (such a complete survey can be found in Cuadra, 1990), but only to present some representative examples and to explain typical features.

6. NUMERIC DATABASES

A database is called numeric when it consists mainly of structured numeric data. Before considering the databases in more detail let us look at the kind of data stored in such a database. In physics we are dealing with numeric values of observables. The data we have to consider usually has the following structure:

1. Measured value
2. Uncertainty
3. Units (SI)
4. Data type
5. Range of validity
6. Method of measurement
7. Combination with other quantities
8. Bibliographic citations

Generally speaking a *datum* can also be any parameter deduced with the help of a model or a theory. It should, however, be noted, that very often some of the fields listed above are empty.

The automatic techniques that are nowadays employed in experiments either in laboratories or outside, for instance in satellite experiments, often generate large amounts of data, sometimes more than can be handled afterwards. These raw or primary data are recorded continually or at specific intervals.

```
L2   ANSWER 12 OF 32
AN   PAS:48318  CONF        SN   K99805312
MT   International ANS Topical Fast Reactor Safety Meeting.
MO   American Nuclear Society (ANS, La Grange Park, IL (USA); Euro-
     pean Nuclear Society, Petit-Lancy (Switzerland); Internatio-
     nal Atomic Energy Agency (IAEA), Vienna (Austria). Internatio-
     nal Working Group on Fast Reactors
ML   Snowbird, UT (USA)
MD   19900812 - 19900816
NTE  Contact: ANS/Idaho Section  International Fast Reactor Safety
     Meeting  attn.: R.W. King  P.O. Box 2528  Idaho Falls, ID 83402
     USA
     Deadlines: for summaries (1500 words): 1 Jan 1990; for complete
     papers with abstracts: 4 May 1990
     Abstracts: (summaries) will not be published
CC   *10P Nuclear reactor technology
      13L Safety engineering
      100 Nuclear power plants
UT   Safety design criteria; public perception/implication of the
     Chernobyl accident; safety experiments/fast reactor testing and
     operational safety experience; designing for passive safety;
     instrumentation and control; safety codes and calculational
     techniques; consequences modeling of accidents/severe accidents/
     containment design and analysis; seismic and external event design
     and accomodation
STA  Entered           Revised
STD  19880519          19891016
```

AN	Accession Number	MT	Meeting Title
MO	Meeting Organizer	ML	Meeting Location
MD	Meeting Date	NTE	Notes
CC	Classification Codes	UT	Uncontrolled Terms
STA	Status	STAD	Status Date

FIG. 9. Retrieval example from CONF.

Later on they are analyzed and interpreted according to the physical quantity involved. Very often this methodology leads to databases containing large volumes of raw data. We shall, however, not consider these in the following because their usage is mostly restricted to the very few specialists carrying out the particular experiments. We shall also not discuss further how to build private numeric files and databases (this has been sketched in the sections before) or go into further details concerning the numerous small home-built data files in physics. As before, we shall only elucidate the publicly available larger databases that are produced in a professional and organized manner.

In physics we are mostly dealing with time- and location-independent data. Therefore, in the following we shall restrict ourselves to the handling and storage of this type of data. In certain subject areas, including astronomy, astrophysics, and geophysics, time series are important. However, that would be a topic of its own.

More often than not we have more than a single experimental result available for a physical quantity. Sometimes these measurements even contradict one another. It is therefore very important to know what must be done in order to store these data in machine-readable form. It is now customary to distinguish between the following forms of data handling and presentation:

- *Tabular lists*. These are simple systematic lists of numerical information.
- *Data compilations*. These are systematic collections of numerical data with some

reorganization for an optimal presentation.

- *Evaluations*. These are collections of numerical data, but with just one recommended best value for one measured quantity. This recommended best value is derived by a critical consideration of available information.

It is obvious that the last form, "evaluations," is of major importance when we are considering storage of data in machine-readable form in a field where we have very often more than one value for one physical quantity.

The structure of numerical databases under consideration can be described in four parts as follows:

1. Bibliographic part—title, authors, bibliographic citation.
2. Information part—for example, substance name, molecular formula, material description, experimental conditions.
3. Data part—for example, melting points, lists of spectrum peaks.
4. Computational part—For example, general program packages, including statistical packages, interpolation routines; specialized program packages, including calculation of thermodynamic properties, spectrum estimation.

This outline shows that a numerical database in physics will contain, aside from the dominant numerical information, also other alphanumeric information (for example, the source where the data came from in the bibliographic part). Very important, sometimes even essential, are also the specialized computer packages integrated in a numerical database. Their purpose is normally to offer a possibility for calculating data not stored on the basis of basic data or of some kind of knowledge base.

The structure of the numeric part is described by matrices (data tables) with m rows and n columns (m and n have any numerical value) where the elements A_{mn} of these matrices contain the numerical information (sometimes also alphanumeric comments). Further details are best illustrated by examples. For this we have selected three: one from the field of crystallography, one from the field of nuclear physics, and one from the field of spectroscopy.

The first database we would like to consider is the Inorganic Crystal Structure Database (ICSD). It contains crystallographic data (crystal structure data) from inorganic compounds:

- reference (authors, journal, year of publication, volume and page)
- chemical name (and phase designation)
- mineral name (and origin)
- chemical formula
- Chemical Abstracts registry number (if available)
- density
- unit-cell dimensions
- space group
- atomic coordinates, site occupation, oxidation state, and temperature factors for each element of the structure
- reliability index (R value)
- remarks concerning conditions of measurement or specialties of the structure not described by the above parameters

ICSD is produced by FIZ Karlsruhe and Gmelin-Institut in cooperation with the University of Bonn (for more details see IUCR, 1987). A typical record is shown in Fig. 10. ICSD contains about 30 000 structures that have been compiled from the open literature. Before the data were entered into ICSD, they were carefully checked both manually and automatically by application of suitable test programs. There are also two specialized program packages integrated into ICSD, one for the calculation of the different angles between the various atoms in the crystal and one for the calculation of distances between specified atoms. Both components can be accessed with the retrieval software without any problems.

As stated above, our next database example has been taken from the field of nuclear physics. In this area the database ENSDF (Evaluated Nuclear Structure Data File), where the printed version is the *Nuclear Data Sheets*, plays a major role. It contains the nuclear structure and radioactive decay characteristics of all isotopes known so far, i.e., the existing energies of the energy levels, the spins, the parities, and the half-lives and all decay energies of the related α, β, and γ transitions. It consists of a data part and bibliographic part, with the data part containing only evaluated data. ENSDF is produced and kept up to date by an international coop-

```
L1    ANSWER 1 OF 4
AN    34308  ICSD
CN    Calcium carbonate
        ***Aragonite***
LSF   Ca C O3
MF    C Ca O3
TI    Refinement of the crystal structure of the aragonite phase of Ca C
      O3
AU    Dickens, B; Bowen, J S
SO    Journal of Research of the National Bureau of Standards, Section
      A.
      Physics and Chemistry. (1971) Vol. 75 p. 27-32; CODEN=JNBAAR
CLP   A=4.9598(5) B=7.9641(9) C=5.7379(6)    unit: Angstrom
      ALPHA=90. BETA=90. GAMMA=90.    unit: Degrees
      NFU  4
      VOL  226.65  Angstrom**3
DEN   2.947(2)  g/cm**3
CSG   PMCN; 62
RVAL  0.04
ST    MINR Compound with mineral name
      TEM  Temperature in Kelvin at which the determination was made (if
      not room temperature)
      TEM 298
LAU   MMM
PRS   OP20
CCLS  MMM; D2H
CSYS  ORT
CSYM  CEN; NPOL
FTYP  ABX3

ATP   Atomic Parameters
```

At	Nr	Ox	Wy	X	Y	Z	SOF
Ca	1	+2	4c	0.2	0.58507(5)	0.25974(6)	
C	1	+4	4c	0.2	0.2386(2)	0.4148(3)	
O	1	-2	4c	0.25	0.0770(2)	0.4043(2)	
O	2	-2	8d	0.4737(2)	0.3196(1)	0.4131(2)	

```
TF   Temperature Factors      unit: Angstrom **2
```

Atom	B(1,1)	B(2,2)	B(3,3)	B(1,2)	B(1,3)	B(2,3)
Ca 1	0.664(9)	0.599(9)	0.601(9)	0.0	0.0	-0.01(1)
C 1	0.75(5)	0.85(5)	0.46(5)	0.0	0.0	0.07(5)
O 1	1.50(5)	0.54(4)	1.04(4)	0.0	0.0	0.03(4)
O 2	0.71(3)	1.03(3)	1.02(3)	-0.32(3)	-0.02(3)	-0.09(3)

```
TFLG 52 At least one temperature factor is implausible but agrees
with the value given in the paper
```

FIG. 10. Data set from ICSD.

eration coordinated by the IAEA. In detail its contents can be characterized by the following figures.

1. Bibliographic part: contains about 120 000 bibliographic citations (from 1910 to 1991).
2. Data part: contains about 920 000 card image-based records. Data sets number about 10 700 of which 24% are adopted levels data sets, 27% are radioactivity data sets, and 49% are nuclear reaction data sets.

The structure of the database is shown in Figs. 11 and 12. An example of a typical data set is presented in Fig. 13.

The next database we would like to consider in this context is the spectroscopic C13 NMR/IR (carbon-13 nuclear magnetic resonance/infrared) database, which is produced jointly by Chemical Concepts and BASF (a large chemical company in Ludwigshafen, Germany) (the on-line version is offered by STN International).

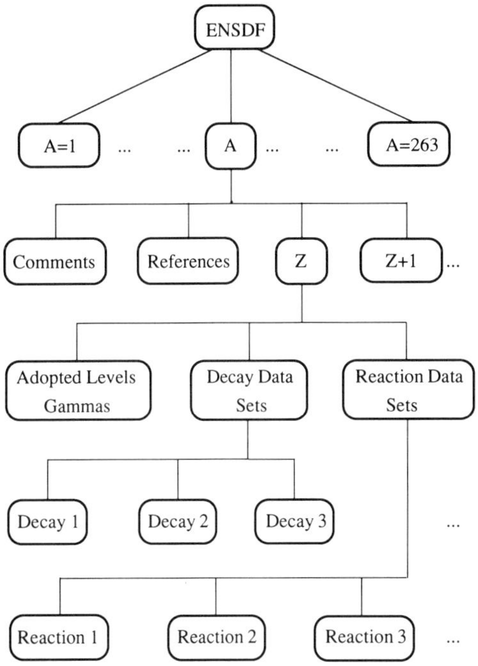

FIG. 11. File organization in ENSDF.

Typical contents	
^{13}C NMR part	IR part
Accession number	Accession number
Molecular formula	Molecular formula
Chemical name	Chemical name
Chemical Abstracts registry number	Chemical Abstracts registry number
Chemical structure (connection tables)	Chemical structure (connection tables)
Molecular weight	Molecular weight
Solvent	Source (sample)
Standard	Sample form
Instrument code	Melting point
Temperature	Peak lists
Conversion shift	Number of peaks
Origin number	Normalization factor
Chemical shift	
Relaxation times	
Bibliographic reference	

A sample for a ^{13}C NMR record is presented in Fig. 14 (see also Luksch, 1990). It should be noted that the database also has a graphical component that allows graphic representation of chemical structures and corresponding spectra. The database contains 102 000 ^{13}C NMR spectra (85 000 chemical structures) and 16 000 IR (15 000 structures). The data are taken from journal articles, spectral catalogs, and unpublished sources. In addition, specialized software packages are connected with the database. These are packages for spectrum estimation, spectrum interpretation, and similarity search. Of these, the most important and interesting one is that for spectrum estimation. It allows the calculation of unknown ^{13}C NMR spectra (chemical shifts) from a known chemical structure. This is based on the correlation between spectral features and the environment of an individual atom in a molecule described by the HOSE and HORD codes (hierarchically ordered spherical description of environment, hierarchically ordered ring description; see Bremser, 1978). A special library with spectral information linked to these HOSE and HORD codes is then built up which allows the derivation of spectra from topological matrices (see Bremser, 1978 and

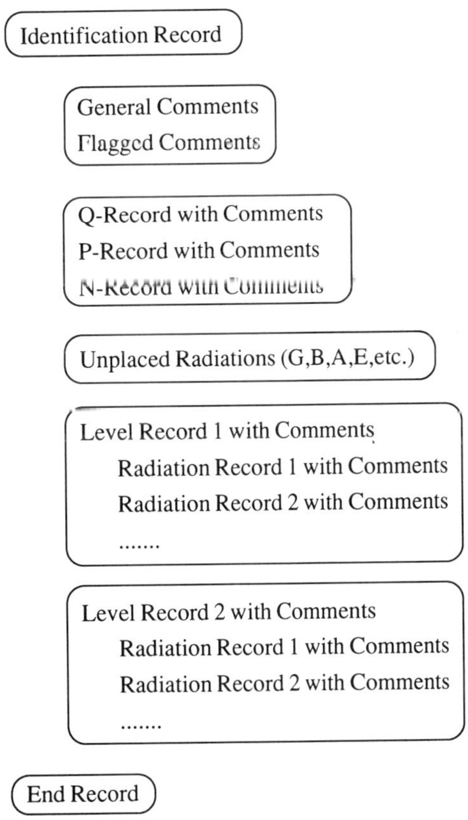

FIG. 12. Structure of a single ENSDF data set. All records are 80-byte card images. The letters Q, P, N, G, B, A, and E, respectively, stand for quantities important in nuclear physics, viz., Q values, parent record, normalization record of emitted radiation, γ, β, and α emissions, and energy data.

```
86RB     86RB IT DECAY (1.017 M)          69SC10,70PA09,67YU01      88NDS      8808
86RB CL             ISOMERIC STATE PRODUCED BY (N,2N) REACTION ON 86RB.
86RB CL             OTHERS: 71RU15, 69SC10, 70PA09, 53SC39, AND 51FL17.
86RB CL J           FROM ADOPTED LEVELS.
86RB  N 0.9817      9 1.00      1.0
86RB CN BR          B- DECAY <0.3% OF TOTAL 86RB IT DECAY FROM ABSENCE OF
86RB2CN             1153G (86SR) (69SC10).
86RB  L 0.0           2-
86RB  L 556.07      186-             1.017 M   3                              M
86RB CL T           FROM 67YU01. OTHER: 1.020 M 17 (69SC10).
86RB  G 556.07      18100         (E4)                 0.01868  100
86RBS G KC=0.01581    $LC=0.00216
86RB3 G EKC=0.0158 15$  ECC=0.0184 15
86RB CG E           WEIGHTED AVERAGE OF 556.03 25 (69SC10) AND 556.1 25 (70PA09
86RB CG EKC,ECC     FROM I(XK)/IG(556) AND ICE/IG(556) (69SC10).
86RB2CG THEORETICAL (68HA52): KC=0.0139 (M3) AND 0.0158 (E4);
86RB3CG                       CC=0.0158 (M3) AND 0.0182 (E4)
```

FIG. 13. Example of an ENSDF data set.

1987; Zupan, 1986). Why is this so important? As we said before, 102 000 spectra are in the database. We have, however, many more compounds existing in nature (about 12 million). Thus in case a spectrum must be interpreted that is not in the database, the corresponding chemical shifts for a postulated chemical structure can be estimated. In a retrieval session it is also very easy to manipulate the structures of the chemical compounds under investigation in order to achieve the greatest possible correspondence between calculated and measured spectra.

Since each of the spectroscopic methods [^{13}C NMR, IR, mass spectroscopy (MS)] has its own specific analysis potential for an unknown chemical structure, it is of great advantage to apply several spectroscopic techniques in parallel (multidimensional spectroscopy) to one unknown compound. The future of spectroscopic databases is therefore devoted to multidimensional spectroscopic databases in which the different sublibraries of specific spectroscopic data such as ^{13}C NMR data, IR data, etc., are linked to one central structure library with connections to the CAS, Beilstein, and Gmelin databases via corresponding registry numbers (see Fig. 15). A typical example is the SpecInfo database originally developed by BASF in Germany and now offered by Chemical Concepts.

These multidimensional databases can be combined with suitable intelligent software to build a multidimensional interpretation system, which is shown schematically in Fig. 16. The flow of information can be described in the following way. The MS, NMR, and IR spectra for the unknown substance are transferred to the system. On the basis of the MS, NMR, and IR subdatabases, chemical structure sets are found by applying similarity searches and statistical analysis of the resulting "hit" list. A structure generator then assembles the substructures and derives structures. MS, NMR, and IR spectra can afterwards be predicted and then fed back with suitable ranking into the interpretative system. Finally, the solution space of the structure gen-

```
CNMR 55459   REG:          MW: 227.0460

NAME: N,N-DIMETHYL-4-NITRO-BENZALHYDRAZONE CHLORIDE
MF: C9 H10 CL1 N3 O2
```

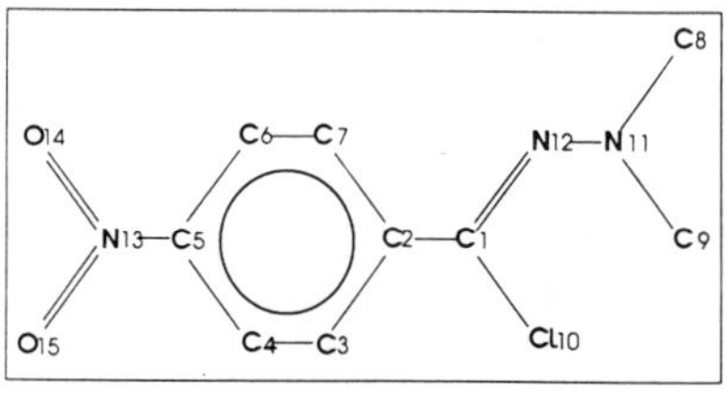


```
M.MUEHLSTAEDT,L.WEBER,Z.CHEM.25,323(1975)
 SO: CDCL3         IN: NOT REPORTED     CO: 0.00 PPM
 ST: TMS           TE: AMBIENT          OR: NEU00038
 DATE: 03.06.86
CHEMICAL SHIFTS (PPM), MULT:
 1  148.20 -      4  127.80 -      7  123.40 -
 2  127.10 -      5  141.40 -      8   46.70 -
 3  123.40 -      6  127.80 -      9   46.70 -
SPECTRUM? (N/Y)*y
```

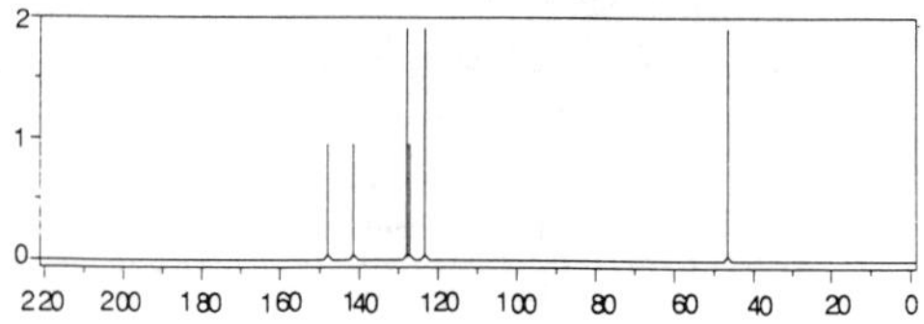

FIG. 14. An example of a retrieved ^{13}C NMR data set.

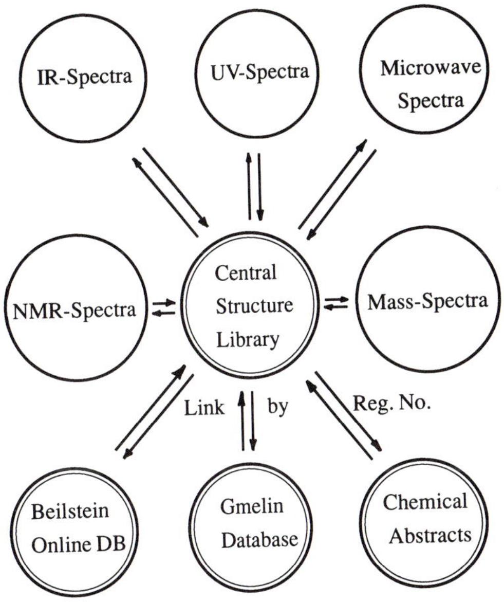

FIG. 15. Spectral data pools with a central chemical structure library.

erator is restricted in a kind of convergence process such that a small number of possible candidates for the structure remain (see Bremser, 1978; Zupan, 1986).

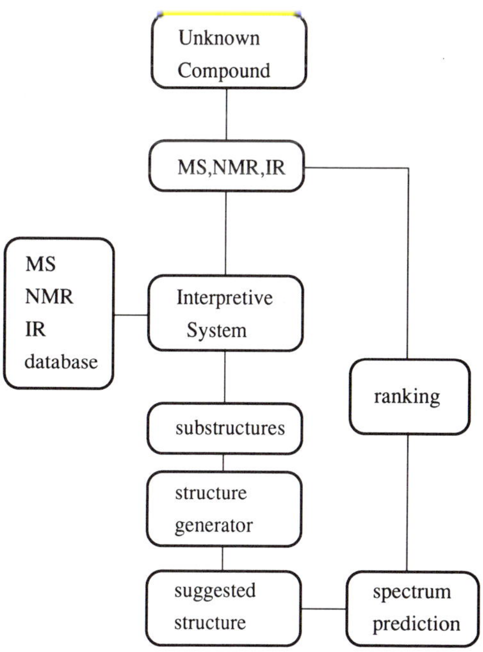

FIG. 16. Organization of a multidimensional spectral database system.

Contrary to the situation we have for the bibliographic databases, the coverage of the whole field of physics by existing numeric databases is very uneven. For example, the fields of crystallography (see Table 1), of nuclear physics (Table 2), of thermophysics and thermodynamics, and of spectroscopy are well covered. In the wide field of solid-state physics, on the other hand, there are hardly any databases. Those that exist contain mostly materials data and strictly speaking do not belong directly to the field of solid-state physics.

Concerning quality criteria there are specific differences between bibliographic and numeric databases. Reliability of data is most important in numeric databases, whereas this point is not so critical in bibliographic databases. In addition, it is important to store data and metadata (information necessary to define data), to refer to the sources of the data, to give information about type of apparatus, experimental conditions etc., and to make available special comments whenever necessary. In many cases the storage of the uncertainties belonging to the data is also of importance. In addition, continuous data maintenance and updating are often essential.

7. ON-LINE SERVICES

Usually, the producers of databases do not make their databases available to the public themselves, but pass these on to database vendors or on-line service centers, the so-

Table 1. The most important crystallographic databases (for more details see IUCR, 1987).

Atomic coordinates
1. Cambridge Structural Database (organic compounds)
2. Inorganic Crystal Structure Data Database (inorganic compounds)
3. Metals Data File (metals and alloys)
4. Protein Data Bank (biological macromolecules)
Unit-cell parameters
NIST Crystal Data Identification File (Organic and inorganic compounds)
Powder patterns
Powder Diffraction File (organic and inorganic compounds)

Table 2. The most important nuclear physics databases (for more details see Behrens and Rittberger, 1985).

1. Nuclear structure and decay data databases:
ENSDF (Evaluated Nuclear Structure Data File)

2. Nuclear reactions (non-neutron) databases:
None

3. Neutron reaction databases:

a. Experimental data:
EXFOR (EXchange FORmat database for experimental nuclear reaction data)

b. Evaluated data:
BROND 90 (USSR Evaluated Neutron Data Library)
ENDF/B-VI (Evaluated Nuclear Data File, Brookhaven, USA)
ENDL-84 (Evaluated Nuclear Data Library, Lawrence Livermore Labs.)
IRDF 90 (International Reactor Dosimetry File, IAEA)
JEF-2 (Joint Evaluated File, Nuclear Energy Agency, NEA)
JENDL-23 (Japanese Evaluated Nuclear Data Library)

called *hosts*. These hosts have suitable mainframes at their disposal where the databases are loaded. By offering also a more or less refined retrieval software they make the databases available to the public. The on-line access for the normal user will be via the usual telecommunication networks. As the word "on-line" implies, this search technique is an interactive process, where user and computer are conversing in real time. For the user the following type of equipment is required (see COMPUTER HARDWARE). First, a terminal or, even better, a microcomputer with ASCII standard and with some suitable communication software is needed. To link the PC (or the terminal), which should have an RS232 port and an asynchronous communication card, to the phone, a *modem*, which stands for modulator/demodulator, is also needed. The modem is a device that converts the digital signals coming out of the PC into the analog language of the telephone system, and vice versa. An important point in this context, however, is the number of characters to be transmitted per second over the line. Usually, the modem speed is measured in baud (bits per second). A 300-baud modem can handle 30 characters per second (approximately five average-length words) and a 2400-baud modem, 240 characters per second (approximately 40 words). Thus it is advantageous to have at least 2400-baud facilities, but today even faster lines and modems are available.

In the first part of the communication link, the connection between the user's PC and the host is made via the ordinary telephone line. Because of cost considerations, the second and major part of the path makes use of special telecommunication networks. These are accessed by making a telephone call to the nearest node; from this the call is automatically switched through the telecommunication networks (packet-switching systems) to the host. At present, one sometimes finds more complicated schemes, viz., a mainframe at the center with telecommunication links to both the host and the user. Of course, the customer has to sign an agreement with the host in order to receive a password that permits access to the databases.

Today we have about 5000 databases publicly offered on line and about 600 hosts (Cuadra, 1990). Such a host may offer from very few to hundreds of databases for all fields of science. For physics the most important public hosts are STN International, DIALOG, BRS, ESA-IRS, Questel, and Data-Star (addresses and more detailed description can be found in Cuadra, 1990). Each of these hosts uses its own retrieval software. This has the consequence that the corresponding search languages are different. In case of STN International, for instance, the name of this language is Messenger. In many cases, however, the necessary commands are similar to a large degree. It is clear that the difference between the various command languages complicates the human–host interaction since a user must learn the language peculiar to each service. In the interaction process that takes place between a human user and the host, there are also other frustrating although often trivial problems. One of these is the log-in procedure to a host computer, which is a trivial but very often cumbersome procedure. It could require typing in a certain number of consecutive digits with restrictions for timing. One typing error normally has the consequence of an invalid call and of the obligation to repeat it. Therefore a large number of on-line searching aid software packages are nowadays available. When installed on the PC they can, for instance, be used for retrieval. Their purpose is to improve the user-friendliness of the host access. These on-line searching aid systems can be classified as (Efthimiadis, 1990)

- gateway systems,
- front ends, or
- intermediary systems.

Gateway software allows auto-log-on (automatic dialing, automatic selection of a communication network, automatic connection to the host computer, automatic transmission of the user's log-in password) and possibly uploading and downloading. Front-end software includes the above features, but goes beyond by simplifying steps of the searching process itself (presearch formulation, editing, uploading of search formulation, downloading, and post-processing). Intermediary systems are often real expert systems and have the aim of replacing human intermediary searchers in some way (query reformulation, query expansion, natural language interface). For more details, see, e.g., Efthimiadis (1990).

From the above it follows that on-line searching aids also simplify the problem of accessing more than one host. In addition, file transfers such as downloading (capturing of data sent by the host computers) and uploading (sending data from the user's computer to the on-line host) are easily performed by making use of online searching aids. This is important because the users very often want to manipulate (post-process) the data they retrieved, which can be done after downloading by application of various packages. Examples for gateway systems are PC Net/Link and iNet; for front-end systems ProSearch, Sci-Mate, STN-Express, STN Personal File System, Molkick, etc.; and for intermediary systems CONIT, NDA, etc.

8. OTHER DATA DISTRIBUTION CHANNELS

An on-line connection between user and host computer is but one of many data distribution channels. Nowadays, quite a few alternative methods of storing and transferring data from one location to another are being used. The possibilities include tape drives, Bernoulli cartridge systems, removable hard disks, write-once optical disks (WORMs), erasable optical disks, floppy disks, and even VHS video tapes.

Tape drives are by far the most popular medium. Data are transferred between mainframes on nine-track, half-inch, open-reel tapes with capacities between 100 and 300 MB depending on reel size and recording density (e.g., 6250 bits per inch). Capacities for systems for use with mini- and microcomputers range from 40 MB to 5 GB. These include cartridges of various makes, such as TK50 and TK70 (VAX computers), DC 6000 QIC, DC 2000 minicartridges, Teac's D/CAS cassettes, and 4-mm digital audio tape (DAT). An overview of the different systems available can be found in Stone (1990).

Because of the predominance of MS-DOS™ and its offspring in the world of personal computers, exchange of smaller amounts of data on standard $5\frac{1}{4}$- or $3\frac{1}{2}$-in. floppy disks is very common. The use of optical disks with their larger storage capacities is increasing (CD-ROMs). WORMs and erasable optical disks are still quite expensive at present, and lack of standardization hampers their use for data exchange.

GLOSSARY

Bibliographic Database (also called Reference Database): A computer-readable file containing literature citations or references to published articles and documents.

Bit: A single linear digit having one of two states, 0 or 1.

BTREE (or Bayer Tree): A way of organizing structured data in a computer in order to minimize the time needed to access it.

Byte: Group of eight bits, used as the smallest computer storage and processing unit. In addition, it is the basis of our character representations.

CAS: Chemical Abstracts Service, P.O. Box 3012, Columbus, Ohio 43210.

CD-ROM: Compact-disk read-only memory. Read-only optical disk for the storage of digitized data.

DBMS or Database Management System: A software system with management and administrative capabilities for maintaining and controlling access to large amounts of structured data on a computer. A DBMS is a DBS with management facilities.

DBS or Database System: A software system supporting structured data on a computer, usually arranged in one or more files.

DDL: Data definition language, used to describe the logical file structures in DBS.

DDM: Data definition module, containing the data definitions for a file in a DBS in original text or compiled form.

Descriptor: A keyword contained in a thesaurus.

Direct Access File: A collection of data in a computer file to which access can be gained directly, i.e., without scanning the data from the start of the file.

DML: Data manipulation language. Command language of a DBMS. Sometimes there is no clear distinction between DDLs and DMLs.

Documentation Unit: The smallest unit of a bibliographic database containing all the data of a literature citation together with the terms of the subject analysis.

ER or Entity–Relationship Model: Conceptual model of the real world used in a DBS containing objects (entities) and their relationships.

Entity: Any object that exists on its own and is distinguishable (e.g., a person, an automobile).

Factual Database: A database containing factual information, mainly of non-numerical nature.

Field: The smallest subdivision of a record and usually contains numeric or alphanumeric information for a single topic.

File: A named organized collection of related data elements stored in a computer system.

File Design: Conceptual and logical design of a DBS.

Front-End Software: Software that is loaded on the user PC in order to improve the user-friendliness of the host access by allowing automatic log-on, prepared search formulations, editing, etc.

Full-Text Database: A database containing the entire text of a document.

Gateway Software: PC software allowing automatic log-in to a host computer.

Hardware: Any piece of physical equipment associated with computer processing or communication networks.

Hashing Algorithm: A function that computes, from a character string serving as key for a file, an integer in the range 0 to some maximum value. This "hash value" is used for directly gaining access to the required data in the file.

Hierarchical Model: Data model structured like an inverted tree. Each item may have one or more descendants, but only one parent.

HORD Codes: Hierarchically ordered ring descriptions. They describe rings and their environments in chemical molecules.

HOSE Codes: Hierarchically ordered spherical description of environments of atoms in chemical molecules.

IMS: IBM's Information Management System.

Index: An ordered list of the contents of one or more fields or parts thereof, each element or key pointing to or identifying the location of the corresponding record.

Indexing (of Documents): The assignment of keywords (descriptors) to bibliographic documents. The term is sometimes also used as a synonym for subject analysis.

Information or Data Retrieval: Transfer of selected information from long-term storage to computer memory for subsequent processing, e.g., for display on a terminal or printer, or downloading to some other device.

Inverted File: A file that contains the addresses or locations or records in which the values for given keys exist. It is arranged for each key in order of record number.

ISAM: Index sequential access method. A method for gaining access to data in a file containing records sorted by key values. A dictionary and a telephone book are everyday examples of index sequential lists. Access to these is by first locating the correct page, i.e., by looking at the entries on the top left or right corners or each page, and then sequentially reading the individual entries on the page.

Link: Connection between two data items or records. A link may be physical, i.e., one item may contain a pointer (or address) to the other, or logical, where two items are connected by fields having identical contents.

Modem: An acronym for modulator/demodulator, the name for a hardware device that interfaces between a telephone line (analog signals) and a terminal or PC (digital pulses).

Network Model: Data model in which the elements have complex (m:n) relationships. Each item may be connected to a number of parents and descendents.

Numerical Database: A database that predominantly contains structured numerical data such as physical properties of substances.

On-Line: Interactive process where the user and the host computer are conversing in real time by a constant connection via a computer terminal or a PC.

On-Line Searching Aids: Software systems providing support for on-line searching.

Packet-Switching Network: A communication network which, in order to transmit a message from one location to another, breaks the message into small units called packets containing the destination address. Through multiple use of a single transmission line the costs are shared and are much lower than with normal telephone lines.

Pointer: A data element containing the address of the object being pointed to. In the programming language C, pointers are 2- or 4-byte quantities that may be used in simple arithmetic operations (addition, subtraction).

Record: A linear logical unit containing data associated with one entity or object of a database.

Relational Model: Data model in which the elements can be ordered linearly like the rows of a table.

RS232: A specification defined by the Electronic Industries Association (EIA) detailing the physical configuration of the serial interface connector between data terminal equipment and data communication equipment.

Subject Analysis: The assignment of terms such as classification codes, keywords (descriptors), and other "value-added" codes to bibliographic citations.

Thesaurus: A controlled vocabulary of keywords to be used for indexing of documents. Entries are ordered hierarchically in narrower (NT) and broader (BT) terms. An entry may be classified as forbidden or point to related (RT) or preferred terms.

WORM: Write once, read many times optical disk for storing digitized information.

List of Works Cited

Bayer, R., McCreight, E. M. (1972), "Organization and Maintenance of Large Ordered Indices," *Acta Informatica* **1**:3, 173–189.

Behrens, H., Rittberger, W. (1985), "Nuclear and Particle Physics," in: D. F. Shaw, (Ed.), *Information Sources in Physics*, London: Butterworths.

Bremser, W. (1978), *Anal. Chim. Acta* **103**, 355–365.

Bremser, W. (1987), *ESN–European Spectroscopy News* **75**, 10–27.

Cuadra (1990), *Directory of Online Databases*, Vol. 11, No. 3, New York: Cuadra/Elsevier.

Date, C. J. (1987), *A Guide to the SQL Standard*, Wokingham: Addison-Wesley.

Efthimiadis, E. N. (1990), *J. Documentation* **46**, 218–262.

Franklin, C. (1989), "Hypertext Defined and Applied", *ONLINE* **13**, No. 3, 37–49.

ICSTI (1991), *International Classification Scheme for Physics*, ICSTI 1991.

IUCR (1987), *Crystallographic Databases*, Bonn: Data Commission of the International Union of Crystallography.

Lockemann, P. C., Schmidt, J. W. (Eds.) (1987), *Datenbank-Handbuch*, Heidelberg: Springer-Verlag.

Luksch, P. (1990), "Spectroscopic Online Databases—A Tool for Structure Elucidation," International Symposium on the Future of Scientific, Technological and Industrial Information Services, IAEA-SM-317, p. 22.

Marchionini, G., Shneiderman, G. (1988), "Finding Facts versus Browsing Knowledge in Hypertext Systems," *IEEE Computer*, **21**(1), 70–80.

Ross, R. G. (1978), *Data Base Systems*, New York: AMACOM.

Salton, G. (1983), *Introduction to Modern Information Retrieval*, New York: McGraw-Hill.

Salton, G. (1989), *Automatic Text Processing*, Reading, MA: Addison-Wesley.

Seymour, J. (1988), "Project 3: Programmable Databases", *PC Magazine* **7**, No. 9, 93–272.

Silberschatz, A. Stonebraker, M, Ullman, J. D. (Eds.) (1990), *Database Systems: Achievements and Opportunities*, Lagunita Report of NSF Invitational Workshop on the Future of DBS Research, Palo Alto, CA.

Stone, M. D. (1990), "QIC 150 Tape Drives: The Rising Star for Backup," *PC Magazine*, **9**, No. 17, 269–338.

Tepel, J. W., Müller, H.-W. (1990), "GAMCAT—A Personal Computer Database on Alpha Particles and Gamma Rays from Radioactive Decay," *Nucl. Instrum. Methods Phys. Res.* **A286**, 443–446.

Ullman, J. D. (1982), *Principles of Database Systems*, Rockville, MD: Computer Science Press.

Zupan, J. (Ed.) (1986), *Computer-Supported Spectroscopic Databases*, Chichester: Ellis Horwood Limited; New York: Wiley.

Further Reading

Date, C. J. (1986), *An Introduction to Database Systems*, Wokingham: Addison-Wesley.

Hall, J. L. (1986), *Online Bibliographic Databases*, London: ASLIB.

Jaschek, C. (1989), *Data in Astronomy*, Cambridge, England: Cambridge University Press.

Nadrachal, J. (Ed.) (1984), *Data Bases and Data Structures in Physics*, Proceedings of the Fifth Summer School on Computational Physics, *Computer Phys. Commun.* **33**, 1–297.

Perry, J. T., Lateer, J. G. (1989), *Understanding ORACLE*, San Francisco: Sybex.

Rossmassler, St. A., Watson, D. G. (1980), *Data Handling for Science and Technology*, Amsterdam: North Holland.

Rumble, J. R., Hampel, V. E. (Eds.) (1984), *Database Management in Science and Technology*, Amsterdam: North Holland.

Salton, G., McGill, M. J. (1983), *Introduction to Modern Information Retrieval*, New York: McGraw-Hill.

Salton, G. (1989), *Automatic Text Processing*, Reading, MA: Addison-Wesley.

Schwarz, St., Watson, D. G. Alvfeldt, O. (1983), *Nonbibliographic Data Banks in Science and Technology*, Paris: ICSU Press.

Shaw, D. F. (1985), *Information Sources in Physics*, London: Butterworths.

Shneiderman, B. Kearsley, G. (1989), *Hypertext Hands-On! An Introduction to a New Way of Organizing and Accessing Information*, Reading, MA: Addison-Wesley.

Registered Trademarks

ADABAS and NATURAL are products of the Software AG, Darmstadt, Germany.

dBASE is produced by Ashton-Tate, Torrance, California.

CLIPPER is produced by the Nantucket Corporation, Los Angeles.

MS-DOS is the trademark of Microsoft Corporation.

ORACLE is produced by the Oracle Corporation, Belmont, California.

PostScript is the trade mark of Adobe Systems, Inc.

COMPUTER GRAPHICS

BRUCE NAYLOR AND ARUN N. NETRAVALI, *AT&T Bell Labs, Murray Hill, New Jersey, U.S.A.*

INTRODUCTION

Computer graphics is the subfield of the computing sciences concerned with the computational creation of visuals, that is, of visually perceived and interpreted information with a predominantly spatial character. This is in contrast to the display of language-based information using symbols (e.g., letters). Visuals may take the form of something as simple as a graph plot of empirically acquired data, in which the geometric setting is used to convey nongeometric information, or it may be as complex as synthesizing realistic images that are not easily distinguishable from photographs, in which the visual component is the essence.

Regardless of the objective, computer graphics uses analytic geometry embedded in various data structures, such as arrays, lists, and graphs, as the basis for representing the basic geometry. This geometry is augmented by a variety of attributes, most notably color, which are appropriate for the intended use. Collectively this forms the *geometric model.* Operations on this model, such as creating a picture of it, are provided by a variety of algorithms.

Since visual perception is a dominant component for much of human experience, and is the sense that receives information at the greatest rate, computer graphics has many application domains. To date, these include visualization of data in business, science, medicine, engineering, architecture, mathematics, and the visual arts, and the simulation and control of processes in manufacturing. Its reliance upon computational representations of geometry, a subject addressed by the subfield of geometric modeling, is shared with other computational fields such as robotics, computer vision, image processing, and computational geometry, as well as with various forms of scientific geometric computation.

Computer graphics began in the early 1960s, most notably with a system called Sketch Pad developed by Ivan Sutherland (1989 Turing Award). Sketch Pad was an interactive design program for electronic circuits that used for its display a variant of an oscilloscope: a CRT in which the electron beam draws the image in a manner analogous to a person drawing with pen and paper. Such a device has been called variously a stroke or vector or random-scan display, and has with rare exceptions been monochromatic.

A flowering of basic research occurred in the field at the University of Utah's Computer Science Department, headed by Ivan Sutherland, during the very late 1960s and throughout much of the 1970s. This work focused on synthesizing more realistic images using TV-based CRTs, called raster-scan displays, that could easily paint a full-color image on the entire screen. Similar work was taking place

1-56081-063-7/92/$5.00+.50

in industry with the creation of flight simulators: mock cockpits used to train aircraft pilots in which the images in the cockpit windows were generated by computer graphics, thus creating a "virtual reality." Raster-scan-based graphics has since become the dominant display method, made possible by greatly reduced costs of memory and much greater processor performance. It is now available on almost all computer systems, including personal computers. Accompanying this technological advance has been a flowering of the field in research, academic, and commercial arenas.

1. GEOMETRIC MODELING

Creating visuals is primarily a spatial activity, and so one must have a computational representation of space. This space is usually taken to be a Euclidean space, and it is almost always two dimensional (2D) or three dimensional (3D). Euclidean space is modeled algebraically either as an affine space, in which parallel lines never meet, or as a projective space, which adds to an affine space the "points at infinity" at which parallel lines meet (and so has the elliptic parallel property). Real affine spaces use the standard Cartesian coordinates to denote points: (x,y) if in 2D, or (x,y,z) if in 3D, where the coordinates are real numbers. Projective spaces add one more coordinate so that the affine point (x,y) corresponds to the projective points (xw,yw,w) for all $w \neq 0$. If $w=0$, then it is a point at infinity in the (x,y) direction. Points are typically represented computationally as one-dimensional (1D) arrays of reals, and a real number is usually represented by "floating point," where a number is represented in a manner comparable to scientific notation.

Geometric objects, such as lines, circles, or polygons, are interpreted as sets of points. These points are usually defined as those satisfying some equations and inequalities. Thus, for instance, a line in 2D is the set of points satisfying the equation $Ax+By+C=0$. A line segment is the subset of a line between two points on the line, say (x_1,y_1) and (x_2,y_2), which determine a set of inequalities, viz., $x_1<x<x_2$ and $y_1<y<y_2$. These expressions can then be used to test if an arbitrary point lies on the line segment by seeing if it satisfies the line equation and the inequalities. It is also possible to generate any number of points known to be on the line segment using these expressions. The data structure for a line segment can be just its two end-points, since they determine the line and bound the segment as well. Analogously, a circle can be represented by its center and radius.

1.1 Affine Transformations

A fundamental class of operations on sets of points consists of those that map each point onto another point to form a one-to-one correspondence between points, while preserving the neighborhoods of the points; that is, the relative closeness of points is preserved. As a consequence, they preserve the topology or "connectedness" of the set, and such transformations are called *homeomorphisms*. The simplest and most important member of this class are the *affine transformations*. These include translation, scaling, and rotation. If one excludes translation from this class, one is left with the class of *linear transformations*, also called linear operators, which has been studied thoroughly in the context of linear algebra on vector spaces. One can represent a 2D point (x,y) as a 2D vector $[x\ y]$ of a 2D vector space, and coordinate axes can be represented by the basis vectors of that space: the x axis by [1 0] and the y axis by [0 1]. Similarly for any dimensionality. Linear transformations are equivalent to a change of the vector basis: each basis vector is mapped to any arbitrary vector. The length of each new basis vector provides scaling along each axis. A rotation is achieved if the new basis vectors remain mutually orthogonal; otherwise, shearing is introduced. Any linear transformation on a d-dimensional space can be represented by a $d \times d$ matrix, whose computational representation is usually a 2D array of reals.

However, a change of vector basis cannot provide translation. While vector addition can do so, the use of only linear transformations can be retained if instead one uses projective space, so that a 2D Euclidean point will be represented by a 3D projective point (xw, yw,w). Most often $w=1$, so that one can think of the original 2D space as simply the $w=1$ plane in the 3D projective space (see Fig. 1). This solves the translation problem by mapping the 2D origin/zero-vector [0 0] to the new w basis vector [0 0 1]. A translation of $[Tx\ Ty]$

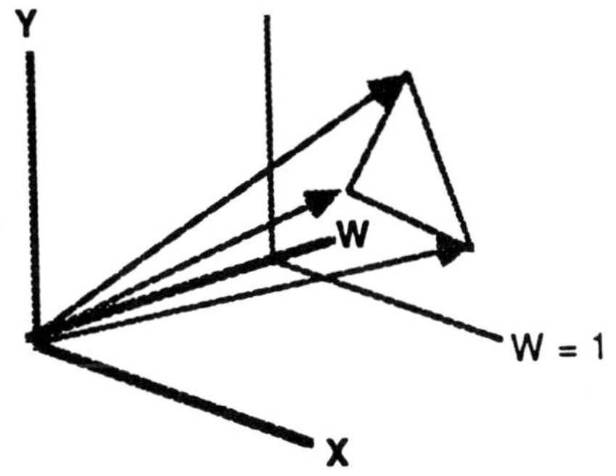

FIG. 1. Homogeneous coordinates.

is achieved by mapping the w basis vector to [Tx Ty 1], while the standard 2D linear transformations affect only the x and y basis vectors. In general, then, affine transformations of a d-space are represented by a restricted class of linear transformations of a $(d+1)$-space.

Projective space is also useful for central projections, e.g., the perspective projection in 3D, as well as for representation of surfaces as rational functions, and is generally preferable to affine space. The coordinates of a projective space are called *homogeneous coordinates,* because a point in d-space is mapped to a line through the origin in $(d+1)$-space. Similarly, lines are mapped to planes that contain the origin; that is, a line $Ax+By+C=0$ is represented by a vector $[A\ B\ C]$, which can be interpreted as the normal of a plane containing the origin. As a result, the zero vector is a solution to the new equation representing each original point or line: thus, the term "homogeneous."

This mapping of a point or line to a set of one greater dimension applies to spaces of any dimension. The dimension-independent versions of point, lines, and planes are called *flats* or *affine subspaces* or *linear varieties.* Therefore, homogeneous coordinates map k-dimensional flats, or simply k-flats, to $(k+1)$-flats that contain the origin. A k-flat can be thought of as a k-dimensional space, say determined by the first k coordinates of a d-space, to which an arbitrary d-dimensional affine transformation has been applied. They are also defined as the intersection of hyperplanes, where a hyperplane is a $(d-1)$-flat defined by the equation $a_1x_1+a_2x_2+\cdots+a_dx_d=a_{d+1}$, i.e., a linear equation in d variables. In 2D a line is a hyperplane, and in 3D a plane is a hyperplane. The intersection of n hyperplanes is in general a $(d\text{-}n)$-flat, $0\leqslant n\leqslant d$.

1.2 Polytopes

Points, line segments, polygons, and polyhedra are the most commonly used geometric sets, and they are respectively the 0D (zero-dimensional), 1D, 2D, and 3D instances of the dimension-independent class known as *polytopes.* These are objects with "flat faces" (Fig. 2). One method of defining polytopes is by a recursive definition in terms of their boundary. Observe that a line segment, which is a 1D polytope, has for its boundary two points, which are 0D polytopes. Similarly, the boundary of the 2D polygon is comprised of a set of the 1D line segments, and the 3D polyhedron is bounded by a set of 2D polygons. In general, a d-polytope is bounded by a set of $(d-1)$-polytopes, which are referred to as the $(d-1)$-faces of the polytope. If one unwinds this recursive definition, it becomes evident that a d-polytope has a boundary comprised of k-dimensional faces for all k, $0\leqslant k<d$.

One can construct a data structure representing a polytope based on this property. Given the 0-faces (vertices), each 1-face (edge) is represented by its two vertices. Each 2-face (polygon) is then a list of 1-faces, and in general each k-face is represented by a list of $(k-1)$-faces. These lists of k-faces are further organized into connected sets, called manifolds. This method of representing each polytope by its boundary is called a *boundary representation,* and there is a one-to-one correspondence (an isomorphism) between connected components of the boundary and elements of the data structure (Fig. 3). This data structure is a graph in the form of a hierarchy

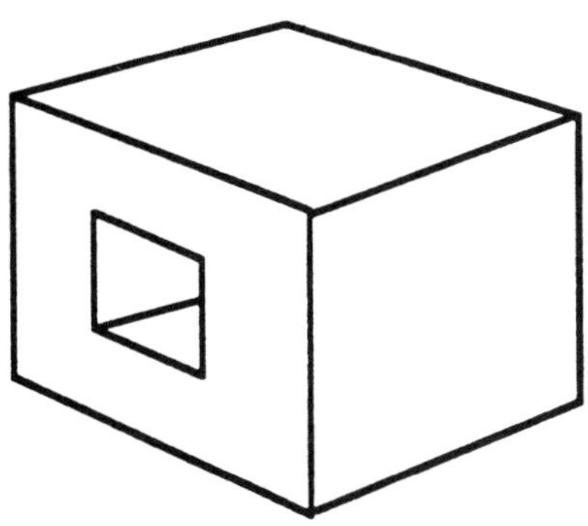

FIG. 2. A polyhedron.

of lists: nodes at one level correspond to all faces of a particular dimension. Arcs in the graph are of two types: between a k-face node and the nodes of its bounding $(k-1)$-manifolds, and between elements of the same manifold. The only purely geometric information is the coordinate values of the vertices; all other information is topological, i.e., information indicating which faces are connected to which other faces. Thus, a boundary representation, also denoted as *b-rep*, is often characterized as being a topology-based representation.

A crucial property of polytopes is that all of the faces are flat; that is, each k-face is contained in a k-flat. This restriction to sets defined in terms of linear equations (hyperplanes) greatly simplifies computations, as the intersection of two flats is another lower-dimensional flat. The fact that any k-face lies in the intersection of $d-k$ hyperplanes can be used as a basis for a representation of polytopes.

First note that a hyperplane h partitions space into two half-spaces, denoted as h^+ and h^-. The intersection of any set of half-spaces is a convex polytope (or the empty set), where a convex set is one for which the intersection of it with a line is a single line segment (or single point). All faces of a convex polytope lie in the hyperplanes that form the half-spaces. However, since a half-space is a convex set, and the intersection of convex sets is always another convex set, this approach can only represent convex polytopes.

The approach can, nonetheless, be extended to represent any polytope, not just convex ones, by allowing both half-spaces of a hyperplane to contain parts of the polytope. This is achieved by recursively subdividing space by the hyperplanes that contain the faces of the polytope. A single local operation is needed: partition a region r by a hyperplane h that intersects r. Usually, h contains a face that lies in r. This binary partitioning of r creates two subregions $r^+ = h^+ \cap r$ and $r^- = h^- \cap r$, where the subregions are open sets, i.e., they do not include h (*open sets* are sets that have had their boundary points removed). This process can be represented by a three-node graph in which the nodes correspond to regions r, r^+, and r^-, and arcs are between r and its two "child" regions (Fig. 4). The right arc can be interpreted as corresponding to the operation $h^+ \cap x$ for any set $x \in r$, and the left arc to $h^- \cap x$. Now the two subregions can be treated just like r; that is, they can be partitioned by hyperplanes containing faces lying in them, and then their children can be partitioned, and so on recursively. The result is a hierarchy of regions, and the coresponding graph is a binary tree in which each node of the tree corresponds to one region, and arcs indicate the relation between parent and child regions of creating the later from the former by partitioning. Such a geometric structure is called a *binary space partitioning tree* or, alternatively, a *partitioning tree* or *bsp tree.*

Generally, the initial region is all of d-space, and a consequence of inducing a partitioning tree structure on d-space is that a disjoint convex decomposition/partitioning is formed by the unpartitioned leaf regions, called *cells.* Convexity arises from the fact that

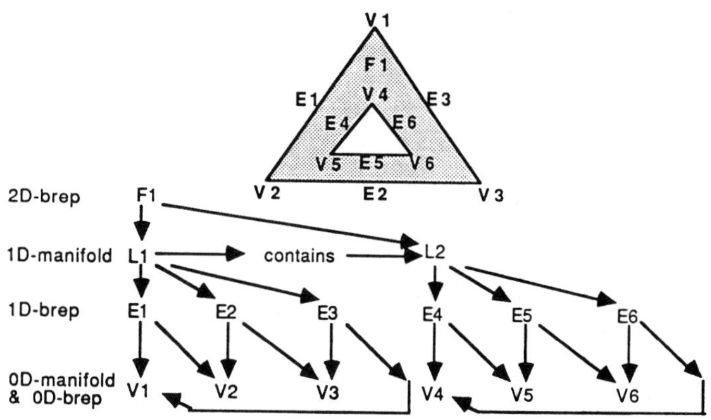

FIG. 3. Boundary representation of a polygon.

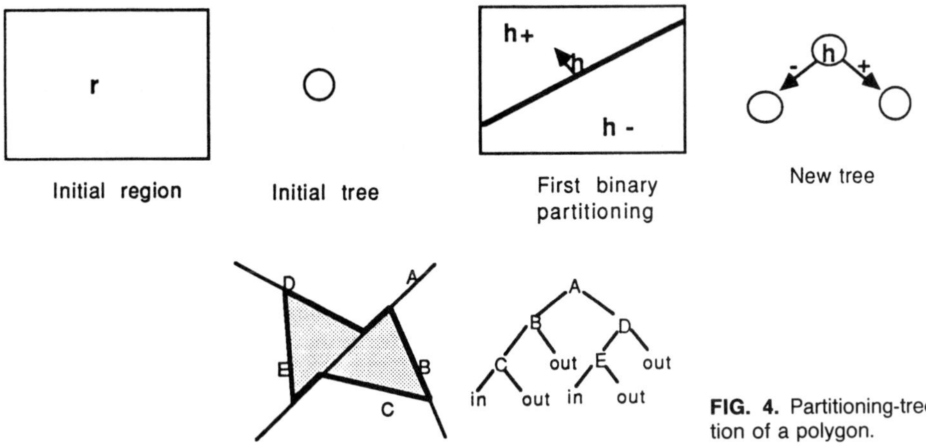

FIG. 4. Partitioning-tree representation of a polygon.

each region *r* of the partitioning tree is created by the intersection of half-spaces corresponding to the arcs in the path of *r* starting at the root of the trees. Since the partitioning tree has been constructed so that all polytope faces lie in the partitioning hyperplanes, none of the boundary lies in any cell, and so each cell with be either entirely in the polytopes's interior or in its exterior. The interior cells then form a convex decomposition of the polytope (and the out-cells form a convex decomposition of the exterior as well). The polytope is defined to be the union of the in-cells, which are open sets, followed by a closure operations that adds boundary points and fills in the cracks between in-cells.

The development of polytopes has implicitly assumed that the flatness of faces is an intrinsic property. However, it is easy to imagine warping the faces of a polytope into curved faces, In fact, systems have been developed that use nonflat polyhedral faces (called *algebraic polytopes,* if the faces lie in polynomials). So flatness need not be considered an intrinsic property of polytopes. However, the idea of a geometric set whose boundary is a set of "smooth" faces, i.e., faces without "interior edges," is intrinsic to the concept of polytopes; that is, polytopes have *piecewise* smooth boundaries. Thus the intrinsic properties of polytopes are that they have smooth faces, and they have a finite number of faces. The former is satisfied through the use of continuous functions, developed in classical mathematics, whose gradient (tangent function) is continuous. The latter is a combinatorial property, which together with the computational setting is an issue addressed by computer science; in particular, it is concerned with the development of representations and efficient algorithms providing operations on those representations. The boundary representation and the partitioning tree are examples of this.

1.3 Nonlinear Sets

While using only linear expressions to define a geometric set simplifies considerably all computational aspects, restricting faces to being flat is too limiting, since many desirable surfaces are not piecewise flat. Defining surfaces that are curved requires using nonlinear functions, such as polynomials of degree greater than 1, or transcendental functions such as sine or cosine or exponential functions, or any easily computable function whose gradient is continuous. Unfortunately, nonlinear functions can be considerably more complicated than linear functions. Because of this, it is very common to use the nonlinear functions to define the model, and then to construct a piecewise linear approximation of the nonlinear sets so that the difficult computations are performed using linear polytopes.

There are two fundamental forms for defining sets: the *characteristic function* that determines whether a point is a member of a set, and the *generating function* that enumerates the points that are members of the set. For sets of integers, the computable sets, or *recursive sets,* are those that have a computable characteristic function, and the semicomputable sets, or *recursively enumerable sets,* are those with a computable generating function. For polynomials, implicit representations provide characteristic functions and parametric functions provide generating functions. The implicit form defines the sets in d-space as $\{\mathbf{x}|\mathbf{f}(\mathbf{x})=0\}$, where $\mathbf{x}$ is a d-dimensional vector and $\mathbf{f}$ is a scalar-valued function. For example, $x^2+y^2-r^2=0$ is the implicit form defining a circle of radius r centered at the origin. The parametric form uses a set of parameters to generate the set $\{\mathbf{x}|\mathbf{g}(\mathbf{t})=\mathbf{x}\}$, where $\mathbf{x}$ is a d-dimensional vector and $\mathbf{t}$ is a vector of dimension $<d$. The dimension of the set is that of t. A parametric form of the circle above using parameter t is $x=r \cos t$ and $y=r \sin t$, for $0 \leqslant t < 2\pi$.

Any set whose characteristic function is given by a polynomial is referred to as an *algebraic set*, and the 1D and 2D algebraic sets are known, respectively, as algebraic curves and algebraic surfaces. The study of algebraic sets is the principal topic in the branch of mathematics known as *algebraic geometry.* While algebraic geometry considers polynomials of any degree, the lower-degree algebraic sets are the most widely used. For example, the second-degree polynomials in 2D define the conics: circle, ellipse, hyperbola, and parabola. In 3D they are the quadric surfaces: sphere, ellipsoid, hyperboloid, paraboloid, and cone. For sets defined by higher-degree polynomials, comparable classifications are not always known; and, unlike first- or second-degree sets, they may intersect themselves, creating singularities, and they may not admit a simple parametrization. In general, the greater the degree of the polynomial, the greater the computation that is involved and also the greater the numerical error due to the floating-point representation of reals. Thus, in practice, only polynomials of relatively low degree are computationally viable if the polynomial is an *irreducible* polynomial. A polynomial is said to be reducible if it is the product of two or more lower-degree polynomials, and irreducible if it is not. Thus $x^2 \equiv x \times x$ is reducible, whereas x^2+y^2 is irreducible. Irreducible polynomials are an important class in that they provide the primitive algebraic sets from which all other algebraic sets can be constructed, and they are referred to as *algebraic varieties.* Planes and spheres are examples of algebraic varieties.

1.3.1 Parametric Curves and Surfaces. Defining a complicated surface using only a single polynomial requires using one of high degree, which as noted above is computationally unattractive. An effective alternative is to define a set piecewise, where each piece is defined by a low-degree polynomial. The pieces then connect to each other in a patchwork manner to form a continuous set. Since each piece is a subset of an algebraic set, it is continuous for all derivatives. But at the places where the pieces join together, the continuity varies according to need. Tangent continuity C^1 is required to ensure that at each join the set does not form a corner, whereas curvature continuity C^2 is used to induce greater smoothness; only rarely are higher degrees of continuity needed.

An important instance of constructing piecewise polynomial curves with smooth joins in the class of curves known as *splines,* which are usually piecewise cubic curves with C^2 continuity at the joins. Piecewise definitions are most often formulated using the parametric rather than the implicit form because the desired piece of the polynomial can be specified by restricting the parametric domain. For example, a set in 3D is defined by the coordinate functions $x=X(t)$, $y=Y(t)$, and $z=Z(t)$, where each coordinate function is a polynomial, and a piece can be defined by limiting the coordinates of t to be in the interval [0 1]. If t is a 1D vector, i.e., a scalar, then the set is a space curve. If t is a 2D vector, a *biparametric* surface is defined. The biparametric surfaces, often denoted as $P(u,v)$, can in fact be interpreted as being a set of parametric curves: holding one parameter constant determines a curve in the other parameter (Fig. 5). This property has given rise to calling these surfaces *cross-product surfaces or tensor product surfaces.*

If one wants to design curves and surfaces interactively, that is, to use a person's visual esthetics to select the shape, the method of shape specification must be simple and intuitive. Specifying the coefficients of polynomials directly does not provide this. Consequently, geometric methods, rather than algebraic ones, have been developed for the interactive design process. A classical approach is *interpolation:* given a set of user-defined points, compute a piecewise polynomial curve or surface that goes through the points and provides the desired degree of continuity at the joins. In fact, splines were developed for exactly this purpose. Interpolation is sufficient if one does not want to adjust

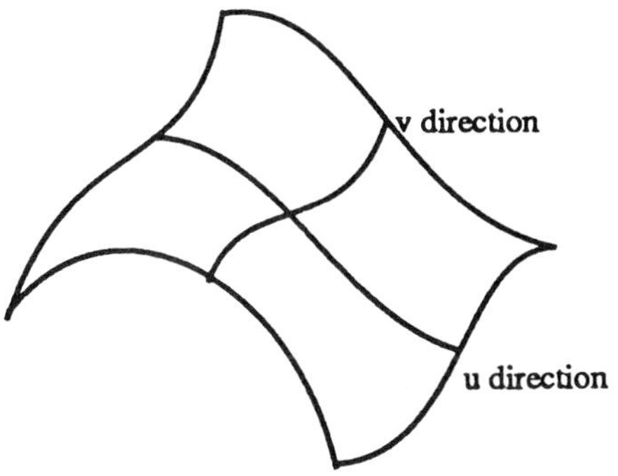

FIG. 5. A parametric surface patch.

the points subsequently, for doing this may cause the shape to change everywhere. With interpolation, any change can have a global effect, whereas the intent is usually to make a local modification, that is, to move only one point and the part of the curve around that point. To circumvent this problem, methods have been developed based on the idea of approximating a set of points rather than interpolating them. The points that are approximated are called *control points* since the designer uses them to control the shape; but the set only lies near the points, not through the control points (in some cases a few of the control points are indeed on the set).

The most well known of these are the Bezier curves and surfaces. A Bezier curve, which is defined by a single polynomial of degree n, is determined by $n+1$ control points, numbered P_0, P_1, ..., P_n (see Fig. 6). To obtain local control, a curve is constructed piecewise from a set of Bezier curves. Thus, only a subset of each Bezier curve is used. The end points of that curve segment are given by the first and last control points, P_0 and P_n. These two control points lie on the curve itself and are the only control points that do so. The tangent line at P_0 is the line through P_0 and P_1, and similarly the tangent line at P_n is the line through P_{n-1} and P_n. An analogous relation holds for higher derivatives at the end points (e.g., the curvature at P_0 is determined by P_1 and P_2), and so there is an intuitive geometric relation between control points and the curve.

The control points taken by themselves as a list of points can be linearly interpolated to form what is called the "control polygon" (a bit of a misnomer, since the first and last control points are not considered "connected," and so it is really a piecewise linear curve). The importance of the control polygon is that the Bezier curve is an approximation of it, and this is why Bezier curves are considered an approximating method. The algebraic analogue to the geometric formulation of Bezier curves is the approximation of functions using the Bernstein basis. Any polynomial of degree n can be represented exactly by the weighted sum of $n+1$ Bernstein basis functions $B_{i,n}(t)$, $0 \leqslant i \leqslant n$, where each is a polynomial of degree n. In particular, $B_{i,n}(t) = C(n,i)t^i(1-t)^{n-i}$ where $C(n,i)$ are binomial coefficients. They have the property that they sum to 1, i.e., $\Sigma_i B_{i,n}(t) = 1$. The control points are used to determine the weights of each basis function; that is, P_0 is the weight for $B_{0,n}$, P_1 for $B_{1,n}$, and so on, so that the curve is $P(t) = \Sigma_i B_{i,n}(t) P_i$. While the algebraic form is important for analysis and computation, it is the geometric characteristics of manipulating control points that has made the Bezier form appealing for geometric design.

When one is constructing a curve by joining several Bezier curve segments, maintaining continuity at the joins requires that the control points of the two segments at a join be geometrically related. For example, tangent continuity between curve segments A and B requires that the control points $A_n = B_0$ and that the tangent line at the join contains control points A_{n-1} and B_1. Constraining control points to achieve continuity is a consequence of the fact that the definition of each segment is independent of every other; and so, only constraining the control points can induce a relation between segments.

An alternative that performs this correlation intrinsically is the *B-spline* method. It retains from Bezier curves the idea that the curve is a piecewise polynomial and that each segment is the weighted sum of $n+1$ basis functions with the weights being the control

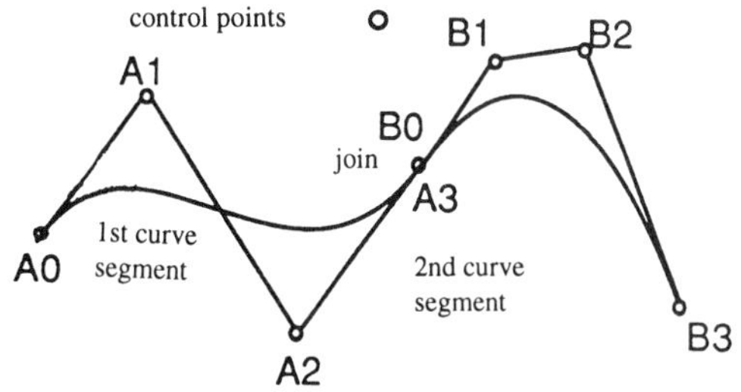

FIG. 6. Cubic Bezier curve with two segments.

points (the unweighted basis functions sum to 1 as well). The difference lies in the fact that a single control point is used as a weight in $n+1$ consecutive segments, and this then provides a means of ensuring continuity at the joins up to C^{n-1}. Thus, a cubic *B*-spline curve is piecewise cubic with C^2 continuity at the joins. Control point P_i is used as a weight in curve segments $S_i, S_{i+1}, \ldots, S_{i+n}$, but it is the weight of a different basis function in each of the segments. Since P_{i+1} is used in segments $S_{i+1}, S_{i+2}, \ldots, S_{i+n+1}$, two consecutive control points contribute to n common segments, which result in the desired automatic correlation between segments to ensure continuity.

Unfortunately, parametric piecewise polynomial curves in 2D of any degree cannot exactly define a conic, which is an important class of curves. However, if the curves are defined in projective space, they indeed can define conics. This is because the perspective projection of a 3D, degree 2, parametric space curve can be exactly a conic. Thus, it is common to use projective space for defining parametric curves, and such curves are called *rational curves,* because the perspective division of $X(t)$ and $Y(t)$ by $Z(t)$ results in a ratio of two polynomials. Analogous reasoning yields rational surfaces.

1.4 Constructive Methods

The methods just described define a shape directly, primarily in terms of its boundary. These direct methods can be supplemented by constructive ones in which two or more of the direct methods are used in conjunction to construct a new shape. The most common such method is the use of Boolean set operations: union, intersection, and both symmetric and asymmetric difference, which correspond to the Boolean functions of "or," "and," "exclusive-or," and "subtraction" (A&~B). Figure 7 illustrates their use in forming a new polygon from set operations on two other polygons. The result can then be used in subsequent set operations, resulting in a hierarchy of shapes. This method of shape creation is called, when in 3D, *constructive solid geometry.*

Another constructive method, called *instancing,* creates multiple copies of a shape, or "instances," by applying a unique affine transformation to the original for each copy. Inherent to this concept is the property that any subsequent change to the original will cause all instances to change; that is, the instances are always defined to be the current value of the original transformed by the current value of the transformation.

A third basic constructive method utilizes the idea of a shape moving in time. For example, if a line segment is moved some distance along a particular direction, all of the points "visited" by the line segment form a quadrilateral. If a line segment is rotated 360° about one of its points, the locus of points swept out is a disk. If instead the center of rotation is not on the segment but is on the line containing the segment, a disk with a hole in it is formed. There are all methods of creating *swept objects.* The technique of moving a shape along some path, i.e., along a curve, is called *translational sweeping,* while rotating a shape about some axis is called *rotational sweeping* (also called, in 3D, generation of surfaces and solids of revolution).

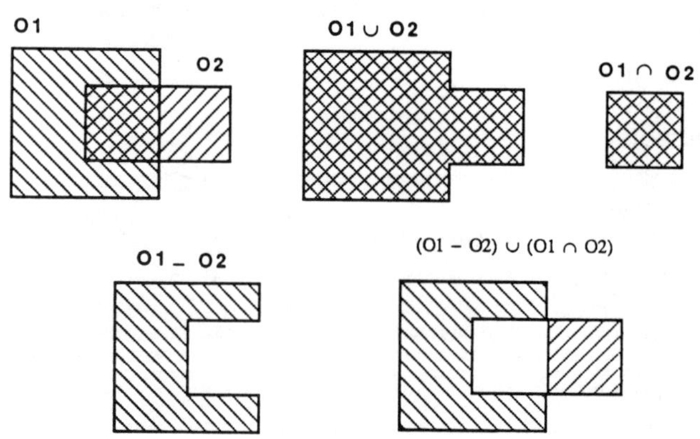

FIG. 7. Constructing sets with set operations.

1.5 Discrete-Space Methods

The methods discussed so far have all made explicit use of the idea that space is continuous. An alternative is to treat space as discrete in the form of a regular lattice of points created by restricting the coordinates to the integers. At each lattice point, an arbitrary set of attributes, like color, can be defined, and a *d*-dimensional space can then be represented computationally by a *d*-dimensional array of attributes. In 3D, each point is referred to as a *voxel* (derived from "volume element"). This representation is typically generated by discretely sampling the analog output of a sensing device, such as a TV camera, which detects some form of electromagnetic energy. Medical devices used for imaging the human body are another example of such devices.

While the array representation is very simple, it can require substantial amounts of memory when compared to typical continuous space representations. An approach to reducing this is to find connected regions of elements that have the same attributes, called *homogeneous regions,* and to represent each such region as one element in the data structure. A standard approach to this is to subdivide space into homogeneous regions using recursive subdivision, as was done by the partitioning tree discussed above. For discrete domains a commonly used method is called the *quadtree* in 2D and the *octree* in 3D. A quadtree is formed by recursively partitioning a square region into four "quadrants" (Fig. 8), and an octree is formed by recursively partitioning a cubic region into eight "octants." During the recursive subdivision, the partitioning terminates whenever a region is produced that is homogeneous; otherwise the region is partitioned. The limit to the subdivision occurs whenever a lattice point is reached, as this is a homogeneous region by definition.

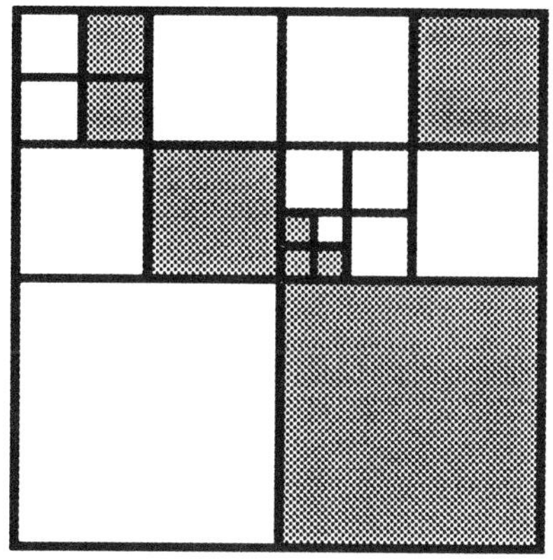

FIG. 8. Quadtree representation.

2. RENDERING

Given a geometric model representing a 3D scene, defined using any of the above techniques, the most important and computationally intensive operation is to synthesize a 2D image of the scene for display with video, film, or print media. This is called *rendering.* The goal is to create an image that will resemble, to some degree, what a person or camera would see if the scene were in some sense real. If the scene is a realizable 3D world, then rendering may try to achieve photorealism; that is, the image generated is indistinguishable from a photograph of a real scene. However, an alternative objective is simply to communicate information of some sort without being concerned with photorealism. This information may have an intrinsic 3D geometric form, such as weather data, or the 3D setting may be used simply as a means of communication, as in the display of statistical data points. But even when photorealism is not an objective, the appropriate rendering algorithms are essentially simplified versions of those needed for photorealism, since the human visual system has evolved for the express purpose of perceiving a 3D physical world.

Rendering algorithms simulate, to varying degrees of accuracy, the physics of the propagation of light through a homogeneous media and the interaction of light with matter, such as reflection, transmission, diffraction, and scattering. However, to do this accurately, one might think that all such propagation and interaction with objects in the scene would need to be simulated at the level of photons. It would, however, be computationally intractable to perform such a simulation. Fortunately this is not necessary since the resolution of the human visual system is limited, and so every detail provided by such a simulation could not be perceived. Therefore, the real challenge in rendering is to develop computationally attractive algorithms that produce realistic-appearing replicas of as large a class of phenomena as possible. Over the last two decades, many such algorithms have been developed that can indeed effectively simulate at reasonable costs a variety of optical effects; among these are reflections, refraction, transparency, shadows, and surface texture.

Synthesizing an image using a geometric model of a 3D scene composed of a variety of objects and sources of illumination requires, in addition, a virtual camera described by its viewing position, orientation, and field of view, as well as an image plane that models the display surface onto which the scene will be projected. Rendering can then be understood as being performed in two distinct steps: *visibility calculations,* which compute the propagation of light through homogeneous media, and *shading,* which performs the computation needed to simulate interaction of light with matter. (While in a number of algorithms these steps may not be very distinct, such a distinction will simplify the presentation.)

For visibility calculations, we consider the light that reaches the viewing position along rays emanating from the object's surfaces. However, objects closer to the viewing position will occlude objects farther away, and this occlusion must be computed. If we then consider rays emanating from the viewer instead of from the surfaces ("antioptic" rays), then any such ray may intersect any number of surfaces, but it will intersect the image plane on which we will create the picture at only one point. From which of these surfaces will the light reach the image plane point? Only the light from the closest surface will, and so this surface point is considered visible. If for all rays emanating from the viewing position the visible surface point is determined, then we will have solved what is called the visible surface or hidden surface problem.

The second component, called *shading* or *local illumination,* computes how incident light is reflected, refracted, and absorbed within a small area on a surface, on the basis of its geometric and optical properties. Models have been developed, based on physics, to cover the spectrum from purely diffuse matte surfaces to purely specular mirrorlike surfaces. Diffuse surfaces are characterized by radiation heat-transfer models (blackbody radiation), and specular surfaces by geometric optics (including refraction and Fresnel effects). Modeling the range between these two extremes is achieved by combining models for the extremal cases.

Since the geometric model is defined in continuous space whereas the image is defined in discrete space as a set of pixels, care must be taken when converting between these two different fundamental forms of space. For example, an inclining line will be represented as a set of pixels that, when examined closely, will resemble a staircase instead of a smooth line. This effect is called *aliasing.* It arises from attempting to represent rates of change (frequencies) that are greater than the spacing of the pixel lattice. Visible artifacts due to aliasing can be reduced by higher sampling rates combined with low-pass filtering; this is called *antialiasing.*

2.1 Visibility Determination

Visibility can be computed in a number of ways. For example, one can directly simulate the propagation of light by constructing a ray from the viewpoint through each image point, determine the objects it intersects, and choose the object that has the smallest distance along the ray. This approach is called *ray tracing.*

An alternative is to transform the model to the coordinate system shown in Fig. 9 in which the viewing position is at the origin and the x and y coordinates correspond to the horizontal and vertical directions of the display, respectively. In addition, the image plane is orthogonal to the z axis and at unit distance from the origin. As a consequence, the z coordinate corresponds to depth. One can then apply a perspective projection to all the points in the model by mapping each point (x,y,z) to $(x/z, y/z, -1/z)$. This has the effect of mapping a ray through the image plane point P to one that is parallel to the z axis but still contains P, and also preserves the depth ordering along the ray. As a consequence of this mapping, the visible surface at image point (x,y) is the surface that has the smallest z coordinate.

2.1.1 Z-Buffer Algorithm. This algorithm is the most commonly used visible surface algorithm. It operates on any set of 3D points in the above post-perspective space whose xy coordinates lie on the pixel lattice. All points with the same xy coordinates project onto the same pixel. Since the visible surface is composed of those points that are closest, we

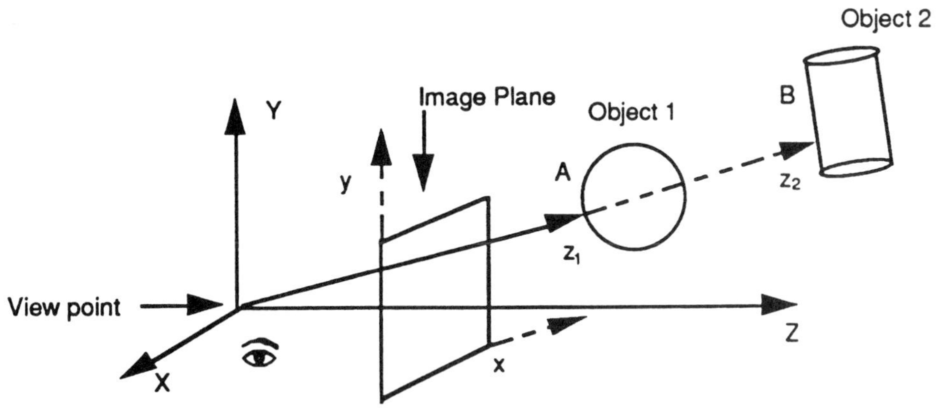

FIG. 9. Visibility: A ray intersects object 1 at distance z_1 and object 2 at z_1, $z_1 < z_2$. Only light from object 1 reaches the image plane.

simply need to sort the points into buckets, one for each pixel denoted as *pixel*[*x*,*y*], while keeping only the point with smallest depth (z value). This can be accomplished by maintaining both a color buffer, *pixel*[*x*,*y*].*color*, and a z buffer, *pixel*[*x*,*y*].*z*, which are set initially to the background color and maximum depth, respectively. Each surface is quantized, one at a time, into a set of 3D points. For a point (x,y,z), if $z<pixel[x,y].z$, then this point is closer than any previous point with those xy coordinates, and its color and z value replace the contents of *pixel*[*x*,*y*]. Otherwise, the point is known to be occluded and so is discarded. After all points have been processed, the contents of the color buffer is the desired image.

2.1.2 Scan-Line Algorithms. Here the image is treated as being composed of a set of horizontal *scan lines*, and it is synthesized one scan line at a time, going from top to bottom. For each scan line, the plane consisting of the viewpoint and the scan line is extended to determine the objects it intersects. If the objects are described by polygons, then such intersections are easy to calculate and this results in a collection of horizontal line segments, called *spans*, one span for each intersected polygon. Visibility calculation can then be carried out on these spans, thereby reducing a 3D problem to a set of 2D problems. These spans are kept in an "active polygon" list sorted left to right by their left end points. Between two consecutive end points, only one polygon can be visible. The algorithm sweeps left to right, computing at each span end point the closest span, and then drawing that span. When a scan line is completed, the next scan line in the top-to-bottom order is processed. This requires updating the active list of polygons by deleting old polygons that do not intersect the new scan line and adding new polygons that do. Polygons remaining in the list have the end points of their spans updated to reflect the change in position between two consecutive scan lines.

2.1.3 Visibility Priority. Another method is based upon visibility priority. If a ray from the view intersects two surfaces A and B such that A is closer to the viewer than B, then A is said to *occlude* B. If it is also the case that A occludes B for every ray that intersects both of them, then A is said to have visibility priority over A. If this fact is known, then it is possible to draw B entirely first, and then draw A "on top of" B; as a consequence the part of A that occludes B will overwrite it. This has been dubbed the *painter's algorithm*, alluding to the occasional practice of painters to overpaint a background with a foreground. If a visibility priority ordering of all surfaces can be generated, then drawing the surfaces "back to front"—that is, lowest priority to highest priority—will produce an image with correct visibility. If one uses the partitioning-tree representation of polyhedra, as discussed in Sec. 1.2, one can generate such a priority ordering by simply traversing the tree in an order determined by the current viewing position.

2.1.4 Ray Tracing. All three of the methods just described are not suitable for computing interobject reflections and transmittance. To accomplish this, ray tracing can be employed (Fig. 10). Here for each pixel on the image plane, a ray is propagated from the viewpoint through the pixel into the object space. Among all the objects that this ray intersects, the object that is the closest to the viewpoint is chosen. From this point of intersection, a second ray is traced toward the light sources. If this secondary ray encounters an object between the surface and the light source, then the intersection point is considered to be in shadow. Rays are also propagated in the directions of reflection and refraction. Such rays may intersect other objects and so generate other reflection and refraction rays, and so on recursively until the ray goes out into empty space or hits a light. This process creates a tree of reflection, refraction, and shadow rays forming a geometric network by which light flows back to the viewer. The intensity and chromaticity of the light is determined by the local illumination models at each surface in the network.

Ray tracing is capable of generating photorealistic images that contain most of the optical effects, except those that can only be modeled using the wave nature of light. There are two principal difficulties, however. As the number of objects in the scene increases, ray-surface intersection calculation increases, particularly if there is a large number of secondary rays. To give an indication of the magnitude of the problem, consider an image (1000×1000) containing a million pixels generated from a scene containing a thousand objects. If there are, say, three secondary rays per primary ray (a very modest estimate), and if each object is represented by a cubic polynomial, then intersection calculations could require computing the real solutions of cubics 4×10^9 times, which is quite formidable.

The second problem for ray tracing arises from the infinitely thin nature of the rays that result in point sampling of the scene. This generates aliasing artifacts. Several techniques have been proposed to overcome these difficulties. Ray-surface intersection can be simplified if surfaces that do not lie near a ray can be excluded. A hierarchy of bounding boxes that enclose objects has been used for this purpose. If a ray does not intersect a box, which is easy to check since boxes are simple, then it does not intersect any object in that box. However, if it intersects the box, then the ray is tested for intersection with each of the smaller boxes contained within this box. This process is continued until the smallest box is reached, at which point the ray is tested for intersection with all the objects in that box.

Another similar technique is to divide the object space into disjoint cells using, for example, a partitioning tree or an octree. A list of all surfaces that intersect a cell is stored with that cell. A ray can then be inserted into the tree to quickly find the cells that it intersects, and the cells can be visited in a near-to-far priority order. Ray-surface intersections are then limited to those surfaces that lie in the cells traversed by the ray, and can be performed in the near-to-far ordering. The first intersection is then the visible surface, and so the rest of the surfaces in the remaining cells need not be tested. Such algorithms

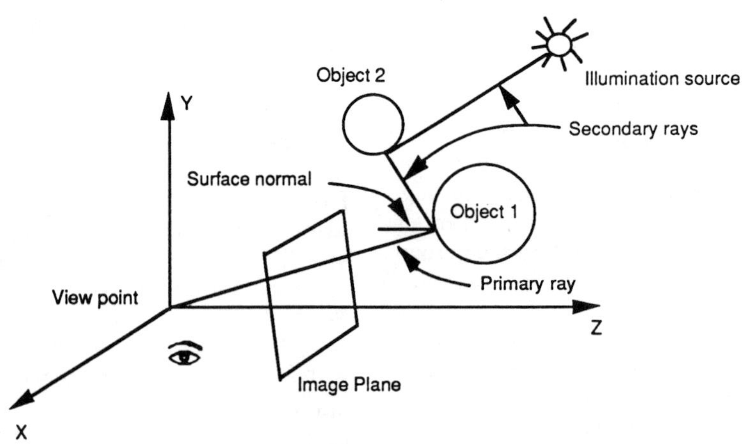

FIG. 10. Primary and secondary rays in ray tracing.

can improve the efficiency of ray-surface intersection calculations by orders of magnitude. But despite these improvements, ray intersection continues to be the principal computational burden for ray-tracing algorithms.

2.2 Illumination

Having determined the point on a surface in the object space that is visible, one must determine the color of light traveling in any particular direction, for example, from a visible surface to the viewer. This requires modeling the interaction between light incident upon a surface with the surface itself. As an example, for reflection, the intensity I in any direction can be modeled as a sum of three components:

$$I = I_d + I_s + I_a,$$

where I_d is diffuse reflection, I_s is specular reflection, and I_a is reflection due to ambient light. The diffuse component is due to light being reflected by a rough surface uniformly in all directions and is treated as being direction independent. Specular reflection corresponds to reflections along the direction of reflection, determined by the angle of incidence and the surface normal. For mirrors this is the only illumination term, but for less perfect reflectors this component is what models highlights. For matte surfaces, the specular component is 0 and reflections are modeled soley by the diffuse term.

I_a is determined by light reflected off of other nearby surfaces. This can be treated simply as being a constant level of direction-independent illumination or it can be computed exactly. *Radiosity,* originally developed for studying radiation heat transfer, can be employed for this. The radiosity of a surface is the total light emanating from it. It consists of all the reflected light and, for light-emitting surfaces (lights), all emitted light. The objects must be subdivided into surface elements that are small enough so that the radiosity term can be treated as being constant over the element. The unknown radiosities are then computed from a system of simultaneous linear equations, one for every pair of these surface elements, expressing the quantity of light transferred between the two elements.

2.3 Antialiasing

Since objects are modeled as having value-continuous, but not tangent-continuous, surfaces, they may have arbitrary sharpness; i.e., they can have edges. Sharpness represents a high rate of change as one moves, say, horizontally across an image, and this rate of change may be thought of in terms of frequency. More exactly, one can in principle perform a Fourier transform on a spatial function to obtain its representation in terms of frequencies. Thus edges correspond to spatial regions with high frequencies since the function changes very rapidly. (If the function is value discontinuous, then there is no bound on the highest frequency.) The displays, being physical systems, are capable of showing only images whose bandwidth is limited. It is, therefore, necessary to carry out low-pass filtering on the signal before it is displayed. If filtering is not done, artifacts such as jagged edges (spatial aliasing) and backward spinning of wagon wheels (temporal aliasing) are clearly seen. However, too much filtering reduces the sharpness of the images, and certain filter shapes can cause other visible artifacts (e.g., ringing). Thus the filter design represents a trade-off between a number of subjective factors such as sharpness and visibility of aliasing artifacts.

Prefiltering a signal by an amount such that the bandwidth of the resulting signal makes the sampling rate a Nyquist rate is a preferred mode of overcoming aliasing. In a number of situations, this can be easily done. However, in ray tracing such prefiltering may require computing a large number of samples by tracing many rays. It is typical, for example, to compute 8 or 16 ray-traced samples for each pixel and then filter them. Another method is to filter the objects directly in 3D space to make them smoother. Unfortunately, such filtering, if properly done, must depend on the direction and distance of the viewpoint and the image plane relative to the object. Moreover, only simple surfaces such as spheres and polygons can at this time be handled, and special effects such as shadows and transparency are quite difficult. A preferred approach is to do adaptive sampling of the image plane. In areas of high detail (or uncertainty) the density of the rays is increased and the location of the rays is randomized. The former method allows more local bandwidth where needed, and the latter makes aliasing noise more random and makes it contain higher frequencies, thus reducing its visibility.

2.4 Surface Color Variations

Visibility calculation and shading described in the last section can, in many cases, create images that look artificial because their surfaces look too smooth. Metallic objects may look fine, but natural objects such as fruits and human skin will not. Realism can be increased in many cases by adding detail, i.e., texture, on the surface of the objects. This is usually done either by adding a prespecified two-dimensional pattern of colors on top of a smooth surface or by making the surface rough. The former can be considered as a mapping of a 2D "image" onto a 3D surface with associated warping of the image, whereas the latter can be modeled as randomization of one or many of the surface attributes, such as the reflection property or surface normal. Since the addition of texture patterns on the object's surface can be thought of as a mapping from texture space onto the object space, it can be done in one of two ways:

1. compositing the mapping from texture space to object space with the mapping from object space to the image plane used in visibility calculation; or
2. utilizing an inverse mapping from the image plane to texture space.

A different approach to varying the color across a surface is to interpolate the color. As mentioned earlier, nonlinear surfaces can be approximated by polygons in order to reduce the complexity of the algorithms dealing with such surfaces. However, the illumination will be discontinuous at polygon edges even though the nonlinear surface would have been continuous in this region. If, however, the color is known at the vertices, then these values can be easily interpolated across the polygon. This guarantees that two polygons that share an edge will have the same color along that edge, and so the image will remain value-continuous at any edge not lying on the silhouette. This method is called *color interpolation* or *Gouraud shading*. A similar idea is to interpolate surface normals defined at the vertices across the polygon and compute the illumination using these normals at each interpolation point. This method is called *normal interpolation* or *Phong shading*, and it allows polygonal models to display highlights.

3. ANIMATION

Computers have been used in creating image sequences showing animation. Techniques of the previous section can be used to create a set of *key frames* in an image sequence. Key frames are those frames where a discontinuity in motion or content occurs during the animation sequence. Computer algorithms are then used to create the "in-between" frames. Smooth trajectories of motion are traditionally used to compute these in-between frames.

In traditional commercial animation, the motion is selected directly by the animator without any explicit reliance upon physics (other than what is contained in the animator's mind). In contrast, *physically based modeling* incorporates the physics of moving bodies explicitly using kinematics or dynamics to determine differential equations for modeling motion. Here the objects, environment, motion equations, and constraints are modeled directly and can be used to generate a significant quantity of the motion in an animation. Fairly sophisticated motion models, such as for the motion used in walking and running by human beings, have been developed.

4. HARDWARE

Computer graphics has from its outset required computing hardware specific to the task of generating images. This arises both from the basic requirement of displaying an image (for example, on a TV screen), and from the performance requirements needed to synthesize the image in an acceptable length of time. For real-time applications, such as simulators for aircraft training, acceptable means 60 images per second, while interactive applications require at least 1–6 images per second. Given the significant performance requirements, hardware designers have turned to creating special-purpose VLSI (very large scale integration) ICs and to designing parallel computer architectures.

A digital image is represented by a 2D array of color values, called *pixels* or *pels*. Each pixel value encodes the color to be drawn at the corresponding point of the display surface. It may be only a single bit if a bilevel black and white display is used; or it may be as many as 24 bits if full color is desired, so that the three primary colors of red, green, and blue (RGB) each have 8 bits, for a total of 24 bits per pixel. Alternatively, one can add one level of indirection and let the value at each pixel be an index into a 24-bit wide *color table*. This allows, for example, a system with only 8 bits per pixel to select, at any one time, any 2^8 colors from a palette of 2^{24} possible colors. This can be used to introduce color into a grey-scale picture (called pseudocolor), as is common with satellite images.

For display on a standard TV monitor (NTSC), a typical resolution is 640 pixels horizontally by 480 pixels vertically for a total of $640 \times 480 = 307{,}200$ pixels displayed at 30 Hz. Higher-resolution monitors are widely used, and are commonly $1280 \times 1024 = 1\,310\,720$ pixels displayed at 60 Hz. Thus, a full-color high-resolution display requires almost 4 Mb of memory accessed at a rate of around 250 Mb/s. Current memory technologies cannot easily provide this video refresh rate in addition to the image-drawing access rate, so special memory chips called VRAMs (video random-access memory) have been designed that provide an additional access mechanism specifically for video refresh. Many of these used in parallel (e.g., 128) can provide the required bandwidth over today's bus technology. The specific block of memory formed from these VRAMs that is used to hold the pixels being drawn on the CRT is called the *frame buffer*, since it holds at any one time a single frame/image of a sequence of frames/images.

Rendering the geometric model at high rates has always relied on specially designed graphics systems. Performing the basic operations of transforming, clipping, and drawing visible polygons, in that order, has led to a number of pipelined architectures, using a variety of hardware implementations. One such design used as many as 16 copies of a graphics VLSI processor to implement the transformation and clipping operations, while an alternative implementation used nine programmable DSPs (digital signal processors) to provide the same functions. Similarly, the drawing of visible polygons (possibly with texture mapping and individual shading calculations for each pixel) has been performed by specially designed VLSI chips (e.g., 20) or programmable DSPs (up to 64). The image computation is partitioned among the k drawing processors so that each is responsible for one out of every k pixels. To obtain a uniform load balancing among the drawing processors, the pixels assigned to a single processor are distributed uniformly throughout the image.

5. APPLICATIONS

Since geometry and images are fundamental components of our experience, it is not surprising that computer graphics has been applied widely. The most common and simplest use is the depiction of statistical data using graphs, bar charts, pie charts, etc., and the creation of illustrations to accompany text, such as the figures in many papers and books. This kind of applying computer graphics carries on activities in the computational domain that were mature long before computers. Since by and large they represent a transference of manual tasks, their computational complexity is not great, and the principal software issues are those dealing with the design of good user interfaces.

Engineering has from the outset been a significant consumer of computer graphics. Electrical engineering uses graphics for designing printed circuit boards on which the IC chips are mounted, for designing the logic diagrams specifying the function of a circuit, and for creating the physical layout of the ICs themselves. All of these, however, are treated as multiple layers of 2D sheets, even though printed circuits and chips are 3D physical objects.

Mechanical, aerospace, civil, chemical, and industrial engineering are concerned with the creation of physical objects that are inherently three-dimensional. To exploit computer graphics fully, interactive design capabilities are essential, as well as nongraphics computations that determine physical properties such as mass, effects of stress, and heat-transfer properties. In certain cases, the designed object can be manufactured directly from the computational representation using numerically controlled milling machines, or, if the target material is plastic, "drawn" directly using lasers.

Closely allied to the needs of these engineering disciplines are those of architecture. In this case, the esthetic dimension is of greater importance than physical properties, and much more of the design is decided by visual criteria as opposed to functional requirements, such as heating and power distribution. Since architecture is concerned with designing spaces for people to be in, having the ability to "walk through" a computer model of a building is considered to be an important capability, a goal yet to be satisfactorily attained. Ideally, the user would be completely enveloped in a *virtual reality* by wearing a head-mounted display. Such a display has separate images for each eye, forming a stereo pair, and everything in the person's field of vision is computer generated. The position of the head is tracked as the person moves, and the images correspond to what would be seen if the environment were physical, thus giving a feel of authenticity to the environment.

Physical scientists are increasingly using computer graphics both to visualize data and to observe simulations of physical processes, which has been dubbed *scientific visualization.* This can provide considerably more insight into the nature of a physical process than just the equations that describe state transitions. It can also provide access to processes whose space-time scale is significantly different from ours. Thus astronomers can see a process entailing millions of light-years develop in a few minutes of computer animation, and similarly for processes at the atomic level involving only a microsecond or less.

Another important area of data visualization is medical imaging. Various sensing devices, such as CAT scan or MRI, produce a set of slices where each slice is a 2D array of values correlated to the kind of tissue at each point in the sample. These slices, when stacked on top of each other, form a 3D array of voxels (volume elements) showing the part of the body that was scanned. Medical diagnosis can then be greatly enhanced if it is possible not only to see a static view of this data, but to manipulate it interactively as well. Providing such capabilities is called *volume rendering.*

Since computer graphics generates pictures, one should expect that it could be of great use to the visual arts. This is indeed the case, and has been since its very beginnings. Initially, much of this work was 2D in character, emulating painting, and in fact many computer paint programs have been written. More recent efforts have included 3D sculptures seen through stereographic viewers. Another visual art form, film, has its computational correlate, namely computer animation, which has been used for both artistic and commercial purposes. Much use has been made in the making of commercials for TV and for traditional graphics, such as the introductions to TV shows (referred to as "flying logos"). More ambitiously, several full-length movies have employed computer graphics to generate an appreciable percentage of the footage, while still more have utilized it for special effects.

GLOSSARY

Affine Subspace: Any subspace defined by the intersection of hyperplanes; *flat* is an alternative name.

Affine Transformations: Transformations that preserve flatness (linearity): translation, rotation, scaling, and shearing.

Algebraic Set: Any set of points that satisfies a polynomial (i.e., its solution set).

Antialiasing: Various filtering operations to remove the artifacts due to discretizing continuous functions, such as those that define curves and surfaces. A discrete image of a line without any filtering appears "jagged" or "staircased."

Binary Space Partitioning Tree: A computational representation of polytopes that partitions space into disjoint sets of convex cells. Each cell is in the interior or the exterior of the polytope.

Boundary Representation: A computational representation of polytopes in which there is a one-to-one correspondence between faces of the polytope and elements of the representation.

Constructive Solid Geometry: Defining 3D solid objects using set operations, affine transformations, and instancing (copying).

Diffuse Reflection: Reflection from a rough or matte surface that is perceived as uniform in all directions within the hemisphere above the surface.

Frame Buffer: A special 2D array of memory containing the image that is displayed on a CRT monitor 30–60 times a second.

Geometric Model: The computational representation of geometric objects and any of their attributes such as color.

Homogeneous Coordinates: Coordinates used in projective space; a normal 2D point (x,y) has in projective space the coordinates (wx,wy,w) for any $w \neq 0$.

Hyperplane: Dimension-independent entity defined by a single linear equation. Examples are a plane in 3D and a line in 2D. Always one dimension less than the embedding space.

Illumination: Modeling the interaction of light with surfaces. Local illumination is concerned with how the light that is incident upon a very small area is reflected, refracted, and absorbed. Global illumination is concerned with the interaction of light among all of the surfaces.

k Flat: A set in d-dimensional space of dimension k, $0 \leqslant k < d$, that can be defined by the intersection of $d-k$ hyperplanes or as all the weighted (affine) combinations of $k+1$ points. For example, a plane is a 2-flat and a line is a 1-flat regardless of the dimensionality of the space.

Nonlinear Sets: Sets defined by nonlinear equations.

Nyquist Rate: The minimum rate at which a continuous function can be sampled and then subsequently reconstructed without introducing errors (aliasing). That rate is twice the highest frequency present in the function.

Parametric Set: A set of points defined by a one-to-one mapping of a parameter t to a point p, where t is a vector whose dimension determines the dimension of the set. The parametric representation of a line through the points p_0 and p_1 is $p = tp_1 + (1-t)p_0$.

Pixels: The points of color out of which an image is composed when represented in discrete space. The pixels lie on a uniformly spaced 2D array.

Polynomial: An expression that is the weighted sum of a set of monomials. A monomial is a constant times any product of coordinate variables where each coordinate is raised to some nonnegative integer power. Example: $x^2y^3 + 3x + 2y^2 - 5$ ("missing" coordinates in a monomial have been raised to the zeroth power).

Polytope: Any set of points with an interior and an exterior, and whose boundary lies entirely in hyperplanes. The boundary is composed of a set of faces. A polyhedron is a 3D polytope, and a polygon is a 2D polytope.

Ray Tracing: Rendering a scene by tracing rays throughout the scene, analogous to geometric optics. The rays originate from lights and viewing positions, and are reflected, refracted, and absorbed by surfaces.

Rendering: Modeling the propagation of light and its interaction with surfaces to generate an image of a geometric model.

Set Operations: Forming the union, intersection, or difference of two sets.

Specular Reflection: Reflection of light from perfectly smooth surfaces; the light is concentrated along one direction, as with a mirror.

Splines: Curves that are comprised of pieces, each defined by a polynomial, in which the pieces fit together to form a smooth curve (tangent or curvature continuous).

Texture Mapping: Defining the color of a surface by a function that varies the color across the surface. A common example is mapping an image onto a surface lying in 3D.

Virtual Reality: Computer-generated sensory input correlated to human motion so that the subjective experience is interpreted as having a realistic quality.

Visibility: Determining for each member of a set of rays that surface the ray intersects first, i.e., what is the visible surface along these rays.

Volume Rendering: Generating images of 3D data that was produced by sampling space using electromagnetic radiation. Common example is medical data generated by scanning a body.

Further Reading

Farin, G. (1988), *Curves and Surfaces for Computer Aided Geometric Design,* San Diego: Academic.

Foley, J., van Dam, A., Feiner, S., Hughes, J. (1990), *Computer Graphics Principles and Practices,* Reading, MA: Addison-Wesley.

Glassner, A. (Ed.) (1989), *An Introduction to Ray Tracing,* San Diego: Academic.

Hearn, D., Baker, P. (1986), *Computer Graphics,* Englewood Cliffs, NJ: Prentice-Hall.

Mantyla, M. (1988), *An Introduction to Solid Modeling,* New York: W. H. Freeman.

COMPUTER HARDWARE

KENNETH H. SLATER, *Semiconductor Engineering Group, Digital Equipment Corporation, Hudson, Massachusetts, U.S.A.*

INTRODUCTION

Computers are now ubiquitous in industrialized society. While the cost, performance, and capacity of computer hardware span several orders of magnitude, the vast majority of machines are stored-program digital computers built of common subsystems. Our focus will be to describe and discuss hardware aspects of these machines, including common contemporary peripheral devices.

Computers have helped fuel the worldwide electronics industry, providing a ready market for semiconductors, especially dynamic memories, which are the key to driving advanced semiconductor technology. Advances in semiconductors have, in turn, helped create the worldwide consumer electronics market. With the advent of the personal computer, computer hardware has become a mass market item, and consumer electronics technology such as compact disks and digital audio tapes are now found in both industrial and personal computers. Over all, the demand for computer hardware has helped create and shape the worldwide electronics economy as we know it today.

Preceded by many mechanical devices and electronic analog computers, the first generation of electronic digital computers was built

1-56081-063-7/92/$5.00+.50

with vacuum tube technology in the 1950s. The advent of the discrete transistor ushered in the second generation of cheaper, smaller, and more reliable machines in the 1960s. The third generation in the 1970s introduced minicomputers, made practical by small-scale integrated circuits. With the advent of large-scale integrated circuits in the 1980s, the personal computer and workstation emerged as the leading products of the fourth generation. The fourth generation also saw the widespread use of high-speed networks and the addition of optical storage devices in addition to traditional magnetic storage media. The fifth generation of machines, still in their infancy with no one architecture having emerged as dominant, will likely take advantage of the low cost and high density of very large-scale integrated circuits and rely on multiple-processor techniques. Different and promising approaches include massively parallel machines, neural networks, systolic arrays, and data-driven machines (see NEURAL NETWORKS; PARALLEL COMPUTING).

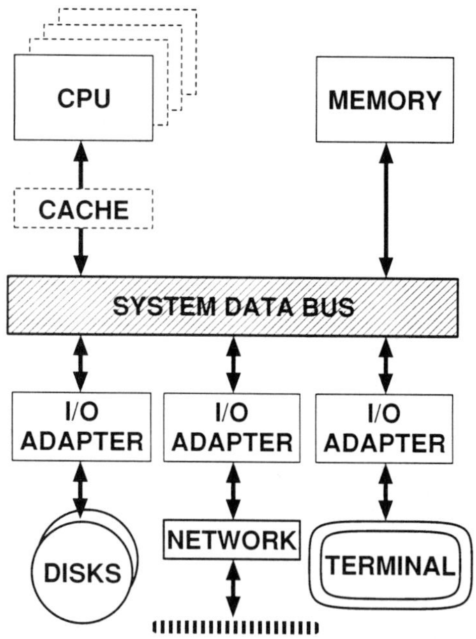

FIG. 1. The major functional blocks of a typical stored-program digital computer. Optional components are shown in dashed lines. Not shown is the possibility of separate memory and I/O busses, used to improve performance by not burdening the CPU memory bus with lower speed I/O traffic.

1. GENERAL ASPECTS

Today's general-purpose digital computers, be they common personal computers or large supercomputers, have many hardware similarities (Fig. 1). They all incorporate one or more central processing units (CPUs) which execute the instruction stream and process data. The CPU retrieves data and instructions from a main memory, and writes modified data back to it. The ability to interpret items in memory as either instructions or data is a key attribute of a stored-program digital computer.

Often the CPU is much faster than the main memory, and a memory cache is employed to match CPU and memory speeds better. Data travel between main memory and the CPU over a pathway called a bus, to which input/output (I/O) adapters (also called peripheral controllers or adapters) are also attached. These controllers connect the CPU and memory system to other devices such as user interfaces, data storage, or communications devices.

Contemporary computer hardware relies on fast and dense electronic integrated circuits for processing logic and main memory. Rigid magnetic or optical disks are used for persistent mass storage, and flexible magnetic media, be they floppy disk* or magnetic tape, for archival storage and data transport. The primary human interface is usually a cathode-ray-tube display, with an electromechanical keyboard and pointing device (such as a mouse or tablet) for human interaction. Despite this common technology, the performance, capacity, and cost of these machines vary widely. The number, size, speed, and configuration of these hardware elements determine the overall characteristics and cost of the computer system.

1.1 System-Wide Characteristics and Metrics

1.1.1 Data Representations. Computer hardware directly interprets only a small number of data representations; more complex data representations are created by software building upon these basic entities.

*Some inexpensive or older personal computers use floppy disks for persistent magnetic storage instead of a hard disk.

The fundamental unit of information in a digital computer is the bit (short for Binary digIT). A bit can have one of two values, 0 or 1, and easily maps to mechanisms in the underlying technologies, such as the presence or absence of electrical charge or the polarity of a magnetic field.

In order to represent data that has more than two values, bits must be aggregated into sets that can represent computer instructions or data. There has been widespread convergence towards representing data in multiples of 8 bits, called a byte, which has one of 256 (2^8) values. Common uses for the byte include

1. representing one of 256 possible characters,
2. representing an unsigned integer in the interval of 0 to 255,
3. representing a two's-complement signed integer in the interval of -128 to $+127$, and
4. representing a simple computer instruction.

Sets of bytes are used to represent larger integers, floating-point data, memory addresses, or complex computer instructions. Common sizes for integer data are 2, 4, and 8 bytes, and floating-point data are typically 4, 8, or 16 bytes in length. While these types are nearly universal, many machines implement additional hardware data types, such as linked lists, character strings, or decimal integers.

1.1.1.1 Data Representation Standards. Character data first became widely standardized with adoption of the American Standard Code for Information Interchange (ASCII), which was based upon a 7-bit character code (ANSI, 1977). While ASCII is still in widespread use, more recent standards issued by the International Standards Organization are beginning to supersede it. ISO Standard 8859 offers 1, 2, or 4 byte-per-character representations (in ISO terminology, an 8-bit grouping of bits is an Octet) which permits simultaneous use of many character sets, a growing requirement in international markets and in desktop publishing. For example, ISO Latin-1 is a one-octet character set in which characters 32 through 126 map directly to the popular 7-bit ASCII set; other standardized single-octet character sets include Arabic, Greek, Cyrillic, and Hebrew. Many Asian character sets, with their rich store of characters, are standardized as two-octet sets, including Katakana, Kanji, Hanzi, and Korean.

Integer data have not been formally standardized, but a *de facto* standard does exist. The vast majority of current computing hardware uses integers sized in multiples of 8-bit bytes (8, 16, 32, 64, and 128 bits being most common) for both signed and unsigned data.

Two's-complement notation for signed integer data has become standard in the industry, being selected over one's-complement and sign-magnitude notations used in some earlier machines. It has the advantage of being able to perform arbitrary signed operations on the same hardware that handles unsigned addition, and it has a unique representation of zero (the other representations have both a positive and negative zero). Given an N-bit word, two's-complement notation can represent numbers in the range of -2^{N-1} to $+2^{N-1}-1$ (for example, an 8-bit byte can represent -128 to $+127$). To convert a number to two's-complement notation, convert the absolute value of the number to an unsigned binary integer. If the original number was negative, then invert (logically complement) each bit and add 1 to the result.

Floating-point data format and floating-point data operations have been standardized through the Institute of Electrical and Electronics Engineers (IEEE) standard 754, commonly called IEEE Floating Point (IEEE, 1985). While many proprietary formats are still in widespread use, largely for historical reasons, the IEEE format is steadily becoming accepted and is supported by most newly designed computer architectures. The standard defines

1. single (32-bit) and double (64-bit) precision basic formats, and single ($\geqslant$ 43-bit) and double ($\geqslant$ 79-bit) precision extended formats (Fig. 2);
2. add, subtract, multiply, divide, square root, remainder, and compare operations;
3. conversions between
   a. integer and floating point formats,
   b. different floating point formats, and
   c. basic floating point formats and decimal strings; and
4. floating-point exceptions and their handling.

1.1.2 Word Width or Word Length. The word width (word length is used interchange-

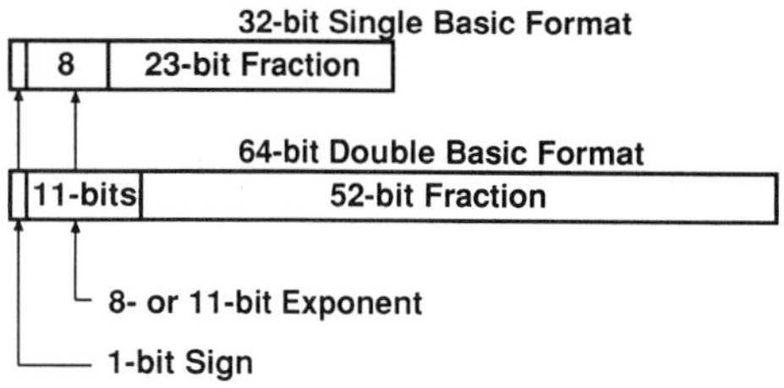

FIG. 2. The sign, exponent, and fraction fields of the IEEE-754 single and double precision basic floating-point data formats. The standard defines the most significant bits of each field to be on the left.

ably) indicates how many bits of data can be processed at once; the wider the word, the higher the performance of a machine compared to a similar machine with a smaller word width. Also, larger word widths provide added precision or range of values in computations directly (the same precision can be achieved by using multiple smaller words and appropriate software; the penalty is reduced speed). Typical word widths in current machines are 8, 16, 32, or 64 bits. Word widths may vary between different parts of the machine; currently popular personal computers based on the 386SX microprocessor have a 32-bit word width in the central processor, a 16-bit word width to memory, and 16- or 8-bit word widths to peripheral devices. Reducing the word width reduces required hardware, reducing hardware costs, but also reduces overall system performance.

1.1.3 Addressing Characteristics. Addressing granularity, or the smallest addressable unit, describes the smallest unit of memory that may be uniquely addressed. Current practice is to address memory at the byte level, which greatly facilitates character processing. This usually does not preclude addressing and accessing larger units, including and up to the word width of the machine.

There are two different conventions for addressing bytes within multiple-byte words (which includes integer, floating-point, and address data): Does the address indicate the most significant bits of the word or the least? Because of a famous paper by Cohen (1981) these two conventions have come to be known as "Big Endian" and "Little Endian." While on an isolated computer either convention works just as well, there are problems to overcome when exchanging programs or data between machines that use different conventions. Both conventions are in widespread use on contemporary machines.

Data alignment is an important, and sometimes restrictive, characteristic of how data may be accessed in memory. Memory subsystems are designed with a natural word width, usually in multiples of 2, 4, 8, or 16 bytes. Access to full words that are aligned to occupy a single word in memory is naturally fast and efficient. Unaligned memory access to full words requires access to two aligned memory locations, selection of the proper bits from each aligned word, and assembly of the bits into the requested unaligned word, as illustrated in Fig. 3. Needless to say, support of unaligned access to memory requires additional hardware and/or software and almost always incurs a performance penalty.

Some computers, particularly the newer reduced instruction set computer (RISC) designs, do not provide any hardware support for unaligned memory access, and if it is supported at all it is usually done with low-level software. Other machines, including several popular complex instruction set computer (CISC) models, provide transparent hardware support for unaligned access, although in most cases aligned data has a noticeable performance advantage.

Total address space is another key characteristic of a machine and is the number of unique addresses that are provided by mem-

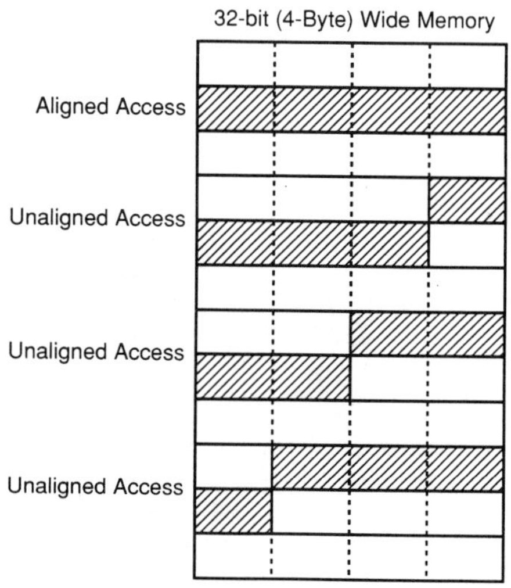

FIG. 3. Aligned and unaligned memory access. Depiction of an aligned memory access of a 32-bit word from a 32-bit byte addressable memory. Also depicted are the three possible cases of unaligned access.

ory. This has direct impact on the data and instruction capacity of the machine. Machines that utilize a 16- to 32-bit address space are very common today, with 48- and 64-bit address machines now arriving on the market. The following table indicates the relationship of the number of bits in the machine address and the number of uniquely addressable items in memory:

Bits in Address	Uniquely Addressable Memory Items
16	65 536 (2^{16})
32	4 294 967 296 (2^{32})
48	281 474 976 710 656 (2^{48})
64	18 446 744 073 709 551 616 (2^{64})

1.1.4 Performance Metrics. "How fast is that computer?" This is a complicated question to answer; even if theoretical or measured performance characteristics of the major subsystems are known, at best they will provide an upper bound of the machine's overall performance. The actual performance of the system is a function of the application to be run, its data, the compilers and operating system being employed, the hardware configuration, and operating environment overhead factors. Benchmark programs have had a particularly difficult time keeping up with the rapid increases in main-memory size and cache sizes; many benchmark programs now fit completely within the machine's high-speed caches and give unrealistically good results (an actual application would likely not be able to fit entirely within the fast cache memory, and would perform significantly slower, because of cache misses which cause reference to the slower main memory). Despite these difficulties, computer users and vendors still like to sum up their system performance in a simple numerical rating.

Millions of instructions per second (MIPS) attempts to quantify the performance of a computer by how many instructions it can execute per second. Computations performed per instruction is not well defined and varies widely over different hardware. Instruction mix and frequency varies widely over different workloads. With these ambiguities, the MIPS number is often ridiculed as the "meaningless indicator of performance." Nonetheless, the MIPS rating of a computer is still often quoted and serves at least as a first-order approximation to relative performance.

Millions of floating-point operations per second (MFLOPS) is used to quantify overall floating-point performance (as MIPS quantifies integer and logical performance). MFLOPS numbers are also widely quoted. As all current computers have comparable sets of floating-point instructions, the use of MFLOPS as a performance indicator between different machines is less problematic than the use of MIPS is for integer and logical performance.

In the personal computer area, clock speed in megahertz is often used as a performance indicator. This has worked well to date as an overall indication of performance between different models of PCs because most PCs are based on a common family of processors and have common hardware configurations; in general the faster the clock runs, the faster the processor will execute instructions and process data, assuming the overall system is reasonably well balanced to operate at that clock speed. Between processor families, however, the number is largely meaningless.

Recently, the Systems Performance Evaluation Cooperative (SPEC) was formed by hardware vendors to promote the use of standardized benchmarks of representative problems of realistic size. A suite of benchmarks which stress different performance aspects and applications of a computer system are provided, with the results summarized by a single figure of merit, the SPECmark.

The SPEC suite consists of four integer-intensive benchmark programs written in C and six floating-point-intensive benchmarks written in FORTRAN. SPECmark numbers are reported for each individual benchmark program, and are combined via a geometric mean to give integer, floating-point, or overall computer system performance. The overall performance of a Digital VAX-11/780™ using VMS™ compilers is defined to be 1.0 SPECmark; in late 1991 the fastest RISC workstations are reporting SPECmarks of about 70, current model VAX processors are about 40, and an Intel™ 50-MHz 486™ is near 18 SPECmarks. See Weicker (1990) for further information on benchmarks and performance measurement.

2. DIGITAL COMPUTER COMPONENTS

2.1 The Central Processor

The central processing unit (CPU) executes a program stored in main memory, manipulating data to achieve a desired calculation. A minimal CPU will have an instruction unit that retrieves instructions from the memory and controls the elements of the CPU to execute the instruction; high-speed temporary storage called registers or accumulators to store intermediate calculations; and an arithmetic/logic unit (ALU) which actually performs the logical or arithmetic operations on the data. The program counter is a register that holds the address of the next instruction in main memory to be executed; a change in program flow (a branch or jump) can be accomplished by modifying its value. A program status word (PSW) register is usually implemented to record specific state information, such as whether the last add instruction resulted in carry or if an interrupt should be generated on a divide by zero operation.

2.1.1 Operation of a Basic CPU. The following example illustrates the operation of a basic CPU, as shown in Fig. 4. A RISC-like machine is used for this example (see Sec. 2.1.2 for a discussion of RISC and CISC machines). The example illustrates the operation of the two instructions:

1. load the contents of a word of memory into a CPU register, and
2. add that register together with a small constant integer and store the results into another register.

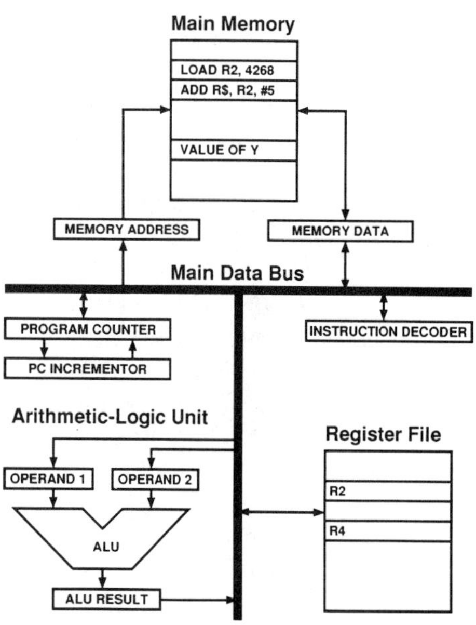

FIG. 4. The CPU used in the example of instruction execution in Sec. 2.1.1.

In a high-level programming language, the programmer might express this operation as $X=Y+5$. After compilation to machine instructions, this statement is comprised of two instructions:

1. Load the value of variable Y, which is stored in main memory at a known address, into a CPU register, and
2. add the value of Y, now in a register, with the integer constant 5 contained in the add instruction itself, and place the result in another CPU register (which holds the value of X in our example).

In summary, we are assuming the following locations of integer data:

Y in main memory at address 4268, temporarily stored in CPU register $R2$.
X stored in CPU register $R4$.
5 a constant, embedded in the add instruction as an immediate operand.

The first instruction is LOAD $R2$, 4268. This instructs the CPU to load the value stored at main memory location 4268 into CPU register $R2$. Our example starts with the program counter register in the CPU containing the address of the example LOAD instruction and proceeds as follows:

1. The CPU uses the value of the program counter to fetch (read) the next instruction from main memory. It sends the value of the program counter to the memory address register, and the main memory returns the instruction to the memory data register. The memory address register and memory data registers are invisible to the programmer, as are several other registers we will encounter in this example, being part of the internal workings of the CPU.
2. The program counter is incremented by the length of an instruction, so that it now holds the address of the next instruction to execute. Typical of many machines, the program counter is incremented with a dedicated piece of hardware to make this

frequent operation efficient.

3. The fetched instruction is moved from the memory data register to the instruction decoder. The instruction is a "load from memory to register" instruction. The decoding circuits separate out two important fields of information within the instruction: The memory address to read from (4268) and the destination register ($R2$).
4. The decoded memory address is used to read the value of the data from memory, by moving it from the instruction decoder to the memory address register. The returned data from memory is temporarily held in a memory data register awaiting storage.
5. The data returned from memory is moved to the specified destination, CPU register $R2$ in the register file (an array of programmer-visible registers).

The second instructions is ADD $R4$, $R2$, #5. This instructs the CPU to add the contents of CPU register $R2$ to the integer constant held in a field of the instruction (in this case with a value of 5) and to place the result in CPU register $R4$.

1. The CPU uses the value of the program counter to fetch (read) the next instruction from main memory.
2. The program counter is incremented by the length of an instruction, so that it now holds the address of the next instruction to execute.
3. The fetched instruction is decoded inside the CPU. It is an "add constant to register" instruction. The decoding circuits separate out three important fields of information within the instruction: The value of the constant (5); the source register ($R2$); and the destination register ($R4$).
4. The value of the constant is sent to the first operand holding register of the arithmetic logic unit (ALU).
5. The value in the source register is sent to the second operand holding register of the ALU.
6. An ADD operation is done by the ALU, and the result is temporarily held in the ALU result register.
7. The value in the ALU result register is stored in the destination CPU register ($R4$).

As the above simple example illustrates, even the simplest instructions require several steps, the movement of data within the machine, and the control of various registers and logic units.

2.1.2 RISC vs CISC. Until the middle of the fourth generation of computers, the richness and complexity of the instruction set understood by the CPU continued to grow. Machines with literally thousands of instructions were designed and are in widespread use. These were later dubbed CISC after RISC appeared in the mid-1980s. RISC computers gain a significant speed advantage over CISC machines by keeping both their CPU and instruction set simple and uniform. They usually employ large numbers of registers to improve performance further. The availability of high-level language optimizing compilers allows programmers to be productive while shielding them from the simple yet often difficult to program underlying machine instructions. The compilers could effectively optimize the small set of fast RISC instructions, while they often could not effectively optimize or even use many of the large set of CISC instructions. RISC machines are now broadly accepted and are the basis for most new designs.

CISC machines are characterized by

1. variable-length instructions formats,
2. a large number of often specialized instructions,
3. a small number of general-purpose registers ($\leqslant 16$),
4. performing arithmetic and logic operations directly to memory, and
5. instructions that have multiple results, such as adding two numbers and incrementing an address of one of the operands in the same instruction (autoincrement)

RISC machines are characterized by

1. fixed-length instructions formats, often only of one basic size,
2. a small number of very general instructions,
3. a large number of general-purpose registers ($\geqslant 32$),
4. arithmetic and logic operations performed only on registers; operands must be loaded and stored from memory with separate instructions, and
5. instructions having only one result with no other side effects (change of machine state).

2.1.3 Pipelining and Super Scalar (Multiple Issue). Pipelining is a technique to improve CPU performance by processing more than one instruction at a time. A typical instruction has several distinct phases of execution: fetch the instruction; decode the instruction; fetch the operands; process the operands; and store the result. In many CPUs, these various phases are performed by distinct pieces of hardware that sit idle during part of the instruction execution cycle. A pipelined CPU will process multiple instructions simultaneously, by overlapping the end of one instruction with the beginning of another. A continuous flow of instructions moving through the CPU "pipeline" overlapped by phase keeps more of the CPU hardware productive, improving performance (Fig. 5). The simplicity of RISC instructions allows pipelines to be built generally with less hardware and to perform more effectively than comparable CISC implementations.

A super scalar, or multiple-issue, machine has several replicated elements, and can process and complete multiple instructions simultaneously. While the pipelined machine processes instructions sequentially with partial overlap, the multiple-issue machine processes instructions in parallel with full overlap (Fig. 5).

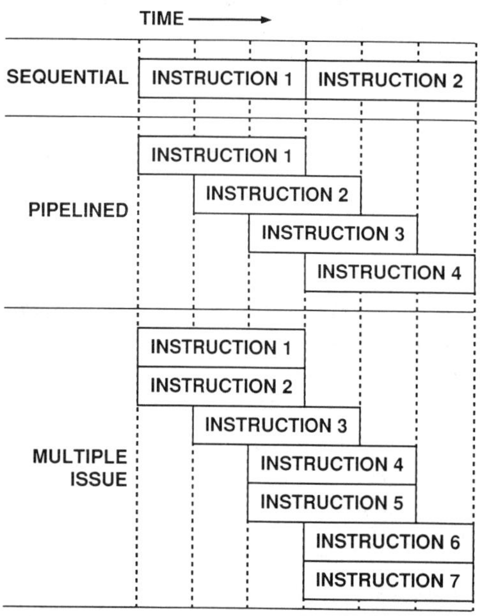

FIG. 5. Instruction execution sequences in sequential, pipelined, and multiple-issue machines. Only one of two possible instructions is being issued in the multiple-issue machine in the second time period, presumably because of a dependence that prevents instruction 4 from being issued with instruction 3.

The standard sequential processor executes only one instruction at a time. In a pipelined machine, instruction execution is overlapped to improve performance, taking advantage of the inherent multiple phases to complete most instructions. Super scalar machines have sufficient additional hardware to process multiple instructions in parallel through multiple pipelines. In both pipelined and super scalar machines, interinstruction data or control dependencies usually limit the theoretical maximum performance, as some instructions must wait for results of previous instruction to become available.

2.1.4 Floating-Point Processors. Floating-point processors are hardware especially designed for floating-point computations; they perform floating operations which the main CPU's arithmetic and logic unit cannot do (it is designed for integer and logical operations). A floating-point processor typically has its own set of registers and control unit, and the main CPU "hands off" entire floating-point instructions to the floating-point processor for rapid execution. Many floating-point units are designed to let the main CPU continue to process other instructions while the floating-point unit completes the floating-point operation independently. Current practice is for most computers to include floating-point processors; low-end personal computers are about the only common machines in which a floating-point processor is optional. When it is not present, floating-point operations can be performed in software by the integer unit at a considerable performance penalty.

2.1.5 Vector Processors. Vector processors are special hardware that can process vectors of data faster than a general-purpose CPU. A vector processor has its own vector register set, with each register holding from 64 to 256 vector elements. A single vector-machine instruction will operate on all elements of a vector or pair of vectors, completing the operation on all vector elements much faster than a scalar processor could complete the operation iteratively. Typical operations would include arithmetic or logical operations on a vector by a constant or between two vectors. Vector processors are usually implemented as pipelined scalar machines which are optimized for vector operations.

2.1.6 Multiprocessors. A multiprocessor is a computer system with multiple CPUs that share a common main memory. Each CPU executes its own set of instructions and operates on data that is possibly shared with other processors. With a properly designed program, a multiprocessor can execute portions of a computation in parallel, achieving a significant performance improvement over a single processor. Because of simultaneous memory access limitations, most multiprocessors in use today have between 2 and 64 CPUs. Programs that wish to exploit the speedup offered by a multiprocessor currently must contain special instructions to control the splitting (and later merging) of parallel instruction streams. Software technology to automate this task fully is still maturing and a programmer is usually required to obtain satisfactory results. The extra development effort required for creating parallel programs, and the lack of parallel programming standards, has limited the availability of programs that take advantage of parallel processors. Multiprocessors are most easily exploited by multitasking operating systems; they can assign each task to a specific processor and effectively use them without special programming effort by their users.

Multiple processors with local main memory avoid the bottleneck of shared memory, but the new bottleneck to performance becomes interprocessor communication. This is the approach taken by massively parallel machines, discussion of which is beyond the scope of this article (see PARALLEL COMPUTING).

2.1.7 Interrupts. Most every processor supports some form of interrupt mechanism. When this mechanism is enabled, the receipt of an interrupt causes the current instruction stream to be suspended and an interrupt service routine program to be executed. The interrupted program can be resumed transparently when the service routine completes. Interrupt hardware usually supports a number of distinct interrupt types, and a priority scheme so that the most important interrupts can preempt ones of lesser importance. Individual interrupts can usually be enabled, which allow them to interrupt the processor, or disabled, in which case they do not effect the processor.

Interrupts allow computer systems to respond to events as they occur without the overhead and possible delay of explicitly checking to see if an event of interest occurred (polling). They are particularly important for computers involved in real-time control of equipment, for efficient interfacing with computer peripherals, for time-keeping and scheduling functions, and for handling unexpected conditions and errors. On pipelined machines, excessive interrupts can adversely affect performance, as the switch in the instruction stream caused by the interrupt causes a delay while the new instruction stream is fetched.

2.2 Main Memory

Computer memory is arranged in a memory hierarchy (Fig. 6) in order to strike a balance between cost, capacity, and performance. The fastest, most expensive memory in the machine is the CPU registers, which have a small total data capacity. Successively larger, less expensive, and slower memory is arranged in order of decreasing speed (and cost) and increasing capacity. A well balanced memory hierarchy is key to good overall system cost and performance. This section will discuss main memory and main memory caches.

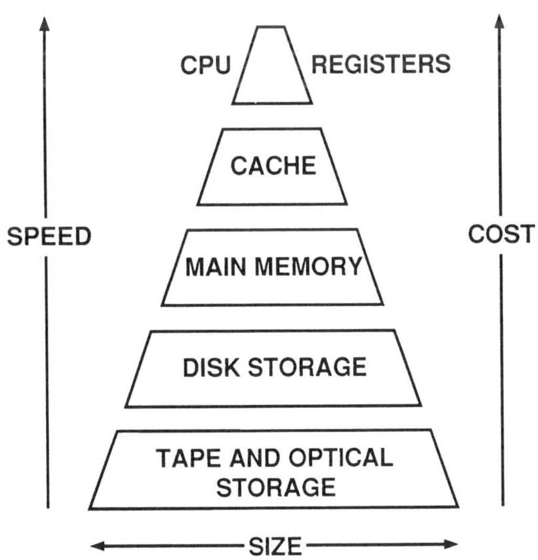

FIG. 6. The relationship between cost, size, and speed in the hierarchy of computer memory systems.

2.2.1 Main Memory Organization and Interleaving. Main memory is random-access: when presented with a memory address, it will directly read or write a given word (or smaller unit if the addressing scheme is so organized). The word width of the memory is usually the word width of the CPU, but designs in which the memory word is twice the size of the CPU to increase performance or half the size to reduce cost are not uncommon. In both cases additional hardware makes the size mismatch invisible to the computer user. Physical memory sizes of computers varies widely: personal computers will have from a few hundred kilobytes up to 16 megabytes; workstations typically have from 4 to a couple of hundred megabytes; larger machines may have hundreds of megabytes or several gigabytes. Memory access times also vary widely, but are generally in the range of 1 to 100 ns. Semiconductor memory density has quadrupled every two to three years throughout the 1980s permitting ever larger main memories and caches, and this trend continues unabated.

Memory systems require a fixed amount of time, called the cycle time, to perform a read or write operation and become ready to perform the next operation. There are usually three or more distinct time periods within the memory cycle:

1. accept an address and operation code from the CPU;
2. accept data (write cycle) or return data (ready cycle); and
3. perform internal operations to complete the cycle, which ensures that the data are properly stored, and to prepare for a new memory cycle.

Interleaving is a technique, similar to pipelining in the CPU, which offsets the cycles of two or more banks (sections) of memory so that the CPU does not have to wait for the entire memory cycle before it can issue a new memory operation.

For example, imagine we had a memory system with two banks or sections, with one section handling even-numbered addresses and the other handling odd-numbered addresses. Further imagine that the third phase of the cycle takes half of the memory cycle time. These memory banks are configured such that the memory cycle of second bank always starts halfway through the cycle time of the first (conversely, the first always starts halfway through the cycle time of the second). With such a scheme, if the processor alternately accesses even and odd memory addresses, it can complete memory transactions at twice the basic memory cycle time. The portion of the memory cycle that does internal operations completes while the processor works with the alternate bank on new memory operation in the early phases of its cycle.

Memory refresh is required for data retention in dynamic semiconductor memories, which are most commonly used in main memories. These devices store information as the presence or absence of electric charge in memory cells of 1 to 4 metal–oxide–semiconductor (MOS) transistors. When a charge is present, it will leak away over time and requires periodic replacement. The memory refresh process periodically reads all memory cells and rewrites them, thereby recharging all previously charged cells. Memory refresh cycles typically need to occur every few milliseconds, while normal memory operations are measured in nanoseconds, and so the overhead is not too large.

Some memory systems provide for protection of sections of memory. Memory protection schemes are commonly used on multiprocessing systems, to prevent independent programs from inadvertently or maliciously reading or altering another program's memory.

2.2.2 Parity and Error-Correcting Code (ECC). To protect against hardware errors in the memory, parity or an error-correcting code (ECC) is used to detect and correct memory errors. In the parity scheme, an additional error-detection bit, the parity bit, is stored with every byte by the memory subsystem. This bit is set by the memory hardware such that the total number of "1"s that are stored is odd (odd parity). Should a read operation result in an even number of ones, a parity error has been detected. This scheme detects single or triple bit errors, but cannot correct errors or detect double bit errors.

ECC memory (error correction code) stores several additional bits with each word of data both to detect memory errors and, in many cases, to correct them. A common scheme is to store 7 ECC bits with each 32-bit word. Such a scheme uses a Hamming code (Hamming,

1950) to create the 7 ECC check bits, which duplicate enough information from the stored 32-bit data to allow all single-bit errors to be corrected and all double-bit errors to be detected (Siewiorek and Swarz, 1982).

2.2.3 Memory Mapping. When the address space available to programs is smaller than the available physical memory (typical of 16-bit machines), techniques are employed to allow access to the larger physical memory within the confines of the smaller program address space.

One popular technique is memory mapping. In this scheme the upper portion of the logical address is used to select one of several mapping registers. The contents of the selected mapping register are used as the upper portion of the physical address while the remaining low bits of the logical address form the low bits of the physical address. In Fig. 7 this is illustrated for a system with a 16-bit logical address space and a 28-bit physical address space. The upper four bits of the logical address are used to select 1 of 16 mapping registers which each provide the upper 16 bits of a physical address. They are concatenated with the lower 12 bits of the logical address to produce a 28-bit physical address. By manipulating the contents of the mapping registers, all of a large physical memory can be accessed.

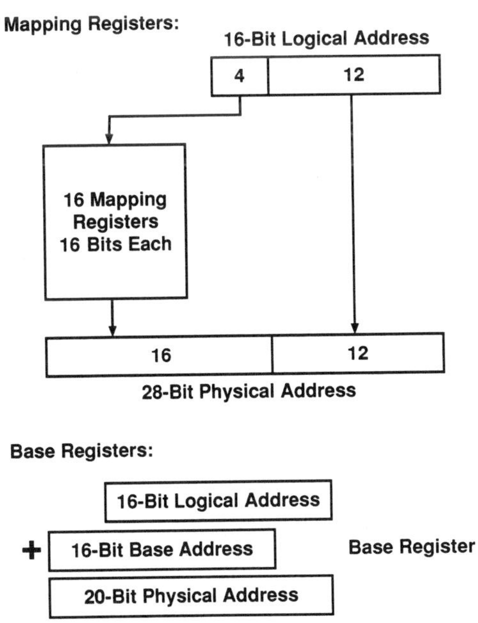

FIG. 7. Typical configurations of mapping registers and base registers.

Another technique is the use of base registers. These registers contain high-order memory addresses, and they are offset and added to the logical address automatically by the hardware. Often the choice of base register is based upon the context of the reference, be it for an instruction, to a stack, or for data. Figure 7 illustrates the scheme used by the 8086 family of processors used in PCs; the 16-bit base register is offset from the program address by 4 bits to produce a 20-bit physical address. There are many variations on these two techniques; see Siewiorek *et al.* (1982) for further examples.

2.2.4 Virtual Memory. Virtual memory is a technique that permits a physical memory to appear to be very much larger than it actually is, giving the illusion of a large, linear address space, which makes programs easier to write and more independent of the physical resources of a particular computer. The illusion of virtual memory is accomplished with additional logic in the memory subsystem, a large backing store which is usually a disk, and supporting software (Hennessy, 1990). Both virtual and physical memory is divided into blocks called pages. Each virtual page has an entry in a section of memory called the page table, which keeps track of the location of a page, whether it is currently resident in physical memory or paged out onto backing store. When a referenced page is resident, the translation of the virtual page address to a physical page is handled entirely through hardware via the page table, with only a very slight delay. However, when a referenced page is not resident, the supporting software is signaled that a page fault has occurred by the memory hardware. This supporting software will then suspend the requesting program, make the requested page resident (which may involve making room by moving another page out to backing store), and then resume the requesting program. The program is not aware that a page fault occurred, except for the fact that some additional time elapsed while the physical memory was being rearranged to satisfy the virtual address reference.

In order to improve virtual memory performance, a cache of recently translated virtual addresses is kept in a translation buffer. When a virtual address is found in the translation buffer, the computer hardware can accomplish the memory access without reference to the page tables, further improving performance.

Virtual memory systems depend on an observed characteristic of most all programs known as locality—that when a reference to memory is made, there is a very high probability that subsequent references will be to a location near the previous ones. Consequently, the probability that a subsequent reference is to a page already in memory is high, and overall system performance will be good. Thrashing is a problem that occurs when a program accesses virtual memory in such a way it is constantly page faulting, usually because it does not exhibit locality. It therefore has a very high page fault rate and consequently poor performance.

2.2.5 Caches. Caches are fast memories located between the main memory and the central processor and used to buffer the CPU from the relatively slow main memory. They maintain a copy of selected items in main memory, and rely on the fact that most programs make repeated references to clustered areas of memory (see discussion of locality in Sec. 2.2.4). Caches are designed to have a high probability of a cache hit (finding the required information in the cache), so that the faster cache can supply the information to the processor without incurring the speed penalty of a main memory reference. When the information is not in the cache, it is fetched from main memory and supplied to both the processor and the cache. Many caches maintain information in blocks larger than one word, and the cache will fill the entire block on the first cache miss, betting that other references will be made to that block.

Multiple-level cache systems are used with large main memories and fast processors, each cache successively smaller and faster. A system with a multiple-level cache would have several trapezoids in Fig. 6 in place of the single one shown.

Cache memory usually uses static semiconductor memories (static RAM). These devices, while significantly less dense then the dynamic RAMs used in main memory, are usually an order of magnitude faster and do not require refresh cycles (each memory cell is active and can maintain its state while the part is powered up).

While reads dominate cache access, information must also be handled on write. A write-through cache writes both the cache and the main memory, slowing the processor. A write-back cache writes only to cache initially, and the main memory is not updated until the cache block is replaced by other information. Other design issues include the lookup (mapping) scheme, block size, total size, and block replacement scheme (Hennessy 1990). Cache coherency is an important issue in multiple processors where several processors may be writing and reading simultaneously. If each processor in a multiple processor has its own cache, there can exist several copies of the same data. Should one processor modify a piece of data, it must ensure that all other copies of the data in other caches or in main memory are either updated or marked as invalid. Various schemes to maintain cache data coherency have been invented, with various performance characteristics and hardware complexities; see Dubois (1990) for approaches to this issue.

2.3 Busses

Busses (see BUS SYSTEMS AND COMPUTER INTERFACES) connect various subsystems together in the computer, moving addresses and data over a well-defined interface. Many computers have just one major bus that connects the CPU, memory, and input/output (I/O) adapters together (Figs. 1 and 4). Faster machines usually employ separate busses for the slower I/O, connecting the I/O busses to the faster CPU memory bus through an adapter or directly to memory.

2.3.1 Standardized Busses. Standardized busses allow peripheral devices to be designed once and used on a variety of machines, and for machines to be expanded after they have been designed and manufactured. The bus of the IBM PC-AT, probably the most used bus in the history of computing, is a *de facto* standard. A few of the most common formally standardized busses include the VME bus (IEEE 1014) and the SCSI bus (ANSI X3.131).

2.4 Input/Output

Input/output, or I/O, is the general term used to describe the movement of data between main memory and any peripheral device, such as disks, tapes, networks, terminals, or printers. Peripheral devices connect to the computer's I/O bus through an adapter. Adapters convert the specialized interfaces of the peripheral to a series of registers that can be read to or written from the I/O bus. Most adapters have some form of buffering, a small local memory that temporarily stores I/O data and helps to overcome processor and peripheral speed differences.

Programmed I/O is the simplest technique, using the central processor to read and write data, peripheral control, and peripheral status to or from the adapter registers and move the data to or from memory. Direct memory access (DMA) logic in the controller allows the CPU to set up a data transfer by writing control information to the adapter registers, with the actual data movement to or from memory being performed by the adapter, freeing the CPU for other work. Some of the most powerful computers employ separate I/O processors (sometimes called channels) that run specialized I/O programs to provide high-speed I/O within minimum impact on the CPU.

3. PERIPHERAL DEVICES

There is a wide variety of peripheral devices for use with computers. We will discuss the main features of the most common ones in this section.

3.1 Persistent Mass Storage

The vast majority of mass storage is electromechanical, in contrast to the purely electronic main memories discussed previously. While access to information is much slower, the data capacity of these devices is much larger, and the cost per bit of information stored is much lower. The stored information is said to be persistent, i.e., the information is retained when the device is powered off.

Fully electronic persistent storage is available, but currently with very low capacity and high cost, such as flash memories used in some portable computers. Vigorous development continues on these technologies and there are predictions that they will have a significant presence later in the decade.

3.1.1 Magnetic Disks—Floppy and Hard. A magnetic disk stores information on a disk-shaped platter coated with a thin magnetic film (see MAGNETIC STORAGE OF INFORMATION). The disk rotates at constant velocity under a read/write transducer called a head. The head is moved over one of many tracks on the surface of the disk, each track occupying a specific radial distance from the center of the disk. The head can detect previous magnetic patterns stored on the disk in a read operation, or write new magnetic patterns to store information. Information is stored in sectors, which is a subsection of a track. A sector will store from a hundred to a few thousand bytes of data, along with overhead bytes that synchronize the read/write circuits, identify the sector, and store error detection and correcting codes.

Disk striping is a technique used to improve performance in situations where files are large and there are multiple disks available. A file is distributed over several disks (members of a stripe set) such that when a section of a file fills an entire track, the next part of the file is found on the next disk in the stripe set. The goal is to eliminate the delay that moving the disk head from track to track causes, by switching the operation to another disk that has prepositioned its heads to the proper track. Switching the data stream among disks is done at electronic speeds, while moving heads is done at mechanical, inertia-limited speeds. Consequently, a stripe set of several disks with good controlling software can significantly outperform single drives in many situations.

A floppy disk is so named because the magnetic disk is made of a thin flexible plastic disk packaged in a flexible or rigid covering. Information may be stored on one or both sides of the disk. Floppy disks come in 8-, $5\frac{1}{4}$-, and $3\frac{1}{2}$-in. diameters with data capacities up to 2.8 megabytes. Floppy disks may be removed from the disk drive, and consequently a single drive may service a library of disks, or the disk may be used to transport data from one computer to another.

A hard disk is made of rigid material, and usually has several disks stacked on a common hub. Heads are ganged together and moved in unison by a single positioner, with each disk surface being serviced by a single head. 14-, 9-, and 8-in. diameter disks are produced usually for use with large computers, while $5\frac{1}{4}$- and $3\frac{1}{2}$-in. units are popular in workstations and PCs. The advent of laptop computers has kept sizes shrinking, and 2-in. and smaller models are now on the market. Disk capacity for $5\frac{1}{4}$-in. and larger units is now above 1 gigabyte and increasing; $3\frac{1}{2}$-in. hard disks can currently store about half that. Removable hard disks are available but declining in popularity; most hard disks today are built for permanent installation with the disks sealed inside the drive.

The key performance criteria for disks are transfer rate and average access time. Transfer rate, in megabytes per second, indicates how fast information may be moved to or from the disk once it is located. Average access time describes how long it takes to access any piece of information on the disk. Access time depends on seek time (how fast the heads can position to a given track) and rotational latency (how long it takes to rotate the desired information under the head).

3.1.2 Optical Disks. Mechanically, optical disks are very similar to magnetic disks, employing a rotating disk and positionable heads. However, the disk surface is designed to store data on the basis of altering the reflectivity or polarization of light, and storing information involves changing these characteristics, usually by heating a small area with a laser pulse (Hamacher, 1984) (see OPTICAL STORAGE). All current optical disk drives have removable media.

Initial optical disks were read-only devices (most drives could not write information on them). CD-ROM is the most popular format, and stores 600 megabytes of data on a $4\frac{3}{4}$-in. diameter compact disk identical in size and technology to compact audio disks. In fact, many newer CD-ROM drives for computers include headphone jacks and volume controls, and will also play audio CDs.

Write-once read-many (WORM) optical disks are available, useful mainly for archival storage. 14-, 8-, and 5-in. units are available with capacities from 600 megabytes to over a gigabyte.

Finally, read-many/write-many devices are becoming available. Capacities and sizes are similar to the CD-ROM and WORM devices.

3.1.3 Magnetic Tape. Magnetic tape, lacking random access, is the slowest mass storage technology currently in use. However, its low cost per bit and large capacity make it a popular archive and transport mechanism. While nine-track reel-to-reel tapes are still in use, a variety of less expensive and more convenient cartridge-based devices are now more popular. The quarter inch cartridge (QIC) is widely used in workstations, and 8-mm video tapes and 4-mm digital audio tapes (DAT) are now being used, often with stack loaders that can automatically change up to 16 tapes. Cartridge capacities of 5 gigabytes are available.

Tapes store data in a thin magnetic film layered on flexible tape. Nine-track and QIC tapes store data linearly along the length of the tape, using one head per track. 8- and 4-mm units use helical scan technology, similar to that used in video cassette recorders, storing bits on diagonal stripes. Helical scan technology is more complex, but permits much higher bit densities.

The useful capacity of all types of magnetic tapes is improved by data compression technology, which uses hardware and/or software to encode the stored data in such a way that it occupies less space on the storage medium. Extremely effective data compression schemes have been devised and it is not unusual to be able to double the raw capacity of a tape for many common data storage situations. The same technology is finding application in portable computer disk drives, where the user is willing to accept the lower performance caused by the compression/decompression process in favor of significantly increased storage capacity.

3.2 Communications and Ports

3.2.1 Serial and Parallel Ports. A port is a defined interface on the computer that can read and/or write digital data to an external device. A parallel port transfers all the bits of a word at once, or in parallel. A serial port transfers the bits one at a time in sequence, at rates that typically range from 300 to 38,000 bits per second (MaNamara, 1988). Most terminals and modems attach to serial ports, while printers may use either.

3.2.2 Modems. A modem (MOdulator/DEModulator) is used to transfer serial data over an audio channel such as a telephone line or radio link. Common modem speeds are 1200, 2400, and 9600 bits per second. Many newer modems have data compression and error correction features. A typical modem attaches to a serial port and to a modular telephone jack, and can dial the telephone under computer control. As all-digital telephone services, such as ISDN, become widely available, the need for modems will be eliminated.

3.2.3 Networks. Networks allow a group of computers to share programs and data easily, and usually provide shared facilities such as file and printer servers. See NETWORKS, COMPUTER for additional information.

3.3 User Interface Devices

3.3.1 Hard-Copy Devices—Printers and Plotters. Hard copy is a term used to refer to printed output. While printers have traditionally focused on character-based documents and plotters drew charts and graphs, the distinction is blurring as newer hard-copy devices, including laser, ink jet, and 24 pin printers can do both. Laser printers use much of the same technology as a plain-paper office copier; the optical input used by the office copies is replaced by a computer-controlled laser to form the image (see XEROGRAPHY). Resolutions of 300 dots per inch (DPI) on $8\frac{1}{2}$ by 11 or 11 by 17 in. sheets are available, and color laser printers are now available. Ink jet printers squirt ink drops out of a nozzle to form dots on the paper. Ink-jet printers are quiet and can print at 150 DPI in both single color and multiple colors. Impact printers, now predominantly using thin wires in the print head to strike an inked ribbon and then the paper, also can produces 150-DPI text and graphics. Multiple-color ribbons offer the ability to produce limited multiple-color output slowly.

3.3.2 Soft-Copy Devices—Terminals and Displays. Computer terminals and displays (see DISPLAY TECHNOLOGY) are used to produce soft-copy output on a screen. They may be character only or mixed characters and graphics, in color or monochrome. The basic character display today displays 80 columns and 24 or 25 rows of characters. Each character may be displayed with a small number of special characteristics, such as an underline, high intensity, or blinking. Color displays usually allow each character to have selected foreground and background colors.

While character displays model the display space in terms of character cells only, a graphics display models the display as a two-dimensional array of picture elements (pixels) which can display both characters and arbitrary graphics. Typical PC displays have 640 by 480 pixel displays, while workstations commonly have 1024 by 768 or 1280 by 1024 pixel displays.

Monochrome displays use one bit per pixel to indicate whether a pixel is on or off. Gray-scale displays use multiple bits per pixel (usually 2, 4, or 8) to represent the brightness of a particular pixel. Color displays also use multiple bits per pixel (up to 24). Frequently, color displays use the value of a pixel to index into a color map, which then selects the displayed color from a very large palette of potential colors. A typical color display uses 8 bits per pixel, so that 256 different colors can be displayed simultaneously. The color map translates 8-bit pixel values to 24-bit color values, allowing the selection of 256 colors from a palette of 2^{24}.

Most terminals use cathode-ray tube (CRT) technology, with diagonal measures of 10 to 19 in. The demand for laptop computers has driven the development of flat-panel display technology. Liquid crystal displays (LCDs) are in widespread use in laptops, with early units offering 640 by 480 notebook-sized gray-scale displays. 16 color units are available at higher cost. With strong worldwide demand for portable high-resolution color LCD displays for both computer and television use, the technology is expected to evolve rapidly to high resolutions and lower cost.

3.3.3 Input Devices. Input devices have remained remarkably stable in an otherwise rapidly changing field. Keyboards still dominate, and the mouse has established itself as the predominant pointing device. A mouse is a palm-sized device that may be rolled across a work surface and transmits relative movement to the computer. Software or hardware then moves a cursor on a display to mimic the motion of the mouse. From one to three buttons on the mouse allow a user to signal some action to software, usually manipulating a displayed object that the cursor is positioned on. A trackball is a partially exposed sphere that a user may roll in place; the relative motion is transmitted to the computer as in the case of a mouse. An advantage of a trackball is that it does not require desk space for motion. Tablets have a rectangular surface over which a puck with buttons is moved (not unlike a mouse); a tablet transmits absolute coordinates to the computer, based upon the position of the puck on the surface of the tablet.

Recent work in handwriting recognition software foreshadows the introduction of so-called "pen-based computers" in the near future, where the keyboard and mouse would be replaced by a stylus which "draws" on a display panel.

3.3.4 Optical Scanners. Optical scanners convert printed or handwritten copy into bit-mapped images, so that they may be manipulated, displayed, or printed by the computer. Hand-held and single-sheet scanners are low in cost and typically produce 300 to 600 dots per inch monochrome or gray-scale images. Color scanners are available at greater cost and at various resolutions. Software is available to convert scanned-in character images to computer character codes, permitting further processing by most existing programs.

4. SUMMARY

Computers continue to evolve, providing ever higher performance and larger capacities at lower costs. Despite this constant evolution, the majority of successful machines over the first four generations have had a common gross structure and similar subsystems, which we have reviewed in this article.

The advent of the personal computer has moved computing into the mass market, and consumer electronics technology such as compact disks and digital audio tape are being applied to solve computing problems. Other mass market technology, such as cameras that use floppy disks as "film," cellular radio networks, handwriting recognition software, and voice recognition systems (Kamel, 1990) all indicate a healthy and vigorous future for computers with even broader application.

ACRONYMS

ALU: Arithmetic logic unit.

ASCII: American standard code for information interchange, a character encoding standard.

CD-ROM: Compact disk—read-only memory.

CISC: Complex instruction set computer.

CPU: Central processing unit.

CRT: Cathode-ray tube.

DAT: Digital audio tape.

DMA: Direct memory access.

ECC: Error correction code.

ISDN: Integrated services digital network, a packet-switching digital telephone service.

MFLOPS: Millions of floating-point operations per second. A popular but nonstandardized figure of merit for floating-point performance.

MIPS: Millions of instructions per second, a popular but nonstandardized figure of merit for integer performance.

PC: (1) Personal computer, (2) program counter.

QIC: Quarter-inch (tape) cartridge.

RAM: Random access memory. While ROM is also random access memory, the term has come to be used almost exclusively to refer to read-write memory based upon charge storage in semiconductor chips.

RISC: Reduced instruction set computer.

ROM: Read-only memory, written once with a fixed set of instructions and/or data can be repeatedly read thereafter.

WORM: Write once read many.

GLOSSARY

Bit: The smallest unit of information represented by a digital computer, which can have one of two values: 0 or 1. From Binary digIT.

Byte: A set of eight bits, commonly used to represent a character. On many computers, the smallest addressable entity.

Core or **Core Memory:** Random-access read-write memory based upon information storage in magnetic cores. Some operating systems refer to any type of main memory as core for historical reasons.

Pixel: A single dot or picture element on a graphic display.

SPECmark: An overall figure of merit of computer performance, based on benchmarks distributed by the Systems Performance Evaluation Cooperative (SPEC).

Word: An addressable collection of bits. In current practice, words are usually a multiple of 1, 2, 4, or 8 bytes (8, 16, 32, or 64 bits, respectively).

List of Works Cited

ANSI (1977), *X3.4-1977 Code for Information Interchange*, New York: American National Standards Institute.

Cohen, D. (1981), *Computer* **14**, 48–54.

Dubois, M., Thakkar, S. (Eds.) (1990), Special Issue on Cache Architectures in Tightly Coupled Multiprocessors, *Computer* **23** (6).

Hammacher, V. C., Vranesci, Z. G., Zaky, S. G. (1984), *Computer Organization*, 2nd ed., New York: McGraw-Hill.

Hamming, R. W. (1950), *Bell Syst. Tech. J.* **29**, 147–160.

Hennessy, J. L., Patterson, D. A. (1990), *Computer Architecture: A Quantitative Approach*, San Mateo, CA: Morgan Kaufmann.

IEEE (1985), *SIGPLAN Notices* **22**, 9–25.

Kamel, R. (Ed.) (1990), Special Issue on Voice in Computing, *Computer* **23** (8).

McNamara, J. E. (1988), *Technical Aspects of Data Communication*, 3rd ed., Bedford, MA: Digital Press.

Siewiorek, D. P., Swarz, R. S. (1982), *The Theory and Practice of Reliable System Design*, Bedford, MA: Digital Press.

Siewiorek, D. P., Bell, C. G., Newell, A. (1982), *Computer Structures: Principles and Examples*, New York: McGraw-Hill.

Weicker, R. P. (1990), *Computer* **23**, 65–75.

Further Reading

Duncan, R. (1990), *Computer* **23**, 5–16. Describes the multitude of parallel computer architectures and classifies them.

Gajski, D. D., *et al.* (Eds.) (1987), *Computer Architecture*, Washington: IEEE Computer Society Press. A tutorial on computer architecture presented as a collection of important papers from 1977 through 1985. Includes sections on microprocessors, RISC machines, and special purpose architectures.

Hennessy, J. L., Patterson, D. A. (1990), *Computer Architecture: A Quantitative Approach*, San Mateo, CA: Morgan Kaufmann. An excellent, easily read textbook that covers all aspects of computer architecture and hardware design, with an emphasis on performance and cost implications. Includes analysis of contemporary RISC machines.

Hennessy, J. L., Juppi, N. P. (1991), *Computer* **24**, 18–29. An article tracing the impact that advances in technology have had on computer architecture and suggesting what the future holds.

Horning, R. J., *et al.* (1991), *IEEE Spectrum* April 1991, 58–68. A good paper on the choices and challenges facing contemporary RISC workstation architects and designers.

Krick, R. F., Dollas, A. (1991), *Computer* **24**, 5–15. Presents a history of advances in instruction sequencing technology and discusses current strategies.

Stone, H. S., Cocke, J. (1991), *Computer* **24**, 30–38. Discusses the impact that high speeds will have on computer design, and the need to consider optical interconnects, greater parallelism, and use of locality to increase performance.

Weicker, R. P. (1990), *Computer* **23**, 65–75. A good overview of popular benchmark programs and the pitfalls of comparative benchmarking.

COMPUTER NETWORKS

See NETWORKS, COMPUTER

COMPUTER PROGRAMMING LANGUAGES

HERBERT J. BERNSTEIN, *Bernstein+Sons, Bellport, New York, U.S.A.*

INTRODUCTION

Computer programming languages are used by people to specify the actions to be taken by computers in solving problems. As the needs of different types of users to solve problems with computers vary, so do the computer programming languages that have been developed. In this article we concentrate on the aspects of computer programming languages of interest to physicists, both for major scientific calculations and for control of experiments. Thus we will give only brief attention to languages and approaches to languages that, while of major importance in other fields, such as computer science, are not critical to working physicists. The so-called imperative or statement-oriented programming languages, which specify sequences of actions to be taken on objects (such as adding a list of numbers, and then dividing by the count of things in the list to get an average), are the major tools of modern physics.

In the space available we cannot fully describe the rich structure of even this small subset of computer programming languages. At best we can give some of the essential terminology, and examples of the most heavily used features. Languages used by people to communicate with people are learned by using them. The same is true of languages used by people to communicate with computers.

Computers have their own internal languages, private jargons of strings of numbers that are difficult for most people to learn and rarely used as programming languages. A single computer may have more than one such language, and different types of computers, even though made by a single manufacturer, may have different languages. The lists of numbers direct the machine to perform certain simple actions. Because the choice of numbers may vary from machine to machine, the number 103344 might be an "instruction"

1-56081-063-7/92/$5.00+.50

(an "order" or "command") to add two numbers stored in locations 33 and 44 on one machine, and an instruction to print the character in location 3310 on another. Such lists of numbers are almost as cryptic as the codes and ciphers used to hide secrets from spies. These "machine languages" require a programmer to specify in full detail what is to be done, and slight conceptual changes can require large portions of a program to be rewritten.

In order to save the effort of writing a new and different program for each machine to solve the same problems, to make computers accessible to more people, and to increase the chance of writing instructions for computers correctly, many programming languages for computers have been created, languages that express the actions to be performed in terms more comprehensible to people, and in ways that allow the same programs to be used on many different types of computers.

There is a large range of problems in physics for which computer languages are used. For problems expressed as formulas and equations, there are languages, such as FORTRAN, that accept formulas in ordinary scientific notation. For problems of process control and control of scientific experiments, languages that allow a finer control over a machine and facilitate management of real-time processes running in parallel, such as C and assembly languages, are used. For many problems it is important to be able to make rapid changes in a program, and the efficiency of execution is not as critical. In such cases, interpretive languages, such as APL, BASIC, and FORTH, that allow "on-the-fly" changes are used.

Some languages are close in style to the operation of computers, consisting of sequences of actions to be taken on objects. Such a style simplifies efficient translation of a problem specification into the internal language of a computer, but may force a strained interpretation of the original problem, increasing the chances of programming errors, and increasing program maintenance costs. To reflect more naturally the demands of theproblems to be solved rather than the demands of computers, languages using other styles of programming have been developed: object-oriented, functional, and logic programming. These offer promise of increased productivity for appropriate problems, but for the classic problems of theoretical and experimental physics, FORTRAN and its statement-oriented, imperative relatives predominate.

We will begin by considering the history of computer programming languages and the differing programming styles that have been developed. Then we will look in some detail at the languages for major scientific calculations, exemplified by FORTRAN, C, and PASCAL, and the additional languages used in control of experiments, exemplified by BASIC and FORTH. For each we will give some of the vocabulary of the languages: the "operators," which are effectively the verbs of the language, and the "data structures" and "data types" that define the nouns and their modifiers.

1. HISTORY

The history of computer programming languages is intimately related to the history of computers themselves (see COMPUTERS; COMPUTER HARDWARE). Many of the restrictions and inconveniences in early languages were reflections of the limitations of the computers of the time. Because our focus here is on languages, rather than computers, we will omit many critical contributions to the design of computers and discuss only some examples of the changes in hardware that had a significant impact on the development of languages.

The first consideration of programming was for Charles Babbage's proposed Analytical Engine by Augusta Ada, Countess of Lovelace, in 1842 (Morrison and Morrison, 1961, pp. 245–295; Hollingdale and Toothill, 1965). The programming was in a direct machine language, with a manual translation from normal mathematical notation. Babbage's machine had a memory, a processor for arithmetic and control, and input-output capabilities. At this early date, the fundamental concepts of operations (actions to be performed), variables (information that may change during the calculation), and constants (numbers, such as e or π, that would not change) were established.

1.1 Plug-Board Programming, Punch Cards, and Paper Tape

In Babbage's time, cardboard cards with holes punched into them were used for the control of looms, and he chose such punch cards as a mechanism for specification of both operations and variables. In 1890, Herman Hollerith invented a series of electromechanical punch-card devices that could do census calculations under the control of a wired-in program of simple actions, such as adding, subtracting, or printing certain columns of numbers. The numbers were stored as holes punched into cardboard cards; hence the name "punch cards." "Programming" such machines consisted of throwing switches and inserting jumper wires into plug boards (like an old-fashioned telephone switchboard). Programs were saved by simply keeping the entire plug board with its wires, and putting in a fresh one for the next program. These electromechanical machines were put to increasingly complex and general tasks, culminating in the design and construction of the Automatic Sequence Controlled Calculator, the Mark I, by Howard Aiken from 1937 to 1944 (Slater, 1987, pp. 81–90). In addition to its internal plug board wiring, it accepted a general program of instructions on paper tape. The operations of the machine were composed from addition, subtraction, multiplication, division, and retrieval of previously computed variables. Data came from three more paper tapes and two stacks of punch cards. The values of 60 constants were set from 1400 rotary switches. Programming the machine was a complex process of preparing tapes, cards, and switch settings. The program on the control paper tape consisted of lines of 24 holes each, broken into three fields, two for operands and one to specify the operation (Bashe *et al.*, 1986).

1.2 Stored Programs and Machine Language Programming

A great simplification of programming came with the creation of digital electronic stored-program computers because a program was stored in memory as numbers. The concept was pioneered during work on Eckert and Mauchly's Electronic Numeric Integrator and Calculator (ENIAC); see Rosen (1969). ENIAC was programmed by wiring, but Wilkes' EDSAC in 1949 fully implemented stored-program computer design. Early internal memories were small, and would have greatly limited program size, so that programs still existed primarily as holes in cards and paper tape external to computers. However, having the ability to store large portions of a program internally made it possible to reuse the same sequence of instructions rapidly without having to reload them for each execution from cards or tape, thereby having efficient "loops" and "subroutines." Commonly used modules of code, e.g., routines to do calculations on floating-point numbers (i.e., real numbers in which the decimal point might be at different places) or complex numbers on machines that did not have built-in hardware for such calculations, were programmed in the native language of the machine, "machine language," and saved for use with many programs. Machine languages settled down to a variety of simple numeric representations of operations of the form

Operation Operand-Address Operand-Address...

where the number of operand addresses (the places from which data was to be fetched, or to which results were to be stored) was determined by the design of the machine and the particular operation being done. Sometimes all the operands were explicitly listed with the instruction, and sometimes one or more operands were implicit in the operation chosen. On some machines the address from which the next instruction would be fetched had to be explicitly given as part of the instruction. In others, that was needed only to create conditionally executed sections of code—loops and subroutines.

If the computer used decimal numbers internally, programming could be a simple transliteration of operations into appropriate decimal digits, and laying out of locations of instructions and operands in memory. This was a complex extension of the problem of seating many people for a large dinner party.

Hardware economies could be found by building computers based on binary numbers (base 2) rather than decimal (base 10), so, for such machines, instructions had to be created in that numbering system. Octal (base 8) notation was sometimes convenient for creating such instructions.

1.3 Assemblers

Translation of names for operations and names of the things that need to be fetched and stored is itself a calculation, one for which the computer itself is well suited. Thus, in the 1950s machine-language programming was replaced for most purposes by assembly-language programming (e.g., the work of Rochester and McClelland; see Bashe *et al.*, 1986, pp. 323 ff.). An assembler translates names for operations, such as "ADD" and "SUB," into the appropriate numbers, and keeps track of which memory locations have already been used for instructions, variables, and constants, so that the programmer might refer to them by name instead of by number. Though each machine had its own assembly language, at least the pieces of code for that machine saved in symbolic form did not have to be rewritten when used with other code that would force changes in addresses.

For example, let us invent a decimal machine that has the following instructions:

01ssdd:

MOV ss [source to] dd [destination]

02ssdd:

ADD ss [source to] dd [destination]

03vvll:

IF vv [variable to test is] NOT-ZERO [then goto location] ll

04ssdd:

READ [a number from source device] ss [and store in] dd

05ssdd:

PRINT [the number in location] ss [on device] dd

060000:

END

and assume that the source of input data is device 5 and the printer is device 6. Then to read in 10 numbers, add them, and print the result, we might write

```
TEN:      DATA      10
NEGONE:   DATA      -1
ZERO:     DATA      0
INPUT:    VARIABLE
ICOUNT:   VARIABLE
ISUM:     VARIABLE
START:    MOV       TEN,ICOUNT
          MOV       ZERO,ISUM
LOOP:     READ      05,INPUT
          ADD       INPUT,ISUM
          ADD       NEGONE,
                     ICOUNT
          IF        ICOUNT NOT-
                     ZERO, LOOP
          PRINT     06,SUM
          END
```

rather than

```
10
-1
0
0
0
0
010004
010205
040503
020305
020104
030408
050506
060000
```

This was a significant improvement in making it easier to write programs. However, programmers still had to write out every instruction in full detail. To save time in writing similar sequences, assemblers were expanded to become macro-assemblers, in that sequences of instructions could be given names.

1.4 Compilers

Assemblers were and still are useful tools, but even with macro facilities, they are still strongly linked to particular machines, not to the problems of primary interest to users of the machines. Translators for languages that do not directly represent machine language, but that require compilation of the meaning of an entire program or major segments of a program into appropriate instructions as large units, are called "compilers." In 1952, Grace Hopper created a more user-oriented computer language called A-O for UNIVAC computers. By 1955, a group at IBM led by John

Backus had designed a language called FORTRAN (for IBM Mathematical Formula Translating System), which would take statements written in ordinary algebraic mathematical notation and translate them into a program in machine language. This language, and its successors, became the workhorses of scientific calculations in the United States, first on the IBM 704 and its descendants, and then on other computers.

In FORTRAN, the example of adding ten numbers, above, might be written as:

```
        ISUM=0
        DO 100 ICOUNT = 1,10
        READ (5,1000) INPUT
100     ISUM=ISUM+INPUT
        WRITE (6,1000) ISUM
1000    FORMAT (I6)
        STOP
        END
```

For business applications, compilers for a "Common Business Oriented Language" (COBOL) based on earlier work by Grace Hopper were developed in 1959, and for such applications the language is still in heavy use. The style is closer to English than the mathematical formulas of FORTRAN, as in

```
ADD MYVALUE TO MYSUM
```

The great success of FORTRAN and COBOL and the increasing power of computers led to attempts to create better languages. Starting in the late 1950s, Peter Naur and others worked on "Algorithmic Language" (ALGOL), intended to provide code more structured and easier to read and to maintain than FORTRAN. Partially because it was oriented toward character sets not well supported on the card punches in the United States, and because FORTRAN was adequately meeting the scientific programming needs of the time on the computers available in the United States, ALGOL did not achieve great popularity in the United States, but did attract a wide following in Europe in the 1960s. By creating PL/I in 1964, IBM attempted to correct some of the limitations of FORTRAN and combine it with all the business-oriented features of COBOL. PL/I is still in use, but has not attracted a large following. ALGOL evolved into two different languages. ALGOL-68 addressed the need for more general data structures, but was hard to read. In 1971 Niklaus Wirth proposed an alternative language, called PASCAL, that proved effective for that purpose, and was simple and clear. Use of ALGOL-68 faded out.

The special needs for computing in the military environment were addressed by Jules Schwartz's JOVIAL ("Jules' Own Version of the International Algebraic Language"), but real-time programming for control computers was done in a mixture of assembler and FORTRAN. This produced a body of code that was hard to debug and maintain as computers changed, and in the mid-1970s development of a language specially suited to programming computers operating in real time and embedded in military hardware was begun. Attempts were made to address all aspects of the cycle of creation and maintenance of software by providing tools needed for effective software engineering in the language, and to learn the lessons of the languages developed up until then. The result, developed primarily by J. Ichbiah and colleagues in 1979, was called ADA, in honor of the Countess of Lovelace; see Booch (1983). We will return to this language in considering software engineering below.

A compiled language called C was designed in 1972 by D. Ritchie (Kernighan and Ritchie, 1978). C provides the user-oriented economy of expression of FORTRAN and PASCAL with the intimate access to hardware features of assemblers, and has replaced a major portion of the use of assemblers for time-critical real-time programming in operating systems and device control code.

1.5 Interpreters

A compiler takes the user source code as a whole and translates it all to machine language before executing any of it. This can imply a long delay between changes in a program and seeing the results, and also causes difficulty in highly dynamic decisions about the use of memory. For applications requiring a more immediate response to code changes or highly dynamic uses of data structures, interpretive languages have been developed. The computer is effectively programmed to simulate a machine closer to the user needs. The first such interpreter, called the Speedcoding Interpreter, was developed by John Backus and others at IBM in 1953. Interpreters have been used for many significant languages since. Examples are APL ("A Programming Language"), developed by Kenneth Iverson in 1962, BASIC, developed by Kemeny and Kurtz in 1965, LISP, developed by John McCarthy in 1962, and FORTH, developed by Charles Moore in 1971.

The languages implemented by use of interpreters often also have compilers to allow for more efficient execution once the programmer is past the debugging stage. Many languages that are more appropriate to compilation because of complex context, such as FORTRAN and C, also have had interpreters created to aid in debugging programs.

1.6 Software Engineering

As an increasing investment was made in programs for use on computers, it became clear that the costs of software development would exceed by far the costs of the computers on which the software would run, and that the major source of failures of computer-based systems would be unreliable software, rather than unreliable hardware. Languages were proposed that more closely reflected the functions to be performed, rather than the internal workings of computers. John McCarthy developed LISP (for List Processing) from 1956 to 1962 which, though it could be used in the step-by-step style of statement-oriented (or imperative) languages, introduced the ability to solve problems simply by evaluating functions made up of yet more functions, perhaps defined recursively. LISP became the basic tool of artificial intelligence research and computerized symbolic algebra. PROLOG, in 1972, introduced an extremely high level of functional programming, permitting general statements of problems to be effectively their code. The development of ALGOL and PASCAL had provided a basis for programming in a hierarchical structure or routines using hierarchical structures of data. These approaches were the beginnings of a serious attempt to convert the art of programming into a process of engineering—creation of a quality product in a predictable time that would perform according to its specifications, and that could be maintained in an operational state over considerable periods of time despite changes in the hardware being used. It was also hoped that an engineering approach would allow the rapid and reliable development of new software products and new versions of existing products in response to new problem specifications.

The most serious effort in this direction, of creating an environment for "software engineering," has been in the ongoing development of ADA, where, because of the large expenditures on software in the military, and the high risks from failures of that software, an engineering approach is desirable. The degree of success of that particular effort is not known as of this writing. It is clear that a highly disciplined and organized approach to creating programs, in which each part of the software is self-documenting and small enough to be thoroughly understood and tested, with the cleanest possible interfaces to other pieces, greatly increases the chances of producing correctly functioning programs in a reasonable length of time.

One of the most significant changes in approach to programming adopted by software engineering has been in the area of "object-oriented" languages. Beginning with work on SMALLTALK from 1971 to 1980, languages supporting this concept allow the creation of objects that respond to messages from other objects by taking appropriate actions, shifting the emphasis in software design from complex sequences of actions to more modularized actions within the mechanisms of objects. Such an approach is particularly appealing when one works with systems requiring many interacting parallel threads of execution. Languages for object-oriented programming provide the ability to define new "abstract data types," in which the user may specify not only the representation of data within memory, but the operations that may be performed on that data, and hide the internals of the representation from processes using the type, so that unneeded dependences on the specifics of a representation will not proliferate.

For more on the history of software engineering, see Ghezzi and Jazayeri (1987).

1.7 Supercomputing

In the major mathematical calculations of physics, the largest and fastest computers of the moment, so-called supercomputers, are pressed into service. The programming languages used are adapted to the special capabilities of those machines. In 1963, Control Data introduced the 6600, a computer capable of 1 million floating-point operations per second and the supercomputer of its day. The machine had a complex and subtle interaction among its operations because some of them happened in parallel, making it difficult to get maximum execution speeds from ordinary FORTRAN compilers. It was common to code critical sections of programs in assembler for speed. Since then, supercomputer hardware has become more complex. The CDC 7600, the Cray-1, CDC Cyber 200 series and NEC SX series, Fujitsu VP series, and Hitachi S-800 series pushed the limits of available hardware speeds for properly organized programs and data. The present generation of machines combines the processing of entire vectors of data in single instructions with the availability of more than one processor to handle a single problem. Preprocessors to FORTRAN, highly optimizing compilers, extensions to FORTRAN, and use of C allow close to optimal execution without having to resort to assembler code, but the programs often have to be extensively redesigned when moved from one machine to another in order to realize maximum speeds.

See Perrott and Zarea-Aliabadi (1986) for more detail on supercomputer languages.

1.8 Parallel Computing

It appears that we may be reaching the limits of speed achievable by a small number of powerful processors, and that the supercomputers of the future will have much greater parallelism. Languages designed for sequential execution of instructions on one processor may not be appropriate for this new generation of parallel computers. Suppose we have a machine with 1000 processors that are able to work in parallel, and we try to convert an existing sequential code to run on our new system. We must ensure that the result of the parallel execution is the same as that of the sequential execution of the program, so we introduce code to synchronize the use of data that must be updated to ensure that it is updated in the desired ordering. If, as a result of this synchronization, we force, say, 1% of the executed instructions to execute on just one processor in total control of the machine, then we can never gain a speed improvement of even 100 times, much less the 1000 times we would like to see from 1000 processors (see PARALLEL COMPUTING).

Many paradigms need to be considered in the design of a language for parallel processing: shared memory versus message passing, tightly coupled versus loosely coupled processors, homogeneous versus heterogeneous processors, many small processors versus a few powerful processors versus many powerful processors. When a large number of processors attempt to share a common memory, mechanisms must be provided to synchronize their use efficiently, or access to a single location may force the processors to "serialize" execution, i.e., to run one after the other, rather than in parallel. When many processors use private memories instead, then careful estimation of the network traffic needed to keep feeding each one of them data is needed. The 1980s have seen an explosion of new languages and extensions of existing languages to handle various types of parallel systems. Of particular interest is the language LINDA (see Ahuja *et al.*, 1986) with data structures based on the concept of Tuple Space. Tuples are ordered sets of values, usually given as a parenthesized list. In a sequential language we would expect to have a fixed address telling us where the tuple is to be found. In the Tuple Space approach, we find the tuple partly by the contents of its fields and partly by its structure, so that the necessary addressing information is carried by the tuple.

The basic operations are "out," which adds a new tuple to Tuple Space, "read," which reads a tuple but leaves it in Tuple Space, and "in," which reads a tuple and removes it from Tuple Space. Trying to "read" or "in" a tuple that is not yet in Tuple Space causes the requester to wait until someone else puts the desired tuple there. This provides the essential interlock to allow effectively simultaneous access to the same data. See Bal *et al.* (1989) for more on the subject of languages for distributed computing.

As with existing supercomputers, there is a certain attractiveness to programming in a variant of FORTRAN, and then having the parallel conversion done by a preprocessor or the language translator. An example is Alliant FX/Series FORTRAN (see Alliant, 1988).

1.9 4GLs, Packages, and Nonprogramming

In the early days of computing, there was a feeling of moving from one generation of computers to another, and from one family of languages to its descendants. With the faster pace of development today, it is hard to distinguish any clear boundaries marking the end of one generation and the start of another. However, the term "4GL," for "Fourth-Generation Language," has come to mean packages and systems of powerful tools for programming that allow someone who is not particularly familiar with the intricacies of programming to solve problems with computers. This includes powerful database query languages, spreadsheet systems, graphics tool boxes, and systems integrated with text processing for the production of complex compound documents. In many cases the need to program is avoided by the use of menus or graphic user interfaces allowing the nonprogrammer to piece together the solution to his problem out of simple known steps and objects. It is to be hoped that the fourth-generation packages will be followed by a "Fifth Generation" of even more powerful systems functioning as expert systems able to assist the user actively in finding the solution to his problems.

With the spread of individual workstations and the decentralization of the management of information, the 1980s have seen the rapidly increasing use of more powerful tools by a wider community of users. Solutions to problems can be pieced together with much greater speed than ever before, but with that speed also comes the risk that the solution, being composed of unexamined black boxes, may not be what was initially intended. There may no longer be a rigorous trace of what was done to achieve that solution, nor even of what it actually is. With powerful tools comes the risk of large mistakes.

For more background on fourth- and fifth-generation languages, see Chorafas (1986).

2. PROGRAMMING LANGUAGE CONCEPTS AND STYLES

The wide variety of uses to which computers may be put has led to the creation of more languages than any one person could learn or use in a lifetime. Indeed, it is common to extend languages, effectively to invent whole new vocabularies, when faced with even modest new problems. Good programmers write extensive libraries of "functions" and "subroutines" that they try to use and further extend for additional applications.

In order to deal with this veritable Tower of Babel, we will present the general concepts of programming languages, with supporting examples from some of the more popular languages used in physics. In any particular application on any particular machine, the specific instructions for use of the dialects of languages used there must be consulted.

Digital stored-program computers provide the means to manipulate numbers and other data by sequences of operations. The data and the memory locations within which they are stored are "objects" on which a computer performs a sequence of "actions." Computer programming languages are tools to perform actions, explicitly or implicitly, on objects. They are designed either to work directly with representations of concepts that derive from the structure and function of digital stored-program computers or to work with representations of concepts that derive from representations of other concepts. In that case, the language translator must translate those other representations into terms appropriate to computers. The objects that must be manipulated are the equivalent of nouns and pronouns in human languages. These objects are called "variables," "constants," "parameters," "arrays," "strings," "pointers," "arguments," etc. in programming languages. There are the actions that may be taken either simply as actions by themselves or on some of these objects, the equivalent of intransitive and transitive verbs in human languages. These actions are called "operations," "functions," "subroutines," "procedures," "handlers," "drivers," etc., in programming languages. The basic elements of a language are enriched by additional descriptive elements that describe various properties of objects and actions, and provide a context that may change the meaning of the objects and the actions. Think of these as descriptive sentences involving adjectives and adverbs. Thus, if we were writing a procedure for cooking a chicken, we might begin with descriptive sentences setting the environment for our procedure:

1. The oven should be at 350 degrees.
2. The chicken should be at room temperature.

Then we might list the actions to be taken on the objects involved:

1. Clean and season the chicken.

2. Put the chicken into the pan.
3. Put the chicken and pan into the oven until 2 hours have gone by.
4. Baste the chicken every 15 minutes.
5. Remove the chicken from the oven.
6. Let the chicken set for 15 minutes.

In this example, we can think of the "chicken," the "pan," and the "oven" all as objects, and "clean," "put . . . into," "baste," "remove," and "let . . . set" as the actions in a programming language.

In programming languages we usually use the name "statement" for sentences of the language. As with human languages, the actions (verbs) and objects (nouns and pronouns) may be in a variety of relative positions, and some of these objects and actions may be understood, rather than explicitly included, in the statement. The three most common orderings of actions and objects are these:

1. Put an action before the objects it affects (prefix notation).
2. Put an action after the objects it affects (postfix notation).
3. Put an action between the objects it affects (infix notation).

For example, if we wished to add the numbers 5 and 6 to get 11, this would be expressed as "+ 5 6" in prefix notation, "5 6 +" in postfix notation, and "5 + 6" in infix notation.

When our emphasis is on the sequence of actions, on doing one thing after another, we consider a statement-oriented, or "imperative" programming, approach. When our emphasis is more on the objects, and we consider the actions as an internal mechanism of the objects that communicate by passing messages among themselves, we are concerned with "object-oriented" programming. Instead of giving the detailed steps involved in chicken cooking, we might define a "chicken cooker" as an object, which might contain a raw chicken or a cooked chicken. We would still have to define somehow the internal mechanisms of the chicken cooker, but we could hide the choice of gas stove versus microwave, basting, etc., from the user of the chicken cooker. If we divorce ourselves entirely from considerations of the order in which actions are to be performed, but consider only the most pure function evaluations, then we are concerned with "functional" programming; and if we refuse to consider actions at all, but only the relationships among objects, we are concerned with "logic" programming. In this last case, we might simply define the relationships between characteristics of a raw chicken and a properly cooked chicken, and leave it to the computer to find a way to transform raw chicken into cooked chicken.

2.1 Imperative Programming

FORTRAN, BASIC, ALGOL, PASCAL, C, and MODULA-2 are examples of languages that support statement-oriented programming, following the approach of our chicken-cooking example above. FORTRAN established the accepted programming style for these languages. BASIC is a simplified, FORTRAN-like language often run interpretively to allow rapid changes and experimentation with a program. ALGOL was a competitor to FORTRAN that allowed a more hierarchical structuring of a program. PASCAL is a descendant of ALGOL with a rich control over data structures and types. C is a more flexible language with less rigid enforcement of data structures and types. MODULA-2 is a descendant of PASCAL that provides support for the concepts of modules and information hiding (see below).

A program in such an imperative programming language will consist of some number of related modules containing some declarations that establish the environment in which the work is to be done (primarily specifying the data structures of the objects to be manipulated) and then a well-defined sequence of steps to perform on those objects. The most powerful high-level action to be taken is the execution of an assignment statement, i.e., a statement that puts new information into an object, usually based on the evaluation of arithmetic expressions, possibly containing references to functions of previously computed data. This statement is a generalization of the simple computer order to "store." It looks like an algebraic equation, but has a different meaning.

Consider the FORTRAN statement

```
X=X**2
```

As an algebraic equation, it would have two solutions, 0 and 1. As a statement in FORTRAN, it is an order to take what is currently stored in the variable X, multiply it by itself, and store it back into X. If X started out as 0.5, it would now become 0.25. It need not be 0 or 1 or any other particular value. Thus, in order to determine the state of an object, what value is stored in a variable or whether any value has been stored at all, we must follow the sequence of actions taken from the beginning and see what has been done to the object. This makes debugging of statement-oriented language programs difficult. Special programs, called debuggers, are used to provide traces of all changes to an object or even of all steps taken.

In order to improve the readability and maintainability of imperative programming language programs, it is desirable to keep small the context of objects that any actions may affect. This is done by breaking up the program into procedures, modules, packages, etc.

A consequence of breaking up a program into small pieces in an imperative programming language is the possibility of "side effects." A procedure may have internal data structures, objects only it knows about, or access to less-than-obviously accessible global data structures. By changes in the state of these hidden objects, different invocations of the same object may produce totally different results. Consider the analogy of an employer dealing with an employee who is having problems outside of work, perhaps with bills that are piling up. On a day when his bank is threatening to foreclose on his house, the employee may simply ignore what his boss tells him and sulk. On a day when he has caught up on his payments, he may perform his job quickly and efficiently. In a global sense, the behavior of the employee is deterministic, but from the point of view of the employer, the employee seems to respond differently at different times to the same inputs. Such variability can be highly desirable when, for example, a procedure for printing pages puts a heading on at the top of each page because it is keeping an internal counter of lines, but in general side effects make it difficult to write reliable software.

2.2 Object-Oriented Programming

In object-oriented programming, we attempt to solve problems by working with objects that can send messages to one another. Those messages may then cause the objects to take some appropriate actions, but, essentially, the actions are hidden, internal mechanisms within the objects. This is a realistic model of the behavior of systems such as trucking companies. Goods delivered to them cause them to dispatch trucks to appropriate destinations. Thus we might define objects called "warehouse" and "restaurant A," "restaurant B," and "restaurant C." The warehouse might contain a variety of inventories of goods, say, "chickens" and "ducks," and allow the operations "arrival" (i.e., store) and "dispatch." The arrival of a truckload of ducks might result in the dispatch of three smaller truckloads of ducks, one to each of the restaurants. The mechanism of computing the quantities of ducks to dispatch to each restaurant need not concern us; it is an internal mechanism of the warehouse, perhaps relating to earlier messages (e.g., standing orders from the restaurants). In a similar way we might define a chicken-cooking machine, which accepts the message that a raw chicken has arrived, and which dispatches a cooked chicken by some unspecified internal means. Other than timing, we would not be concerned that the chicken cooker used a gas oven or a microwave, just that the arrival of a raw chicken resulted in the production of a cooked chicken.

An early example of a computer system operating in a similar manner was the CDC 6600 of 1963, a complex of 11 processors communicating via messages left in mail boxes. This approach was used to provide a highly efficient operating system, but was not made available for user programs. SIMULA-67 introduced the concept of "class," which allowed a data structure to be associated with the code that would manipulate that data structure. MODULA-2, SMALLTALK, and C++ provided similar modularization of "objects."

The important distinction from the norms of simple imperative languages is that in, say, FORTRAN, one could organize data and create routines to manipulate those data. The entire mechanism of how the data would be manipulated normally remained visible, so that a programmer could and often did build assumptions about the internals of that manipulation into the programs that used it. Then when any change was made at the lower level, say to improve the efficiency of the routines handling the data structure, or to add new features to the data structure, every piece of code that used those routines had to be examined, and perhaps changed. In an object-oriented language, by associating the manipulations of the data with the data structures themselves, the internal mechanisms can be hidden from the higher-level routines, and the effects of changes within the lower level routines can be kept purely internal.

However, there can still be side effects in an object-oriented language. Indeed, because of the hidden nature of actions within objects, side effects may become the dominant problem-solving mode. Also, even though there is a considerable relaxation of concern about strict sequential execution of actions on objects by allowing many messages to be outstanding at one time, there is still a strong preference for solving problems by performing actions in some sequence on objects.

2.3 Functional Programming

In functional programming (or "applicative" programming) problems are solved by pure function evaluations, without any possibility of side effects. The concepts began with work of Church on the "Lambda Calculus" in the 1930s and found a viable programming language expression in McCarthy's LISP. Iverson's APL, while not a purely applicative language (nor is LISP for that matter), allows and encourages programming in that style. ML is a more recent example of a functional programming language.

In such a language, the procedures that express functions themselves are valid data, which will be evaluated in some manner consistent with their definition, but not in a strictly determined order. Thus if we were doing a problem in surveying, and were using some lengths and the sine, cosine, and tangent of an angle to compute the boundaries of a field, we would simply write the equations for each of the boundaries in terms of the known quantities, and leave it to the computer to decide on the actual order in which to compute the sine versus the cosine versus the tangent. This functional style can be combined with concepts from other programming styles to achieve a richer language. For example, the abstract data types of object-oriented programming, with their hidden internal mechanisms for performing required actions, can be viewed as valid data types in a functional programming language and the actions recast simply as operators on data to be used in functions without side effects.

Languages for symbolic manipulation of equations are often based on one of the functional programming languages, such as LISP, since they are oriented toward evaluation of formulas.

Because the order of execution is not of concern in functional programming, this programming style may have significant application in efficient use of parallel processors.

For more background on functional programming, see Hudak (1989).

2.4 Logic Programming

We may move still one step further in trying to focus on the application rather than on the operation of computer hardware and consider only the relationships among objects of interest rather than even the implicit actions of functions. This style is called logic programming.

PROLOG was developed by A. Colmerauer in 1972. It deals with the specification of problems on a high level, as statements about objects and the relationships among them, and then allows the user to ask questions about those relationships, making the logical inferences from what is known to what has been asked. This leads to the term "logic programming." PROLOG is an appropriate language for work in artificial intelligence. However, it may be argued that any application of a computer to any problem is an exercise in artificial intelligence, and that the normal human specification of a problem in terms of objects and their relationships could be addressed directly by PROLOG without further programming.

In PROLOG, if we wish to state as a fact that "the child is father to the man," we would write

```
father (child,man).
```

In general we state a fact by putting the relationship first, before a parenthesized list of objects associated by that relationship. Questions about relationships are posed by placing "?-" before the hypothetical relationship. If it can be inferred from the known facts, PROLOG answers "yes"; otherwise "no."

If an object is named starting with a capital letter, then it is treated as a variable that may stand for known or unknown objects. In

PROLOG, when the object for which the variable stands is known it is called "instantiated." When asked a question involving an uninstantiated variable, PROLOG tries to find a match to facts that will allow it to instantiate the variable. The answer may be accepted or rejected by the user. If rejected, the attempts to match continue.

PROLOG uses a comma between facts to indicate "and," and ":-" to show that a fact depends on other facts (making a "rule"), so that we could say "Sally loves men who love dancing and dining" as

```
loves(sally,Men):- loves(Men,dancing),
  loves(Men,dining).
```

PROLOG can work with numbers (integers or real) as well as symbolic identifiers. It has the arithmetic operators + (add), − (subtract), * (multiply), / (divide), and mod (take the remainder), and the relational operators > (greater than), < (less than), >= (greater than or equal to), =< (less than or equal to), and =\= (not equal to). For more detail on programming in PROLOG, see Clocksin and Mellish (1987).

2.5 Concurrent Programming

With all concerns about evaluation as well as order of evaluation taken out of the hands of the programmer, a logic programming language has the maximum freedom to adapt the solution of a problem to parallel and distributed execution. By adopting the view that the system is in open-ended continuous operation, reacting to an environment, rather than transforming input data into output data, and by adopting the attitude that partial problem solutions are interesting, logic languages such as PROLOG have been changed into "concurrent logic" programming languages, one of which is CONCURRENT PROLOG. The approach blends naturally with object-oriented programming style in which objects send messages to other objects, and perhaps delegate them (passing them on) to yet other objects to handle when they are not understood.

These languages are not a particular improvement over imperative languages such as FORTRAN for straight numerical algorithms, since they may execute orders of magnitude more slowly, and not even produce the expected result. They are better suited to solving the challenging problems of making computers able to make humanlike decisions in reasonable times, to reason, and to display artificial intelligence. The role of such languages would then not be in the execution of a numeric algorithm but in helping with the design of the algorithm, or the design of an experiment on the basis of an expert system.

For more on this subject, see Shapiro (1989).

3. LANGUAGES FOR MAJOR SCIENTIFIC CALCULATIONS

When a problem pushes computer resources to their limits, languages that will make optimal use of those resources are needed. This is most easily achieved with languages whose structures correspond to the structures of computers, in terms of sequences of actions to be taken on appropriate data. The first and most enduring of these languages is FORTRAN. It is an evolving language with many dialects and variants, but (with minor modifications) routines written to do numeric calculations in 1959 could still be used in 1989, and most likely will still be useful in 2019. For some things, however, even with the best optimization, FORTRAN may not allow a machine to achieve maximum performance. In earlier days, an assembler would have been used to allow instructions to be written that would extract that last bit of efficiency. Today, this would more commonly be done in a FORTRAN-like language such as C.

We will also look at ALGOL and its descendant, PASCAL, which avoids some of the clumsiness of FORTRAN and which was a popular competitor to FORTRAN for a while. Many important algorithms were published in ALGOL, and an ability to read the language is important. PASCAL is used extensively to handle complex data structures.

FORTRAN is evolving and picking up features from PASCAL and other languages. It is not clear whether the resulting language will be able to retain the performance capabilities of the simpler, existing language.

If one looks carefully at FORTRAN, C, and PASCAL, it is obvious that almost anything one might wish to program could be programmed in any of the three, and with almost equal ease. Indeed, aside from the "syntactic noise" of different statement delimiters, programs in the three languages can be hard to distinguish, and the same well-written codes can be, and often are, transliterated from one to another of the languages to good effect. The significant differences among these languages lie in two areas: ability to help detect errors in writing a program, and what we may call "depth of context."

In any language it is possible to write a program that is expressed correctly in that language, but that does not do what we had intended. PASCAL provides very strong cross-checks to make such errors less common than in FORTRAN or C. However, it is precisely those cross-checks that often make it difficult to use PASCAL for applications that must efficiently push the limits of a computer, as in device drivers for experiment control, which would be more likely to be written in FORTRAN and C.

C, however, shares common ground with PASCAL in allowing a depth of context lacking in FORTRAN. In FORTRAN, the environment of definitions within which we work is usually on one level. Everything about the definition of a name in FORTRAN as a particular type of data structure is visible within the routine manipulating it, because in most dialects of FORTRAN it is not possible to nest complete routines, with their definitions, within other routines. Such nesting can be done for both procedure and data structure definitions in PASCAL and for data structure definitions in C. This allows some of the definitions being used to be brought much closer to the point where they will be used, allowing a disciplined programmer to keep the context that someone reading the code needs to understand to a page or two, but also allows an undisciplined programmer to create a complex hierarchy of contexts, within the depths of which one can get lost.

The difference between approaches is similar to differences between English and German. In normal English, sentences work on a small number of levels of structure; while in German, a single sentence may go on for a paragraph, with complex nested subordinate clauses, and verbs gathered far away from the rest of the sentence. FORTRAN encourages a style familiar to English speakers, while C and PASCAL encourage a more Germanic style of coding. It must be emphasized, however, that such stylistic biases need have little effect on most programs, since the common practice now is to write most programs as sets of independently compiled routines, each of which is so small as not to be likely to make use of many levels of internal structure, providing a depth of context in the chain of external calls, rather than in internal blocks. This approach can be followed in any of these languages, and produces code that is easy to move from one language to another.

3.1 FORTRAN

The family of FORTRAN languages has a simple, imperative style. These languages work with data as integer, real, double-precision real, complex, or logical, and arrays of up to three or more subscripts of such data types. They form expressions by applying the arithmetic operators + (addition), − (subtraction or negation), * (multiplication), / (division), or ** (exponentiation) to numeric constants, variables, array elements (elements of vectors or matrices), and function calls (values returned by other pieces of code that evaluate the value of a function of some given arguments), and assign the values of those expressions to variables or array elements.

Though many variations are allowed on different systems, the following rather restrictive approach is valid on almost all systems. This rigid format is relaxed in many modern dialects, to make typing of programs from terminals simpler and to make it easier to read programs.

A program consists of statements.

Each statement is formatted as a sequence of fixed-length lines (originally intended as the images of 80-column punch cards), with four fixed fields:

Columns 1–5 may contain a number (a nonzero integer in some machine-dependent range, but rarely restricted to less than 32767) to be used as a statement label or may indicate a comment by a "C" in column 1.

Column 6 may contain a continuation mark (not blank, and not the digit zero), indicating that the line on which it appears contains a continuation of the statement that was on the prior line.

Columns 7–71 contain the body of the statement itself.

Columns 72–80 may contain an optional comment (often used for a line sequence number).

The only valid characters that may be used are blanks, the uppercase letters, the digits 0 through 9, the special symbols "=", "+", "−", "*", "/", "(", ")", ",", ".", "$", and, on all except the oldest systems, the single quote " '". Most newer systems now allow the use of lowercase letters and additional special symbols, but a program written in the limited "FORTRAN character set" is portable from machine to machine.

Blanks and leading zeros are not significant in statement numbers, and, with minor exceptions, blanks are not significant within a statement and may be introduced or removed as needed to make a program more readable.

A program consists of its main-line code (the "main program") and whatever subroutines and functions it may need to perform its functions. Each such routine consists of a list of statements, the first group of which (the declarations and other nonexecutable statements) establish the environment within which the following executable statements will work.

The main program is distinguished from the other routines by not having a special header, or, in some dialects, starting with a "PROGRAM" statement. Thus, the general structure of a FORTRAN program would be

PROGRAM name
declarations and other nonexecutable statements
executable statements
END
SUBROUTINE name (argument,argument,...,argument)
declarations and other nonexecutable statements
executable statements
END
FUNCTION name (argument,argument, ...,argument)
declarations and other nonexecutable statements
executable statements
END
⋮

continuing with as many subroutines and functions as may be needed in any order.

Logically, these routines may be intended to fit into a multilevel hierarchical structure, but visually there are only two levels: the main program and the other routines. This simple structure may cause visual confusion in large systems with many routines.

Variables in FORTRAN are named with one to six letters and/or digits, but the name of a variable must start with a letter. (More than six characters are allowed in a variable name in most modern dialects.) If the first letter of the name of a variable is "I", "J", "K", "L", "M", or "N", it is assumed to be an integer unless declared otherwise. It is assumed to be (single precision) real for the other letters. Use of variables of other types requires an explicit declaration. The implicit typing of the most commonly used data by the first letter of its name is a distinctive feature of FORTRAN, significantly reducing the need for declarations of variables, but also removing an opportunity for the compiler to check the declared use of a variable against the ways in which it is used by a program.

Arrays are indicated by a name formed according to the rules for variable names followed by a parenthesized list of up to three (or in some dialects more) integer indices. The set of indices picks which element of an array is of interest, as with XX(1) or XX(3) to select the first or third elements of one-dimensional array XX, or ZZ(3,4) to pick the fourth element of the third row of a two-dimensional array. Functions are referenced in the same way as arrays, except that the parenthesized list contains arguments of the function. The arguments are used in some internal calculation of the resulting "return value" of the function and, unlike array indices, need not be integers. The objects upon which the language acts may be constants, variables, array elements, the values returned by function calls, and, via special input-output statements, values found in files of data stored on disks, tapes, or other external media.

The constants and variables are of type INTEGER (whole numbers), type REAL (numbers with decimal points, usually floating-point numbers in scientific notation), type DOUBLE PRECISION (real numbers of higher precision, often double the precision, than those of type REAL), type COMPLEX, and type LOGICAL (having values ".TRUE." or ".FALSE.")

There is also a way to handle characters. Early FORTRAN compilers supported character strings in a very limited way, given as "nnHaaaaaaaaaa", where the "nn" was a count of the characters "aaaaaaaaaa" that followed. Newer compilers, since the mid-1970s, have supported a type CHARACTER, quantities that may have values that are character strings of a known length, given as "'aaaaaaa'", i.e., strings delimited by single quotation marks.

In FORTRAN, files of data are an important construct, but the language does not handle them like other data types. Considering all types except files, the actions that may be taken are the following:

Assignment. A statement of the form destination = expression, where the destination is a variable or array element, first computes the value of the expression and then stores it into the location implied by the variable or array element. If the expression is of a different type than the destination, a conversion of type is done before the value is stored. Thus, the statement

```
I=X+2
```

is a directive to fetch the contents of the variable "X", add the constant "2", and store the result converted by truncation to an integer into the variable "I". If "X" had contained the value 3.6, then, following the execution of this assignment statement, "I" would contain the value 5.

Arithmetic Operators. "+" for addition, "−" for unary negation or subtraction, "*" for multiplication, "/" for division, and "**" for exponentiation. The values being combined may be numbers of type integer, real, double precision real, or complex. In the strictest versions of FORTRAN, the objects being added (or subtracted or multiplied, divided or exponentiated) must be of the same type (or at least compatible types as with complex multiplied by real or real taken to an integer power), but more general mixed-mode addition is allowed in newer variants.

Relational Operators. For integers, reals, and double precision reals, under the same compatible mode restrictions as subtraction, numbers may be compared to produce a type logical result as follows:

".GT.": greater than
".GE.": greater than or equal to
".LT.": less than
".LE.": less than or equal to
".EQ.": exactly equal to
".NE.": not exactly equal to

Caution is needed in comparing type REAL values, since the reduced internal precision of calculations on a computer may make tests for exact equality fail unexpectedly.

Logical Operators. Expressions of type logical may be combined using the operators ".NOT.", ".AND.", and ".OR.". Some dialects define additional logical operators, but these three are sufficient to allow the most general logical inferences by forming expressions [e.g., .NOT.(A .AND. B) is the "nor" of A and B].

Expressions are formed by taking sequences of constants, variables, array elements, and function references separated by operators and grouped by balanced pairs of parentheses. Evaluation works from the innermost level of function calls and parenthesized groups out, subject to a hierarchy of operators. First all the arguments of function references within the expression are evaluated, and then the functions are evaluated on those arguments. Then parenthesized expressions are evaluated from the innermost level out. Operators are applied in the following order of precedence (highest on the list first):

"**"
"*" and "/"
"+" and "−"
".GT.", ".GE.", ".LT.", ".LE.", ".EQ.", and ".NE."
".NOT."
".AND."
".OR."

so that X/Y/Z+D−E+F is the same as (X/(Y*Z))+(D+F)−E, with the compiler having the right to reorder the evaluation to an algebraically equivalent expression and to evaluate common subexpressions fewer times than they actually appear.

In addition to the assignment statement, FORTRAN includes the following executable statements that control the sequence of execution:

The GOTO statement. Transfers control to the indicated statement, as in "GOTO nnnn", where "nnnn" is a statement label in the same routine, which will be the next statement to be executed. There are two additional GOTO statements, the "computed" GOTO and the "assigned" GOTO, which allow the choice of destination to be decided within the program while running, instead of being chosen beforehand.

The arithmetic IF statement. Transfers control conditionally, as in "IF (expression) nneg, nzero,npos", where the numeric value of "expression" is computed, and then control passes to the statement with label "nneg" if the value of the expression is negative, the statement with label "nzero" if the result is zero, and to the statement with label "npos" if the value of the expression is nonzero positive.

The logical IF statement. Executes a statement conditionally, as in "IF (expression) statement", where the logical expression is evaluated, and the statement is executed if the result is .TRUE.

The block IF statement. Similar to the logical IF statement, except, of the form:

```
IF (expression) THEN
   statements to execute if the expression
     is .TRUE.
ELSE
   statements to execute if the statement
     is .FALSE.
ENDIF
```

allowing most GOTOs to be avoided.

The DO statement. Executes a block of code repetitively, incrementing a control variable on each pass. Originally of the form

```
          DO nnnn ivar = ilow,ihi,istep
             statements
nnnn last statement of the loop
```

where the variable "ivar" would start at the value "ilow", and be incremented by the value "istep" (by 1 if "istep" was not specified) until it reached or exceeded the value "ihi". The statements between the DO statement and the statement with statement label "nnnn" would be executed for each value assigned to "ivar" between "ilow" and "ihi". In earlier dialects the statements would be executed at least once (for ivar equal to ilow), even if ihi was less than ilow. This is not done in more modern dialects.

In some dialects this is now written as

```
DO ivar = ilow,ihi,istep
   statements
ENDDO
```

avoiding the need to use a statement label.

The CONTINUE statement. A do-nothing statement convenient for attaching statement labels for the end of DO-loops.

The CALL statement. Transfers control to a subroutine, as in "CALL name(arg1,arg2, arg3,...)" or just "CALL name".

The RETURN statement. Returns from a subroutine to the statement just after the CALL.

The STOP statement. Terminates execution of the program.

The PAUSE statement. Temporarily suspends execution of the program. This statement is not appropriate in a batch environment.

Files are manipulated by a variety of statements, such as "OPEN", "CLOSE", "READ", "WRITE", "REWIND", and "BACKSPACE". Files may include complex formatting of data to and from external representations, either using separate "FORMAT" statements to hold the descriptions of the various fields or including appropriate character strings in READ and WRITE statements. The environment within which the executable statements are executed is established by the nonexecutable statements:

The COMMON statement. Declares that the variables or arrays listed should be made available to other routines listing the same COMMON statement. Otherwise variables and arrays are local to each routine. There may be more than one common area, distinguished by using labels conforming to the rules for variable names. The names for the same variables in different routines need not be the same. The shared use is determined by their physical position within the list of variables, and their types.

The EQUIVALENCE statement. Declares that particular variables or arrays will share the same storage locations. Originally intended to economize on memory when it was a scarce resource, it has come to be used as a way to view the same bit patterns as being of different types.

The DIMENSION and TYPE statements. These statements declare the types of both variables and arrays, and the size of arrays (by listing the maximum values of each of the indices) or, in the case of the DIMENSION statement, just the size of arrays.

FORTRAN includes a large library of built-in functions, which the programmer may replace or add to with his own routines.

For more on FORTRAN, see ANSI Committee No. X3J3 (1989).

3.2 C

C is a language that even more closely reflects the capabilities of the machine being used than does FORTRAN, allowing bit manipulations and use of internal registers so that an assembler need not be used. C normally produces well-optimized code. The syntax avoids much of the card-oriented clumsiness of FORTRAN, using some of the style of ALGOL-like languages (see below).

A C program consists of statements terminated by semicolons, made into nested blocks by "{" and "}" (the equivalent of "begin" and "end" in ALGOL, described below). The objects that may be manipulated in C are built from three precisions of integer, which may also be unsigned for use as bit patterns, from two precisions of real, and from data of type "character." The types of all variables must be declared. Arrays are declared by using brackets with the range of the index (the quantity that selects the array element desired), e.g., "int xx[10];", which has essentially the same meaning as the FORTRAN "DIMENSION xx(10)". Users may define types. Complex data structures may be created by listing the field names with their types in braces in the "struct" declaration. A "union" allows each field to be mapped to the same locations, acting somewhat like an "EQUIVALENCE" statement in FORTRAN. Bit fields can be given field names in structures.

In FORTRAN, most access to data was explicitly by variable name, but a limited ability to point to particular data items in a larger set of data items was implicitly provided by array indices. C allows for the explicit declaration of pointers to objects (including pointers to pointers, etc.) and extraction of the machine address of objects, and types of objects may be explicitly ignored. This flexibility helps in allowing C programs to be used in place of assembler code, but means that a valid C compiler may provide little protection against programming errors.

C names variables as strings of letters and digits, beginning with a letter, but, unlike FORTRAN, accepts both upper- and lower-case letters, and (in most dialects) treats them as distinct. Only the first eight characters of a C variable name are significant in some dialects.

C describes the objects it manipulates in terms of "storage class" and type. Storage class declarations allow the programmer to direct that a variable will claim storage only during the invocation of the block that uses it ("automatic"), or that the variable will have a life extending past that of any particular invocation of the block using it ("static"), or that the variable has a life external to the entire program ("external"). Variables may also be declared as being best stored in fast machine registers, if possible ("register").

The types are

1. short int, int, long int. The representation of three integer types, intended to provide access to the integer sizes natural to the machine.
2. unsigned int. The representation of an integer without a sign bit, obeying the arithmetic rules of "2's complement arithmetic" commonly used for manipulating machine addresses.
3. float. The representation of a real number. The format of type float constants is similar to that used in FORTRAN, except that all such constants are double precision.
4. double. The representation of a possibly higher precision real number than provided by type float.
5. char. The representation of characters appropriate to the machine. A character constant is given as a single character in single quote marks or by quoting special sequences starting with a backslash:

\n	newline (or line-feed)
\t	horizontal tab
\b	backspace
\r	carriage return
\f	form feed
\\	backslash
\'	single quote
\"	double quote
\nnn	the bit pattern given by the 1–3 octal digits nnn.

Character strings are given as "array of char" strings in double quote marks. A zero character ('\0') is appended to string constants to help programs scan for the end of a string.

C also provides an enumerated data type, in the style of PASCAL (see below).

C has FORTRAN- and ALGOL-like expression evaluation, with the addition of a combinedassignment and increment or decrement (a concept from ALGOL-68). The operator "=" is used for assignment. It is just another operator in C, not a special statement as in FORTRAN or PASCAL. The form "xx += 1" is the same as "xx = xx + 1", and "xx −= 1" is the same as "xx = xx −1". Other appropriate operators may be combined with assignment in the same way.

Another special feature of C is the use of "++xx" to mean increment xx before using it, and "xx++" to mean return the value of xx, and then increment it.

C has an even more complex order of precedence for the actions of its operators than does FORTRAN. Parentheses are recommended.

The operators available in C are

1. Unary operators. Indirection ("*") to retrieve the value pointed to by the object, address of ("&") to obtain a pointer to the object, arithmetic negation ("−"), logical negation ("!"), ones' complement ("~"), increment ("++"), decrement ("−−"), type cast (specified by a type-name in parentheses), size in bytes ("sizeof").
2. Arithmetic operators. The operator "*" is used for multiplication as in FORTRAN; the operator "/" is used for division as in FORTRAN; the operator "%" is used to get the remainder; the operator "+" indicates addition; the operator "−" indicates subtraction.
3. Shift operators. Left shift with zero fill ("<<"), right shift (">>"). The right shift has zero fill for unsigned expressions, but may be an arithmetic right shift for signed expressions.
4. Relational operators. Less than ("<"), greater than (">"), less than or equal to ("<="), greater than or equal to (">="), equal to ("=="), not equal to ("!=").
5. Bit-by-bit operations. And ("&"), exclusive or ("^"), inclusive or ("|").
6. Logical operations. And ("&&"), or ("||").
7. Conditionals. The question mark and colon are used to form an if-then-else operation, as in

   e1 ? e2 : e3

   which returns e2 if e1 is nonzero (i.e., true), and returns e3 otherwise.
8. Assignment operations. Simple assignment ("="), other operations (+, −, *, /, %, >>, <<, &, ^, |) combined with assignment in the form "op"= [e.g., x1+=x2 is equivalent to x1 = (x1) + x2].
9. Comma operation. The comma causes left-to-right evaluation of its arguments and returns the rightmost value.

Operators are applied in the following order of precedence (highest on the list first):

unary operations: * & − ! ~ ++ −− type-cast sizeof

multiplicative operations: * / %

additive operations: + −

shift operations: << >>

ordered relational operations: < > <= >=

equality relational operations: == !=

bit-by-bit and: &

bit-by-bit exclusive or: ^

bit-by-bit inclusive or: |
logical and: &&

logical or: ||

conditional: ... ? ... : ...

assignment = += −= *= /= %= >>= <<= &= ^= |=

comma

Control of the sequence of execution is provided by the following statements allowing compound statements, grouped by braces "{" and "}" with declarations private to the block of statements given after the opening brace, and statements similar to the FORTRAN "GOTO", "LOGICAL IF", and "DO", in this case "goto", "if", and "while", "do...while", and "for" statements.

Additional statements that improve the clarity of C code over FORTRAN in some cases are the following:

1. The "switch" statement. This statement executes one of several statements selected by the value of an expression. It is of the form

```
switch(expression){
   case c1 : s1;
   case c2 : s2;
      ...
   case cn : sn;
   default : sd
};
```

where c1, c2, ..., cn are integer constant expressions. The expression is evaluated and control transferred to the statement listed after the matching constant expression, or to the default if no match is made.

2. The "break" statement. This statement terminates the innermost repeated execution statement or switch statement, avoiding the need for some goto statements.
3. The "continue" statement. This statement ends the current pass of the innermost repeated execution statement, but allows subsequent passes, if any.
4. The "return" statement. This statement is used to return a value from a function and continue execution in the statement where the call was made.

C includes a preprocessor that translates lines beginning with the symbol "#" to allow replacement of identifiers by strings ("#define"), inclusion of lines from other source files ("#include"), and conditional compilation of lines ("#if", "#ifdef", "#ifndef").

Because of its intimate connection with major operating systems development, there is a very extensive library of functions for use with C programs for file manipulations and other useful activities. The libraries used in support of other languages often are written in C.

For more on the C language, see Kernighan and Ritchie (1978).

3.3 ALGOL and PASCAL

ALGOL is a FORTRAN-like language with a less card-oriented, more readable syntax. The major difference from FORTRAN is structuring into nested blocks, allowing a hierarchy of procedure declarations, and a more rigid requirement that the types of all variables be specified, rather than allowing the default integer and real declarations of FORTRAN. The current descendant of ALGOL in heaviest use is PASCAL.

Let us first look briefly at ALGOL syntax. A program is composed of statements terminated by semicolons, organized into at least one BEGIN-END block:

```
BEGIN
   statements declaring the types of variables
   executable statements and procedures
   comments
END
```

Procedures follow the same structure as the outer level after a statement of the form

```
PROCEDURE name;
```

immediately followed by declarations of variables to be used only in that procedure.

On any level the types of the objects to be used are declared by statements beginning "REAL," "INTEGER," or "BOOLEAN," possibly modified by the word "ARRAY" to indicate an array. In FORTRAN each array index implicitly starts with 1. In C indexing starts with 0. In ALGOL, a lower and an upper value for each index are given to allow indexing to start from other values.

The actions that can be taken on these objects are similar to those allowed in FORTRAN, but slightly different symbols are used, depending on the characters available on the particular computer. The FORTRAN "**" for exponentiation can be replaced by an up-arrow (often printing as "^"), and the relational operators by their usual mathematical symbols; the assignment operator is indicated by ":=" to distinguish it from the test for equality. Expression evaluation is similar in algebraic style.

Labels are symbolic, rather than numeric as in FORTRAN, and delimited by a ":", so that no special field needs to be reserved for them. There are the equivalent of the FORTRAN logical IF, the block IF, the GOTO, and the computed GOTO in ALGOL.

Only minimal file handling was specified in the language and varied from system to system until later languages, such as PASCAL.

PASCAL (see Jensen and Wirth, 1974; Wilson and Addyman, 1982) is sometimes referred to as "FORTRAN with semicolons," mixing some of the style of FORTRAN with that of ALGOL. Programs are written as sequences of statements in an imperative style, and formulas and equations translate into PASCAL similarly to the way it is done in FORTRAN and ALGOL. However, PASCAL provides a more natural set of data structures and control structures than was available in FORTRAN at the time PASCAL was introduced, which allowed the writing of cleaner, more comprehensible code, especially when dealing with problems that involved more than simple iteration of the evaluation of mathematical formulas.

The objects that can be used in PASCAL are constants or variables, and variables must have a "type," i.e., the attributes of the values and the legitimate range of values that may be assumed by a variable must be specified. The language starts by defining "simple" types, then provides "structure" types that may be built out of simple types, and "pointer" types that facilitate linking of data together. The four simple types are as follows:

integer-type. A subset of the whole numbers (..., −3, −2, −1, 0, 1, 2, ...).

real-type. A subset of the real numbers (such as 3.141 59, −6 483 221., 0.000 03, etc.).

char-type. The characters that can be handled on that machine, including the upper- and lowercase letters, and the decimal digits, with alphabetic and numeric ordering preserved.

Boolean-type. The logical values "true" and "false."

Except for the real-type, the simple types can all be ordered by integer values, and are called "ordinal-types."

The user can create new types. One can simply enumerate the values that may be taken on by a new type in their appropriate ordering ("enumerated-types"), or pick a range of values out of an existing type ("subrange-types"), or make more complex structures out of existing types. Structure-types may be packed for storage efficiency, or unpacked.

The possible "structure-types" are as follows:

array-type. Vectors ("tuples") of existing types specified by an index set. Because the elements of an array-type may themselves be of array type, matrices may be represented, with the usual comma notation to separate the indices.

record-type. A structure consisting of groups of named fields, where (unlike array elements, each of which must be of the same type) fields may be of different types. This ability to mix types allows one to, for example, create strings of characters with a field of integer-type giving the length and a field or array of char-type holding the characters. Unlike arrays, however, the value of a field identifier cannot be computed. It is just a name. Different record structures ("variants") may be used in a given record-type, determined by the value of a tag field used to select the variant to be used.

set-type. A set composed of elements all of the same type (the "base-type"), which must be an ordinal type. A PASCAL implementation imposes a limit on the cardinality of the base type, usually somewhere between 48 and 256.

file-type. A structure consisting of a sequence of values of a given component type. Such structures provide PASCAL with an input-output mechanism.

In order to facilitate working with lists that may grow and shrink, pointer types are defined:

pointer-type: Either "nil" (i.e., not pointing to anything) or having a value that identifies (points to) data of a type (the "domain-type") associated with the pointer.

This rich structure of types allows the user to build up complex data structures as the fundamental objects of the language. Objects and actions are used to create expressions. PASCAL does not have the same finely detailed precedence of actions seen in FORTRAN, and requires more use of parentheses to avoid ambiguities. The objects and expressions are used in invocations of functions (that return values) and procedures (that do not return values) and in assignment statements. Control over the sequence of execution of statements is provided by a variety of control structures, similar to those discussed for FORTRAN and C. The language allows portions of code and appropriate data structure to be packaged as procedures and functions, and for them to recursively call themselves.

A PASCAL program to add ten numbers and print the result might be written as:

```
program addnum (input,output);
var myinput,count,sum:integer;
begin
   sum := 0;
   for count := 1 to 10 do
   begin
      readln (myinput);
      sum := sum + myinput;
   end;
    writeln (sum)
end.
```

As with our other examples of this problem, we have ignored the niceties of what the program could (or should) do if there are not ten numbers to read, or if there is an error in reading.

For more on PASCAL, see Wilson and Addyman (1982) and Feuer and Gehani (1982).

3.4 Libraries and Language Extensions

Each application of computers and computer languages tends to develop a specialized set of language extensions and libraries appropriate to problems commonly encountered. There are extensive and well-tested libraries of mathematical routines for solving linear equations, differential equations, etc. By imposing a style of programming, of use of data structures, and file formats, these libraries extend the base languages into jargons appropriate to their applications. To pick one example among many, consider the "Basic Linear Algebra Subprograms" (BLAS) by Dongarra *et al.* (1988). This package provides routines for the manipulation of matrices and vectors, using a clear consistent style. The first letter of a routine specifies the precision ("S" for single precision real, "D" for double precision, etc.). The next two letters classify the kind of matrices being handled ("GE" for general, "HE" for Hermitian, "TR" for triangular, etc.), and the next two the operation ("MV" for matrix-vector product, etc.). Storage conventions are specified for triangular matrices.

By adhering to such conventions, the user not only gains the ability to use the routines in the library involved, but also the ability more efficiently to produce new programs that achieve clarity and maintainability by following known conventions; i.e., by speaking a known language.

4. LANGUAGES FOR THE LABORATORY ENVIRONMENT

Computers have become an integral part of the laboratory environment. They control the operation of instruments, move the data from instruments to storage, analyze the data, compare it to models, display it, and archive it. There is a wide range of specialized languages, packages, and systems for this purpose. In some cases the computer aspects are so well integrated with a set of instruments that the user need not be concerned with programming. The system functions as a black box. In other cases, detailed programming is required. In the latter cases, the choice of an appropriate language may be forced by considerations of what languages, if any, may be available for the computers involved. However, when the choice is unconstrained by such questions, typical choices are FORTRAN, PASCAL, C, Assembler, BASIC, and FORTH.

What drives such choices are questions of time—either of the computer or of the person using it. Compared to the demands of some modern experiment designs, computers are slow and can have difficulty moving all the data involved without falling seriously behind. In such cases, highly efficient code, or at least code of highly predictable timing, may be required. Languages oriented toward the automatic management of complex data structures can place demands on the computer for activities such as garbage collection, for which there simply may not be time. In the most extreme cases, use of C or an assembler that can give the maximum control over use of machine resources may be required. Since FORTRAN has a direct and static mapping of its data structures to machine implementation, and has benefitted from decades of effort at optimization of the code it produces, it can often provide almost the same efficiency. PASCAL has been used as a cleaner, simpler language for such applications, but in time-critical PASCAL work has not been as popular as mixtures of FORTRAN with C and assembler.

If, however, the primary consideration is time of the person, then interpretive languages such as BASIC and FORTH become attractive. The speed lost in interpretation with BASIC can be recovered by taking the final version and compiling it. FORTH manages to achieve good speeds by providing what is called threaded code (see Sec. 4.2) and by allowing embedded assembler code use.

The creation of device control code in FORTRAN or PASCAL is difficult because the languages themselves do not provide intimate access to hardware, nor is the timing of code execution easily predicted. They also do not deal directly with asynchronous parallel processes. They do, however, provide a comfortable environment for the creation of data structures, management of external files, and numeric data analysis. Thus a common structure of laboratory code is an outer layer of control logic in FORTRAN or PASCAL, using time-critical services provided by C or assembler subroutines.

Efficient communication among the pieces is essential. For large streams of data, the structure most commonly used is the "circular buffer" or "circular queue." In this structure, two pointers are used, each advancing from the bottom of the structure to the top, and then returning to the bottom to scan the structure again. One of these pointers designates the next location into which data are to be stored, and the other the next location from which data are to be taken. As the pointers chase each other, the gap between them may grow and shrink to deal with the vagaries of timing of the arrival of data and its processing, thereby compensating for the unknown timing of the higher-level language code.

When timing is less critical or ease of modification is the main issue, an interpretive language such as BASIC or FORTH may be used for laboratory work.

4.1 BASIC

There are many dialects of BASIC. The language was developed by Kemeny and Kurtz as a simplified FORTRAN and ALGOL derivative appropriate for students. It is most often used as an interpretive language on time-sharing systems and small computers. Programs consist of sequences of numbered lines. The variables used may be restricted to single-letter names, but it is more common to allow at least a letter and a digit or, in most recent versions, full strings of letters and digits starting with a letter. The types of variables may be either type real or type string, and most current versions allow additional types. The types are usually designated by an extra symbol appended to the name of the variable:

real single precision	!	(or nothing as the default)
integer	%	
double-precision real	#	
string	$	
long integer	&	(on some machines)

Arrays are indicated by a variable name followed by subscript indices in parentheses, and are declared by a "DIM" statement.

Expression evaluation is FORTRAN-like, using the following operations:

1. Arithmetic operations. Addition ("+"), subtraction ("−"), multiplication ("*"), division ("/"), exponentiation ("^").
2. Relational operations. Less than ("<"), less than or equal to ("<="), greater than (">"), greater than or equal to ("=>"), equal to ("="), not equal to ("<>").
3. Logical operations. "AND", "OR", "NOT".

The flow of program control is keyed to the numbers put on each line:

The "GOTO" statement transfers control to the designated line number.

The "GOSUB" statement transfers control to the designated line number. That line and following lines are executed as a subroutine to be executed until a "RETURN" statement. The "RETURN" statement transfers control back to the place immediately after the "GOSUB".

The "FOR" statement causes repeated execution of the following statements in statement number order until a matching "NEXT" statement.

The "IF" statement, of the form "IF expression THEN statement" or "IF expression THEN statement ELSE statement," evaluates the logical expression and executes the statement after the "THEN" if the expression is true.

The assignment statement is as in FORTRAN, except some versions require the word "LET" before the statement.

Files are manipulated by INPUT and PRINT statements.

For device control, many versions of BASIC add an "INP" statement to read device registers and an "OUT" statement to store new values into device registers.

In BASIC the example of adding 10 numbers might read

```
100  SUM = 0
110  FOR COUNT%= 1 TO 10
120     INPUT MYINP
130     SUM = SUM + MYINP
140  NEXT COUNT%
150  PRINT SUM
160  END
```

There are compiled, batch versions of BASIC, some of which avoid the need for statement numbers on every line and which allow the use of symbolic statement labels.

For more on BASIC, see Nolan (1974).

4.2 FORTH

FORTH is a simple stack-oriented, postfix notation language, with a highly structured programming style oriented toward use of small computers. A typical FORTH program is built up from function definitions that are used to create higher-level function definitions. The highest level created is invoked as the solution to the problem. Most implementations of the language make use of a programming technique called "threaded code," which is particularly effective in achieving a compromise between the ease of modification of an interpretive language, and the efficiency of execution of a compiled language.

In threaded code, we create a virtual machine for our application as a set of subroutine calls, and use lists of those calls as the instructions in the virtual machine. Suppose we start with a machine that has only integer arithmetic. Let us define a routine to do floating-point addition, and store it at, say, location 100, and let us define a routine to print a floating-point number and store it at location 200. To keep our machine simple, let us address all the objects we care about by a pointer, called a "stack pointer," which acts like a spring-loaded pile of plates at a buffet. We can only see the top item, the item to which we currently point. We can remove that item ("pop" it off the stack) by changing the pointer by one data object, and place a new item on top ("push" it onto the stack) by changing the pointer in the other direction. Then we can keep our new language simple, by having the addition routine add the two items at the top of the stack, removing both and replacing them by the sum, and having the print routine print what is on top of the stack. A routine to add and print then would just be the sequence 100, 200. This is a compact representation of a complex action. This routine can itself be stored and used as an extension to the language.

For more on FORTH, see Brodie (1981).

5. LANGUAGES FOR GRAPHICS

Computers are a useful tool for capturing, drawing, and modifying images. Their use in such applications increases as raw computing power and storage capacity increase, allowing manipulation of increasingly complex images.

Languages for computer graphics are implemented as extensions to more general languages. The objects to be manipulated are images and the pieces of images, such as lines and circles, as well as the devices that draw and display images, such as pens and graphics screens. A language for graphics must deal both with what is being drawn and the way in which it is being drawn.

The most elementary view is that of a pen being dragged along a piece of paper, as in the language LOGO (Abelson, 1982). The paradigm in LOGO is that of a turtle pointed in a direction, carrying a pen, which may be up or down. The pen is placed down by the command "PENDOWN" and raised by the command "PENUP". A stroke is made in the current direction or its reverse by the commands "FORWARD nnn" and "BACK nnn", where nnn is the length of the stroke. The direction of strokes is changed by "RIGHT ddd" and "LEFT ddd", which specify a rotation in degrees.

Similar minimal graphics can and have been added as libraries or language extensions of most procedural languages, providing at least a vector-drawing capability. However, management of images as objects is simplified when issues of scaling and presentation location are separated from the object definitions themselves by the establishment of windows and viewports. Three-dimensional graphics in particular benefits from separation of the definition of the object from the mechanism of viewing the object, so that orientation, hidden-line removal, ray tracing, and other considerations in looking at the object do not require redefinition of the object. There are many contending approaches to presentation of images, including GKS and PHIGS, with rapid changes being driven by rapid advances in the ability of computers to present increasingly detailed images in real time.

For more information, see COMPUTER GRAPHICS.

6. LANGUAGES FOR DATA RETRIEVAL

Computers manage information. They provide the tools to scan large amounts of information and extract what is needed. Data retrieval works with objects that are identified by their contents, rather than by their physical location. The contents may include explicit keys to help in the search or may stand by themselves.

The paradigm now most used for database languages is one in which data are considered as a set of tables, the columns of which identify attributes of data, and the rows of which are the data items of interest. Some column, or set of columns, serves as the key to identify uniquely the data items to be retrieved.

The operators used are union (combining the rows of two tables with matching columns to form one with the combined information from both), intersection (combining to retain only the common information from both), difference (combining to obtain information in the first table that is not in the second), Cartesian product (combining tables with different columns to make all possible combinations), select (extracting selected rows from a table), project (extracting selected columns from a table), inner join (combining tables by matching entries in a common column), outer join (like inner join, but including rows with no match), and divide (extracting particular rows and columns on the basis of a match in other columns).

A popular language for data retrieval is SQL (Structured Query Language). The typical command in this language has the form

SELECT	the names of columns
FROM	the name of a table
WHERE	the criteria to be satisfied in selecting
ORDER BY	the order of presentation of the columns

The criteria may include relational, arithmetic, and pattern-matching operators.

There are some difficulties in applying this paradigm to data that have search criteria based on considerations other than positions of values within tables. For example, a search on the shape of particular images, or on the symmetries of lattices involving experimental error, may strain this paradigm, and may require the explicit creation of inverted indices organized on the basis of the anticipated search.

For more information, see COMPUTER DATA BASES.

7. SUMMARY

In order to save the effort of writing a new and different program for each machine to solve the same problems, to make computers accessible to more people, and to increase the chance of writing instructions for computers correctly, many programming languages for computers have been created, languages that express the actions to be performed in terms more comprehensible to people, and in ways that allow the same programs to be used on many different types of computers. Some languages are close in style to the operation of computers, consisting of sequences of actions to be taken on objects. Such a style leads to efficient translation of a problem specification into the internal language of a computer, but may force a strained interpretation of the original problem, increasing the chances of programming errors, and increasing program maintenance costs. To reflect more naturally the demands of the problems to be solved, rather than the demands of computers, languages using other styles of programming have been developed: object-oriented, functional, and logic programming. These offer promise of increased productivity for appropriate problems, but for the classic problems of theoretical and experimental physics, FORTRAN and its statement-oriented, imperative relatives predominate.

ACKNOWLEDGMENTS

We wish to thank Frances C. Bernstein, Daniel J. Bernstein, and Lawrence C. Andrews for their helpful comments and suggestions.

GLOSSARY

Assembler: A translator for a symbolic representation of a computer machine language.

Compiler: A translator for a computer programming language that converts an entire program, or major segments of it, as a unit to machine language.

Constant: Datum that does not change.

Function: A self-contained portion of code that returns a single value computed from the parameters given in invoking the function.

Functional Programming: Programming in terms of function evaluations.

Imperative Programming: Programming in terms of a sequence of actions to be taken.

Infix Notation: Programming notation in which an operator is given between the objects upon which it acts.

Instruction: An elementary action that a computer may take, or that a computer programming language defines.

Interpreter: A translator for a computer programming language that converts small portions of a program to machine language immediately.

Logic Programming: Programming in terms of the relationships among objects.

Main Program: The outermost layer of code, which starts and controls the execution of code to solve a problem. It may invoke other routines to solve the problem, in which case control would eventually normally return to the main program, or it may pass on control to some other main program, without the expectation of the return of control.

Object-Oriented Programming: Programming in terms of objects that take actions on the basis of messages passed to them.

Operator: An elementary action on objects defined in a computer language. If there is only one object, the operator is called "unary"; if there are two, "binary," etc.

Order: An elementary action a computer may take; an instruction.

Prefix Notation: Programming notation in which an operator is given before the objects upon which it acts.

Procedure: A routine. See below.

Postfix Notation: Programming notation in which an operator is given after the objects upon which it acts.

Routine: A self-contained portion of code used by other code. It may be a main program, a function returning a value, or a subroutine performing an arbitrary calculation. It may invoke other routines.

Side Effect: A nonobvious change to the state of data structures resulting from the invocation of a function, subroutine, or other procedure.

Statement: The specification of an action to be taken (an executable statement) or the environment (a declaration) in which actions are to be taken in a programming language. It is usually built upon objects and actions, and used in turn to build routines.

Subprogram: A subroutine. See below.

Subroutine: A self-contained portion of code that performs a calculation that may depend on parameters given in invoking the function.

Variable: An identifier for data that may change in the course of execution of a program.

List of Works Cited

Abelson, H. (1982), *Byte* **7**(8), 88–115.

Ahuja, S., Carriero, N., Gelernter, D. (1986), *Computer* **19**, 26–34.

Alliant (1988), *FX/Fortran Programmer's Handbook*, Littleton: Alliant Computer Systems Corp.

ANSI Committee No. X3J3 (1989), *American National Standard for Information Systems Programming Language Fortran S8 (X3.9–198x), version 112*, Washington, DC: Computer and Business Equipment Manufacturers Association (CBEMA).

Bal, H. E., Steiner, J. G., Tanenbaum, A. S. (1989), *ACM Comput. Surveys* **21**, 261–322.

Bashe, C. J., Johnson, L. R., Palmer, J. H., Pugh, E. W. (1986), *IBM's Early Computers*, Cambridge, MA: MIT Press.

Booch, G. (1983), *Software Engineering with Ada*, Menlo Park: Benjamin/Cummings.

Brodie, L. (1981), *Starting FORTH*, Englewood Cliffs: Prentice-Hall.

Chorafas, D. N. (1986), *Fourth and Fifth Generation Programming Languages*, Vols. 1 and 2. New York: McGraw-Hill.

Clocksin, W. F., Mellish, C. S. (1987), *Programming in Prolog*, 3rd, revised and extended ed. New York: Springer-Verlag.

Dongarra, J. J., Du Croz, J., Hammarling, S., Hanson, R. J. (1988), *ACM Trans. Math. Software* **14**, 1–17.

Feuer, A. R., Gehani, N. H. (1982), *ACM Comput. Surveys* **14**, 73–92.

Ghezzi, C., Jazayeri, M. (1987), *Programming Language Concepts*, 2nd ed. New York: Wiley.

Hollingdale, S. H., Tootill, G. C. (1965), *Electronic Computers*, Harmondsworth: Penguin.

Hudak, P. (1989), *ACM Comput. Surveys* **21**, 359–411.

Jensen, K., Wirth, N. (1974), *Pascal User Manual and Report*, 2nd ed. New York: Springer-Verlag.

Kernighan, B. W., Ritchie, D. M. (1978), *The C*

Programming Language, Englewood Cliffs: Prentice-Hall.

Morrison, P., Morrison, E. (Eds.) (1961), *Charles Babbage and His Calculating Engines*, New York: Dover.

Nolan, R. L. (1974), *Introduction to Computing through the BASIC Language*, New York: Holt, Rinehart and Winston.

Perrott, R. H., Zarea-Aliabadi (1986), *ACM Comput. Surveys* **18**, 5–22.

Rosen, S. (1969), *ACM Comput. Surveys* **1**, 7–36.

Shapiro, E. (1989), *ACM Comput. Surveys* **21**, 412–510.

Slater, R. (1987), *Portraits in Silicon*, Cambridge: The MIT Press.

Tenenbaum, A. M., Augenstein, M. J. (1981), *Data Structures Using Pascal*, Englewood Cliffs: Prentice-Hall.

Wilson, I. R., Addyman, A. M. (1982), *A Practical Introduction to Pascal*, 2nd ed. (including the Pascal Standard). New York: Springer-Verlag.

Further Reading

Dijkstra, E. W. (1976), *A Discipline of Programming*, Englewood Cliffs: Prentice-Hall.

Knuth, D. E. (1973), *The Art of Computer Programming*, 2nd ed., Vol. 1: *Fundamental Algorithms*, Reading, MA: Addison-Wesley.

Knuth, D. E. (1981), *The Art of Computer Programming*, 2nd ed., Vol. 2: *Seminumerical Algorithms*, Reading, MA: Addison-Wesley.

Knuth, D. E. (1973), *The Art of Computer Programming*, Vol. 3: *Sorting and Searching*, Reading, MA: Addison-Wesley.

March, S. T. (ed.) (1989), *ACM Computing Surveys: Special Issue on Programming Language Paradigms*, New York: Association for Computing Machinery.

Saib, S. H., Fritz, R. E. (eds.) (1983), *The ADA Programming Language: A Tutorial*, New York: IEEE.

COMPUTERS

John W. Carr, III, *Department of Computer and Information Science, University of Pennsylvania, Philadelphia, Pennsylvania, U.S.A.*

1-56081-063-7/92/$5.00+.50

INTRODUCTION

The description of the structure and continuing growth of a new, rapidly growing science, technology, and industry such as computer usage is difficult within an encyclopedia's pages. These three elements interact within an area that is oriented to

1. intellectual effort,
2. information handling over wide geographic areas,
3. computation, and
4. equipment design and sale.

The reader, like the users of such machines, faces a rush of technological and economic details that continuously changes. A static description here will surely be altered continuously in the future. This article will, however, attempt to emphasize those elements which will remain invariant or change slowly as time progresses.

Classically, electronic computers may be classified as analog, using continuous voltages to represent (usually numerically encoded) data, and digital, in which analog pulses are used to represent bits (binary digits). Each bit can be assigned alternative values of 0 or 1 (absence or presence of a pulse). A string of bits can be used to represent many different informational elements by encoding such elements into so-called data structures. The basic element in such a data structure is usually an 8-bit byte or a 32-bit word.

These can be used to represent integers and floating-point approximations to real numbers. In addition, strings of bytes or words can be used to represent alphabetic and other symbolic characters, as well as pictures. With further interconnection into sets of strings, one can represent sentences, equations, and the internal graphical structure of pictures as well. Video images can be represented by digital numerical pixels, or picture elements. Much of computer science involves theory and practice of manipulation and transformation of such elements.

In this discussion, we will concentrate on electronic stored-program digital computers, that class which has gained dominance in modern society. Through success in a competitive market place of vendors, programmers, and users, along with strong support from governmental organizations, digital computers as described herein are at present dominant in their range of applications. Several possible replacement technologies will be mentioned only as a possible prediction of future directions of both research, development, and production.

Digital computers are complex switching devices, often generically labeled hardware, operating upon

1. data, furnished from outside the computers, by either humans or other machines, and
2. programs, labeled software, created and furnished, up until now, outside the computers, almost completely by human programmers.

Data input into computers must be collected by hand or by machine, and communicated on paper or by electronics. Data collection and assemblage must be considered as outside the range of this discussion. Data output is on paper, on cathode-ray tube or alternative electronic presentation devices, or by audio sounds or video, television-like graphical images.

Programming is now, in most cases, a labor-intensive human-performed operation. The process is marked by possible errors in the finished product. In addition, one needs continuing so-called maintenance of the created sequences of instructions over time, since the program itself may attempt to solve a real-world changing problem.

Programming in the most common case consists of the generation of a series of control operations. These, intermixed with data transformation operations, direct the hardware through a complex path. The end result of the process is transformation of the input data to produce output. The output data are required to satisfy some specified goal—to yield the solution of the problem.

In this survey a top-down approach will be presented, which emphasizes the fundamental structure of the elaborate computational system that has been developed.

1. PARADIGMS FOR COMPUTER USAGE

Because of the interposition of elaborate programs between the user and the computer itself, many users interact using paradigms in which the computer is disguised as a special purpose device, or machine. Among the external interfaces that have been set up by which users may interact with computers are the following:

1. The classical file cabinet, with files which can be viewed by "popping them up" on a cathode-ray tube (CRT) screen. Various "windows" operating systems allow the user to point at or name the wanted files and act upon them, opening and closing windows into different sets of text and data.
2. The classical matrix-like spread sheet, in which the user can name or point at entries, describe them in terms of equations, enter parameters, and operate on rows, columns, individual elements, or the entire matrix, with the computer automatically updating the entries.
3. The classical drafting procedure, in which the CRT face acts like the drawing board, and the user operates on various lines and surfaces with various pseudodrawing tools, including simulated pens, paint brushes, erasers, etc., actually consisting of electronically interacting artifacts. Such devices can be used to animate synthetic pictures in time.
4. The classical video screen, with operations to merge, delete, overwrite, recolor, etc., frames from external sources.
5. The classical notebook, with pen-like devices being used to point to and write onto an interactive representation, which changes and operates the computer inside a notebook-sized container.
6. The classical library card catalog, where the user can search for the desired information, which is presented on the CRT screen. Such information systems (with or without the actual card image) are called databases. As yet the card paradigm is incomplete, in that more often special languages are used to search out the information.
7. The classical typewriter, extended to the word-processor and desktop print shop, where copy is edited on the CRT face and then output using various sized type fonts intermixed with drawings and photograph, on laser printers.
8. The classical teletype system, extended into electronic mail or e-mail, such that users with keyboards can mail text or other information, including fax-like material, to other computer users on a worldwide net.
9. The classical bulletin board in which users, using e-mail, can post documents of their own choosing in a formal or informal exchange of information throughout a group.
10. Cooperative exchange of equipment, in which one user is able, by means of network interconnection of computers, to use programs and equipment located elsewhere geographically or administratively.

Still generally unavailable are widespread voice-mail integration of the telephone network paradigm into these systems, or video conferencing. However, as technology provides wider communications bandwidths, these appear to be coming. So-called video-tex systems, mainly in France and elsewhere overseas, have provided rudimentary picture-information network systems, like libraries or information desks. It has thus become apparent that almost any human information behavior, network, or system is a possible paradigmatic model for computer application.

Many of these paradigms have made it easy, as time went on, for nonusers of computers to learn to solve a restricted but important class of their own problems. However, it is still very difficult, in many cases, to move spread sheets onto video screens, images onto e-mail, etc., pictures and graphs into reports. This interconnection of the paradigmatic systems remains a partially achieved task.

Each time a paradigm becomes useful and important, it may need to be connected to previous such structures, and the sets of these interfaces grow markedly, often geometrically. Behind these user–interface paradigms, of course, flows a history that deals with actual equipment and the complicated programs that run upon it. As these change, the implementation of the paradigms for users must also, in many cases, be reworked.

2. HISTORY OF COMPUTER DEVELOPMENT

Historically, computer equipment arose from the vacuum tube technology of World War II radar. At that time fabrication was both expensive and unreliable. Assembly of individual discrete parts into large one-of-a-kind systems required detailed handwork (such as wiring) by human beings, which raised costs and produced many fabrication errors. These became known as hardware "bugs"; the term was carried over into software or programming.

The first electronic digital computer to be publicly operational was the Electronic Numerical Integrator and Calculator (ENIAC) (shown in Fig. 1), at the University of Pennsylvania in 1946. Other early systems were soon operational in laboratories of U.S. security agencies, the Atomic Energy Commission, and the National Bureau of Standards. Research projects were started in many British, Continental, and Soviet governmental laboratories, and in other universities in the U. S., Canada, Great Britain, and elsewhere. By the mid-1950s commercial equipments, pioneered by such firms as Univac, IBM, NCR, and others in the U.S., and Ferranti, ICL, Bull, Siemens, and others on the Continent, began to make their appearance in extensions of problem solutions previously attempted with classical punched-card equipment.

Costs of these electronic systems began to decrease and reliability to improve in the 1960s. The use of discrete transistors as the switching devices still required them to be interconnected by hand labor at high cost. Only after the labor-intensive circuitry interconnection methods for such devices were replaced by printed circuits and circuit boards produced by machine operations did the cost of digital computer hardware begin to fall.

Mass production of computers came only after 1970 with the invention and development of the integrated circuit (IC). This combined automated production of transistors and wiring on the same silicon base (or substrate). Using photolithographic-like technology, firms such as Texas Instruments, Intel, Motorola, and their Japanese competitors NEC, Toshiba, and others began to produce small versions of centralized computers.

These so-called microprocessors were produced using a different kind of high technology assembly line, in which solid-state physics and chemistry replaced mechanics as the underlying science guiding design and fabrication. This mass production of the central kernel of the digital computer had been paralleled by mass reproduction of software and data input, using, first, punched cards, then magnetic tape and the portable "floppy diskette" of magnetic material. The remaining economic bottleneck to truly efficient mass production and distribution remained in the design and fabrication start-up costs for both equipment and programs.

Design costs for integrated circuits began to drop with the development of computer-aided design (CAD) tools for integrated circuits. As quantities of an individual IC unit, or chip, began to increase, fabrication costs came down. Analogous design techniques for programs, called computer-aided software engineering (CASE), have not yet proved as successful. With both techniques, beginning about 1980, a mass market for both dealer and direct-mail sale of desktop computers began.

Continuing improvements were made in the distributed software. Programs are now sold using mass-published and mass-marketed copies, at relatively low costs with respect to the hardware, on small, inexpensive floppy diskettes. These fabrication and distribution advances now make individual computer interaction a potential for every technologically sophisticated worker in the modern

FIG. 1. The electronic numerical integrator and calculator (ENIAC), first designed and completed in 1946 at the Moore School of Electrical Engineering of the University of Pennsylvania, and used at the Ballistic Research Laboratories of the Aberdeen Proving Ground in the post-World War II era. (Courtesy of Moore School of Electrical Engineering, University of Pennsylvania.)

world, and for many for whom their use will be routinely integrated into their jobs.

2.1 General Structure of Computer Usage

Users should, it appears, view solution of problems of any sort in light of the entire computational system which has become economically available. Such problems may be computational or data processing, real-time control, or database searching. A useful model is to view the process of computation from the top down, or outside in. One starts with the problem and works inward towards program and computer, rather than starting with the detailed technology of the system and working outward.

Of primary importance is the problem, or class of problems, to be solved. Some type of applications program, a solution procedure, must be selected by the user. Such a program is the highest level performer in the hierarchy of program, operating system, and computer. The program (or programs) may have already been in operation, available for purchase, or may need to be written from the beginning. The program must interact with at least two other systems elements:

1. an operating system, or control program that performs the housekeeping for the applications program, on
2. a specific computer, or integrated system of computers, that accepts the language of the applications program, as integrated with the operating system program.

2.1.1 Operating System and Language Interactions. The operating system serves as a buffer between the applications program and the actual machine hardware, the computer itself. It manages the specific idiosyncracies of the input–output elements of the computer, and of the secondary mass storage devices such as magnetic disks. The applications program communicates with the operating system in a language which that system can interpret. In general, the applications program does not have to communicate directly with the hardware. For this reason, an operating system has the potential of offering a unique interface to many applications programs. Yet at the same time it may be tailored to control the hardware of many different computer versions.

Similarly, the language in which an applications program is written may be passed through a compiler–translator which outputs machine code for a particular machine. Different versions of this compiler–translator may produce resulting machine codes that are compatible with different machine hardware. Both operating systems and compiler–translators offer the possibility for the introduction of standardization, in which the programming languages, operating systems languages, or actual computer hardware is defined in a standard fashion. This allows ease of use, design, and manufacture in the interaction between these various elements.

The potential user, having decided on an applications program to solve the proposed problem, must decide on an operating system to accept the applications program, and a computer itself to accept the operating system. In many cases, the latter two are already decided upon as a result of previous analyses and decisions. This may restrict the available applications programs for solution purposes.

In some cases, if no solutions program exists as yet, or none can be obtained or modified to suit the need, someone—user or applications programmer—must write a program to interact with the operating system and at the same time solve the problem. In today's complicated computational systems, a professional programmer will most often be needed here, although for certain problems, so-called higher-level languages or graphical interfaces may allow problem-oriented users to create programs.

2.2 Changes in System Technology Simplifying Usage

In the early 1970s, computers were mainly used through the good offices of professional programmers, who wrote programs under the guidance of noncomputer specialists, fed in data, and monitored and produced the results. Three technological advances changed this situation: time sharing, integrated circuits and cheaper memory systems, and interactive graphical computing.

1. "Time sharing" was developed as a method by which many terminals (CRTs with circuitry) were connected to a large central main-frame computer, and shared its machine cycles among them. These terminals were originally mechanical teletypes, later electronic displays with keyboards. The central computer responded to individual users at these terminals so that in appearance, at least, each user seemed to have control of the central machine. The latter was actually sharing its operation sequences among the many terminals. This change allowed (particularly important) interactive usage of central machines from keyboards at remote sites. Such interactive time sharing included both computing and programming from the terminals. This sped up a previously much slower cycle based on punched cards, not keyboard strokes, as input. Time sharing also required design and development of networks connected between terminals and computers. Terminals themselves soon began to contain microprocessors, and thus became smaller local computers. The original time-sharing networks became computer-to-computer networks, expanding to many-computer interconnections called local area networks (LANs).
2. The development of the integrated circuit described above, along with the parallel development of computer-aided design, began to produce small digital computers. Such personal computers (PCs) based on these microprocessors became available for a few thousand dollars. Speeds continued to increase with decrease in size of the integrated microcircuits. Costs and size of the resulting computers decreased. The combined functions of the resulting computers also increased markedly. Low-cost, small-

size computers (see Fig. 2) began to appear not only in homes but also in small and large businesses. These—combined into networks—began to replace or extend the original far more costly main frames, as the physically large and more expensive original designs were called. By 1991, microprocessor chips with half a million to a million transistors were being produced in large quantities. These were sold at lower and lower costs, with former large-size circuitry and interconnections moving onto the chips themselves. At the same time, the cost of computer memory, mainly dynamic random access memory (DRAM), was coming down because of decreased transistor and wiring sizes on the IC chips. The result was the production and sale of large numbers of these cheaper microcomputers, which could be used in the same way as the time-sharing terminal, but without the need for the central main-frame computer. The sharp distinction between large main-frame computer and microcomputer began to blur with the expansion of personal computer performance and the development of so-called "work stations" with power equivalent to that of previous large computers. It is possible as of 1992 to obtain a computer that can comfortably be held on one's lap, yet performs at levels comparable to those of desktop computers (see Fig. 3).

3. Interactive graphical computing was bolstered by addition of electronic screen-pointing devices for the cathode-ray tube terminal. The "mouse," first developed by Xerox, was commercialized successfully by the firm Apple Computer. Its availability allowed the Macintosh computer to remain a firm competitor of IBM's ubiquitous PC as both began to enter the offices of major businesses internationally. Only after so-called windowing systems began to appear on the work stations and PC systems (about 1990) did they begin to compete in effective nonprogrammer user interaction. Hardware was thus combined with elaborate software, to produce what became known as graphical user interfaces (GUIs) (Sec. 4.3.3.2). These began to replace previous programmer-oriented, rather than user-oriented, interactive interfaces. Newer systems thus place the emphasis on making the interaction between user and computer less complex and more "user friendly." The paradigms for users mentioned above have sped up the so-called learning curve on the part of novice and nonprofessional programmer users. For the first time the computer user rather than professional programmer is more nearly in control.

2.2.1 Changes in Software Technology. Technology in software design and "construction" has lagged behind that of hardware. This is probably because the problems involved in such creation are more difficult and have often been ignored or passed on to the pro-

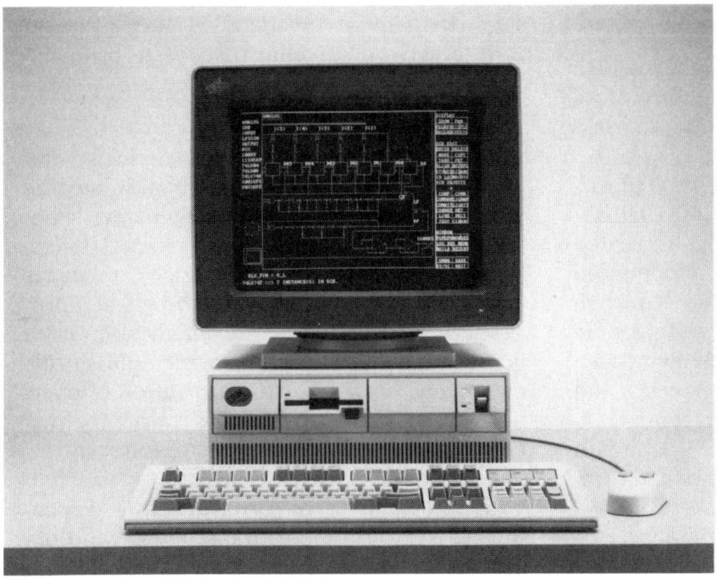

FIG. 2. A personal computer, rated about 100 times as fast as the original microprocessor-based computers, containing up to 8 megabytes of primary storage, a hard disk drive with up to 200 megabytes of secondary storage, and a powerful graphical display and high-resolution laser printer. (Courtesy of IBM Corporation.)

FIG. 3. A laptop computer, weighing about 6 lb, and containing up to 8 megabytes of storage and a $3\frac{1}{2}$-in. floppy diskette drive. Such computers perform at speeds and with programs comparable to desktop computers, which are also microprocessor oriented. (Courtesy of Compaq Computer Corporation.)

gramming creators by the hardware designers. Until now, there has been little computer-aided design for the generation of software. Here again handcrafting of program sequences has most often been the rule. The classical procedure is diagrammed in Fig. 4.

Many software tools, however, are now in use to aid applications programmers in creating and running their problem solution program. Among the more important of these are

1. applications programming languages which made the sequence of instructions more nearly that of the problem solver and less that of the machine itself;
2. translators from such languages to the machine language itself, which included assemblers that assemble previously created and tested sequences (code) into unified programs, and automatic interfaces through the operating system (OS) to such input–output devices as terminals, printers, telephone lines, and mass storage devices; and
3. operating systems, which originally replaced the behavior patterns of the human operators of the original main-frame computers by automatic control programs performing equivalently. Later operating systems changed many decision-making procedures from the human programmer to the system itself at run time. They also provided standard interfaces that applications programmers could use to guarantee correct performance. Thus the OS began to perform more and more "housekeeping" tasks for the applications program and programmer. The computer operator more and more became an onlooker rather than participant at the detailed level, but an interactive controlling force at the problem-solving level, made at what has become a problem-solving console.

These changes led to new efficiencies—and constraints. As certain languages became widely used, they began to become "standards" for a growing group of users. Applications programs written in such languages could be passed back and forth between proprietary hardware machines, and often even across machine families. To accomplish this effectively, however, standardization of operating systems beyond such proprietary machine families also became necessary.

2.2.2 Changes in Programming Procedures. In addition to the language and operational "utilities" described above, certain

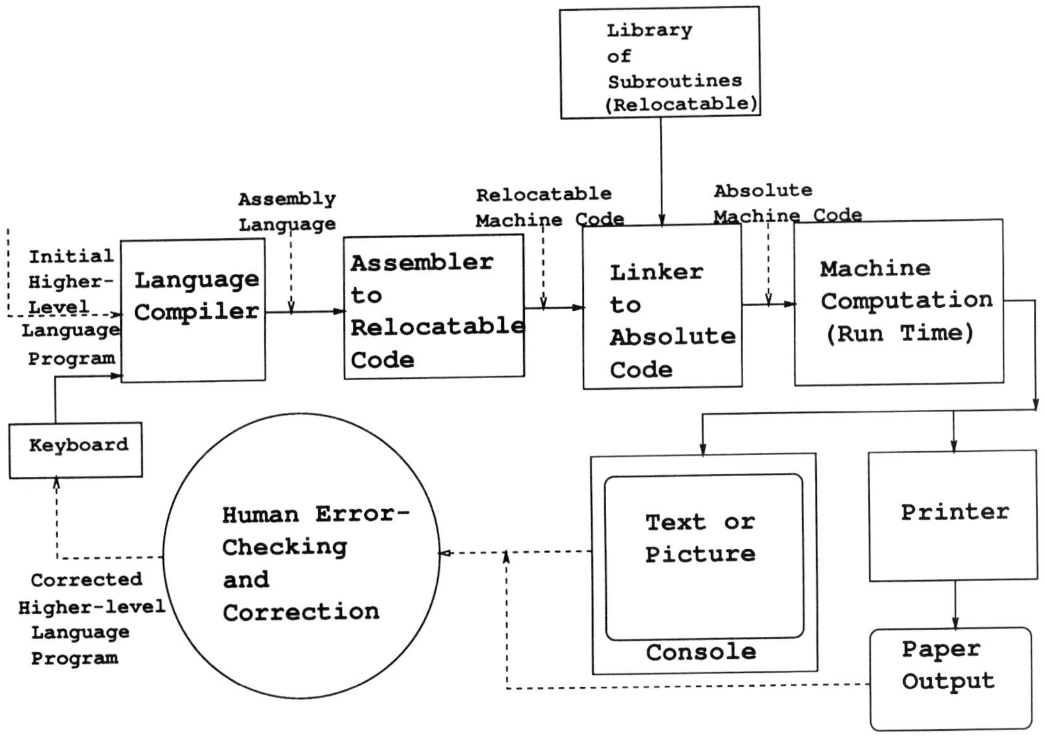

FIG. 4. A diagram of the classical programming procedure, originally with teletype or punched cards as input and only printed paper as output. Now interactive keyboards and cathode-ray tube displays speed up the feedback in correcting errors and changing performance of programs.

fundamental strategies for producing computer programs more nearly automatically began to arise. Some of these were highly complex, such as the tactics of artificial intelligence (*q.v.*) and expert systems. These proposed to copy the behavior of human mental processes to aid in solving real-life problems. In certain areas, such as governmental intelligence and certain medical diagnostics, claims were made that such programs produced results beyond the behavior of human experts.

The concept of "objects," as developed with object-oriented programming, created classes of programs and data structures that passed along historical traits by a process of "inheritance," and abstracted out many programming details so that programs might be used over a broad range of analogous applications. This type of programming, combined with special graphical user interfaces, appeared to produce an interactive design process that benefitted from rapid interaction between programmer and programming system.

More immediately applicable computer-aided design procedures for programs have been collected under the title of computer-aided software engineering (CASE). Here the new programming processes of the graphical user interfaces, operating systems, and interactive programming are combined to aid in the generation of alternative versions of certain standard applications programs. CASE has most often been applied to replace the generation of simple, but nevertheless expensive, programs in commercial and data processing applications, previously produced by hand coding. The results are intended to be user-friendly programs. The generated programs themselves interact with nonprogrammer users to create specialized data handlers and transformers.

The resulting programming process is now a mixture of both labor-intensive human analysis and design, combined with machine aids that attempt to guarantee ease of maintenance of the resulting code. Some systems keep a complete record of the historical changes in large programs, combined with histories of certain tests and their results. As software generation has become a more nearly engineering process (as opposed to art form), the introduction of engineering design and test procedures, similar to those of integrated circuit hardware design, increases.

Software costs continue to grow as a major economic factor in the use of computers. Until adaptive program generation techniques begin to produce correct programs automatically, human programming will remain a major constraint on computer utilization.

3. THE CLASSICAL SEQUENTIAL COMPUTER PARADIGM

This section describes the original stored-program computer and gives both abstract and concrete descriptions of how it has been extended with memory and input–output, becoming the central element in local area networks, client server systems, etc. Later, how it has been replicated into parallel computer systems will be discussed.

The simplest model for a digital computer, now only an approximation to the smallest portions of very elaborate systems, is that of the stored-program digital computer. Often credited to Burks, Goldstine, and von Neumann—who published first—it was first proposed by the creators of the ENIAC, Eckert and Mauchly, at the University of Pennsylvania in the mid-1940s. It is useful to compare these basic ideas with the structure of the most popular present design, the IBM Personal Computer.

Parts of such present-day popular microcomputers copy the separated-memory model of the "Harvard machine" of the 1940s, with separate program storage [now in read-only memory (ROM) in the IBM-PC basic input-output system (BIOS)] and separate data storage (in the PC, always in DRAM).

Most of the systems today, however, store other program instructions and data intermixed in the same storage medium, usually DRAM, following the model of the Pennsylvania designers and von Neumann. In the discussion that follows, the classical model is slightly modified to adapt to changes in structure and perception since it was originally proposed. The structure of such a computer can be viewed, as with most such artifacts, by means of both a behavioral and an architectural model.

3.1 The Control and Data Units

The behavioral model provides an idealized, abstract description of how a single digital computer behaves, described in terms of two units, each formally so-called finite-state machines—"FSMs." The behavioral units are pictured in Fig. 5.

Two separate units, control and data, act in a time-interleaved fashion in this model. Each unit has inputs, outputs, and internal states, drawn in each case from a finite set of alternatives. Passage of information between the two units, as described in Fig. 5, is performed in a "ping-pong" fashion, starting with original initial address and data states inserted from outside.

First the control unit uses its initial starting address to generate or retrieve an instruction, which becomes the control unit output. This becomes the data unit input, and is then performed by the data unit, transforming the internal data in the data unit. At the same time, an output is sent back from the data unit to the control unit. This output becomes the control unit input. The control unit uses this input to choose a new address, and the cycle is repeated, beginning at the first word of this paragraph.

3.1.1 Data Unit. In more detail, the data unit holds the data state (consisting of many bits representing numerical and linguistic or structural information). It possesses arithmetic and information-transforming operations which change the data states. The data state is transformed at each step or cycle of machine operation, as a function of the input data operation coming from the control unit. Its outputs are test results sent back to the control unit as the result of test operation inputs.

3.1.2 Control Unit. Outputs from the control unit are operations sent to the data unit, which perform specific transformations on the data states, or test them (in present microcomputers, usually as "flags") for properties such as sign, overflow, etc., generating outputs which are sent back to the control unit, usually of "true" or "false" nature.

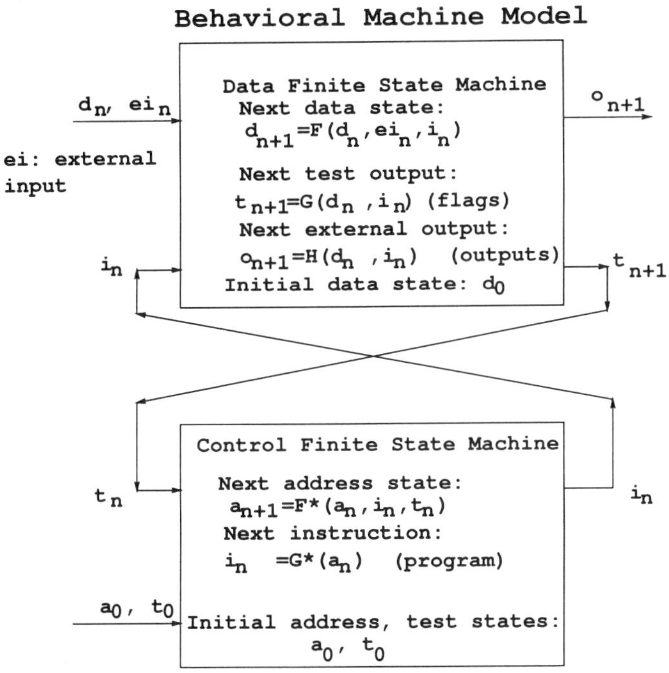

FIG. 5. A behavioral diagram of the classical sequential computer, described in the text. A data unit transfers data upon input of instructions from the control unit, which changes its address as the result of test information from the data unit, sending instructions on as a function of the address. The latter is the program.

The control unit has its own internal state, usually called the address state, which changes at each cycle of the joint system. Ordinarily, the succeeding address is obtained in integer succession, with addresses represented as integers. However, results of test operations on data change addresses out of the regular integer succession sequence. It is the latter result which allows "transfer of control" and "loop control" between instructions in the control unit.

Some operations sent to the data unit are more complicated, such as test operations which also transform data states (such transformations are often termed "side effects" for operations or programs that are not logically clean).

3.1.3 Initial Conditions. The user of such a system (which may be a human, or another program such as the operating system) must furnish

1. the initial address state (which is where the program is to be started) in the control unit,
2. the initial data state (which generally consists of all of the initial numbers, integers, and other parameters which specify the exact problem to be solved),
3. the next address function, which takes data outputs and the previous address, and decides on the new or next address. This function, in the mathematical sense, must be supplied as a part of the computer program from outside this system, and
4. the next-data function, which takes control outputs (operations to be performed) and the present data state, and generates a new data state. This function is hardware, wired into the system and generally not changeable from outside.

3.1.4 Continuing Operation. Once these four elements have been set (either at hardware design time or program and data entry time), starting the computer initiates the cycle of successive (address, data) changes until a "halt" instruction is reached, which leaves the control unit in an address with no output instruction. In actual systems, there generally are no halt instructions. Once one program terminates, it supplies the initial address and data to another program (which might be some other program controlling it, for example).

3.2 A Corresponding Structural Model

Corresponding to the behavioral description given above, which only describes elements being manipulated as abstract, coagulated units, one can produce a structural model, which is usually what the actual hardware designer works from. One such model, with four units interconnected in a more elaborate fashion, is given in Fig. 6. Parts of these units may belong to one or the other of the control and data units above. Engineering and hardware realities dictate a more elaborate representation.

The four units of the present usual hardware configuration are usually connected by one (or sometimes more than one) common electronic highway or "bus." Some systems have separate instruction and data buses; others have special buses to match different speeds of information transfers of the units that follow:

1. Memory or storage unit: In "stored-program machines" the address-instruction function (or program) in the control unit is stored as instructions in positions given by machine addresses (usually implemented in the integrated-circuit DRAM). In addition, data states are stored in addressable positions, usually as short (8-, 16-, or 32-bit) sequences, as are the instructions.
2. Control unit: In this, the control or address state is represented as a changeable stored integer, along with the hardware next-address function. This latter function changes addresses depending upon test results of flags, or individual bits developed as the result of test instructions.
3. Arithmetic–logical unit (ALU): In this unit, data states, with changes occurring in words in the memory, are transformed according to the definition of the arithmetic or other data operations, wired in and defined by the machine's designers. Key data information is generally stored in accumulators or registers (analogous to the accumulators on the original desk calculators) and transformed by operations on one or more of these, with results fed back to the same or a different accumulator.
4. Input unit: This unit brings in new data from outside sources such as keyboards, telephone lines through modems, analog-to-digital converters bringing in audio and video information, and floppy diskettes and hard disks with programs and data. These are stored in registers or memory to change the computer data state or instruction program.
5. Output unit: This unit passes information, including data and control commands (instructions), to effector output devices such as printers, cathode-ray tube and video displays, digital-to-analog converters, telephone lines, and the floppy diskettes and hard disks.

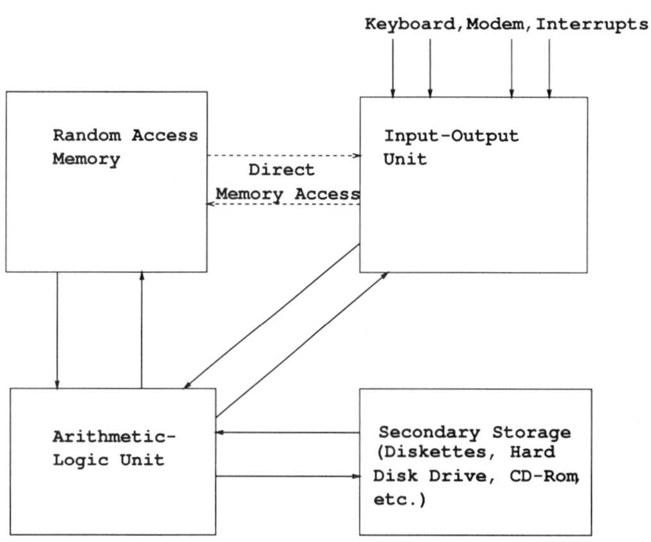

FIG. 6. The classical four-unit structural model of a sequential computer, with arithmetic-logic unit, storage unit, and input and output units to communicate with the outside world.

One can also consider the floppy diskettes and hard disks, along with compact optical disks and magnetic tape units, as representing a unit:

1(a) Secondary storage unit or memory, in which a mechanical device backs up the initial electronic random access memory (RAM).

3.3 Methods of Speeding Up Performance

Hardware designers, in efforts to speed up performance often slowed by mechanical or lower-speed electronic information transfers, use buffering techniques and parallel memory transfer as expediters. Most common is the so-called "cache" memory, a smaller, more expensive high-speed memory that holds most-often or most-recently used instructions and data. In many cases, sophisticated compilers and assemblers allow these artifacts to be "transparent" to the user-programmer at some level, although quite complicated algorithms may be necessary to achieve best performance.

3.3.1 Use of Static Memory. At the present time, the memory or storage units are most often DRAM. These are mass-produced integrated circuits whose time of information access has continually decreased during the past ten years. At the same time, the number of memory units (bytes or words) per chip has moved upwards the megabyte range. With increase in the density of the storage transistors on the chip, costs have come down. These units are called "dynamic" because they do not hold their information without constant refreshing—reading and writing back the storage contents. Alternative, higher speed but higher priced, storage elements are the static RAM, which do not need this refreshing process. Other memories (as used in the PC BIOS) are used with permanent program or data storage, ROM. These contain information (data or instructions) stored at fabrication time, or later changed only by substitution of new chips.

3.3.2 Combined Control and Arithmetic-Logic Unit. The control unit and ALU have from the time of the first microprocessors been combined on a single integrated circuit chip, or silicon element. Transistors are used to perform the switching functions necessary to change addresses and data. More recently, as more transistors became available on a single chip, further input and output units of increasing complexity have been added to the central structures. All of these units contain smaller portions of their own memory, usually fabricated as high-speed local static RAM.

The overall speed of both memory and the other units is dependent on the speed of action of individual transistors. This is usually described in terms of a "clock" or high-speed counter, which synchronizes storage and shifting of pulses between different elements on the chip. Present commercial microprocessor clock rates range from 20 to 100 MHz, or millions of cycles per second. This means that the smallest time duration for local information transfer might be in the neighborhood of 10^{-8} s. (Actual arithmetic operations, which require more complex pulse manipulation, would be longer in duration.)

4. ACTUAL IMPLEMENTATION OF HARDWARE CONFIGURATION

The model for actual hardware design is not of the single sequential computer, but, as on the integrated circuit (IC) chip, of a set of parallel subsystems made up of transistors mounted on a common substrate. On the chip, the silicon substrate is the physical common element that allows electronic interaction by transistor switching. The parallel subsystems interconnected on this substrate can interact both synchronously and asynchronously.

A synchronous system is locked into a common clock, which coordinates the timing of the many subsystems in the overall systems. Common actions—moving or complementing pulses—are performed at (more or less) the same time when a clock pulse, propagated all over, triggers local transistor switching.

Most simple ICs are controlled synchronously; the upper physical size of a synchronous system becomes dependent on the maximum delay in clock pulse propagation from one point of the chip to another (called "clock skew"). An asynchronous system is one in which a local action may have to wait on the completion of a more remote operation elsewhere in the system. Such systems contain inherent delays, obviously, as well as possible blocking conditions (two subsystems waiting on each other to complete each one's present task, for example). There exist some asynchronous circuits at the IC level (the classical "flip-flop" circuit is an example), but most asynchronous interaction comes at higher levels, between chips, computers, independent memory units, etc.

4.1 A Typical PC Architecture

A typical computer of the PC variety will have a synchronous, clock-controlled microprocessor as its central processing unit (CPU). The CPU, however, will change its control asynchronously upon being interrupted by such input–output units as keyboard, hard-disk controller, and telephone modulator–demodulator (modem). In addition, if other so-called coprocessors are present (such as a high-speed floating-point arithmetic unit), the microprocessor will asynchronously pass information back and forth at arbitrary, irregular intervals dependent only on the size of different parameters and corresponding calculations. Sometimes this "coprocessing" technique moves information so as to by-pass the CPU, such as in the direct memory access (DMA) process which captures a bus (hardware highway) for a time, in order to move information back and forth between storage and input–output.

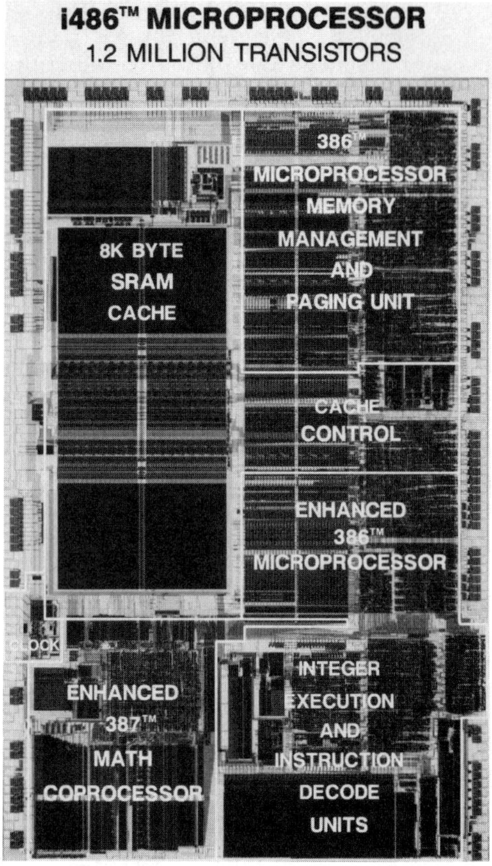

FIG. 7. An Intel 80486 integrated-circuit microprocessor, showing the detailed wiring and transistors that make up this state-of-the-art complex instruction-set computer (CISC). (Courtesy of Intel Corporation.)

4.2 Standard Microprocessor Families

Large-scale main frames and minicomputers more often have had structures which were later copied in microprocessors, and so it is useful to look at microprocessors as the most prominent examples (at least in manufactured quantity) of single processors. In practice, microprocessors have over time come to be fabricated in large quantities. In this process, the number of different hardware architectures has come to be narrowed down to families, usually created by one manufacturer, with upward compatibility of programs. Original systems, such as the Intel 8080, were later expanded into more complicated 8086, and then 80186, 80286, 80386, and 80486 (Fig. 7) systems, over a period of about ten years. A comparable family developed by Motorola started with a very simple 6800 microprocessor and expanded upward through the 68000 series. There are other families for less standard purposes (e.g., digital signal processing, by Texas Instruments), but quantity occurrence in the marketplace has often been restricted to a few families.

4.2.1 Changes Within a Family. Successive numbered versions of such a processor family usually consist of chips in which added architectural features are adjoined at each stage. In addition, basic clock rates increase as transistor rectangle size decreases with improved fabrication technology. "Upward compatibility" of programs attempts to guarantee that programs written for previous versions in the family will perform on later versions, thus allowing older programs to be used with newer processors. Through the use of interfacing software such as operating systems and compilers, this is often accomplished.

Among the architectural features which vary are those which improve speed of computation and decrease cost of fabrication. The central limiting factor in such a process is often the bus or joint information communication path on the chip (see BUS SYSTEMS AND COMPUTER INTERFACES). Such a bus appears abstractly not only in hardware on a chip, but at a higher circuit-board level, and abstractly in the software operating system as well.

4.3 Hardware and Software Buses

Abstractly, the common highway or bus of the microprocessor chip, from which subsystems are connected asynchronously, has a parallel in the common interconnection structure of the software operating systems. Just as subcircuits are designed independently and then connected on a chip, subprograms (often called subroutines) are designed independently with standard behaviors, and then interconnected to form a working program. Such programs can be "utility" programs, used to perform housekeeping tasks for the overall computer system, or parts or wholes of application programs created to solve specific problems.

The standard hardware bus will have a set of wires, divided into subsets transmitting data, instructions, and control information back and forth between the asynchronous (and sometimes synchronized) subunits. The addition of a new electronic subunit consists of attaching its inputs and outputs to the appropriate data, instruction, and control wires of the bus, so that those common bus signals can be manipulated by the new subunit. Thus, a special memory, input, or output system can be mounted on a printed circuit board and plugged into the bus to expand the capability of the original system. Obviously, along with the new circuitry, in most cases appropriate changes and additions to the central microprocessor program instructions will also be necessary in order to synchronize with the control signals to the newly added unit.

4.3.1 The Central Hardware "Motherboard". Those parts of the computer hardware which are fixed at fabrication time (e.g., of the standard PC) are mounted on what is called a motherboard. This contains wired connections between sockets into which the IC chips are to be inserted. These standard chip sockets are usually mounted on a printed circuit board, along with a series of open sockets connected to the bus. Such sockets on board or bus may be filled at manufacturing time, or later by the vendor or actual user. The standard chip sockets allow the possible potential substitution, e.g., of a later microprocessor in a family series for an earlier slower version. This is a substitution at the chip level. The bus sockets, on the other hand, allow the interconnection of extra boards, fitted with appropriate pins, into the bus. Such extra boards, often from a different manufacturer than that of the original motherboard or of the entire system itself, allow an extension of the memory, control, or operating capability of the original motherboard configuration.

Thus, the flexibility of the motherboard with its sockets along the standard bus allows many different designers and fabricators to add new or different subsystems to the original system, provided they follow standard socket and bus specifications.

These generally have been set up with standard signal, timing, and electrical characteristics, by specific manufacturers or groups of manufacturers, or by IEEE or other standards committees. Such buses, unless accepted widely, tend to again become proprietary in nature, locking in users for a period of time to a restricted set of sources of hardware modifications and extensions. The hardware bus is thus the standard interface by which disparate subsystems can interact.

With these ideas of flexible socketing, memory and microprocessors can be extended or replaced, at least within the electrical and physical limits of the bus system itself. Many of the newer paradigms for computer users, including audio, video, and telephone interaction, have been achieved by development of special-purpose boards able to be plugged into available sockets on a bus. Sequential and parallel data transmission, and digital-to-analog converters (as well as the inverse a-to-d devices), were first added to original personal computer configurations by special boards. Powerful graphics systems, themselves often containing microprocessors, can be added into standard buses, provided accompanying programs are made available. Later, many such functions have been moved down to the central microprocessor ICs themselves, since these have been growing continuously in available functionality.

4.3.2 Standard Buses. Present standard buses in the PC world include ISA (for Industry Standard Architecture; used in the original PC), EISA, an extension, and the Micro Standard Architecture of IBM's PS-2 newer configuration. Newer standards generally have a wider, faster bus to fit the later-developed microprocessors and memory that are to be combined on the motherboard configuration.

Other standard buses used in larger, faster work stations (which have grown out of proprietary buses), such as Multibus, VME bus, etc., were originally designed around a particular microprocessor family, but later in some cases were able to be used with other equipment. The larger minicomputers and mainframe computers also have their standard buses, to allow "third party" vendors to design and furnish extensions to the original architecture.

4.3.3 Operating Systems as Software Buses and Interactive Interfaces. A similar commercial-industrial method of furnishing software extensions to computer control systems uses the operating system as a "software bus," which allows the "socketing" of new programs and data structures into the overall computer software structure. Corresponding to the wires of the hardware bus are parameters in the programs which are to be "plugged in" to the operating system's data storage. Standard transfer of control procedures correspond to the control wires of the hardware bus. Since CPUs most often contain only one processor at a time, classical operating systems have asynchronous-like features only when multiprogramming (performance of two programs on the same single processor) is being carried out.

4.3.3.1 Combined Hardware and Software Extensions. Standard operating system definitions allow separate hardware producers or vendors to specify extensions to the software. The results are intended to operate correctly, if they correspond to the system constraints. Among the best known of operating systems was the original OS-360 for the IBM-360 series, which allowed many applications to be successfully integrated into this IBM main-frame series of computers. For the PC systems, Microsoft's disk operating system became such a standard that all application programs for PCs, for many years, have used the MS-DOS rules for interaction with that operating system. More recently, the Windows program for the same computer family has belatedly introduced a GUI standard. The power of the Apple Macintosh computers was in the existence of a combined GUI and operating system many years earlier, using the original "mouse" screen-pointing concept pioneered by Xerox, and now available on Windows.

Broader standards for windowing systems, e.g., X-Windows, are appearing as GUIs for faster work-station systems made by Sun, Hewlett-Packard, and others. These systems, like many recent computers now being marketed, have as their common extended software "bus-like" structure the UNIX system. This operating system, originally developed at Bell Laboratories, is becoming, more and more, the OS of choice for hardware manufacturers, applications program vendors, and users. Its combination with a successful common standard GUI may become the future overall standard for programs generated by different originators.

UNIX was originally designed for word processing and text manipulation at Bell Laboratories, but its use has been markedly extended over about 20 years. It contains the software equivalent of many of the hardware techniques implanted on chips or motherboards, including software coprocessors, splitting of processes into parallel tasks, and complete control of files, following the file-cabinet paradigm. It works with many interacting terminals dealing with a central system, and allows mutual interactive operation on common files, under appropriate controls. Only recently have memory prices decreased enough so that UNIX systems appeared economic on PC platforms.

4.3.3.2 Graphic User Interfaces. GUIs like the Macintosh clipboard–file-cabinet paradigm extend user interaction to graphical presentations which allow the mouse pointer to manipulate files by pointing to "icons" or "buttons" on the screen (Fig. 8), rather than requiring lengthy key-stroke sequences as with UNIX or the PC-oriented DOS. Microsoft's windows 3.0 extends the Macintosh behavior by directly linking various programs through

FIG. 8. A graphical user interface, in use with a "mouse" to point to icons, "buttons," and specific windows, text, and pictures on the screen. This point-and-click process bypasses much of earlier character-oriented interfaces on microprocessor operating systems. (Courtesy of Apple Corporation.)

dynamic data exchange, which will cause a chart from a spread sheet, embedded in a word-processing program, to change as individual cells in the spread sheet are changed. This is an example of computational paradigms being interconnected.

4.3.4 Asynchronous Hardware–Software Interaction. The asynchronous behavior of hardware systems obviously causes a need for appropriate response by programs. Two methods of having CPUs such as microprocessors (and much larger machines) respond to the real-time (often mechanical) behavior of the outside system are as follows:

1. "Polling": The CPU synchronously, at regular intervals, looks at electronic control signals from other units, and responds by changing its internal software control to respond to that unit with highest priority to be serviced. This of course places a burden on the CPU, which must spend a portion of its time in checking on its environment. (This is indeed analogous to a "poll" of a population to learn each individual's situation.)
2. "Interrupt processing": The CPU only responds when interrupted by an external unit (analogous to a telephone ring), and then transfers its internal control to software which responds to the particular situation. This system is much faster and much more efficient, but also requires more complex software and computer operation capability on the part of the ALU and peripherals.

This programming behavior is, of course, very similar to the behavior of interconnecting hardware devices interacting in a "master-slave" relationship or requiring a "handshake" procedure to guarantee that both subsystems have the same common control information. Many of the tactics of circuits have found their way into programs, and conversely.

4.4 From Integrated Circuit to Area Network

4.4.1 The Integrated-Circuit Chip as a Similar System. Because integrated-circuit chips need efficiency both in space and speed, the structure of a chip follows many of the characteristics of the larger systems, particularly since entire microcomputers are now being moved onto single chips. Individual chip configurations are, in general, synchronous with respect to the fundamental clock cycle frequency, which times basic transfers between transistor–capacitor storage elements.

Larger subsystems on many chips act relatively independently and asynchronously, waiting on completion of actions by other subsystems before starting on a local task. Such interactions often contain feedback that corresponds to rudimentary hand shaking. This is the more elaborate interaction between higher-level subsystems to guarantee that mutual tasks can be continued because local subgoals have been reached.

Communication on a two-dimensional planar chip is of course constrained by that planarity, somewhat relaxed by the availability of multiple levels of wire crossing.Thus, not every subsystem can easily connect directly to every other, and the concept of bus is forced upon the designer, as a standard pathway with standard interfaces that allow independent subdesign.

4.4.2 Design of Integrated Circuits. IC chips were originally designed by "hand," with drawings made of the many levels of silicon and metallic compounds that were to be superimposed. The entry of programming technology, particularly in the case of graphics, began a continuing development of "silicon compilers" (see COMPUTER-AIDED DESIGN IN ELECTRONICS). These latter were oriented towards linguistic or GUI descriptions of the circuits, which were to be translated into the final sequences of rectangles to be deposited on the silicon substrate. The most sophisticated of such programming systems were aimed at creating application-specific integrated circuits (ASICs). Among the many applications that have been moved from software onto hardware circuit boards and finally onto ASIC chips are graphics, telephone modems, spelling checkers, games, disk controllers, and many other key "bottleneck" processors where lower cost or higher speed is absolutely necessary to make such behavior economically feasible. These similarities in structure, creation, and economics of various levels of computer software and hardware emphasize the abstract (although certainly not concrete) similarities of these varied forms of information storage and manipulation.

4.4.3 A Higher Level of Area Networks. Once low-cost computers began to appear on many desks, it became apparent that their interconnection had the potential for mutual information processing and control among many users. Microcomputers and work stations are limited to the number of common interactive users that they can accept; they cannot carry out such interactions and at the same time perform requisite calculations. From this need and opportunity came the LAN concept (Fig. 9), which allows machines to call on storage, programs, and even program performance from and by other machines. The development of such commercial standards as that of Ethernet and corresponding packet-switching networks first produced LANs, and then began their interconnection into wide-area networks.

At present, LANs are usually configured out of a collection of PCs or work stations, acting asynchronously except for bus message communication, which is synchronized to a standard cyclical system. This means that LAN designs are constrained, like the structure of chips, as well as the structure of PC interrupt processing, to those design tactics that guarantee the absence of deadlocks and other fatal interactions. One can thus note that from smaller to large systems, the same rules of synchronous and asynchronous interaction apply:

1. individual transistor circuits;
2. interactive subsystems on a chip;
3. chip sets in a cooperating system, such as on a PC motherboard;
4. interactive multiprocessing systems;
5. common-bus parallel machines;
6. many-processor locally connected massively parallel machines; and
7. LANs.

It is this potentially recursive, replicative nature of the hardware and software artifacts that have been created that allows individual humans the possibility of understanding their behavior. At some near future stage, however, it may be only through embedded computer descriptions within themselves that humans will be able to maintain enough knowledge to make satisfactory use of such systems.

4.5 The Philosophy of Interchangeable Parts

The sometimes chaotic growth of all these various combined hardware and software systems illustrates a growing organization of their structure. Under the constraints of technology and the economic-sociological environment of the marketplace, a process classically

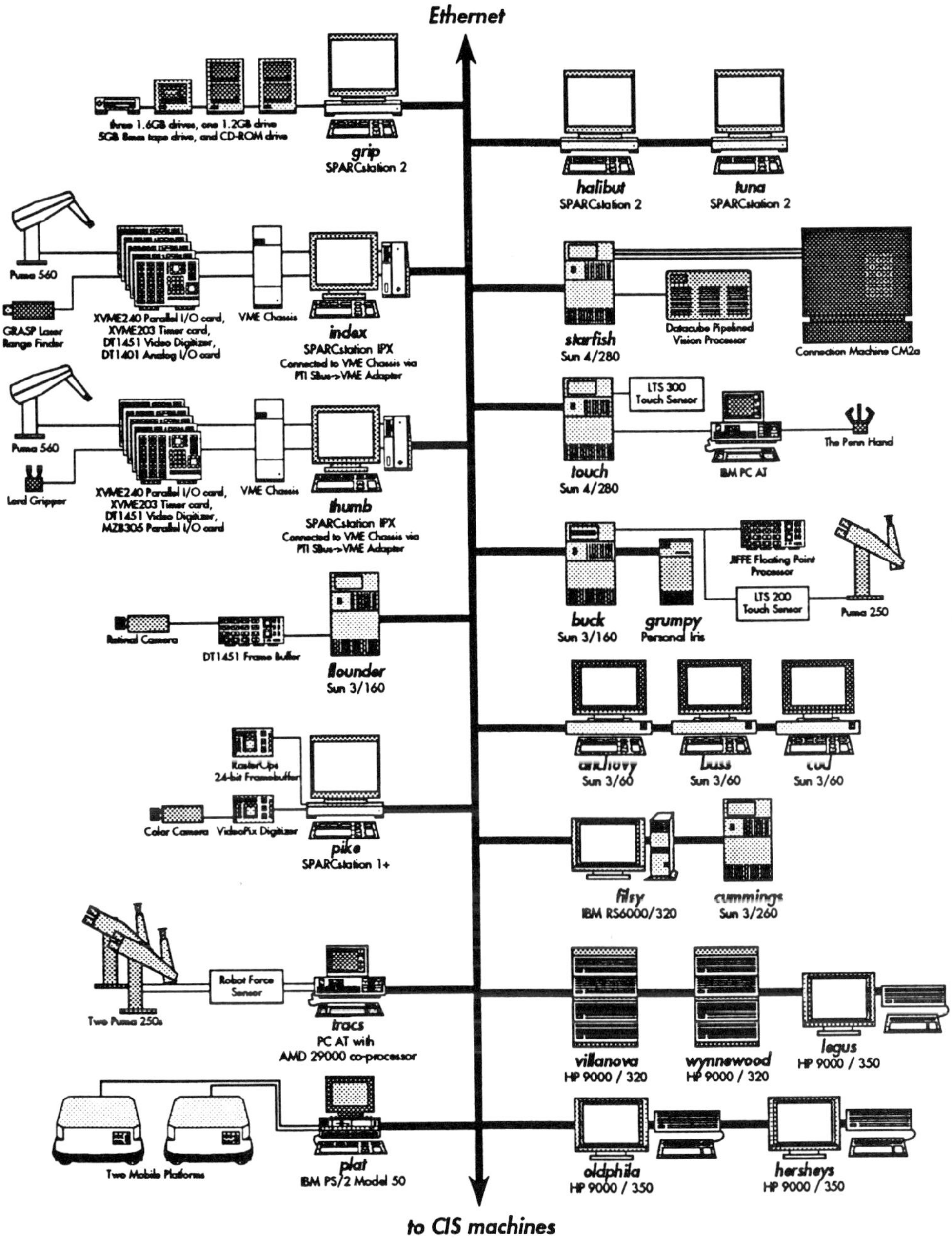

FIG. 9. Diagram of a local-area network (LAN), showing the asynchronous interaction between semi-independent processors which nevertheless may call on each other for data, program, or actual execution of program and data in combination at a remote site. (Courtesy of GRASP Laboratory, Dept. of Computer Science, University of Pennsylvania.)

akin to the "interchangeable parts" systems of 19th century New England has been developing.Standards are beginning to guarantee that individual subpieces of the overall system are replaceable, sometimes by software, sometimes by hardware. As long as the interfaces are well-defined as to interconnection syntax and behavioral semantics, individual designer groups can work on improving subsystems without interfering with the remainder of the system.

The evolution of computer systems thus can be viewed as a continued design process, making use of improvements and changes in technology as increased physical information is discovered, obeying flexible but locally fixed laws of interconnection.

5. THE LOGICAL EQUIVALENCE OF SOFTWARE AND HARDWARE

Through the systems optimization, or at least improvement, process that goes on alternatively with new hardware and software design, runs a common thread. It may be stated as a somewhat constrained principle: Computer software and computer circuitry are logically equivalent; at any time an example of one may be replaced by an equivalently behaving example of the other.

This is certainly true in that any circuit can be simulated by a program. Such simulators have become increasingly important in the CAD process for circuits, where simulators are used to test integrated circuit behavior. The hardware design language VHDL (a very high-level hardware description language) is an example of a language for creating such programs which simulate hardware. In the other direction, only programs that, intuitively speaking, do not modify themselves can be replaced by circuitry, since at present circuitry is not able to modify itself.

With the coming of low-cost, high-speed ASICs, these units are designed rapidly through CAD procedures. Portions of operating systems and application programs are beginning to be replaced by individual chips or subportions of larger systems on a chip. This process seems to indicate that the flow of innovation in digital computers will move from application and utility programs, able to be easily modified and improved. Such programs, relatively expensive to create, use, and maintain, will be transformed to chips, chip sets, and portions of larger microprocessor systems. These hardware systems will run much faster than the original programs. Primitive examples, mentioned elsewhere, include many cases where microprocessor programs were later reimplemented as peripheral processors in hardware.

Replication of design and fabrication costs is a chief factor contributing to overall systems cost. Such movement from software to hardware will probably prove most effective in applications in parallel to produce hardware parallel processes. The chief immediate application of such replication in this fashion, of course, is in the development of parallel processors consisting of many interconnected microprocessors (or microprocessor-like subsystems) performing the same or different operations at the same time. Such hardware will be described further below.

6. MODERN TECHNOLOGY

6.1 Applications Programs

The structure of applications programs has evolved towards a concept of "user friendly." A definition of this term might initially be that the program does not require from the user any specific knowledge of programming skills or operating system or computer structure. In addition, the program, through its GUI, should guide the user through the computational process by asking questions. For example, it can request responses by simple user actions (motions of the pointer-mouse, single "hot-key" keyboard responses). These replace the elaborate and arcane sequences of keystrokes originally required in many of the first applications programs, tailored to the programming technology of that time. At its most straightforward, the interactive program might offer to the user an elaborate version of "20 questions" in which the program is the questioner, working to discover the true desire of the user.

Such programs may be centered around a single (or more than one) paradigm or conceptual programming structure. The GUI, which likens the terminal screen to a notebook or blackboard, is one such presentation. The spread sheet, a matrix-like representation in which the user enters numbers and/or letter sequences to be processed by standard actions (preprogrammed or under user-control) is another. Standard word-processing programs offer a visual structure not unlike the pages of a manuscript. The more a user can call on previous experience, the easier use of such a system can be.

Many applications programs (like some operating system interfaces) offer menus, either in printed characters or in pictures on a graphical user interface. These constrain and simplify the choices that a user has when interacting. The system may offer the ability to browse through the menus, or additionally, through the text, pictures, or data that are being processed—or the program itself. The system should offer complete information and preventive editing so that incorrect or out-of-range data does not cause the program to "crash" (end up with insufficient information to recover automatically or manually, or to continue).

Applications programs for more complicated problems may require even more complicated interfaces. These should produce graphical results posed at high speed for the user to assimilate and respond to. Computer user and system thus become a pair of joint co-processors working interactively to solve a specific problem or class of problems. This process can of course be extended to many computers and many machines, provided an understanding of the needed interactions can be placed into the hardware and software.

6.2 RISC vs CISC

The advent of CAD for computers themselves, in the form of silicon compilers and layout programs, has produced basic changes. These are changing the output of the silicon chip assembly lines in such fabrication firms as Intel, Motorola, Texas Instruments, AMD, and National Semiconductors in the United States, and Siemens, NEC, Toshiba, and others overseas.

Because of the increased ease of designing and fabricating logical structures, a new overall systems approach including both hardware and software is now being undertaken. Most indicative of this process is the development of new reduced instruction-set computer (RISC) devices. System architects analyze the entire system structure consisting of software (program, compiler, operating system) and hardware (central microprocessor and peripheral chip set). The resulting optimization, originally carried out by researchers at Stanford University, the University of California at Berkeley, and IBM, among others, produced a different design from that of the classical main-frame computer and microprocessor—a class usually called complex instruction-(CISC) (for example, the Intel 80486 shown in Fig. 7). (It is the instructions themselves that are reduced or complex, not the set.)

In such RISC machines, much of the behavior of a complex instruction in a classical CISC machine is generated by the RISC machine's compiler at compile time, rather than by the hardware designer far earlier at the instruction design time itself. The combination of an optimized compilation sequence of simplified instructions is claimed (and often verified in actual hardware designs) to be much faster than a possible shorter sequence of the more complicated complex instructions. The inclusion of pipelined instruction operation and storage caches yields an advantage, in many cases, to simplified RISC instructions as compared to their CISC complex counterparts. Since the cost of memory has decreased with integrated circuit costs, a larger program instruction memory is now available, and thus brings down overall RISC cost.

One major efficiency of RISC designs (which may be possible to introduce into more classical designs such as the Intel 80x86—x=null, 1,2,...)—is the introduction of parallel performance inside individual operations themselves, so that additions and multiplications, e.g., can be performed simultaneously on different pairs of operands. As the cost of memory decreases, compilers can use more storage for instructions, translating and unwinding loops so that the product (memory access/instruction) times (number of instructions) continues to decrease.

Although RISC designs have been used in some main-frame and/or minicomputers, their most publicized appearance has been in microprocessors, which have become more powerful as their speeds encroach upon those of past minicomputers. RISC designs are competing markedly with the more classical Intel and Motorola CISC chips, both for performance and in the market place as parts of so-called work stations. The Intel 80x86 family, e.g., has served as the central processor in the family of IBM PCs and their clones (duplicate copies). Partially because of its speeds, this chip sequence in time has come to constitute an estimated 80% of the microprocessor sales market for personal computers, as the clock frequency, directly related to the overall performance rate, has increased to about 50 MHz (early 1992).

On the other hand, comparisons of performance of a recent RISC-based Hewlett-Packard work station, using the manufacturer's own RISC chip, generate claims of speeds "up to ten times faster" than the most recent, highest-speed Intel CISC microprocessor, the

80486. Machines based on this and similar RISC designs, by HP, MIPS, Sun, and IBM, and even by Intel and Motorola, have produced work stations fundamentally different from the PC architecture of the 80*x*86 family, and much faster. Part of the difference is masked by differing compiler technologies, and by different operating systems, as previously discussed.

Even so, the coming of these new architectures seems to presage further overall system optimizations that may change the software–hardware combination more markedly than before.

The future of mass-produced and -marketed machines (almost everything, in the near future, but the largest computers) will therefore depend on conditions of production and marketing among the large international companies that now increasingly dominate integrated hardware–software systems, as well as the larger number of entrepreneurial smaller organizations that appear to effect distribution.

Just as important, perhaps, will be subsequent software–hardware optimizations that may produce further marked efficiencies, not predictable before the integrated-circuit design and fabrication revolution. These may produce classes of designs as yet untried and perhaps not yet thought of.

6.3 Artificial Intelligence

Beginning in the mid-1950s, researchers began using the term "artificial intelligence" (*q.v.*) to describe potential computers that would behave like intelligent beings—e.g., humans. Competition for research funds has moved emphasis back and forth in this area between mechanical translation, game playing, so-called planning, and later "expert systems." Portions of work in image-processing analysis appeared to fall into this area.

6.3.1 Division of Artificial Intelligence into Problem Solving and Learning. Early efforts seemed to divide the area of study into applications to problem solving on the one hand, and efforts to create learning systems on the other. One of the key methods, now appearing in some commercial and governmental applications, was the attempt to copy a human expert's capabilities into a corresponding computer program running on a machine.

Applications have included copying of such human activities as geologic exploration techniques, testing and tasting of soup, recognition and discovery of terrorists and their activities, medical diagnosis, and a host of other less exotic human behaviors. Special programming procedures, built around a list processing language (LISP) or PROLOG (a logic programming language), bypassed some of the difficulties of the more classical machine-oriented and numerically oriented languages. This was in an attempt to move computer behavior much further into the logical area. So-called knowledge engineers produced "protocols" of medical experts' diagnosis investigations, in order to translate them into computer programs that would produce a similar behavior.

Most of these systems, independent of language, were based on a (condition, action) paradigm, in which the presence of a particular condition in the case at hand (whether medical, geological, or whatever) caused the computer to produce a corresponding information-processing action, whether inside its memory, or output to the observer for possible outside action. The priority of which conditions to give immediate action to, whether to perform them in parallel, etc., in some ways copied the structure of certain computers themselves.

The computer-expert system and the external user were once again a co-processing system, each responding to the other's input by changing state and yielding output. The design of a controlled interactive sequence of (condition, action) pairs thus became a first approximation at generating an intelligent computer program, whose instructions were made up of these individual responses.

Although the knowledge engineer might be considered to be learning from the expert whose protocols he recorded, the need for systems that modify themselves to produce better performance is still there. Such "adaptive systems" have been developed to aid various control systems for aircraft and factories to improve their performance, but these in general have depended upon classical statistical techniques—after all, a type of learning to predict behavior.

A new engineering-oriented theory, the "fuzzy sets" of Zadeh, attempts to introduce enough vagueness into the measurement of the external world that the computer can produce results that might be considered qualitative rather than quantitative. Results from such systems appear to remain experimental.

6.3.2 Neural Nets. The most sophisticated proposed technique for fine-grained intellectual behavior is based on an attempt to copy brain behavior at some detailed level. Both programs and hardware have been designed to copy individual neuron action, with self-adaptation through feedback to the artificial "synapse-like" input elements in individual units. Actual neural net machines, with variable control, have been designed and constructed using special ICs (see NEURAL NETWORKS). Compiler–assemblers are to interconnect substructures both initially, and perhaps later dynamically. Neural net programs so far have been successful in certain classes of simple tasks, but their competitiveness in practice still remains to be seen. They do offer a potential common method of applying learning techniques which otherwise appear markedly different in different problems to be attacked.

The most impressive examples of human-like behavior have perhaps been with chess-playing machines, which in certain cases have been able to defeat very competent human players. However, neither learning nor complex logical expert systems are deeply involved with such machines. Instead, they make use of the brute power of massive integrated circuit techniques to search widely and deeply within the tree of possible moves available to the machine and its opponent. A simple but powerful intellectual heuristic technique is able to limit this search by ignoring certain branches in which it would be of no value to continue the search. For the moment, the expert behavior of human chess masters does not appear to have been captured.

6.4 Creation and Standardization of Languages

A major factor in the development and usage of computers has been the creation of a mass of computer languages (see COMPUTER PROGRAMMING LANGUAGES) to specify the behavior of a specific machine or class of machines.

Two basic techniques of responding to computer languages, which are usually input through a programmer's interface, involve the level of the system which responds to the language:

1. Compilation-translation: This term describes the replacement of external language strings, understandable by the human programmer, by equivalently behaving computer language strings immediately interpretable as programmed instructions by the target computer. Such compilers are one-time performers, which do their translation job, and then retire from the system and leave the computer to interact with its own language translated strings.
2. Interpretation: This term describes a process in which a program (interpreter) stored permanently inside the computer continuously intercepts, interprets, and acts upon the external language strings, dynamically, responding element by element to the external language.

Obviously interpreters are more powerful than compilers, since they include the compiler capability, and are always resident, and thus contain the potential of changing the linguistic program strings with time—adaptively. Correspondingly, compiler-translators, because they look at the external language only once, yield a much faster overall system performance.

The classical example of a language that is compiled is FORTRAN, created by IBM programmers in the early 1960s. This is a formula translator that translates numerical formulas, appropriately modified at the programming input stage, into the languages of various target machines. FORTRAN has proved both effective and long-lasting; it is still being used as a standard language for certain classes of numerical analysis problems even with massively parallel computer systems, in a modified form to include control of the synchrony and asynchrony of individual computers performing concurrently.

An alternative example of an interpretive language is that of BASIC, first created at Dartmouth University at about the same time. BASIC proved particularly useful for interactive computing, which arose at the same time, because the need for recall of a compiler was obviated. It was adopted, because of its relative simplicity in usage (first designed for students), as the original programming language for the IBM PC, embedded in its ROM as a fundamental artifact purchased with the machine. Because of its interpretative slowness, however, BASIC does not compete in large problems, and does not appear as a widespread language of use, except in certain special cases.

Languages evolve as a function of classes of problems being solved, and classes of users or programmers. LISP and PROLOG, oriented towards functional and logical descriptions of problem solution techniques, came out of mathematical studies almost completely independent of computers. On the other hand, the language C, developed at Bell Laboratories, was conceived as a human-oriented programming language nevertheless allowing close interaction with actual machine instructions. Many important non-numerical applications, such as operating systems and utility programs for newer machines, are written in C. A sample is shown in Fig. 10.

An important step forward in computer language structure was the development of command languages (in classical computer command form) that were recursive in the formal mathematical recursive-function behavior that allows functions to call on themselves. From LISP, which had this characteristic, came the algorithmic language (ALGOL) languages, which for the first time had well-studied plans for such recursive behavioral implementation.

A family of computers, first created by Burroughs, and later adapted by microprocessor designers, used a so-called "push-down stack" to store data and addresses automatically on a "last-in, first-out" basis. Without these continually developed hardware capabilities, programs that call on other programs and each other recursively would be much more costly to implement. The extension of this initial structure to more elaborate register "context switching" is purported to produce much of the power of the new RISC architectures.

University users, in attempting to codify the fundamental structure of the more pragmatic commercial languages and simplify the ALGOL complexity, invented other languages such as PASCAL and MODULA, with which many students wrote their first programs.

The U.S. Government, a large user of programs and programmers, demanded standardization of its data processing procedures and produced (with other outside organizations) a common business oriented language (COBOL). The Department of Defense went further to produce an extremely complex language, ADA (named after an English woman regarded as the first computer programmer). This has elaborate certification procedures to guarantee that the results of translation, compilation, and performance would be the same for any machine on which it was to be run. Complex language such as COBOL and ADA combine behavior of both translators and interpreters. They perform certain transformations initially, and then the system interacts dynamically from time to time at run time.

Similar languages, such as VHDL for integrated circuit and chip description, were formally codified and have become part of the design process for microprocessors, ASICs, and other assorted hardware boards and systems.

In fact, one can view this expansion of languages as describing a second level of paradigm, in which the computer is viewed as an instrument to which a programmer communicates, talking about how to solve a restricted class of problems, but in many cases using a very complicated set of nouns (data structures) and verbs (commands).

While higher-level languages have offered tremendous power to the programmer in developing more elaborate data structures and transformations, they have not solved the problem of guaranteeing correctness of the program to produce a specific result. This is the goal of the logic languages (of which PROLOG is a somewhat modified offshoot) which attempt to describe the goals desired and from them produce commands to achieve those goals. Most such languages have achieved only research performance, although Japanese commercial plans indicate use of PROLOG.

The world of computer languages is Darwinian-like in its evolutionary behavior. The power and number of users, as well as investment in previous programs in a language, may guarantee its survival. IBM's imprimatur produced FORTRAN's longevity, but failed to preserve a more elaborate in-house language, PL-1, in the external world. LISP, a powerful and classical artificial intelligence language, failed to produce a much-wanted hardware interpretive commercial machine to run LISP

```
/* A single address, accumulator, machine simulator.  Program first
reads a program into memory and then executes it.  opcodes are:

        00      halt
        10      divide
        20      multiply
        30      subtract
        40      add
        50      load
        60      store
        70      branch on greater of equal
        80      branch always
        90      print value at memory address
*/
#include <stdio.h>
int memory[1000];                /* program memory */
int pc;                          /* program counter */
int acc;                         /* the accumulator */

main() {
  int number;                    /* number argument */
  int run = 1;                   /* run flag */
  int sym;                       /* value read in */

  while (scanf("%d", &sym) != EOF) /* read in program */
    memory[pc++] = sym;
  pc = memory[--pc];             /* starting address */

  while (run)
    switch (memory[pc++]) {      /* decode instruction */
    case 00:                     /* halt */
      run = 0; break;
    case 10:                     /* div */
      acc /= memory[memory[pc++]]; break;

    case 20:                     /* mul */
      acc *= memory[memory[pc++]]; break;

    case 30:                     /* sub */
      acc -= memory[memory[pc++]]; break;

    case 40:                     /* add */
      acc += memory[memory[pc++]]; break;

    case 50:                     /* load */
      acc = memory[memory[pc++]]; break;

    case 60:                     /* store */
      memory[memory[pc++]] = acc; break;

    case 70:                     /* bge, branch greater or equal */
      if (acc >= 0)
        pc =  memory[pc];
      else
        pc++;
      break;

    case 80:                     /* ba, branch always */
      pc =  memory[pc++]; break;
    case 90:                     /* print */
      printf("%d\n", memory[memory[pc++]]); break;
    }
}
Richard (Lou) Paul           Department of Computer and Information Science
email: lou@cis.upenn.edu     University of Pennsylvania
```

FIG. 10. A program written in C, one of the most prevalent languages at present in use by programmers developing applications programs for commercial marketing. (Courtesy of Professor Richard Paul, University of Pennsylvania.)

programs directly. (This would have speeded up productivity many times using this language. Nevertheless, users apparently could not support the costs of the equipment involved.) A very powerful language, APL, once supported by IBM and popular in Canada, the land of its creator, has almost disappeared despite its power in certain classes of vector computations.

The disappearance of such once-popular languages, like APL and PL-1, and the rise of the use of C, demonstrate how the constraints of technology, programmer acceptance, compiler complexity, and user demand on the part of government and industry produce survival of the "fittest" languages only in terms of very complex, far from only technical, measures of effectiveness.

The rise of C as a programming language, similar to, but less sophisticated theoretically and technically than, university-oriented PASCAL, shows the technical and market forces at work in language adoption and usage. The language C is closer to machine language than almost any other higher-level language, and contains operators that approximate many machine operations in the standard CISC microprocessors. Its close association with UNIX at Bell Laboratories and University of California, Berkeley, also gave a wide impetus to its usage. Moreover, it is a programmer's language, not forced upon its users by educational curricula.

Specialized languages include those for such user paradigms as the database search process. Designers here created so-called "fourth generation languages," such as SQL, successful in this area, but not extended more widely. The interest of the Japanese in interacting with potential European users of Japanese computers has boosted the potential of PROLOG, originally created and nurtured in France and England. Japanese "fifth generation computers," supported in development by their Ministry of International Trade, may use this logic language based on nonimperative structures similar to declarative sentences. On the other hand, the most powerful new Japanese supercomputers may competitively use FORTRAN, because of the nature of available programs.

A powerful, potentially useful class of languages are those first based on SMALLTALK, another product of the Xerox Park Research Center (which also developed the "mouse" and portions of the IC design systems). SMALLTALK is labeled an object-oriented language, in which the objects are generalized data structures and operations on each class of object are associated with that class. Object classes have a hierarchical hereditary structure in which descendants have the same characteristics as their predecessors, but with potentially more detailed structure and corresponding operations.

Object-oriented languages apparently owe their initial concepts, among others, to mathematical category theory, in which structures are essentially viewed in terms of pointing arrows describing abstract interactions. Despite what might be considered an arcane beginning, pragmatic object-oriented languages such as SMALLTALK and C++, a C derivative extension, are beginning to be used with claims of markedly increased programmer productivity. The abstract view of elements, as opposed to the details of the more classical languages, offers an alternative top-down view.

In the university research community, so-called "typed languages," again corresponding to mathematical structures in abstract algebra, allow powerful but correct manipulation of complex data structures. Such languages have not yet gained a foothold in any commercial or production sense.

Certain small groups of users, such as astronomers, have adapted special languages that prove easy to use and exchange (in this case FORTH, a "growing machine" language, which provides semantic clarity and abbreviational simplicity by sequences of Polish suffix operators, using interpreters).

Throughout this Babel-like set of language developments runs the pragmatic thread of appealing jointly to a body of users on one hand and to a specific computer architecture on the other. Nevertheless, up until very recently, the desire for control of the result by the user population has generally kept resulting programs, in almost any language, highly labor intensive in generation.

Clashes between esthetic formal theory and pragmatic user acceptance generally have found the latter winning out. The coming of the GUI, with its amplification of user performance, has begun to produce acceptance of automatic programming generation, triggered from such graphical interactive interfaces, by nonprogrammer users who wish to produce programs.

The clash here becomes one between the classical programmer, who desired freedom to create at will, on the one hand, and the more casual problem solver, who is willing to accept constraints on what can be communicated in return for more speed and power in producing results. Cheaper computation would appear to increase the potential for more constraints and less creative freedom. Among such constraint systems now are syntax-directed compilers, which only allow correct language strings to be typed. Also there are strong restrictions, in many newer systems, on what commands are allowed, and on use of replication features that manipulate previously written code, just as hardware designers use previously designed subsystems.

In general, this process of automatic generation of computer programs, still unsuccessful, illustrates the paradox that relates the two philosophical extremes: freedom and constraint. Machine languages offer almost absolute freedom to the programmer (with the sometimes crude constraints of the actual hardware). Only very crude syntax is required. As languages become higher level, stricter syntax is required, but semantics becomes more widely defined. The user is constrained as to what program primitives can be used. But these can be used, hopefully error-free, freely, and at very low cost. Just how the Darwinian process of language evolution will proceed in this situation is still to be seen.

6.5 Creation and Standardization of Operating Systems

It has been noted above that the operating system is abstractly the equivalent of the hardware bus inside various computer families, such as that of the PC. Its construction allows addition of subroutines and applications programs into slot-like positions, in a standard process, so that the set of programmed operations can be extended in a growing fashion.

The origin of the operating system, perhaps first developed on the Digital Equipment Corporation's model PDP-8, was the result of applying a classical expert system technique. One observes the sequence of interaction protocols of an operator seated at the console. These sequences, translated into a program, allowed replacement of the expert, in this case the operator, by a computer program that performed the same actions much more quickly and effectively.

The development of the operating system OS-360, for IBM's family of similarly named machines, powerfully demonstrated the potential of the operating system as the integrating factor for a family of computers. Its use continued over many users, and down through tens of years of time. The OS-360 system, however, did not have the interactive power that would allow humans to program and solve problems easily on line.

The development of the operating system Multics by the government-sponsored Project MAC at Bell Labs and MIT first proved the large-scale potential of interactive computing. Honeywell and General Electric used this design to produce the first commercially successful time-sharing operating system. The specialization of Multics to UNIX by workers at Bell Laboratories produced a system originally emphasizing processing of strings using terminals connected to a minicomputer. Over almost 20 years the efforts of the telephone company, AT&T, as well as other firms and universities, such as University of California, Berkeley, produced adaptations that showed success.

UNIX is now expanded to a broader range of equipment, including larger machines as well. Despite its requirements for large amounts of storage, improvement in hardware performance has moved it closer to becoming a possible standard for even personal computers. Since many work stations and larger computers now also use UNIX, this will allow passage of programs and data back and forth electronically between computers from different vendors. Users of one computer can perform portions of their programs on more powerful computers in the same network, with data from elsewhere in a network file system. Such capabilities require interactive network utility programs tailored to fit the cooperating operating systems at both ends.

The development of similar small operating systems on microprocessors, often restricted UNIX-like versions, culminated with Microsoft's disk operating system (DOS) on the IBM PC. This became a de facto standard for those operating systems running on that common de facto standard personal computer, the IBM PC.

The embedding of operating systems in graphical interfaces was first developed by Xerox and commercialized by Apple with the Macintosh. The Macintosh used the paradigm of the desk top, the clipboard, and the file cabinet. The transporting of this technology to the PCs by Microsoft recently produced a powerful, popular GUI, Microsoft's Windows 3.0, as an extension or replacement of that vendor's ubiquitous MS-DOS.

Other technological attempts to improve operating systems include placement of operating systems on ROMs or IC chips. The BIOS on the PC is one such example. Special operating systems have also been developed for real-time systems controlling man-made artifacts, such as aircraft, rockets, automobiles, and factory systems.

At a certain level, the operating system may be viewed as replacing the hardware computer itself as a machine paradigm, with its own operations, data structures, and interfaces. Portions of it may obviously move into the hardware itself. Probably more of the applications programs operations will become codified within the operating system, in a classical process of movement from human information and behavior into automated instructions in the machine.

7. SEQUENTIAL VS PARALLEL PROCESSING

With the coming of low-cost microprocessors, fabricated using VLSI technology, construction of massively parallel-processing systems became possible (see PARALLEL COMPUTING). These consist of large numbers of simpler subprocessors. Originally, parallel systems contained few subprocessors and were very expensive. With new technology, designs have grown in size from tens to thousands of subprocessors, connected in arrays that attempt to pass information and control from element to element in as efficient a fashion as possible.

Such systems require elaborate extensions of previous sequential-programming technology, its algorithms, routines, and languages. An entirely new set of control instructions, many borrowed from parallel hardware design procedures, had to be added to classical languages such as FORTRAN. Alternatively, new languages had to be developed. Investigations into which classical sequential algorithms could be easily and efficiently extended into parallel computation have been necessary. For those problems where this extension technique would not work, entirely new parallel algorithms were needed to make use of the new experimental parallel programming structures.

7.1 Classes of Parallel Equipment

In general, an individual subprocessor in an array is some type of sequential processor. Since each individual subprocessor in an array has both program and data, it is straightforward to classify parallel processing systems according to program (instruction) and data. If all of the basic subprocessors scan the identical program stream (and the same instruction) at once, the system is termed single instruction (SI). Otherwise, if different subprocessor control units scan and interpret different instructions, the term multiple instruction (MI) is used. If all of the data units act upon the same data element, the system is called a single-data (SD) parallel system. Otherwise, with different data being processed in different subprocessors, the label is multiple data (MD).

Thus, to a first approximation, such systems may be classified into four obvious categories: SISD, SIMD, MISD, and MIMD.

SISD machines are classical von Neumann–Eckert–Mauchly digital computers such as the first stored-program computers of the early 1950s. Recently, designers have come up with possible extensions to produce very long instruction word (VLIW) instruction processing. Such processing is now making its way into RISC microprocessor, as well as into single high-performance processor, design. VLIW instructions perform different operations, such as multiplication and addition (included in one instruction), in parallel on local data sets which are made independent of one another through pipelining, vectorizing, or other tactics.

SIMD computers most often operate on vectors of many individual data elements, with the same instruction in parallel. Such architectures very early became important in large numerical calculations able to be expressed in vector-like form. Searches of many databases in parallel, for the same information, also falls into this category. Such problems replace sequential many-input calculations, repeated in time, by the same calculation sequence performed by duplicate processors over space. With these, there are local problems whenever programs produce different conditional transfers for different data inputs. This obviously no longer permits easy use of the same instruction simultaneously in every subprocessor.

7.1.1 MISD and MIMD Systems. MISD systems involve solving many problems with one data stream. A computer with multiple searches on a single data-base stream is an example of such a processor type. Another technique is the use of pipelining, which takes sequential instructions and performs them concurrently on successive iterates in a sequential data stream. The most sophisticated of such systems might well be termed MIMD systems, as described below.

Most important in the future will probably be MIMD systems, which can support multiple instruction streams by pipelining and parallel processing. In the latter case, the interconnection structure by which information, both instructions and data, is passed between processors becomes important, and can be used for further classification. MIMD systems can, e.g., use a bus, or common information channel, to pass information at lower cost but lower overall speed. Most lower-cost recent commercial systems, such as the Sequent and Encore computers, originally used standard microprocessors as the processors connected with common memory, along a bus.

More complicated MIMD systems, such as *n*Cube, and the Hypercube and later versions from Intel, connect each memory-equipped processor to its *n* nearest neighbors, to form an abstract logical hypercube (see Fig. 11). Messages are sent between the node subprocessors in the hypercube to synchronize the concurrent computations.

Hierarchical direct connections include such geometrical forms as trees, pyramids, and prisms. Crossbar machines (in which every processor is connected to every other), are overly expensive, so far, except for small systems. Marked changes in cost of such communication between subprocessors would be necessary before complete crossbar systems became commercially viable.

A most important characteristic of a parallel processing system is its granularity, defined in terms of the size of unit of work allocated to individual processors. Until recently, the most powerful systems were termed "coarse-grained," such as Cray, IBM, and previous ETA systems, with four to ten very high-speed processors, each individually constructed, generally not using VLSI fabrication. Such systems have computing power ranging up to thousands of megaflops. ("Flops" stands for "floating-point operations per sec-

FIG. 11. A multiple-microprocessor parallel machine with potential for more than 1000 individual separate processing units connected to "nearest neighbors" in a cube-like "hypercube" array. (Courtesy of Intel Corporation.)

ond;" mega and giga are the standard prefixes for 10^6 and 10^9.)

The relative cost performance of such coarse-grained processors, compared with alternatives using many-times replicated VLSI microprocessors, appears to be decreasing through multifabrication economies of scale. These "supercomputers," therefore, may tend to be replaced by the massively parallel newer designs.

7.1.2 Cost of New Large Systems. Current 1991 prices for Cray's smallest to largest computers range from $300,000 to $30 million, according to published reports. The largest of these, the Y-MP C-90, is announced as having a central processing unit that can perform at the rate of one gigaflops. For the top price unit, there will be 16 such processors, each program-compatible with previous systems.

New examples of massively parallel designs have been announced recently. An Intel announcement of a new massively parallel processor (Paragon XP/S) with up to 4000 internal processors, gives a range in cost from $2 to $55 million. This system uses standard Intel 860XP RISC-like processors in each node, and is claimed to generate 5 to 300 gigaflops. Such parallel systems with so many internal subprocessors would tend to fall into the medium-grained category, based on the necessary detailed work each subprocessor might do. An Alliant Computer announcement gives price ranges of $2.5 million for a 100-processor system to $16 million for an 800-processor system.

Among other medium-grained commercial systems would be those using tens of Intel 80*x*86 or competitive RISC processors connected along a bus (Sequent, Encore, and other less powerful commercial versions). So-called fine-grained systems were built around simpler and cheaper basic units. The Connection Machine, built in hypercube interconnection with a special chip as subprocessor, is an SIMD machine which replicates the same computation in every one of its thousands of elements.

7.2 Relative Serial and Parallel Efficiencies

Simulation and direct experimentation on existing equipments is being used to test classes of problems according to a criterion of "efficiency": the ratio of actual performance of processors, interacting in an inefficient manner due to delays in waiting for the results of other computations, to that performance possible with perfect undelayed response from all other processors at the same time. [A sequential machine, which does not wait for outside results, has an efficiency of 1, for one processor. An n-element parallel processor with perfect efficiency of 1 would have an efficiency ratio of $n \times$(n-machine parallel eff.)/(sequential eff.), with a maximum value of n.]

Certain types of near-linear partial differential equations are solved on key MIMD machines with a very high efficiency, near 1, and efficiency ratio close to n. Nonlinear partial differential equations, on the other hand, which require more elaborate communication (therefore larger communication delays) when solved on parallel processors, have a much lower efficiency ratio.

7.2.1 Methods of Improving Performance. Methods of improving performance of such complex solution procedures in parallel include the attempted use of Monte Carlo (probabilistic) analyses, easily replicated in parallel. Recent conference reports on high-performance computing include examples of this and other new tactics. These develop parallelized solution processes, alternatives to sequential classical procedures that are not easily extended.

In any case, of course, even with low efficiency of usage of parallel architectures, any marked improvement means that solutions to otherwise impossible problems may be obtainable in a finite time. Most of the improvements in productivity in analysis and design of large physical and engineering systems will probably depend on improved usage of such massive architectural computer complexes. Two recent descriptions of experience at Sandia Laboratories and IBM Research make the claim that massively parallel MIMD architectures are becoming more commercially competitive and application effective.

7.2.2 Examples of Problem Solution Implementations. Among the applications whose efficiencies have been studied for MIMD shared and distributed memory architectures is the use of "farm parallelism," with many workers doing the same task, applied to the modeling of gas-phase reactive systems. The problem is broken into several elementary chemical reactions, each investigated separately and concurrently. Many classical events (so-called trajectories) are calculated concurrently. Experimental results on an IBM-3090 multiprocessor (6 subprocessors) and an *n*Cube 2 with 128 nodes were carried out for 4096 trajectories. Efficiencies of over 90% were obtained with several alternative grains of parallelism.

7.3 Areas in Which Parallel Computers Have Succeeded

Other numerical calculations recently described, where parallel computers improved performance, include Navier–Stokes equations (IBM-3090-VF, 6 processors); simulation of laminar–turbulent transition (Cray 2); structure of semiconductor strained superlattices (IBM-3090VF); 3D shallow-water models (Alliant FX/4, 4 processors); and Poisson problem using domain decomposition (IBM-3090VF, Alliant FX/80). Again, future effectiveness of such architectures will depend upon growth in effectiveness of the overall complex of programming languages, compilers, and operating systems, as applied to such parallel performance.

8. SPECIAL-PURPOSE EQUIPMENT STRUCTURES

Because of the increasing ease of design using silicon (or alternative technologies, of which gallium arsenide is the only remote future contender, at the present time), many special-purpose devices for replacement or extension of standard classical machines have been tried; some have been successful. Almost all retained the stored program.

Among past entries not able to be successful commercially were special-purpose database machines. Among those which have proved successful is the graphics image creation machine originally developed at Stanford University, and mapped into silicon by Silicon Graphics. This forms a powerful chip set that has enabled computer graphics presentations on work stations to be markedly sped up at a large decrease in price over previous techniques using the central processing unit.

8.1 Simplified Microprocessor Controllers and Coprocessors

Special simplified controllers have been developed for controlling portions of a microcomputer system. For example, a separate microprocessor, of simplified design, often controls the microcomputer keyboard or the hard-disk drive. Even more simplified special-purpose computers, often without stored programs, control interrupt processing, real-time processing, or analog-to-digital and digital-to-analog conversion outside of the central microprocessor.

The original microprocessors had too few transistors to carry out floating-point arithmetic operations. Therefore, arithmetic coprocessors, on a separate chip, accepted operational commands and data from the central microprocessor. This chip performed the actual multiplications and divisions for scientific and engineering applications programs that required high-speed arithmetic. Successful co-processors of this type were originally off-chip (such as the 80*x*87 floating-point processor for the Intel 80*x*86 series). With growing transistor count on chips, many of these original special-purpose co-processors have since been incorporated into the newer designs which allow for far more functions on the same chip.

8.1.1 Real-Time Applications. Special-purpose real-time microprocessor controllers, such as the Motorola 6802, have become integral parts of automotive systems, acting to control antilock brake systems and to monitor engine behavior. Other classes of specialized digital computers, mass produced in silicon, have found mass applications. Large areas of technology, previously based on analog electronic voltages, are now being manipulated with digital devices. For example, the telephone industry has spawned the digital signal processor, especially made to handle digitized audio signals. High-definition television (HDTV), which has competing analog and digital proposed systems, will probably end up almost completely digital in character, with digital computers as the central controlling and processing elements.

8.1.2 Future Control Devices. Such devices are example of mass-produced, low-cost, stored-program computers, interacting through sensors inputting signals from the outside world. Their output, through effectors, can be used to change and control the outside world. The ability to change their internal behavior, either from outside, or by means of their own program, through tactics of control theory, artificial intelligence, or other heuristic techniques, offers the capability of small, local devices, independently responding, where needed. Their applications to experiment in many areas are still ahead.

8.2 Digital Signal Processors

With its capabilities, an important section of digital computer usage, not ordinarily known to the standard user, is that of the digital signal processor (DSP). These specially designed digital computers are particularly oriented to handle streams of signals in such devices as cellular phones, private branch exchanges, central telephone exchanges, industrial controllers, telephone answering machines, motor controls, multimedia applications, and such internal digital computer mechanisms as hard-disk drives.

The DSPs themselves are now being produced by use of customized design procedures that allow inclusion of new and specialized logical functions on a central computer chip. An example is the Texas Instruments TMS320 series of chips, which allows customized inte gration of analog-to-digital and digital-to-analog converters on the central integrated circuit. Such input–output devices translate from classical audio and other signal levels into digital samples, and conversely.

Such combinations are now being fabricated in low-power CMOS technology, with 32-bit floating point arithmetic units included on the same chip. Unit prices are quoted at "$1/Mflops in 1992," with 27-Mflops units currently being shipped.

The integration of a standard central unit (here the DSP itself, oriented to handling streams of numbers very efficiently), with peripheral elements designed in at the silicon level, produces a chip specific to the application at hand. This is an example of the fundamental logical equivalence of software and hardware and of the continuing movement of functions, originally implemented as programs, into the design of the central silicon chips themselves.

Almost any electronic consumer product, from automotive brake and audio systems to modems and telefacsimile devices, will probably use something like a DSP as its central control element. Most users of standard commercial computers and networks will not directly interface with DSPs. Nevertheless, their existence offers the capability of extension of real-time networks into a broad range of data collection and control procedures, far beyond the standard artifacts most often thought of as computers. As applications arise, other analogous designs, still with stored programs, may extend the range of classical computer systems even further.

9. COMPUTER ECONOMICS IN 1991

9.1 Overall Economics of Computers

Early usage of digital computers entailed high cost of equipment and low productivity due to individual human programming. Financial support was originally economically centered on governmental or large corporation funding.

Moreover, within the United States and Western Europe, construction and distribution of the machines themselves was controlled by means of proprietary systems. These were specially designed so as to prevent easy passage of computer programs from one computer manufacturer's machine to that of another manufacturer. A commercial or scientific user, therefore, was locked into its manufacturer's equipment and programs. Most scientists or engineers involved with large calculations, until recently, at least, may have experienced this situation.

A similar proprietary situation began with the software, although it too generated a conflict: proprietary restrictions vs ease of circulation. On one side, computer programs are easily reproducible using computers themselves (like books with photocopying). On the other, their purchase has become surrounded with complex licensing procedures, copyrights, and other attempts by the vendors to restrict unauthorized reproduction. Just how exactly innovation will be encouraged or discouraged by such processes is still to be worked out, since proprietary families of programs obviously become tied into operating systems, for example.

9.2 Standards Organization

To allow crossover of programs from proprietary hardware or software family to other families, a standardization process began to be developed for hardware, languages, and operating systems. Such standardization mitigates against the restrictions of proprietary hardware, operating systems, and applications programs, all of which aid individual manufacturers in protecting their market share. There is thus a conflict between the forces of commercial hardware and software producers, and the users. The former want to control certain parts of the overall system, to optimize market share and profit. The users want some optimal method of producing the resulting computation at low overall cost in time and money. This clash provides the key continuing dialectic for computers today. Much of the ongoing changes, in particular the effort at standardization, are attempts at resolving this conflict.

9.2.1 International Economic Conflicts. Among the hardware devices that came to be fabricated in quantity, one became an international *cause célèbre* rivalling that over imported automobiles. DRAMs became a cheap but powerful method of storing programs and data during processing. Without them, the PC would have been impossible. In the United States, only the largest manufacturers such as IBM and AT&T were able to compete with their own in-house versions of these devices.

Overseas, this part of the computer technology market became economically dominated by Japanese integrated circuit manufacturers, often accused of "dumping" products below cost onto the American market. This alleged oligopoly and its future effect on production of U.S. computers (mirrored by alleged U.S. control of microprocessor production) became a prime "political football" in U.S.–Japanese economic relations. This economic conflict emphasizes the rapid rise of computers to a most important level in the international economy. The DRAM's marketing as a commodity product demonstrates the importance of the standardization process for hardware, here built around the concept of interchangeable electronic parts.

With standardization, the sociological structure of machine usage began to form around standardized technological artifacts: standard hardware, standard languages, standard operating systems, and standard graphical user interfaces. Behind these standards, the hardware and software being distributed began to be commodity products, sold on price and effectiveness, not the proprietary products sold on the historical necessity of the user not losing the investment in past hardware and software. Standardization began when U.S. and European governmental organizations, such as the Department of Defense (DOD), and industrial user or vendor consortia began to negotiate common standards. Among classically important ones were the ALGOL standards, the International Standards Organization agreements on the open systems interface, and the standards on the FORTRAN, COBOL, and ADA languages, the latter two required for U.S. governmental program procurement. Such standards increased the probability that hardware and software from different origins would nevertheless interact with each other successfully.

9.2.2 Hardware Becomes a Commodity Product. As a result of this effort, at the end of the 1980s, the economics of computer manufacture, marketing, and usage began to change markedly. Before, the use of proprietary hardware (the IBM series of IBM-360 and subsequent mainframes) or of proprietary operating systems (such as the VAX-VMS system of Digital Equipment Corporation) locked users into a continuing sequence of same-vendor equipments.

During the last years, much of the computer industry has been forced out of this pattern. A lead story in a September 1991 Wall Street Journal, entitled "Computers Become A Kind of Commodity, To Dismay of Makers," indicates the extent of this change. This report, as well as somewhat more technical reports in trade journals, lists the changes:

1. Customers, except for the buyers of the larger machines, are buying computers as if they were commodities, attempting to buy as cheaply as possible.
2. Profit margins have dropped; sales channels have been transformed away from retailers towards mail order distribution and mass purchasing. The result of decreased margins has forced virtually every computer manufacturer to try to "reinvent itself." Some manufacturers (e.g., Wang)

no longer sell their own computers, only software for another manufacturer. New alliances, such as that between once arch-rivals Apple and IBM, are springing up.

3. With the loss of proprietary computer system monopolies and the coming of "open systems," the sale of "clones" (or near-exact duplicates), such as for the IBM PC, brings new competition. New quasimonopolies, for the moment, have arisen in software (such as the Microsoft DOS PC operating system and its Windows derivative). In hardware, Intel, the maker of the system 8086 microprocessors and derivatives, may also be momentarily considered a quasimonopoly, despite the beginning fabrication of duplicates by competitors.
4. The coming of UNIX as a standard operating system offers a new level of open system and commodity product structure. UNIX has now become the standard operating system of computers at the work station level above the IBM PC and its clones. For much of the past, implementation of full-scale UNIX on PC-level machines was impossible because of RAM and hard-disk storage limitations, but recent pricing decreases offer storage capabilities that make UNIX competitive. However, the complexities of the system, without appropriate graphical interfaces such as the open-system X-Windows, do not make it a system for the casual user.

The result predicted is that the number of vertically integrated hardware vendors will decrease markedly, replaced by a smaller number of fabricators in the U.S. and overseas. Consulting and integration companies will become more important. Software companies, with proprietary programs, may replicate the past domination by proprietary hardware systems—until their products, too, become commodity products.

All this, of course, does not take into consideration very much what foreign (most likely Asian) competitors in both areas may do. It also assumes that computers will merge "gracefully" into all areas of modern industrial society, which is far from likely. Nevertheless, such critiques seem to indicate that, whereas the technological future of computers might be considered predictable, their place in the economy may shift drastically due to such technological changes themselves.

New paradigms for computer users that will be designed to hide the actual software–hardware structure will continue to be invented. The beginning capability for computer statistical, neural net, or other "learning" will begin to grow if small practical commercial attempts prove successful. The size of computers, decreasing from main frame to desktop to laptop to notebook size and below, makes new applications suddenly feasible. The interactive one-person, portable computer, connected by cellular phone networks to e-mail and giant database servers, appears almost within the present technological-economic range.

There thus appears to be a dichotomy in future usage: On the one hand, smaller and smaller physical computers will become more tailored to individual users, with these individual software–hardware systems adapting themselves to the needs and idiosyncracies of the individual human with whom they are in synergy. On the other hand, such devices (and others) will also serve as end organs of input–output to giant client-server network systems. These will provide knowledge, computational, and retrieval services for varied forms of information, with many alternative methods of presentation. Interaction with other humans on this network will be rapid and powerful. Problems of misuse of such systems are among the most important not yet posed.

9.3 Fundamental Changes in Scientific Economics

One major effect of the computer revolution on scientists and engineers has been the disappearance of many "wet laboratories." These classical laboratories, where experiments were carried out on the actual physical, chemical, or mechanical devices, have in many cases been replaced by digital computer computations to predict comparable behavior. Wherever the theory is trusted, stored-program computers, with programs to solve problems based on that theory, should often be able to produce predictable results far less expensively.

On the frontiers of a new field, experiments obviously are necessary before any theory can be created or verified. Nevertheless, many highly sophisticated experimental laboratories, such as wind tunnels, are no longer being duplicated in quantity. Predictive behavior of supersonic planes, in most cases, are the results of elaborate computer calculations. The software component of such classical organizations as Bell Laboratories has become an increasingly important financial factor. Originally, the experimental laboratories were often most expensive; they are still present, but economically may no longer dominate. After that the hardware costs of the computer equipment, so often first supported by the government in national laboratories, became a marked burden. Finally, in time sequence, the software development for such computer hardware has become increasingly expensive. The process of prediction and verification of experiment is thus intricately entwined, both technologically and economically, with use, design, and development of both hardware and software. In certain cases, the computer has become the experiment.

GLOSSARY

Adaptive System: A system which changes its behavior as a function of its input.

Analog: Method of representing information by continuously varying physical quantities.

Arithmetic–Logic Unit (ALU): A portion of a computer which performs arithmetic calculations and also makes logical decisions based on the arithmetic results.

Application-Specific Integrated Circuit (ASIC): Usually an IC designed especially for some specific problem solution, as opposed to general-purpose microprocessor ICs.

Basic Input–Output System (BIOS): That part of the operating system concerned with handling the details of input and output procedures; usually stored in ROM.

Bus: A (usually parallel) set of wires connecting many units together in a common pathway for data or instruction passage.

Cache: A small, high-speed storage unit between RAM and the CPU or microprocessor's register storage, that requires far less time to access, and thus allows the computer to run much faster.

Computer-Aided Software Engineering (CASE): An area of study and development of ways to produce effective programs semiautomatically or automatically.

Central Processing Unit (CPU): That part of a computer which interprets the program and carries out instructions.

Chip: A technical term for the usually rectangular piece of silicon substrate onto which integrated circuits are deposited and fabricated as a microelectronic device.

Complex Instruction-Set Computer (CISC): One in which individual operations carry out elaborate data processing. Such designs are alleged by RISC protagonists to be much less efficient, slower, and more costly.

Clone: A "duplicate copy," usually marketed, designed so as to duplicate the performance of the original design. For computers, most recent clones are copies of the IBM PC.

Coarse-Grained: A parallel processing system with each subprocessor working on large subsets of a programming procedure at the same time.

Compiler-Translator: A service program that translates from a higher-level language to the basic machine code language.

Computer Operator: A technician who ordinarily controls the sequence of operations of a computer, originally manually.

Converter: Electronic circuit to transform analog signal representation to digital, and conversely.

Digital: Method of representation of information by variables that can take on only a limited set of discrete values.

Digital Signal Processor: A special type of microprocessor designed specifically to manipulate streams of digital representations, usually of external analog signals.

Dynamic Random-Access Memory (DRAM): Storage device which must continually be refreshed to retain information.

Expert Systems: Programs which attempt to copy the effective behavior of human experts.

Fine-Grained Processing System: A parallel processing system with each subprocessor working on a very small portion of a programming procedure at the same time.

Finite-State Machine: A formal representation of mechanistic behavior describing the overall flow of performance in terms of inputs, outputs, internal states, and functions relating their local behavior.

Flag: Single binary digit indicating the existence of a particular situation—usually the recent occurrence of a particular event, such as an addition operation with overflow.

Floating Point: Numerical encoding in terms of powers of 2 or 10 with fractional coefficients, to allow computation without the overflow obtainable with fixed-length fixed-point numerical representation.

Fuzzy Set: Nontechnically, an imprecisely defined set which allows imprecise but hopefully useful reasoning to be carried out on its behavior.

Graphical User Interface (GUI): A pictorial presentation, usually on a cathode-ray tube, of the user's range of choices for the next step, allowing much easier programming and operation of a computer.

Handshake: A two-way passage of signals between two concurrent portions of an interacting system, that allows synchronization of their behavior.

Hardware: Generic term for any or all computer electronic equipment.

Hot Key: A key on the computer or terminal keyboard that is programmed to cause a sequence of operations to occur that would otherwise require a more elaborate sequence of key strokes.

Icon: A pictorial representation of a file or program, on a GUI.

Interpreter: A continuously resident service program that translates higher-level language on line at run time, performing actions based on the semantics of the linguistic elements.

Interrupt: An external signal from a peripheral to the CPU or microprocessor stating that it needs some type of service.

Local Area Network (LAN): An interconnection of computers usually within a contiguous geographical area.

Machine Code: The simplified code of the actual machine hardware, which translates at run time into actual electronic switching actions.

Macintosh: A very successful personal computer with a file and clipboard paradigm for its CRT presentation, using a mouse.

Main Frame: Classical term for a physically very large expensive CPU with all its peripherals.

Menu: A listing, as in a restaurant, for options—programs, data, etc.—that a programmer may choose to have the computer perform or operate on.

Microprocessor: A CPU fabricated, usually from silicon, with very small transistors and a very small chip substrate.

Multiple Instruction, Multiple Data (MIMD): Description of parallel processing systems that allow many individual processors to work simultaneously, in one program, on different data with different instructions.

Motherboard: A large printed circuit board, constituting the basic physical structure onto which a computer's electronic elements would be mounted.

Mouse: Small hand-held device used to move a cursor around the screen.

MS-DOS: A small operating system originally applied to the IBM PC as PC-DOS; often called simply DOS. (From Microsoft Disk Operating Systems.)

Neural Net: A detailed interconnection of processors with the aim of producing behavior comparable to that of human brain neurons or neuron groups.

Object Oriented Programming (OOP): The use of object-oriented languages to manipulate classes of data structures with common properties, including the partial ordering of inheritance.

Operating System: Control program that performs housekeeping for applications programs, including file storage and retrieval and input–output, also offering simplified language for control.

Parallel Processor: A digital computer made up of more than one subprocessor, which performs more than one instruction at the same time.

PC: Personal computer—a computer, first available about 1978, which usually interacted with only one person.

Pipeline: A processing structure taking sequential information, such as numerical elements in a vector, and processing more than one at a time, to speed up arithmetic or data manipulation.

Polling: Cyclic consultation of peripherals by a CPU or microprocessor.

Protocol: In nondiplomatic usage, the detailed recording of experiments, human interactions, or computer behavior, for example.

Pushdown Stack: A hardware or software structure allowing "last-in, first-out" behavior of storage into it, usually broadened to include even more elaborate storage procedures.

Random-Access Memory (RAM): An electronic read-and-write memory in which every stored element can be consulted equally easily.

Read-Only Memory (ROM): An electronically unchangeable memory that can be consulted at very high speed.

Reduced Instruction-Set Computer (RISC): A computer based on a set of simplified instructions; claimed because of its instructions' simplicity to perform far more effectively than previous designs.

Run Time: That interval when the computer is actually performing a program, as opposed to compile time, when the compiler is operating, or other preliminary intervals not associated with actual performance of the solution program.

Sequential Processor: A digital computer which performs operations only in sequence, not several at the same time.

Silicon Compiler: A compiler–translator whose target language is a linguistic description of an actual chip-fabrication plot.

Software: Generic term for any and all types of programs.

Spread Sheet: A computer program that presents a diagram ruled horizontally and vertically into which numbers, equations, and other information may be entered, to be operated upon by very general operations (such as matrix-vector multiplication, for example). The term "spread sheet" comes from the original paper equivalent.

Static Random-Access Memory (SRAM): Random-access memory that retains its contents without the need of repeated refreshing.

Stored Program: Class of computer where the problem solution procedure is stored in a changeable memory.

Time Sharing: A computer usage method by which many users at terminals share the performance of a powerful central processor.

UNIX: An operating system originally developed by Bell Laboratories about 1970.

User-Friendly: An adjective describing hardware or software which is easy for a nontechnically oriented user to interact with.

Utility Program: A program to aid in understanding and manipulating the contents of storage and behavior of the machine, often after nonpredicted behavior.

Very High-Level Description Language (VHDL): A language for describing computer architecture along with more detailed circuitry description down to integrated circuits.

Video-Tex: An interactive computer-driven public system for interrogating data bases such as telephone books, airline schedules, etc.

Very Long Instruction Word (VLIW): Describing a type of processor that allows concurrent behavior in a single instruction, and therefore speeds up each instruction.

Windows: Rectangular CRT areas in which different data or instruction sets can be presented, so that the user can deal with more than one group of information on the screen at the same time.

ACRONYMS

ALU: Arithmetic-Logical Unit
ASCII: American Standard Code for Information Exchange
ASIC: Application-Specific Integrated Circuit
BIOS: Basic Input–Output System
CAD: Computer-Aided Design
CASE: Computer-Aided Software Engineering
CISC: Complex Instruction-Set Computer
CPU: Central Processing Unit
CRT: Cathode-Ray Tube
DMA: Direct Memory Access
DOS: Disk Operating System
DRAM: Dynamic Random-Access Memory
DSP: Digital Signal Processor
GUI: Graphical User Interface
LAN: Local Area Network
MIMD: Multiple Instruction, Multiple Data
MISD: Multiple Instruction, Single Data
PC: Personal Computer
RAM: Random-access memory
RISC: Reduced Instruction-Set Computer
ROM: Read-Only Memory
SIMD: Single Instruction, Multiple Data
SISD: Single Instruction, Single Data
SQL: Standard Query Language
SRAM: Static Random-Access memory
VHDL: Very High-Level Development Language
VLIW: Very Long Instruction Word

Further Reading

Bach, M. J. (1986), *The Design of the UNIX Operating System*, Englewood Cliffs: Prentice Hall.

Barkakati, N. (1991), *X Window System Programming*, Carmel, IN: SAMS.

Birnes, W. J., ed. (1989), *McGraw-Hill Personal Computer Programming Encyclopedia: Languages and Operating Systems*, New York: McGraw-Hill.

Brinch-Hansen, P. (1977), *The Architecture of Concurrent Programs*, Englewood Cliffs: Prentice Hall.

Camposano, R., Wolf, W. (1991), *High-Level VLSI Synthesis*, New York: Kluwer Academic.

Curry, D. (1991), *Using C on the UNIX System (A Nutshell Handbook)*, Sebastopol, CA: O'Reilly and Associates.

DeCegama, A. L. (1989), *The Technology of Parallel Processing*, Vol. 1, *Parallel Processing Architectures and VLSI Hardware*, Englewood Cliffs: Prentice Hall.

Denker, J. ed. (1986), *Neural Networks for Computing*, New York: American Institute of Physics.

Durand, M., El Dabaghi, F., eds. (1991), *High Performance Computing II, Proceedings of the Second Symposium on High Performance Computing, Montpellier 1991*, Amsterdam: North Holland.

Goldberg, A. (1984), *SMALLTALK-80, The Interactive Programming Environment*, Reading MA: Addison-Wesley.

Huber, J. P., Rosneck, M. W. (1991), *Successful ASIC Design the First Time Through*, New York: Van Nostrand Reinhold.

Hwang, K., Briggs, F. A. (1984), *Computer Architecture and Parallel Processing*, New York: McGraw-Hill.

Iannucci, R. A. (1990), *Parallel Machines: Parallel Machine Languages*, Boston: Kluwer Academic.

Katevenis, M. G. H. (1985), *Reduced Instruction Set Computer Architecture for VLSI*, Cambridge, MA: MIT Press.

Kochan, S. G., Wood, P. H. (1985), *UNIX Shell Programming*, Indianapolis: Hayden.

Lipsett, R., Schaefer, C., Ussery, C. (1989), *VHDL: Hardware Description and Design*, Boston: Kluwer Academic.

Mead, C., Conway, L. (1980), *Introduction to VLSI Systems*, Reading, MA: Addison-Wesley.

Patterson, D. A., Hennessy, J. L. (1990), *Computer Architecture: A Quantitative Approach*, San Mateo, CA: Morgan Kaufman.

Peek, J. (1991), *MH & xmh*, Sebastopol, CA: O'Reilly and Associates.

Reiss, L. (1987), *Introduction to Local Area Networks with Microcomputer Experiments*, Englewood Cliffs: Prentice Hall.

Rubin, S. M. (1987), *Computer Aids for VLSI Design*, Reading MA: Addison-Wesley.

Stevens, (1989), *Unix Networking Programming*, Englewood Cliffs: Prentice Hall.

Tabak, D. (1990), *RISC Systems*, New York: Wiley.

Veronis, A. (1991), *Survey of Advanced Microprocessors*, New York: Van Nostrand Reinhold.

Wechsler, H., ed. (1990), *Neural Networks for Perceptrons*, Vol. 2. New York: Academic.

Wiener, R., Pinson, L. J. (1988), *An Introduction to Object-Oriented Programming and C++*, Reading, MA: Addison-Wesley.

COMPUTERS IN PHYSICS

See PHYSICS APPLICATION OF COMPUTERS

CONDUCTION IN ONE AND TWO DIMENSIONS

See ELECTRON STATES IN ZERO, ONE, AND TWO DIMENSIONS

CONDUCTIVITY, SURFACE

See INSULATORS AND SEMICONDUCTORS, SURFACE CONDUCTIVITY IN METALS AND ALLOYS, SURFACE CONDUCTIVITY IN

CONFORMATIONAL ANALYSIS

S. SCOTT ZIMMERMAN, *Department of Chemistry, Brigham Young University, Provo, Utah, U.S.A.*

INTRODUCTION

Conformational analysis is the study of the three-dimensional structure of molecules. It specifically deals with spatial arrangements that result from rotations about single bonds between the atoms in molecules (Carey, 1987). Conformational analysis is a major field of physics, chemistry, and biology because the conformation of a molecule has a significant effect on its physical and chemical properties and on its biological activity (Orville-Thomas, 1974).

The following are the objectives of this article:

Provide an overview of conformational analysis, including a complete definition of conformational analysis, a discussion of its importance, and a description of how scientists determine the conformation of a molecule.

Describe some of the major experimental approaches to conformational analysis, including x-ray diffraction, nuclear magnetic resonance (NMR), ultraviolet spectroscopy (UV), infrared spectroscopy (IR), circular dichroism (CD), and optical rotatory dispersion (ORD). This article will not describe all the physical techniques employed for conformational analysis. For a more complete description of experimental approaches, consult the list of Further Reading at the end of this article.

Describe the major theoretical (computational) approaches to conformational analysis, including *ab initio*, semiempirical, and empirical methods, which are part of the more general procedures of conformational analysis called molecular modeling, molecular mechanics, and molecular dynamics.

The example systems described in this article are restricted to organic (carbon-containing) compounds: simple hydrocarbons, peptides, carbon-based polymers, and biological polymers, with emphasis on biological applications. This article does not treat inorganic compounds, which often do not have rotatable single bonds and therefore occur in only one conformation. The reader should understand, however, that conformational analysis of inorganic compounds is also an important field of chemistry and physics. Fortunately, the principles and techniques discussed in this article also apply to inorganic compounds.

1-56081-063-7/92/$5.00+.50

1. OVERVIEW OF CONFORMATIONAL ANALYSIS

Organic and biological molecules are made up of atoms joined together with covalent bonds. Most carbon-carbon, carbon-nitrogen, and carbon-oxygen single bonds can readily rotate at room temperature, whereas double and triple bonds have severely restricted rotations. A rotation of more than about 15° in a double or triple bond is extremely rare. Therefore, conformational analysis deals primarily with the different shapes of a molecule as a result of rotations about single bonds or about partially double bonds.

1.1 The Goals of Conformational Analysis

The major goals of conformational analysis can be expressed in terms of the following questions (see Cantor and Schimmel, 1980a; Zimmerman, 1985):

What is the three-dimensional shape of the molecule? The immediate aim of experimental and theoretical conformational analysis is to know the relative locations in space of all the atoms in the molecule being studied.

How does the shape of the molecule change in response to environmental changes? For example, what are the effects of temperature, pressure, and solvent? How does the conformation in the solid state differ from the conformation of the molecule in solution? What are the effects of small ions and other solutes on the conformation of the molecule in solution?

What is the relationship between the conformation of the molecule and its physical, chemical, and biological properties? This question addresses the important structure-function relationship in organic and biological molecules. The question is also important in materials science and polymer science, since the development of new materials depends on an understanding of the conformational properties of the molecules involved.

Can theoretical methods be used to predict the conformation of a molecule? One of the major goals of the modern field of molecular modeling, for example, is to use computational methods to predict the conformational properties of organic and biological molecules.

Given the conformational characteristics of a molecule with known biological activity, can other molecules be designed and synthesized having similar physical and biological properties? This question is the basis for drug design in modern pharmaceutical companies. If a certain molecule, for example, has desirable properties but also has some undesirable side effects, a related molecule can possibly be designed to enhance the desirable effects and diminish the undesirable effects.

1.2 Dihedral Angles

The conformation of a molecule is described completely by the *X*, *Y*, and *Z* Cartesian coordinates of all the atoms in the molecule. Since the conformation is independent of the location of the molecule in space, the number of independent variables is $3N-3$, where N is the number of atoms.

Often the molecule is treated as having fixed bond lengths and bond angles, and therefore the conformation can be described as a set of angles of rotation (called *dihedral angles* or *torsional angles*) about the single bonds. Because of bond stretching and bending, the bond lengths and angles between atoms are in fact not fixed but fluctuate about equilibrium locations. The concept and use of dihedral angles is still valid because the equilibrium locations are essentially constant.

Any dihedral angle θ is defined by the position of four atoms: two central atoms (see atoms 2 and 3 in Fig. 1) connected by the covalent bond whose torsional angle is being described, and two flanking atoms (see atoms 1 and 4 in Fig. 1), each attached to one of the

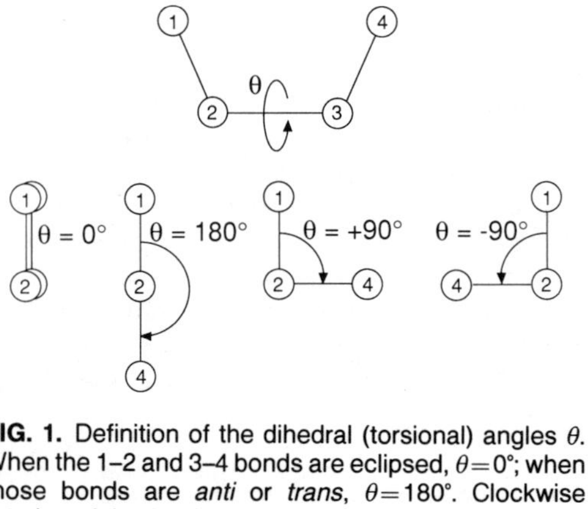

FIG. 1. Definition of the dihedral (torsional) angles θ. When the 1–2 and 3–4 bonds are eclipsed, $\theta=0°$; when those bonds are *anti* or *trans*, $\theta=180°$. Clockwise rotation of the 3–4 bond relative to the 1–2 bond gives positive values of θ; counterclockwise rotation gives negative values of θ.

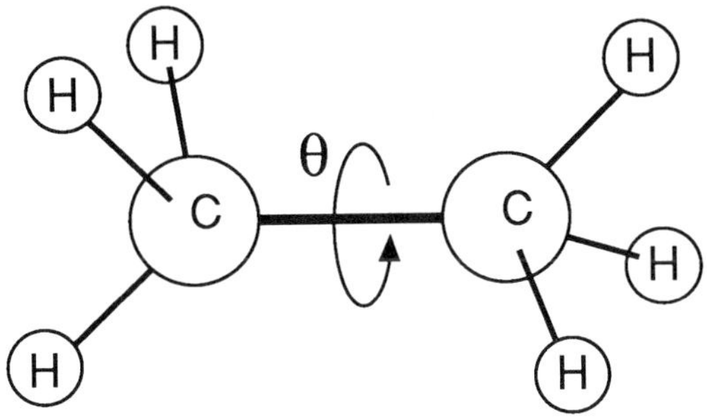

FIG. 2. Ball-and-stick model of ethane.

central atoms. If the four atoms are in the same plane with the two flanking atoms on the *same* side of the bond under consideration, then (by definition) the dihedral angle θ is zero (that is, $\theta=0°$). When atoms 1, 2, and 3 are kept stationary and atom 4 is moved by a clockwise rotation about the bond between atoms 2 and 3, the value of θ increases from 0° to 180°. When the bond is rotated counterclockwise, the value of θ decreases from 0° to $-180°$. When all four atoms are in the same plane but on *opposite* sides of the bond under consideration, the dihedral angle θ equals $\pm 180°$ (IUPAC-IUB, 1970). See Fig. 1.

1.2.1 The Dihedral Angle in Ethane and Butane. One of simplest organic compounds that can exist in different conformations is ethane, CH_3–CH_3. Figure 2 shows a ball-and-stick representation of the structure of ethane. Since the three hydrogen atoms on each carbon are equivalent, the four atoms used to specify the dihedral angle θ (for rotations about the C–C bond) include the two carbons and any one of the three hydrogens on each of the two carbons. The shape of the molecule with a specific θ value can be represented in several ways. The three most common are wedge-and-dash drawings, sawhorse drawings, and Newman projections. Figure 3 shows ethane in all three of these representations with $\theta=60°$, called the *staggered* conformation. Figure 4 shows ethane in the three representations with $\theta=0°$, called the *eclipsed* conformation.

Although all values of θ from $-180°$ to $+180°$ for rotations about the C–C bond in

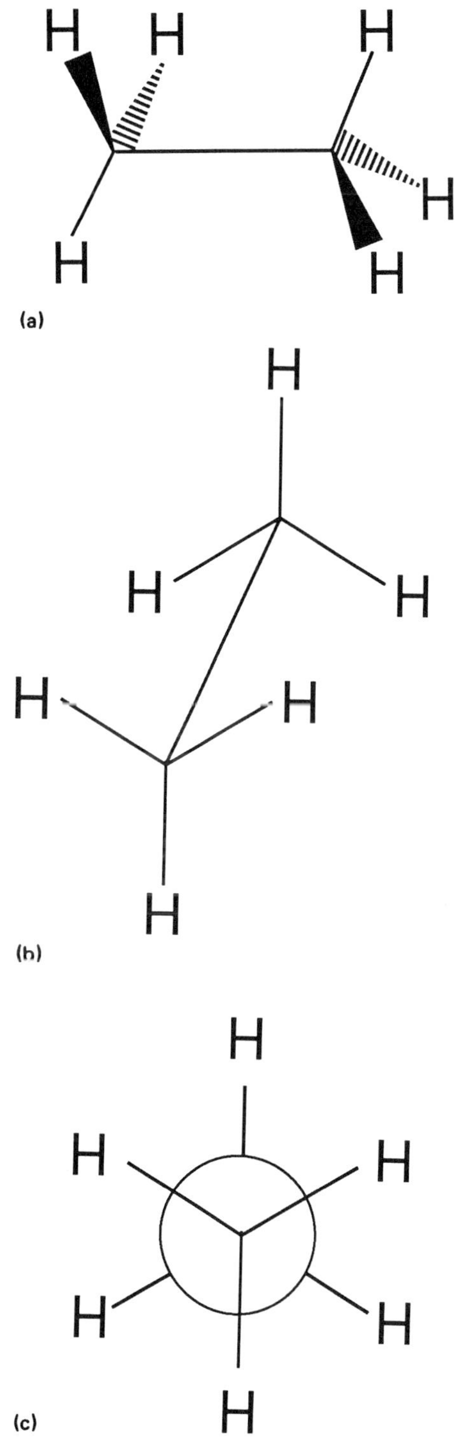

FIG. 3. Models of ethane with $\theta=60°$. (a) Wedge-and-dash model. (b) Sawhorse model. (c) Newman projection.

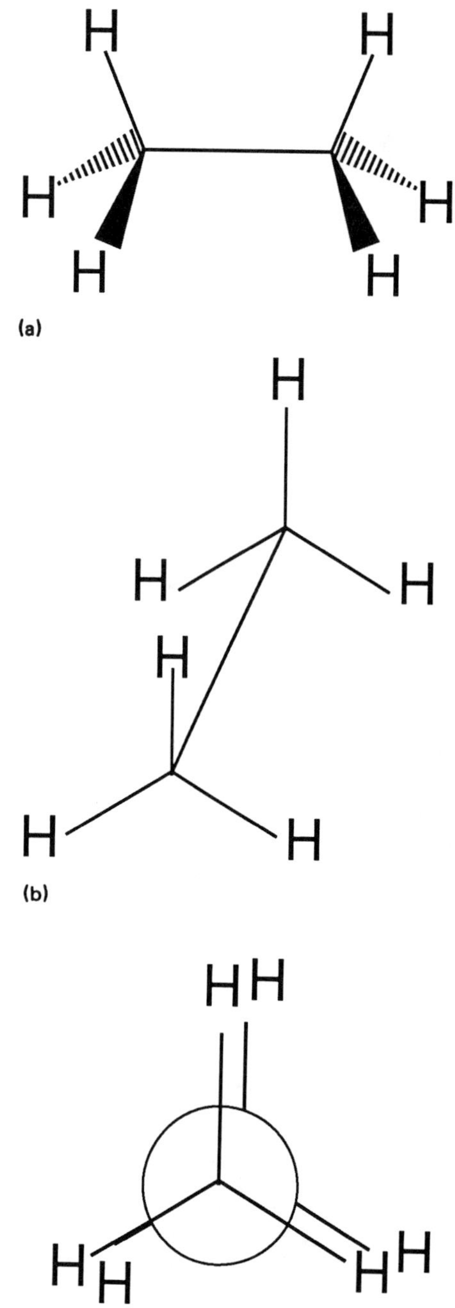

FIG. 4. Models of ethane with $\theta=0°$. (a) Wedge-and-dash model. (b) Sawhorse model. (c) Newman projection.

ethane are possible, all values of θ are not equally probable, and therefore rotation about the C–C bond is not completely free. The diagram in Fig. 5, derived from experimental

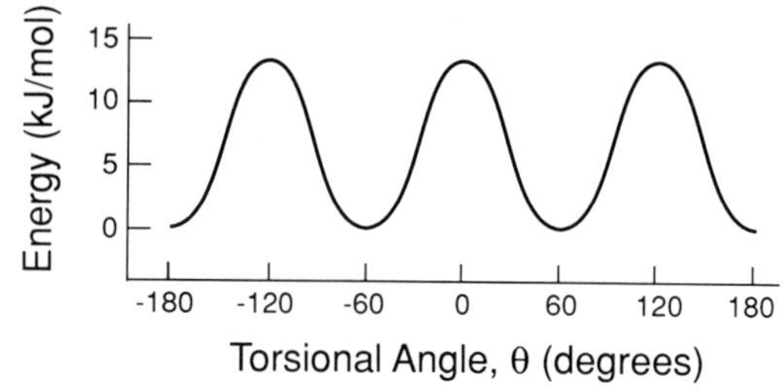

FIG. 5. The potential (conformational) energy of ethane as a function of the torsional angle θ.

and theoretical conformational analysis, shows that the potential energy is a minimum at $\theta=+60°$, $-60°$, and 180°. These three values of θ are, in fact, the same conformation, which is the most stable (lowest-energy) form of ethane. The potential energy is a maximum (13 kJ) at $\theta=0°$ and $\pm 120°$. These values of θ also yield the same conformation, which is the least stable form of ethane.

The conformational properties of organic compounds are further illustrated by examining the rotations about the central C–C bond of butane, CH_3–CH_2–CH_2–CH_3. This molecule can exist in two different stable conformations, one called *gauche* (with $\theta=\pm 60°$) and one called *anti* (with $\theta=180°$). Although both the gauche and anti conformations are stable—that is, they both occur in minimum-energy conformations—anti is the *global* (overall) minimum, having a potential energy of 3.3 kJ lower than the gauche. The potential-energy diagram of butane is shown in Fig. 6.

1.2.2 The Dihedral Angles in Peptides. Conformational analysis of peptides and proteins is a major field of biophysical chemistry

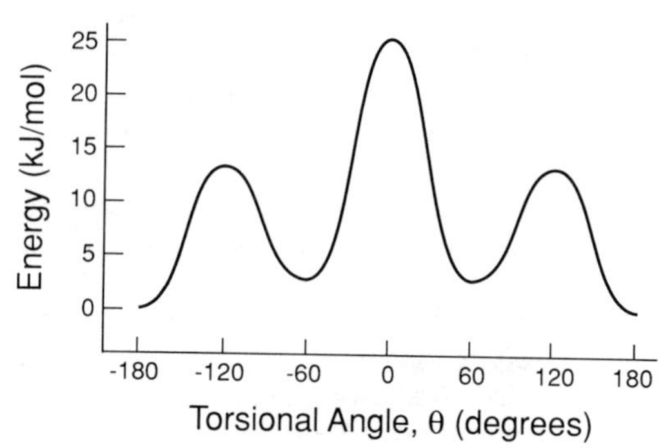

FIG. 6. The potential (conformational) energy of the central C–C bond of butane as a function of the torsional angle θ.

because the shape of a protein is the major factor in determining its biological function.

Peptides and proteins are composed of amino acids joined together with peptide bonds. Naturally occurring amino acids have the general structure shown in Fig. 7(a). Two amino acids joined together by a peptide bond are generally called dipeptides [Fig. 7(b)]; three amino acids joined together with two peptide bonds are generally called tripeptides; and so forth. Long, unbranched chains of amino acids joined together with peptide bonds are called polypeptides. If the polypeptide is a naturally occurring biological molecule, it is called a protein.

The dihedral angles that dictate the conformation of peptides and proteins are ϕ, ψ, ω, and χ (see Fig. 8). Angle ϕ is the dihedral angle described by the four atoms C′–N–C^{α}–C′ that is, rotation about the N–C^{α} bond. (C^{α} is the α carbon, adjacent to the carboxyl carbon in an amino acid. C′ is the carbonyl carbon in a peptide unit.) Angle ψ is described by the atoms N–C^{α}–C′–N, that is, rotation about the C^{α}–C′ bond. Angle ω is described by the atoms C^{α}–C′–N–C^{α}, that is, rotation about the C′–N or peptide bond. The peptide bond has partial double-bond character because of the overlap

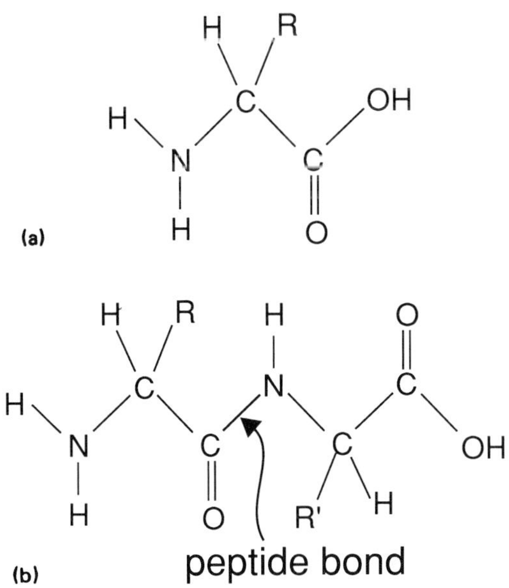

FIG. 7. (a) The structure of an amino acid (shown here in its un-ionized form). (b) The structure of a "dipeptide." R represents an organic substituent that defines a specific amino acid. For example, R is $-CH_3$ in the amino acid alanine.

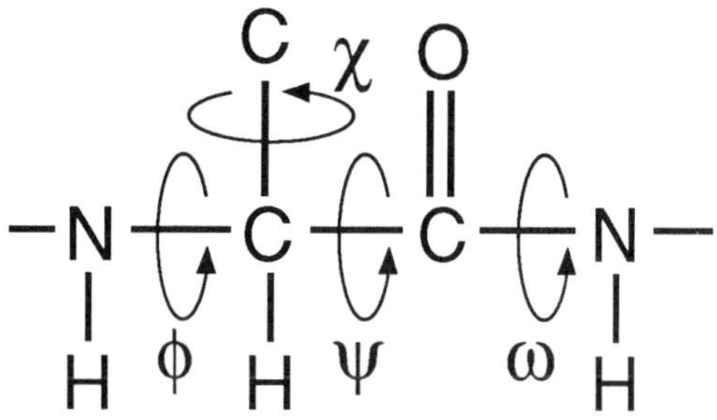

FIG. 8. The Greek letter designations of the angles of rotation in peptides and proteins.

of the unpaired electrons on the nitrogen with the π-electrons of the C═O double bond. Since the peptide bond has partial double-bond character, ω is usually found to have the value 180° (planar trans) or, on rarer occasions, the value 0° (planar cis). Angle χ is a rotation about a side-chain bond (see Fig. 8).

A theoretical analysis of the molecule N-acetyl-N′-methylalanineamide (Fig. 9), a model of an amino acid unit within a polypeptide or protein, yields the contour plot (Fig. 10) of the conformational potential energy as a function of the dihedral angles ϕ and ψ, with ω and χ being fixed at 180° and 60°, respectively (see Zimmerman *et al.*, 1977). (A conformational-energy contour plot is a three-dimensional potential-energy curve analogous to the two-dimensional curves shown in Figs. 5 and 6.) The diagram shows lines of constant energy at increments of 1 kcal mol^{-1} (4.186 kJ mol^{-1}) above the global minimum at about $\phi=-80°$, $\psi=80°$ (marked with a +). For simplicity, lines for conformational energies greater than 9 kcal mol^{-1} (40 kJ mol^{-1}) above the global minimum have been omitted from the plot. This large blank space indicates disallowed conformations in which the ϕ-ψ values cause

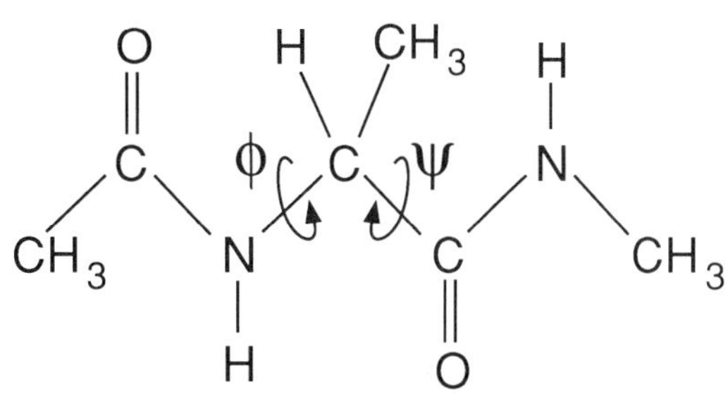

FIG. 9. The structure of N-acetyl-N′-methylalanineamide, a model of a single alanine residue in a peptide or protein.

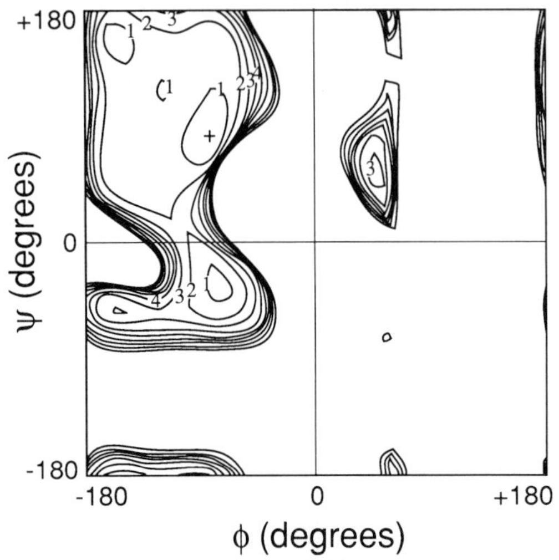

FIG. 10. The conformational energy contour map of the alanine residue (*N*-acetyl-*N*′-methylalanineamide). The + marks the global minimum. Lines of constant energy are drawn for 1–9 kcal mol^{-1} above the global minimum.

overlaps between one or more pairs of atoms in the peptide molecule.

Figure 11 shows the conformation of the global minima of N-acetyl-N′-methylalanineamide. This conformation ($\phi=-80°$, $\psi=80°$) is called C_7^{ax} because the hydrogen bond forms a seven-member ring (or cyclic) structure and the alanine side chain is approximately in the same plane as (that is, axial to) the seven-membered ring.

The above discussion of butane, ethane, and peptides demonstrates the nature of conformational analysis and lays the foundation for the discussions of experimental and theoretical techniques in conformational analysis.

2. EXPERIMENTAL TECHNIQUES

Chemists and physicists have developed a wide range of experimental techniques for analyzing the conformation of organic and biological molecules. Each technique has its own special set of advantages and disadvantages, depending on the type and amount of information that the technique provides, the purity and size of sample required to perform the experiment, the ease of varying environmental conditions, the effort required to prepare samples and obtain results, and so forth (Cantor and Schimmel, 1980b; see MOLECULAR SPECTROSCOPY). In many cases, two or more techniques are required to yield the desired conformational data.

The following is a brief description of some of the most popular experimental techniques in conformational analysis.

2.1 X-Ray Diffraction

X-ray diffraction is a powerful technique for determining the precise location of the atoms in a molecule (see X-RAY DIFFRACTION). The technique is based upon three principles:

1. Electrons scatter x rays such that the amplitude of the scattered wave is proportional to the electron density of the sample;
2. the scattered waves recombine constructively or destructively; and
3. the way in which the scattered waves combine (interfere) depends upon the distances between the reflecting planes and the wavelength of the electromagnetic radiation.

For constructive interference to occur (that is, for the reflected waves to be reinforced), the equation

$$2d\,\sin\theta=n\lambda \qquad (1)$$

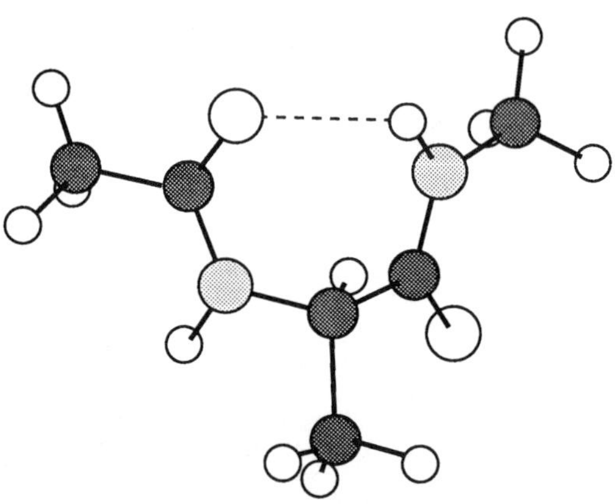

FIG. 11. The conformation of the alanine residue (N-acetyl-N′-methylalanineamide) in its global minimum. The dotted line represents a hydrogen bond. This conformation is called C_7 because the hydrogen bond forms a cyclic group of seven atoms. The large white atoms are oxygen; the small white atoms are hydrogen; the light gray atoms are nitrogen; and the dark gray atoms are carbon. The backbone atoms of alanine begin with the first nitrogen from the left and proceed right to the C^{α} atom and then to the C′ (carbonyl) atom.

must be satisfied, where d is the perpendicular spacing between the lattice planes in the crystal, θ is the angle between the incident (or the reflected) beam and the plane of the reflecting particles, n is an integer, and λ is the wavelength. This is called the Bragg law.

Since bond lengths in molecules are $\sim 1.5 \times 10^{-10}$ m (~ 15 nm), the wavelengths of the radiation must also be in the range of ~ 15 nm. That is, the diffracting electromagnetic radiation must be x rays, which have wavelengths of 10^{-12}–10^{-7} m.

To determine the locations of the atoms within a molecule using x-ray diffraction requires these general steps:

1. Purifying and crystallizing the sample.
2. Measuring the diffraction angles and intensities using an x-ray diffractometer.
3. Calculating electron-density maps based upon the data gathered from the diffractometer and analyzed using Fourier transforms (see FOURIER AND OTHER MATHEMATICAL TRANSFORMS).
4. Relating the location of the atoms within the molecules to the electron-density maps.

The major advantage of x-ray diffraction is that the precise location of all the atoms (except hydrogen) within the sample molecule can be obtained. The conformation of the molecule in the crystalline state is therefore known unequivocally. The major disadvantages of the technique are that

1. producing a crystal of sufficient purity and size can be difficult or even impossible, especially for large molecules;
2. gathering and analyzing data can be extremely time consuming (although modern diffractometers and computers have removed much of the tedium from the process); and
3. determining the dynamics of conformational change in response to environmental changes is usually impossible.

An x-ray structure represents the single most stable conformation of the molecule based upon intermolecular and intramolecular interactions. Molecules in the gas phase or in solution, on the other hand, can often exist in (and interconvert among) many different conformations. Therefore, the x-ray structure may have little relationship to the physical and chemical properties of the molecule in solution.

2.2 Nuclear Magnetic Resonance

Nuclear magnetic resonance (NMR) is one of the most popular tools for analyzing the conformation of organic and biological molecules (see NUCLEAR MAGNETIC RESONANCE; MAGNETIC RESONANCES, DETERMINATION OF STRUCTURES BY). This popularity results from three factors (Freifelder, 1982):

1. the theory of NMR is sufficiently developed to determine—at least in principle—the arrangement of all the atoms in a molecule from the NMR spectrum;
2. hydrogen atoms—whose electron densities are too low for observation by x-ray diffraction—can be located using NMR; and
3. different types of nuclei—such as ${}^{1}_{1}$H, ${}^{13}_{6}$C, ${}^{15}_{7}$N,—can be examined separately.

Moreover, NMR is effective in analyzing conformations in solution and in measuring conformational changes in response to environmental changes. For species with molecular weights <500, NMR can usually yield the locations of all the atoms in the molecule. For larger species, where determining the precise locations of all the atoms may be impossible, NMR can still yield much valuable information about the conformation, including hydrogen-bonding schemes, interatomic distances between certain atoms, and binding by other species.

The principles of NMR can be understood from the quantum-mechanical description of atomic nuclei (see ELECTRONIC STRUCTURE OF ATOMS). Consider, for example, the nucleus of a hydrogen atom. Its proton possesses two possible "spin" quantum numbers (angular momentum eigenvalues): $+1/2$ and $-1/2$. Each of these spin states gives rise to a magnetic field whose strength and direction is expressed as the magnetic moment μ. When a proton is placed in an external magnetic field of strength H_0, the magnetic moment of the proton can assume one of two angles relative to the direction of H_0: θ (aligned with) or $180-\theta$ (aligned against the external magnetic field). The difference in potential energy between these two magnetic states is $2\mu H_0 \sin\theta$.

When a proton absorbs this amount of energy—which lies within the radio-frequency range of the electromagnetic spectrum—the proton undergoes a transition between the two magnetic states. The proton is said to "flip" from one orientation to the other.

What has been stated about protons (the nuclei of hydrogen atoms ${}^{1}_{1}H$) is also true of other atomic nuclei, most notably ${}^{13}_{6}C$, ${}^{15}_{7}N$, ${}^{19}_{9}F$, and ${}^{31}_{15}P$. All have a quantum-mechanical spin number I of $\frac{1}{2}$, which gives rise to two nuclear energy levels in the presence of an external magnetic field.

If all nuclei exposed to the same applied magnetic field underwent a transition at the same frequency, NMR would have little value in determining molecular conformations. In reality, however, when a molecule is placed in an applied magnetic field, the individual nuclei within the molecule do not see the same magnetic field. The effective magnetic field, H_{eff}, seen by each nucleus differs from the applied magnetic field H_0, because each nucleus in a molecule is surrounded by electrons that also generate a magnetic field of their own, thus "shielding" the nucleus to some degree from the applied magnetic field. Moreover, the effective magnetic field seen by nucleus is influenced by the magnetic field of nearby nuclei. Thus, the observed resonance frequency of a particular nucleus depends on its environment. This is why NMR spectroscopy is valuable in conformational analysis; each nucleus of the same type has a resonance frequency that depends on the conformation of the molecule.

The major advantages of NMR are that the technique can

1. yield data from ensembles of conformation (although they are sometimes difficult to understand);
2. provide information about conformational dynamics;
3. yield spectra of samples in the solid, liquid, or gas states, including aqueous and nonaqueous solutions; and
4. provide a considerable amount of conformational information, including precise locations of atoms in small molecules.

The major disadvantages are that

1. the signals from water can sometimes interfere with the sample signal from proton (${}^{1}H$) NMR (although taking the spectrum in D_2O or using high-resolution Fourier-transform NMR instruments can usually overcome this problem);
2. the spectra from dilute or impure samples are difficult to obtain; and
3. the spectra of large, complex molecules are difficult (and sometimes impossible) to interpret.

2.3 Ultraviolet Absorption Spectroscopy

Ultraviolet (UV) absorption spectroscopy has long been used as an analytical technique for identifying compounds and determining concentrations, but it also has value in analyzing conformational properties of compounds. For example, the UV absorption spectrum of a polypeptide can be used to demonstrate the presence of α-helices, β-pleated sheets, or nonregular ("random coil") structure, and to analyze helix-to-coil transitions.

UV spectroscopy cannot match x-ray crystallography or NMR spectroscopy for the amount of detailed information obtained, but UV spectroscopy has fewer experimental constraints. The major advantage of UV spectroscopy over most other popular techniques is that spectra can be obtained from only partially purified samples, from dilute solutions, and from small samples, with relatively inexpensive instruments. Moreover, changes in conformation as a function of change in temperature, pressure, or other environmental conditions are easily monitored with UV spectroscopy.

UV spectroscopy is based upon transitions between different electronic energy levels of the electrons upon absorbing UV light (electromagnetic radiation with a wavelength below 400 nm). For example, the amide unit in peptides and other molecules undergoes the so-called n-π^* transition at ~220 nm, the π_1-π^* transition at ~190 nm, and the π_2-π^* transition at ~165 nm. The exact values of these transitions depend upon the specific type of molecule and on its conformation. In the synthetic polypeptide poly-L-lysine, for example, the location of the maximum in the absorption spectrum is at 191 nm for the α-helix, 193 nm for the random coil, and 196 nm for a β-pleated sheet. Although these changes are small, they can be measured accurately. Moreover, the differences in the amount of absorption are more dramatic: the three types of conformation have molar absorptivities of 4300, 7100, and 7800, respectively (Gratzer, 1967).

UV spectroscopy, therefore, is a powerful technique in conformational analysis when the spectra of different conformations have been well characterized and the goal of the analysis is to determine the general conformation of the sample or to study its conformational changes.

2.4 Infrared Spectroscopy

Infrared spectroscopy (IR) is best known as a tool for determining structures of organic compounds, but it can also be used as a tool in conformational analysis. For example, O–H and H–H groups produce different IR absorption bands depending on whether or not they are involved in a hydrogen bond (see MOLECULAR SPECTROSCOPY).

IR spectroscopy is based upon the absorption of electromagnetic radiation by transitions in the vibrational levels (due to bond stretching and bending) in a molecule. The energies of these transitions are generally in the range of infrared radiation, which has a wavelength range of 10^{-6}–10^{-4} m (1–100 μm). IR spectra usually cover the range 2.5–16 μm, which corresponds to the wave-number range of 400–625 cm^{-1}. (Traditionally, IR absorption is plotted as a function of wave numbers rather than wavelengths.)

One of the major disadvantages of traditional IR spectroscopy is that the absorption band for H_2O (and in some cases even D_2O) occurs in the same wave-number range as the main absorption bands of many organic and biological molecules, greatly diminishing the value of IR spectroscopy in analyzing conformations in aqueous solutions. With the advent of Fourier-transform infrared spectroscopy (FT-IR), the technique has greatly increased in its usefulness as a method of conformational analysis.

2.5 Circular Dichroism and Optical Rotatory Dispersion

One of the most useful techniques for analyzing the conformations and the conformational changes of asymmetric (or optically active) compounds, particularly biopolymers, is circular dichroism (CD) and the related technique, optical rotatory dispersion (ORD). In concept, CD and ORD provide the same information, but in practice, CD has become the method of choice because its spectra show greater sensitivity to conformation.

When plane-polarized light passes through a sample of a chiral compound, the light is changed in two ways:

1. The maximal amplitude is no longer confined to a plane but rather traces out an ellipse. The resulting ellipticity of the light is defined as the arc tangent of the ratio of the minor axis to the major axis of the ellipse. CD is a measure of the ellipticity caused by the chiral compound.
2. The orientation of the ellipse may be rotated. In other words, the plane of polarized light is rotated. Optical rotatory dispersion is a plot of the amount of rotation as a function of wavelength.

Consider an alternative way of looking at ellipticity. Plane-polarized light can be resolved into two circularly polarized components of equal amplitude. The sum of the two components at all points results in the original plane-polarized wave. In an optically active medium, the left circularly polarized light (A_L) and the right circularly polarized light (A_R) are absorbed differently. This is called circular dichroism. When the two opposite circularly polarized waves are combined, the result is an ellipse, as shown in Fig. 12. Thus, CD is equivalent to ellipticity. A CD spectrum is a plot of the ellipticity θ (or specific ellipticity, $[\theta]$) as a function of wavelength.

The major advantages of CD and ORD are that the spectra are sensitive to conformational changes and that the environmental conditions are easily modified. Thus, conformational changes in biopolymers are easily monitored using ORD and CD. For example, Fig. 13 shows the CD spectra of poly-L-lysine (a model polypeptide) in α-helix, β-pleated sheet, and random coil conformations (Greenfield and Fasman, 1969). In the figure, $[\theta]$ is

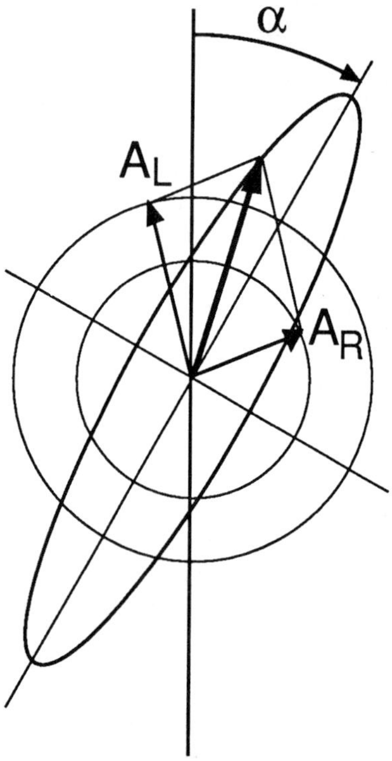

FIG. 12. Circular dichroism and optical rotation. The circular dichroism (or ellipticity) is proportional to the difference between the amplitude of the left circularly polarized wave A_L and of the right circularly polarized wave A_R. The optical rotation is given by α.

the specific ellipticity, and the wavelength λ is taken in the ultraviolet region where the peptide unit absorbs light. Notice the huge difference in the spectra for the three different types of conformation. CD can likewise be used to analyze the general conformation of proteins, carbohydrates, and nucleic acids, as well as other kinds of molecules.

2.6 Other Experimental Techniques

The five experimental techniques discussed above are the most popular methods for analyzing the conformation of molecules, but they are by no means the only methods. Other popular experimental techniques include fluorescence spectroscopy, ultracentrifugation (and other hydrodynamic techniques), light scattering, Raman spectroscopy, Mössbauer spectroscopy, electron paramagnetic resonance spectroscopy, and so forth. For a general description of these techniques, see Cantor and Schimmel (1980b) and Freifelder (1982).

3. THEORETICAL TECHNIQUES

Theoretical (computational) techniques are valuable in the conformational analysis of organic and biological molecules because the results of a theoretical analysis can

1. provide a model or framework to aid in choosing proper experiments;
2. help explain experimental results; and
3. provide information about conformational properties of molecules when experiments are difficult or impossible to perform.

The major objectives of theoretical analysis are

1. to calculate the conformational properties of a molecule given its covalent structure;

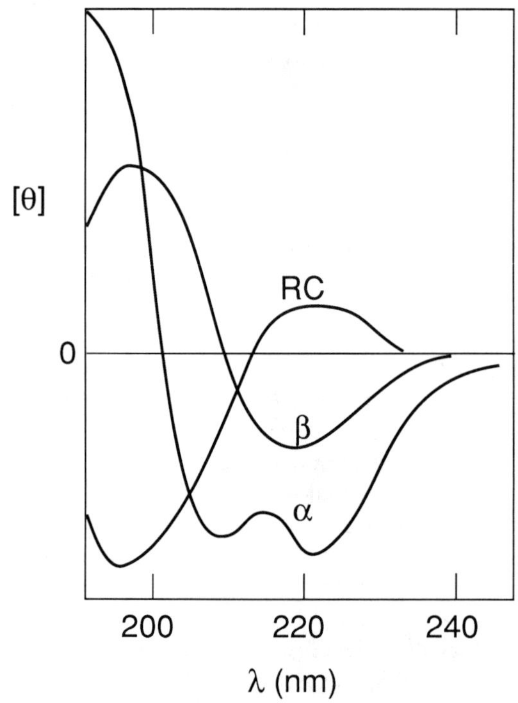

FIG. 13. The circular dichroism spectrum of poly-L-lysine in three different conformations: α-helix, β-pleated sheet, and random coil (RC). The conformational changes are induced by changing the *p*H. From Greenfield and Fasman (1969).

2. to determine the effects of the environmental conditions on the conformation; and
3. to relate the calculated conformation of a molecule to its physical properties and its biological activity.

3.1 Empirical Methods

Empirical methods of conformational analysis involve empirical equations (that is, equations based on experiment or observation without regard to fundamental theory) that relate the positions of the atoms to the conformational energy, and involve parameters used in the empirical equations. In a so-called molecular mechanics system, computer software employs the equations and parameters to calculate the conformational energy, minimize the total conformational energy for locating local and global minima, and display on a computer monitor a graphic representation of the conformations.

A commonly used empirical equation (Momany *et al.*, 1975; Feller *et al.*, 1989) is the so-called Lennard-Jones "6-12" potential:

$$U_{\mathrm{NB}}=F\epsilon\left(\frac{r_g}{r_{ij}}\right)^{12}-2.0\epsilon\left(\frac{r_g}{r_{ij}}\right)^{6}, \qquad (2)$$

where r_g is the equilibrium distance between the atoms i and j, ϵ is the energy at the equilibrium distance, and r_{ij} is the distance between atoms i and j. The positive (repulsive) term contains r^{12}, and the negative (attractive) term contains r^6. The attractive term is the London dispersion force (London, 1937). The F is an adjustment factor taken as 1.0 for 1–5 interactions (those between atomic pairs separated by at least four bonds) and as 0.5 for 1–4 interactions (those between atomic pairs separated by three bonds). The subscript NB stands for "nonbonded," meaning the energy of interaction between atoms that are not bonded to each other. The values of r_g and ϵ depend on the particular types of atoms involved in the interaction.

Along with the Lennard-Jones 6-12 potential, conformational energy calculations also require an equation to take into account the electrostatic interactions. For example, the most common equation for determining electrostatic energy between two atoms is

$$U_{\mathrm{el}}=kq_iq_j/Dr_{ij}, \qquad (3)$$

where k is a constant to ensure that the energy has the correct units, q_i and q_j are the partial charges on atoms i and j, D is the dielectric constant, and r_{ij} is the distance between atoms i and j.

For certain types of bonds (particularly double bonds), the nonbonded and electrostatic energies do not completely describe the restricted rotation observed for the bond. Therefore, torsional energy is also used in some procedures to determine the conformational energy:

$$U_{\mathrm{tor}}=\frac{U_0}{2}(1\pm\cos n\theta), \qquad (4)$$

where U_0 is the maximum torsional energy, θ is the torsional angle, and n is the symmetry of the barrier.

The total conformational energy of a molecule in a particular conformation is the sum of nonbonded energies U_{NB} and electrostatic energies U_{el} for all i,j pairwise interactions and of all the torsional energies U_{tor}. This procedure was used to calculate the total conformational energy at each ϕ, ψ point of the conformational energy contour map shown in Fig. 10. It can also be used to analyze the conformational properties of larger peptides and other biological and organic compounds. Because the calculations require large amounts of computer time, the procedure does not permit the complete conformational analysis of proteins, nucleic acids, carbohydrates, or other macromolecules.

In many approaches to molecular mechanics, the bond lengths and bond angles are also considered variable. Appropriate potential-energy functions include not only the interactions between the nonbonded atoms, but also between the bonded atoms (Burkert and Allinger, 1982).

3.2 Quantum-Mechanical Approaches

Pullman (1976) has argued that empirical energy functions "suffer from two obvious drawbacks. In the first place, whatever the practical justification for the partitioning of the total potential energy into a series of components, the procedure involves necessarily an element of arbitrariness and, possibly, incompleteness. In fact, most authors take into consideration different selected components only and frequently different sets of them in different publications. Secondly, the fundamental formulas and parameters used to define the various components are far from being well established and differ, often appreciably, from one author to another."

The conclusion that Pullman and others draw from these observations is that only quantum-mechanical methods, not empirical methods, have the ability to account for the subtle quantum-mechanical effects that may be part of the conformational properties of a molecule.

These arguments certainly have some merit. Quantum mechanics has indeed contributed to the field of theoretical conformational analysis (see QUANTUM MECHANICS). But in actual practice, quantum-mechanical methods also involve a certain degree of arbitrariness, require broad approximations, and involve huge amounts of computer time (Zimmerman, 1985). Both quantum-mechanical and empirical methods are valuable in addressing the problems of molecular conformations, since each approach brings to the problem its own strengths and weaknesses.

The most popular semiempirical molecular-orbital methods used for conformational analysis include the extended Hückel (EHT) method (Hoffman, 1963), the complete neglect of differential overlap (CNDO/2) methods (Pople and Segal, 1966), and the all-valence electron molecular-orbital (perturbative configuration interaction using localized orbitals) method (Diner *et al.*, 1969) (see ELECTRONIC STRUCTURE OF ATOMS AND MOLECULES).

EHT is a molecular-orbital treatment in which the interactions between electrons in different orbitals are ignored. CNDO/2 assumes the atomic orbitals are spherical when electron repulsions are evaluated. Directionality of *p* orbitals is included, however, through the so-called one-electron resonance integrals. CNDO calculations are aimed primarily at reproducing electronic properties, such as dipole moment (Clark, 1985).

Perturbative configuration interaction using localized orbitals goes beyond CNDO/2 by including a refined molecular-orbital all-valence electron procedure that incorporates an appreciable fraction of the correlation energy (Maigret *et al.*, 1971).

Ab initio methods (see QUANTUM MECHANICS) are also used in theoretical conformational energy analysis (e.g., see Shipman and Christofferson, 1973; Peters and Peters, 1981).

3.3 Applying Theoretical Techniques

Whether using empirical, semiempirical (molecular-orbital), or *ab initio* theoretical methods, the theoretician utilizes the following general steps in determining the conformational properties of a molecule:

1. Selecting a set of initial trial conformations, either randomly (e.g., using a Monte Carlo method; see Li and Scheraga, 1987) or systematically (e.g., by stepping through all of conformational space).
2. Calculating the total conformational energy of the molecule in each of the trial conformations. If the theoretician selects and calculates enough trial conformations, the results will yield the most favored (lowest-energy) conformation or will provide an ensemble of conformational states on which to base a statistical-mechanical analysis.
3. Minimizing the total conformational energy for each of the trial conformations to yield local and global minima. If this method yields a large number of minima, the set of minima may constitute the partition function for a statistical-mechanical analysis of the conformational properties. If minimization yields a small number of minima, inspection of those minima may provide all the necessary information about the conformational properties of the molecule.

Applying these techniques to peptide molecules, as an example, has yielded important information about the structure-function relationship in biological systems (see Zimmerman, 1985).

GLOSSARY

α-Helix: A polypeptide or protein conformation in which the backbone chain forms a right-handed screw with 3.6 amino acid residues per turn and a pitch (rise per turn) of 54 nm.

β-Pleated Sheet: A polypeptide or protein conformation in which the backbone chains are extended and hydrogen bonded to adjacent chains.

Conformation: The overall three-dimensional shape of a molecule.

Contour Plot: A plot of the conformational energy of a molecule as a function of two variable dihedral angles. Each contour line represents conformations of equal energy.

Dihedral Angle: The rotational angle of a bond, defined such that the angle is 0° when the two bonds flanking the torsional bond are eclipsed when looking along the bond and 180° when the two flanking bonds are *anti* (or *trans*) to each other.

Global Minimum: The conformation of the lowest energy of a molecule.

Local Minimum: A conformation found at a minimum-energy point (not necessarily the lowest, or global, minimum).

Molecular Modeling: A technique that employs computer graphics and calculations to study molecular conformations and dynamics.

Molecular Mechanics: A method for calculating the conformations of molecules based upon empirical energy functions.

Random Coil: A statistical ensemble of polymer conformations in which the backbone of the molecule has no order.

Torsional Angle: See dihedral angle.

List of Works Cited

Burkert, U., Allinger, N. L. (1982), *Molecular Mechanics*, Washington, DC: American Chemical Society.

Cantor, C. R., Schimmel, P. R. (1980a), *Biophysical Chemistry. Part I. The Conformation of Biological Macromolecules*, San Francisco: Freeman.

Cantor, C. R., Schimmel, P. R. (1980b), *Biophysical Chemistry. Part II. Techniques for the Study of Biological Structure and Function*, San Francisco: Freeman.

Carey, F. A. (1987), *Organic Chemistry*, New York: McGraw-Hill.

Clark, T. (1985), *A Handbook of Computational Chemistry*, New York: Wiley.

Diner, S., Malrieu, J. P., Jordan, F., Gilbert, M. (1969), *Theor. Chim. Acta* **15**, 100–110.

Feller, D. C., Delmoe, E. F., Zimmerman, S. S. (1989), *Comput. Chem.* **13**, 337–341.

Freifelder, D. (1982), *Physical Biochemistry*, 2nd ed. San Francisco, Freeman.

Gratzer, W. B. (1967), in G. D. Fasman (Ed.), *Poly-α-amino Acids*, New York: Marcel Dekker, pp. 177–238.

Greenfield, N., Fasman, G. D. (1969), *Biochemistry* **8**, 4108–4116.

Hoffman, R. (1963), *J. Chem. Phys.* **39**, 1397–1412.

International Union of Pure and Applied Chemistry–International Union of Biology Commission on Biochemical Nomenclature (1970), *J. Mol. Biol.* **52**, 1–17.

Li, Z., Scheraga, H. A. (1987), *Proc. Natl. Acad. Sci. U.S.A.* **84**, 6611–6615.

London, F. (1937), *Trans. Faraday Soc.* **33**, 8–26.

Maigret, B., Perahia, D., Pullman, B. (1971), *Biopolymers* **10**, 1649–1660.

Momany, F. A., McGuire, R. F., Burgess, A. W., Scheraga, H. A. (1975), *J. Phys. Chem.* **79**, 2361–2381.

Orville-Thomas, W. J. (1974), in: W. J. Orville-Thomas (Ed.), *Internal Rotation in Molecules*, New York: Wiley.

Peters, D., Peters, J. (1981), *Int. J. Quantum Chem.* **19**, 1121–1132.

Pople, J. A., Segal, G. A. (1966), *J. Chem. Phys.* **44**, 3289–3296.

Pullman, B. (1976), in: B. Pullman (Ed.), *Quantum Mechanics of Molecular Conformations*, New York: Wiley.

Shipman, L. L., Christofferson, R. E. (1973), *J. Am. Chem. Soc.* **95**, 1408–1416.

Zimmerman, S. S. (1985), in: V. J. Hruby (Ed.), *The Peptides: Analysis, Synthesis, Biology*, Vol. 7, *Conformation in Biology and Drug Design*, Orlando, FL: Academic.

Zimmerman, S. S., Pottle, M. S., Némethy, G., Scheraga, H. A. (1977), *Macromolecules* **10**, 1–9.

Further Reading

Derome, A. E. (1987), *Modern NMR Techniques for Chemistry Research*, New York: Pergamon.

Glass, R. S. (1988), *Conformational Analysis of Medium Sized Heterocycles*, New York: VCH.

Glusker, J. P., Trueblood, K. N. (1985), *Crystal Structure Analysis: A Primer*, Oxford: Oxford University.

Marchand, A. P. Takeuchi, Y. (Eds.) (1986), *Applications of NMR Spectroscopy to Problems in Stereochemistry and Conformational Analysis*, Deerfield Beach, FL: VCH.

Mason, S. F. (1982), *Molecular Optical Activity and the Chiral Discriminations*, New York: Cambridge University.

Rasmussen, K. (1985), *Potential Energy Functions in Conformational Analysis*, New York: Springer-Verlag.

Stout, G. H. Jensen, L. H. (1989), *X-ray Structure Determination*, New York: Wiley.

CONSERVATION LAWS AND SYMMETRIES

See SYMMETRIES AND CONSERVATION LAWS

CONSTANTS, FUNDAMENTAL*

BARRY N. TAYLOR *National Institute of Standards and Technology, Gaithersburg, Maryland, U.S.A.*

INTRODUCTION

This article deals with the fundamental physical constants of nature such as the speed of light in vacuum c, the mass of the electron m_e, the Newtonian gravitational constant G, the elementary charge e, and the Planck, Avogadro, Faraday, Rydberg (for infinite mass), fine-structure, Josephson, and von Klitzing constants h, N_A, F, R_∞, α, K_J, and R_K, respectively. Such constants are nature's true invariants in contrast to, say, the freezing point of gallium or the time required for the Earth to revolve once about its axis. Thus, they have a unique place in science and technology and have even acquired a certain mystique. No more cogent explanation of why this is so could likely be written than the following portions of the prefaces to the proceedings of two international conferences devoted to precision measurement and fundamental constants (Langenberg and Taylor, 1971; Taylor and Phillips, 1984):

*Contribution of the National Institute of Standards and Technology, not subject to copyright in the United States.

1-56081-063-7/92/$5.00+.50

Precision measurement plays an essential and crucial role in the development of all science and technology. Scientific theories cannot be tested, instruments and machines cannot be designed and built, and even the daily routine of the market place cannot proceed without precise quantitative measurement. The demand for precision and accuracy is greatest at the frontiers of science and technology. Here our increasing understanding of the fundamental nature of the universe and our increasing ability to put our knowledge to practical use depends on continuing refinement of our theories and of the technology used in testing theories against reality.

In all of our fundamental theories there appear a certain few parameters which characterize the fundamental particles and interactions we find in nature. These are the fundamental physical constants. The precision determination of the numerical values of these constants has long been and remains one of the principal objectives of experimental science. This is not because there is any intrinsic virtue in accumulating lists of ever longer numbers, but because the fundamental constants are the quantitative links between our most basic theories and the physical reality we wish them to describe. Our theories must stand or fall according to their ability to make quantitative predictions which agree with experimental observations to the maximum accuracy possible. The progress of our understanding of the physical world is therefore very much intertwined with the advance of the art of precision measurement and its application to the determination of the fundamental physical constants. In recent years there have been notable advances in both theory and experiment relevant to the fundamental constants. New technologies have made possible new levels of precision: Computers have made possible more complex calculations. The laser and the maser have found application in many different experiments. The discovery of the Josephson effects in superconductors and ··· [the quantum Hall effect in semiconductors] has provided new ways to determine several important fundamental constants. There has been steady progress in precision measurements of the properties of fundamental particles and simple atomic systems. Current theoretical and experimental efforts in this field span almost the whole of physics, from elementary particle theory to experimental solid-state physics.
To this ··· [must be added] the idea that because of the interrelationships which exist among the fundamental constants and related quantities, for example, the proton gyromagnetic ratio, the Faraday, Avogadro, and fine-structure constants, $2e/h$ via the Josephson effect, the absolute ohm and ampere, the quantized Hall resistance, the electron and muon anomalous magnetic moments, the proton to electron mass ratio, etc., it is appropriate, indeed even necessary, for the solid-state theorist to be aware of the work of the electrical metrologist; the atomic mass spectroscopist of the work of the precision electrochemist; and the quantum electrodynamics theorist of the work of the experimental semiconductor physicist. The fundamental constants of nature and closely related precision measurements are truly the common meeting ground for many of the disciplines of science and technology.

It should be clear from the above that the fundamental constants field is so broad that any one article can only touch upon a few selected topics. Nevertheless, the aspects chosen to be covered in this article are believed to be among the most important and together should provide the reader a reasonable picture of the field, especially the interplay between theory and experiment. However, the decision to be broad in scope means that no one topic is covered in great depth. In particular, no detailed descriptions of how specific constants are actually measured in the laboratory are given. It is left to the reader to peruse the original papers in order to understand fully the difficulty of determining the value of a constant to an accuracy of, say, 1 part in 100 million (1×10^{-8}). Short of visiting the laboratories of experimenters carrying out fundamental constant determinations, this is the only way to gain an appreciation of their ingenious measurement techniques, their exquisite apparatus, and the great care with which they perform their measurements. Indeed, it is not uncommon for a fundamental constant experiment to last 10 to 15 years!

The main body of the article (Sec. 1) begins with a brief discussion of the "origins" of the fundamental constants, that is, how they might be usefully categorized; and a lengthier discussion of "the romance of the next decimal place," or why the fundamental constants are important and why their determination to ever greater levels of accuracy can have a profound effect on physics. This is done through the late-1960s example of the determination of the quotient $2e/h$ via the Josephson effect (JE) in superconductors, the derivation of a value of the fine-structure constant α from it and other constants, and the use of the resulting value of α to test the quantum electrodynamic (QED) theory of the ground-state hyperfine splitting of the hydrogen atom. A similar example still under active investigation is also presented in which a value of α obtained from the quantum Hall effect (QHE) is used to test the QED theory of the electron magnetic moment anomaly a_e.

Section 2 is devoted to a discussion of how sets of "best" or recommended values of the fundamental constants are derived and is based on the example of the 1986 Committee on Data for Science and Technology (CODATA) least-squares adjustment of the constants, the most recent internationally accepted evaluation currently available. Included in this section are recommended values for a comprehensive group of constants as taken from the 1986 CODATA set.

Section 3 demonstrates the importance of the fundamental constants to technology by reviewing the 1 January 1990 introduction of new practical representations of the volt and ohm as defined in the International System of Units (SI). These representations, based on the JE and QHE, and internationally adopted values of the Josephson and von Klitzing constants K_J and R_K, are now being used worldwide in electrical metrology.

Section 4 summarizes the principal advances made in the fundamental constants field since the 1986 CODATA adjustment was completed and discusses their implications for both the 1986 set of recommended values and the next least-squares adjustment. Section 5 again demonstrates the importance of the fundamental constants to technology, this time by exploring the possible future role of the constants in replacing the kilogram, the only SI base unit that is still defined in terms of a material artifact. Section 6 briefly discusses the question of why the constants have the particular values they do and their possible variation with time. Section 7, which concludes the main body of the article, looks ahead in an attempt to see where the fundamental constants field might be heading over the next decade.

Before beginning Sec. 1, it is perhaps useful to comment briefly on the history of the fundamental constants field. Few measurements of the constants existed prior to the start of the 20th century because the modern era in physics did not begin until then; quantum theory, relativity, and atomic and nuclear physics all emerged after 1900. The two notable exceptions are c and G which were first measured in the 18th century. Extraordinary advances in determining the values of the constants were made immediately after World War II because of the significant advances made in electronics, microwaves, and other technologies during the war. The period starting in about 1970 has also been especially productive because of the wide introduction of highly sophisticated digital electronic instruments, lasers, computers, and cryogenic techniques. To a considerable extent the accuracy with which the fundamental constants have been known at any given epoch has depended upon the level of sophistication of the measuring instruments and apparatus available at the time. But it is not only improvements in measurement technology that have led to more accurate values of the constants. The discovery of new phenomena, such as the JE and QHE, that provide new and better methods for determining certain constants have also led to significant advances in the field.

Measurements of the constants and the frontiers of measurement science or metrology, especially the development of new reference standards for the units of measurement, have always been closely linked, but in recent years the distinctions between them have all but disappeared. Perhaps the most striking example of this is the determination of c in the early 1980s based on extremely difficult measurements of the frequencies and wavelengths of several different laser radiations. As a direct consequence of this work, the definition of the SI unit of length, the meter, which was based on a particular optical transition in the ^{86}Kr atom, was replaced in 1983 with one based on an adopted value for the speed of light (Giacomo 1984; Hudson, 1984). As will be discussed in Sec. 5, if current efforts to determine h or N_A with a relative uncertainty of about 1×10^{-8} are successful, it is quite possible that this scenario will be repeated with regard to the kilogram in the first decade of the 21st century.

1. "ORIGINS" AND IMPORTANCE OF THE CONSTANTS

1.1 Origins of the Constants

Three distinct "sources" of the fundamental constants may be identified. The first is the physical theories that have been developed to describe nature's basic laws, for example, Maxwell's theory of electromagnetism, Einstein's theories of relativity, quantum mechanics, and QED, and their application as appropriate to atomic systems. c, G, h, α, and R_∞ fall within this category. These particular constants (see beginning of Introduction) are of great importance because c is the velocity of all electromagnetic radiation or photons (in vacuum) and the limiting velocity of all energy and matter; G determines the strength of the gravitational force between bodies; h is the fundamental constant of quantum mechanics and determines the energy of a quantum of radiation of a given frequency; R_∞ sets the scale of the energy levels of atoms; and α determines the strength of the electromagnetic force that governs how electrically charged elementary particles and electromagnetic radiation interact.

The second source is the fundamental particles such as the electron, proton, and neutron. The elementary charge e, the rest mass of the electron m_e, the rest masses of the proton and neutron m_p and m_n, the magnetic moment of the electron in units of the Bohr magneton μ_e/μ_B, where $\mu_B = eh/4\pi m_e$, and the magnetic moments of the proton and neutron in units of the nuclear magneton μ_p/μ_N and μ_n/μ_N, where $\mu_N = eh/4\pi m_p$, are all examples of quantities that characterize a basic property of nature's elementary building blocks.

The third source is conversion factors. Although not true fundamental constants like those in the first two categories, these quantities nevertheless play a critical role in the fundamental constants field since knowledge of their values is essential to many fundamental constant determinations. Examples include the local value of the gravitational acceleration g, which relates mass to force; the ratio of the old x-ray unit of length or kilo-x-unit to the angstrom Å (1 Å = 0.1 nm); and the ratio of various so-called as-maintained electrical units to their International System of Units (SI) definitions, for example, Ω_{NIST}/Ω, where Ω_{NIST} is the National Institute of Standards and Technology (NIST) as-maintained ohm defined (prior to January 1, 1990) in terms of the mean resistance of a particular group of wire-wound standard resistors kept in an oil bath at 25 °C, and Ω is the SI unit of resistance, the ohm.

It should be noted that $\alpha = \mu_0 ce^2/2h$ and $R_\infty = m_e c\alpha^2/2h$ are examples of constants that are actually combinations of other quantities, but are considered fundamental constants in their own right since the combination always appears in theoretical equations in the same way. (Here $\mu_0 = 4\pi \times 10^{-7}$ N/A^2, exactly, is the permeability of vacuum.)

1.2 Importance of the Constants

There are at least four reasons why the fundamental physical constants play a critical role in science and technology and must thus be known to as high an accuracy as possible. First (see Introduction), accurate values of the constants are required for the critical comparison of theory with experiment, and it is only such comparisons that enable our understanding of the physical world to advance. A closely related idea is that by comparing the numerical values of the same fundamental constants obtained from experiments in the different fields of physics, the self-consistency of the basic theories of physics themselves can be tested.

Second, determining the fundamental constants to ever greater levels of accuracy fosters the development of state-of-the-art measurement methods that may have wide application. Determining the next decimal place is never trivial and usually requires an entirely new measurement technology. An example is the determination of the Avogadro constant N_A, which necessitated the development of techniques to measure the lattice spacing of pure, single crystals of silicon in meters with a relative uncertainty of one-tenth of a part per million or 0.1 ppm. (Note that 1 ppm$=1 \times 10^{-6}$ and throughout this article, all uncertainties are one-standard-deviation estimates.) The end result has been the extension of our ability to make accurate measurements of length in SI units to the picometer or 10^{-12} m range.

Third, the fundamental constants are the obvious key to the development of a reproducible and invariant system of measurement units, a major goal of measurement science or metrology. If our measurement system could be based entirely on fundamental constants, there would be no need for artifacts such as the platinum-iridium cylinder kept in a vault at the International Bureau of Weights and Measures (BIPM) that defines the SI unit of mass, the kilogram.

Finally, values of the fundamental constants are required for computations and measurements throughout science and technology—for example, the calculation of the excited states of atoms and molecules that play an important role in air pollution and nuclear fusion.

1.2.1 QED and the Josephson Effect Determination of 2*e*/*h*. As an example of the significant role improved measurements of fundamental constants can play in increasing our understanding of physical theory, consider the determination of the quotient $2e/h$ using the JE in weakly coupled superconductors that was first reported in the late 1960s (Parker, Taylor, and Langenberg, 1967). When a Josephson tunnel junction—for example, two thin films of lead separated by a 1-nm-thick thermally grown oxide layer—is irradiated with microwave radiation of frequency f, its current versus voltage curve exhibits current steps at highly precise quantized Josephson voltages U_J (Barone and Paternò, 1982). The voltage of the nth step, $U_J(n)$, where n is an integer, is related to f by

$$U_J(n)=nf/K_J, \tag{1}$$

where K_J is the Josephson constant.

The theory of the JE predicts, and the experimentally observed universality of Eq. (1) is consistent with the prediction, that $K_J=2e/h$. It therefore follows from Eq. (1) that a measurement of f and U_J for the nth step will yield a value of $2e/h$. The result reported in 1969 (Parker *et al.*, 1969) using X-band microwaves (8–12 GHz) was

$$K_J=2e/h=4.835\,976(12) \times 10^{14}\ \mathrm{Hz/V_{NIST}}\ (2.4\ \mathrm{ppm}), \tag{2}$$

where V_{NIST} is the unit of voltage maintained at the U.S. National Institute of Standards and Technology (NIST) and at the time was based on the mean electromotive force (emf) of a group of electrochemical standard cells. [In Eq. (2), the first number in parentheses is the one-standard-deviation uncertainty in the last digits of the quoted value; the second is its fractional equivalent in ppm.] The difference between this value when expressed in SI units (Hz/V) and the previous best value, that resulting from the 1963 least-squares adjustment of the fundamental constants (Cohen and DuMond, 1965), was (35 ± 10) ppm. As shall shortly be seen, the cause of this rather severe discrepancy was the use in the 1963 adjustment of a value of α derived from the early 1950s measurement of the fine-structure splitting of the deuterium atom (energy difference between the $2\,^2P_{3/2}$ and $2\,^2P_{1/2}$ atomic energy levels), which subsequently turned out to be incorrect.

Because QED, the relativistic quantum field theory that describes the interaction of charged particles and electromagnetic radiation or photons, is one of the most precise and important theories of modern physics (Kinoshita, 1990), it is essential to see how well it can withstand experimental tests. The Josephson $2e/h$ determination is therefore significant mainly because a reliable indirect value of α independent of QED can be derived from it, and this value can in turn be used to compare QED and experiment critically and unambiguously. This is in marked contrast to the situation that existed prior to 1967 when no such value was available and tests of QED were mainly checks of internal consistency.

The equation relating α to $2e/h$ may be written as

$$\alpha=\left[\frac{4R_\infty(\Omega_{NIST}/\Omega)\gamma_p'(\mathrm{lo})_{NIST}}{c(\mu_p'/\mu_B)(2e/h)_{NIST}}\right]^{1/2}, \tag{3}$$

where as before Ω_{NIST}/Ω is the ratio of the NIST as-maintained ohm to the SI ohm; μ_p'/μ_B is the magnetic moment of the proton in units of the Bohr magneton (throughout, the prime means that the protons are not free but in a spherical sample of pure H_2O at 25 °C); and $\gamma_p'(\mathrm{lo})_{NIST}$ is the gyromagnetic ratio of the proton obtained by the so-called low (magnetic) field method and is measured in terms of the NIST as-maintained unit of current $A_{NIST}=V_{NIST}/\Omega_{NIST}$. [Note that the subscript NIST may be replaced by the more general LAB (for laboratory); and that $\gamma'_p=\omega'_p/B=4\pi\mu'_p/h$, where ω'_p is the proton nuclear magnetic resonance or spin-flip angular frequency in the magnetic field (magnetic flux density) B. In the low-field method, B is established via a precision solenoid of known dimensions carrying a current known in terms of A_{LAB}.] The value of α derived from Eq. (3) in 1969 (Taylor, Parker, and Langenberg, 1969) using the best data available was

$$1/\alpha=137.036\,08(26)(1.9\text{ ppm}). \tag{4}$$

Included among the quantities that require an accurate value of α for comparing theory and experiment are the g factors of the electron and muon; the energy levels in hydrogen-like atoms, especially the $n=2$ Lamb shift ($2\,^2S_{1/2}-2\,^2P_{1/2}$ splitting); and the ground-state hyperfine splitting in hydrogen, muonium (an electron bound to a positive muon or μ^+e^- atom), and positronium (a bound electron-positron pair or e^+e^- atom). Of particular interest in the late 1960s was the hydrogen hyperfine splitting ν_{Hhfs}. This transition frequency, which corresponds to essentially the energy difference between a hydrogen atom in which the electron and proton spins are aligned and one in which they are in opposite directions, can be determined experimentally to the extraordinary accuracy of 1 part in 10^{12} using the hydrogen maser. The theoretical QED equation for ν_{Hhfs} is

$$\begin{aligned}\nu_{\text{Hhfs}}(\text{theor})=(16/3)cR_\infty(\mu_p/\mu_B)\\ \times(1+m_e/m_p)^{-3}\\ \times\alpha^2[1+\text{H.O.C.}],\end{aligned} \tag{5}$$

where H.O.C. means higher-order corrections $\ll 1$. In contrast to the 1×10^{-12} relative uncertainty of $\nu_{\text{Hhfs}}(\text{expt})$, Eq. (5) is limited to an accuracy of a few ppm because of the difficulty in calculating some of the higher order corrections. [Even in the late 1960s, c, R_∞, μ_p/μ_B, and m_e/m_p were sufficiently well known that their total contribution to the uncertainty of $\nu_{\text{Hhfs}}(\text{theor})$ was considerably less than 1 ppm.]

The most uncertain H.O.C. in Eq. (5) is $\delta_N^{(2)}$, the proton polarizability contribution, which arises from the various excited states or internal structure of the proton. In the late 1960s, theoretical calculations predicted $\delta_N^{(2)}=(0\pm5)$ ppm. This was in conflict with what was implied by the value of α accepted at that time, which, as noted earlier, was derived from a measurement of the fine-structure splitting in deuterium. When the deuterium value of α was used to calculate $\nu_{\text{Hhfs}}(\text{theor})$ from Eq. (5), when this result was compared to $\nu_{\text{Hhfs}}(\text{expt})$ as obtained from the hydrogen maser, and when their difference was assumed to arise solely from the existence of a polarizability term, it was found that $\delta_N^{(2)}=(43\pm9)$ ppm. This meant that the probability for $\delta_N^{(2)}$ to be as small as predicted by direct calculation was only 1 in 20 000, an obvious inconsistency. In contrast, when the value of α derived from the JE measurement of $2e/h$ [Eq. (4)] was used in place of the deuterium value, it was found that $\delta_N^{(2)}=(2.5\pm4.0)$ ppm, in keeping with the theoretical calculations. Hence the JE value of α removed a discrepancy that during the 1960s was termed one of the major unsolved problems of QED. This example further illustrates the overall unity of science—a low-temperature, solid-state physics experiment has provided information about the excited states of the proton—as well as how precise measurements of the fundamental constants can illuminate apparent inconsistencies in our physical description of nature.

1.2.2 QED and the Quantum Hall Effect Determination of h/e^2. Another example of a critical test of QED, but one that is still under active experimental and theoretical investigation in the 1990s, involves the QHE and the magnetic moment anomaly of the electron $a_e=g_e/2-1=\mu_e/\mu_B-1$, where g_e is the g factor of the electron.

The current best experimental value for a_e,

$$a_e(\text{expt})=1\,159\,652\,188.4(4.3)\times10^{-12}(0.0037\text{ ppm}), \tag{6}$$

is that obtained from microwave resonance measurements on a single electron stored in a Penning trap (van Dyck *et al.*, 1987). The QED theoretical expression for a_e may be written as (Kinoshita and Lindquist, 1990)

$$a_e(\text{theor})=C_1(\alpha/\pi)+C_2(\alpha/\pi)^2+C_3(\alpha/\pi)^3+C_4(\alpha/\pi)^4+\cdots+\delta a_e, \tag{7}$$

with $C_1=\frac{1}{2}$, $C_2=-0.328\ 478\ 965...$, $C_3=1.176\ 11(42)$, $C_4=1.434(138)$, and $\delta a_e=4.46\times10^{-12}$. The total uncertainty in a_e(theor) assuming α is exactly known is 5.8 parts in 10^9 or 0.0058 ppm, 0.0046 ppm from C_3 and 0.0035 ppm from C_4, and arises from the uncertainties of various numerical integrations. To test QED by comparing a_e(theor) with a_e(expt) to the full accuracies of a_e(theor) and a_e(expt) requires, therefore, a value of α having an uncertainty significantly less than the 0.0069-ppm uncertainty of the value of α obtained by equating Eqs. (6) and (7), say 0.002–0.003 ppm. Such a value is not yet available, but one with only ten times this uncertainty has already been obtained from the QHE.

The QHE is characteristic of a completely quantized two-dimensional electron gas (Prange and Girvin, 1987). Such a gas may be realized in a high-mobility semiconductor device such as a silicon metal-oxide-semiconductor field-effect transistor (MOSFET) or GaAs-$Al_xGa_{1-x}As$ heterojunction of standard Hall bar geometry in an applied magnetic flux density B of the order of 10 T and cooled to about 1 K. Hence, like the JE, the QHE is a low-temperature, solid-state physics phenomenon.

For a fixed current I (typically 10–50 μA) through the device, there are regions in the curve of Hall voltage U_H versus gate voltage for a MOSFET, or of U_H vs B for a heterojunction, where U_H remains constant as either the gate voltage or B is varied. These regions of constant U_H are termed quantum Hall plateaus. In the limit of zero dissipation (zero voltage drop) in the direction of current flow, the Hall voltage-to-current quotient $U_H(i)/I$ or Hall resistance $R_H(i)$ of the ith plateau, where i is an integer (we consider only the integral QHE), is quantized and given by

$$R_H(i)=U_H(i)/I=R_K/i, \tag{8}$$

where R_K is the von Klitzing constant (after the discoverer of the QHE).

The theory of the QHE predicts, and the experimentally observed universality of Eq. (8) is consistent with the prediction, that

$$R_K=h/e^2=\mu_0 c/2\alpha. \tag{9}$$

Since in the SI $\mu_0=4\pi\times10^{-7}$ N/A^2 exactly, and $c=299\ 792\ 458$ m/s exactly as a result of the 1983 redefinition of the meter in terms of the speed of light, a measurement of R_K in SI units (i.e., ohms) with a given uncertainty will yield a value of the fine-structure constant α with the same uncertainty.

In practice, R_K is measured in terms of Ω_{LAB} [indicated as $(R_K)_{LAB}$] so that the conversion factor Ω_{LAB}/Ω must be determined in a separate experiment using an apparatus known as a calculable cross capacitor in which the unknown resistance of a reference resistor is compared with the known impedance of the capacitor. Nevertheless, the QHE has already yielded a value of α with an uncertainty of 0.024 ppm (Shields, Dziuba, and Layer, 1989). When used to compare a_e(theor) with a_e(expt), it gives a difference of a_e(expt) $-a_e$(theor) $=\Delta a_e=(0.041\pm0.025)$ ppm. Since the 0.041-ppm difference is only 1.7 times the 0.025-ppm standard deviation of the difference, Δa_e is within statistically acceptable limits. This agreement confirms QED to the unprecedented level of 4 parts in 10^{11} through the near equality of $g_e(\text{expt})=2[1+a_e(\text{expt})]$ and $g_e(\text{theor})=2[1+a_e(\text{theor})]$, since QED actually predicts the value of the entire g factor.

2. OBTAINING "BEST VALUES" OF THE CONSTANTS

2.1 Relationships Among the Constants

Equation (3), Eq. (6) together with Eq. (7), and Eq. (9) can all provide values of α. However, because in practice a value of Ω_{LAB}/Ω is required to obtain α from Eq. (9) as well as from Eq. (3), the two values will not be completely independent. Moreover, $2e/h$ (in SI units) can be obtained from the relation

$$2e/h=\left[\frac{16R_\infty(m_p/m_e)N_A}{\mu_0 c^2 M_p \alpha}\right]^{1/2}, \tag{10}$$

where M_p is the molar mass (atomic weight in units of kg/mol) of the proton; and γ'_p(lo) from the equation

$$\gamma'_p(\text{lo})_{LAB}=\frac{F_{LAB}K^2_{A-LAB}(\mu'_p/\mu_B)(m_p/m_e)}{M_p}, \tag{11}$$

where F_{LAB} (Faraday constant) is measured in terms of a particular laboratory's as-maintained ampere A_{LAB}, and $K_{A\text{-}LAB}=A_{LAB}/A=(V_{LAB}/\Omega_{LAB})/(V/\Omega)=(V_{LAB}/V)/(\Omega_{LAB}/\Omega)=K_{V\text{-}LAB}/K_{\Omega\text{-}LAB}$. Further, the Faraday constant is related to the Avogadro constant by $F=N_A e$, and $K_{A\text{-LAB}}$ may be obtained from direct measurements using a so-called ampere force balance as well as from the equation

$$K^2_{A\text{-LAB}}=\frac{\gamma'_p(\text{lo})_{\text{LAB}}}{\gamma'_p(\text{hi})_{\text{LAB}}}, \qquad (12)$$

where $\gamma'_p(\text{hi})$ is the gyromagnetic ratio of the proton as obtained by the so-called high (magnetic) field method. (The magnetic flux density B is determined by measuring the force it exerts on a conductor of known dimensions carrying a current known in terms of A_{LAB}.)

It should now be clear that complex relationships can exist among groups of fundamental constants and conversion factors, and that a particular constant may be obtained either by direct measurement or indirectly by appropriately combining other directly measured constants. If the direct and indirect values have similar accuracy, both must be considered in order to arrive at a best value for the quantity. However, each of the various routes that can be followed to a particular constant, both direct and indirect, provides a somewhat different numerical value. The best way to handle this situation is by the mathematical technique known as least squares.

2.2 Least-Squares Adjustments of the Constants

The least-squares method furnishes a well-defined procedure for calculating best "compromise" values of the constants from all of the available data. It takes into account all possible routes to a particular constant and yields a single value for each constant by weighting the different routes according to their relative uncertainties. The weights themselves are obtained from the *a priori* uncertainties assigned to the individual direct measurements that constitute the original set of data.

The least-squares study of the constants was pioneered in the late 1920s by R. T. Birge and continued by others, notably J. W. M. DuMond and E. R. Cohen. The three most recent reviews were those of B. N. Taylor, W. H. Parker, and D. N. Langenberg in 1969, based on their JE determination of $2e/h$; and of Cohen and Taylor in 1973 and 1986, both carried out under the auspices of The Committee on Data for Science and Technology (CODATA). The 1986 adjustment is the most recent one available, and the recommended set of constants resulting from it are those officially adopted by CODATA for international use.

2.3 The 1986 CODATA Adjustment

In general, the data entering a least-squares adjustment of the constants are divided into two groups:

1. the more precise data or auxiliary constants and
2. the stochastic input data.

The auxiliary constants are either defined quantities such as μ_0 and c with no uncertainty, or constants such as R_∞ and M_p with assigned uncertainties sufficiently small in comparison with the uncertainties assigned the stochastic input data with which they are associated in the adjustment that they can be taken as exact (i.e., their values are not subject to adjustment in contrast to the stochastic data). The auxiliary constants used in the 1986 adjustment, none of which had an assigned uncertainty greater than 0.02 ppm, included μ_0, c, R_∞, M_p, m_p/m_e, g_e, μ_e/μ_p, and the g factor of the muon $g_\mu=2(1+a_\mu)=2\mu_\mu/(eh/4\pi m_\mu)$, where a_μ is the magnetic moment anomaly of the muon and μ_μ is its magnetic moment.

Also included in the auxiliary constant category were the representations of the volt, ohm, and ampere in terms of which all electrical-unit dependent stochastic data were expressed: $V_{76\text{-BI}}$, Ω_{BI85}, and $A_{\text{BI85}}=V_{76\text{-BI}}/\Omega_{\text{BI85}}$, respectively. Here $V_{76\text{-BI}}=(K_{J\text{-}72}/K_J)V=K_V\,V$ is the representation of the volt (i.e., "practical unit" of voltage or as-maintained volt) realized by the JE at the BIPM with $K_{J\text{-}72}=483\,594$ GHz/V exactly (equivalent to defining $K_J=483\,594$ GHz/$V_{76\text{-BI}}$ exactly). This value of K_J was suggested by the Consultative Committee for Electricity (CCE) of the International Committee for Weights and Measures (CIPM) in 1972 (see Sec. 3) and was to be used by every national standards laboratory that based its national representation of the volt on the JE. The practical or as-maintained unit of resistance Ω_{BI85} is the BIPM representation of the ohm (prior to January 1, 1990) based on the mean resistance of a particular group of wire-wound standard resistors kept at BIPM in an oil bath at 20 °C and as they existed on January 1, 1985.

To carry out a least-squares adjustment, a subset of M constants is chosen in terms of which all of the stochastic input data can be individually and independently expressed—if necessary, with the aid of the auxiliary constants. It is actually the M constants constituting this subset, termed the adjustable constants or unknowns, that are directly subject to adjustment. In the 1986 evaluation there were $M=5$ unknowns and these were taken to be α^{-1}; the conversion factor $K_V=V_{76\text{-BI}}/V$; the conversion factor $K_\Omega=\Omega_{BI85}/\Omega$; d_{220}, the (220) repeat distance or lattice spacing of a perfect crystal of pure silicon at 22.5 °C and in vacuum; and μ_μ/μ_p, the ratio of the magnetic moment of the muon to that of the proton.

An equation which relates a stochastic input datum to the M adjustable constants (with the aid of the auxiliary constants if necessary) is known as an observational equation. Thus, if there are N items of stochastic input data, there are N observational equations. However, the observational equations for data items of the same type are of the same form. The weight w_i of each stochastic datum and hence of each observational equation is related to the datum's *a priori* assigned uncertainty s_i by $w_i=1/s_i^2$. Consequently, if one value of a datum for a particular quantity has half the uncertainty of another, it carries four times as much weight in the adjustment. Taking into account the weight assigned each observational equation and with the aid of a computer, the least-squares adjusted values of the M unknowns may be readily obtained. "Best values" in the least-squares sense for these M adjustable constants, with their variances and covariances, are thus the immediate output of the adjustment. Optimal values in the least-squares sense for most constants of interest not directly subject to adjustment may then be calculated from the M adjustable constants with the aid of the auxiliary constants. (But this does not apply to the auxiliary constants themselves since, for the purpose of the least-squares adjustment, they are assumed to be exactly known.)

In contrast to the auxiliary constants of the 1986 adjustment for which no uncertainty exceeded 0.02 ppm, the uncertainties assigned to the 38 items of stochastic input data initially considered in the 1986 adjustment were in the range 0.065–9.7 ppm. These 38 items, of 12 distinct types with the number of items of each type ranging from one to six, were:

1. $K_\Omega=\Omega_{BI85}/\Omega$: five values with uncertainties of 0.11 to 0.36 ppm.
2. $K_A=A_{BI85}/A=(V_{76\text{-BI}}/\Omega_{BI85})/(V/\Omega)=K_V/K_\Omega$: six values with uncertainties of 4.1 to 9.7 ppm.
3. $K_V=V_{76\text{-BI}}/V$: two values with uncertainties of 0.60 and 2.4 ppm.
4. F_{BI85}: one value in units of A_{BI85} s/mol $=C_{BI85}$/mol with an uncertainty of 1.33 ppm.
5. $\gamma_p'(\text{lo})_{BI85}$: six values in units of s^{-1}/T_{BI85} with uncertainties of 0.24 to 3.25 ppm.
6. $\gamma_p'(\text{hi})_{BI85}$: four values in units of C_{BI85}/kg with uncertainties of 1.0 to 5.4 ppm.
7. d_{220}: two values in units of m with uncertainties of 0.10 and 0.23 ppm.
8. $V_m(\text{Si})=M(\text{Si})/\rho(\text{Si})$: one value of the mean molar volume of pure single-crystal silicon at 22.5 °C and in vacuum in units of m^3/mol with an uncertainty of 1.15 ppm. [M(Si) is the mean molar mass of silicon (units of kg/mol) and ρ(Si) is its density (units of kg/m^3).]
9. $(R_K)_{BI85}$: six values in units of Ω_{BI85} with uncertainties of 0.12 to 0.22 ppm.
10. α^{-1}: two values with uncertainties of 0.065 and 0.60 ppm.
11. μ_μ/μ_p: two values with uncertainties of 0.36 and 0.53 ppm.
12. ν_{Mhfs}: one value of the ground-state hyperfine splitting of muonium in units of Hz with an uncertainty of 0.14 ppm taking into account the theoretical expression which relates ν_{Mhfs} to α^{-1} and μ_μ/μ_p.

Because new results that can influence a least-squares adjustment of the constants are reported continually, it is always difficult to choose an optimal time at which to carry out a new adjustment and to revise the recommended values of the constants. In the case of the 1986 adjustment, all data available up to 1 January 1986 were considered for inclusion, with the recognition that any additional changes to the 1973 CODATA recommended values that might result by taking into account more recent data would be much smaller than the changes resulting from the data available prior to that date. Indeed, the improved measurements of R_K and Ω_{LAB}/Ω, and the improved measurement of a_e(expt) and calculation of a_e(theor), which have yielded values of

α with uncertainties of 0.024 and 0.0069 ppm, respectively (see Sec. 1.2.2), were not available by January 1, 1986. Their impact and that of other more recent results on the 1986 recommended values are discussed in Sec. 5.

The quantity d_{220} was introduced as an unknown in the 1986 adjustment, and the quantities d_{220} and $V_m(\mathrm{Si}) = M(\mathrm{Si})/\rho(\mathrm{Si})$ were treated as separate items of stochastic input data, because the two available experimental values of d_{220} were in gross disagreement. If they had not been discrepant, then their weighted mean could have been used in conjunction with the single available value of $V_m(\mathrm{Si})$ to obtain a single value of N_A through the relation $N_A = 8V_m(\mathrm{Si})/(d_{220}\sqrt{8})^3$. This result could then have been included in the adjustment directly and there would have been no need to introduce d_{220} as an unknown—the adjustment could have been carried out in terms of the four unknowns α^{-1}, K_V, K_Ω, and μ_μ/μ_p.

2.3.1 Uncertainty Assignment and Discrepant Data. Because the weight of each stochastic datum included in a least-squares adjustment of the constants is taken as $w_i = 1/s_i^2$, critical analysis of the input data and deciding what *a priori* uncertainty s_i should be assigned to each measurement is the main problem in successfully carrying out an adjustment. The uncertainty problem is made especially difficult because in most experiments sufficient data are taken to reduce the uncertainty from random effects to negligible amounts, and the final uncertainty assigned the experiment is determined primarily from estimates of the uncertainties associated with systematic effects. Because such effects frequently arise from phenomenona that the experimenter knows little about, the estimation of the uncertainties that properly characterize them is often subjective and based on educated guesses at best.

Another problem area is deciding how to handle "discrepant" data, that is, measurements that differ from each other by statistically significant amounts in comparison with their assigned uncertainties s_i. Such data cannot be included in an adjustment uncritically because the inconsistencies imply either incorrect uncertainty assignments or the presence of unknown measurement errors.

When confronted with such a situation, the reviewer can in general either include the inconsistent data, but only after expanding (increasing) their s_i so that they are no longer discrepant; or decide, on as sound a theoretical and experimental basis as possible, which of the inconsistent data are least reliable, and discard them. These two approaches clearly imply that different reviewers might treat the same data differently, thereby obtaining somewhat different sets of best values.

Motivated by the all too frequent occurrence of large changes in the recommended values of many constants from one adjustment to the next, new algorithms for handling discrepant data in least-squares adjustments of the constants were developed during the years between the 1973 and 1986 adjustments. These new algorithms, which added increased objectivity to expanding *a priori* uncertainties and discarding inconsistent data, were used extensively by Cohen and Taylor in carrying out the 1986 adjustment. Indeed, with their aid, the initial group of 38 items of stochastic input data was reduced to 22 items by deleting those that were either highly inconsistent with the remaining data or had assigned uncertainties so large that they carried negligible weight.

2.3.2 Recommended Values of the Constants. Table 1 gives the 1986 CODATA recommended values and one-standard-deviation uncertainties of a number of fundamental physical constants. As noted in Sec. 2.3, the direct output of a least-squares adjustment is best values in the least-squares sense for the adjustable constants or unknowns and their variances and covariances. For the 1986 adjustment, this means for the five unknowns α^{-1}, K_V, K_Ω, $d_{220}(\mathrm{Si})$, and μ_μ/μ_p. Thus, the vast majority of the quantities listed in Table 1 were calculated from these five unknowns and auxiliary constants as appropriate.

2.3.3 Comparison of the 1986 and 1973 Adjustments. The 1986 CODATA adjustment represents a major advance over its 1973 counterpart; the uncertainties of many recommended values have been reduced by roughly an order of magnitude through the enormous advances made throughout the precision measurement-fundamental constants field between 1973 and 1986. This can be seen from column 4 of Table 2, which compares the recommended values and uncertainties of a selected group of constants as obtained from the two adjustments. It is also clear from the last

Table 1. Recommended values of selected physical constants from the 1986 CODATA least-squares adjustment by Cohen and Taylor (1987). The digits in parentheses are the one-standard-deviation uncertainty in the last digit(s) of the given value.

Quantity	Symbol	Value	Units	Relative uncertainty (ppm)
Speed of light in vacuum	c	299 792 458	$m\ s^{-1}$	(exact)
Permeability of vacuum	μ_0	$4\pi \times 10^{-7}$	$N\ A^{-2}$	
		$=12.566\ 370\ 614\cdots$	$10^{-7}\ N\ A^{-2}$	(exact)
Permittivity of vacuum, $1/\mu_0 c^2$	ϵ_0	$8.854\ 187\ 817\cdots$	$10^{-12}\ F\ m^{-1}$	(exact)
Newtonian constant of gravitation	G	6.672 59(85)	$10^{-11}\ m^3\ kg^{-1}\ s^{-2}$	128
Planck constant	h	6.626 075 5(40)	10^{-34} J s	0.60
$h/2\pi$	$\hbar$	1.054 572 66(63)	10^{-34} J s	0.60
Elementary charge	e	1.602 177 33(49)	10^{-19} C	0.30
Magnetic flux quantum, $h/2e$	Φ_0	2.067 834 61(61)	10^{-15} Wb	0.30
Electron mass	m_e	9.109 389 7(54)	10^{-31} kg	0.59
Proton mass	m_p	1.672 623 1(10)	10^{-27} kg	0.59
Proton-electron mass ratio	m_p/m_e	1 836.152 701(37)		0.020
Fine-structure constant, $\mu_0 ce^2/2h$	α	7.297 353 08(33)	10^{-3}	0.045
Inverse fine-structure constant	α^{-1}	137.035 989 5(61)		0.045
Rydberg constant, $m_e c\alpha^2/2h$	R_∞	10 973 731.534(13)	m^{-1}	0.0012
Avogadro constant	N_A, L	6.022 136 7(36)	$10^{23}\ mol^{-1}$	0.59
Faraday constant, $N_A e$	F	96 485.309(29)	$C\ mol^{-1}$	0.30
Molar gas constant	R	8.314 510(70)	$J\ mol^{-1}\ K^{-1}$	8.4
Boltzmann constant, R/N_A	k	1.380 658(12)	$10^{-23}\ J\ K^{-1}$	8.5
Stefan-Boltzmann constant, $(\pi^2/60)k^4/\hbar^3 c^2$	σ	5.670 51(19)	$10^{-8}\ W\ m^{-2}\ K^{-4}$	34
Josephson frequency-voltage quotient	$2e/h$	4.835 976 7(14)	$10^{14}\ Hz\ V^{-1}$	0.30
Quantized Hall resistance, $h/e^2 = \mu_0 c/2\alpha$	R_H	25 812.805 6(12)	Ω	0.045
Bohr magneton, $e\hbar/2m_e$	μ_B	9.274 015 4(31)	$10^{-24}\ J\ T^{-1}$	0.34
Nuclear magneton, $e\hbar/2m_p$	μ_N	5.050 786 6(17)	$10^{-27}\ J\ T^{-1}$	0.34
Bohr radius, $\alpha/4\pi R_\infty$	a_0	0.529 177 249(24)	10^{-10} m	0.045
Compton wavelength, $h/m_e c$	λ_C	2.426 310 58(22)	10^{-12} m	0.089
Electron magnetic moment	μ_e	928.477 01(31)	$10^{-26}\ J\ T^{-1}$	0.34
in Bohr magnetons	μ_e/μ_B	1.001 159 652 193(10)		1×10^{-5}
Electron magnetic moment anomaly, $\mu_e/\mu_B - 1$	a_e	1.159 652 193(10)	10^{-3}	0.0086
Electron g factor, $2(1+a_e)$	g_e	2.002 319 304 386(20)		1×10^{-5}
Muon-electron mass ratio	m_μ/m_e	206.768 262(30)		0.15
Muon magnetic moment anomaly, $[\mu_\mu/(e\hbar/2m_\mu)]-1$	a_μ	1.165 923 0(84)	10^{-3}	7.2
Muon g factor, $2(1+a_\mu)$	g_μ	2.002 331 846(17)		0.0084
Muon-proton magnetic moment ratio	μ_μ/μ_p	3.183 345 47(47)		0.15
Proton magnetic moment	μ_p	1.410 607 61(47)	$10^{-26}\ J\ T^{-1}$	0.34
in Bohr magnetons	μ_p/μ_B	1.521 032 202(15)	10^{-3}	0.010
in nuclear magnetons	μ_p/μ_N	2.792 847 386(63)		0.023
Proton gyromagnetic ratio	γ_p	26 752.212 8(81)	$10^4\ s^{-1}\ T^{-1}$	0.30
	$\gamma_p/2\pi$	42.577 469(13)	$MHz\ T^{-1}$	0.30
Molar volume (ideal gas), RT/p $T=273.15$ K, $p=101\ 325$ Pa	V_m	0.022 414 10(19)	$m^3\ mol^{-1}$	8.4
		Non-SI units used with SI		
Electron volt, $(e/C)J=\{e\}J$	eV	1.602 177 33(49)	10^{-19} J	0.30
(Unified) atomic mass unit, $1\ u = m_u = \frac{1}{12}m(^{12}C)$	u	1.660 540 2(10)	10^{-27} kg	0.59

Table 2. Comparison of the recommended values and uncertainties of a selected group of constants as obtained from the 1986 and 1973 CODATA least-squares adjustments of the constants.[a]

Quantity	1 std. dev. uncertainty of 1986 recommended value, σ_{86}, in ppm	1 std. dev. uncertainty of 1973 recommended value, σ_{73}, in ppm	Ratio of σ_{73} to σ_{86}	Change in 1973 recommended value implied by 1986 recommended value, δ, in ppm; and corresponding number of standard deviations, δ/σ_{73}	
				δ	δ/σ_{73}
α^{-1}	0.045	0.82	18	−0.37	−0.5
e	0.30	2.9	9.7	−7.4	−2.6
h	0.60	5.4	9.0	−15.2	−2.8
m_e	0.59	5.1	8.6	−15.8	−3.1
N_A	0.59	5.1	8.6	15.2	3.0
m_p/m_e	0.020	0.38	19	0.64	1.7
F	0.30	2.8	9.3	7.8	2.8
$2e/h$, K_J	0.30	2.6	8.7	7.8	3.0
R	8.4	31	3.7	12	0.4
k	8.5	32	3.8	−3	−0.1
σ	34	125	3.7	34	0.3

[a]Cohen and Taylor, 1987; Cohen and Taylor, 1973.

column of Table 2 that unexpectedly large changes have occurred in the 1973 recommended values of several of these constants (i.e., a change which is large relative to the uncertainty assigned the 1973 value). These changes are a direct consequence of the 7.8-ppm decrease in the quantity K_V from 1973 to 1986 and the high correlation between K_V and the calculated values of e, h, m_e, N_A, and F. Since $2e/h=K_{J\text{-}72}/K_V$, the 1986 value of K_V also implies that the value of $2e/h$ adopted by the CCE in 1972, which was believed to be consistent with the SI value and which most national standards laboratories adopted to define and maintain their national representation of the volt, is actually 7.59 ppm smaller than the SI value. As will be discussed in Sec. 3, this unsatisfactory situation was rectified starting on January 1, 1990.

The large change in K_V between 1973 and 1986 would have been avoided if two determinations of F that seemed to be discrepant with the remaining data had not been deleted in the 1973 adjustment. In retrospect, the disagreement was comparatively mild. Fortunately, there are no similar disagreements in the 1986 adjustment; the measurements that were deleted were so discrepant that they obviously could not be correct, or of such low weight that if retained the adjusted values of the five unknowns would change negligibly. Thus, it is unlikely that any reasonable alternative evaluation of the data considered in the 1986 least-squares adjustment could lead to significant changes in the 1986 recommended values.

3. NEW JOSEPHSON AND QUANTUM HALL EFFECT REPRESENTATIONS OF THE VOLT AND OHM

The fundamental constants have a key role to play in the establishment of reproducible and invariant units of measurement, as was pointed out in Sec. 1.2. Exemplifying this role is the world-wide introduction on January 1, 1990 of new practical units of voltage and resistance based on the JE and QHE, the topic of this portion of the article.

3.1 The SI and Representations of the Volt and Ohm

The SI is the system of units used throughout the world for expressing the results of physical and chemical measurements. The three SI *base* mechanical units are the meter (m), kilogram (kg), and second (s) (Taylor, 1991b). The *derived* SI unit of force, the newton (N), is then defined in terms of these three base mechanical units by 1 N=1 m kg s^{-2}. The SI base unit of current, the ampere (A), is defined as that constant current which, if maintained in two straight parallel conductors of infinite length, of negligible circular cross section, and placed 1 m apart in vacuum, would produce between these conductors a force equal to 2×10^{-7} N per meter of length. This definition leads to $\mu_0=4\pi\times10^{-7}$ N A^{-2} exactly.

The SI derived unit of electric potential difference, the volt (V), and the SI derived unit of electric resistance, the ohm (Ω), are then defined in terms of the three base mechanical units and the base electric unit, the ampere. Formally, 1 V=1 m^2 kg s^{-3} A^{-1} and 1 Ω=1 m^2 kg s^{-3} A^{-2}. In practice, the volt may be realized using a force balance in which an electrostatic or electromagnetic force is compared with a mechanical force; and the ohm may be realized by comparing the impedance of a calculable capacitor or inductor with that of a resistor. However, because both experiments are exceedingly difficult and time consuming to carry out and are of limited accuracy, it has become necessary to establish practical or as-maintained units of voltage and resistance, that is, *representations* of the volt and ohm, that provide better long-term repeatability and constancy than the direct realizations of the SI units themselves.

Historically, representations of the volt have been based on large groups of electrochemical standard cells. However, in 1972 (Taylor, 1987), the CCE suggested that the national standards laboratories base their national representations of the volt on the JE in order to avoid the variation with time and severe temperature dependence of the emf of such cells. In particular, the CCE proposed, and the CIPM subsequently approved the proposal, that the national standards laboratories adopt K_{J-72}=483 594 GHz/V, exactly, as the conventional value of the Josephson constant for use in realizing and maintaining an accurate and stable national representation of the volt by means of the JE. But while most national laboratories accepted this value, three did not. Moreover, as pointed out in Sec. 2.3.3, the 1986 CODATA adjustment shows that K_{J-72} is nearly 8 ppm smaller than the SI value, implying that the national representations of the volt of those countries that use it are about 8 ppm smaller than the SI unit.

During the late 1970s and early 1980s, the CCE also became increasingly concerned that, because most national standards laboratories based their representation of the ohm on the mean resistance of a particular group of precision wire-wound standard resistors, and because these artifact standards age, the various national representations of the ohm differed significantly from each other and from the ohm, and some were drifting excessively. Although in principle a calculable cross capacitor can now be used to realize the ohm with an uncertainty of <0.1 ppm, it is a difficult experiment to perform routinely; only one laboratory in the world has had such an apparatus in continuous operation since the method was first developed in the early 1960s. Consequently, electrical metrologists enthusiastically welcomed von Klitzing's 1980 discovery of the QHE since it promised to provide a method for basing a representation of the ohm on fundamental constants in much the same manner as the JE, discovered 19 years earlier in 1961, provided a method for basing a representation of the volt on fundamental constants.

3.2 The Conventional Values K_{J-90} and R_{K-90}

To address the problems associated with national representations of the volt and ohm discussed above, the CCE in 1986 (Giacomo, 1987) established the CCE Working Group on the Josephson Effect. The CCE charged the Working Group to propose a new value of K_J consistent with the SI value based upon all relevant data that became available by 15 June 1988. Similarly, recognizing the rapid advances made in understanding the QHE since its comparatively recent discovery in 1980, the CCE established the Working Group on the Quantum Hall Effect. The CCE charged the Working Group

1. to propose to the CCE, based upon all relevant data that became available by 15 June 1988, a value of R_K consistent with the SI value for use in maintaining an accurate and stable national representation of the ohm by means of the QHE; and
2. to develop detailed guidelines for the proper use of the QHE to realize reliably such a representation.

Further, the CCE stated its intention to meet in 1988 with a view to recommending that the proposed values of K_J and R_K come into effect on 1 January 1990. These values would be used by *all* those national standards laboratories (and others) that base their representations of the volt on the JE and that choose to base their representations of the ohm on the QHE. These proposals of the CCE were subsequently approved by the CIPM and by the 18th General Conference on Weights and Measures (CGPM).

In response to the CCE's directives, each Working Group reviewed and analyzed the values of K_J and R_K that were available by the cutoff date and derived a recommended value of K_J and of R_K (Taylor and Witt, 1989). The Working Group on the JE considered ten values of K_J, some of which were based on experimental data included for use in the 1986 CODATA least-squares adjustment but most of which had become available since the latter's 1 January 1986 cutoff date. (The CCE's 15 June 1988 cutoff date stimulated the completion of much experimental work relevant to deriving values of K_J and R_K, or at least the reporting of first results from ongoing experiments.) Similarly, the Working Group on the QHE considered 13 values of R_K, only one of which was based on experimental data available for use in the 1986 CODATA adjustment.

The values finally recommended by the working groups, which were subsequently approved by the CCE and CIPM at their respective 1988 meetings (Quinn, 1989) and which went into effect worldwide as planned on 1 January 1990, are

$$K_{J\text{-}90}=483\ 597.9\ \text{GHz/V exactly,} \tag{13a}$$

$$R_{K\text{-}90}=25\ 812.807\ \Omega\ \text{exactly.} \tag{13b}$$

The respective CCE Working Groups estimated that $K_{J\text{-}90}$ is consistent with the SI value of K_J to within a conservatively assigned one-standard-deviation uncertainty of 0.2 GHz/V or relative uncertainty of 0.4 ppm, and that $R_{K\text{-}90}$ is consistent with the SI value of R_K to within a conservatively assigned one-standard-deviation uncertainty of 0.005 Ω or relative uncertainty of 0.2 ppm. This implies that an ideal representation of the volt based on the JE and $K_{J\text{-}90}$ should be consistent with the SI volt to within 0.4 ppm or 0.4 μV, and that an ideal representation of the ohm based on the QHE and $R_{K\text{-}90}$ should be consistent with the SI ohm to within 0.2 ppm or 0.2 μΩ.

Since the experimental uncertainties associated with realizing representations of the volt and of the ohm based on the JE and QHE lie typically in the range 0.01 to 0.2 ppm, all national representations of the volt and of the ohm should now be equivalent to within a few parts in 10^7, and consistent with the SI volt and ohm to within about 5 parts in 10^7. Thus the fundamental constants of nature have led to a very practical result: the significant improvement of the international uniformity of electrical measurements and their consistency with the SI. This will no doubt be of major benefit to science, commerce, and industry throughout the world for many years to come.

4. IMPLICATIONS OF RECENT ADVANCES IN THE FUNDAMENTAL CONSTANTS FIELD

Many advances have been made in the fundamental constants field since the completion of the 1986 CODATA least-squares adjustment, as was clear in Sec. 3. In this section, the principal new results are summarized and their implications for both the 1986 CODATA recommended values and the next least-squares adjustment are discussed (Taylor and Cohen, 1990). The new results lead to values of the constants that in general have uncertainties 5 to 7 times smaller than the uncertainties assigned the 1986 values. However, the changes in the values themselves are less than twice the 1986 one-standard-deviation uncertainties and thus are not highly significant. Although much new data has become available since 1986, three new results dominate the analysis: a value of h obtained from a realization of the watt; a value of α^{-1} obtained from a_e; and a value of the molar gas constant R obtained from a new measurement of the speed of sound in argon.

4.1 New Results Obtained Since 1986

The discussion of the new results and their impact will follow closely that of the 1986 CODATA adjustment, Sec. 2.3. However, no new recommended values for any constants are given. One reason is that because the recommended values of a least-squares adjustment are correlated, the new results cannot be readily incorporated in the 1986 table of recommended values; to do so would require nothing less than a new adjustment. More important, although the new results would lead to significant reductions in the uncertainties of many of the 1986 recommended values, it is inappropriate to replace the 1986 set prematurely. First, it takes considerable time for a new set of recommended values to diffuse throughout all of science and technology; handbooks, textbooks, encyclopedias, and other reference works are not revised yearly. Second, the 1986 values adequately serve the needs of the vast majority of users—those few users who require the most up-to-date and accurate values of a particular constant can consult the primary literature. On this basis, approximately eight years between adjustments would seem reasonable.

Because the uncertainties of the auxiliary constants in a least-squares adjustment are generally 10–20 times less than the uncertainties of the stochastic input data, as might be expected, the new results for some of the auxiliary constants used in the 1986 adjustment have little impact on the vast majority of the 1986 recommended values. Further, it is unlikely that any quantity in the auxiliary constant category in the 1986 adjustment will become a stochastic input datum in the next adjustment.

In the 1986 adjustment, electric unit-dependent quantities such as F and γ'_p were expressed in terms of the practical units $V_{76\text{-}BI}$ and Ω_{BI85}, and the dimensionless ratios $K_V = V_{76\text{-}BI}/V$ and $K_\Omega = \Omega_{BI85}/\Omega$ were taken as two of the five adjustable constants or unknowns. With the introduction on 1 January 1990 of the new volt and ohm representations based on the JE and QHE and the conventional values $K_{J\text{-}90}$ and $R_{K\text{-}90}$ (see Sec. 3), it is now natural to express electric unit dependent quantities in terms of the new representations, namely, V_{90} and Ω_{90} defined by $K_J = 483\ 597.9\ \text{GHz}/V_{90}$ and $R_K = 25\ 812.807\ \Omega_{90}$. This is equivalent to defining $V_{90} = (K_{J\text{-}90}/K_J)V$ and $\Omega_{90} = (R_K/R_{K\text{-}90})\Omega$. The relationships between units also imply $A_{90} = V_{90}/\Omega_{90}$, $W_{90} = V_{90}A_{90} = V_{90}^2/\Omega_{90}$, and $C_{90} = A_{90}$ s.

For the least-squares analysis of the data, K_Ω and K_V may again be taken as unknowns but defined as $K_\Omega = \Omega_{90}/\Omega$ and $K_V = V_{90}/V$. The only other unknown required for the analysis is μ_μ/μ_p. Thus the data may be analyzed using the three unknowns K_Ω, K_V, and μ_μ/μ_p compared with the five unknowns used in 1986: α^{-1}, K_V, K_Ω, d_{220}, and μ_μ/μ_p. This reduction has been made possible by

1. following the CCE Working Group on the QHE and including only measurements of $(R_K)_{LAB}$ that are directly tied to a realization of the ohm based on a calculable cross capacitor so that the measured quantity is R_K in (SI) ohms; and
2. the elimination of the discrepancy between the two values of d_{220} available in 1986, by the discovery of a systematic error in one of the values, plus the completion of an independent measurement of the molar volume of silicon $V_m(\text{Si}) = M(\text{Si})/\rho(\text{Si})$ by the laboratory whose d_{220} result was retained in the 1986 adjustment.

As a consequence of these latter advances, two completely independent and consistent directly measured values of N_A with uncertainties of 1.10 and 1.19 ppm are now available.

The 12 distinct types of data initially considered in the 1986 least-squares adjustment were $K_\Omega = \Omega_{BI85}/\Omega$, $K_A = A_{BI85}/A$, $K_V = V_{76\text{-}BI}/V$, F_{BI85}, $\gamma'_p(\text{lo})_{BI85}$, $\gamma'_p(\text{hi})_{BI85}$, d_{220}, $V_m(\text{Si})$, $(R_K)_{BI85}$, α^{-1}, μ_μ/μ_p, and ν_{Mhfs}, although all values of K_A were eventually eliminated. As just discussed, d_{220} and $V_m(\text{Si})$ may now also be eliminated and replaced by N_A, and measurements of $(R_K)_{LAB}$ without a closely tied calculable cross-capacitor ohm realization are no longer considered useful. To these remaining 9 data types must be added 2 new ones, thereby bringing the total to 11: $K_W = W_{90}/W$ obtained from a realization of the watt via a moving-coil balance; and γ'_p in SI units (s^{-1}/T) obtained from the relation $\gamma'_p = [\gamma'_p(\text{lo})_{LAB} \times \gamma'_p(\text{hi})_{LAB}]^{1/2}$, where the measurements of $\gamma'_p(\text{lo})_{LAB}$ and $\gamma'_p(\text{hi})_{LAB}$ are carried out in the same laboratory using the same laboratory electrical units. Two sets of such measurements are treated in this way rather than as separate values of $\gamma'_p(\text{lo})_{90}$ and $\gamma'_p(\text{hi})_{90}$ because of the difficulty of reexpressing them in terms of the new volt and ohm representations, V_{90} and Ω_{90}.

There are 20 items of stochastic data distributed over the 11 data types that are appropriate for consideration, as follows:

1. $K_\Omega = \Omega_{90}/\Omega = R_K/R_{K\text{-}90}$: three values with uncertainties of 0.066, 0.054, and 0.024 ppm.
2. $K_W = W_{90}/W$: two values with uncertainties of 0.136 and 1.33 ppm.
3. $K_V = V_{90}/V = K_{J\text{-}90}/K_J$: two values, each with an uncertainty of 0.27 ppm.

4. F_{90}: one value in units of C_{90}/mol with an uncertainty of 1.33 ppm. It is the same result used in the 1986 adjustment but reexpressed in terms of C_{90}.
5. $\gamma'_p(\mathrm{lo})_{90}$: two values in units of s^{-1}/T_{90} with uncertainties of 0.11 and 0.36 ppm.
6. $\gamma'_p(\mathrm{hi})_{90}$: one value in units of C_{90}/kg with an uncertainty of 1.01 ppm. This is the same result used in the 1986 adjustment but reexpressed in terms of C_{90}.
7. γ'_p: two values in units of s^{-1}/T with uncertainties of 0.80 and 0.86 ppm obtained from $\gamma'_p(\mathrm{lo})_{\mathrm{LAB}}$ and $\gamma'_p(\mathrm{hi})_{\mathrm{LAB}}$ as discussed above.
8. N_A: two values in units of mol^{-1} with uncertainties of 1.10 and 1.19 ppm as discussed above.
9. α^{-1}: two values with uncertainties of 0.0069 and 0.20 ppm. The first was derived from a_e via Eqs. (6) and (7); the second from a measurement of h/m_n.
10. μ_μ/μ_p: two values with uncertainties of 0.36 and 0.53 ppm. These were also used in the 1986 adjustment.
11. ν_{Mhfs}: one value in units of Hz with an uncertainty of 0.14 ppm; it is the same value used in the 1986 adjustment.

Surprisingly, the measurement of $K_W = W_{90}/W$ is actually a measurement of the Planck constant. This follows from the fact that $W_{90} = V_{90}^2/\Omega_{90}$, $W = V/\Omega$, and thus $K_W = K_V^2/K_\Omega$; and $K_V = K_{J\text{-}90}/K_J$, $K_\Omega = R_K/R_{K\text{-}90}$, $K_J = 2e/h$, and $R_K = h/e^2$. These relations lead to $h = 4/K_J^2 R_K$ and hence

$$h = 4K_W/K_{J\text{-}90}^2 R_{K\text{-}90}. \tag{14}$$

Since $K_{J\text{-}90}$ and $R_{K\text{-}90}$ are defined constants with no uncertainty, a measurement of K_W (i.e., a measurement of W_{90} in watts or realization of the SI watt) with a given relative uncertainty yields a value of h with the same relative uncertainty.

It is also surprising that of the 20 data items listed above, only 5 were available at the time of the 1986 CODATA adjustment. Moreover, the value of the molar gas constant R used in the 1986 adjustment and from which were derived the Boltzmann constant k, Stefan-Boltzmann constant σ, and other thermophysical constants has been replaced by a new value of R with a relative uncertainty of 1.7 ppm, nearly 5 times smaller than the 8.4 ppm of the 1986 value. A comparable and thus significant reduction in the uncertainties of the 1986 recommended values that are dependent upon R, such as k and σ, may be anticipated. All of this new data shows how rapidly the fundamental constants field advances.

A third surprise is that of the 20 data items listed above, 2 play a dominate role because their uncertainties are significantly smaller than those of the other items of stochastic data. These are the value of K_W (and thus h) with 0.136-ppm uncertainty and the value of α^{-1} from a_e with 0.0069 ppm uncertainty. In fact, essentially these same two data items dominated the analysis of the two CCE Working Groups. Taking into account the new value of R, the situation described in the opening paragraph of Sec. 4 obtains.

4.2 Impact on 1986 Recommended Values

As indicated in Sec. 4.1, the currently available items of stochastic data may be analyzed using the least-squares technique taking as the adjustable constants K_Ω, K_V, and μ_μ/μ_p. The analysis shows (Taylor and Cohen, 1990) that the value of $\gamma'_p(\mathrm{lo})_{90}$ with 0.36 ppm uncertainty and the value of γ'_p with 0.80 ppm uncertainty are both highly discrepant and should be discarded. Nevertheless, the results from the least-squares adjustment of the remaining 18 data items plus the new value of R lead to changes in most of the 1986 CODATA recommended values and their uncertainties. Such changes for a selected but representative group of constants are given in Table 3 and, along with those for some additional constants, are depicted in Fig. 1. The changes in the 1986 CODATA recommended values are significant in comparison with the uncertainties of the 1986 values but not disturbingly so; shifts in the range 1.5–1.8 standard deviations are typical (see the last column of Table 3). In fact, in light of recent past adjustments, changes of less than two standard deviations may be considered a welcome sight. In contrast, the reductions in the uncertainties of the 1986 recommended values are major and typically in the range 4.7 to 6.9 (see column 4 of Table 3).

The values of $R_K = h/e^2 = \mu_0 c\alpha^{-1}/2$ and $K_J = 2e/h$ resulting from this 18 data-item adjustment imply that $R_{K\text{-}90}$ exceeds R_K by only (0.0285 ± 0.0064) ppm and that $K_{J\text{-}90}$ is only (0.026 ± 0.062) ppm smaller than K_J. This excellent agreement between the conventional and SI values of the von Klitzing and

Table 3. Changes in the 1986 CODATA recommended values and uncertainties of a selected group of constants implied by the data currently available.

Quantity	1 std. dev. uncertainty of 1986 recommended value, σ_{86}, in ppm	1 std. dev. uncertainty of new value, σ, in ppm	Ratio of σ_{86} to σ	Change in 1986 recommended value implied by current data, δ, in ppm; and corresponding number of standard deviations, δ/σ_{86}	
				δ	δ/σ_{86}
α^{-1}, R_K	0.045	0.0064	6.9	0.024	0.5
e	0.30	0.062	4.9	−0.52	−1.7
h	0.60	0.12	4.8	−1.02	−1.7
m_e	0.59	0.12	4.7	−0.97	−1.6
N_A	0.59	0.12	4.7	0.97	1.6
F	0.30	0.064	4.7	0.45	1.5
$2e/h$, K_J	0.30	0.062	4.8	0.50	1.7
m_μ/m_e	0.15	0.13	1.2	0.00	0.0
R	8.4	1.7	4.9	−4.7	−0.6
k	8.5	1.7	4.9	−5.7	−0.7
σ	34	6.9	4.9	−20	−0.6

Josephson constants as implied by the currently available data is not surprising inasmuch as the dominant values K_W and $\alpha^{-1}(a_e)$ were also available to the CCE Working Groups and played a major role in their selection of $R_{K\text{-}90}$ and $K_{J\text{-}90}$.

4.3 Impact on Next Adjustment

The strong dependence of the changes in the 1986 recommended values and their uncertainties on just three new results has significant implications for the timing of the next CODATA least-squares adjustment. While the large reduction in the 1986 uncertainties arising from these three results provides some motivation for carrying out a new adjustment sooner rather than later, their dominant role is a caution sign; it is highly desirable to obtain corroboration of each of these three results at a comparable level of uncertainty before a new set of recommended values is introduced. It is thus premature to decide when the next least-squares adjustment should be carried out, even though a new set of values could in principle be introduced today with uncertainties considerably smaller than those of the 1986 set. Clearly the progress made in obtaining additional data will strongly influence the timing of the next adjustment.

5. POSSIBLE ROLE OF THE FUNDAMENTAL CONSTANTS IN REPLACING THE KILOGRAM

Like Sec. 3, this section will demonstrate the important role that the fundamental constants can play in practical metrology.

5.1 Problems with Present Kilogram

The unit of mass in the SI, the kilogram, is the only SI base unit still defined in terms of a material artifact; it is equal to the mass of the international prototype of the kilogram, a 90% platinum, 10% iridium (by weight) right circular cylinder about 39 mm in height and diameter. Established by the CGPM in 1889, it is rarely used for fear of damage.

The problems with the present material artifact kilogram may be summarized as follows (Quinn, 1991; Taylor, 1991a):

1. can be damaged or even destroyed;
2. cannot be used routinely for fear of wear;
3. available in only one laboratory;
4. ages in an unknown manner, perhaps as much as 75 μg in the last 100 years (0.075 ppm or 7.5 parts in $10^8 = 7.5\times10^{-8}$); and
5. not well defined because it accumulates foreign material and is difficult to clean reproducibly.

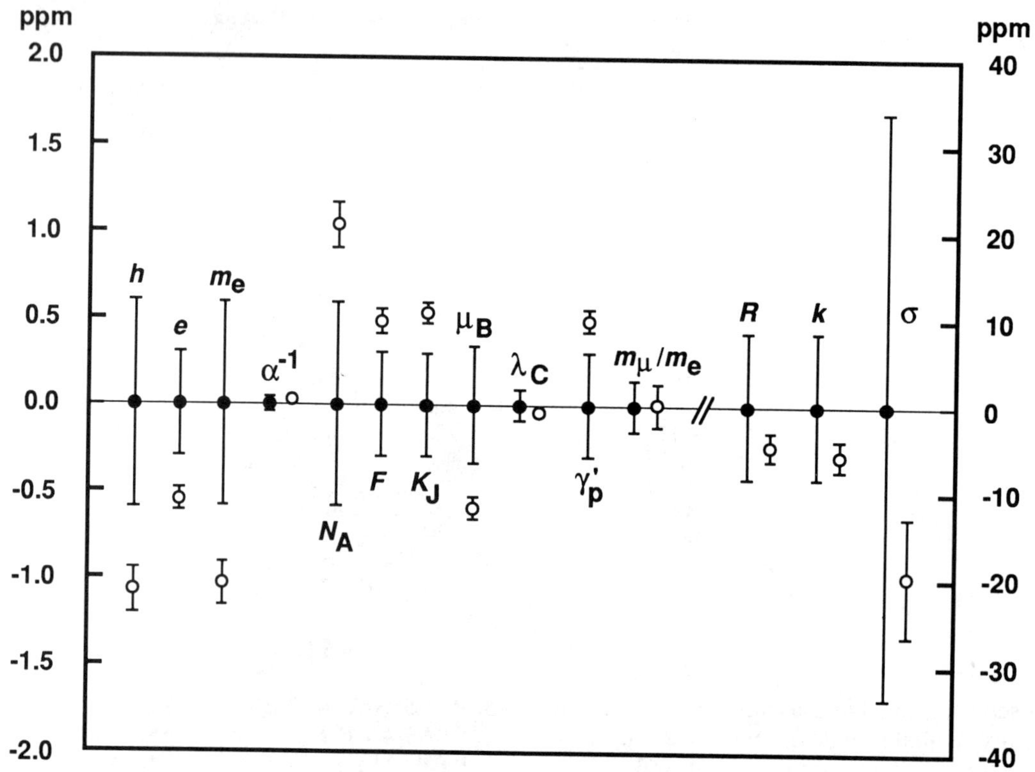

FIG. 1. Graphical display of the changes in the 1986 CODATA recommended values and one-standard-deviation uncertainties of a representative group of constants as implied by the data currently available. The filled circles with vertical error bars on the 0-ppm line are the 1986 values and uncertainties. The open circles with vertical error bars indicate the new values and their one-standard-deviation uncertainties relative to the 1986 values. (The scale at the right is for R, k, and σ; $\lambda_C = h/m_e c$ is the Compton wavelength of the electron.)

Using the best available balances and under the proper experimental conditions, Pt-Ir kilogram standards can be compared with an uncertainty of 0.001 μg (0.001 ppm or 1×10^{-9}) or even less. However, it is generally agreed that a new definition of the kilogram based on a constant of nature, such as the mass of a stable elementary particle or atom, that could be realized in practice with a relative uncertainty of the order of 1×10^{-8} at the 1-kg level is all that is required to replace the current definition. The reason is that the uncertainties associated with the cleaning and aging of the international prototype, the air buoyancy correction required when masses of different densities are compared, and the scaling of masses up and down from the 1-kg level limit practical mass measurements to a relative uncertainty of a few parts in 10^8 at best.

5.2 An Alternative Based on Fundamental Constants

One approach to eliminating the artifact kilogram as the definition of the SI unit of mass is to replace it with a definition based on the mass of an atom, for example, carbon-12 (^{12}C). A possible definition might read: *The kilogram is the unit of mass; it is equal to the mass of* 5.018×10^{25} *free* ^{12}*C atoms at rest.* To avoid introducing a discontinuity in the SI mass unit, the actual number $n(^{12}\mathrm{C})$ given in the definition would, to the fullest possible extent, be taken equal to the number of free ^{12}C atoms at rest whose total mass equals the mass of the international kilogram prototype. Thus

$$n(^{12}\mathrm{C}) = (1\ \mathrm{kg})/m(^{12}\mathrm{C}) = (1\ \mathrm{kg})N_A/M(^{12}\mathrm{C}), \qquad (15)$$

where $m(^{12}C)$ is the mass in kg of a free ^{12}C atom at rest; N_A is the Avogadro constant in units of mol^{-1}; and $M(^{12}C)$ is the molar mass of ^{12}C in units of kg/mol and from the definition of the base unit mole in the SI is exactly 0.012 kg/mol. It follows from Eq. (15) that once the new definition is implemented [i.e., once an exact value for $n(^{12}C)$ is adopted], N_A and the atomic mass unit $u=m_u=m(^{12}C)/12$ will become exactly defined constants.

Cyclotron-resonance mass-ratio measurements of single charged particles and ions cooled and trapped in a Penning trap have reached the point where the molar masses of most particles or atoms of interest, such as the electron, proton, or ^1H, are now known or soon will be known with a relative uncertainty of $(1–3)\times10^{-9}$. Similarly, the 20×10^{-9} relative uncertainty of the reported value for the proton-electron mass ratio, $m_p/m_e=M_p/M_e$, should soon be reduced by a factor of 10 or so. These small uncertainties imply that if the mass m of any elementary particle or atom can be obtained in kilograms with a relative uncertainty of 1×10^{-8}, then $m(^{12}C)$ and hence $n(^{12}C)$ will be known with essentially the same uncertainty. Further, Eq. (15) implies that determining N_A with a relative uncertainty of 1×10^{-8} will lead to a value of $n(^{12}C)$ with the same uncertainty. Consequently, one need only consider the determination of N_A and, since the electron is the particle whose mass can be obtained most accurately at present, the determination of m_e.

5.3 Determining N_A and m_e

It was pointed out in Sec. 2.1 that values for constants may be obtained by direct measurement or indirectly from appropriate combinations of other directly measured quantities. As touched upon in Sec. 2.3, N_A may be determined *directly* by the x-ray crystal density (XRCD) method in which one measures the molar mass, density, and unit-cell volume of a material such as silicon that can be prepared in the form of extremely pure and nearly perfect single crystals. For silicon, a cubic crystal with eight atoms per unit cell of edge length $a = d_{220}\sqrt{8}$ and volume $v=a^3$, the basic relation is $N_A=8M(\mathrm{Si})/\rho(\mathrm{Si})\times(d220\sqrt{8})^3$. N_A may be obtained *indirectly* by a number of different routes but since $m_e=M_e/N_A=M_p/[N_A(m_p/m_e)]$, any indirect measurement of N_A may be viewed as an indirect measurement of m_e.

Since m_e cannot yet be measured directly (one cannot yet "weigh" electrons), only indirect measurements (other than via N_A) need be considered. Based on the equations $R_\infty=m_e c\alpha^2/2h$, $\alpha=\mu_0 ce^2/2h$, and relations previously discussed, one can show that

$$m_e=\frac{8R_\infty\alpha^{-2}K_W}{cR_{K\text{-}90}K_{J\text{-}90}^2} \tag{16a}$$

$$=\frac{16R_\infty\alpha^{-1}K_V^2}{\mu_0c^2K_{J\text{-}90}^2} \tag{16b}$$

$$=\frac{4\mu_0R_\infty\alpha^{-3}K_A^2}{R_{K\text{-}90}^2K_{J\text{-}90}^2}, \tag{16c}$$

where K_W, K_V, and K_A have the same meaning as they had throughout all of Sec. 4, namely, $K_W=W_{90}/W$, $K_V=V_{90}/V$, and $K_A=A_{90}/A$.

Now μ_0, c, $R_{K\text{-}90}$, and $K_{J\text{-}90}$ are exact constants; R_∞ is known with a relative uncertainty that is currently limited by the practical realization of the meter at visible wavelengths or about 2×10^{-10}; and α^{-1} as obtained from a_e has a relative uncertainty of 6.9×10^{-9} and it should eventually be possible to reduce this to $(1–2)\times10^{-9}$. Consequently, realization of the watt, volt, or ampere (i.e., determination of K_W, K_V, or K_A) with a relative uncertainty of about 1×10^{-8} could lead to a new definition of the SI unit of mass.

One can also show that

$$m_e=\frac{2(\mu_p'/\mu_B)}{R_{K\text{-}90}K_{J\text{-}90}\gamma_p'(\mathrm{hi})_{90}} \tag{17a}$$

$$=\frac{2M_p}{R_{K\text{-}90}K_{J\text{-}90}(m_p/m_e)F_{90}}, \tag{17b}$$

where again the various quantities entering these relations have the same meaning as in previous sections. The current relative uncertainty of μ_p'/μ_B is 11×10^{-9} and can likely be significantly reduced if need be. The current relative uncertainty of m_p/m_e is 20×10^{-9}, and, as noted in Sec. 5.2, this will eventually be reduced by a factor of 10 or more. Consequently, a determination of either $\gamma_p'(\mathrm{hi})_{90}$ or F_{90} with an uncertainty of about 1×10^{-8} can also lead to a value of m_e, and thus $m(^{12}C)$ and

$n(^{12}C)$, having essentially the same uncertainty and therefore to a new definition of the kilogram.

It should be recognized that Eqs. (16) and (17) are potentially useful as a basis for replacing the kilogram only because it is possible to calibrate laboratory voltage standards in terms of V_{90} with a relative uncertainty of about 1×10^{-9} using a Josephson-array voltage standard, and laboratory resistance standards in terms of Ω_{90} with a similar uncertainty using the QHE and all-cryogenic techniques.

5.4 Practical Use of New Kilogram Definition

Suppose that one of the above approaches eventually yields a value of m_e with the requisite relative uncertainty of about 1×10^{-8} and the kilogram is redefined in terms of $n(^{12}C)$. How would the newly defined kilogram be used in practice? It should be recognized that present-day platinum-iridium national prototype kilograms and the prototype kilogram itself would not be discarded. Rather, they would serve as "flywheels," that is, as a means of preserving the unit of mass between realizations based on the new definition. Their periodic calibration would be performed by carrying out, for example, a watt-realization experiment or an Avogadro-constant determination in terms of the mass standard to be calibrated, taking as its value that currently assigned to it. Any deviation of the measured value of m_e and thus $n(^{12}C)$ from its adopted value would be attributed to a shift in the mass standard. A new value would then be assigned to the standard, thereby recalibrating it. Only a few laboratories in the world would need to have the apparatus necessary to carry out such experiments since the results could be readily transferred between laboratories by intercomparing mass standards. If more accurate values of relevant fundamental constants (e.g., R_∞, α^{-1}) became available, the adopted value of $n(^{12}C)$ would not be changed. Rather, one would take into account any change in $n(^{12}C)$ implied by the improved values of the constants by assigning new values to mass standards. Of course, before any new kilogram definition is adopted, the international prototype would be measured periodically (i.e., monitored) in terms of the proposed new definition to minimize introducing a discontinuity in the SI unit of mass.

6. "NUMEROLOGY" AND "TIME VARIATIONS" OF THE CONSTANTS

No article on the fundamental physical constants would be complete without at least a brief discussion of why the constants have the values they do, and whether they are in fact really constant. For example, why does the ratio of the rest mass of the proton to that of the electron happen to be 1836.15...? Why does α^{-1} happen to equal 137.035...? Why not other, perhaps simpler, values? Three rather different points of view have evolved about the constants, especially their dimensionless combinations.

One viewpoint is that the numbers are not at all arbitrary but can be calculated from some as yet unknown basic theory (or theories) in much the same way that the g factors of the electron and muon can be calculated from QED, or even in the way that the ratio of the circumference of a circle to its diameter need not be experimentally determined but can be calculated to arbitrary accuracy from mathematics. For example, in the late 1960s an expression was derived for α involving only simple integers and π, based on the volumes of certain bounded spaces associated with the invariance group of a relativistic quantum wave equation, and giving a value for α in excellent agreement with the recommended value at the time (Wyler, 1969). However, the physical basis for the derivation was not at all clear and it is safe to say that there does not yet exist a physically meaningful "derivation" of an accurate value for a fundamental constant, or dimensionless ratio or combination of constants.

The second viewpoint is not aimed at explaining the values of the physical constants but simply notes that if they were terribly different from their currently observed values, we would not be here to measure them—life as we know it on earth, and perhaps the existence of the universe itself, depends on complex physical processes that require the constants to have their observed values. This is the idea behind the so-called weak anthropic principle (or WAP) (Barrow and Tipler, 1988). The strong anthropic principle (or SAP) goes even further by speculating that the fundamental constants and laws of physics must be such that life and the universe as we know it can exist.

An extreme statement of the third point of view is that the fundamental constants change with time as the universe evolves and therefore the currently observed values are not particularly significant. This idea had its origins in 1937 when Dirac noted that the ratio of the electric to gravitational forces between the electron and proton is of the order 10^{40}, very nearly equal to the age of the universe expressed in an appropriate atomic time unit, for example, the time required for light to travel a distance equal to the classical electron radius. He speculated that either the equality is accidental or it indicates a causal relationship between electromagnetism, gravitation, and cosmology. If such a relationship is in fact true, then since the universe is continually aging, the equality would remain unchanged only if one or more of the fundamental constants entering the two ratios changed—for example, the Newtonian gravitational constant G or elementary charge e.

These speculations have stimulated a considerable amount of theoretical and experimental work over the last 50 years, the latter often involving astronomical or geophysical observations. As a result, stringent limits (albeit frequently model dependent) have been placed on the possible time dependence of a number of physical constants. For example, it has been shown that h changes less than 2 parts in 10^{12} per year; that G changes less than 3 parts in 10^{11} per year; and that α changes less than 5 parts in 10^{18} per year. Clearly, for a long time to come, the changes in the values of the fundamental constants caused by man's ineptitude in measuring them will dominate any changes that might be caused by Nature's laws!

7. FUTURE TRENDS

The fundamental-constants field is ever evolving; new experiments are undertaken with each major new advance in measurement science. Indeed, roughly speaking, our knowledge of the values of the fundamental constants has improved an order of magnitude each decade during the last half of the 20th century. However, the question naturally arises as to whether this sort of improvement can continue indefinitely. Will phenomena and measuring techniques yet to be discovered lead to continued reductions in uncertainty of this magnitude into the 21st century?

While it is always difficult to predict what will happen in this field in the future, a number of experiments currently under way can provide clues. For example, a new experiment to determine both μ_μ/μ_p and ν_{Mhfs} is being undertaken which, together with improvements in the QED-dependent theoretical expression for ν_{Mhfs}, could yield a value of α with an uncertainty of about 0.01 ppm. While this value would depend on QED, comparing it with the value of α derived from a_e(expt) and a_e(theor), which has an uncertainty of 0.0069 ppm, would still provide a highly useful check on the internal consistency of QED.

A somewhat related experiment is aimed at determining a_μ with significantly improved accuracy using a new muon storage ring. Coupled with improvements in the QED-theoretical expression for a_μ and the completion of associated experiments needed to improve the calculation of the so-called hadronic contribution to a_μ, a comparison of a_μ(expt) and a_μ(theor) will test predictions of the "Standard Model" of particle physics, which describes the electromagnetic, weak, and strong forces through which the elementary particles interact. A reliable, QED-independent value of α is also required for this comparison.

Another area of considerable promise (alluded to in Sec. 5.2) involves measurements of the mass ratios of the same or different atoms in various ionization states by measuring the cyclotron resonance frequencies of single ions stored in a Penning trap. Potential accuracies (in the ratios including m_p/m_e) approaching a few parts in 10^{12} seem feasible and would lead to a several orders of magnitude improvement in the molar masses of atoms as well as a new means for measuring atomic binding energies. The method might even allow the determination of the Lamb shift in the ground state of completely ionized heavy atoms such as uranium. As noted earlier, comparing experimental and theoretical values of the Lamb shift is yet another way to test QED critically.

Then there are all the various experiments discussed in Sec. 5 dealing with replacing the kilogram. Among those determinations being actively pursued or considered and their approximate target relative uncertainties are these: R_∞, 1×10^{-12}; $\alpha^{-1}(a_e)$, 1×10^{-9}; N_A by the XRCD method, 1×10^{-8}; h by a moving-coil balance realization of the watt, 5×10^{-9}; h by a levitated-mass realization of the joule, 1×10^{-8}; F_{90} by counting and weighing ions, 1×10^{-8}. To these should be added $\gamma_p'(\text{lo})_{90}$, 3×10^{-8}, which would yield a value of α^{-1} with a relative uncertainty of 1×10^{-8}. Whether some or all of these very ambitious goals are reached remains to be seen. Given the typical 10–15-year time span of a fundamental constant experiment, it will likely be necessary to wait until after the turn of the century to find out how successful the researchers have been. One thing is certain, however. Scientists will continue their efforts to advance our knowledge of the fundamental constants unabated as long as they believe there is even a slim chance of success. The reasons were given in the Introduction and throughout Sec. 1 but are best summarized in the following closing quotation by a famous American physicist of the early decades of this century:

"Why should one wish to make measurements with ever increasing precision? Because the whole history of physics proves that a new discovery is quite likely to be found lurking in the next decimal place" (F. K. Richtmyer, 1932).

GLOSSARY

***a_e*:** Electron magnetic moment anomaly; $a_e=g_e/2-1=\mu_e/\mu_B-1$, where g_e is the g factor of the electron, μ_e is the magnetic moment of the electron, and μ_B is the Bohr magneton: $\mu_B=eh/4\pi m_e$, where e is the elementary charge, h is the Planck constant, and m_e is the electron mass.

BIPM: Bureau International des Poids et Mesures (International Bureau of Weights and Measures), located in Sèvres, France, a suburb of Paris. BIPM is the international standards laboratory.

CCE: Comité Consultatif d'Électricité (Consultative Committee for Electricity), one of the eight advisory committees to the CIPM.

CGPM: Conférence Général des Poids et Mesures (General Conference on Weights and Measures), the highest international metrological authority established by the Convention du Mètre (Treaty of the Meter) signed in 1875; it is a diplomatic organization.

CIPM: Comité International des Poids and Mesures (International Committee for Weights and Measures), one of the organizations established by the Convention du Mètre (Treaty of the Meter) signed in 1875. The CIPM supervises the BIPM and is under the authority of the CGPM.

CODATA: Committee on Data for Science and Technology, an interdisciplinary committee of the International Council of Scientific Unions. It seeks to improve the compilation, critical evaluation, storage, and retrieval of data of importance to science and technology.

JE: Abbreviation for Josephson effect, a solid-state, low-temperature physics phenomenon characteristic of weakly coupled superconductors cooled below their transition temperatures. An example is two thin films of superconducting lead separated by a thermally grown oxide layer about 1 nm thick.

***K_J*:** Josephson constant characteristic of the Josephson effect and assumed equal to $2e/h$, where e is the elementary charge and h is the Planck constant.

***$K_{J\text{-}90}$*:** Conventional value of the Josephson constant adopted by the CIPM for international use in basing a representation of the volt on the Josephson effect starting 1 January 1990; $K_{J\text{-}90}=483\ 597.9$ GHz/V exactly.

Muonium: Hydrogenlike atom composed of a positive muon and an electron (μ^+e^- atom).

NIST: National Institute of Standards and Technology, the national standards laboratory of the United States (called the National Bureau of Standards or NBS prior to 23 August 1988).

QED: Quantum electrodynamics, the relativistic quantum field theory of the interaction of charged particles and electromagnetic radiation (photons). QED is an amalgamation of Maxwell's theory of electromagnetism, quantum mechanics, and Einstein's special theory of relativity.

QHE: Abbreviation for quantum Hall effect, a solid-state, low-temperature physics phenomenon characteristic of a two-dimensional electron gas. The latter may be realized in a high-mobility semiconductor device such as a silicon metal-oxide-semiconductor field-effect transistor (MOSFET) or GaAs-$Al_xGa_{1-x}As$ heterojunction cooled to about 1 K. The applied magnetic field (magnetic flux density) B must be of order 10 T.

$\boldsymbol{R_K}$**:** von Klitzing constant characteristic of the quantum Hall effect and assumed equal to h/e^2, where h is the Planck constant and e is the elementary charge.

$\boldsymbol{R_{K\text{-}90}}$**:** Conventional value of the von Klitzing constant adopted by the CIPM for international use in basing a representation of the ohm on the quantum Hall effect starting 1 January 1990; $R_{K\text{-}90}=25\ 812.807\ \Omega$ exactly.

SI: Système International d'Unités (International System of Units), the modernized metric system with the seven base units meter (m), kilogram (kg), second (s), ampere (A), kelvin (K), mole (mol), and candela (cd); and numerous derived units such as the hertz (Hz), joule (J), Newton (N), watt (W), volt (V), and ohm (Ω).

$\boldsymbol{\alpha}$**:** Fine-structure constant, $\simeq 1/137$. α is the coupling constant or the measure of the strength of the electromagnetic force that governs how electrically charged elementary particles (e.g., electron, muon) and light (photons) interact.

$\boldsymbol{\alpha^{-1}}$**:** Inverse fine-structure constant; $\alpha^{-1} = 1/\alpha \simeq 137$.

List of Works Cited

Barone, A., Paternò, G. (1982), *Physics and Applications of the Josephson Effects*, New York: Wiley.

Barrow, J. D., Tipler, F. J. (1988), *The Anthropic Cosmological Principle*, Oxford: Oxford University Press.

Birge, R. T. (1929), *Phys. Rev. Suppl.* **1**, 1–73.

Cohen, E. R., Du Mond, J. W. M. (1965), *Rev. Mod. Phys.* **37**, 537–594.

Cohen, E. R., Taylor, B. N. (1973), *J. Phys. Chem. Ref. Data* **2**, 663–734.

Cohen, E. R., Taylor, B. N. (1987), *Rev. Mod. Phys.* **59**, 1121–1148.

Dirac, P. A. M. (1937), *Nature* **137**, 323.

van Dyck, Jr., R. S., Schwinberg, P. B., Dehmelt, H. G. (1987), *Phys. Rev. Lett.* **59**, 26–29.

Giacomo, P. (1984), *Metrologia* **20**, 25–30.

Giacomo, P. (1987), *Metrologia* **24**, 45–51.

Hudson, R. P. (1984), *Metrologia* **19**, 163–177.

Kinoshita, T. (Ed.) (1990), *Quantum Electrodynamics*, Singapore: World Scientific.

Kinoshita, T., Lindquist, W. B. (1990), *Phys. Rev. D* **42**, 636–655.

Langenberg, D. N., Taylor, B. N. (Eds.) (1971), *Precision Measurement and Fundamental Constants*, National Bureau of Standards (U.S.), Special Publication 343, Washington, D.C.: U.S. GPO.

Parker, W. H., Taylor, B. N., Langenberg, D. N. (1967), *Phys. Rev. Lett.* **18**, 287–291.

Parker, W. H., Langenberg, D. N., Denenstein, A., Taylor, B. N. (1969), *Phys. Rev.* **177**, 639–664.

Prange, R. E., Girvin, S. M. (Eds.) (1987), *The Quantum Hall Effect*, New York: Springer.

Quinn, T. J. (1989), *Metrologia*, **26**, 69–74.

Quinn, T. J. (1991), *IEEE Trans. Instrum. Meas.* **IM-40**, 81–85.

Richtmyer, R. K. (1932), *Science* **75**, 1–5.

Shields, J. Q., Dziuba, R. F., Layer, H. P. (1989), *IEEE Trans. Instrum. Meas.* **IM-38**, 249–251.

Taylor, B. N., Parker, W. H., Langenberg, D. N. (1969), *Rev. Mod. Phys.* **41**, 375–496.

Taylor, B. N., Phillips, W. D. (Eds.) (1984), *Precision Measurement and Fundamental Constants II*, National Bureau of Standards (U.S.), Special Publication 617, Washington, D.C.: U.S. GPO.

Taylor, B. N. (1987), *IEEE Trans. Instrum. Meas.* **IM-36**, 659–664.

Taylor, B. N., Witt, T. J. (1989), *Metrologia* **26**, 47–62.

Taylor, B. N., Cohen, E. R. (1990), *J. Res. Natl. Inst. Stand. Technol.* **95**, 497–523.

Taylor, B. N. (1991a), *IEEE Trans. Instrum. Meas.* **IM-40**, 86–91.

Taylor, B. N. (Ed.) (1991b), *The International System of Units (SI)*, National Institute of Standard Technology, Special Publication 330, Washington, D.C.: U.S. GPO.

Wyler, A. (1969), *C.R. Acad. Sci. Ser. A* **269**, 743–745; (1971), **271**, 186–188.

Further Reading

Cohen, E. R., Du Mond, J. W. M. (1965), *Rev. Mod. Phys.* **37**, 537–594.

Cohen, E. R., Taylor, B. N. (1973), *J. Phys. Chem. Ref. Data* **2**, 663–734.

Cohen, E. R., Taylor, B. N. (1986), *Rev. Mod. Phys.* **59**, 1121–1148.

Cutler, P. H., Lucas, A. A. (Eds.) (1983), *Quantum Metrology and Fundamental Constants*, NATO ASI Series, Series B: *Physics*, Vol. 98, New York: Plenum.

Ferro-Milone, A., Giacomo, P., Leschiutta, S. (Eds.) (1980), *Metrology and Fundamental Constants*, Proceedings of the International School of Physics "Enrico Fermi," Course LXVIII, Amsterdam: North Holland.

Kinoshita, T. (Ed.) (1990), *Quantum Electrodynamics*, Singapore: World Scientific.

Langenberg, D. N., Taylor, B. N. (Eds.) (1971), *Precision Measurement and Fundmental Constants*, Natl. Bur. Stand. (U.S.), Spec. Publ. 343, Washington, D.C.: U.S. GPO.

Petley, B. W. (1988), *The Fundamental Physical Constants and the Frontier of Measurement*, Bristol: Adam Hilger.

Taylor, B. N. (1991), *IEEE Trans. Instrum. Meas.* **IM-40**, 86–91.

Taylor, B. N., Cohen, E. R. (1990), *J. Res. Natl. Inst. Stand. Technol.* **95**, 497–523.

Taylor, B. N., Cohen, E. R. (1991), *Phys. Lett. A* **153**, 308–312.

Taylor, B. N., Parker, W. H., Langenberg, D. N. (1969), *The Fundamental Constants and Quantum Electrodynamics*, New York: Academic; also published in *Rev. Mod. Phys.* **41**, 375–496.

Taylor, B. N., Phillips, W. D. (Eds.) (1984), *Precision Measurement and Fundamental Constants II*, National Bureau of Standards (U.S.), Special Publication 617, Washington, D.C.: U.S. GPO.

Taylor, B. N., Witt, T. J. (1989), *Metrologia.* **26**, 47–62.

COPPER

MARY W. COVINGTON,* *Consultant*

NATHAN L. CHURCH AND WILLIAM H. DRESHER, *International Copper Association, New York, U.S.A.*

BERNARD R. COOPER, *Dept. of Physics, West Virginia University, Morgantown, West Virginia, U.S.A.*

INTRODUCTION

Copper and its alloys are used when the application requires high electrical and thermal conductivity, formability, ease of joining, corrosion resistance, and moderate to high strength. In addition to its traditional uses in building construction, electrical and electronic applications, industrial machinery, transportation equipment, and consumer products, new uses for copper are constantly being found. Some of these are in solar heating, electric vehicles, superconductors (both high- and low-temperature types), fire safety systems, and even pharmaceuticals.

We first discuss the basic electronic and physical behavior of copper. All of the valuable functional properties of copper find their foundation in this electronic behavior. As an example, we describe the features of the electronic structure that lead to the high electrical conductivity of copper.

The useful properties of copper in practice are determined not only by the ideal behavior based on the fundamental electronic and physical behavior, but also by the modifications of that behavior brought about by processing and the presence of impurities. In Sec. 2, we therefore discuss the impurity content characteristic of high-conductivity copper and then summarize some of the key physical and mechanical properties of copper. After doing this, we discuss the practical aspects affecting the utilization of copper: resources and raw materials, production, alloy terminology, and standards. In Sec. 6, we describe the utilization of copper in a variety of applications before providing a brief summary in the final section.

1. BASIC ELECTRONIC AND PHYSICAL BEHAVIOR

Copper (atomic number $Z=29$) is in subgroup 1*B* of the periodic table of the elements. The copper atom contains 2, 8, and 18 electrons in its inner shells and 1 electron in its outer shell. The cuprous ion (Cu^{+}) is formed by loss of the first electron and the cupric ion (Cu^{2+}) is formed by loss of a second electron. The atomic weight is 63.54. There are two stable isotopes, ^{68}Cu and ^{65}Cu. The specific gravity is 8.89 and the melting point is 1083 °C.

*Deceased.

1-56081-063-7/92/$5.00+.50

The most important attribute of the electronic behavior of copper is that it behaves like a monovalent metal such as the alkali metals, but with behavior modified by the existence of the filled *d*-shell states at an energy about 2 eV below the unfilled *s*-like states that provide the monovalent behavior. The two most conspicuous electronic-behavior-based properties of copper, its high electrical conductivity and its distinctively reddish "copper" color, follow as consequences of these features of the electronic structure.

To envision the relationship of electrical conductivity, optical behavior, and other properties to the electronic structure of copper we have to consider the first Brillouin zone for the face-centered-cubic (fcc) structure, as shown in Fig. 1, and the nature of the Fermi surface encompassing the filled electronic states within that zone (Callaway, 1964). For a perfect three-dimensional lattice, there are three primitive translations, $\mathbf{a}_1,\mathbf{a}_2,\mathbf{a}_3$, which define the periodicity of the lattice. The requirement that the potential acting on an electron moving through a solid be identical at any two points $\mathbf{r}$ and $\mathbf{r}'$ connected by any linear combination of multiples of these primitive translation vectors leads to the requirement that the electronic wave functions obey Bloch's theorem and are labeled by a crystal-momentum quantum number $\mathbf{k}$, which is specified in terms of primitive reciprocal lattice vectors $\mathbf{b}_1,\mathbf{b}_2,\mathbf{b}_3$. Thus,

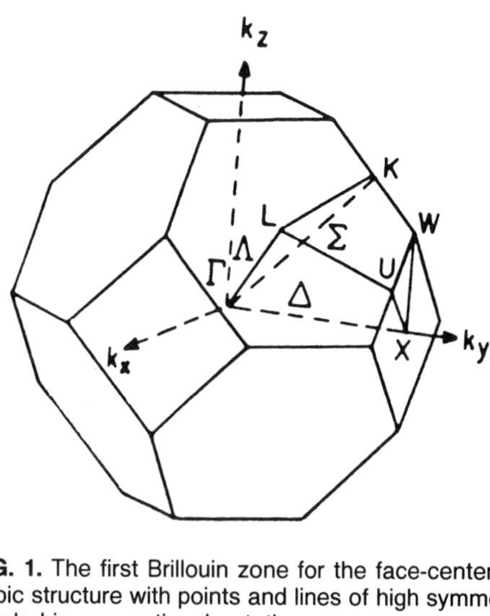

FIG. 1. The first Brillouin zone for the face-centered-cubic structure with points and lines of high symmetry labeled in conventional notation.

$$\mathbf{k}=2\pi\left(\frac{k_1\mathbf{b}_1}{N_1}+\frac{k_2\mathbf{b}_2}{N_2}+\frac{k_3\mathbf{b}_3}{N_3}\right), \tag{1}$$

where k_1, k_2, k_3 and N_1, N_2, N_3 are integers with $k_1,k_2,k_3 < N_1,N_2,N_3$, and

$$\mathbf{b}_1=\frac{\mathbf{a}_2\times\mathbf{a}_3}{\mathbf{a}_1\cdot(\mathbf{a}_2\times\mathbf{a}_3)}, \tag{2}$$

with $\mathbf{b}_2$ and $\mathbf{b}_3$ obtained by cyclic permutation of the indices of the $\mathbf{a}$'s.

The *reciprocal-lattice vectors* $\mathbf{K}_m$, which are linear combinations of integral multiples of $\mathbf{b}_1$, $\mathbf{b}_2$, and $\mathbf{b}_3$, define a periodic lattice in reciprocal (or *lattice-momentum*, $\mathbf{k}$) space; The part of $\mathbf{k}$ space closer to the origin than to any nonzero $\mathbf{K}_m$ is called the *first Brillouin zone*, and is constructed by setting up planes that perpendicularly bisect the lines connecting the origin to all $\mathbf{K}_m$, and then taking the volume about the origin enclosed by these intersecting planes. Figure 1 shows the first Brillouin zone for the face-centered-cubic structure with conventional labeling of points and lines of high symmetry. Since the product $N_1N_2N_3$ is a very large number essentially corresponding to the number of copper atoms in a macroscopic sample ($\sim 10^{24}$), and k_1, k_2, k_3 are integers ranging from zero up to N_1, N_2, N_3, the $\mathbf{k}$ form a very finely spaced semicontinuum that in practice can be treated as a continuous variation of lattice momentum across the first Brillouin zone.

At each value of lattice momentum $\mathbf{k}$ in the first Brillouin zone there are a number of discrete electron energy states (E_m); and as $\mathbf{k}$ varies across the Brillouin zone, the set of curves given by the variation of the E_m with $\mathbf{k}$ gives the electronic energy band structure (Callaway, 1964). Because of the Fermi statistics obeyed by electrons, the first Brillouin zone can hold two electrons per real-space crystal lattice site per band. At absolute zero temperature the band states will be filled with electrons up to an energy level, the Fermi energy (E_F), such that all the electrons are accounted for. For a metal there are one or more partially filled bands. Thus there is a surface in reciprocal space separating filled from empty states. This surface is the *Fermi surface*. Since E_F is typically several electron volts and 1 eV is the energy equivalent of a temperature of 11 606 K, there is only a very slight blurring of the Fermi surface at finite temperatures. The properties of a metal are dominated either by the behavior of the electrons at or near the Fermi surface or by transitions between filled electron states below the Fermi energy and unfilled states above the Fermi energy. For example, the electrical conductivity is determined by the situation that, under the influence of a dc electric field, the only electrons that can respond are those at the Fermi surface, since they can be displaced into adjacent empty states with different lattice momenta. On the other hand, an optical-frequency electric field can cause transitions between filled states below the Fermi energy and unfilled states above E_F, and either or both states can be well separated in energy from E_F.

The simplest example of a Fermi surface is the spherical surface that occurs for a noninteracting electron gas, and the Fermi surface given by band theory for the monovalent alkali metals is quite close to spherical. For copper, as shown in Fig. 2, the Fermi surface is also basically spherical, but with distinctive *necks* contacting the hexagonal <111> faces of the first Brillouin zone (Segall, 1962). This behavior has its origin in the outer electron configuration of atomic copper and the way in which these outer electrons interact in the solid to form the energy bands. There is one valence electron state per copper atom, corresponding to the atomic 4*s* state; and it is these states that basically form the partially filled band, the filling of which determines the Fermi energy and Fermi surface. If this were the only effect determining the Fermi surface, it would be very similar to those of the alkali metals, quite close to spherical. However, the filled 3*d* states affect the electronic structure and consequent properties of copper in a very important way.

The 10 3*d* electrons per copper atom form filled bands in the solid and do not give a Fermi surface; however, these 3*d* states interact with the 4*s* states causing changed behavior (Segall, 1962) of the 4*s* band as shown in Fig. 3. The 4*s* band is "bent" upward by this interaction, so that the Fermi surface intercepts the hexagonal faces of the Brillouin zone with a ring-shaped intersection centered around the <111> direction in **k**-space. Thus it is the 4*s*-3*d* interaction (mixing) that causes the *necks* in the copper Fermi surface shown in Fig. 2.

The distinctive reddish color of copper corresponds to the sharp onset in frequency of strong interband optical transitions between filled states at the upper part of the 3*d* band (which has a relatively flat variation of energy with **k**) and empty conduction-band 4*s*-like

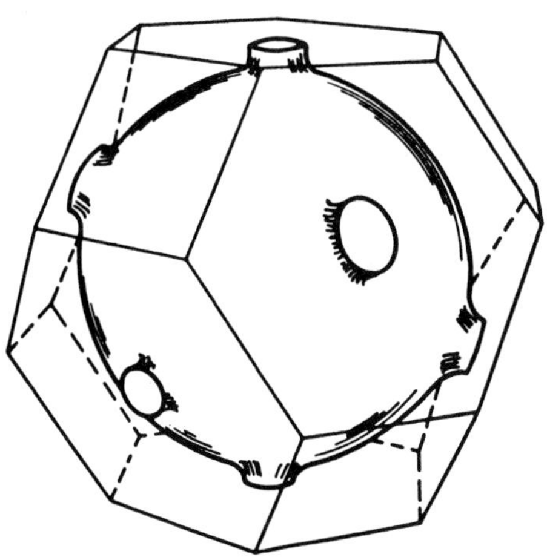

FIG. 2. A sketch of the Fermi surface of copper. (After Segall, 1962.)

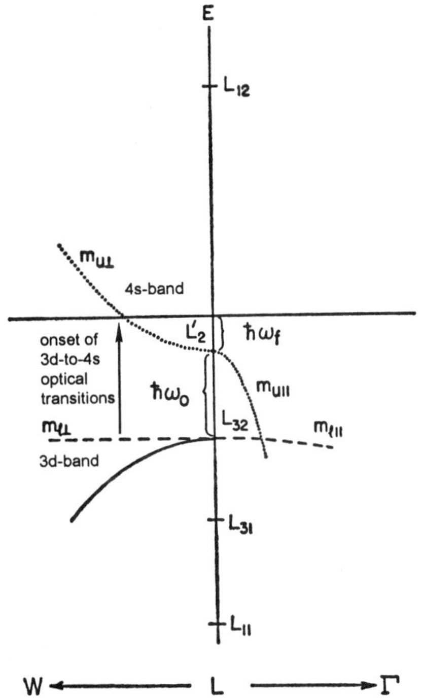

FIG. 3. The band structure near *L* in the first Brillouin zone for the noble metals. (After Cooper *et al.*, 1965.)

states (Ehrenreich and Philipp, 1962; Cooper *et al.*, 1965). As can be seen from Fig. 3, these transitions become allowed only at an energy given by the energy difference between the 3*d* and 4*s* bands at the **k**-space locations where the 4*s* band cuts the Fermi energy E_F; and it is the associated sharp onset of strong interband optical transitions (indicated by the arrow in Fig. 3) that dominates the optical behavior of copper.

The high electrical conductivity of copper also arises from the modification of its free-electron-like (alkali-metal–like) monovalent behavior caused by the filled 3*d* states. The filled *d* shells cause (see Ziman, 1960, Chap. IX) the compressibility of copper to be much smaller than that of the alkali metals. Since the electrical resisitivity of copper at room temperature and above is dominated by scattering of electrons by thermal vibrations, the substantial decrease in amplitude of these vibrations, corresponding to the decreased compressibility, gives rise to the increased electrical conductivity.

2. CHARACTERIZATION AND PROPERTIES

The useful properties of copper in practice are determined not only by the ideal behavior based on the fundamental electronic and physical behavior described in Sec. 2, but in the modification of that behavior brought about by processing and by the presence of impurities. For example, we can consider such effects on the electrical conductivity. High-purity copper's electrical conductivity is slightly over 100% IACS (as based on a volume electrical resistivity of 0.0168 $\mu\Omega$ m). This value refers to the International Annealed Copper Standard (IACS) established in 1913 and is based on a copper standard defined as 100%, being given by 100 times the reciprocal ratio to the volume resistivity of annealed copper at 20 °C as measured at that time (0.017 241 $\mu\Omega$ m). Subsequent improvements in refining have raised the actual value of copper to slightly over 100% IACS. Electrical conductivity is temperature- and processing-sensitive; for copper it is 800% IACS at −240 °C and 38% IACS at 425 °C. Although copper does not become superconductive, its conductivity decreases markedly as temperature approaches absolute zero. Cold working of annealed copper to about 90% reduction in area results in a slight drop in electrical conductivity of 2–3% IACS. Conductivity is also affected by impurities or alloying elements.

Impurity content characteristics of high-conductivity copper are given in Table 1, while Table 2 summarizes some of the key physical and mechanical properties of such high-purity copper. Below we briefly summarize a number of the useful and interesting characteristic properties of copper.

2.1 Fatigue

Fatigue strength, usually specified at a life of 10^8 cycles, is enhanced by fine grain size and the absence of nonmetallic inclusions. The techniques used to strengthen coppers (see Sec. 2.3.1) are also effective in increasing resistance to fatigue.

2.2 Magnetic Properties

Copper and most of its alloys are nonmagnetic. Copper is a diamagnetic material and has a mass susceptibility (χ) of -0.086×10^{-6} cgs units per gram at room temperature. This is equivalent to a gram-atomic susceptibility of -5.1×10^{-6}. Susceptibility varies with increasing temperature; for example, it is −0.085 at 300 °C, −0.082 at 600 °C, and −0.079 at 900 °C.

2.3 Formability

Copper and its alloys are readily hot worked by all of the usual processes, and may be fabricated into end-use forms by rolling, bending, drawing, stretching, and coining. All operations that deform the metal result in work hardening to various extents depending on chemical composition and previous annealing or fabricating history.

Casting to shape, using sand or permanent molds, is a widely used process for copper and many of its alloys, including many specifically designed for this purpose.

2.3.1 Strengthening. In addition to cold working, copper alloys can be metallurgically strengthened by solid solution hardening, phase transformations, precipitation hardening (including spinodal decomposition), and grain size refinement.

Table 1. Chemical composition of some coppers.

Composition	C10100 Oxygen-free Electronic[a]	C10200 Oxygen-free[a]	C11000 Electrolytic Tough Pitch[b]
Copper (including silver) (min %)	99.99	99.95	99.90
Trace element (ppm max)			
Antimony	4	—	—
Arsenic	5	—	—
Bismuth	1	—	—
Cadmium	1	—	—
Iron	10	—	—
Manganese	0.5	—	—
Mercury	1	—	—
Nickel	10	—	—
Oxygen	5	10	Variable
Phosphorus	3	—	—
Selenium	3	—	—
Silver	25	—	—
Sulfur	15	—	—
Tellurium	2	—	—
Tin	2	—	—
Zinc	1	—	—

[a]ASTM B 170 Standard Specification for oxygen-free electrolytic copper-refinery shapes.
[b]Impurity elements vary according to process.

Solid solution hardening is achieved by short-range order, chemical interaction, lattice misfit, electrical and valence interaction, or a combination of these mechanisms.

Precipitation hardening is used primarily for chromium- and beryllium-containing coppers. The latter are the strongest generally available copper alloys, exceeding 1100 MPa tensile strength compared to 700–965 MPa reached by other forms of treatment.

Grain size refinement hardens copper alloys while retaining a high degree of ductility and toughness. A low-temperature anneal following cold working produces grain refinement in wrought material such as strip or wire.

2.4 Corrosion and Oxidation

Copper corrodes at low rates in unpolluted air, water, and deaerated nonoxidizing acids. It is not resistant to oxidizing acids, aerated nonoxidizing acids, oxidizing heavy-metal salts, sulfur, hydrogen sulfide, ammonia, and some sulfur and ammonia compounds, the latter of which forms highly soluble complexes with copper.

Copper and its alloys resist corrosion because of their low driving force toward the oxidized state and because their corrosion products form adherent protective films in some media. Under most outdoor conditions, the film eventually forms a green patina by reacting with moisture and oxides of sulfur in the air to form an artificial mineral, brochantite [$Cu_2SO_4 \cdot 3Cu(OH)_3$]. The actual color of the patina, ranging from blue-green to yellow- or brown-green will vary with the atmospheric environment to which the copper is exposed. The presence of sulfides in the atmosphere tends to blacken the patina and the presence of acids tends to dissolve it. Hence, the status of the patina on copper roofs, etc., is a good indicator of the degree of air pollution present.

Pure copper oxidizes in air at temperatures of approximately 250 °C and a film of Cu_2O is formed. Up to 100 °C, the oxide film grows linearly with the logarithm of time. From 100 to 1025 °C, the scale consists of an inner layer of Cu_2O and an outer layer of CuO. Oxidation behavior of alloys depends on alloying elements.

Hydrogen embrittlement at high temperatures occurs when hydrogen diffuses into copper and reacts with oxygen to form water vapor under high pressure. Subsequent defor-

Table 2. Physical and mechanical properties of oxygen-free copper C10100.

Property	English units	Metric units
Melting point	1981 °F	1083 °C
Electrical resistivity (annealed)[a]	10.3 Ω (circ mil/ft) @ 68 °F	1.71 μΩ cm @ 20 °C
Electrical conductivity (annealed)[b]	101% IACS @ 68 °F	0.586 MS cm @ 20 °C
Thermal conductivity	226 Btu/ft^2 ft h °F @ 68 °F	0.934 cal/cm^2 cm s °C @ 20 °C
Coefficient of thermal expansion	9.4×10^{-6}/°F (68–212 °F) 9.6×10^{-6}/°F (68–392 °F) 9.8×10^{-6}/°F (68–572 °F)	17.0×10^{-6}/°C (20–100 °C) 17.3×10^{-6}/°C (20–200 °C) 17.7×10^{-6}/°C (20–300 °C)
Specific heat (thermal capacity)	0.092 Btu/lb °F @ 68 °F	0.092 cal/g °C @ 20 °C
Density	0.323 lb/in.^{3} @ 68 °F	8.94 g/cm^3 @ 20 °C
Specific gravity	8.94	8.94
Modulus of elasticity (tension)	17 000 ksi	117 000 MPa
Modulus of rigidity	6,400 ksi	44 000 MPa
Tensile strength	42 ksi	290 MPa
Yield strength, 0.1% offset	36 ksi	248 MPa
Fatigue strength[c,d]	17 ksi	117 MPa
Elongation, % in 2 in.	14	14

[a]Maximum resistivity of Certified Grade.
[b]Minimum conductivity of Certified Grade (volume basis). Mechanical properties are for flat products, one-half hard temper.
[c]Rod, drawn to 35% reduction in area.
[d]At 100 million cycles.

mation results in cracking. Embrittlement is most common in oxygen-containing copper, typically C11000 (numbering system designates chemical composition, see Sec. 6.1), electrolytic tough pitch copper commonly used for applications where the highest electrical conductivity is not a requirement. Embrittlement is avoided by the use of oxygen-free or deoxidized grades.

2.5 Joining

Copper is easily soldered and brazed and can be welded. Mild fluxes or cleaning in a diluted acid bath is used to remove tarnish films. Commercial solders of lead plus tin in the range of 35–60% achieves a superior metallurgical bond, although such solders are prohibited in many areas for use in potable-water plumbing systems because, under certain conditions, the lead can leach into water. For plumbing use, antimony-tin, silver-tin, and copper-antimony solders are available. Copper is also easily brazed for higher-temperature applications. The welding processes most used for copper are resistance welding, gas-metal-arc (GMAW), and gas-tungsten-arc (GTAW). Electron-beam welding can be used for precision welds.

The brass alloys can be soldered, brazed, or welded. The other copper alloys are generally brazed or welded.

3. RESOURCES AND RAW MATERIALS

For at least 6000 years, copper has been one of the most important materials in the development of civilization and culture. The earliest copper objects were made by beating and hammering native copper, a form of metallic copper found mixed with copper-bearing ore. The earliest known copper object to have been melted and cast, a mace head dated at 5000 BC, was found at Can Hasan in Anatolia (now Turkey). Archaeological evidence shows that copper ore was first smelted between 4000 and 3500 BC, thus marking the beginning of the science of metallurgy. Current theory is that the smelting process evolved from observations of the reduction of copper minerals in the glaze used in pottery. Thus, the first smelting furnaces were undoubtedly pottery kilns. In 1965 archaeologists excavated the earliest furnace designed specifically for smelting copper at Timna in the southern Sinai in Israel. The furnace was dated at 3200 BC. Charcoal had been the reducing agent and iron oxide the fluxing agent.

By 3000 BC, silver, tin, and lead were being used, and copper began to be alloyed with these other metals. Alloying with tin produced bronze, a hard alloy used for tools, plows, weapons, armor, and decorative objects. Brass, an alloy of copper and zinc, has been in use for almost 2000 years. The first intentional use of brass was by the Romans, who used it for coins, ornaments, and decorative objects.

There was a thriving trade in bronze by 1500 BC, which gave this era the name of the Bronze Age. In the ancient world, copper was mined in Cyprus and other Middle Eastern sites; however, since there are no known tin deposits at the bronze-making centers of this region, the tin had to have been transported to the region. Evidence exists that it could have come from as far away as Cornwall, where it was transported from Penzance by the Phoenicians—the great metal traders of that era. Since the tin had to have come from a distant site, it is clear that the Bronze Age could not have started until a reliable network of trading contacts had been established.

By 1100 BC, iron had replaced bronze in most functional uses. The Iron Age came about both because of iron's wider availability and because swords made from iron were proven to be superior to those of bronze when the Hittites conquered Assyria and captured Babylon in 1300 BC.

The modern history of copper begins with the discovery of electricity, and electrical applications continue to be one of its principal uses. Copper ranks third in metal consumption after steel and aluminum. Copper is mined in about 58 countries, with the major producing countries accounting for nearly 95% of world production. The major producing countries, in order of production in 1990, were Chile, United States, Soviet Union, Canada, Zambia, Poland, China, Zaire, Australia, Peru, and Mexico. The leading copper refining countries in 1990 were United States, Soviet Union, Chile, Japan, Germany, China, Canada, Zambia, Belgium, and Poland. World mine production was 9 100 500 metric tons in 1990. Of this amount, the United States produced 1 565 100 metric tons. Despite copper's long history, about 75% of all copper ever mined has been consumed since World War II. Copper ore reserves in the United States and the world are considered more than adequate to meet cumulative demand in the foreseeable future.

4. PRODUCTION

In present-day mining operations, large quantities of low-grade ore are crushed and loaded into trucks or conveyor belts for delivery to the ore treatment plant. In the United States, the ore averages about 0.6% copper, the richer values having long since been mined out. The ore is treated by grinding and froth flotation to produce a beneficiated product containing about 25% copper. These concentrates are reduced to the metallic state by smelting to produce copper matte containing up to about 60% copper. The matte is converted by selective oxidation to blister copper containing 98.5% or more copper. This intermediate product is refined electrolytically into copper cathodes containing 99.9% or more copper.

In recent years, more and more mines are leaching waste dumps and oxide ores for recovery of copper. Sulfuric acid, recovered from smelter off-gases as part of air-pollution control, is used as the leachant. Acid liquors containing copper are purified by solvent extraction, and the copper is recovered, in the form of cathodes, by electrowinning.

Cathodes from both the electrolytic refining process and the electrowinning process are, for the most part, melted, cast, and rolled into wire rod—a precursor for drawing into wire. The remainder is melted and cast into cakes or billets for conversion into semifabricated forms by manufacturers of wrought copper and copper alloys. Approximately 75% of all newly mined copper and 50% of all copper (including scrap) is made into wire and cable by wire mills. Most of the remainder is made into sheet, strip, plate, rod, bar, pipe, tube, wire, forgings, and extrusions by brass mills. A small amount is used in castings by foundries or manufactured into powder. Certain mill products, such as wire and most tubular products, are used without further fabrication. Most flat-rolled products, bar, rod, mechanical wire, and forgings go through multiple forming, finishing, machining, and assembly operations before their ultimate use in a manufactured product.

Free-cutting brass rod, which exhibits outstanding machinability and good corrosion resistance, and brass strip, which has high strength, corrosion resistance, excellent formability, and good electrical properties, constitute more than three-fourths of the total tonnage of copper alloys shipped from U.S. brass mills. Other alloy types of major commercial significance include tin bronzes (phosphor bronzes), which are noted for their excellent cold-forming behavior and strength; tin brasses, known for outstanding corrosion resistance; copper nickels, which are strong and particularly resistant to seawater; nickel silvers, which combine a silvery appearance with good formability and corrosion resistance; beryllium coppers, which provide outstanding strength when hardened; and aluminum bronzes, which have high strength along with good resistance to oxidation, chemical attack, and mechanical abrasion.

Foundries use prealloyed ingot, scrap, and virgin metal as raw materials. Their chief products are shaped castings for many different industrial and consumer goods, the most important of which are plumbing products and industrial valves. Centrifugal- and continuous-cast products find major application as bearings, cylinders, and other symmetrical components. Powder plants produce powder and flake for further fabrication into powder metallurgy parts, chiefly small sintered bronze bushings for bearings.

In 1990, the United States consumed 3 064 000 metric tons of copper mill and foundry products. The total supply of copper and its alloys (wire mill, brass mill, foundry, and powder products) to U.S. domestic users totaled 6643 million pounds (metal content including alloying ingredients). Of this, the amounts to the five major end-use markets were (in millions of pounds) building construction, 2627; electrical and electronic, 1723; industrial machinery and equipment, 923; transportation equipment, 746; and consumer and general products, 624.

A large proportion of copper supply is provided by scrap. This recycled material is responsible for about 45% of the total copper supply in the United States.

In 1990, the United States was self-sufficient in copper, exporting 400 000 metric tons of concentrates and refined copper more than it imported, and exporting 4400 metric tons of scrap more than the 95 600 metric tons of copper contained in imported semifabricated products it imported. Imports of copper metal, in the form of primary refinery shapes or mill products, were about 7% of U.S. total supply.

5. COPPER AND ALLOY TERMINOLOGY AND STANDARDS

To ensure orderly identification of specific wrought and cast copper chemical compositions, a six-digit Unified Numbering System is used in the United States and Canada. This system has been adopted by North American specifying bodies such as the American Society for Testing and Materials (ASTM), Society for Automotive Engineers (SAE), and various federal organizations in specifying standards for copper and copper alloy products.

Besides the ASTM standards used in the United States and Canada, other national specifications are BS in the United Kingdom and DIN in Germany.

In the designation system, numbers from C10000 through C79999 denote wrought alloys. Cast alloys are numbered from C80000 through C99999. Within these two categories, the compositions are grouped into the following families of copper and copper alloys:

Coppers (C10000–C15999 and C80000–C81299): Metal that has a designated minimum copper content of 99.3% or higher.

High-Copper Alloys (C16000–C19999 and C81300–C82999): For the wrought products, these are alloys with designated copper contents less than 99.3% but more than 96% that do not fall into any other copper alloy group. The cast high-copper alloys have designated copper contents in excess of 94%, to which silver may be added for special properties.

Brasses: These alloys contain zinc as the principal alloying element with or without other designated alloying elements such as iron, aluminum, nickel, and silicon. The wrought alloys comprise three main families of brasses: copper-zinc alloys (C20500–C28599); copper-zinc-lead alloys (leaded brasses) (C31200–C38599); and copper-zinc-tin alloys (tin brasses) (C40400–C49999).

The cast alloys comprise four main families of brasses: copper-tin-zinc alloys (red, semi-red, and yellow brasses) (C83300–C85899); "manganese bronze" alloys (high-strength yellow brasses)(C86100–C86399); leaded "manganese bronze" alloys (leaded high-strength yellow brasses)(C86400–C86899); and copper-zinc-silicon alloys (silicon brasses and bronzes) (C87300–C87999).

Bronzes: Broadly speaking, bronzes are copper alloys in which the major alloying element is not zinc or nickel. Originally "bronze" described alloys with tin as the only or principal alloying element. Today, the term is generally used not by itself but with a modifying adjective. For wrought alloys, there are four main families of bronzes: copper-tin-phosphorus alloys (phosphor bronzes)(C50100–C52499); copper-tin-lead phosphorus alloys (leaded phosphor bronzes) (C53200–C54899); copper-aluminum alloys (aluminum bronzes) (C60600–C64499); and copper-silicon alloys (silicon bronzes) (C64700–C66199).

The cast alloys have four main families of bronzes: copper-tin alloys (tin bronzes) (C90200–C91799); copper-tin-lead alloys (leaded and high-leaded tin bronzes) (C92200–C94599); copper-tin-nickel alloys (nickel-tin bronzes) (C94700—C94999); and copper-aluminum alloys (aluminum bronzes) (C95200–C95900).

The family of alloys known as "manganese bronzes," in which zinc is the major alloying element, is included in the brasses, above.

Copper-Nickels (C70100–C71999 and C96200–C96999): These are alloys with nickel as the principal alloying element, with or without other designated alloying elements.

Copper-Nickel-Zinc Alloys (C73000–C79999 and C97300–C97999): Known commonly as "nickel silvers," these are alloys that contain nickel and zinc as the principal and secondary alloying elements, with or without other designated elements. They contain no silver.

Leaded Coppers (C98200–C98999): These comprise a series of cast alloys of copper with 20% or more lead, with usually a small amount of silver present, but without tin or zinc.

Special Alloys (C99300–C99999): Alloys whose chemical compositions do not fall into any of the above categories.

6. APPLICATIONS

Copper and its alloys are widely used commercially because of their excellent electrical and thermal conductivity, resistance to corrosion, and ease of fabrication, with good strength and fatigue resistance. Pure copper is used for electrical applications such as wire and cable, bus bars, electrical contacts, and any part required to conduct electricity. Copper and certain brasses, bronzes, and copper nickels are used for automobile radiators, heat exchangers, home heating systems, solar-energy panels and other applications requiring high thermal conductivity. Copper, brass, some bronzes, and copper nickel are used for corrosion-resistant pipe, valves, and fittings in potable-water, process-water, and other aqueous fluid systems. The copper-nickel and bronze alloys are most noted for their application in marine systems. Copper and the brass and bronze alloys are widely used in architecture, for both functional and decorative purposes, in such applications as roofing and guttering, locks and hinges, railings, and window and door mullons.

Copper chemicals, such as copper sulfate, carbonate, oxychloride, and oxides, are used in a wide range of applications: as an algicide in water treatment; as a micronutrient in chicken and cattle feed and on copper-deficient agricultural soil; as a fungicide and bactericide on crops grown in tropical areas; as a wood preservative; in antifouling marine paint; and as oxidation catalysts in petroleum product and organic chemical manufacture.

Copper and copper alloys are used in a number of newly developed technologies. For example, copper is an important material in the use of semiconductors and microcircuitry, where it is used in lead frames and in printed circuit boards. It is used as the stabilizer in superconducting cables. Here the superconducting alloy filaments are embedded in a matrix of pure copper. In this role, the copper acts as the electricity conductor when the filaments are not in their superconducting state and acts as a heat transfer agent to cool the filaments to the temperature at which they are superconducting. In the conventional, low-temperature system, the superconducting filaments are niobium-titanium or niobium-tin alloy. In the new high-temperature system, the superconducting filaments are ceramics that may themselves contain copper oxide.

Copper has also made possible the use of a number of new technologies, and its use is growing as a result. For example, in telecommunications the FAX machine would not be possible without copper wire, and increasing use of FAX for information transmission has greatly increased the use of copper telecommunications cable. Likewise, more and more automation and instrumentation is being used in automobiles. This has been made possible by copper wire and, as a result, has significantly increased the use of copper in an average automobile, now standing at approximately 45 lb per vehicle.

New uses are continually being developed for copper and its alloys and compounds. Among these are applications that take advantage of its bacteriostatic properties such as for clean surfaces in hospitals and public buildings; as a biofouling- resistant surface on ships and other marine structures; in fire safety sprinkler systems and gas distribution systems; and for pharmaceuticals used in diseases where tissue damage is involved.

7. SUMMARY

Copper has been one of the most important materials in the development of civilization, and even today is a significant factor in the development and application of new technologies. The modern growth of the industry is linked to the widespread use of electricity and the continual development of new ways to use electricity. Copper's high electrical and thermal conductivity, formability, and corrosion resistance account for its major applications today.

List of Works Cited

Callaway, J. (1964), *Energy Band Theory*, New York: Academic.

Cooper, B. R., Ehrenreich, H., Philipp, H. R. (1965), *Phys. Rev.* **138**, A494–A507.

Ehrenreich, H., Philipp, H. R. (1962), *Phys. Rev.* **128**, 1622–1629.

Segall, B. (1962), *Phys. Rev.* **125**, 109–122.

Ziman, J. M. (1960), *Electrons and Holes: The Theory of Transport Phenomena in Solids*, Oxford: Oxford University Press.

Further Reading

AMAX Copper, Inc. (1974), *OFHC Copper: A Survey of Properties & Applications*, New York: AMAX Copper, Inc.

American Society for Metals (1979), *Metals Handbook*, Vol. 2: *Properties and Selection: Nonferrous Alloys and Pure Metals*, Materials Park, OH: ASM.

American Society for Testing and Materials (1991), *Anual Book of ASTM Standards*, Vol. 02.01: *Copper and Copper Alloys*, Philadelphia: ASTM.

Butts, A. (1954), *Copper, The Science and Technology of the Metal, Its Alloys and Compounds*, New York: Reinhold.

Copper Development Association Inc. (1975), *Low-Temperature Properties of Copper and Copper Alloys*, Application Data Sheet No.104/5, New York: Copper Development Association Inc.

Copper Development Association Inc. (1990), *Standard Designations for Copper and Copper Alloys*, New York: Copper Development Association Inc.

International Copper Association, Ltd. (1973), *Cryogenic Properties of Copper* (wall chart), New York: International Copper Association.

Nonferrous Metal Works of the World, 5th ed. (1989), New York: Metal Bulletin Inc. (Lists world copper and brass suppliers by country.)

Touloukian, Y. S. (1967), *Thermophysical Properties of High Temperature Solid Materials*, New York: Macmillan.

U. S. Bureau of Mines (1985), Copper chapter in *Mineral Facts and Problems*, Bull. No. 576, Washington, DC: U.S. GPO.

U. S. Bureau of Mines, *Minerals Yearbook*, Published annually, Washington, DC: U.S. GPO.

Ziman, T. M. (Ed.) (1969), *The Physics of Metals*, Vol. 1: *Electrons*, Cambridge, UK: Cambridge University Press.

COSMOLOGY

See ASTROPHYSICS
GALAXIES AND COSMOLOGY

CRITICAL PHENOMENA

CYRIL DOMB, *Department of Physics, Bar-Ilan University, Ramat-Gan, Israel*

INTRODUCTION

If the thermodynamic properties of matter in equilibrium are surveyed, it will be found that they fall into two groups: those that are smooth and continuous, and those that have sharp discontinuities. As examples of the first group we may cite the properties of ideal or nearly ideal gases (energy, entropy, specific heat, and equation of state), of ideal or nearly ideal solids, of ideal or nearly ideal mixtures of gases or solids, paramagnetism and diamagnetism, and properties of electrons and phonons in normal metals. The second group is usually associated with phase transitions of various types: liquid–vapor equilibrium and the critical point, the melting of solids, phase separation in liquid or solid mixtures or solutions, order–disorder transitions in alloys, ferromagnetism, antiferromagnetism, superconductivity, and λ-point anomalies (e.g., liquid helium).

Standard statistical mechanics (see STATISTICAL MECHANICS, CLASSICAL; STATISTICAL MECHANICS, QUANTUM) can deal with the first group with relative ease. The macroscopic body under consideration can be decomposed into microscopic constituents (molecules, harmonic oscillators, spins, and electrons) that can to a good approximation be treated as noninteracting. Of course there must be interactions present to ensure that thermal equi-

1-56081-063-7/92/$5.00+.50

librium is attained, but in describing the behavior of the macroscopic bodies these interactions can be ignored. The energy levels of the macroscopic body can be easily related to the energy levels of the microscopic constituents, and with the aid of some elementary combinatorial analysis, a partition function can be constructed from which the equilibrium behavior of the macroscopic bodies can be derived. Elementary mathematical considerations show that the thermodynamic properties will be continuous.

In general, the discontinuities described in the properties of the second group above originate in *interactions* between the microscopic constituents, and the problem of how to take such interactions into account in calculating a partition function will be central to our discussions. Much of the mathematics involved in a proper treatment of interactions is specialized and sophisticated. The major aim of the present article will be to convey the results and significance of the calculations, which, we believe, can be understood by the general physicist even if the detailed mathematical techniques are beyond him or her.

When the interactions between the constituents are weak, PERTURBATION METHODS (*q.v.*) can be used, but as long as only a finite number of terms are considered, the continuity of the thermodynamic properties will not be destroyed. Discontinuous behavior can be introduced only by taking the perturbation series to infinity so that the mathematical character of the partition function is changed. In fact, the problem to be tackled in dealing with phase transitions is the "strong interaction" problem in which the interactions can no longer be treated as a small perturbation, but play a dominant role in the calculations and in the resulting physical properties.

We will be largely concerned with behavior near critical points of fluids, magnets, and solutions, and with various types of λ-point transition. The past two decades have seen great progress in the theoretical understanding of how this behavior depends on the nature of the intermolecular interactions, e.g., their symmetry and range. We shall endeavor to describe this progress and bring the reader up to the frontiers of current research.

Our treatment will follow the historical development of the subject, and we shall divide this history into four periods:

1869–1944	The classical period
1944–1965	The Onsager revolution
1965–1971	Reconciliation of Onsager with van der Waals
1971–	Renormalization group

In the first period the term classical is used in the sense of "well established" and not with any relationship to quantum theory. The subject starts with the introduction of the term "critical point" by Thomas Andrews in his Bakerian Lecture of 1869 to the Royal Society. Andrews had used the term before in 1863, but the exposition of the true nature of the critical point of a fluid took place in the Bakerian Lecture in which he discussed in detail the experimental behavior of the isotherms of carbon dioxide. A few years later van der Waals provided a theoretical explanation of Andrews's results. The classical period is characterized by the interplay of experiment and theory, and the introduction of new concepts that seemed adequate to provide a qualitative and quantitative description of the phenomena under consideration. Developments that took place in different areas of physics seemed to have many features in common and led eventually to the idea of a unified description of all phenomena involving critical points or λ points.

The second period was initiated by a mathematical *tour de force*, the exact solution by Onsager of a realistic two-dimensional model, giving results that were in basic disagreement with the classical calculations. This period was characterized by further exact calculations of the Onsager kind and also by "down-to-earth" numerical calculations that made extensive use of computers; in addition, new experimental techniques were exploited to furnish more accurate data than had been available previously. As a result the classical theories were completely discredited in relation to quantitative predictions of critical behavior. However, no coherent new pattern was formulated to put in its place.

This led Uhlenbeck to suggest at a conference on "Phenomena in the Neighbourhood of Critical Points" at the National Bureau of Standards in April 1965 that the most important task facing workers in the field was to reconcile Onsager with van der Waals. Indeed, shortly after this, a wider framework emerged into which the new theoretical and experimental results seemed to fit. During the period 1965–1971 this framework was explored in more detail, and the pattern of behavior was summarized most succinctly in the hypothesis of universality. However, the pattern was obtained empirically — there was no theoretical explanation.

The latest phase supplied the theoretical understanding that has made the field "respectable." It was initiated in 1971 by Kenneth Wilson, who suggested a brilliant application of the renormalization group, a concept taken from field theory, to behavior near a critical point. The burst of research activity stimulated by Wilson's ideas has opened up many new avenues of exploration; Wilson's work received rapid recognition and he received a number of important awards culminating in the Nobel prize for physics in 1982. The recognition accorded Wilson is also a recognition of the field of critical phenomena and of the work of all those who helped build the structural foundation on which the renormalization group could base itself.

1. THE CLASSICAL PERIOD

1.1 Liquid–Gas Critical Point

The relation between gases and liquids was the subject of much attention during the first half of the 19th century. It was realized quite early that the liquid phase ceased to exist above a certain temperature, but the general assumption was that it disappeared into the gaseous phase. Thus Faraday talked of the "disliquefying point," while Mendelev used the term "absolute boiling temperature" for the point at which the latent heat of evaporation becomes zero.

In addition to careful and accurate experiments on carbon dioxide described in the Bakerian Lecture of 1869, Thomas Andrews introduced a new concept of symmetry between the liquid and gaseous phases; the two phases merged at the critical point into one fluid phase, "but if any one should ask whether it is now in the gaseous or liquid state, the question does not, I believe, admit a positive reply." He emphasized this feature in the title of his lecture "On the Continuity of the Gaseous and Liquid States of Matter," and pointed out how it was possible, by suitable choice of path, to pass from the liquid to the gaseous phase without any discontinuity. Figure 1 represents the isotherms of carbon dioxide taken from Andrews's Bakerian Lecture (1869). The difference of behavior below and above the critical temperature is manifest. Andrews estimated the critical temperature at 30.92 °C.

Only four years elapsed before van der Waals used the newly developing ideas on the kinetic theory of gases to give a plausible theoretical explanation of Andrews's experimental data. Van der Waals assumed that a gas is made up of molecules with a hard core and a long-range mutual attraction. The range of the attractive forces was assumed to be long compared to the mean free path, and they give rise to a negative "internal pressure." To calculate its value van der Waals made use of the virial theorem that Clausius had introduced in 1870 only a few years before, which states that

$$\left\langle \sum_i \tfrac{1}{2} m v_i^2 \right\rangle = -\tfrac{1}{2} \left\langle \sum_i (x_i X_i + y_i Y_i + z_i Z_i) \right\rangle , \qquad (1)$$

where v_i is the velocity, (x_i,y_i,z_i) the coordinates, and (X_i,Y_i,Z_i) the force on the ith molecule. He found that

$$P_{\text{internal}} \propto 1/v^2, \qquad (2)$$

where $v=V/N$ is the volume per molecule. For the hard core he made the simplest assumption that the available volume is reduced from v to $v-b$. Hence the equation he put forward was

$$P = P_{\text{internal}} + RT/(v-b), \qquad (3)$$

$$(P+a/v^2)(v-b) = RT. \qquad (4)$$

Van der Waals was a doctoral student, and the above ideas were put forward in his thesis in Dutch. Their importance was quickly recognized by Maxwell, who reviewed the thesis in *Nature* in 1874, and in a lecture to the Chemical Society in 1875 he wrote as follows:

"The molecular theory of the continuity of the liquid and gaseous states forms the subject of an exceedingly ingenious thesis by Mr. Johannes Diderick van der Waals, a graduate of Leyden. There are certain points in which I think he has fallen into mathematical errors, and his final result is certainly not a complete

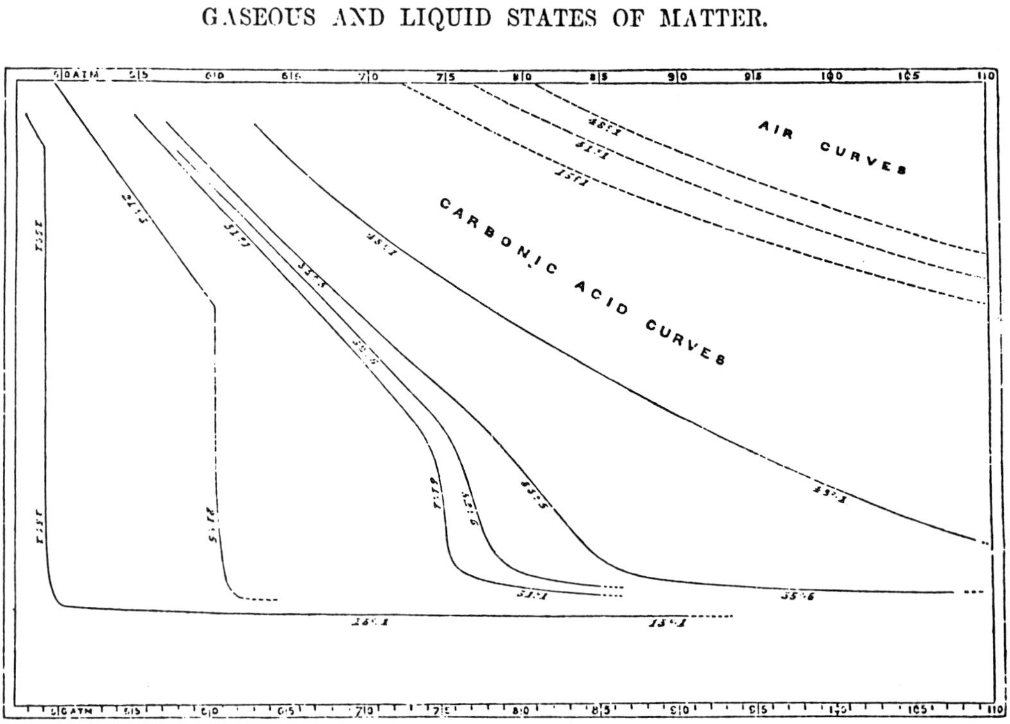

FIG. 1. Isotherms of carbon dioxide taken from Andrews's Bakerian Lecture to the Royal Society in 1869. Andrews's estimate for the critical temperature was 30.92 °C.

expression for the interaction of real molecules, but his attack on this difficult question is so able and so brave, that it cannot fail to give a notable impulse to molecular science. It has certainly directed the attention of more than one inquirer to the study of the Low-Dutch language in which it is written."

The form of isotherms given by van der Waals's equation (4) is shown in Fig. 2. In order to compare them with the experimental results of Andrews, it was necessary to decide where to draw the horizontal line that represents the liquid and gas in equilibrium. Here Maxwell put forward his "equal area" construction, which is illustrated in Fig. 3.

Boltzmann too was impressed by van der Waals's results, and although he said that "unfortunately, van der Waals had to abandon mathematical rigour at a certain point in order to carry out his calculation," he was nevertheless impressed by the practical value of his theory.

It is significant that both Maxwell and Boltzmann were reasonably happy with the attractive a/v^2 term; their main anxiety was the repulsive term.

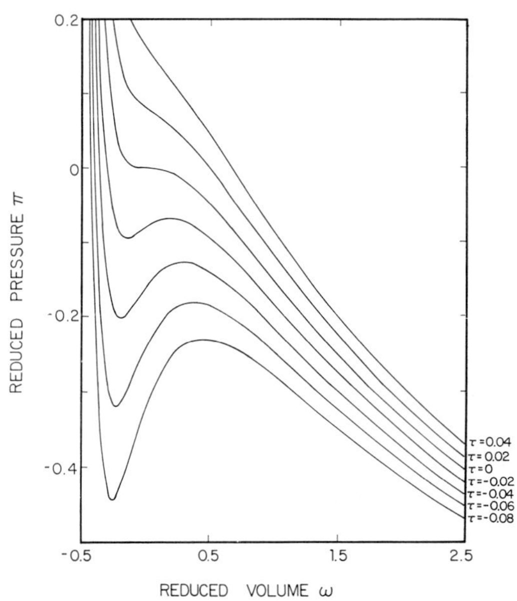

FIG. 2. Isotherms corresponding to van der Waals's equation (4). Reduced pressure $\pi(=P/P_c)$ is plotted against reduced volume $\omega(=V/V_c)$.

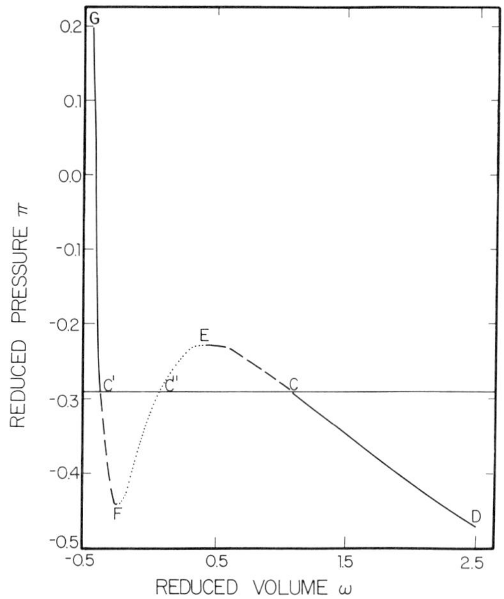

FIG. 3. Maxwell's "equal area" construction to determine the horizontal part of the isotherm in van der Waals's equation. Area $C'FC''$ = Area $C''EC$. The construction was proposed in a lecture to the Chemical Society in 1875.

The new concept of "internal pressure" that van der Waals introduced was to bear fruit 30 years later in a completely different area of physics.

1.2 Magnet–Fluid Analogy: Curie Point of a Ferromagnet

The fact that a magnet loses its magnetic power at high temperatures was noted by Gilbert in his famous treatise "De Magnete" in 1600. A more detailed quantitative investigation started in the 19th century involving Faraday in the 1830s, Barrett in 1874, Bauer in 1880, and Hopkinson in 1889, who was the first to introduce the term "critical temperature" for "the temperature at which the magnetism disappears." But the definitive paper for magnets comparable to that of Andrews for fluids was written by Curie in 1895 before he started his more famous investigations of radioactivity.

One of the most interesting new ideas put forward in this paper is the analogy between magnets and fluids. Taking pressure P as the analog of magnetic field and density ρ as the analog of magnetization, Curie points out the close similarity between the P-ρ and M-H isotherms (Fig. 4). The paramagnetic state at high temperatures corresponds to the gaseous phase and the ferromagnetic state at low temperatures to the liquid phase. Curie says that this analogy could be used to suggest new and useful experiments, and he poses the question whether there exists a precisely defined critical point with associated critical constants for a ferromagnet analogous to a fluid.

It was this analogy that led Pierre Weiss in 1907 to postulate his *molecular field* hypothesis, in which he assumed that the mutual interactions between the molecules could be replaced by a uniform field nM proportional to the magnetization in the same direction. He states, "One may give to nM the name *Internal Field* to mark the analogy with the internal pressure of van der Waals."

For the magnetic susceptibility of a gas, Curie had discovered experimentally the inverse temperature dependence

$$\chi = C/T, \tag{5}$$

and Langevin had applied the techniques of statistical mechanics to explain this result theoretically (see PARAMAGNETISM). For an ideal gas of molecules each having magnetic moment μ_0, Langevin derived the magnetic equation of state (analogous to $PV=RT$ for an ideal gas)

$$M = L(\mu_0 H/kT), \tag{6}$$

where $L(x)$ is the Langevin function $\coth x - x^{-1}$.

For a ferromagnet Weiss put forward the simple modification

$$M = L[\mu_0(H+H_{\rm int})/kT] \quad (H_{\rm int}=nM), \tag{7}$$

and this led to far reaching conclusions. There is indeed a sharply defined critical point analogous to that for a fluid; Weiss and Kamerlingh Onnes later termed this the *Curie point* in memory of Curie, who had been killed in an accident in 1906. Below the Curie temperature there is a nonzero spontaneous magnetization; in the paramagnetic state above the Curie temperature T_c, Curie's relation (5) for the susceptibility is modified to

$$\chi = C/(T-T_c), \tag{8}$$

and this is usually called the Curie–Weiss law.

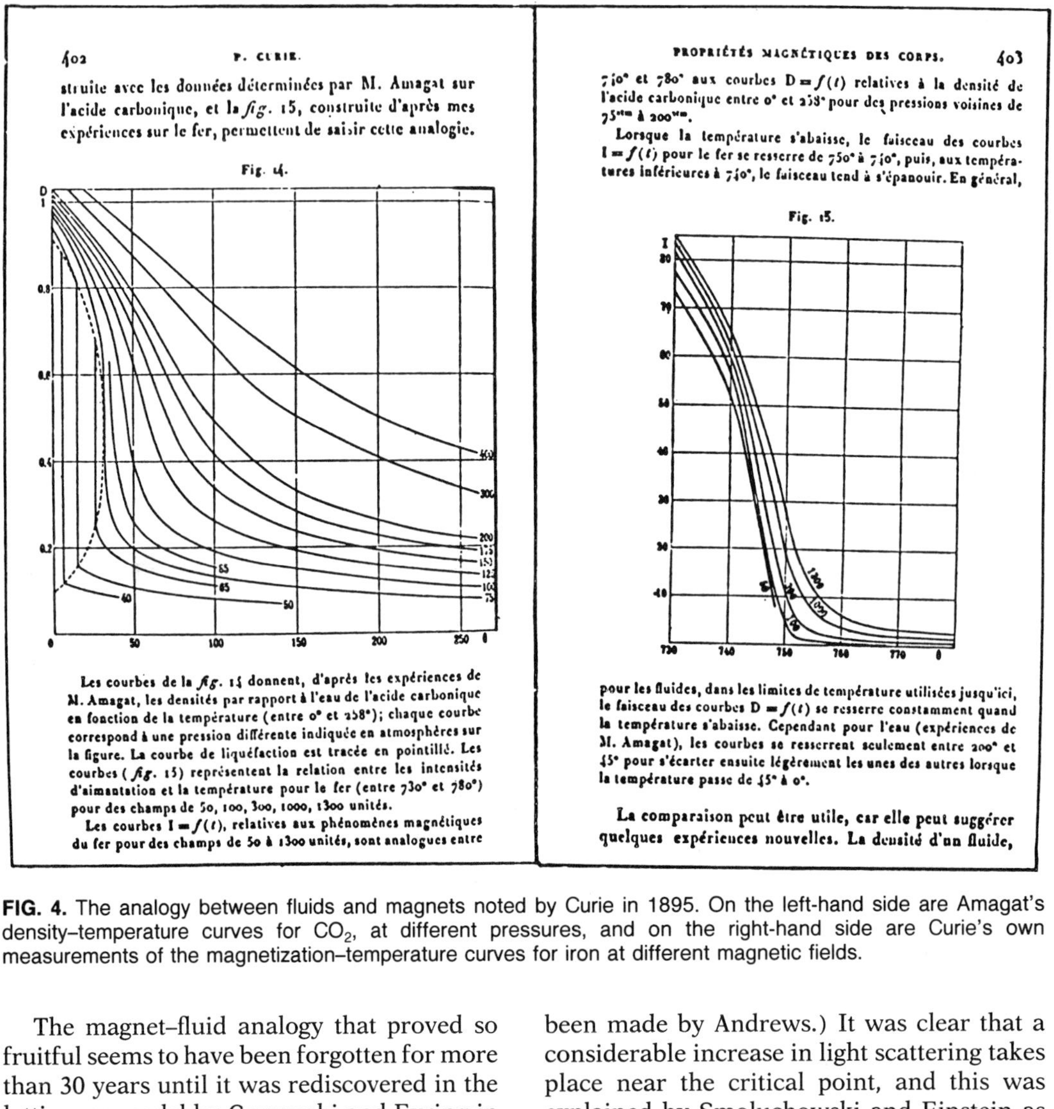

402 P. CURIE.

struite avec les données déterminées par M. Amagat sur l'acide carbonique, et la *fig.* 15, construite d'après mes expériences sur le fer, permettent de saisir cette analogie.

Fig. 14.

Les courbes de la *fig.* 14 donnent, d'après les expériences de M. Amagat, les densités par rapport à l'eau de l'acide carbonique en fonction de la température (entre 0° et 258°); chaque courbe correspond à une pression différente indiquée en atmosphères sur la figure. La courbe de liquéfaction est tracée en pointillé. Les courbes (*fig.* 15) représentent la relation entre les intensités d'aimantation et la température pour le fer (entre 730° et 780°) pour des champs de 50, 100, 300, 1000, 1300 unités.

Les courbes $I = f(t)$, relatives aux phénomènes magnétiques du fer pour des champs de 50 à 1300 unités, sont analogues entre

PROPRIÉTÉS MAGNÉTIQUES DES CORPS. 403

740° et 780° aux courbes $D = f(t)$ relatives à la densité de l'acide carbonique entre 0° et 258° pour des pressions voisines de 75^{atm} à 200^{atm}.

Lorsque la température s'abaisse, le faisceau des courbes $I = f(t)$ pour le fer se resserre de 750° à 740°, puis, aux températures inférieures à 740°, le faisceau tend à s'épanouir. En général,

Fig. 15.

pour les fluides, dans les limites de température utilisées jusqu'ici, le faisceau des courbes $D = f(t)$ se resserre constamment quand la température s'abaisse. Cependant pour l'eau (expériences de M. Amagat), les courbes se resserrent seulement entre 200° et 45° pour s'écarter ensuite légèrement les unes des autres lorsque la température passe de 45° à 0°.

La comparaison peut être utile, car elle peut suggérer quelques expériences nouvelles. La densité d'un fluide,

FIG. 4. The analogy between fluids and magnets noted by Curie in 1895. On the left-hand side are Amagat's density–temperature curves for CO_2, at different pressures, and on the right-hand side are Curie's own measurements of the magnetization–temperature curves for iron at different magnetic fields.

The magnet–fluid analogy that proved so fruitful seems to have been forgotten for more than 30 years until it was rediscovered in the lattice-gas model by Cernuschi and Eyring in 1939. It was fruitful again in subsequent developments after 1944.

1.3 Microscopic Critical Behavior of Fluids

The experiments of Andrews were concerned with the thermodynamic properties of bulk fluids. An experimental phenomenon that excited theoretical interest at the beginning of the 20th century was critical opalescence, in which a colorless transparent fluid suddenly becomes opaque in a narrow region of temperatures near T_c; the onset of opacity is accompanied by dramatic changes of color. (These experimental results had been known much earlier, an initial observation having been made by Andrews.) It was clear that a considerable increase in light scattering takes place near the critical point, and this was explained by Smoluchowski and Einstein as arising from large fluctuations in density as follows.

In 1910 Einstein showed how Boltzmann's formula relating the probability of any state of a system to its entropy (see STATISTICAL MECHANICS, CLASSICAL) could be used to calculate fluctuations of thermodynamic quantities. For the density he derived the formula

$$\langle \Delta\rho^2 \rangle = (\rho^2 kT/V) K_T, \tag{9}$$

where K_T is the isothermal compressibility,

$$K_T = \frac{1}{\rho}\left(\frac{\partial\rho}{\partial P}\right)_T = -\frac{1}{V}\left(\frac{\partial V}{\partial P}\right)_T. \tag{10}$$

Assuming that a change in density is accompanied by a change of refractive index according to the Lorentz–Lorenz law

$$(\epsilon-1)/(\epsilon-2)=A\rho \tag{11}$$

and that scattering takes place randomly, Einstein came to the conclusion that the scattering of light of wavelength λ should be proportional to K_T/λ^4. Since van der Waals's equation leads to an infinite value of K_T at T_c, this seemed to provide a satisfactory explanation of critical opalescence.

It was remarkably perceptive of Ornstein and Zernike (1914, 1916) to note an inconsistency in the treatment arising from the assumption that the fluctuations in all elements of volume are independent of one another. In fact, they pointed out that there must be a correlation between different elements that increases indefinitely in range as the critical point is approached.

To treat this correlation, Ornstein and Zernike introduced a basic new function, afterwards called the "pair distribution function," which has played a central role in the theory of liquids ever since. Let $\nu(d\mathbf{r})$ be a random variable representing the number of molecules in a volume $d\mathbf{r}$ centered at $\mathbf{r}$. Since $d\mathbf{r}$ is very small, the probability of occupation by more than one particle being of order $d\mathbf{r}^2$ can be neglected. Hence we can write

$$\langle \nu(d\mathbf{r})\rangle = n_1(\mathbf{r})d\mathbf{r}, \tag{12}$$

where $n_1(\mathbf{r})$ is the density, which Ornstein and Zernike took to be a constant ρ. Similarly for the correlation between particles at points $\mathbf{r}_1$ and $\mathbf{r}_2$,

$$\langle \nu(d\mathbf{r}_1)\nu(d\mathbf{r}_2)\rangle = n_2(\mathbf{r}_1,\mathbf{r}_2)d\mathbf{r}_1 d\mathbf{r}_2, \tag{13}$$

and for a homogeneous isotropic fluid $n_2(\mathbf{r}_1,\mathbf{r}_2)$ is of the form $n_2(r)$ $(r=|\mathbf{r}_1-\mathbf{r}_2|)$.

Ornstein and Zernike introduced the function

$$g(r)=n_2(r)/\rho^2, \tag{14}$$

which approaches 1 at large distances where the correlation becomes negligible. They used the fluctuation relation (9) to derive a fundamental identity between K_T and $g(r)$. But they also obtained an integral equation for $g(r)$ that they were able to solve explicitly.

The basic physical idea behind Ornstein and Zernike's treatment was to differentiate between the direct influence of molecular interactions, which should be short ranged and was represented by a function $f(r)$, and the correlation between densities, represented by $g(r)$ above, which should become long ranged as the critical temperature is approached. The integral equation relates $g(r)$ and $f(r)$.

Their calculations suggested that the correlations fall off asymptotically as

$$\frac{1}{r}\exp(-\kappa r), \tag{15}$$

where the value of κ can be determined from van der Waals's equation [$\kappa \sim (T - T_c)^{1/2}$]. Their detailed conclusions about light scattering differed from those of Einstein; near the critical temperature the λ^{-4} Rayleigh dependence on wavelength ceases to be valid, and there is a "whitening" of the scattered light. By temperature T_c the wavelength dependence has become of the form λ^{-2}.

1.4 Critical Behavior of Binary Alloys

The early years of the 20th century saw the development of x-ray diffraction as a powerful tool in the investigation of crystal structures, and compounds such as NaCl were found to have a regular ordered structure. However, the ionic bond between Na and Cl is so strong that no significant disordering occurs when the temperature is raised; the crystal melts before it disorders.

In 1919 Tammann suggested that similar ordering might occur in metallic alloys, and this was demonstrated experimentally a few years later (Bain, 1923; Johansson and Linde, 1925) by the existence of superlattice lines in the x-ray diffraction pattern of copper–gold alloys. However, for such systems there were soon indications that significant disordering takes place with an increase in temperature and that this is accompanied by an anomalous specific heat.

The mathematical description of the disordering process is usually associated with the names of Bragg and Williams. In their classic paper in 1934 they introduced a parameter S to characterize the *degree of order* and used Boltzmann's principle to calculate its behavior as a function of temperature. They found a pattern closely analogous to the Weiss theory of ferromagnetism, with S falling rapidly to zero at a critical temperature T_c and remaining zero for $T > T_c$. In fact, for a binary alloy with equal concentrations of constituents the equation derived for S was

$$S = \tanh(ST_c/T), \tag{16}$$

which is of the same form as Eq. (7) with $H=0$ and $\tanh x$ replacing the Langevin function $L(x)$.

In 1935 Bragg and Williams wrote a second paper elaborating their ideas. Because of the analogy with ferromagnetism they called T_c the *Curie temperature* of the alloy; S was now termed *long-distance order* to differentiate it from another parameter that had been introduced by Bethe to characterize the *short-range order* that persists above T_c. But they also apologized in this paper for having ignored the work of other investigators, notably Gorsky, Borelius, and Dehlinger, who had been thinking along similar lines.

The important new concept of long-range order is clearly described in the second paper, and a significant distinction is drawn between long-range order and long-range forces. It is stated clearly that short-range forces can give rise to long-range order, a conclusion exactly parallel to that of Ornstein and Zernicke for fluids described above, of which Bragg and Williams were apparently unaware.

1.5 Universality of Second-Order Transitions (Landau)

It was L. D. Landau (1937) who first attempted to provide a unified description for all second-order transitions. In addition to the phenomena described above, experimental evidence was accumulating about specific heat anomalies that were termed λ-point transitions in liquid helium, ammonium chloride, and a number of other substances, and the superconducting transition was also in this category. In 1933 Ehrenfest had introduced a thermodynamic classification of higher-order transitions, but there were difficulties associated with his treatment, and we shall avoid describing λ-point transitions as second-order phase transitions although this terminology is common.

Landau generalized the idea introduced in the theory of alloys to all systems manifesting λ-point transitions. He suggested that for every such system one must identify an *order parameter* analogous to the long-range order in an alloy, which would be zero on the high-temperature side of the transition and nonzero on the low-temperature side. He emphasized the important role that symmetry plays in phase transitions, and suggested that the important features of behavior in the vicinity of a λ point could be determined by expanding the free energy in a power series as a function of the order parameter η. In the ferromagnetic transition, the order parameter is the spontaneous magnetization, and in the fluid, the density difference between liquid and gas; Landau argued that for reasons of symmetry the form of the expansion of the Gibbs function would be

$$\phi(P,T,\eta) = \phi_0(P,T) + A(P,T)\eta^2 + B(P,T)\eta^4 + \cdots, \tag{17}$$

and the Curie temperature would correspond to

$$A(P,T) = 0, \quad B(p,T) > 0. \tag{18}$$

Above the Curie temperature $A > 0$ and the solution corresponding to a minimum of ϕ is $\eta = 0$; this is the phase of higher symmetry. Below the Curie temperature the minimum of ϕ corresponds to a nonzero value of η given by

$$\eta^2 = -A/2B, \tag{19}$$

and this is the phase of lower symmetry.

Landau's theory enabled one to understand why all the systems discussed previously had the same essential pattern of critical behavior; even though equations of state such as Eqs. (4) and (7) look very different, they both conform to Eq. (18). Critical exponents are the same for all λ-point transitions, and in later terminology the transitions would be described as universal.

The behavior of typical thermodynamic quantities for a ferromagnet near T_c were as follows ($t = T/T_c - 1$):

spontaneous magnetization	$M_0 \sim (-t)^{1/2}$	$(t<0)$,	
initial susceptibility	$\chi_0 \sim t^{-1}$	$(t<0)$,	
critical isotherm	$H \sim M^3$	$(t=0)$,	(20)
derivative of susceptibility	$\frac{d\chi_0}{dH} \sim t^{-4}$	$(t>0)$.	

The specific heat tended to one value C_- on approaching the transition from a lower temperature, and to another C_+ from above. The transcription to appropriate thermodynamic variables can be readily made for other systems.

Landau developed an alternative approach to that of Ornstein and Zernike to deal with critical fluctuations. His method was to seek a Taylor expansion of the excess free energy arising from variations in density (for a magnetic system it would arise from variations in magnetization, and in other systems from variations in the appropriate order parameter). We can divide the total Helmholtz free energy ΔF into contributions from different spatial regions,

$$\Delta F = \int \Delta a(\mathbf{r}) dV, \tag{21}$$

where $a(T,\rho)$ is the local free energy expressed in terms of the density. $\Delta a(\mathbf{r})$ depends on $\delta\rho(\mathbf{r})$ and combinations of the derivatives of ρ, which are spherically symmetrical. In an expansion of $a(\mathbf{r})$ the first derivatives of ρ can enter only as $(\text{grad}\,\rho)^2$ and the second derivatives as $\nabla^2\rho$. Terms in $\nabla^2\rho$ can be integrated to give a surface contribution that can be ignored and to give a bulk contribution of the form $(\text{grad}\,\rho)^2$. We are thus led in the first order to an expansion of the form

$$\Delta a(\mathbf{r}) = \tfrac{1}{2} c \delta\rho^2 + \tfrac{1}{2} b (\nabla\rho)^2 + \cdots, \tag{22}$$

where $c = (1/\rho)(\partial P/\partial\rho)_T$ is zero at T_c, and b must be greater than 0 for $\Delta a(\mathbf{r})$ to correspond to a minimum at T_c.

If $\delta\rho$ is broken down into Fourier components, then ΔF is a sum of contributions from independent normal modes. The absence of interaction between the modes is due to the truncation of expansion (22) at the square term. Assigning an average energy kT to each mode, Landau was led to the universal formula (15) for the pair correlation function.

The approximation of stopping with the gradient terms in the expansion in Eq. (22) is reasonable for long-wavelength fluctuations which are responsible for critical behavior. However, since c is zero at T_c it is reasonable to assume that a higher-order term $d(\delta\rho)^4$ is needed to get a valid picture in the critical region. This was the suggestion of Landau and Ginzburg for the order parameter of the superconducting transition. Its adoption complicates the above treatment very considerably and has far-reaching consequences, as we shall see in Sec. 5.4.

2. ISING MODEL: EXACT SOLUTIONS

2.1 Ising Model

In 1925 W. Lenz, who had been looking for a simple model that might serve to explain ferromagnetism, put forward the following suggestion to his graduate student E. Ising. Assume that each atom possesses a spin and hence a magnetic moment μ_0, which can orient either parallel or antiparallel to an external magnetic field H. There is an interaction between nearest-neighbor spins in the lattice, parallel spins having an interaction $-J$ and antiparallel spins $+J$ $(J>0)$. Lenz hoped that the model might manifest a nonzero spontaneous magnetization, i.e., that as $H \to 0$ the ratio

$$(N_2 - N_1)/(N_2 + N_1) \tag{23}$$

might tend to a nonzero value. Here N_1 and N_2 are the number of spins antiparallel and parallel to the field H.

Ising was able to solve the problem in one dimension only, where the statistical problem is quite simple, and the solution did not possess a spontaneous magnetization at any temperature above zero. He concluded wrongly that the model would not give rise to a spontaneous magnetization in higher dimensions. In 1936 Peierls was the first to demonstrate convincingly that for a two-dimensional Ising model, at sufficiently low but nonzero temperatures, the ratio (23) does not tend to a nonzero value as $H \to 0$.

The Hamiltonian for the Ising model can be conveniently written in the form

$$\mathcal{H} = -J \sum_{\langle i,j \rangle} \sigma_i \sigma_j - \mu_0 H \sum_i \sigma_i, \qquad (24)$$

where the sum i is taken over all N spins of the lattice and $\langle i, j \rangle$ over all of the nearest-neighbor pairs of the lattice, counting each pair once only. Each σ_i takes on the values ± 1 according as the spin i is pointing in the direction 1 or 2. Ising considered only a model of spin $\frac{1}{2}$; the natural generalization to spin s is

$$\mathcal{H} = -\frac{J}{s^2} \sum_{\langle i,j \rangle} s_{zi} s_{zj} - \frac{\mu_0 H}{s} \sum_i s_{zi}, \qquad (25)$$

where s_{zi} are standard quantum spins taking on the values $s, (s-1), \ldots, -(s-1), -s$. The normalization is chosen to keep the maximum interaction between nearest-neighbor spins, and the maximum interaction with the external field, independent of s. The term Ising model without qualification usually refers to the model of spin $\frac{1}{2}$; it is sometimes called the simple Ising model for greater clarity.

Let us now consider the partition function (see STATISTICAL MECHANICS, QUANTUM) for the Ising model. Any configuration of spins characterized by the numbers N_1 and N_2 of spins pointing in directions 1 and 2, and N_{11}, N_{12}, and N_{22} different types of nearest-neighbor pairs, will have energy

$$\mu_0 H(N_1 - N_2) + J(N_{12} - N_{11} - N_{22}). \qquad (26)$$

It is easy to establish the relations

$$2N_{11} + N_{22} = qN_1, \quad 4N_{12} + N_{22} = qN_2, \qquad (27)$$

by enumerating all q bonds that originate in each atom of the lattice (q being the coordination number of the lattice).

For convenience, to avoid surface effects, we adopt the usual solid-state physics device of cyclic boundary conditions, so that all lattice sites are equivalent. Also we know that

$$N_1 + N_2 = N; \qquad (28)$$

hence Eq. (26) can be written

$$-(\mu_0 H + \tfrac{1}{2}qJ)N + 2\mu_0 H N_1 + 2JN_{12}. \qquad (29)$$

If $J > 0$ the lowest-energy state corresponds to $N_1 = 0$, $N_{12} = 0$, i.e., all the spins are aligned in direction 2. Of the five numbers introduced only two are free, and these are conveniently taken to be N_1 and N_{12}.

The major difficulty in the calculation of the partition function is the combinatorial problem of determining how many states there are on a lattice with a given (N_1, N_{12}). If we denote this by $g(N; N_1, N_{12})$, the partition function can be written as

$$Z_N^I = \exp(-\beta\mathcal{H})$$
$$= y^{-N/2} z^{-qN/4} \sum_{N_1, N_{12}} g(N; N_1, N_{12}) y^{N_1} z^{N_{12}}$$
$$[y = \exp(-2\beta\mu_0 H), \quad z = \exp(-2\beta J),$$
$$\beta = 1/kT]. \qquad (30)$$

We now consider the problem of an alloy consisting of N_A atoms of type A and N_B atoms of type B on the same lattice ($N_A + N_B = N$). Since the alloy has a fixed composition, N_A and N_B remain fixed, in contrast to N_1 and N_2 above. But since N_A and N_B represent particular values of N_1 and N_2, relations (27) are valid for N_A, N_B, N_{AA}, N_{AB}, and N_{BB}. To calculate the energy of a given configuration it is reasonable to use a model taking account only of nearest-neighbor interactions, since interatomic forces decrease rapidly with distance. Let us denote the interactions of AA, AB, and BB pairs of neighbors by $-\epsilon_{AA}$, $-\epsilon_{AB}$, and $-\epsilon_{BB}$, respectively. Consider a configuration in which there are N_{AA}, N_{AB}, and N_{BB} such pairs. The energy of the configuration is

$$w = -N_{AA}\epsilon_{AA} - N_{AB}\epsilon_{AB} - N_{BB}\epsilon_{BB}$$
$$= -\tfrac{1}{2}q(N_A\epsilon_{AA} + N_B\epsilon_{BB}) - N_{AB}\epsilon$$
$$[\epsilon = \epsilon_{AB} - \tfrac{1}{2}(\epsilon_{AA} + \epsilon_{BB})]. \qquad (31)$$

If $\epsilon < 0$ the lowest-energy state corresponds to the minimum of N_{AB}, i.e., to two separated phases of A and B atoms. This gives rise to the analog of an Ising ferromagnet in the realm of alloys, a *solid solution*.

The partition function for the solid solution is, from Eq. (30),

$$Z_N^s = z_A^{(1/2)qN_A} z_B^{(1/2)qN_B} \sum_{N_{AB}} g(N;N_A,N_{AB}) z^{N_{AB}}$$

$$[z_A = \exp(\beta\epsilon_{AA}),\ z_B = \exp(\beta\epsilon_{BB}),\ z = \exp(\beta\epsilon)]. \tag{32}$$

On comparing Eqs. (32) and (30) we see that, except for the ground-state terms, the partition function for the alloy is obtained by picking out the coefficient of y^{N_A} in the partition function for the magnet. In statistical-mechanical terminology (see STATISTICAL MECHANICS, QUANTUM) the partition function for the magnet corresponds to the *grand partition function* for the alloy. Writing

$$\Lambda_N(y,z) = \sum_{N_1,N_{12}} g(N;N_1,N_{12}) y^{N_1} z^{N_{12}}, \tag{33}$$

the behavior of both the ferromagnet and the alloy can be calculated from a knowledge of $\Lambda_N(y,z)$.

In Sec. 1.4 we referred to order–disorder transitions in binary alloys that form superlattices at low temperatures. This corresponds to $\epsilon > 0$ in Eq. (31) which gives rise to a lowest-energy state with N_{AB} a maximum. The magnetic analog with $J < 0$ is an Ising model of an antiferromagnet in which the lowest-energy state corresponds to alternating up and down spins. Both of these problems are likewise determined by $\Lambda_N(y,z)$ defined in Eq. (33) but for a different range of values of the variable z.

The introduction of a specific model (25) enables us to mention an important alternative method of deriving the classical result. If we focus attention on the field acting on a given spin as a result of interactions with its neighbors, this field will fluctuate as the orientation of the neighbors changes. The exact solution must take these fluctuations into account exactly. But we can approximate by replacing the fluctuating field by its mean value, and from this a self-consistent equation can be derived for the magnetization. This is equivalent to the classical result, and is usually called the *mean-field approximation.* We would expect from the statistical considerations that it would be a reasonable approximation when the number of interacting neighbors is sufficiently large.

2.2 The Lattice-Gas Model

In 1939 Cernuschi and Eyring realized that the Ising model could readily be adapted to provide a simple model of liquid–gas equilibrium. If, in the solid-solution model of the previous section, type *A* atoms are replaced by the molecules of the fluid, and type *B* atoms by holes, i.e., unoccupied cells, the possibility of varying the number of holes allows for changes in free volume, one of the essential features of a liquid. Any two neighboring occupied cells have an attractive interaction; the fact that an occupied cell cannot accept another molecule provides a crude representation of a hard-core repulsion, and the model thus represents the two basic features of intermolecular forces. The concentration of holes is not fixed but allowed to vary, and this is taken into account by means of the grand partition function, as indicated in the previous section. The interactions ϵ_{AB} and ϵ_{BB} are zero, ϵ_{AA} represents the intermolecular attraction, and ϵ defined in Eq. (31) is equal to $-\frac{1}{2}\epsilon_{AA}$.

For this model the magnet–fluid analogy discussed in Sec. 1.2 is formulated precisely, and the thermodynamic quantities associated with fluids, such as pressure and density, can be given exact magnetic analogs. All of the classical results discussed in Sec. 1 can be related to a simple function of two variables $\Lambda_N(y,z)$ defined by Eq. (33), and we can understand the similarities in experimental behavior of the various systems. However, when we come to examine the critical behavior of the function $\Lambda_N(y,z)$ we shall find that it differs very significantly from the Landau pattern described in Sec. 1.5.

2.3 Transfer Matrix

For a one-dimensional assembly of systems each having t possible energy levels, a general treatment is possible that reduces the evaluation of the partition function to the determination of the largest eigenvalue of a matrix. Let $U(x_i,x_{i+1})$ denote the energy between any two systems x_i and x_{i+1}, each x_i taking t possible values. For simplicity we shall deal with a ring of systems in which the Nth is connected cyclically to the first. The partition function for N systems is then given by

$$Z_N = \sum_{x_i} \exp\{-\beta[U(x_1,x_2)+U(x_1,x_2)+\cdots +U(x_N,x_1)]\}$$
$$= \sum_{i,j,\ldots,k=1}^{t} V_{ij}V_{jk}\cdots V_{kl}V_{li}, \qquad (34)$$

where $V_{ij}=\exp(-\beta U_{ij})$ corresponds to the interaction between neighboring systems in states i and j. The elements V_{ij} thus form a $t\times t$ matrix V and the continued product (34) will readily be identified as the trace of V^N. Thus if $\lambda_1,\lambda_2,\ldots,\lambda_t$ are the eigenvalues of V, then

$$Z_N=\lambda_1^N+\lambda_2^N+\cdots+\lambda_t^N. \qquad (35)$$

For sufficiently large N only the largest eigenvalue, λ_1, is significant, and we may identify Z_N with λ_1^N. It should be noted that if the interaction U_{ij} is symmetric in i and j, the matrix V is symmetric. Also the ith diagonal element of V^N is proportional to the probability that a particular system is in state i.

The above method of treating one-dimensional assemblies was suggested independently in 1941 by Kramers and Wannier, by Lassettre and Howe, and by Montroll. It constituted the first preliminary step towards modern theories of critical behavior, focusing attention on the exact calculation of the partition function. The matrix V is usually called the *transfer matrix.*

For the one-dimensional simple Ising model $t=2$, and the eigenvalues in Eq. (35) can readily be calculated:

$$\left.\begin{matrix}\lambda_1\\ \lambda_2\end{matrix}\right\}=\tfrac{1}{2}(1+y)\pm[(1-y)^2+4yz^2]^{1/2}. \qquad (36)$$

This is the solution found by Ising in 1925. For large N we can ignore λ_2, and since λ_1 is a smooth analytic function of (y,z) there are no singularities. There is no spontaneous magnetism for $T>0$; the only manifestation of singular behavior occurs at $T=0$ where the susceptibility takes on the value $\beta\mu_0^2\exp(\beta J)$, which becomes infinite.

The method outlined above can be applied to a finite strip of width l, and the number of different states t to be taken into account for the transfer matrix is then 2^l. Similarly, for a three-dimensional block of dimensions $l\times m$ the transfer matrix is of order 2^{lm}. As long as the transfer matrix is finite and the largest eigenvalue is nondegenerate, it is easy to show that it involves no singularities. To achieve singular behavior the matrices must be infinite; hence for the Ising model only systems infinite in at least two dimensions can manifest singular critical behavior. Such dimensional considerations were completely absent in the classical Landau theory (Sec. 1.5).

2.4 Onsager's Solution for the Rectangular Lattice

In 1936 Peierls had demonstrated rigorously that the Ising model on the quadratic lattice in two dimensions does indeed possess a spontaneous magnetization. In 1944 Onsager published an exact solution for this model in zero magnetic field, which exercised a revolutionary impact on the whole subject, since its properties differed essentially from those of the classical theories. This will be seen from the following brief summary:

1. *The specific heat* manifests a striking singularity at the Curie point. It is logarithmically infinite on both sides of the Curie temperature, and is shown in Fig. 5 for the symmetric net. For temperatures near to T_c the following approximation may be used:

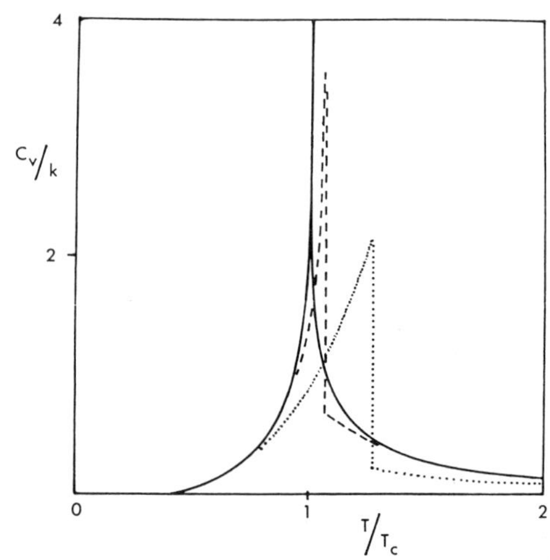

FIG. 5. Onsager's classic calculation (1944) of the specific heat of the spin-$\frac{1}{2}$ Ising model on a simple quadratic lattice. The solid curve represents the true specific heat, which is logarithmically infinite. The dotted curve represents a closed-form approximation due to Bethe, and the dashed curve an improved closed-form approximation due to Kramers and Wannier.

$$C_v/NK \cong (2/\pi)(\ln \cot \pi/8)^2 \times (K_1 - 1 - \pi/4)$$

$$\cong 0.4945 \ln |T - T_c|$$

$$[K_1 = \ln(2^{1/2}|K - K_c|),\ K = \beta J]. \quad (37)$$

It should be noted that the specific-heat curve differed substantially from any curves that had been observed experimentally, particularly in regard to the magnitude of the "tail" above the Curie temperature. This can be seen more clearly from the critical value of the entropy, which shows that the entropy change above T_c ($=0.387$) is greater than that below T_c ($=0.306$).

For an asymmetric net with interactions J, J' in the principal directions, the logarithmic nature of the specific-heat anomaly is unchanged. As one of the interactions J' becomes small the Curie temperature moves to zero, but this movement is very slow; if the interaction J' is J/ν, then the reduction in T_c is of order $\ln \nu$ for large ν. The specific-heat curve for an asymmetric net in which $J' = J/100$ was calculated by Onsager, and is shown in Fig. 6 for comparison with the symmetric net, and with the specific heat of a one-dimensional chain ($J'=0$). The abscissa is $2kT/(J+J')$ so that the areas under all the curves are the same. It will be seen that the large asymmetry reduces the Curie temperature by a factor of 2 only. Also there is an essential difference below the Curie temperature between a two-dimensional net with one small interaction and a one-dimensional chain.

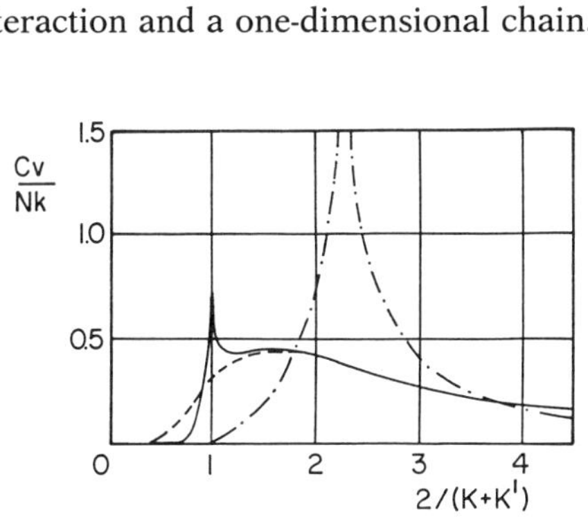

FIG. 6. Onsager's calculation of the specific heat of an asymmetric rectangular lattice with a ratio of interactions 100:1. The logarithmic infinity is retained in contrast to the specific heat of the linear chain (dashed line). The central curve (–·–) corresponds to the symmetric quadratic lattice.

2. *Other thermodynamic quantities.* The internal energy E has a vertical tangent on both sides of the Curie point. Near the Curie point it is of the form

$$A_0 + B_0(T - T_c)\ln|T - T_c|. \quad (38)$$

The free energy F involves an integral of the internal energy and therefore has infinite curvature at the Curie point. All thermodynamic quantities have a singularity at T_c, so that *no Taylor expansion of the type envisaged by Landau* (Sec. 1.5) *is possible.*

3. *The specific heat of a finite crystal.* The specific heat of a finite crystal of n rows is finite at all temperatures, since the partition function is an analytic function of the temperature, but a maximum occurs near the transition point of the infinite crystal. Onsager showed that for large n the maximum specific heat is given approximately by

$$C_v/Nk = 0.4945 \ln n + 0.1879. \quad (39)$$

4. *Correlations.* In Sec. 1.3 we described the classical theory of correlations between the molecules of a fluid as formulated by Ornstein and Zernike. By analogy, for a magnetic system we introduce the correlation between spins. For the simple Ising model we define

$$\langle \sigma_0, \sigma_{\mathbf{r}} \rangle = (\Sigma \sigma_0 \sigma_{\mathbf{r}} \exp(-\beta \mathcal{H}))/Z_N^I, \quad (40)$$

which plays a role analogous to $g(r)$ in Eq. (14). The calculation of correlations is more sophisticated than that of thermodynamic quantities, and a number of important exact results were derived by Kaufman and Onsager in 1949. Of particular significance was their conclusion that at T_c the correlations decay with distance as $r^{-1/4}$. In his original paper Onsager had calculated the mean range over which correlation persists for $T > T_c$ as $\sim (T - T_c)^{-1}$. For the classical theory [Eq. (15)] the corresponding behavior was r^{-1} and $(T - T_c)^{-1/2}$.

5. *Long-range order and spontaneous magnetization.* It can be readily shown that for the simple Ising model the long-range order and spontaneous magnetization are effectively identical. Onsager in 1947 and Yang in 1951 concluded independently that its critical behavior is $\sim (T_c - T)^{1/8}$, whereas classical theory gave $(T_c - T)^{1/2}$ [Eq. (20)].

2.5 Other Exact Results

For some time after 1944 hopes ran high that a solution of the three-dimensional Ising model might soon be forthcoming. But as a variety of different methods were advanced to rederive the Onsager solution, none of which could be extended to three dimensions, the realization spread that the three-dimensional model is a problem of exceptional difficulty.

However, solutions were derived for a variety of other two-dimensional lattices, e.g., the triangular and honeycomb. The analytic character of the solution was individual to each lattice. But the critical behavior followed the same pattern as Onsager's solution and indicated that properties like the coordination number of the lattice have little relevance to critical behavior.

In 1952 Yang and Lee proved that the function $\Lambda_N(y,z)$ in Eq. (33) is analytic for all values of y except for $y=1$ ($H=0$) and $T<T_c$. This important result established the pattern of singularities and showed rigorously that the lattice-gas model is capable of accounting for the pattern of discontinuities in the isotherms representing liquid–gas equilibrium that were described in Sec. 1.

It was therefore of great importance to look for alternative methods that could provide reliable information on three-dimensional models and on alternative forms of interaction. The most useful source of such information in the initial period following Onsager was the development of series expansions at high and low temperatures. Further important information was extracted from Onsager's solution, and other exact solutions were discovered corresponding to unrealistic interactions but nevertheless providing insight into the relationship between critical behavior and intermolecular forces. The cluster integral theory of fluids, which had been one of the triumphant theoretical developments of the late 1930s, had seemingly come up against a barrier in its attempt to explain critical behavior. Onsager's work aroused interest in the far more primitive lattice-gas model, which is isomorphic with the Ising model. And finally the exact results for the two-dimensional Ising model and the new theoretical conjectures that began to emerge for other models provided a challenge to experimentalists to look for systems in nature that might be expected to conform to the theoretical models, and to provide accurate measurements of their critical behavior.

3. THREE-DIMENSIONAL MODELS

3.1 The Challenge of Onsager's Solution

Onsager's solution demonstrated clearly that the classical treatment was incorrect for the two-dimensional Ising model, and there was no valid reason for thinking that it would be correct for a three-dimensional model. But the thermodynamic functions arising from Onsager's solution differed radically from experimental results. The spontaneous magnetization, which comes to zero with an eighth power law, was much steeper than anything that had been observed in magnetic materials. The specific heat has a large tail that did not accord with measurements of order–disorder and λ-point transitions, and the logarithmic singularity symmetric on both sides of T_c seemed strangely unreal. Was the disagreement with experiment due to the dimensionality of the model, the neglect of forces other than nearest neighbor, or perhaps the inadequacy of the Ising interaction?

3.2 More General Models

In 1928 Heisenberg suggested that the origin of the large interactions giving rise to ferromagnetism lay in the quantum-mechanical exchange forces, which can be represented by a vector coupling of spins,

$$J\mathbf{s}_i\cdot\mathbf{s}_i=J(s_{xi}s_{xj}+s_{yi}s_{yj}+s_{zi}s_{zi}). \tag{41}$$

Here s_x,s_y,s_z are quantum-mechanical noncommuting operators representing the components of spin. Because of the noncommutation of the spin operators the quantum-mechanical Heisenberg model presents special difficulties in calculation. If we allow s to become infinite, we derive the *classical Heisenberg model* with an interaction

$$J\boldsymbol{\sigma}_i\cdot\boldsymbol{\sigma}_j, \tag{42}$$

where $\boldsymbol{\sigma}_i\boldsymbol{\sigma}_j$ are unit vectors that take on all possible orientations.

The Heisenberg model has spherical spin symmetry. A generalized Ising model of spins s has been defined in Eq. (25) by the interaction of

$$Js_{zi}s_{zj}, \tag{43}$$

where each s_z can take on $2s+1$ values. The Ising model has anisotropic spin interaction.

It is possible to combine the above interactions into a more general Heisenberg–Ising interaction

$$J_1 s_{xi}s_{xj}+J_2 s_{yi}s_{yj}+J_3 s_{zi}s_{zj}. \tag{44}$$

The particular case $J_3=0$, $J_1=J_2$ is called the x-y model; it has two-dimensional circular spin symmetry. When $s\to\infty$ it leads to a classical Heisenberg model with two-dimensional spins.

We shall see in Sec. 6 that the spin dimension plays an important role in modern theory and is usually denoted by n. Confining attention to isotropic spin interactions, the Ising model corresponds to $n=1$, the x-y model to $n=2$, and the Heisenberg model to $n=3$.

3.3 Series Expansions

During the two decades that followed the publication of Onsager's solution, the critical properties of model systems had to be established on an individual basis — each model and each property entailed a separate calculation. The most useful tool turned out to be the generation of lengthy perturbation series expansions at high and low temperatures.

The idea of generating lengthy series expansions and correlating the coefficients with critical behavior was first advanced by Domb (1969). Onsager's solution applies only in zero field, and the first target was to explore the properties of the model in a nonzero field.

The method was developed with the help of Rushbrooke and Wakefield at Oxford and Sykes, Fisher, and other members of the Theoretical Physics Department at King's College, London, and applied to a number of problems. A few general remarks are necessary to clarify the basis of the method. In normal statistical work it is hazardous to attempt to extrapolate asymptotic behavior from a finite number of terms. However, in the case of a ferromagnet one could anticipate on physical grounds a pattern of expected critical behavior, and use the coefficients of the series to obtain the best fit to parameters such as critical temperature and critical exponents in the assumed asymptotic form for the coefficients. Moreover, the methods of analysis and fitting could be tested on Onsager's exact solution to see how well they worked and to get an idea of the error to be expected. The method proved capable of furnishing important reliable information, and as numerical techniques improved, its use became widespread.

The actual generation of terms required a close familiarity with graph theory, and the programming of sophisticated graphical enumeration problems on computers. There are basically three different types of series expansion.

1. *High-temperature expansions* are derived by expanding the partition function in powers of β:

$$\langle\exp(-\beta\mathcal{H})\rangle=1-\beta\langle\mathcal{H}\rangle+\frac{\beta^2}{2!}\langle\mathcal{H}\rangle-\cdots+\frac{(-\beta)^n}{n!}\langle\mathcal{H}^r\rangle-\ldots, \tag{45}$$

where $\langle\ \rangle$ denotes the classical average or quantum-mechanical trace.

Those familiar with statistics will recognize Eq. (45) as the analog of a *moment-generating function* and will doubtless remember that it is more convenient to take logarithms and evaluate the *cumulant generating function*

$$\ln\langle e^{-\beta\mathcal{H}}\rangle=-\beta\langle\mathcal{H}\rangle+\frac{\beta^2}{2!}(\langle\mathcal{H}^2\rangle-\langle\mathcal{H}\rangle^2)-\frac{\beta^3}{3}(\langle\mathcal{H}^3\rangle-3\langle\mathcal{H}\rangle\langle\mathcal{H}^2\rangle+2\langle\mathcal{H}\rangle^3)+\ldots. \tag{46}$$

Such expansions were developed for all of the models described in the previous section (Sec. 3.2). For the Ising model of spin $\frac{1}{2}$ simplifications arise, and the expansions could be taken to considerable length. The terms of the expansion for the free energy in zero field, and the magnetic susceptibility and its field derivatives, are usually positive in sign. These expansions were the most useful from which to obtain information about critical behavior.

2. *Density expansions* use the variable y [$=\exp(-2\beta\mu_0 H)$, Eq. (30)] or its equivalent, and since one expects asymptotically that

$$\Lambda_N(y,z) \sim [\Lambda(y,z)]^N, \quad (47)$$

they give rise to an expansion of the form

$$\ln \Lambda(y,z) = yg_1(z) + y^2 g_2(z) + \cdots + y^r g_r(z) + \cdots . \quad (48)$$

For a lattice-gas model these expansions are equivalent to the well-known Mayer cluster integral expansions that were developed in the late 1930s.

In principle, such expansions could be determined for all the models of the previous section, but calculations were practicable only for the Ising model. The expansions start from a state of complete alignment of the spins in a large magnetic field, and terms of Eq. (48) correspond to groups of 1,2,3,... overturned spins.

3. *Low–temperature expansions* have proved practicable only for the Ising model. They use z [$=\exp(-\beta J)$, Eq. (30)] as expansion variable, the coefficients being functions of y. They are simply a rearrangement of the density expansions (48).

As a typical example of a high-temperature series we reproduce the expansion of the initial magnetic susceptibility χ_0 for the simple Ising model on the fcc lattice:

$$\chi_0/\beta\mu^2 = 1 + 12w + 132w^2 + 1404w^3 + 14\,652w^4 + 151\,116w^5 + 1\,546\,322w^6 + 15\,734\,460w^7 + 59\,425\,580w^8 + \ldots \quad (w = \tanh \beta J). \quad (49)$$

Before discussing the analysis of the terms we make a few observations on power series. Consider a function $f(z)$ defined by

$$f(z) = \sum_{n=0}^{\infty} a_n z^n. \quad (50)$$

Then, if

$$\lim_{n\to\infty} |a_n|^{1/n} \quad (51)$$

exists and is equal to $1/z_c$, the series converges for $|z| < z_c$. We can then write

$$|a_n| \sim \phi(n)/z_c^n, \quad (52)$$

where

$$\lim[\phi(n)]^{1/n} = 1. \quad (53)$$

There is always a singularity on the circle $|z| = z_c$. If all the a_n are consistent in sign, then the dominant singularity lies on the positive real axis. Replacing z by $-z$, we see that if the a_n alternate regularly the dominant singularity lies on the negative real axis. More irregular alternations indicate dominant singularities in the complex plane. If a_n is real, these must occur in complex pairs $(1/z_c)\exp(\pm i\sigma)$. For a single pair we should expect

$$a_n \sim [\phi(n)/z_c^n] \cos n\sigma.$$

If σ is a simple fraction, this gives rise to cyclic behavior; otherwise it is more random.

If the a_n are all known *exactly*, we can (in principle) continue the function analytically across the whole plane. Asymptotic values of a_n determine the behavior near to the dominant singularity. A dominant singularity on the positive real axis corresponds to a positive temperature and can reasonably be identified with the Curie point. Hence we see that series of terms consistent in sign are particularly useful since a numerical analysis of the a_n provides direct information about critical behavior.

In view of Eq. (53) a simple assumption for the asymptotic behavior of $\phi(n)$ is

$$\phi(n) \sim An^g, \quad (54)$$

and to test whether this assumption is in accord with the numerical data, it is convenient to plot a_n/a_{n-1} for series (49) as a function of $1/n$. If assumption (54) is valid, then

$$a_n/a_{n-1} \sim (1/w_c)(1+g/n), \quad (55)$$

and hence a straight line should be obtained whose intersection with $1/n=0$ determines w_c and whose slope determines g. Such *ratio plots* are reproduced in Fig. 7 for the susceptibility and specific heat, and it will be seen that they approach an asymptotic linear value quite rapidly. Both lead to the same value of $1/w_c$, as one might expect on physical grounds.

In contrast to high-temperature series, low-temperature series for the simple Ising model are inconsistent in sign and irregular in behavior for many common lattices. For example, the spontaneous magnetization m_0 for the fcc lattice can be put in the form

$$m_0/\mu_0 = 1 - 2u^6 - u^{11}(24 - 26u + 48u^4 + 252u^5$$

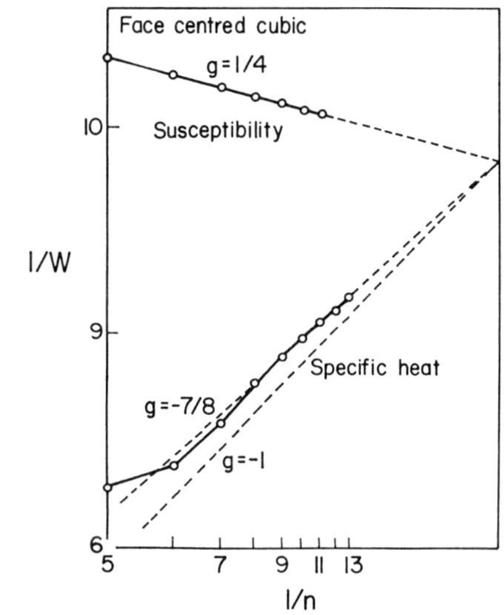

FIG. 7. An example of the use of series expansions to determine critical behavior. The susceptibility and specific heat are expressed as high-temperature expansions $\Sigma_n a_n w^n$ where $w=\tanh J/kT$. The ratios a_n/a_{n-1} of successive terms in the series expansion are plotted as a function of $1/n$, and from this an estimate can be obtained of their asymptotic behavior for large n. The ratios approach linearity, the extrapolated lines meeting the axis $1/n=0$ at $w=w_c$, the critical value, and the slopes give the critical exponents. If $a_n \sim n^{g-1}$, the corresponding thermodynamic quantity has a singularity of the form $(1-T_c/T)^{-g}$. *From this plot the estimates* $\gamma \cong 5/4$, $\alpha \cong 1/8$ *were derived (based on work by M. F. Sykes and his collaborators in* 1967 *at King's College London).*

$$-720u^6+438u^7+192u^8+984u^9 +1008u^{10}-12\,924u^{11} +19\,536u^{12}+\cdots)\ (u=z^2), \qquad (56)$$

and it is difficult to detect any pattern in the coefficients.

A breakthrough in dealing with series of this kind occurred in 1961 when Baker introduced the Padé approximant into the field. This piece of mathematics, which had lain dormant since the end of the 19th century, enables *all* the singularities of a function defined by a power series to be studied; the unphysical singularities can be ignored, and attention can be focused on the singularity of physical interest.

Density series were of importance in attempting to find the *m-H* behavior at T_c. A power series in y is obtained from Eq. (48), the coefficients being evaluated at the critical value, z_c, of z estimated from high-temperature expansions.

3.4 Critical Exponents

Assumption (54) corresponds to a branch-point singularity at T_c. We therefore define a critical exponent (or index) θ of a thermodynamic quantity Ψ by the relation

$$\theta = \lim_{t\to 0} \frac{\ln \Psi}{\ln t}, \quad \Psi \sim At^{\theta}. \qquad (57)$$

This definition ignores logarithmic factors. In a more refined analysis one might need to modify Eq. (57) to

$$\Psi \sim At^{\theta}(\ln t)^{\phi}. \qquad (58)$$

Fortunately, in the practical application to critical phenomena in two and three dimensions, Eq. (57) is usually sufficient, although in four and more dimensions Eq. (58) is often needed. Also, when $\theta=0$ it is usual to specify whether the behavior is discontinuous (e.g., classical specific heat), or logarithmic (e.g., specific heat in the Onsager solution).

The notation for magnetic critical exponents developed empirically and the standardization provided by Fisher in 1965 at the Washington Conference on "Phenomena in the Neighbourhood of Critical Points" received wide acceptance. Table 1 defines the commonly used exponents and lists their classical values.

The Ising model served as a *trailblazer* in unravelling the general pattern of behavior of critical exponents. For the simple two-dimensional model α, α', β, ν, and η were known from the exact calculations quoted in Sec. 2. The remaining exponents were estimated from series expansions (e.g., $\gamma=7/4$, $\delta=15$) and were later shown to be exact.

In general, high-temperature series expansions relating to the initial susceptibility χ_0 were the most useful, and as the exponent γ was evaluated for different lattices and different models, a number of general conjectures were advanced as follows:

1. Critical exponents for a given model depend on dimension and not on lattice structure.
2. Critical exponents for the Ising model are independent of the spin s.
3. High-temperature critical exponents are

Table 1. Magnetic critical exponents ($t=T/T_c-1$, $t'=-t$).

Exponent	Quantity	Definition	Curve	Classical Value
$T>T_c$				
α	Specific heat	$C_m \sim t^{-\alpha}$	$m=0$	0 (discontinuous)
γ	Initial susceptibility	$\chi_0 \sim t^{-\gamma}$	$m=0$	1
ν	Coherence length $\sim \delta$	$\kappa \sim t^{\nu}$	$m=0$	$\frac{1}{2}$
2Δ	Derivatives of susceptibility	$d^2\chi_0/dH^2 \sim t^{-\gamma-2\Delta}$	$m=0$	3
$T<T_c$				
α'	Specific heat	$C \sim t'^{-\alpha'}$	$m=0$	0 (discontinuous)
β	Spontaneous magnetization	$m_0 \sim t'^{\beta}$	$H \to 0+$	$\frac{1}{2}$
γ'	Initial susceptibility	$\chi_0 \sim t'^{-\gamma'}$	$m=m_0$	1
ν'	Coherence length	$\kappa \sim t'^{\nu'}$	$m=m_0$	$\frac{1}{2}$
Δ'	Derivatives of susceptibility	$d\chi_0/dH \sim t'^{-\gamma'-\Delta'}$	$m=m_0$	$\frac{3}{2}$
$T=T_c$				
δ	Field vs magnetization	$H \sim m^{\delta}$		3
η	Pair correlation function	$\Gamma(\Gamma,0) \sim r^{d-2+\eta}$	$m=0$	0

ordered in a one-to-one manner with the symmetry of the magnetization in the ground state.

These conjectures represented important steps towards the hypothesis of universality (Sec. 4.3). The significance of the last conjecture can be illustrated by reference to the anisotropic Heisenberg model (44) with the following interaction:

$$J\sigma_{zi}\sigma_{zj}+J'(\sigma_{xi}\sigma_{xj}+\sigma_{yi}\sigma_{xy}). \qquad (59)$$

If $J'=0$, the model reduces to the simple Ising model in three dimensions for which γ had been estimated at 1.25, whereas if J' grows to J, the model becomes classical Heisenberg with corresponding estimate 1.38. What happens when $0<J'<J$? Conjecture 3 suggests that the change takes place discontinuously at $J'=J$, and that $\gamma \cong 1.25$ when $0<J'<J$. Likewise if $J=0$ the model is classical x-y with $\gamma \cong 1.32$; conjecture 3 suggests that if $J'>J$, γ retains the value 1.32.

Critical exponents for the three-dimensional Ising model lay between those for the two-dimensional model and those of classical theory. Thus, as the dimensionality increased, classical values were approached. However, increase in spin dimension generated a move away from the classical results. In keeping with this it was shown rigorously that the two-dimensional classical Heisenberg model has no ordered state except at $T=0$ and is in this respect similar to the *one-dimensional* Ising model.

3.5 Spherical and Gaussian Models

In 1952 Berlin and Kac discussed two new models for which they were able to calculate an exact solution in three dimensions in a nonzero field. The Gaussian model consists of independent spins with a nearest-neighbor Ising interaction, the magnitude of each spin being governed by a probability distribution

$$(2\pi)^{-1/2}\exp(-\sigma^2/2)d\sigma. \qquad (60)$$

The spherical model allows the spins to take any value subject to the overall restriction

$$\sigma_1^2+\sigma_1^2+\cdots+\sigma_N^2=N. \qquad (61)$$

Neither model is realized by any known physical system, and the solutions exhibit certain marked unphysical features.

For many years the models were regarded as of mathematical interest only. The discovery by Stanley in 1968 that the spherical model can be regarded as the limit of the classical vector spin model in which the *spin* dimension becomes infinite secured an important place for the model in the general pattern of critical behavior. The Gaussian model found its place as a first step in Wilson's application of the renormalization group (Sec. 5.4).

Like the Heisenberg model, the spherical model does not manifest critical behavior in two dimensions. In three dimensions Kac and

Berlin found that $\gamma=2$, and because of Stanley's result, the variation of γ as a function of spin dimension could be understood, starting with 1.25 for $n=1$, 1.32 for $n=2$, and 1.38 for $n=3$, and reaching 2 for $n=\infty$.

An important contribution of the spherical model to our understanding of critical behavior arose from the calculation by Joyce in 1966 of the properties of the model when the dimension $d\geqslant4$. Joyce found that with the exception of logarithmic factors, all critical exponents attain their classical values when $d=4$ and retain them for higher dimensions. Joyce also calculated the properties of the model for long-range forces of the form $J(r)\sim 1/r^{d+\sigma}$ and found that for a sufficiently small σ, exponents attain classical values (even when $d=1$). In such a case a large number of neighbors effectively contribute to the interaction, so that the mean-field approximation discussed at the end of Sec. 2.1 gives the correct answer.

We shall see that these results are of great help in constructing an overall pattern of critical behavior.

3.6 Systems with Weak Long-Range Forces

In 1959 Kac introduced a one-dimensional continuum model with long-range forces that seemed to offer the possibility of exact treatment. The molecules interact with a hard-core repulsion of length b, and an exponential attraction $\lambda e^{-\lambda r}$. For this model he showed that the partition function can be derived as the maximum eigenvalue of a linear integral equation with a positive-definite Hilbert–Schmidt kernel.

The calculations were pursued by Kac, Uhlenbeck, and Hemmer in 1963. They first showed that for any finite value of λ the partition function is continuous, and there is no phase transition. However, in the limit of $\lambda\to0$ there is a phase transition, and it can be evaluated in closed form. It gives rise, in fact, to the van der Waals equation [Eq. (4)] with the Maxwell construction automatically included. At the end of their paper, Kac, Uhlenbeck, and Hemmer discussed the three-dimensional model and advanced arguments to show that in this case also the attractive and repulsive parts of the potential make independent contributions to the free energy. In three dimensions the hard-sphere partition function is not known in closed form, but Monte Carlo estimates are available.

Calculations of a similar kind for the one-dimensional Ising model led to a Weiss type of solution [Eq. (7)]. In fact, it seemed that a mean-field solution will always be obtained for a model with a weak long-range attractive force of the form

$$J(r)=\lim_{\lambda\to0}\lambda^{d}e^{-\lambda r}. \tag{62}$$

Classical theory was thus partially reinstated—it is a valid theory for long-range forces, but in practice intermolecular forces in nature are usually short ranged.

The work of Joyce on the spherical model indicated that a mean-field solution should result for models with forces of the form $J(r)\sim1/r^{d+\sigma}$ for a sufficiently small σ.

3.7 Experimental Results

The experimental investigation of the behavior of a substance near a critical point is a matter of considerable difficulty. Very accurate temperature measurements are needed since small changes of temperature are accompanied by large differences in equilibrium values. Substances must be very pure; a small concentration of impurity can exercise a significant influence. As T_c is approached the time needed to reach equilibrium becomes large, so that a long time scale is needed for the performance of experiments. For fluids gravity must be taken into account. The theoretical values of several of the thermodynamic properties investigated become infinite at T_c. But such an infinity applies only to an unbounded sample; in practice the finite size of the sample becomes important since it provides a bound on the values that can be obtained. In solids this involves a consideration of grain boundaries.

The first suggestion that experimental results for real fluids deviated from the van der Waals picture came from Verschaeffelt at the end of the 19th century. Unfortunately his work was ignored by his contemporaries. In 1945 Guggenheim undertook a critical analysis of experimental data on the coexistence curves of a number of gases. According to van der Waals theory these gases should obey a law of corresponding states, i.e., if reduced units T/T_c and ρ/ρ_c are used (where ρ_c and T_c are the density and temperature, respectively, at the critical point), the coexistence curves for the different gases should fall on a single universal curve. Guggenheim found that the data provided good support for such a law of corresponding states (Fig. 8), but the form of the curve was cubic,

$$\Delta\rho \sim (-t)^{1/3},$$

rather than quadratic as required by van der Waals theory.

In 1954 Habgood and Schneider published precise measurements of the isotherms of xenon near the critical point. They found that the critical isotherm is much flatter than the cubic curve predicted by van der Waals (Fig. 9), and they suggested that the third and fourth derivatives of P with respect to ρ are zero at the critical point.

The theoretical developments described in Sec. 3.4 stimulated a new wave of precise experimental measurements, making use of new techniques and new magnetic materials to attain a much greater accuracy than had been available previously. We shall quote a few representative examples. In 1957 Fairbank, Buckingham, and Kellers published the result of a new investigation into the nature of the λ point of liquid ^{4}He. Improved low-temperature techniques enabled them to get very close to the λ point, and domains and

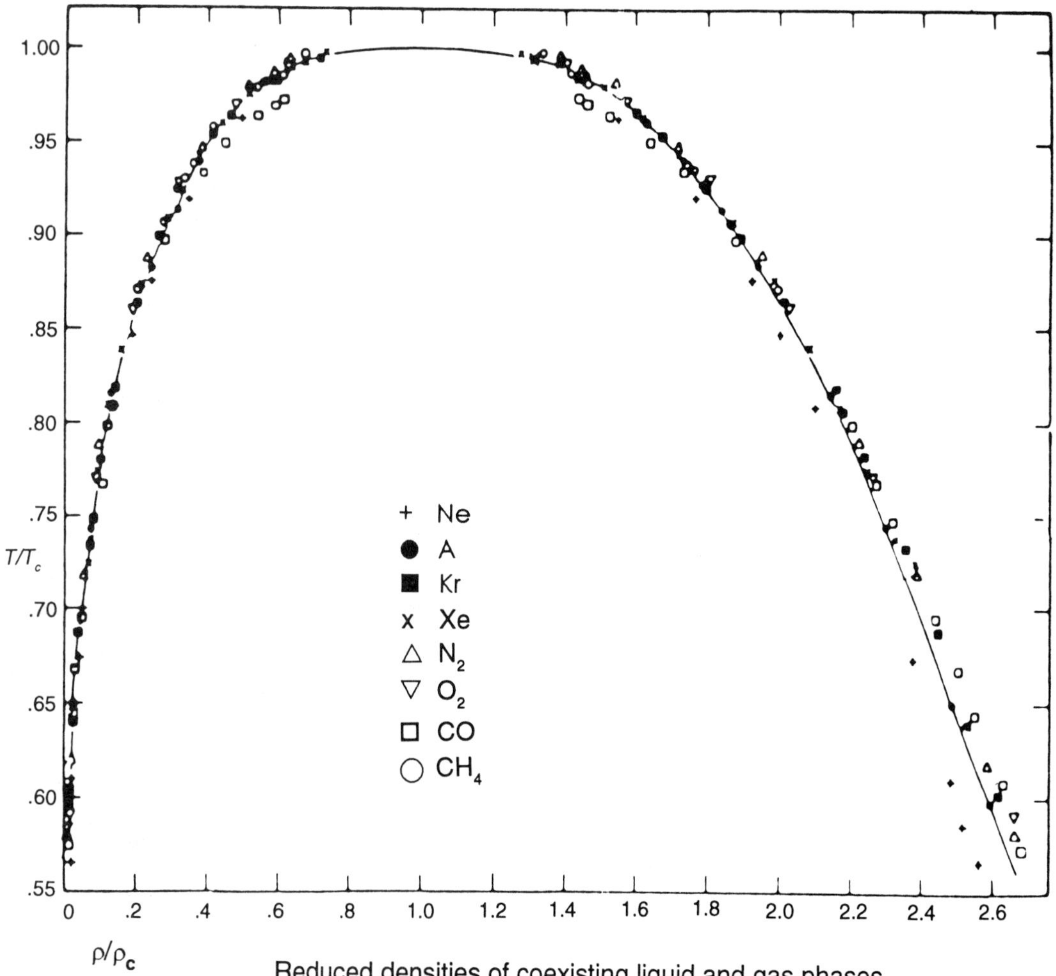

FIG. 8. Reduced densities of coexisting liquid and gas phases (T/T_c vs ρ/ρ_c). The data fit a law of corresponding states, but the resulting curve is cubic rather than quadratic as given by van der Waals's equation (after Guggenheim, 1945).

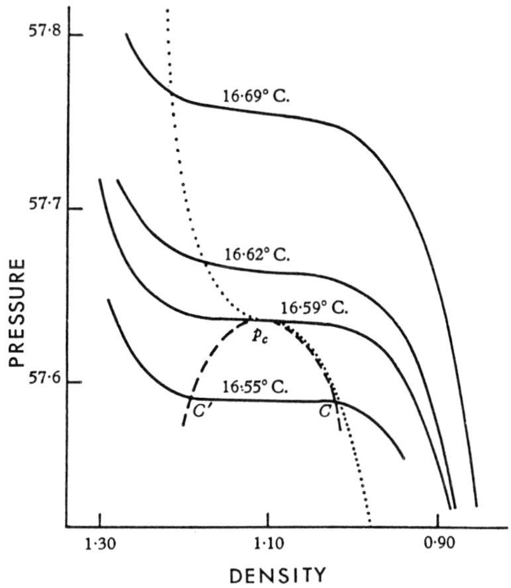

FIG. 9. Isotherms of xenon near the critical point. The dashed line marks the region of coexistent phases. The dotted line is the critical isotherm according to van der Waals's equation to be contrasted with the measured 16.59 °C isotherm (after Habgood and Schneider, 1954).

grain boundaries that limit the growth of specific heats of crystals are absent in liquid helium. They concluded that the specific heat is logarithmically infinite on both sides of the λ point (Fig. 10) and is in many respects similar to the Onsager specific heat depicted in Fig. 5. Of course they did not suggest that there was any basic reason for the similarity between a three-dimensional quantum fluid and a two-dimensional Ising ferromagnet, but at least they demonstrated that logarithmically infinite specific heats are a practical reality.

Conventional ferromagnets like iron, nickel, and cobalt have high Curie temperatures, and accurate measurements of critical behavior are difficult. However, new ferromagnets were discovered, such as europium sulfide ($T_c \sim 16.50K$), which are much more amenable to accurate measurements, and nuclear magnetic resonance can be used to measure the fraction of spins oriented in a particular direction at a given temperature and hence to deduce the spontaneous magnetization. Figure 11 is an example of such a measurement by Heller and Benedek in 1965 from which they deduced that the cube of the spontaneous magnetization is very nearly linear.

It is difficult to find a magnet that conforms satisfactorily to the Ising interaction. However, we have seen in Sec. 2.1 that an order–disorder transition in a binary alloy can also provide a realization of an Ising model. β-brass (CuZn) was chosen by Als-Nielsen and Dietrich for neutron-scattering experiments to test the theoretical predictions for the three-dimensional Ising model, and they estimated for the critical exponents $\beta = 0.30 \pm 0.005$ and $\gamma = 1.25 \pm 0.02$. These were in good agreement with the predictions of series expansions.

The new experimental results seemed to fit well to the theoretical calculations, even though the theoretical models were simple and crude,

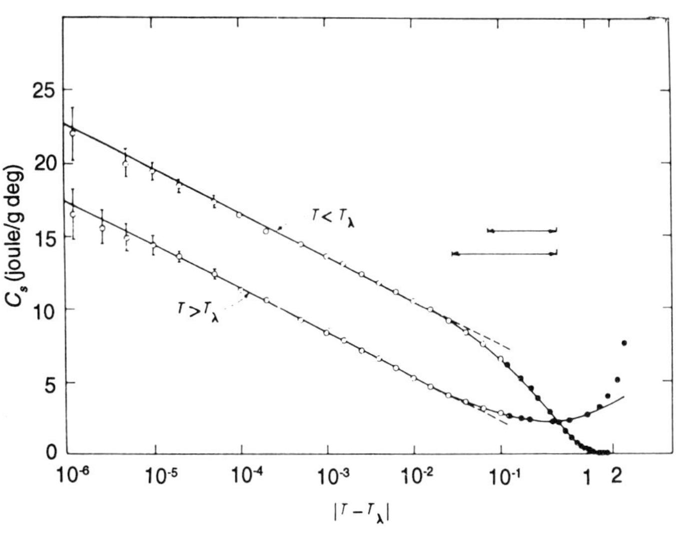

FIG. 10. Specific heat of liquid ^{4}He near the λ point. Accurate measurements by Fairbank, Buckingham, and Kellers in 1957 indicated a logarithmic curve of the type derived by Onsager for the Ising model.

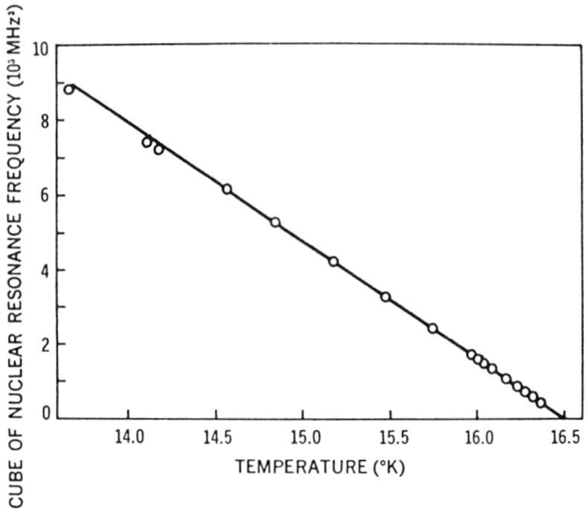

FIG. 11. Temperature dependence of the cube of the nuclear frequency of ^{153}Eu in EuS between 13.60 and 16.33 K. In this range the law $m_0 \sim Dt'^{1/3}$ is quite accurately obeyed (after Heller and Benedek, 1965).

and did not adequately map the true physical interactions. It seemed as if critical behavior was insensitive to the details of the interaction mechanism.

3.8 Conclusion

In this section we have described the chain of theoretical and experimental work set in motion by Onsager's solution. Agreement between theory and experiment was gratifying, but there was no coherent framework to replace the classical theory. In the next section we shall describe how such a framework was constructed.

4. SCALING AND UNIVERSALITY

4.1 Reconciliation

The first international conference devoted specifically to critical phenomenon took place in April 1965 in Washington at the National Bureau of Standards. It could reasonably be termed the founding conference of critical phenomena, since for the first time all the different strands of the subject were woven into a coherent fabric, and the major obstacles to progress were clearly delineated.

The keynote address was given by G. E. Uhlenbeck, and it is particularly fascinating now to quote the actual words of one of his concluding paragraphs:

"If there is such an universal, but non-classical behaviour, then there must be a universal explanation which means that it should be largely independent of the nature of the forces. The only corner where this can come from is I think the fact that the forces are not long range. The Onsager solution gives I think a strong hint. It may well be so that away from the critical points the classical theories give a good enough description but that they fail close to the critical point where the substance remembers so to say Onsager. I think that to show something like this is the central theoretical problem. One can call it the reconciliation of Onsager with van der Waals."

Uhlenbeck was looking for a new type of universality for short-range forces that would replace the universality of the classical theories, which are valid only for very-long-range forces.

The first requirement in effecting the reconciliation was a coherent description of the "nonclassical universality" analogous to the van der Waals description of classical behavior. A preliminary attempt in this direction was the "droplet model" introduced by Essam and Fisher in 1963 and elaborated by Fisher a few years later. The model attempted to find an empirical approximation to $g_r(z)$ in Eq. (48) by analogy with droplet formation in a condensing gas. The resulting partition function depended on only two parameters and led to a relation between critical exponents,

$$\alpha' + 2\beta + \gamma' = 2, \tag{63}$$

which was satisfied by the exact exponents for the two-dimensional Ising model. Rushbrooke noted that from thermodynamics alone one could derive an inequality

$$\alpha' + 2\beta + \gamma' \leqslant 2. \tag{64}$$

Rushbrooke's observation triggered off thermodynamic investigations that led to other inequalities of a similar kind. However, the heuristic droplet model had a number of unphysical features, and could not be regarded as providing a representation of the equation of state over the total range of the variables.

A satisfactory form of a nonclassical equation of state valid in the critical region emerged within a few months of the above conference.

It was suggested independently by three different groups: Widom (1965) in the U.S., who searched for a generalization of van der Waals's equation that could accommodate nonclassical exponents; Domb and Hunter (1965) in the U.K., as a result of an analysis of the behavior of series expansions of higher derivatives with respect to magnetic field at the critical point; and Patashinskii and Pokrovskii (1966) in the U.S.S.R., who considered the behavior of multiple correlations near the critical point.

Widom used fluid language in developing his ideas. Because of the exact symmetry $H \to -H$, $m \to -m$ in magnetic systems we shall find it convenient to reproduce the results in magnetic language. The transition to fluids is achieved by replacing H by $p-p_c$ and m by $\rho - \rho_c$.

4.2 Equation of State in the Critical Region: Scaling

The critical magnetic equation of state was derived from high-temperature series expansions of the form Eq. (45) and involved the plausible assumption (which was afterwards checked numerically) that the critical behavior of the rth field derivative of the magnetic susceptibility is of the form

$$\frac{d^r \chi_0}{dH^r} \sim t^{-\gamma - r\Delta}, \qquad (65)$$

where Δ was termed the gap exponent. This led to an equation of state of the form

$$m = t^{\beta} U(Ht^{-\Delta}). \qquad (66)$$

However, since one wishes to use the equation for positive and negative values of t, the form (66) is not very convenient, and Griffiths reformulated the equation so as to involve t only linearly,

$$H = m^{\delta} h(tm^{-1/\beta}). \qquad (67)$$

Equation (67) is correct only for positive H and m. By symmetry H and m can both be reversed without making any physical difference to the system. Therefore the general equation, valid for all m, H is

$$H = m|m|^{\delta-1} h(t|m|^{-1/\beta}). \qquad (68)$$

It is convenient to use Eq. (67) with the symmetry assumption latently understood.

There are two characteristic features of Eq. (67). First, all critical exponents depend on only two parameters, β and δ, and this is the basis for relations like Eq. (63) that are indeed satisfied exactly. The critical exponents of other thermodynamic quantities can be readily derived, and are reproduced in Table 2.

Second, critical data satisfy a *scaling relation*, i.e., if $Hm^{-\delta}$ is plotted against $tm^{-1/\beta}$, the two-dimensional data will fall on a single curve $h(x)$. As a result of the Yang–Lee theorem quoted in Sec. 2.5, $h(x)$ is analytic and the variable x ranges from $-x_0$, the zero of $h(x)$ (which corresponds to a spontaneous magnetization), to $x=\infty$. For a mean-field theory $h(x)$ is linear; as the dimension decreases and the spin dimension increases, $h(x)$ becomes more curved. $h(x)$ is termed the *scaling function.*

Patashinskii and Pokrovskii related susceptibility derivatives to correlation functions and were therefore able to relate the correlation exponent ν to Δ:

$$2\Delta = \gamma + d\nu. \qquad (69)$$

Such a relation was usually termed a *length scaling* relation as opposed to *thermodynamic scaling* relations of type (63).

These scaling relations were satisfied by the known exact exponents for the two-dimensional Ising model and for the spherical model (Table 2). For the three-dimensional Ising model, numerical estimates supported thermodynamic scaling, but there were small but persistent discrepancies in the length scaling relations, which will be discussed subsequently (Sec. 5.5). They were subject to experimental test for a large number of systems of fluids and magnets, and the results were very encouraging.

Similarly, the hypothesis of a scaling function $h(x)$ was tested for fluids and magnets, and a typical result using data for ^{4}He is reproduced in Fig. 12. It will be seen that the data fit well to a single curve.

4.3 Universality

Equation (67) says nothing about possible values of β, δ, and $h(x)$. We have already mentioned empirical information accumulating about regularities in the pattern of critical exponents. These were collected together in the *hypotheses of universality* put forward by

Table 2. Scaling relations between critical exponents.

Thermodynamic quantity	Exponent	Simple Ising model in two-dimensions	Mean field	Spherical model in three dimensions	Spherical model—long-range forces[a]
Spontaneous magnetization	β	$\frac{1}{8}$	$\frac{1}{2}$	$\frac{1}{2}$	$\frac{1}{2}$
Magnetization vs field at T_c	δ	15	3	5	$(d+\sigma)/(d-\sigma)$
High-temperature susceptibility	$\gamma=\beta(\delta-1)$	$\frac{7}{4}$	1	2	$\sigma/(d-\sigma)$
High-temperature gap	$\Delta=\beta\delta$	$\frac{15}{8}$	$\frac{3}{2}$	$\frac{5}{2}$	$(d+\sigma)/2(d-\sigma)$
High-temperature specific heat	$2-\beta(\delta+1)$	0	0	-1	$(d-2\sigma)/(d-\sigma)$
Low-temperature susceptibility	$\gamma'=\gamma$	$\frac{7}{4}$	1	...	...
Low-temperature gap	$\Delta'=\Delta$	$\frac{15}{8}$	$\frac{3}{2}$	...	...
Low-temperature specific heat	$\alpha'=\alpha$	0	0	...	...
		Length scaling			
High-temperature coherence length	$\nu=\beta(\delta+1)/d$	1	$\frac{1}{2}$	1	$1/(d-\sigma)$
Low-temperature coherence length	$\nu'=\nu$	1	$\frac{1}{2}$	...	...
Spin pair correlation at T_c	$\eta=d(\delta-1)/\delta+1$	$\frac{1}{4}$	0	0	$2-\sigma$

[a]Force $J(r)\sim r^{-(d+\sigma)}$ in d dimensions ($1<d/\sigma<2$, $0<\sigma<\min\{2,d\}$).

Kadanoff in 1970 at a summer school on critical phenomena; an analogous *smoothness postulate* was advanced independently by Griffiths. This nonclassical universality is more sophisticated than had been envisaged by Uhlenbeck; different classes are defined by the space dimension d and the spin dimension n, and within a given class critical behavior is universal. If a third parameter σ is introduced to take account of the range of the intermolecular forces, both classical and nonclassical behavior can be taken into account in a wider pattern of universality. Once d, n, and σ have been specified, the exponents δ, β, and the function $h(x)$ are determined, and the behavior is smooth and universal within the class. Changes in critical exponents with accompanying discontinuities occur in the *crossover* between one universality class and another (e.g., in the transition from a two-dimensional to a three-dimensional system by the introduction of a new interaction), and we shall now discuss this crossover behavior in more detail.

4.4 Crossover Behavior

It was noted by Hankey and Stanley in 1972 that scaling laws are closely related to homogeneous functions. A homogeneous function of degree p in two variables u,v is defined by the relation

$$g(\lambda u,\lambda v)=\lambda^p g(u,v) \qquad (70)$$

for all λ. But if we use Eq. (70) with the value $\lambda=1/u$, we deduce that

$$g(u,v)=u^P g(1,v/u)=u^P G(v/u). \qquad (71)$$

Hence a homogeneous function of two variables scales to a function of a single variable.

To take account of the types of exponent arising in relations like (66) or (67), Hankey and Stanley introduced the concept of a *generalized homogeneous function* that satisfies the relation

$$g(\lambda^a u,\lambda^b v)=\lambda g(u,v). \qquad (72)$$

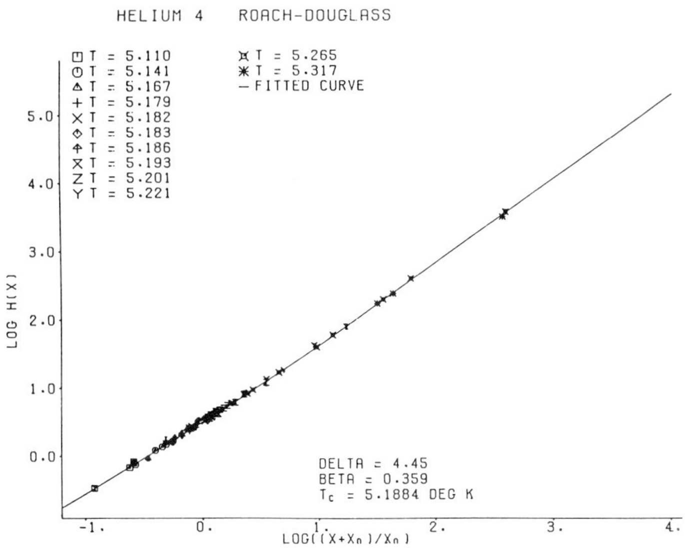

FIG. 12. Scaling relation for the fluid ^{4}He. Experimental data of Roach (1968) fitted by Vicentini-Missoni (1971).

A standard homogeneous function (70) corresponds to $a=b$. Taking λ to be $u^{-1/a}$ we find as a generalization of Eq. (71)

$$g(u,v)=u^{-1/a}g(1,v/u^{b/a})=u^{-1/a}G(v/u^{b/a}), \tag{73}$$

and this is of the same form as Eqs. (66) or (67). From (67), for example, we have

$$a=1/\delta, \quad b=-1/\beta\delta. \tag{74}$$

We have seen in Sec. 3.4 that critical exponents for a given model depend on dimension and not on lattice structure. If we consider a Hamiltonian of the form

$$\mathscr{H}=-J\sum_{\mathbf{i},\mathbf{j}}\sigma_{\mathbf{i}}\sigma_{\mathbf{j}}-gJ\sum_{\lambda,\mu}\sigma_\lambda\sigma_\mu=\mathscr{H}_0+g\mathscr{H}_1 \tag{75}$$

for a simple-cubic lattice, where **i**,**j** refer to nearest neighbors in horizontal planes, and λ,μ to nearest neighbors in the vertical direction, then for $g=0$ the behavior will be two-dimensional in character, but for $g\neq 0$ it will be three-dimensional. How does the transition or *crossover* depend on the parameter g when g is small? We shall investigate this example in detail as typical of the crossover from one universality class to another.

For the Hamiltonian (75), by analogy with Eq. (66), we postulate for the singular part of the free energy a generalized homogeneous function

$$f(g,t)=t^{2-\alpha_2}V(gt^{-\phi}), \tag{76}$$

where the subscript 2 is to remind us that we are dealing with the two-dimensional critical exponent, and ϕ is the new *crossover* exponent analogous to the gap exponent Δ in Eq. (66).

In presence of a magnetic field H, Eq. (76) would generalize to

$$f(g,H,t)=t^{2-\alpha_2}V(gt^{-\phi},Ht^{-\Delta_2}). \tag{77}$$

By differentiating Eq. (77) we conclude that thermodynamic quantities corresponding to the derivatives of the free energy with respect to H satisfy scaling relations similar to Eq. (76); for example, for the initial susceptibility we can write

$$\chi_0(g,t)\cong t^{-\gamma_2}Z(gt^{-\phi}). \tag{78}$$

However, there is an important difference between the scaling relations with respect to g and those with respect to H of Eqs. (66) and (67). For the latter the transition disappears when $H\neq 0$, whereas for the former a new transition point $T_c(g)$ appears with new critical exponents. For this to be possible the scaling function $Z(x)$ in Eq. (78) must have a singularity at some point x_c, and the singular behavior must be such as to enable the susceptibility when $g\neq 0$ to correspond to the three-dimensional Ising model:

$$\chi_0(g,\dot t)\cong A(g)\dot t^{-\gamma_3} \quad [\dot t=T/T_c(g)-1]. \tag{79}$$

From this we first deduce that the new critical point is given by $gt^{-\phi}=x_c$, or

$$t_g = T_c(g)/T_c(0) - 1 = g/x_c^{1/\phi}. \quad (80)$$

To divide the crossover region more precisely into areas of temperature where the behavior is two-dimensional in character and those where it is three-dimensional, for a given g, we require knowledge of the scaling function $Z(x)$. But the general pattern of these areas can readily be outlined. Let us fix arbitrarily a percentage deviation p from two-dimensional behavior that we regard as significant, and let x_A be the value of x for which $Z(x_A)$ deviates from $Z(0)$ by p%. Then for $0 < x < x_A$ we can reasonably consider the behavior as two dimensional, i.e.,

$$t > (g/x_A)^{1/\phi}. \quad (81)$$

Let us now determine x_B at which the deviation of $Z(x)$ from its singular behavior is less than p%. For $x_B < x < x_c$ we can reasonably consider the behavior as three dimensional, i.e., for

$$(g/x_B)^{1/\phi} > t > (g/x_c)^{1/\phi}. \quad (82)$$

4.5 Kadanoff's Treatment

The progress described at the beginning of this section depended on an empirical and pragmatic approach. In a key paper published in 1966 Kadanoff tried to provide a theoretical basis for understanding the scaling properties that had been discovered in critical behavior.

Kadanoff noted that near T_c the coherence length ξ becomes very large, and hence it is possible to find a length L large compared with the lattice spacing but small compared with ξ. He then considered replacing the interaction of individual spins by the interaction of blocks of L^d spins. One might perhaps expect that in each block the spins would be nearly all up or nearly all down, and the original Ising model of spin σ and interaction J could then be replaced by a new model with block spin $\tilde{\sigma}$ and interaction $\tilde{J}$. If the new block spin model was effectively the same as the original model, the free energies would be related by

$$f(h,t) = L^{-d} f(\tilde{h},\tilde{t}) \quad (h = \beta H). \quad (83)$$

But how would $\tilde{h}$ and $\tilde{t}$ be related to h and t? Kadanoff suggested that

$$\tilde{h} = L^x h, \quad \tilde{t} = L^y t. \quad (84)$$

It is then easy to show that all the exponents (including ν) can be expressed in terms of x and y, that relations (63) and (69) are satisfied, and that a scaling-type equation of state (66) results.

Relation (69) is satisfied by the Ising model in two dimensions, but, as we have noted, numerical calculations for the three-dimensional model showed a small but persistent discrepancy. Also, there are several features of Kadanoff's argument that do not stand up to critical examination. However, his ideas stimulated the next major theoretical advance that completed the reconciliation sought by Uhlenbeck.

5. RENORMALIZATION GROUP

5.1 Background Ideas

The renormalization group (RG) was developed in the 1950s in connection with field theory (see PARTICLES, ELEMENTARY; QUANTUM ELECTRODYNAMICS), but it had no apparent practical consequences and had not been taken very seriously. Wilson saw that the theory was capable of providing an understanding of universality and a framework for the detailed calculation of critical behavior.

The RG does not produce exact solutions of the Onsager type, and its application involves quite drastic approximations, but it endeavors to retain all the essential physical characteristics of the problem near T_c, while rejecting those features that are not significant. No reliable method has been found of assessing directly the magnitude of the error at any stage of approximation of an RG treatment, and comparison with the results of exact solutions and series expansion estimates has played, and continues to play, a key role in the successful application of the RG.

Before discussing the RG approach we shall outline (following Wilson) the nature of the problem to be tackled in trying to account for critical behavior. We shall then give a qualitative account of how the RG deals with it.

If we consider the behavior of a fluid such as water, far from the critical point there are microscopic density fluctuations on an atomic scale (wavelengths ~1 Å). If the temperature and pressure are increased towards the critical point, fluctuations become important at longer wavelengths. Sufficiently close to T_c and P_c there are fluctuations on the scale of 1000–10 000 Å that scatter ordinary light and give rise to critical opalescence, making the water look milky. However, the microscopic fluctuations have not decreased in magnitude. Close to the critical point there are fluctuations at all wavelengths from 1 Å up to the correlation length, which tends to infinity at the critical point. The essence of the problem is how to deal with an assembly with so many scales of length.

The aim of the RG approach is to reduce in a systematic way the enormous number of degrees of freedom of the system, eliminating those that are irrelevant and do not contribute to critical behavior, and focusing attention on those which are relevant. A transformation is set up in the parameter space of the Hamiltonian,

$$\mathcal{H}'=\mathrm{R}\{\mathcal{H}\}, \tag{85}$$

which preserves the dimensionality and symmetries of the system and reduces the correlation length by a factor $b>1$ and the number of degrees of freedom from N to $N'=N/b^d$. The transformation is chosen so that the partition function is preserved:

$$Z_{N'}(\mathcal{H}')=Z_N(\mathcal{H}). \tag{86}$$

A very wide choice of transformations is possible that can be characterized in *real space* or in *momentum space*. In real space the transformation usually follows the Kadanoff idea of introducing block-spin variables or eliminating some of the spins by partial summation (decimation transformation). However, the edge of the block is not long compared with the lattice spacing, but contains one or two lattice spacings; also the Kadanoff picture must be amplified, and the relation between the original spin and the block spin must be described precisely. In momentum space the aim is usually to eliminate high-momentum variables.

The transformation is iterated:

$$\mathcal{H}'=\mathrm{R}[\mathcal{H}],\ \mathcal{H}''=\mathrm{R}[\mathcal{H}'],\dots . \tag{87}$$

A brief reference to the behavior of standard elementary iterative processes of the form

$$x_{n+1}=f(x_n) \tag{88}$$

will help to clarify the next step. It can easily be shown in Eq. (88) that the x_n tend to a limit x^* (fixed point), which is a solution of the equation

$$x=f(x). \tag{89}$$

If there are several solutions of this equation some may be *stable* (any deviation leads back to the fixed point) and others *unstable* with the opposite property. It can be shown easily that the fixed point is stable if $|f'(x^*)|<1$ and unstable if $|f'(x^*)|>1$. If $f'(x^*)=1$ the fixed point is *marginal*, and we must proceed to higher derivation of $f(x)$ at x^*.

For iterative processes in s variables we can write Eq. (88) in the more general form

$$x_{(n+1)i}=f_i(x_{nj})\quad (i,j=1,2,\dots,s), \tag{90}$$

and fixed points x_i^* will be determined by solutions of the s equations

$$x_i=f_i(x_j). \tag{91}$$

The behavior at the fixed point is now governed by the eigenvalues of the matrix $|\partial f_i/\partial x_j|$ at x_i^*. If they are all less than 1, the fixed point is stable; if they are all greater than 1, it is unstable. But there are now other possibilities of partial stability; if t are >1 and $s-t$ are <1, we can describe the fixed point as t-unstable. In the theory of critical phenomena we are concerned with t-unstable fixed points with $t=1$, 2, and perhaps 3.

The characteristic of a stable fixed point is that for a certain range of variables the solution is independent of the starting point. Each stable fixed point $x_i(r)^*$ has a *range of attraction*, and for any starting value within this range the limit will be $x_i(r)^*$. It is thus assumed in Eq. (87) that convergence takes place in the parameter space of the Hamiltonians to a fixed-point Hamiltonian, satisfying

$$\mathcal{H}^*=\mathrm{R}[\mathcal{H}^*], \tag{92}$$

and that critical properties are determined by behavior near the fixed point. Each fixed point has its own range of attraction, and all Hamiltonians within this range will have the same pattern of critical behavior (universality). All such Hamiltonians lie on the *surface of criticality* in the parameter space of Hamiltonians. But different fixed points in the parameter space give rise to different universality classes.

5.2 RG Derivation of Critical Behavior

Consider the partition function

$$Z_N=\langle\exp(-\beta\mathscr{H})\rangle, \tag{93}$$

which is characterized by a number of coupling constants $K^{(1)},K^{(2)},K^{(3)},\ldots$ that depend on $T,H,\ldots$ and can be conveniently written as a vector $\mathbf{K}$. Suppose that the correlation length associated with this Hamiltonian is $\xi(K)$. We now apply a block-spin or decimation transformation R_b to obtain a new set of parameters $\mathbf{K}'$ satisfying

$$\mathbf{K}'=\mathbf{f}(\mathbf{K}), \tag{94}$$

where the function will usually be nonlinear. Since we have reduced the unit of length by b we have the relation

$$\xi(\mathbf{K}')=b^{-1}\xi(\mathbf{K}), \tag{95}$$

and because of Eq. (86) we have for the partition function per particle $Z=(Z_N)^{1/N}$

$$[Z(\mathbf{K}')]^{N'}=[Z(\mathbf{K})]^{N} \tag{96}$$

and hence for the free energy per particle

$$\psi(K')=b^{d}\psi(\mathbf{K}). \tag{97}$$

Applying the transformation successively as in Eq. (87) in the form

$$\mathbf{K}_{n+1}=\mathbf{f}(\mathbf{K}_n), \tag{98}$$

we find that

$$\xi(\mathbf{K}_{n+1})=b^{-1}\xi(\mathbf{K}_n)=\cdots b^{-n}\xi(\mathbf{K}_0). \tag{99}$$

Suppose now that $\mathbf{K}_n$ approaches a fixed point $\mathbf{K}^*$ satisfying

$$\mathbf{K}^*=f(\mathbf{K}^*). \tag{100}$$

It then follows from Eq. (99) that if we start at $\mathbf{K}_0$, then

$$\xi(\mathbf{K}_0)=b^{n}\xi(\mathbf{K}_{n+1}), \tag{101}$$

so that if $\xi(K^*)$ is not zero the correlation of the original system must be infinite, i.e., $\mathbf{K}_0$ must correspond to the critical point.

To obtain critical behavior near $\mathbf{K}^*$ we write

$$\mathbf{K}_n=\mathbf{K}^*+\mathbf{k}_n, \tag{102}$$

where $\mathbf{k}_n$ is small. Applying this to Eq. (98) we can develop an expansion and retain only linear terms, obtaining

$$\mathbf{K}_{n+1}=\mathsf{L}\mathbf{k}_n, \tag{103}$$

where relations (103) are linear so that L is a matrix. Let $\mathbf{l}_i$ be the eigenvectors and Λ_i the eigenvalues, satisfying

$$\mathsf{L}\mathbf{l}_i=\Lambda_i\mathbf{l}_i. \tag{104}$$

Expand $\mathbf{k}_{n_0}$ in the form

$$\mathbf{k}_{n_0}=\sum_i u_i^{(0)}\mathbf{l}_i \tag{105}$$

where n_0 is sufficiently large for Eq. (103) to be valid. We then have

$$\mathbf{k}_{n+n_0}=\sum_i \Lambda_i^n u_i^{(0)}\mathbf{l}_i. \tag{106}$$

Any eigenvalue Λ_i whose modulus exceeds 1 will grow in importance (*relevant*); one whose modulus is less than 1 will be *irrelevant*; an eigenvalue for which $|\Lambda_i|=1$ is called *marginal*. The $u_i^{(0)}$ are called *scaling fields* (and we can drop the superscript); for example, u_1 is the temperature t, u_2 the magnetic field H, etc.

From Eq. (99) we have the relation for the correlation function

$$\xi(u_1,u_2,\ldots)=b^n\xi(\Lambda_1^n u_1,\Lambda_2^n u_2,\ldots) \tag{107}$$

while the corresponding relation for the free energy follows from Eq. (97),

$$\psi(u_1,u_2,\ldots)=b^{-n}\psi(\Lambda_1^n u_1,\Lambda_2^n u_2,\ldots). \tag{108}$$

When $H=0$ and the only relevant field is t we find from Eq. (107)

$$\xi(t,0,\ldots)=b^n\xi(\Lambda_1^n t)\cong b^n(\Lambda_1^n t)^{-\nu}. \tag{109}$$

Hence we find that

$$b\Lambda_1^{-\nu}=1 \quad\text{or}\quad \nu=\ln b/\ln\Lambda_1. \tag{110}$$

For the free energy we define scaling indices y_T and y_H by

$$\Lambda_1=b^{dy_T},\quad \Lambda_2=b^{dy_H}, \tag{111}$$

and we then have, for all λ,

$$\psi(t,H)=b^{-dn}\psi(b^{ndy_T}t,b^{ndy_H}H), \tag{112}$$

i.e.,

$$\psi(\lambda^{y_T}t,\lambda^{y_H}H)=\lambda\psi(t,H) \tag{113}$$

where $\lambda=b^{dn}$.

The free energy is a *generalized homogeneous function* (Sec. 4.4). Take λ to satisfy

$$\lambda^{y_T}t=1,\quad \lambda=t^{-1/y_T}, \tag{114}$$

and we have

$$\psi(t,H)=t^{2-\alpha}K(Ht^{-\Delta}), \tag{115}$$

where

$$2-\alpha=y_T^{-1}, \quad \Delta=y_H/y_T, \quad K(u)=\psi(1,u), \tag{116}$$

i.e., we have derived a relation analogous to Eq. (66). Finally we have the additional relation, from Eqs. (110) and (111),

$$\nu^{-1}=dy_T. \tag{117}$$

Hence *all* exponents are determined by y_T and y_H in agreement with Kadanoff, and in support of length scaling.

The extension to crossover behavior (Sec. 4.4) is immediate. One needs to use a scaling field g and to introduce a scaling index y_g satisfying a relation corresponding to Eq. (111), and this will lead to a free energy of the form (77).

5.3 Real-Space Renormalization

To give a specific illustration of the above ideas for real-space renormalization, we take an example used by Wilson in relation to the simple Ising model on the two-dimensional square lattice of lattice spacing a. The block spin relates to a 2×2 cell of the lattice that contains 9 spins. The value of the block spin is determined by a simple majority rule; if 5 or more of the 9 spins point upwards, the block spin has value $+1$, otherwise it has value -1. It is a straightforward matter with the help of a computer to transform any configuration of the original lattice to one of the transformed lattice. The 2^9 possible configurations of a cell in the original lattice are reduced to 2 in the transformed lattice, the value of b above is 3, and the new lattice has spacing $3a$. Speaking roughly, we can say that the transformation averages out fluctuations in spin direction whose scale is smaller than the block size $3a$. It is as if one looks at the lattice through an out-of-focus lens that blurs the smaller features but does not affect the larger ones.

When the transformation is applied for a second time, fluctuations in spin direction whose scale is smaller than 3^2a will have been averaged out. After n applications, fluctuations whose scale is smaller than 3na will have been averaged out. Eventually fluctuations at all scales up to the correlation length are averaged out.

Block spins do not have the same couplings as the spins in the original configuration lattice. In the original formulation, there is one nearest-neighbor interaction parameter. The first stage of the block spin transformation introduces a finite but substantial number of more distant neighbors and multispin couplings. If the second state is carried out exactly, a new round of coupling parameters will be introduced; if this is continued indefinitely an infinite number of coupling parameters will be required. An infinity of parameters is needed to furnish an exact solution. Wilson's crucial conjecture was that all but a finite and relatively small number of coupling parameters can be ignored without impairing the physical description of behavior near the critical point. He later tested this conjecture by performing a calculation for the simple Ising model on the simple quadratic lattice using a decimation transformation (developed by Kadanoff) and retaining 217 parameters! His conclusion is as follows:

"The dominant interaction at the fixed point is the nearest-neighbor coupling. For accurate calculations one must include many more; but for a qualitative picture the nearest-neighbor constant K plus perhaps the next-nearest-neighbor coupling L should be enough. The remaining couplings are at least a factor of 5 smaller than L.... Thus the old idea of Kadanoff that there would be effective nearest-neighbor Ising models for block spins is very close to the truth."

5.4 Momentum-Space Renormalization

For the use of the RG in momentum space it proved convenient to employ a continuous-spin adaptation of the Ising model. This enabled contact to be made with the Gaussian model (Sec. 3.5) which gives valid results for critical exponents when $d>4$; on replacing the lattice structure by a continuum (as for a fluid) the theory could be merged with the Landau–Ginzburg development (Sec. 1.5) and with the field theories with which the RG originated.

The partition function for the simple Ising model (24) can be written in the alternative form

$$Z^I=\int_{-\infty}^{\infty} d\sigma_1 2\delta(\sigma_1^2-1)$$

$$\times \exp\left(K \sum_{\mathbf{l},\mathbf{m}} \sigma_{\mathbf{l}}\sigma_{\mathbf{m}} + \beta\mu_0 H \sum_{\mathbf{l}} \sigma_{\mathbf{l}} \right) \quad (118)$$

with $\sigma_{\mathbf{l}}$ a continuous variable. Wilson replaced the function $2\delta(\sigma_{\mathbf{l}}^2-1)$ by the continuous function

$$w(\sigma_{\mathbf{l}}) = \exp(-\tfrac{1}{2}c\sigma_{\mathbf{l}}^2 - u\sigma_{\mathbf{l}}^4), \quad (119)$$

where u is a positive number. When $c=-4u$, $w(\sigma_{\mathbf{l}})$ has the form

$$w(\sigma_{\mathbf{l}}) = \exp[-u(\sigma_{\mathbf{l}}^2-1)^2+u], \quad (120)$$

which reproduces the simple Ising model when $u \to \infty$. For finite positive u, $w(\sigma_{\mathbf{l}})$ then has maxima at $\sigma_{\mathbf{l}} = \pm 1$, and if u is large, the probability function is sharply peaked at $\sigma_{\mathbf{l}} = \pm 1$. When $u=0$ Eq. (119) the Gaussian model is retrieved for positive c.

Following the idea of universality classes enunciated in Sec. 4.3 one might expect that any weighting function of the form (120) would belong to the Ising universality class. Wilson made the more daring suggestion that as long as $u>0$ *any* weighting function of the form (119) will correspond to the Ising universality class, i.e., that the only value of the parameters (c,u) giving rise to a change in behavior is $u=0$.

The transition to a continuum-space model can be achieved on replacing the $-\sigma_{\mathbf{l}}\sigma_{\mathbf{m}}$ in Eq. (118) by

$$\tfrac{1}{2}(\sigma_{\mathbf{l}}-\sigma_{\mathbf{m}})^2. \quad (121)$$

Since the lattice spacing is small and not expected to exercise a significant influence on critical behavior, the discrete σ is replaced by the continuous function $\sigma(x,y,z)$ and becomes

$$\tfrac{1}{2}(\nabla\sigma)^2. \quad (122)$$

To operate in momentum space it is convenient to introduce the Fourier-transformed spin variables

$$\hat{\sigma}_{\mathbf{q}} = \sum_{\mathbf{l}} \exp(i\mathbf{q}\cdot\mathbf{l})\sigma_{\mathbf{l}}. \quad (123)$$

For a lattice structure of spacing a, the momenta $\mathbf{q}$ are essentially restricted by $|q| \leqslant \pi/a$ (neglecting the shape of the Brillouin zone which, as one might expect, proves irrelevant). This cutoff is characteristic of the physics of condensed matter, and contrasts with quantum field theory where there is normally no natural cutoff and the momenta must be taken to infinity; the latter is therefore represented by zero lattice spacing. The RG transformation R of Eq. (85) is defined by averaging out all the $\hat{\sigma}_{\mathbf{q}}$ variables satisfying

$$b^{-1} < |q|a/\pi \leqslant 1. \quad (124)$$

This corresponds as before to a spatial rescaling factor b.

The first practical application by Wilson of the RG was in momentum space taking $b=2$. Severe approximations were required in evaluating the integrals to put the results into a form from which iterative recursion relations could be derived. They were solved numerically and gave fair values for the critical exponents.

In a subsequent paper by Wilson and Fisher the key suggestion was advanced that the appropriate expansion parameter for the critical exponent is $\epsilon=4-d$ treated as a continuous parameter, and expansions to the first order in ϵ were derived for a number of models. Wilson then generalized the treatment to the n-vector classical Heisenberg model, using for the exponent in Eq. (119)

$$-\tfrac{1}{2}c_0 \sum_{i=1}^{n} \sigma_i^2 - u_0\left(\sum_{i=1}^{n} \sigma_i^2 \right)^2. \quad (125)$$

For this model he used Feynman graph methods of field theory (see QUANTUM ELECTRODYNAMICS) to derive the following expansions:

$$\gamma = 1 + \frac{n+2}{2(n+8)}\epsilon + \frac{(n+2)(n^2+22n+52)}{4(n+8)^3}\epsilon^2 + O(\epsilon^3), \quad (126)$$

$$\eta = \frac{n+2}{2(n+8)^2}\epsilon^2 + \frac{n+2}{2(n+8)^2}\left(\frac{6(3n+14)}{(n+8)^2} - \frac{1}{4} \right)\epsilon^3 + O(\epsilon^4), \quad (127)$$

$$\phi = 1 + \frac{n\epsilon}{2(n+8)} + \frac{\epsilon^2(n^3+24n^2+68n)}{4(n+8)^3}\epsilon^2 + O(\epsilon^3). \quad (128)$$

Even though ϵ is assumed to be a small parameter, substituting $\epsilon=1$ gave results that compared surprisingly well with the series expansion estimates.

This seminal work initiated a new diagrammatic approach to the calculation of critical exponents, and powerful field-theoretical techniques were used to extend the ϵ expansion to higher orders, and to improve the accuracy of the critical exponent estimates.

5.5 Critical Exponents: Corrections to Scaling

Using the above techniques Nickel added several terms to the ϵ expansion, but the series was found to be divergent. However, Nickel and Lipatov showed how to calculate the asymptotic behavior of the general term in the expansion, and this enabled a Borel transformation to be used to convert the series into a convergent one. From this Le Guillou and Zinn-Justin in 1977 made the accurate estimates of the exponents of the simple Ising model with $\epsilon=1$ as follows: $\gamma=1.240, \beta=0.325, \nu=0.630$. Series expansions had favored $\gamma=1.25$, $\beta=0.3125$, $\nu=0.64$. The differences, although relatively small, were beyond the expected error bounds. Of more fundamental importance was the failure of the second set of exponents to satisfy the length scaling (or hyperscaling) relation (69).

The key to the resolution of this discrepancy lay in the form of the correction terms to scaling. It is latently assumed in the ratio analysis, Eq. (55), that there is no *confluent singularity*, i.e., that the critical behavior of the thermodynamic function under consideration is of the form

$$t^{-h}G(t), \tag{129}$$

where $G(t)$ is analytic. However, Wegener showed in 1972 that the RG leads to the conclusion that there is a correction term

$$t^{-h-\Delta_1}G_1(t), \tag{130}$$

where Δ_1 need not be an integer. Moreover, it was possible to derive an ϵ expansion for Δ_1, and for $\epsilon=1$ this gave $\Delta_1 \cong \frac{1}{2}$.

In 1980 the discrepancy was resolved in a paper by Nickel presented at the International Conference on Statistical Mechanics in Edmonton, Alberta. Nickel added a significant number of new terms to the series expansion for the susceptibility of the bcc lattice and showed that with a correction term of the form (130) it was possible to achieve agreement between the series expansion and RG estimates with acceptable error bounds.

It is now generally assumed that the most accurate estimates of critical exponents in three dimensions are provided by the ϵ expansion. The results shown in Table 3 are taken from the final 1980 paper of Le Guillou and Zinn-Justin.

5.6 Conclusion

It is very gratifying that such good agreement should finally have been achieved between the two most accurate methods of estimating critical behavior. As regards the contributions to progress in the RG, real-space renormalization has proved useful for illustrating the general principles but has not proved very accurate in three dimensions. Momentum-space renormalization has surprisingly provided the most accurate estimates of three-dimensional critical behavior to date.

6. MISCELLANEOUS ADDITIONAL TOPICS

Although we have described in detail the mainstream developments in critical phenomena, a number of related topics have been passed over for lack of space. We summarize them briefly in this section.

Exact solutions have played a role of special importance in critical phenomena. We have seen that the modern era was initiated by Onsager's exact solution, which revealed the defects of the classical theories and served as a test for new techniques of calculation. For about 20 years after Onsager's work there were no new solutions for realistic interactions that differed in a substantial way from

Table 3. ϵ-expansion estimates of critical exponents, from Le Guillou and Zinn-Justin (1980).

n	γ	ν	η	β	α	Δ_1
0	1.1615±0.0020	0.588±0.0015	0.027±0.004	0.302 ±0.0015	0.236±0.0045	0.470±0.025
1	1.241 ±0.0020	0.630±0.0015	0.031±0.004	0.325 ±0.0015	0.110±0.0045	0.498±0.020
2	1.316 ±0.0025	0.669±0.0020	0.033±0.004	0.3455 ±0.0020	−0.007±0.006	0.522±0.018
3	1.386 ±0.0040	0.705±0.0030	0.033±0.004	0.03645±0.0025	−0.115±0.009	0.550±0.016

that of Onsager. Then in 1967 Lieb produced a variety of new solutions for two-dimensional ferroelectric models. The critical behavior followed a completely different pattern from the Ising model, and it seemed that new ground was being broken.

Lieb was followed by Baxter, whose solution of another ferroelectric model, the eight-vertex model, gave rise to exponents that vary continuously with the strength of the interaction. This was a major challenge to the idea of universality, since the model is equivalent to an Ising model with a two- and four-spin interaction. The issue was resolved by an analysis by Kadanoff and Wegener in 1971, who indicated that the model possesses a special symmetry that gives rise to continuously varying exponents; this particular feature would not be expected to arise in normal physical systems.

Baxter subsequently derived a number of other new solutions and as a by-product extended very significantly exact series expansions for two-dimensional models which he was unable to solve exactly. (For more details of this exact work, see Baxter, 1982).

An interesting point that led to fruitful ideas is the behavior of two-dimensional models when $n=2$. Peculiarities in this model were noted some years ago, and their investigation by Kosterlitz and Thouless in 1973 has suggested a new type of critical phenomenon related to *topological order*. Experimental work has already confirmed some of the theoretical predictions.

Other problems in condensed-matter physics that have been profoundly influenced by progress in critical phenomena are the shape and size of polymer chains in solution (see, e.g., de Gennes, 1979), and percolation processes (see, e.g., Stauffer 1985). The model of a self-avoiding walk on a lattice (see, e.g., Domb, 1969) had for many years been regarded as a useful representation of a polymer chain with a realistic hard-core "excluded volume." It had been noted empirically that the properties of such self-avoiding walks paralleled those of the Ising model. The precise connection was revealed by de Gennes in 1972, who noted that the self-avoiding-walk configurations correspond to a ferromagnetic model with spin dimension n when n is allowed to take the value zero (see Table 3).

A theoretical model that burst into prominence after being ignored for nearly two decades is the Potts model. This was first suggested in 1952 as a generalization of the Ising model having q orientations but only two different energies of interaction; the critical point could be located exactly. At the Kyoto Conference on Statistical Mechanics in 1968, Kasteleyn and Fortuin showed that percolation processes associated with random mixtures that have applications to dilute magnets, amorphous semiconductors, and a number of other problems in condensed-matter physics correspond to the Potts model with $q=1$! A number of physical systems have been identified for which the Potts model is a reasonable representation, and the difference in symmetry between the n-vector ferromagnetic model and q-component Potts model gives rise to important differences in critical behavior. The literature on percolation and the Potts model has escalated sharply in the past few years.

GLOSSARY

Coherence Length: The range over which the behavior of atoms is significantly correlated.

Critical Exponents: Parameters that characterize the singular behavior of thermodynamic functions near a critical point.

Crossover: A change from one specific pattern of critical behavior to another.

Curie Temperature (Point): The temperature at which a ferromagnet loses its intrinsic magnetization.

Gaussian Model: A model of ferromagnetism in which the values of the atomic spins follow a Gaussian probability distribution.

Heisenberg Model: A model of ferromagnetism introduced by Heisenberg in 1928 with a quantum interaction between atomic spins.

Ising Model: A model of ferromagnetism suggested by Lenz to his student Ising in 1925 in which each spin has two values, up or down.

Lambda-Point Transition: A phase transition with no *latent* heat but with a *specific heat anomaly* in the shape of the Greek letter λ.

Lattice-Gas Model: A lattice model of a fluid in which each lattice cell is either occupied or unoccupied by a molecule.

Long-Range Order: Correlation of molecules or spins that extends to infinite range.

Mean-Field Solution: An approximation in the theory of critical behavior in which a fluctuating quantity is replaced by its average value.

Opalescence: Changes in color and appearance arising from light scattering.

Order Parameter: A key thermodynamic quantity identified by Landau that characterizes the critical behavior of a system.

Partition Function: A function constructed from the *microscopic* properties of molecules of a system from which *macroscopic* behavior can be derived.

Relevant (or Irrelevant) Fields: Interactions that do (or do not) significantly affect critical behavior.

Scaling Relations: Relations arising from a simplified equation of state in which two-dimensional data scale onto a single curve.

Universality: Systems of widely different origins whose critical behavior follows the same standard pattern.

List of Works Cited

Andrews, T. (1869), *Philos. Trans. R. Soc. London* **159**, 575–591.

Bain, E. C. (1923), *Trans. AIME* **68**, 625–641.

Baxter, R. J. (1982), *Exactly Solved Models in Statistical Mechanics*, New York: Academic.

Bragg, W. L., Williams, E. J. (1934), *Proc. R. Soc. London Ser. A* **145**, 699–730.

Bragg, W. L., Williams, E. J. (1935), *Proc. R. Soc. London Ser. A* **151**, 540–566.

Curie, P. (1895), *Ann. Phys. Chim.* **5**, 289–405.

DeGennes, P. G. (1979), *Scaling Concepts in Polymer Physics*, Ithaca, NY: Cornell University Press.

Domb, C. (1969), *Adv. Chem. Phys.* **15**, 229–259.

Domb, C., Hunter, D. L. (1965), *Proc. Phys. Soc. London* **86**, 1147–1151.

Guggenheim, E. A. (1945), *J. Chem. Phys.* **13**, 253–261.

Habgood, H. W., Schneider, W. G. (1954), *Can. J. Chem.* **32**, 98–112.

Heller, P., Benedek, G. (1965), *Phys. Rev. Lett.* **14**, 71–74.

Johansson, C. H., Linde, J. O. (1925), *Ann. Phys. (Leipzig)* **78**, 439–460.

Landau, L. D. (1937), *Phys. Z. Soviet Union* **11**, 26, 545. [English Translation in: D. ter Haar (Ed.) (1965), *Collected Papers of L. D. Landau*, London: Pergamon, 193–216].

Onsager, L. (1944), *Phys. Rev.* **65**, 117–149.

Ornstein, L. S., Zernike, F. (1914), *Proc. Akad. Sci. Amsterdam* **17**, 792–806.

Ornstein, L. S., Zernike, F. (1916), *Proc. Akad. Sci. Amsterdam* **18**, 1520–1527.

Patashinskii, A. Z., Pokrovskii, V. L. (1966), *Sov. Phys. JETP* **23**, 292–297.

Roach, P. R. (1968), *Phys. Rev.* **170**, 213–223.

Tammann, G. (1919), *Z. Anorg. Chem.* **107**, 1–239.

Vicentini-Missoni, M. (1971), in C. Domb and M. S. Green (Eds.), *Phase Transitions and Critical Phenomena*, Vol. 2, New York: Academic, Chap. 2.

Weiss, P. (1907), *J. Phys. (Paris)* **6**, 661–690.

Widom, B. (1965), *J. Chem. Phys.* **43**, 3898–3905.

For detailed references of other papers mentioned in the text see

Domb, C. (1985), *Contemp. Phys.* **26**, 49–72.

Further Reading

Review Articles

De Jongh, L. J., Miedema, A. R. (1974) *Adv. Phys.* **23**, 1–269.

Domb, C. (1960) *Adv. Phys.* **9**, 149–361.

Fisher, M. E. (1967), *Rep. Prog. Phys.* **30**, 615–730.

Fisher, M. E. (1974), *Rev. Mod. Phys.* **45**, 597–616.

Heller, P. (1967), *Rep. Prog. Phys.* **30**, 731–826.

Wilson, K. G., Kogut, J. (1974), *Phys. Rep.* **12C**, 75–200.

Books

Baxter, R. J. (1982), *Exactly Solved Models in Statistical Mechanics*, New York: Academic.

De Gennes, P. G. (1979), *Scaling Concepts in Polymer Physics*, Ithaca, NY: Cornell University Press.

Domb, C. (1992), *The Critical Point*, London: Taylor and Francis, in press.

Domb, C., Green, M. S. (Eds.) (1972–6), *Phase Transitions and Critical Phenomena*, Vols. 1–6, New York: Academic.

Domb, C., Lebowitz, J. L. (Eds.) (1983–9), *Phase Transitions and Critical Phenomena*, Vols. 7–13, New York: Academic.

Ma, S. K. (1976), *Modern Theory of Critical Phenomena*, Reading, MA: Benjamin.

Stanley, H. E. (1971), *Phase Transitions and Critical Phenomena*, London: Oxford University Press.

Stauffer, D. (1985), *Introduction to Percolation Theory*, London: Taylor and Francis.

CRYOGENICS

HANS M. BOZLER AND CHRISTOPHER M. GOULD, *Department of Physics, University of Southern California, Los Angeles, California, U.S.A.*

INTRODUCTION

Cryogenics is the field of science and engineering that deals with the production of low temperatures. This subject includes the techniques needed to obtain low temperatures, the methods to measure low temperatures, and the study of the physics of matter at these temperatures. A quest for obtaining low temperatures followed the proposal of an absolute zero temperature by Guillaume Amontons in 1702. Progress accelerated after the middle 1800s when a series of experiments succeeded in liquefying the materials that were gases at room temperature. These were liquefied in the descending order of their boiling points—oxygen, nitrogen, hydrogen, and finally helium. In this article, we will briefly review the sorts of physical changes that take place in matter at cryogenic temperatures. Then we will discuss the techniques used to obtain these temperatures. After this we will introduce the very important subject of measuring temperature. Since the range of

1-56081-063-7/92/$5.00+.50

temperature varies over a number of orders of magnitude, we will frequently refer to logarithmic graphs as shown in Fig. 1. Here, each factor of 10 in absolute temperature is one division on the scale.

1. WHAT HAPPENS AT LOW TEMPERATURES?

The understanding of absolute temperature, first postulated by Guillaume Amontons, was incorporated by William Thomson (Lord Kelvin), James P. Joule, Rudolf Clausius, and several other thermodynamicists into the laws of thermodynamics during the mid-1800s (Mendelssohn, 1977). The absolute temperature is zero when all thermal energy has been extracted from matter. Thus when a material is at zero temperature, only the quantum-mechanical zero-point motions remain.

As a material is cooled from room temperature many changes take place. Liquids freeze, and solids become stiffer and more brittle. Pure metals become less resistive, and many metals become superconductors. All of the gases that exist at room temperature eventually liquefy, and most of these also freeze into solids. At ordinary pressures, helium (the common isotope ^{4}He and the rare isotope ^{3}He) remains a liquid at absolute zero in temperature.

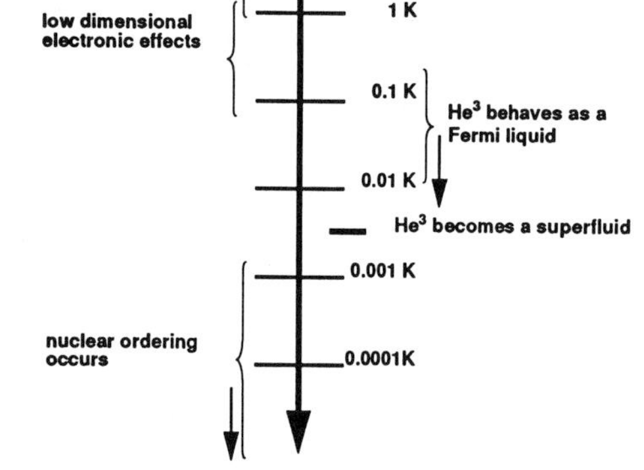

FIG. 1. Logarithmic temperature scale indicating significant events as the temperature is lowered from room temperature. The temperatures are on the kelvin scale.

The process of reducing temperature lowers the energy in materials. This is the energy that causes the random motions. For this reason, as the temperature is reduced, interactions between molecules, atoms, and even nuclei of neighboring atoms progressively become more important. We can substitute an energy scale, with ϵ being a typical single-atom or -molecule energy, for the temperature scale in Fig. 1 by using

$$\begin{aligned}\epsilon &= k_B T \\ &= (1.38\times10^{-23}\ \mathrm{J/K})T \\ &= (8.63\times10^{-5}\ \mathrm{eV/K})T. \end{aligned} \tag{1}$$

As the temperature (or energy) is reduced below the interaction energy, important physical changes are likely to occur. These energies are small and become tiny when the temperatures are reduced. For this reason, cryogenics can be thought of as the physics of small energy.

A second way of thinking about low temperatures is to consider the entropy that is contained within materials (Kittel and Kroemer, 1980; Zemansky and Dittman, 1981). Entropy is a measure of disorder or randomness. The classical definition of entropy is written in terms of the number of ways that one can microscopically rearrange a system consistent with whatever physical constraints exist (e.g., temperature, energy, pressure or volume, magnetic field or magnetization). In usual systems, this number of rearrangements is a very large number, but the entropy is the logarithm of this number. If we use S for the entropy and Γ for the number of combinations we have

$$S = k_B \log \Gamma. \tag{2}$$

When the thermal energy, as shown in Eq. (1), is smaller than interactions, the system goes into its ground (lowest-energy) state. The system becomes highly organized. Thus when a system is highly organized, the number of possible arrangements of particles Γ is small and S approaches zero. This corresponds to the low-temperature state of a system.

The universe has a background temperature of about 3 K due to the residue of radiation from the "big bang." Current measurements show this temperature to be 2.735 K (Mather, 1990). For this reason, there are no naturally occurring regions anywhere in the universe that have a temperature below this 2.7 K. Thus, with cryogenics, we are capable of producing unique states of matter (below 2.7 K) that are not duplicated in the natural world. By contrast, in other frontiers of technology (high energy, high pressure, high temperature, high magnetic fields, low pressure or high vacuum, etc.) manmade efforts are inevitably exceeded by nature.

At very low temperatures (below 1 K) the usual vibrations (phonons) in most materials become insignificant. Properties of materials are dominated by the interactions of electrons and magnetic spins, and eventually as the temperature is reduced, the magnetism of the nuclei becomes important. (Zemansky and Dittman, 1981; White, 1979; Lounasmaa, 1974).

2. PRODUCTION OF LOW TEMPERATURES

In 1877 L. Cailletet expanded high-pressure oxygen, which produced a fine mist of liquid (Mendelssohn, 1977). Simultaneously, R. Pictet had produced a similar result in another laboratory. After this, the race was on to liquefy all of the remaining gases. Sir James Dewar was the first to liquefy hydrogen in 1898; but, in the race to liquefy helium (the last natural gas), he was beaten out by Kamerlingh Onnes, who succeeded in 1908. Soon after this time, temperatures near 1 K were achieved.

As a by-product of the highly competitive effort to be the first to liquefy gases, several inventors worked to improve cryogenic refrigerators. The first patent for an expansion engine was given to Solvay in 1887. Hampson in England and Linde in Germany each filed a patent in 1895 for a process using a compressor, a counterflow heat exchanger, and a Joule–Thomson expansion valve to produce large quantities of liquid oxygen and nitrogen. Shortly after this, Claude filed a patent in France for a refrigerator that placed an expansion engine before the Joule–Thomson expansion valve in the Hampson–Linde refrigerator. An expansion engine greatly improves the refrigerator by letting the gas do work on a moving piston and thus lose heat energy. This staged approach to refrigeration is characteristic of low-temperature refrigerators.

Magnetic cooling was first proposed independently by William Francis Giauque and Peter Debye in 1926 (Mendelssohn, 1977). Giauque succeeded in reaching 0.25 K by 1933.

Currently, the production of low temperatures follows the techniques developed by the pioneers that we have mentioned above. Heat energy is extracted from materials with various methods of refrigeration depending on which methods efficiently remove entropy in the temperature range where the refrigerator is operating. Figure 2 shows, on a logarithmic temperature scale, the approximate temperature ranges in which these various techniques function best. As the temperature is reduced, the preferred method of cooling changes. For this reason, devices that produce or operate at low temperatures (cryostats) frequently employ a series of cooling techniques. This cascade approach is used in state-of-the-art cry-

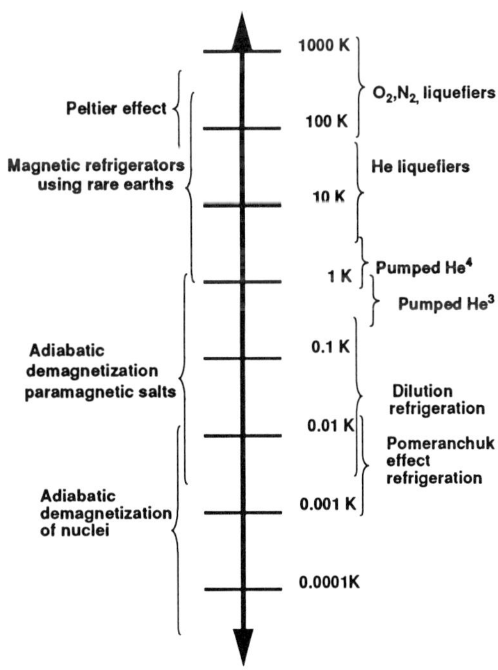

FIG. 2. Logarithmic temperature scale indicating the ranges of temperature in which various refrigeration techniques work most efficiently. Many of these techniques are discussed in the text, with the exception of Peltier-effect refrigerators and the Pomeranchuk effect.

ostats for obtaining microkelvin temperatures. These cryostats start with liquid helium at 4.2 K (a result of a cascade of mechanical coolers), then use an evaporation refrigerator to obtain 1 K; this is followed by a dilution refrigerator that can get to less than 10 mK. Next a nuclear cooling stage gets the temperature to well below 1 mK. Sometimes, a second nuclear cooling stage is added that can obtain a final (nuclear) temperature in the nanokelvin range.

2.1 Thermal Insulation

The production of low temperatures requires that we remove heat energy. Our ability to lower the temperature is ultimately limited by the amount of heat that leaks back in. For this reason, any low-temperature system needs to be surrounded by thermal insulation. This insulation can be in the form of materials that have low thermal conductivity, or a vacuum. Heat flows in three primary ways. First, there is diffusive conduction, which is usually high in pure metals and crystals and low in glasses and composite materials such as ceramics, plastics, and fibers (White, 1979; Powell and Childs, 1972). A second mode of heat flow is convection, in which heat is carried by a gas or liquid. A temperature inversion (hot region below the cold region) generally induces convection so that a fluid moves from a low-density hot region to a high-density cold region. Most gases are actually poor thermal conductors. However, in open regions, they can start a convective flow that transports much more heat than conduction. For this reason, many forms of insulation such as Styrofoam rely on trapping air in small cavities. The third common mode of heat flow is blackbody radiation. The heat flow due to radiation depends on the emissivity of the material and is proportional to T^4 (Stefan's law) (Kittel and Kroemer, 1980).

Containers for cryogenic fluids and equipment are referred to as Dewar flasks—named for James Dewar. Dewars rely on vacuum and reflective surfaces to minimize heat flow. In Fig. 3, we illustrate two types of Dewars. The first is the original vacuum flask. Here the walls are made from glass or stainless steel. Glass Dewars have a shiny metal film added to the walls to reduce radiation. (These Dewars are the same as the household Thermos™ bottles for holding coffee or cold drinks.) The second Dewar is an improved version that uses superinsulation in addition to the vacuum. Superinsulation is made of many layers of Mylar (polyester) film coated with a thin film of aluminum. The superinsulation then acts as a very efficient radiation shield since, going from higher temperature, each layer radiates less then the one outside. Also, residual gas in the vacuum space is trapped between the layers of the superinsulation. In modern Dewars, the neck, in which there is a rapid change in the temperature, is frequently made from an epoxy and fiberglass composite. For efficient storage of cryogenic liquids, the neck is usually kept small since much of the heat leaking into the Dewar is conducted down the neck. Dewars can be made in many shapes ranging from the simple cylindrical shapes shown here to large tank cars.

Many low-temperature devices use "heat switches," which are capable of turning on and off heat flow (White, 1978; Lounasmaa, 1974). Heat switches are used when there is a need to have parts of the cryogenic apparatus in close thermal contact at some times, and well isolated at other times. For example, a

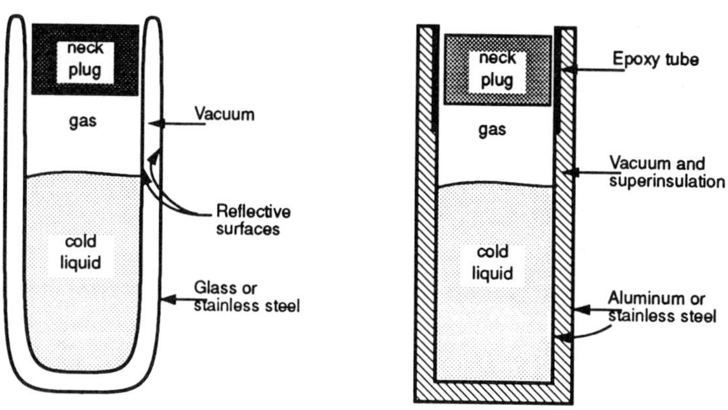

FIG. 3. Typical arrangements of Dewar flasks. The one on the left is close to the original design and uses a vacuum and highly reflective surfaces to insulate the liquid from heat coming through the walls. The walls are made from a poor heat conductor so that the top region, or neck, conducts very little heat. The Dewar on the right uses a more modern construction with the neck constructed of an epoxy tube (usually with fiberglass as a filler) and the walls filled with superinsulation.

heat switch can be used to allow cooling and later isolation of a material, the heat capacity of which is being measured. Such a switch can be a mechanical switch, or a system using a gas that can be later evacuated to exchange heat. At temperatures below 1 K, heat switches are most frequently made from pure superconducting metals such as Pb, Sn, In, and Al. (See the inset in Fig. 14.) When these metals are pure, their heat conductivity is very high in the "normal" state but very low in the superconducting state. A superconducting heat switch is made by surrounding the superconductor with a variable-field magnet. When the magnetic field is raised above the superconductor's critical field, the conductor becomes normal and switches from being a poor conductor of heat to a very good conductor of heat.

2.2 Cryogenic Refrigeration Using Fluids

All large-scale cryogenic refrigerators, and most small-scale refrigerators, use heat engines that have either a liquid, a gas, or a mixture of a liquid and gas as the refrigerant. These engines use the expansion of a compressed fluid as the cooling process. A few heat engines use magnetic or thermoelectric cooling. Graham Walker in his book on cryogenic heat engines (Walker, 1983) has coined the term "cryocoolers" to describe heat engines intended to produce cryogenic temperatures. In this section, we will briefly discuss three processes that can be used singly or in combinations to produce low temperatures. In practice, there are many variations of these processes named after different inventors. The three processes we discuss here illustrate the essential features that exist in all of these refrigerators.

In general, heat engines are devices that use heat to produce mechanical work. Examples of these would be steam turbines or gasoline and diesel engines. In Fig. 4 we have a schematic drawing of two heat engines, one running forward using high-temperature heat to produce work, and one running in reverse using work to remove heat. A forward-running heat engine is characterized by its efficiency ϵ in converting the high-temperature heat Q_h (that's what you pay for) into work W_{out}:

$$\epsilon = W_{out}/Q_h. \tag{3}$$

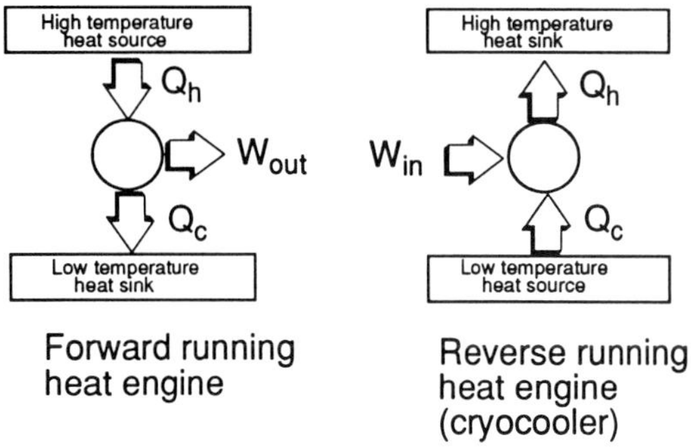

FIG. 4. Schematic diagrams for continuously operating heat engines. The forward-running heat engine uses heat Q_h to produce work W_{out} and rejects heat Q_c. The reverse-running heat engine uses work W_{in} to remove heat Q_c from a low temperature and adds heat Q_h to the high-temperature heat sink.

The first law of thermodynamics (energy conservation) requires that $W_{out} = Q_h - Q_c$. The change in entropy due to heat flow is given by $\Delta S = Q/T$, which means that proportionately more entropy is carried away by the heat carried to the low-temperature reservoir. The second law of thermodynamics requires that the net entropy of the heat engine cannot decrease. Thus there is a maximum efficiency which is equal to $1 - T_c/T_h$ and can never exceed 1.

A cryocooler is a reverse-running heat engine, which can be characterized by the ratio of the extracted heat Q_c to the input work W_{in}. This ratio is called the coefficient of performance C,

$$C = Q_c/W_{in} \tag{4}$$

Here, the second law of thermodynamics shows that the maximum value for C is $T_c/(T_h - T_c)$, which can exceed the value 1 (which occurs in common refrigeration devices such as room air conditions), while the first law requires that $W_{in} = Q_h - Q_c$. (Note that the directions of heat flow and work in the cryocooler are reversed.)

An ideal heat engine follows a Carnot cycle. The Carnot cycle has four parts, which are illustrated in a TS plane in Fig. 5. This figure shows the heat engine going through a series of constant-temperature ($1 \rightarrow 2, 3 \rightarrow 4$) processes and constant-entropy processes ($2 \rightarrow 3, 4 \rightarrow 1$). These correspond to combinations of heating or cooling and compressing or expanding the fluid. In the Carnot cycle, the thermodynamic system plotted is usually a

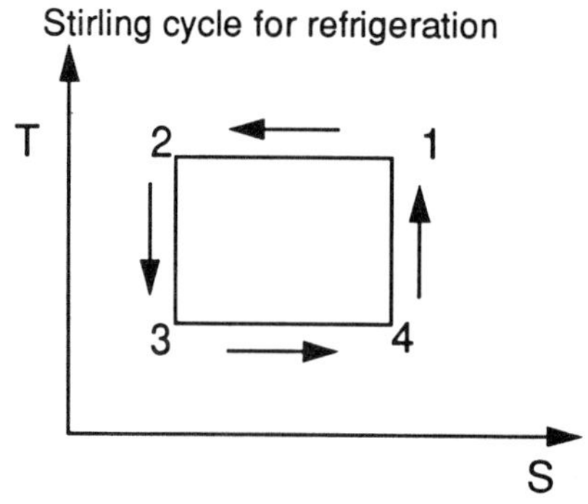

FIG. 5. The Stirling cycle as represented in a *TS* plane. The arrows indicate the direction taken when a Stirling engine is operating as a refrigerator.

single substance—often a gas. However, obtaining the heat switching necessary to have constant-entropy processes in a single substance is impractical.

2.2.1 Stirling Cryocooler. A Stirling heat engine uses a process approximating a Carnot cycle. In contrast to the Carnot engine, the Stirling engine uses two "working substances" that exchange heat between themselves so that the composite system follows a Carnot path. Figure 5 represents the ideal path for the refrigerator, but a real Stirling refrigerator relies on mechanical devices that only approximate the path. Also, friction and thermal gradients reduce the performance. Thus, the real coefficient of performance is less than $T_c/(T_h-T_c)$. Frequently, this cycle is shown in a pressure–volume (*PV*) plane; however, this presentation can be deceptive since a Stirling engine relies on a reversible heat source to produce adiabatic (constant *S*) changes in the temperature.

A still idealized model of a Stirling refrigerator using a gas is shown in Fig. 6. Here step $1\rightarrow2$ is accomplished by compressing the gas on the right (hot) side and simultaneously removing heat Q_h through a heat exchanger (not shown) to the high-temperature reservoir. The next step, $2\rightarrow3$, involves moving both pistons simultaneously with the gas delivering heat to the regenerator. Step $3\rightarrow4$ is the isothermal expansion of the left (cold) cylinder. Now, heat from the low-temperature side Q_c is absorbed.

In a real Stirling engine, the reversible heat source is called a regenerator and is typically made from an assembly of finely divided metal in the form of strips, wires, or tubes. This provides a large heat capacity, efficient heat exchange, and low heat conduction between the hot and cold ends. An actual Stirling engine only approximates the steps described in Fig. 6, since the mechanical linkages to produce these exact piston motions would be unnecessarily complex. On the other hand, smooth oscillatory motion of the pistons does not deteriorate the efficiency much. Stirling engines have a significant advantage in that they do not require valves, which makes them highly reliable. They are among the most efficient cryocoolers, and modern versions of these are used in both large-scale and small lightweight systems. Most modern military aircraft have Stirling cryocoolers on board to cool their infrared detection systems.

2.2.2 Joule–Brayton Recuperative Refrigerator. A Joule–Brayton recuperative cryocooler is very similar to a Stirling engine except in the manner in which the gas is precooled. The Joule–Brayton cryocooler uses a compressor, a counterflow (recuperative) heat exchanger, and an expander. A schematic diagram for this type of refrigerator is shown in Fig. 7. First the gas is compressed, which heats it up. The high-temperature heat exchanger brings the gas back to the temperature of the surroundings. The expander is a reversed compressor. Compressed gas is allowed to expand while doing work on the piston. The walls of the expander are poor heat conductors so that the process is adiabatic and nearly reversible. Thus the work done by the expanding gas on the piston in the expander removes internal energy from the gas, which results in a temperature drop in gas. After passing through the cold region, the returning gas flows through a counterflow heat exchanger which precools the incoming gas.

2.2.3 Joule–Thomson Refrigerator. The third refrigeration technique, the Joule–Thomson expansion, differs from the expander explained above because in this case the gas does no external work. Instead of the gas doing work on a piston, it does work on itself. This is the procedure that is most commonly used in air conditioning, home refrigerators, and freezers, where the working fluid is Freon 12 (a chlorofluorocarbon) or ammonia. A Joule–Thomson refrigerator uses a compressor to feed gas through a narrow

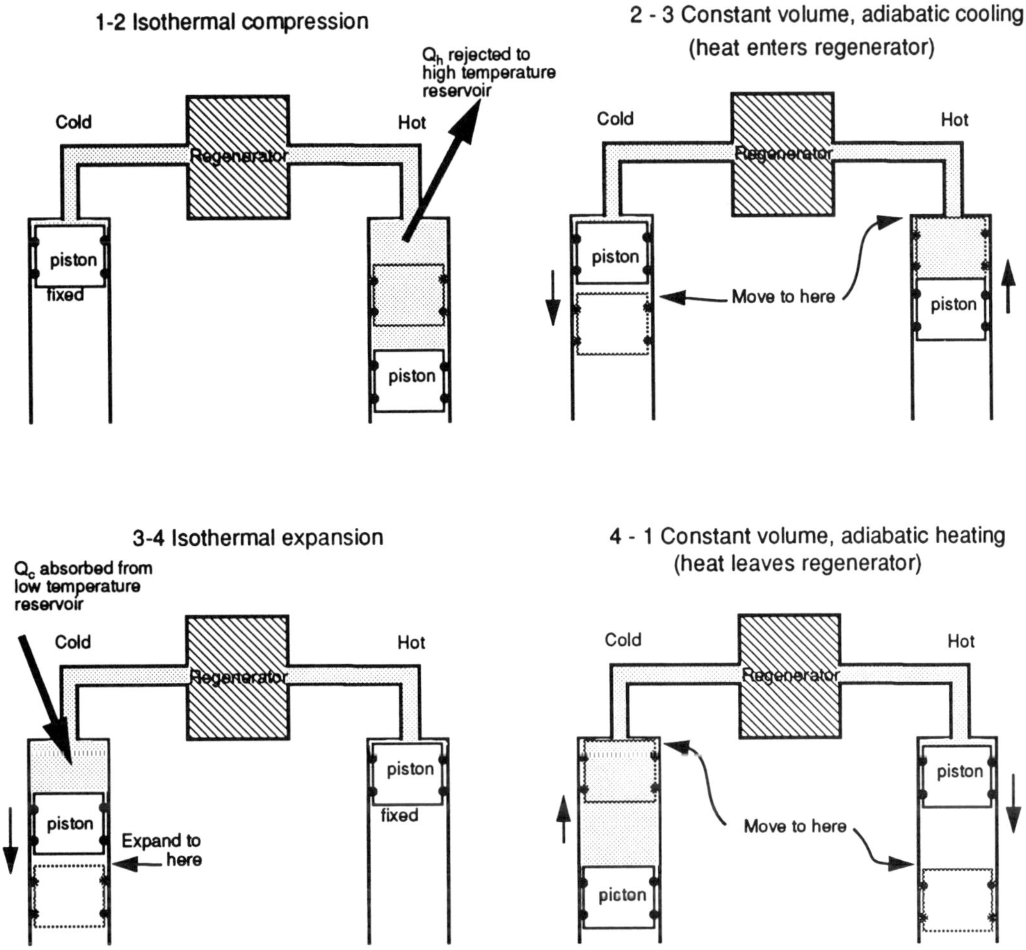

FIG. 6. An idealized model for a Stirling engine. The pistons are connected to external mechanical devices to achieve the correct sequencing of the motion.

restriction. This narrow restriction can be made with a piece of fine capillary tubing or a fine porous material. In many cryogenic refrigerators, the restriction can be adjusted to fine tune the performance of the refrigerator. Thus this restriction is often called a Joule–Thomson valve. We can easily show that a thermodynamic potential called the enthalpy is the same before and after the gas passes through the restriction. The enthalpy is defined by $H=U+PV$, where H is the enthalpy and U is the internal energy. In general, a constant-enthalpy expansion can cause either cooling or warming depending on the sign of the Joule–Thomson coefficient $\mu \equiv (\partial T/\partial P)_H$. In general, gases have a negative μ at high temperatures and warm when expanded, but have an inversion point where μ goes through zero and becomes positive at lower temperatures. For gases at temperatures below the Joule–Thomson inversion point, the result is a reduction in temperature. The maximum inversion temperatures, which also depend on pressure, for N_2 and O_2 are about 621 and 764 K—above room temperature—while the inversion temperatures for H_2 and He are about 202 and 40 K.

Although Joule–Thomson expanders can reduce the temperature of a gas at any starting temperature below the inversion point, they are efficient refrigerators and liquefiers only if the gas going into the Joule–Thomson valve can be precooled to significantly lower temperatures. As a rule of thumb, starting the

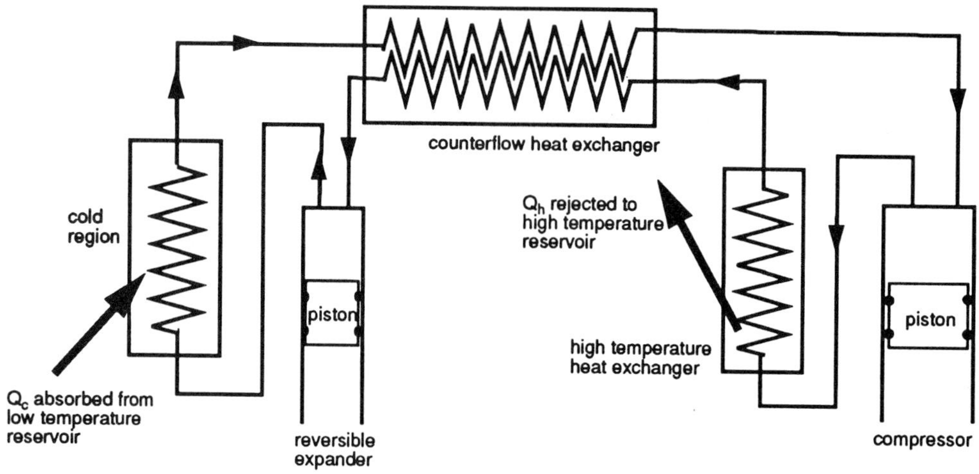

FIG. 7. A schematic diagram of Joule–Brayton recuperative cryocooler. Heat is removed from the gas by allowing it to do work on the piston in the reversible expander.

expansion near the critical temperature of the liquid greatly increases the efficiency of the refrigerator.

A Hampson–Linde cycle (named for its inventors) uses a Joule–Thomson expansion and a recuperative heat exchanger. Figure 8 shows a schematic diagram of this cycle, and the path taken by the gas in a temperature–entropy diagram is shown in Fig. 9. The numbers in the *ST* diagram refer to the points in this Joule–Thomson refrigerator.

$1 \rightarrow 2$ is a compression from the low-pressure side of the system at P_1 to the high-pressure side at P_2. This compression raises the temperature of the gas above the ambient (room) temperature, and as the gas is fed through the high-temperature heat exchanger, the temperature is returned to the ambient temperature, $2 \rightarrow 3$. Then the gas passes through the counterflow heat exchanger, where it is cooled by the returning gas, to point 4. The Joule–Thomson expansion, $4 \rightarrow 5$, follows a line of constant enthalpy and ends in the two-phase region. Thus part of the gas liquefies and can be removed with new make-up gas coming in at the intake of the compressor, or alternatively, the system can be used as a refrigerator using the latent heat of the liquid

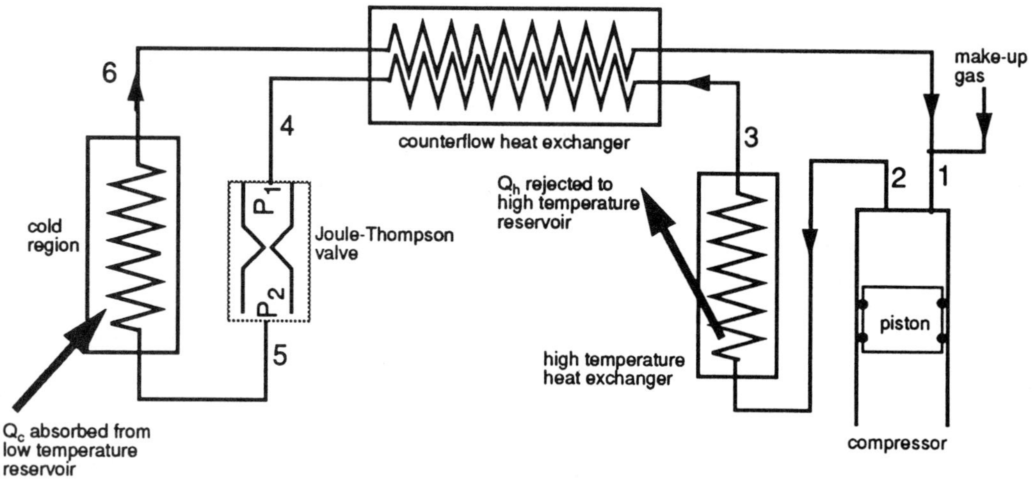

FIG. 8. Joule–Thompson expansion refrigerator. Here, the compressed gas is first returned to close to room temperature and then further cooled by returning gas in the counterflow heat exchanger. The Joule–Thompson expansion results in a liquid–gas mixture if the inlet temperature to the Joule–Thompson valve is sufficiently low.

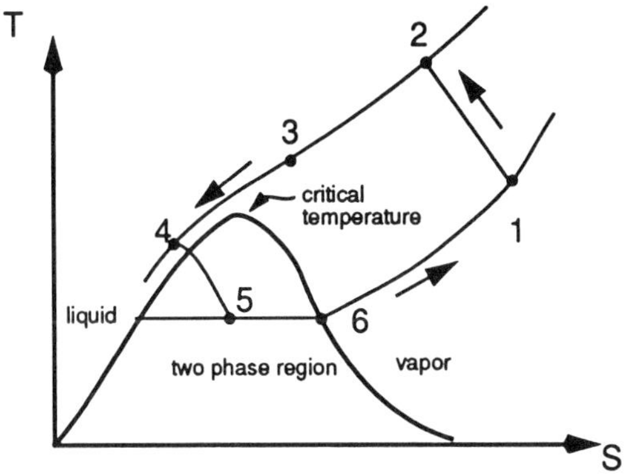

FIG. 9. A temperature-vs-entropy diagram for a typical Joule–Thompson refrigerator. The numbered points in the diagram correspond to points in Fig. 8.

in the two-phase region, 5→6. If the precool temperature (corresponding to point 4) is not sufficiently close to the two-phase region so that the Joule–Thomson expansion does not reach the two-phase region, this system still operates as a refrigerator, but at a greatly reduced efficiency.

By itself, a Joule–Thomson refrigerator can be used for the liquefaction of air and other gases with the exception of hydrogen and helium. However, this process is usually coupled with one or more other expansion engines. For example, Collins developed a helium liquefier during the late 1940s that used two Joule–Brayton type expanders followed by a Joule–Thomson expansion stage. This three-stage liquefier is now common to low-temperature laboratories throughout the world.

2.2.4 Evaporative Coolers—1 K, 0.3 K. Apparatus that requires temperatures below 4 K usually use liquid helium (^{4}He, which boils at 4.2 K at atmospheric pressure) as a starting point. In most low-temperature experiments that use liquid helium, the liquid is produced elsewhere and then transferred via an insulated line to a Dewar. The process of producing lower temperatures involves adding stages to the refrigeration process. As a first stage in cooling below 4.2 K, a simple evaporative cooler can be used.

Figure 10 illustrates a simple pumped-^{4}He refrigerator. The main Dewar holds a supply of ^{4}He at atmospheric pressure. Inside the Dewar, there is a leak-tight container ("can") whose high vacuum acts as insulator. Inside the vacuum can, a container holding liquid ^{4}He is connected to a vacuum pump, usually located outside the Dewar. A small amount of liquid ^{4}He is allowed to trickle into the "1-K pot" through a tiny capillary. The pump reduces the pressure in the 1-K pot causing the liquid ^{4}He to boil, removing the latent heat of vaporization and cooling the remaining liquid. As the temperature is reduced, the ^{4}He vapor pressure is reduced, as shown on the left-hand side of Fig. 10. The lower vapor pressure in turn lowers the amount of ^{4}He that can be pumped away so that eventually a steady-state temperature is obtained that depends on the vapor pressure and the size of the vacuum pump. With a good vacuum pump and fast pumping lines, temperatures near 1 K (more frequently nearer 1.2 K) can be obtained this way.

^{3}He is a rare isotope of helium containing one less neutron than the common isotope ^{4}He. Isotopically pure ^{3}He has significantly different properties from ^{4}He at low temperatures. We have included a sketch of the ^{3}He vapor-pressure curve in Fig. 10. Since ^{3}He has a higher vapor pressure, we can reach lower temperatures (about 0.3 K) by substituting ^{3}He for ^{4}He in the cryostat. Of course, in that case we are very careful to recover the ^{3}He from the back of the vacuum pump since it is very expensive.

2.2.5 Dilution Refrigeration. Most cryostats that reach temperatures below 0.5 K use a dilution refrigerator as the next cooling stage below the 1-K pot. A dilution refrigerator uses a mixture of ^{4}He and ^{3}He. This mixture separates into two phases below 0.86 K. At low temperatures, one phase (the "concentrated" phase) is nearly pure ^{3}He and the other phase is mostly ^{4}He with about 6.4% ^{3}He (the "dilute" phase). When concentrated ^{3}He is diluted, latent heat is absorbed in a way very much analogous to the vaporization of a liquid (such as the ^{4}He discussed above). A dilution refrigerator can operate either by having ^{3}He circulate with ^{4}He remaining stationary or with ^{4}He circulating. Dilution refrigerators with circulating ^{3}He are the most common form of these refrigerators.

The essential parts of this type of refrigerator are shown in Fig. 11. The entire low-temperature portion of the refrigerator is

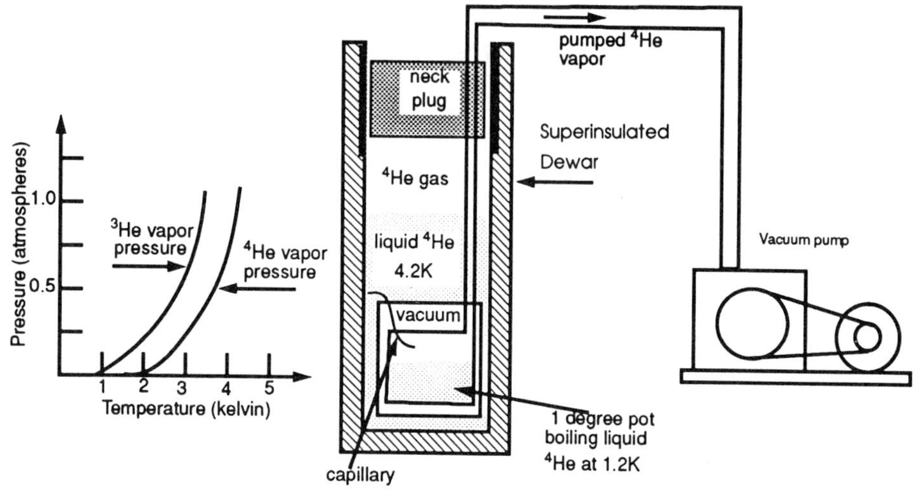

FIG. 10. A schematic diagram showing a Dewar with a 1-K refrigerator inside. The vacuum pump removes helium gas from the 1-K pot, causing cooling. The 1-K pot is refilled by a small capillary connected to the helium in the main Dewar. The inset on the left shows approximate values for the vapor pressures of ^{4}He and ^{3}He. When ^{3}He is used for refrigeration, a second ^{3}He pot is placed below the 1-K pot and the ^{3}He vapor is returned to the ^{3}He pot after passing through a series of heat exchangers.

surrounded by a vacuum can that is immersed in liquid ^{4}He in the same configuration as is shown in the ^{4}He refrigerator. With the dilution refrigerator, the cooling takes place in the mixing chamber, which has both concentrated and dilute phases in coexistence. The ^{3}He is moved through the refrigerator by pumping on the still, which is held near 0.8 K. In that temperature range, the vapor pressure of the ^{3}He is much greater than that of the remaining ^{4}He *even though the concentration of ^{3}He is small.* For this reason, the pump removes from the still almost pure ^{3}He, which is recovered and then returned. The continuous removal of ^{3}He from the still causes a depletion of ^{3}He in the mixing chamber, which in turn causes the concentrated ^{3}He to dilute continuously into the ^{4}He. This process is similar to evaporation, in which the concentrated ^{3}He is the liquid phase and ^{3}He in the dilute phase is analogous to a vapor. (Here the "vapor" phase lies below the "liquid" phase since liquid ^{4}He is denser than ^{3}He.) The temperature of the mixing chamber depends on how much heat it is removing, the size of the pumping system, and the efficiency of the heat exchangers that cool the returning ^{3}He. Dilution refrigerators with large pumping systems and very efficient heat exchangers are capable of reaching temperatures below 0.005 K (5 mK) at the mixing chamber, and some have even reached 2 mK.

Since the ^{3}He in both the concentrated and dilute phases is a Fermi liquid, a closer analogy can be drawn with the Peltier effect (Ebner and Edwards, 1971) in which current through a thermocouple junction cools (or heats). However, Peltier refrigerators do not function well much below room temperature because of the Joule (Ohmic) heating due to the electrical resistance of the wires.

2.3 Magnetic Refrigerators

2.3.1 Introduction to Spin Entropy and Adiabatic Demagnetization. Magnetic refrigerators take advantage of the heat energy associated with the magnetization of materials. These refrigerators are most easily understood in terms of entropy, with the heat required to raise the entropy being given by $\Delta Q = T\Delta S$. In the simplest case, a magnetic material can be thought of as a collection of small magnetic moments. With no magnetic field, the orientation of the spins is random. In a magnetic field, the individual moments have an energy given by

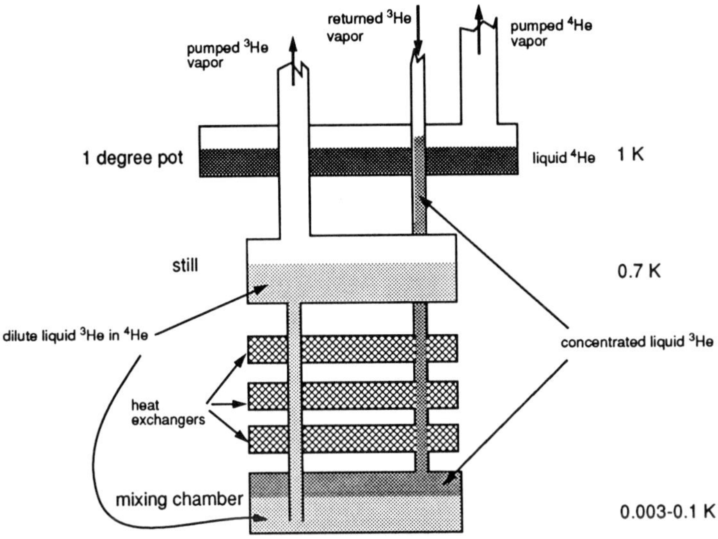

FIG. 11. A schematic diagram of a dilution refrigerator. The parts shown are placed inside a vacuum can that is inside a Dewar similar to that in Fig. 10. The still is connected to an external vacuum pump that removes mostly ^{3}He from the dilute liquid ^{3}He mixture. The 1-K pot acts as a precooler to condense the returning ^{3}He vapor. The heat exchangers further cool the returning ^{3}He. The mixing chamber removes the latent heat of mixing, which causes the mixing chamber to cool. Typical values for the temperatures of the various stages are indicated on the right.

$$\epsilon = -\mathbf{m}\cdot\mathbf{B}, \qquad (5)$$

where **m** is the magnetic moment and **B** is the magnetic field. At high temperatures, where the thermal energy $k_B T$ is greater than ϵ, the orientation of the spins is almost random. In this situation, pictured in Fig. 12(a), the number of possible rearrangements of the spins having the same total magnetization is large. Thus according to Eq. (2), the entropy at high temperatures is large. At lower temperatures where $\epsilon \cong k_B T$, the individual moments are partially aligned, as shown in Fig. 12(b), and the entropy is reduced. At very low temperatures, the spins are almost completely aligned by the field, as shown in Fig. 12(c). Here there remain very few rearrangements of the spins consistent with the same magnetization and so the entropy is near zero. Figure 13 shows a sketch of the entropy versus temperature at two magnetic fields. For isolated spins, the entropy reduction is a function only of the ratio B/T.

2.3.2 Demagnetization of Paramagnetic Materials. Figure 13 shows a typical demagnetization cycle. In step 1→2 the material is magnetized in a large magnetic field. An external refrigerator connected to the sample through a "heat switch" removes the entropy. The magnetic material is then thermally isolated and the field reduced in step 2→3. Ideally, there is no heat flow, so that this step is adiabatic and we see that the temperature is reduced. Heat flowing into the material eventually warms it up, as indicated by step 3→4. Finally the magnetic field can be increased to complete the cycle, making a magnetic heat engine; however, most practical magnetic refrigerators act as "one-shot" systems only using steps 1→2 and 2→3.

Many materials can be used to make a magnetic refrigerator. The choice of material depends on the size of the magnetic moment and the strength of the residual interactions between the magnetic moments. If the magnetic moment is small, the entropy reduction during the magnetization is small, and the refrigerator will not be effective. Interactions between magnetic moments cause magnetic ordering (either ferromagnetic or antiferromagnetic), which reduces the entropy and makes the entropy independent of the magnetic field. Thus, very little magnetic cooling can occur at temperatures below which magnetic ordering occurs.

At high temperatures Gd is frequently used for magnetic refrigeration. Below 1 K a number of rare-earth salts are used. The most common of these in low-temperature laboratories is CMN (cerium magnesium nitrate). This salt orders ferromagnetically near 2 mK (0.002 K), and this sets the minimum temperature that it can achieve.

2.3.3 Nuclear Demagnetization. Nuclear demagnetization follows the same process as that for paramagnetic materials except that the magnetic moments that are used are the nuclear moments. While these moments are much smaller, the interactions between the

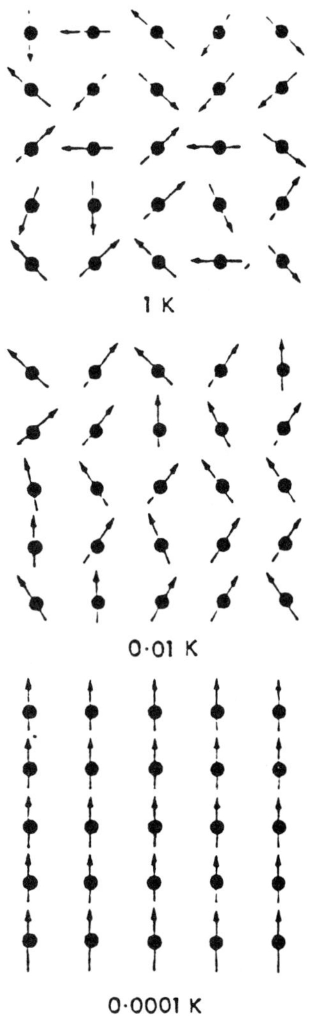

FIG. 12. Arrangements of magnetic spins at a high, intermediate, and low temperatures. The high-temperature arrangement is nearly random with a small total magnetization. There are many arrangements of spins with small magnetization causing a high entropy. At the intermediate temperature, the magnetic energy is approximately the same as $k_B T$. At low temperatures, the spins are aligned with a higher magnetization, but fewer possibilities. Thus the entropy is low. (Mendelssohn, 1977, p. 168.)

moments are also small. A nuclear demagnetization apparatus typically uses a dilution refrigerator to precool a sample such as copper in a high magnetic field (8 T or more) to temperatures near 10 mK. Sometimes, two stages of nuclear demagnetization are used with the first stage providing the precooling for the second. The entropy reduction may be less than 10% of the total entropy. Nevertheless, these refrigerators are capable of reaching extremely low temperatures because these systems, which have small interactions, do not order (ferromagnetically or antiferromagnetically) until the temperature is much smaller than in the case of electron paramagnetic materials. Nuclear demagnetization cryostats hold all of the current low-temperature records, as we will discuss in the last section.

A diagram of a nuclear demagnetization cryostat is shown in Fig. 14. The materials that can be used for nuclear demagnetization include In, Cu, Ag, and even Tl and Pt. Cu is used in the greatest number of nuclear demagnetization cryostats by far.

There is a group of metal alloys with rare earths that have a hybrid form of magnetic moment through which the atomic polarization enhances the field at the nucleus, which gives a larger effective moment. This is called Van Vleck enhancement of the nuclear moment, and these alloys can be very effective for nuclear demagnetization (Andres, 1978). The most effective of these Van Vleck alloys is $PrNi_5$, which can be used to reach temperatures in the range of 0.5 mK.

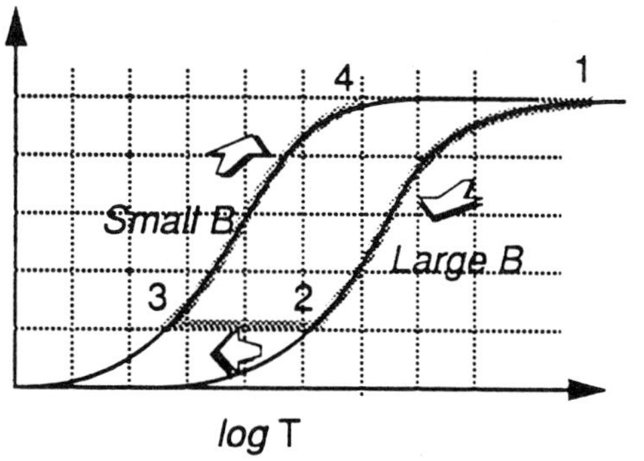

FIG. 13 The path taken in entropy-vs-temperature space for a magnetic refrigerator. The magnetic material is precooled in large magnetic fields (1→2), then demagnetized while it is thermally isolated (2→3). Heat is removed from the surroundings at low field (3→4), and finally the field is raised which increases the temperature (4→1).

3. MEASUREMENT OF LOW TEMPERATURES

3.1 Temperature Scales

3.1.1 The Absolute Temperature Scale. The definition of the true thermodynamic temperature is subtle and annoyingly unsatisfying. The intuitive sense of temperature, being what registers on a thermometer when it has come into thermal equilibrium with the material to be measured, is, however satisfying, insufficient to determine uniquely an absolute universal scale. A correct definition of temperature is that it is the quantity that connects entropy changes and heat exchanges in reversible processes:

$$dS = đQ/T. \tag{6}$$

Equivalently, $1/T$ is the integrating factor that converts differential heat exchanges into a perfect differential, so that thereby entropy is a state variable.

Obviously, every practical thermometer does not immediately refer to this definition, but each does have a connection to it through whatever chain of calibrations establishes its accuracy.

The definition implicitly defines $T=0$ at the absolute zero of temperature, so that all that is required for a complete scale is an overall scale factor, set by defining one special temperature. That special place is chosen by in-

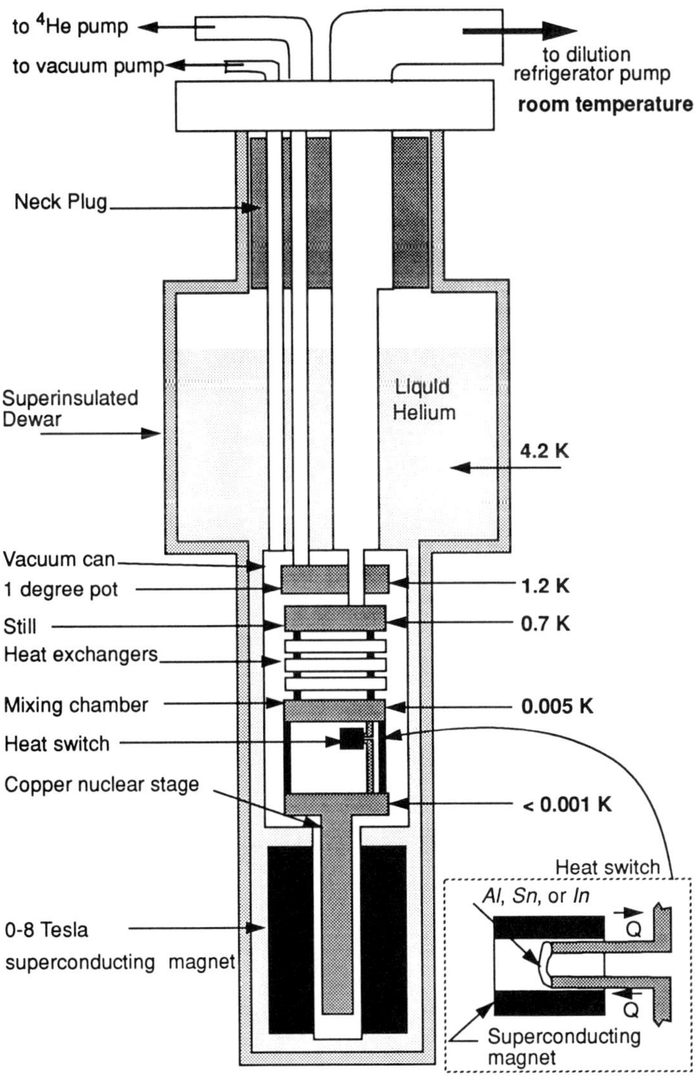

FIG. 14. Nuclear demagnetization. This cryostat combines the features of several refrigeration processes. The main components are indicated on the left. Three pumping lines are connected to external pumps and gas recovery systems as shown. The bold printed temperatures indicated along the right-hand side of the figure are typical operating temperatures of the various stages in the cryostat. Liquid ^{4}He is delivered to the main Dewar. A 1-K pot maintains 1.2 K. The dilution refrigerator provides precooling for the nuclear demagnetization stage, which is connected by a heat switch. When the demagnetization stage is isolated from the dilution refrigerator, the main magnet field is lowered, lowering the last stage to submillidegree temperatures. The inset shows a heat switch using a superconducting bridge between copper rods.

ternational convention to be the triple point of water (at which ice, water, and water vapor are simultaneously in equilibrium, $P=4.579$ Torr$\approx$6.1 mbar) and defined to be $T=273.16$ K, where K stands for the international unit of temperature, the kelvin (not degree kelvin). The Kelvin scale is used principally in scientific work, particularly at low temperatures.

3.1.2 Other Temperature Scales. There are three other temperature scales in common usage today. Each was originally defined by choosing two special points to define two temperatures. Each is thus linearly related to the Kelvin temperature scale. Closely connected to the Kelvin temperature scale is the Celsius scale, with temperature changes numerically equal in each. Today the definition is $T(°\mathrm{C})=T(\mathrm{K})-273.15$. The original definition of this scale used the equilibrium temperature of melting ice at 1 atm as 0 °C, and the normal boiling point of water (where the pressure is 1 atm) as 100 °C. This exact 100 °C interval was preserved (to the then existing level of accuracy of experiments) by the specific choice of Kelvin scale factor given by choosing the triple point of water temperature given above. Accounting for the negative slope of the water–ice coexistence curve, the temperature of the normal melting point is about 0.003 °C. The Celsius scale, having zero closer to ordinary human conditions than the Kelvin scale, is the most commonly used temperature scale in the world, and is used in most scientific work near room temperature.

An earlier scale is the Fahrenheit scale, originally defined to be 0 °F at the lowest temperature then available (ca. 1700, with a salt–ice mixture) and 12 °F (carrying over a proposal by Isaac Newton) at normal human body temperature, later multiplied by eight to become 96 °F. (Nonlinearities in early thermometers led this number to differ from the now standard value 98.6 °F.) This scale is currently defined in terms of the Celsius scale as $T(°\mathrm{F})=\frac{9}{5}T(°\mathrm{C})+32$. It is the one most commonly used in the United States, but is (very) slowly being displaced in favor of the Celsius scale.

Finally, analogous to the relationship between the Kelvin and Celsius scales, in which the size of a unit temperature change is the same but with a zero offset, is the relationship between the Fahrenheit and Rankine scales. The Rankine scale is an absolute Fahrenheit scale and defined equivalently by either $T(°\mathrm{R})=T(°\mathrm{F})+459.67$, or $T(°\mathrm{R})=\frac{9}{5}T(\mathrm{K})$.

3.1.3 International Temperature Scale of 1990 (ITS-90). The current international temperature scale, ITS-90, was established at the 17th session of the Consultative Committee on Thermometry (CCT) in September 1989 (Preston–Thomas, 1990). The CCT is one of eight Consultative Committees of the International Committee of Weights and Measures, which is, in turn, a committee of the General Conference of Weights and Measures, the authority, by international agreement, for the International System of Units (SI). The CCT is composed of representatives of national standards laboratories throughout the world and of other laboratories concerned with the accurate definition of temperature.

ITS-90 establishes temperatures down to 0.65 K. It also defines overlapping ranges and the functions to be used in calibrating the various thermometers used in each range. (See Fig. 15.) From 0.65 to 5.0 K, temperature is defined as a specified function of the vapor pressure of each of the helium isotopes in separate subranges. From 3.0 to 24.5561 K and from 13.8033 to 1234.93 K the scale is defined in terms of, respectively, a constant-volume gas thermometer using each of the helium isotopes and a platinum resistance thermometer. Up to room temperature each

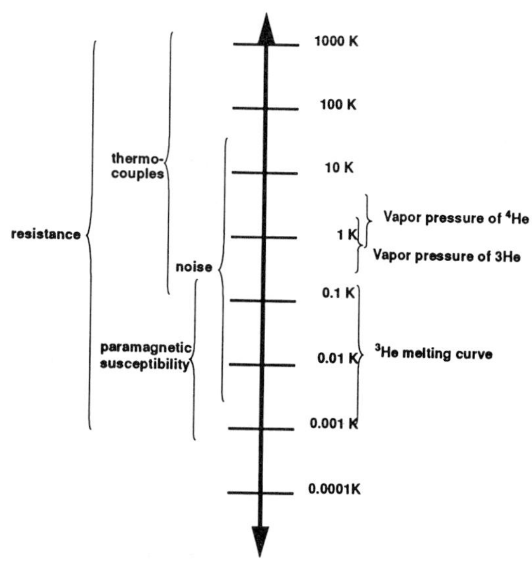

FIG. 15. A logarithmic temperature scale indicating the various types of thermometry that are used.

Table 1. Summary of classes of thermometers for use at cryogenic temperatures.

Class	Theoretical foundation	Examples	Usability	Range	Accuracy	Sample-to-sample reproducibility
Primary	Rigorous, absolute	Ideal gas law, 2nd law, nuclear orientation	Generally poor	Limited	High	Excellent
Secondary	Strong	NMR, noise	Poor to good	Limited	Needs calibration point	Good
Tertiary	Weak	Resistance, diode voltage, thermocouples	Good	High	Needs calibration curve	Poor
Transfer standards	None	Vapor pressure of ^3He and ^4He, solidification pressure of ^3He	Poor	Limited	High	Excellent
Transfer standards	None	Triple points, superconducting transitions	Poor	None	High	Excellent

thermometer is calibrated at the triple points of hydrogen (with equilibrium ortho–para ratio), neon, oxygen, argon, mercury, and water at 13.8033, 24.5561, 54.3584, 83.8058, 234.3156, and 273.16 K, respectively.

ITS-90 superseded the previous temperature scale, known as ITPS-67(75), by extending its lower bound below the triple point of H_2 and changing temperatures below the triple point of H_2O by a mean amount of 10 mK. It also provides better agreement with thermodynamic values than the 1976 Provisional 0.5 K to 30 K Temperature Scale (EPT-76).

3.2 Classes of Thermometers

All thermometers operate by measuring some physical property of a material that changes with changing temperature. Every thermometer can be classified on the basis of its ability to define the absolute temperature. The classification scheme given here is consistent with general use, although there is no universally recognized definition. Examples of each class of thermometer will be given in the following section. These classes are summarized in Table I.

A primary thermometer is one that is capable of defining the temperature by itself, without reference to any other device. It must therefore be grounded in a rigorous thermodynamic principle. For example, a gas thermometer measuring the pressure P of a fixed amount of gas and comparing it to the pressure when in equilibrium with water at its triple point, P_{tr}, would define temperature as

$$T=273.16 \lim_{P_{tr}\to 0} \left(\frac{P}{P_{tr}}\right), \qquad (7)$$

where the limiting process successively reduces the amount of gas in the volume to eliminate the effect of intermolecular interactions. The reliance on the ideal gas law and the experimental production of an ideal gas (through the limiting process) makes this a good primary thermometer, although in practice it is rather limited. In general, the reliance on thermodynamic principles renders primary thermometers independent of the working substance, so that reproducibility from sample to sample (and, more importantly, from laboratory to laboratory) is excellent.

A secondary thermometer is one in which an absolute temperature can, in principle, be measured (at least in some approximation) but for which technical difficulties prevent absolute calibration. For example, the nuclear susceptibility in platinum (far from any ordering temperature) to a high accuracy obeys a Curie law with a known coefficient. The difficulty, however, is that an absolute calibration would require an accurate knowledge of the coupling of the platinum nuclei to the thermometer's electronics. Usually the process of determining this coupling is foregone in favor of calibrating at some known point. In principle, a single calibration point is sufficient to define a secondary thermometer's entire absolute scale.

All other thermometers without simple connection to fundamental thermodynamic principles, which includes almost all thermometers in use, are classified as tertiary. Certainly the physics of their operation is well known, but difficulties in a first-principles calculation prevent their being classified as secondary. For example, the voltage developed by a thermocouple arises from the temperature-dependent chemical potential of the two involved materials, but this is exceedingly difficult to evaluate even for simple metals, let alone the metal alloys usually used in practical thermocouples.

Finally, we mention the role of transfer standards and fixed points. These provide an excellent mechanism to transfer temperature scales between laboratories. These are exclusively associated with phase transitions (fixed points simply being the intersection of at least two phase transitions) and are thus not calculable from first principles without a known equation of state, which we do not have. These transitions can be used either directly, as in the case of ^{3}He melting-curve thermometers in the low-millikelvin regime, or indirectly, to calibrate other, potentially more convenient, thermometers such as by using the pressure-dependent superfluid transition temperature in ^{3}He, $T_c(P)$, which establishes a reliable scale from 0.9 to 2.5 mK (Greywall, 1986).

3.3 Practical Thermometers

A good thermometer should have high sensitivity, short internal thermalization time constants (so that the thermometer itself will have a well-defined temperature), good thermal contact with the system, and negligible total heat capacity so as not to load thermally the system it is measuring (which may be achieved simply by making the thermometer physically small), and should introduce negligible heat to the system through its operation. It should furthermore be insensitive to all environmental influences other than temperature. A good thermometer for one measurement may be a poor one under different conditions. We summarize here some of the thermometers used in cryogenic measurements. The ordering of these thermometers generally follows their useful temperature range, in decreasing order.

3.3.1 Constant-Volume Gas Thermometer. As mentioned above, the pressure of a fixed amount of an approximately ideal gas can be used, although corrections for nonideality are important. This thermometer can be used until the pressure becomes so small that thermomolecular corrections become significant. This limits its practical range to that above several kelvins. For precision metrology there is also a significant problem associated with adsorption of the gas on its container's walls. Even for the noble gases, the binding energy for initial layers to surfaces is 100 K or more, which can significantly affect the inferred molar volume if the surface-to-volume ratio is not negligible.

3.3.2 Thermocouples. Thermocouples have a long history of importance, being used in the definition of the Kelvin scale until the advent of the current scale, ITS-90. They have the advantage of simplicity, but the disadvantage of relatively small voltage signals which vanish as $T \to 0$. Below 10 K they are generally not useful, although special Kondo-enhanced thermocouples (see HEAVY-FERMION PHENOMENA) have been demonstrated to work in a limited range around 10 mK.

3.3.3 Cryogenic-Liquid Vapor Pressure. The vapor pressure of cryogenic liquids can be used as a thermometer. The ITS-90 is, in fact, *defined* in terms of the helium liquid vapor pressures from 0.65 to 5.0 K. This technique has a range limited from above by the liquids's critical point and from below by the rapidly vanishing vapor pressure, decreasing as $\exp(-\Delta/k_BT)$, where Δ is a constant (at low temperatures) proportional to the latent heat of vaporization. Even using ^{3}He, with its small latent heat of vaporization, the lowest measurable temperatures with this technique are only 0.3 K.

3.3.4 Resistance Thermometers. One of the simplest tertiary thermometers to use is a resistance device. Any material with a temperature-dependent resistance can be used. Typical materials are carbon (including ordinary carbon composition resistors), germanium, platinum, and numerous alloys. All resistors exhibit some degree of magnetoresistance so that calibrations are distorted by modest (few teslas) magnetic fields, but most resistors retain sensitivity to thermal changes, which may be sufficient in a given application. Off-the-shelf carbon resistors are convenient, but vary in their characteristics from batch to batch and may show significant irreproducibility during the first few thermal cyclings.

The low-temperature limit is set by different effects for different materials. For carbon resistors, the resistance of which climbs with decreasing temperature, the low-temperature limit is set by the maximum resistance measurable without significant interference from spurious effects, such as leakage resistance. A typical maximum resistance is 1 MΩ. For commercial thermometers the low-temperature limit is around 1 K, although uncalibrated carbon resistors routinely used in ultralow-temperature laboratories can be used down to a few millikelvins. Carbon resistors generally also have a problem with internal thermalization in the millikelvin regime because they are poor thermal conductors.

Platinum resistors decrease in value with decreasing temperature. Their practical low-temperature limit is about 10 K, which arises from their vanishing temperature coefficient at low temperatures and the power dissipation from the temperature-independent part of the resistance.

3.3.5 Diode-Voltage Thermometers. Closely connected to resistance thermometers are diode thermometers. The difference between diodes and resistors is only in the diode's nonlinear current–voltage relationship. For a fixed current, as is used conventionally, the two are identical. Driven with 10 μA forward current, silicon diodes yield a sensitivity of 2–3 mV/K from room temperature down to about 30 K when the carriers begin to freeze out and the sensitivity increases to 50–100 mV/K. GaAs diodes have a similar sensitivity at high temperatures, but this sensitivity decreases below about 20 K. The low-temperature limit is about 1 K, set by the rather large power dissipation of 10 μW.

3.3.6 Capacitance Thermometers. Capacitance thermometers use the temperature dependence of the dielectric susceptibility of ferroelectric (and other ordered) crystals. They have the advantage of insensitivity to magnetic fields and are used principally to regulate temperatures as magnetic fields are applied. Capacitance thermometers utilizing $SrTiO_3$ exhibit a nonmonotonic temperature dependence, peaking around 65 K. They also suffer from irreproducibility when thermally cycled.

3.3.7 Ferromagnetic and Antiferromagnetic Frequency Shifts. The resonance frequency of nuclei in solids depend sensitively upon the internal magnetic fields. Therefore there have been a number of proposals to measure temperature via nuclear frequency shifts in magnetically ordered materials. By choosing the appropriate ferromagnet or antiferromagnet, essentially any temperature interval can be covered. The disadvantages of this technique are the sensitivity to sample preparation and the narrow temperature interval over which this technique has high sensitivity.

3.3.8 Electron Paramagnetic Susceptibility. The magnetic susceptibility of electrons in paramagnetic salts provides a convenient thermometer at low temperatures. For example, by choice of a material with a suitable magnetic ordering temperature (and by use of a large measuring field), temperature resolution of 1 nK near the λ point of ^{4}He can be achieved (Chui and Lipa, 1984; Steinberg and Ahlers, 1983; Marek, 1987). Below 1 K a convenient material is cerous magnesium nitrate (CMN), which has a large susceptibility, but a relatively low ordering temperature of 1.8 mK. In the low-millikelvin regime it is much more common to dilute the magnetic Ce ions with nonmagnetic La in a 1:19 ratio. This material is conventionally named LCMN. The temperature dependence of the susceptibility of LCMN is accurately given by a Curie–Weiss form, $\chi \propto 1/(T-\Delta)$, down to 0.3 mK, although this needs to be checked for each sample, given that the randomness of the La-for-Ce substitution means that there must exist CMN regions which are ordered.

3.3.9 Nuclear Susceptibility. Analogous to the magnetic susceptibility of electrons, the nuclear susceptibility also makes a convenient thermometer that is very common in ultralow-temperature laboratories. Because of the much smaller magnetic moments of nuclei relative to electrons, the susceptibility is smaller, pushing down the upper temperature limit of usability to a few tens of millikelvin. Also because of the reduced moments, the nuclear interactions are smaller than electronic ones, suppressing the ordering temperatures and pushing down the low-temperature limit of usability. As long as the temperature is significantly above 1 μK, well above the nuclear ordering temperatures, the susceptibility is accurately described by a Curie form, $\chi \propto 1/T$. For purposes of measuring temperatures in the low-millikelvin regime, the most commonly used material is platinum, which has a spin-lattice relaxation time of about (30 mK)/T s. As usually implemented, pulsed NMR spectrometers show pulse-to-pulse amplitude reproducibility of only about 1%, requiring considerable averaging for high accuracy.

3.3.10 ^{3}He Melting Pressure. An interesting thermometer used in several ultralow-temperature laboratories uses the melting pressure of ^{3}He directly as a thermometer. All that is needed to measure temperature is an accurate calibration of a pressure gauge. Zero offsets can be eliminated by adoption of the A transition of superfluid ^{3}He as the reference point. The melting curve is roughly independent of magnetic field for small fields ($B \leqslant 0.5$ T). This thermometer suffers from a rather large thermal load, but this is not significant for experiments studying superfluid ^{3}He or using it in an auxiliary role. This thermometer loses sensitivity at the low end below about 1 mK, where the solid orders magnetically. It is also insensitive at temperatures approaching the minimum in the melting curve around 0.3 K.

3.3.11 Noise Thermometry. Thermal noise (known as Johnson noise or Nyquist noise) is a limiting factor in resistance thermometry. However, this noise can itself be used as a thermometer directly. The maximum available noise power per unit bandwidth of any device is k_BT. In the case of a pure resistor of value R, at temperature T, thermal fluctuations of local charge density produce an effective voltage source in series with the resistor with mean square amplitude in a frequency interval Δf of $V^2 = 4k_BTR\Delta f$. The fundamental difficulty at low temperatures is efficient coupling of the noise into a detector that itself has a noise temperature small compared to that being measured, for which several solutions have been attempted. With commercial rf superconducting quantum-interference devices (SQUIDs), detector noise temperatures of order 1 mK have been demonstrated in noise thermometers. In principle, this thermometer could be a primary thermometer, but difficulties in precisely specifying the system transfer function over the entire bandwidth for which noise contributes power has limited devices to date to secondary status, requiring one calibration point to fix the temperature.

3.3.12 Nuclear Orientation. Nuclear orientation thermometry is a primary thermometer in that all of its parameters can be specified by studies of radioactive decay modes. The fundamental idea is that γ rays associated with a chain of radioactive decays are not emitted isotropically, but rather have a calculable probability of being emitted at any specified angle with respect to an axis easily defined by an applied magnetic field. The radiation pattern is isotropic at high temperatures and is maximally anisotropic at $T=0$. By measuring the anisotropy one can deduce the temperature. This thermometer suffers from self-heating associated with the nuclear decay by-products. Reducing the activity of the source to reduce the heating increases the time required to count decays. Furthermore, the accuracy of this device is not high, limited by instrumental difficulties to about 1%. Although there is no shortage of nuclei and hosts to consider, it appears unlikely that nuclear orientation will be a useful primary thermometer below several millikelvin.

3.3.13 Quasiparticle Damping Thermometry. A unique thermometer developed at the University of Lancaster, United Kingdom, measures the damping of a wire vibrating in liquid ^{3}He. At the lowest temperatures, the damping arises from individual quasiparticles being scattered from the wire. Since the number of quasiparticles as a function of temperature is known, this can be a good secondary thermometer. At temperatures above about 500 μK the damping becomes dominated by ordinary hydrodynamic viscosity, but this device is still useful as a tertiary thermometer. Since the ^{3}He itself is the thermometer, this avoids problems of thermal contact between ^{3}He and some other thermometer. This device has been used to measure the lowest temperatures in ^{3}He and has been used in dilute mixtures with some attendant controversy.

3.3.14 Nuclear Quadrupole Thermometry (NQT). A speculative technique being investigated by several laboratories uses a small Zeeman splitting (see ZEEMAN AND STARK

EFFECTS) of nuclear quadrupole levels to establish a primary thermometer. Nuclear resonance measures transitions between different quadrupole levels, with slightly shifted frequencies arising from the Zeeman fine structure. The ratio of the spectral weights of the different transitions depends upon relative populations of different energy levels E_i, which are determined by a Boltzmann factor $\exp(-E_i/k_B T)$. The possibility of such a thermometer has been demonstrated using a variety of nuclei at temperatures from 10 mK to 1 K.

3.3.15 Second-Law Thermometry. Finally, at the coldest temperatures yet reached, none of the above thermometers is reliable. Fortunately, the second law of thermodynamics can be used directly. The basic idea is to measure some thermometric property (such as magnetization) from which we can deduce the entropy while injecting a known amount of heat $đQ = TdS$. The temperature is then immediately deducible.

4. HOW LOW CAN WE GO?

The second and third laws of thermodynamics state that the entropy of any system at zero temperature is zero, but the efficiency of removing entropy also approaches zero at zero temperatures. For this reason, $T=0$ stands as one of Nature's ultimate limits, much like the speed of light in a different context.

4.1 Negative Temperatures

In any discussion of low-temperature limits, it is natural to consider the possibility of exceeding the ultimate limit. While it is impossible to achieve negative temperatures solely by exchange of heat between reservoirs, it is possible to produce negative temperatures by nonthermal means.

In a discussion of *low*-temperature limits, it may appear odd to discuss negative temperatures because, simply put, negative temperatures are hotter than positive temperatures. However, the theoretical considerations involved in producing negative temperatures are identical to those in producing the current ultralow-temperature limits.

The range of systems in which negative temperatures can be achieved is quite limited. Consider an ensemble of two-state systems with energy levels $E_i = \pm\Delta/2$, as shown in Fig. 16, such as a spin-$\frac{1}{2}$ system in a magnetic field where $\Delta = \mu B$. In thermal equilibrium, the probability of level i being occupied is proportional to the Boltzmann factor $\exp(-E_i/k_B T)$. At ordinary (positive) temperatures this is consistent with the usual perception that lower-energy levels are more populated than higher-energy levels, as shown in Fig. 16(a).

If, however, by some means, such as magnetic resonance or rapidly reversing the field direction, the upper level can be made more populated, then the effective temperature describing this situation would be negative, as shown in Fig. 16(b). In fact, exactly this procedure of rapidly (quasiadiabatically) reversing the field direction has demonstrated negative temperatures in silver nuclei and their ferromagnetic tendency in the *minus* 10-nK temperature range.

What was required for this description to work? The obvious requirement was that the spectrum of energy levels needed to be bounded from above. Otherwise, at negative temperatures, the levels at infinitely high energy would be infinitely occupied. This is why a classical gas with translational kinetic energy (unbounded above) cannot exhibit negative temperatures. Since energy levels are

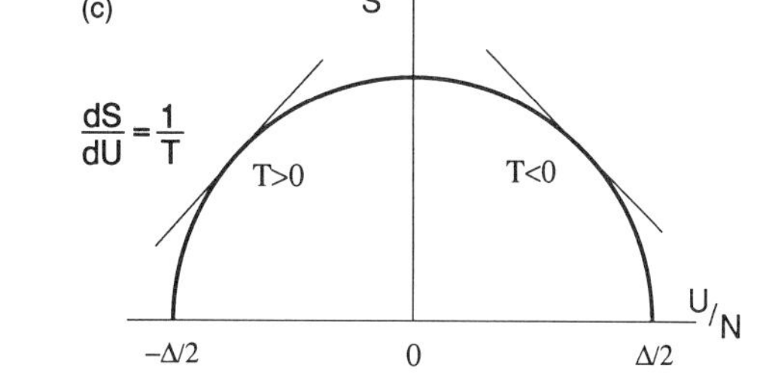

FIG. 16. Negative temperatures. The relative occupation probability of different energy levels in thermal equilibrium defines a temperature which can be (a) positive or (b) negative. In (c) the same problem is viewed from the perspective of the entropy, suggesting an alternative scale of temperature.

always bounded below, there is no converse argument against positive temperatures.

A more subtle requirement is that the system itself must be in thermal equilibrium. That is, there is a hierarchy of characteristic times that must be properly ordered. The characteristic time for the energy in the system to distribute itself thermally must be short compared to the time scale on which temperature is being defined by the experimental measurement, which in turn must be short compared to the time scale on which external influences change the system's temperature. This is analogous to the requirements given earlier for a thermometer to measure accurately the temperature of its environment.

The requirement of fast internal equilibration is somewhat obscured in the two-state example because any ratio of occupation probabilities uniquely defines some temperature, $N(E_1)/N(E_2)=\exp[(E_2-E_1)/k_BT]$. In an N-state problem, however, there are $N-1$ independent ratios of occupation probabilities, all of which must be derivable from one unique temperature if that quantity is to make sense. Given an arbitrary nonthermal disturbance of population levels, these $N-1$ ratios will not yield a unique temperature during an initial time interval before energy can be exchanged within the system.

An entirely equivalent description of negative temperatures is given in Fig. 16(c), where the entropy of the ensemble is shown as a function of its total internal energy. The derivative of this function is, by definition, the inverse of the temperature. The evolution of the temperature as a function of increasing energy can be read off easily. When the energy is minimum, the slope dS/dU is infinite, so that $T=0$. As the energy increases, the (positive) slope decreases until it reaches 0, at which point $T=+\infty$. Further increases in the energy cause the slope to decrease so that the temperature continuously evolves from $-\infty$ to -0, at which point the energy is a maximum.

This leads naturally to the frequent suggestion for an alternative scale of temperatures given by the negative slope $-dS/dU=-1/T$. This scale, called the "minus beta" scale ($-\beta=-1/k_BT$) or names equally unpoetic, satisfies the intuitive sense that it increases with increasing energy, and has the further advantages of correctly ordering negative and positive temperatures, and providing a ready answer to the question of why absolute zero is unattainable: you cannot reach infinity.

4.2 Present Ultralow-Temperature Limits

Finally, we summarize the current state of the art in achieving low temperatures. A logarithmic temperature scale showing the present limits of low-temperature technology is shown in Fig. 17. This is a dynamic field, and by the time this article appears in print some of the records listed here will be out of date. Nonetheless, here is a snapshot of the lowest records known to us as of August 1991.

The reason that there is more than one record is that for different systems there are different tradeoffs. For example, when trying to reach the lowest temperatures in a nuclear system, it is advantageous to leave the electrons at an elevated temperature to act as a larger buffer against heat leaks from the outside. The thermometers used at the end of the temperature scale here are also working at their limits, so that some revision in the reported temperatures is to be expected.

The lowest temperature achievable in a continuous mode is approximately 2 mK (Frossati, 1978). Short of mechanical failure or conscious decision to stop, this temperature could be maintained indefinitely. All of the remaining records listed use one or more stages of nuclear refrigeration, and so are intrinsically limited in duration, though, as mentioned earlier, this duration can be on the order of weeks at temperatures only fractionally above the low limit.

The lowest temperature at which a metal is known to become superconducting is 325 μK for rhodium (Buchal *et al.*, 1983) as discovered in the laboratory of F. Pobell, now at the University of Bayreuth in Germany.

In trying to establish a low temperature, both a low specific heat and low thermal boundary resistance are desirable. Dilute solutions of ^{3}He in ^{4}He do not have either of these advantages. The lowest measured temperature in these dilute mixtures is about 220 μK (Owers-Bradley *et al.*, 1983), although this measurement was limited by thermal contact to the thermometer so that the actual temperature may well be considerably lower.

Moving to ^{3}He lowers the boundary resistance, but increases the problem of high spe-

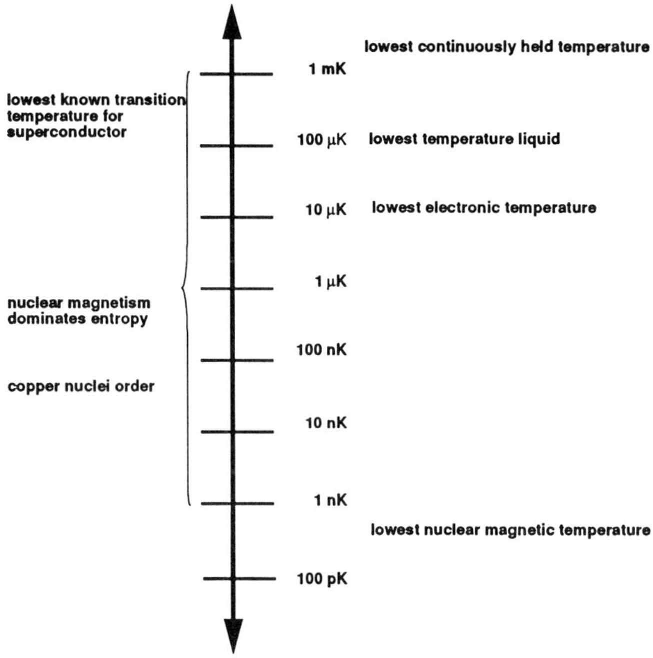

FIG. 17. A logarithmic temperature scale showing the approximate locations of the lowest temperatures that have been achieved to date and some of the new physical phenomena observed in the submillidegree temperature regime.

cific heat. In superfluid ^{3}He-*A*, with a specific heat varying as T^3, the current record is about 140 μK, while in superfluid ^{3}He-*B*, with specific heat decreasing faster as $\exp(-\Delta/k_BT)$, the record is just under 100 μK (Carney *et al.*, 1989). Both of these records were established at the University of Lancaster in the United Kingdom, and were measured via the quasiparticle damping of a vibrating wire immersed in ^{3}He-*B*. Much earlier measurements in Lancaster using a platinum NMR thermometer showed that ^{3}He-*B* could be cooled to about 110 μK.

Solid ^{3}He, with its much lower heat capacity for T less than 1 mK, has been cooled to about 40 μK (Takano *et al.*, 1985; Mamiya *et al.*, 1990).

Moving to metals considerably reduces the specific heat, and consequently gives the lowest achievable temperatures. Both of the groups in Lancaster and in Bayreuth have demonstrated temperatures in copper of 12 μK as measured by a platinum NMR thermometer in good thermal (i.e., electrical) contact with the copper (Gloos *et al.*, 1988).

Finally, we consider the temperature of the nuclear system in a metal. Here we gain the advantage of measuring the temperature of the actual refrigerant and not that of a material cooled by it. At the very lowest temperatures, the time constants for nuclear thermalization with the electronic system in the same block of metal can be made quite long, on the order of hours. Since the internal thermalization time within the nuclear system is much shorter, typically milliseconds, it makes good sense to define separate temperatures for the nuclear and electronic systems, even though they are physically inseparable (cf. the discussion above, regarding negative temperatures).

In copper, the nuclear ordering temperature is 58 nK. Since most of the nuclear entropy is removed in the vicinity of the transition, its cooling power rapidly diminishes, limiting the lowest temperature in copper nuclei to 30 nK, established at the Helsinki University of Technology in Finland (Huiku *et al.*, 1986). (At the same time, it should be noted that the electronic system in the same sample is maintained at about 50 μK, i.e., about a thousand times hotter.)

Silver, with much weaker nuclear interactions than copper, allows a much lower temperature to be reached. Using silver, the Helsinki group established the current nuclear low-temperature record of 560 pK (Hakonen *et al.*, 1991), 9 orders of magnitude below the bottom end of the International Temperature Scale of 1990.

ACKNOWLEDGMENTS

We wish to thank W. J. Gully, G. R. Pickett, and P. J. Hakonen for useful comments on this manuscript. Much of the research by the authors has been supported by the National Science Foundation Low Temperature Program.

GLOSSARY

Antiferromagnetism: A state of magnetic order in which alternate individual spins have opposite orientations, resulting in a reduced total magnetization.

Blackbody Radiation: Electromagnetic radiation coming from a heat source.

Carnot Cycle: A thermodynamically reversible closed cycle consisting of two adiabatic (see entropy) and two isothermal processes.

Compressor: A machine that uses mechanical energy to do work on a gas by reducing its volume.

Convective Flow: A spontaneously circulating flow caused by a gravitational instability associated with a hot, low-density fluid placed below a cold, higher-density fluid.

Counterflow Heat Exchanger: A device that allows heat to flow between two fluids that are moving in opposite directions.

Dewar Flask: Originally, a double-walled glass container with a vacuum between the walls and reflecting surfaces that provides thermal insulation. Currently applies to any well thermally insulated container. This term is often shortened to Dewar.

Dilution Refrigerator: A refrigerator that obtains cooling by diluting ^{3}He into a mixture of ^{3}He and ^{4}He.

Enthalpy: A thermodynamic potential obtained by subtracting the mechanical term ($-pV$) from the internal energy, i.e., $H=U+pV$.

Entropy: The thermodynamic variable that measures the quality of thermal energy. From a statistical-mechanics view, entropy measures the degree of disorder in a system.

Evaporative Cooler: A simple refrigerator that cools by evaporating a liquid.

Expansion Engine: The reverse of a compressor. Here, a high-pressure gas does work on a moving piston, thereby removing thermal energy.

Ferromagnetism: A state of magnetic order in which the individual spins have the same direction, resulting in a high total magnetization.

Heat Switch: A device that can interrupt heat flow. These are often made from superconducting metals, where the thermal conductivity is high in the normal state and low in the superconducting state. Other heat switches have been made from movable mechanical contacts and gas-filled containers.

Joule–Thomson Coefficient: The rate of cooling or heating caused by adiabatic (Joule–Thomson) expansion of a gas.

Joule–Thomson Expansion: An adiabatic expansion of gas through a restriction such as a small tube, a porous plug, or an adjustable valve (Joule–Thomson valve). This is a constant-enthalpy process.

Joule–Thomson Inversion Point: A temperature point in a phase diagram where the Joule–Thomson coefficient changes sign. These points depend on pressure and volume, and thus form a line in a phase diagram.

Latent Heat: The heat of vaporization or melting. This corresponds to the change in enthalpy in going from one phase to another.

Magnetic Ordering: An increase in the organization of magnetic spins as temperature is lowered. The magnetic order can be either ferromagnetic, ferrimagnetic, or antiferromagnetic and is accompanied by a reduction in the entropy.

Magnetic Refrigerator: A refrigerator that uses a magnetic material as the working substance and a magnetic field to provide the external work.

Nuclear Demagnetization: A refrigeration technique that uses the magnetic moments of nuclei at the magnetic material. (See Magnetic Refrigerator.)

Paramagnetic Material: A material that has no magnetic order.

Phonons: Lattice vibrations in a solid. Formally, phonons form a basis set to describe lattice deformations and are thus pseudoparticles.

Recuperative Refrigerator: A refrigerator that regains cooling by the use of a counterflow heat exchanger to precool returning gas.

Regenerator: Formally, an ideal heat source and sink. In heat engines, a regenerator temporarily stores heat during part of a cycle. An efficient regenerator can maintain a thermal gradient so that heat can enter and leave at a series of different temperatures.

Stirling Engine: A heat engine that follows the Stirling cycle, which is equivalent to a Carnot cycle using a gas plus a regenerator. Often, any heat engine that uses a regenerator is referred to as a Stirling engine.

Superinsulation: A highly effective cryogenic insulation made from multiple layers of aluminized polyester film. It operates in a vacuum. Frequently, the layers are spaced by a web of plastic fiber. This insulation works as an effective radiation shield because of the layers of reflective aluminum. The polyester and other plastics are poor thermal conductors. Also, the multiple layers trap any residual gas atoms or molecules.

Working Substance: The part of a heat engine that undergoes mechanical (pressure, volume, magnetization, etc.) and thermal (temperature, entropy) changes during a cycle. Typically, this is a gas, liquid, or magnetic material.

Zero-Point Motion: The minimum motion of atoms and molecules about their equilibrium position that is required by the wave nature of these particles. The energy associated with this motion cannot be removed even at $T=0$, and thus is not part of the thermal energy.

List of Works Cited

Andres, K. (1978), *Cryogenics* **18**, 473–477.

Buchal, Ch., Pobell, F., Mueller, R. M., Kubota, M., Owers-Bradley, J. R. (1983), *Phys. Rev. Lett.* **50**, 64–67.

Carney, J. P., Guenault, A. M., Pickett, G. R., Spencer, G. F. (1989) *Phys. Rev. Lett.* **62**, 3042–3045.

Chui, T. C. P., Lipa, J. A. (1984) in: U. Eckern, A. Schmid, W. Weber, H. Wühl (Eds.) *Proceedings of 17th International Conference on Low Temperature Physics* (LT17) (North-Holland, 1984), p. 931–932.

Ebner, C., Edwards, D. O. (1971), *Phys. Rep.* **2C**, 77–154.

Frossati, G. (1978), *J. Phys. (Paris), Colloq.* **39**, C6-1578–1589.

Gloos, K., Smeibidl, P., Kennedy, C., Singsaas, A., Sekowski, P., Mueller, R. M., Pobell, F. (1988), *J. Low Temp. Phys.* **73**, 101–136.

Greywall, D. S. (1986), *Phys. Rev. B* **33**, 7520–7538.

Hakonen, P. J., S. Yin, and K. K. Nummila (1991), *Europhys. Lett.* **15**, 677–682.

Huiku, M. T., Jyrkkio, T. A., Kyynäräinen, J. M., Loponen, M. T., Lounasmaa, O. V., Oja, A. S., (1986), *J. Low Temp. Phys.* **62**, 433–487.

Kittel, C., Kroemer, H. (1980), *Thermal Physics*, 2nd ed., New York: Freeman.

Mamiya, T., *et al.* (1990), *Physica B* **165 & 166**, 837–838.

Marek, D. (1987) in *Proceedings of the 18th International Conference on Low Temperature Physics* (LT18), *Jpn. J. Appl. Phys.* **26**, Suppl. 3, 1683–1684.

Mather, J. C., *et al.* (1990), *Astrophys. J. Lett.* **354**, L37–L40.

Mendelssohn, K. (1977), *The Quest for Absolute Zero*, 2nd ed. New York: Wiley.

Owers-Bradley, J. R., Chocholacs, H., Mueller,R. M., Buchal, Ch., Kubota, M., Pobell, F. (1983), *Phys. Rev. Lett.* **51**, 2120–2123.

Powell, L. R., Childs, G. E. (1972), in: B. H. Billigs *et al.* (Eds.), *American Institute of Physics Handbook*, New York: McGraw-Hill, p. **4**-142.

Preston-Thomas, H. (1990), "The International Temperature Scale of 1990," *Metrologia* **27**, 3–10.

Steinberg, V., Ahlers, G. (1983), *J. Low Temp. Phys.* **53**, 255–283.

Takano, Y., *et al.*, (1985) *Phys. Rev. Lett.* **55**, 1490–1493.

Walker, Graham (1983), *Cryocoolers*, New York: Plenum.

White, Guy K. (1979), *Experimental Techniques in Low Temperature Physics*, 3rd ed., New York: Oxford University Press.

Zemansky, M. W., Dittman, R. H. (1981), *Heat and Thermodynamics*, 6th ed., New York: McGraw-Hill.

Further Reading

Hudson, R. P. (1972), *Principles and Application of Magnetic Cooling*, Amsterdam: North-Holland.

Mendelssohn, K. (1977), *The Quest for Absolute Zero*, 2nd ed., New York: Wiley. An interesting history of the efforts to reach ever lower temperatures and to understand physics that is revealed at low temperatures.

Richardson, Robert C., Smith, Eric (1988), *Experimental Techniques in Condensed Matter Physics at Low Temperatures*, Reading, MA: Addison-Wesley.

Temperature, Its Measurement and Control in Science and Industry, Pittsburgh: Instrument Society of America. An episodic overview of the state of the art in articles focusing on specific issues by the workers themselves. Volume 5 was published in 1982.

Tilley, D. R., Tilley, J. (1990), *Superfluidity and Superconductivity*, 3rd ed., Bristol: Adam Hilger.

Wilkes, J. and Betts, D. D. (1987), *An Introduction to Liquid Helium* 2nd. ed., Oxford: Clarendon.

CRYSTAL DEFECTS

See IMPERFECTIONS IN CRYSTALS

CRYSTAL GROWTH

ERICH SCHÖNHERR, *Max-Planck-Institut für Festkörperforschung, Stuttgart, Germany*

1-56081-063-7/92/$5.00+.50

INTRODUCTION

Single crystals consist of a highly ordered atomic assembly. They can be produced in highest purity or with defined impurity (dopant) concentrations to have various important properties of scientific and technological interest. For these reasons crystals are produced in two scales: large crystals (e.g., 1.5-m-long silicon crystals) grown at industrial sites for technical applications and smaller crystals at laboratories for research (e.g., 10^{-3} m for structure analysis and 10^{-2} m for optical investigations).

Extensive study of crystal growth from solutions started in the late 19th century. Historical reviews are given by van Hook (1961), Buckley (1951), and Bohm (1985). The essential crystal-growth techniques which are used at the present time were developed between 1900 and 1960. In the last 30 years these methods have been enormously improved by the realization of control of heat, mass flow, and crystal form through electronic instrumentation. In recent years, experimental investigations of crystal-growth systems in space together with extensive numerical investigations of heat and mass flow have given sufficiently quantitative analyses of some crystal-growth systems that their state of art is now transforming to a state of science. The invention dates of the most important techniques are listed in Table 1.

In this article, the fundamentals of various growth methods of bulk crystals are emphasized with the goal of giving a general insight into the various procedures and their functions. Hence the common properties of different techniques are identified to improve understanding of particular procedures. In

Table 1. Dates of most important inventions of crystal growth methods.

Method	Year	Reference	Comments
Verneuil	1902	Verneuil (1902)	Flame fusion of ruby
Hydrothermal technique	1905	Spezia (1905)	Hydrothermal synthesis of quartz
	1943	Nacken (1950)	Start of industrial quartz production
Solution	1910	Krüger and Finke (1910)	Two tanks technique
Bridgman	1914	Tammann (1914)	Metal crystals in vertical tubes
	1924	Obreimov and Schubnikov (1924)	Metal crystals in sealed ampoules with cone
	1925	Bridgman (1925)	Metal crystals in sealed ampoules with capillary end
	1936	Stockbarger (1936)	LiF in graphite crucibles, controlled environment
Czochralski	1915	Nacken (1915)	Salol crystals from melt through top seeding
	1918	Czochralski (1918)	Pulling of crystalline wires from melt
	1926	Kyropoulos (1926)	Alkali halides from melt through top seeding
Pyrolysis	1922	Koref (1922)	W deposition on hot wire through WCl_6 decomposition
Chemical transport	1923	van Arkel (1923)	W transport through $WCl_6 = W + 3Cl_2$
Zone melting	1952	Pfann (1952)	Horizontal zone refining of Ge
	1953	Keck and Golay (1953)	Float zoning of Si
Pizzarello	1954	Pizzarello (1954)	PbS crystals through sublimation
Diamond synthesis	1955	Bundy, Hall, Strong, Wentorf (1955)	Diamond synthesis from Ni solvent under high pressure

addition, the advantages, disadvantages, and economic aspects of each method are pointed out. The basic physical processes of crystal growth, namely, mass transfer, heat transfer, incorporation of crystal species, and impurity segregation are undoubtedly rather complex and are therefore presented only by very simple quantitative relations. Crystallization is not considered that which is commercially applied to the production of sugar, salt, and fertilizer as reviewed by Mullin (1985). Also, electrocrystallization is ignored which is described, e.g., by Bockris and Razumney (1967).

Section 1 deals with physical and technical principles of crystal growth to the extent that they are needed to comprehend at least qualitatively the crystal-growth procedures, which consist of a controlled change of a mobile phase into a solid. The conduction of the crystal species through the fluid phase to the crystal interface by a suitable temperature field appears as the most important principle of crystal-growth processing in a monocomponent system. In a multicomponent system, change of the composition is an additional way of processing.

In Sec. 2 the basic components which are common to most crystal-growth systems are outlined. The components are essentially heat sources, fluid containers, and particular crystal environments.

The growth methods are described in Sec. 3 and are classified by the kind of mobile phase: crystal growth from the vapor phase in Sec. 3.1, from the melt in Sec. 3.2, from the solution in Sec. 3.3, and from the solid phase in Sec. 3.4. The particular techniques of each category are essentially characterized by technical aspects such as support, attachment, and geometry of fluid and crystal. Additional categories are obtained from physicochemical aspects such as the change of a temperature distribution or composition of the fluid phase (solution).

Since thin films are grown by the same physicochemical principles as those of bulk crystals, layer deposition techniques are briefly discussed in Sec. 4.

Some essential points of crystal characterization such as growth form, impurity segregation, and point, line, and two-dimensional defects are discussed in Sec. 5.

Section 6 closes the article and gives some examples of sophisticated applications of crystal-growth methods.

1. PHYSICAL AND TECHNICAL PRINCIPLES

1.1 Driving Force, Undercooling, and Supersaturation

From the view of thermodynamics, crystal growth consists of the change of a fluid phase into a solid phase which represents the crystal. The transition requires a driving force which is given by the difference in chemical potential $\delta\mu = \mu_f - \mu_c < 0$ between the species in the fluid (f) and crystalline (c) phase. Measurable thermodynamic quantities such as temperature T, vapor pressure p, and concentration C are preferred for the description of practical crystal growth. In particular,

1. the undercooling $T_m - T_c = T_m\delta\mu/H_m$ is used for a melt monosystem where T_m is the melting point and H_m the heat of fusion;
2. the supersaturation $p_f - p_c$ with $\ln(p_f/p_c) = -\delta\mu/kT$ for a vapor monosystem; and
3. the supersaturation $C_f - C_c$ with $\ln(C_f/C_c) = -\delta\mu/kT$ for a solution.

1.2 Formation of an Initial Crystal

The formation of an initial crystal is for all self-seeding techniques an important procedure. The creation of only one nucleus can be achieved by a sufficiently slow undercooling or supersaturation of the fluid in a cone or capillary. When the particular seeding area is idealized by a small circular cavity, the formation of a single seed can be explained by the thermodynamic approaches described by Sigsbee (1969) and Walton (1969). For an arbitrary cylindrical cluster of height h, the Gibbs energy of formation G_i results in

$$G_i = ivG_v + iv(\sigma_{12} + \sigma_{13} - \sigma_{23})/h + 2(\pi v h)^{1/2}\sigma_E\, i^{1/2}, \qquad (1)$$

where i is the number of cluster members; v the solid volume per monomer; G_v (<0) the difference in Gibbs energy between solid and fluid (driving force); σ the surface Gibbs interfacial energy among cluster (1), fluid (2), and wall (3); and σ_E the edge Gibbs energy. The energy G_i assumes a maximum value G_i^* at

$$i^* = -\pi h \sigma_E^2 v^{-1}[G_v + (\sigma_{12} + \sigma_{13} - \sigma_{23})/h]^{-2}.$$

The size i^* is denoted as critical since clusters smaller than i^* are unstable while those greater than i^* form growing nuclei. An increased possibility of nucleation exists for the cavity since the edge free energy σ_E is reduced through the reaction with the wall. The preferred nucleation is schematically illustrated in Fig. 1. Curve (1) represents the derivative of the interfacial energy of G_i as a function of the size i for a cluster on a plane crucible wall. Curve (2) shows the derivative for that cluster in a cavity of matching radius. For the actual cavity where the radius is in accordance with the cluster size i_c^* a nucleus is formed as soon as the driving force reaches νG_{vc}. The formation of other nuclei is unlikely since clusters of improbable sizes larger than i^* are required.

1.3 Crystal Growth

Crystal growth occurs through the transport of crystal species from the fluid phase to the crystal interface followed by incorporation of these species into the crystal. The incorporation is accompanied by chemical reactions at the surface if the fluid species differ from the structure units of the crystal, e.g., in the cases of chemical transport and associative and dissociative sublimation. In the case of a multicomponent system additional chemical reactions can happen in the fluid. When the reactions in the fluid can be ignored the three processes

1. mass transfer,
2. chemical reactions at the interface, and
3. incorporation

are sequentially coupled so that one of these determines the growth rate.

1.3.1 Mass Transfer of Crystal Species and their Incorporation. A quantitative description of the motion of fluid species is very complex since four conservation laws have to be considered—namely, conservation of momentum (Navier-Stokes), chemical species (transport), mass (continuity), and energy (heat flow). General treatments are presented by Bird *et al.* (1960), Hershey (1973), and Peube (1977). In general, mass transport controls the growth rate through volume diffusion for most multi-component systems. In comparison to a purely monocomponent fluid, here diffusive fluxes, caused by thermal (Soret effect) or concentration gradients, are superposed to the convective flux of the species as discussed by Wilcox (1971), Faktor and Garrett (1974), and Rosenberger (1979). A one-dimensional steady-state approximation for the two-component nonviscous fluid leads to the velocity

$$V_{cf}=(D/\delta)\ln[(C_t-C_{cf})/(C_t-C_f)] \quad (2)$$

of the crystal species in the fluid, where D is the binary diffusion coefficient and C_t the total concentration, assumed to be spatially constant. The concentrations C_{cf} and C_f are those of the crystal species at the interface and at the

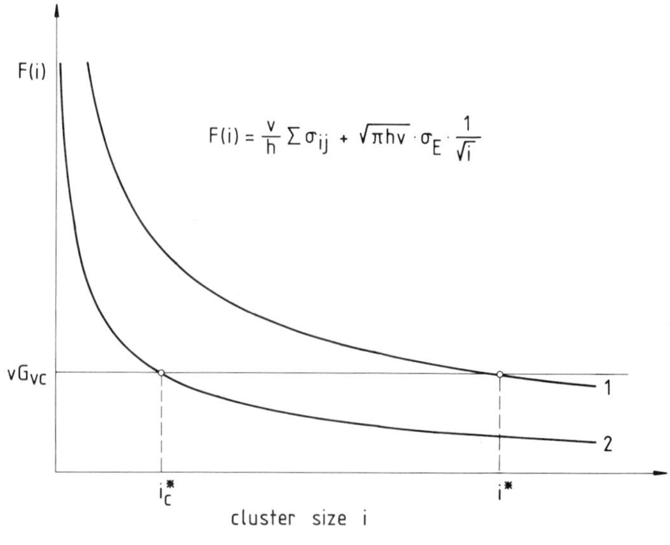

FIG. 1. Formation of an initial crystal. (See text.)

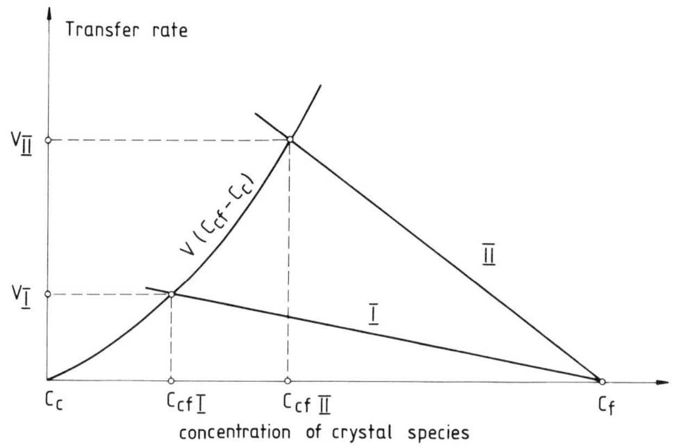

FIG. 2. Stationary crystal growth at the rates V_I and V_{II} for two different transport conditions. The mass transfer rates I and II equal the incorporation rate at $C_{cf\,I}$ and $C_{cf\,II}$, respectively. In both cases the saturation concentration difference C_f-C_c is the same.

effective distance δ (see Sec. 5.2.1) from the interface, respectively. The concentration distribution, and thus the thickness δ, are strongly affected by convections resulting, e.g., from stirring, buoyancy, or tension gradients of wall-free fluid surfaces (Marangoni convection); therefore, the same total driving force C_f-C_c, where C_c is the saturation concentration of the crystal, can lead to different growth rates as illustrated in Fig. 2. Curve $V(C_{cf}-C_c)$ represents the surface reactions which are driven by the supersaturation $C_{cf}-C_c$. Curve I decribes the mass transfer for a convectively undisturbed flux. The intersection of the two curves determines the growth rate V_I. By pumping the fluid at a high speed towards the crystal, the effective diffusion zone δ becomes small resulting in curve II and thus in an increased growth rate V_{II}.

The growth is denoted as interfacial kinetics controlled if $C_{cf}-C_c \gg C_f-C_{cf}$ and mass transfer controlled if $C_f-C_{cf} \gg C_{cf}-C_c$. The growth rate V_c and fluid mass velocity V_{cf}, normal to the interface, are linked by the conservation relation

$$\rho_c V_c = \rho_{cf} V_{cf}\,, \qquad (3)$$

where ρ_c and ρ_{cf} are the densities of crystal and fluid, respectively, as shown by Wilcox (1971).

The interfacial processes are mainly affected by the surface structure, i.e., by the number of available bonds per approaching crystal-growth unit. The actual structure is between atomically rough and atomically smooth as schematically shown in Fig. 3. In the first case nearly every surface site has sufficient free bonds to bind an incoming growth unit, so that the growth occurs continuously normal to that interface. Rough interfaces achieve the highest incorporation rates and may appear close to the melting point or at high supersaturation as described by Jackson (1975) and Bennema and Gilmer (1973).

In most cases smooth plane as-grown surfaces are observed which run parallel to (low-

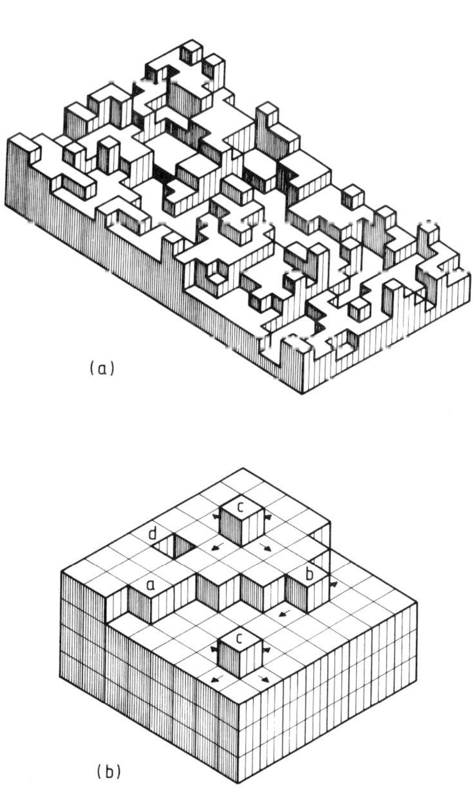

FIG. 3. Crystal surface models. (a) Rough surface. (b) Kossel-Stranski surface (Kossel, 1927; Stranski, 1928) characterized by some particularities: An atom is in a kink position (*Halbkristallage*) at *a*, at a ledge at *b*, and adsorbed at *c*. A hole (*Lochkeim*) is formed at *d*.

index) crystallographic faces (singular faces) of low surface energy. They grow layer by layer through the capture of growth units in kinks along monomolecular steps (ledges). The energy of kink formation is very low, about a tenth of the heat of evaporation, so that the growth rate is mainly determined by the creation of steps. Sources of ledges are screw dislocations (BCF theory) as discussed by Burton, Cabrera, and Frank (1951), Strickland-Constable (1968), Mutaftschiev (1980), and Chernov (1984) and two-dimensional seeds as described by Voronkov (1983). Screw dislocations continuously provide the surface with self-perpetuating spiral steps. In contrast to the free energy of a two-dimensional seed, the free energy of the first ledge winding around a screw dislocation is so low that the growth along the spiral step occurs at nearly zero driving force without observable critical behavior. In cases of growth from vapors and solutions, the kinks are mainly fed from adsorbed species so that the ledge spreading is strongly affected by the average surface diffusion length and desorption probability of the adsorbed species. At low supersaturations the distance between steps is much greater than the diffusion path and the growth rate becomes proportional to the square of the driving force. At high supersaturations the step density of the spirals becomes high and the growth rate linearly proportional to the driving force. When, in addition, heterogeneous reactions are involved the surface processes become very complex as described by Jones and Shaw (1975).

Surface nucleation models, which explain the growth of perfect crystals, lead to exponential, linear, or more complex relations between growth rate and driving force as, e.g., discussed by Brice (1973).

1.3.2 Mass Transport through Chemical Reactions. Chemical vapor transport is extensively discussed by Schäfer (1964) and Faktor and Garrett (1974). Transport of a solid in the vapor phase requires the presence of a volatile product of the solid. In the simplest case the volatile product is formed by the reaction of the solid with a gas (transporting agent), e.g., $M(s)+X_2(g)=MX_2(g)$, where M is a divalent metal, X a halogen, and MX_2 the volatile reaction product. When thermal diffusion and any kind of sublimation is excluded, the one-dimensional stationary flux J_i of the vapor species i is expressed by

$$J_i=\frac{D}{RTd}\,\frac{p^*(d)p_i(0)-p_i(d)p^*(0)}{p^*(d)-p^*(0)}\times\ln\frac{p^*(d)}{p^*(0)}, \tag{4}$$

where T is the average temperature between crystal (at the location 0) and source (at d). The average diffusion coefficient is given by

$$D=D_0(T/T_0)^n(p_0/p) \tag{5}$$

with $T_0=273$ K, $p_0=1$ Pa, $p=\Sigma p_i$ (total pressure), and $1.5\leqslant n\leqslant 1.8$. The standard diffusion coefficient D_0 varies between 1.2×10^{-4} and 0.01×10^{-4} m^2/s. The mass velocity of the total vapor is

$$V_t=(D/d)\ln[p^*(d)/p^*(0)].$$

The stoichiometric average pressure, $p^*=\Sigma\alpha_j p_j$, follows from the condition of insolubility ($J_j=0$) of the transporting agent j in the solid. The coefficients α_j are the stoichiometric ones of the gaseous transport agent compounds j. Transport of species i occurs from hot to cold places for

$$p^*(d)p_i(0)>p_i(d)p^*(0)$$

and from cold to hot for

$$p^*(d)p_i(0)<p_i(d)p^*(0).$$

In the example $M(s)+X_2(g)=MX_2(g)$ one has $D=D_1=D_2$, $p_1=p_{X2}$, $p_2=p_{MX2}$, and $p^*=p_{X2}$.

1.3.3 Heat Transfer into the Crystal. The balance of heat at the crystal interface leads to a quantitative description of crystal growth. In the case of growth in one dimension x the crystal-growth rate becomes

$$V_c=[k_c(dT_c/dx)-k_f(dT_f/dx)]/(\rho_c H_c), \tag{6}$$

where k_c and k_f are the heat conductivities, dT_c/dx and dT_f/dx the temperature gradients of crystal (c) and fluid (f), ρ_c the crystal density, and H_c the heat of phase transition. When A is the cross section of crystal and fluid, $Ak_f(dT_f/dx)$ is the heat transfer rate from the fluid into the crystal, and $Ak_c(dT_c/dx)$ the heat removal rate from the crystal. Most of the crystal-growth procedures consist of controlling these rates. A complete heat-flow description of a crystal fluid system leads to the Stefan problem as described, e.g., by Parker (1970) and Crank (1988).

1.3.4 Growth Rates from Theoretical Models. Maximum crystal-growth rates for a monocomponent system can be estimated from basic physical relations.

1. For the melt, from the Wilson-Frenkel relation

$$V=a\alpha\,\exp\left(-\frac{H_0}{RT_c}\right)\exp\left(-\frac{H_m}{RT_m}\right)\times\left[1-\exp\left(-\frac{H_m}{R}\frac{T_m-T_c}{T_m-T_c}\right)\right], \qquad (7)$$

where a is the interatomic distance (10^{-10} m), α the lattice vibration frequency (10^{13} Hz), H_0 the activation enthalpy of self-diffusion in the melt as discussed by Jackson (1975), H_m the heat of fusion, R the gas constant, T_m the melting point temperature, and T_c the temperature of the crystal.

2. For the vapor, from the Hertz-Knudsen relation

$$V=\frac{M_f}{\rho_c(2\pi M_f RT_c)^{1/2}}(p_f-p_c), \qquad (8)$$

if every atom is incorporated which hits the (rough) interface. In this expression, M_f is the molar mass of the vapor, ρ_c the density of the crystal, p_f the vapor pressure in front of the crystal interface, and p_c the saturation pressure of the crystal.

3. For the vapor in a cylindrical ampoule with interfacial equilibrium, from the Hagen-Poiseuille relation

$$V=\frac{M_f a_d^2}{16\rho_c\eta LRT}(p_f^2-p_c^2), \qquad (9)$$

where η is the dynamic viscosity, a_d the ampoule diameter, p_f the vapor pressure of the source, L the distance between crystal and source, and T the average temperature of crystal and source.

4. For the vapor with an additional inert component A, from

$$V=\frac{M_{fc}D}{\rho_c LRT}\ln\left(\frac{p_t-p_c}{p_t-p_{fc}}\right), \qquad (10)$$

where M_{fc} is molar mass of the crystal species of the vapor phase, D the binary diffusion coefficient defined in Eq. (5), $p_t=p+p_A$ the constant total pressure, p_{fc} the partial vapor pressure of the crystal vapor species at the distance L in front of the crystal, and T an average temperature.

Growth rates described by the relations (7)–(10) are shown in Fig. 4.

1.4 Principles of Crystal-Growth Procedures

1.4.1 Crystal Growth in Monocomponent Systems. In monocomponent systems, crystal growth is generally forced by adjusting a temperature distribution across the interface in such a way that heat is continuously flowing from the hotter fluid into the colder crystal. The growth rate is essentially determined by the rate of heat flow according to Eq. (6). Continuous growth in one direction toward the fluid is maintained by two principal techniques:

1. A temporary constant temperature distribution is slowly shifted toward the fluid by moving the crystal or in a few cases by the moving of the heater.

2. The container and heater are kept in a fixed

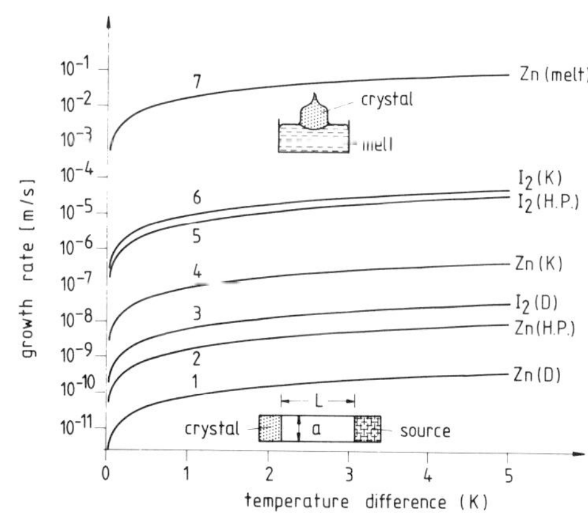

FIG. 4. Various growth rates from theoretical models for different monocomponent systems as a function of undercooling. *Curve 7:* Growth of Zn from the melt according to Eq. (7), H_0 = 21.31 kJ/Mol (from Ramachandrarao, 1978). *Curves 1–6:* Crystal growth rates of Zn and I_2 from vapor phase as a function of temperature difference between source and crystal. The parameters are T_c = 410 °C, $D_0=10^{-5}$ m²/s for Zn, and T_c=40 °C, $D_0=5\times10^{-6}$ m²/s for I_2. The common parameters are L=0.15 m, $a=2\times10^{-2}$ m, η = 2×10^{-4} P, and P_A = 133 Pa. Curves 1 and 3 represent Zn and I_2 crystal growth by diffusion through argon according to Eq. (10). Curves 2 and 5 represent crystal growth of Zn and I_2 by a pure viscous flow according to Eq. (9). Curves 4 and 6 represent Zn and I_2 incorporation according to Eq. (8).

position while the temperature of the heater is slowly lowered.

In a few cases crystals are isothermally grown from supercooled melts by extremely slow cooling rates as described by Scheffen-Lauenroth *et al.* (1981).

The shape or the temperature distribution of the fluid near the interface affects the stability of the interface, particularly if the fluid contains impurities or dopants. The possibility of parasitic nucleation is reduced by a strong normal temperature gradient through superheating the fluid in front of the interface.

1.4.2 Crystal Growth in Multicomponent Systems. In principle, the methods for crystal growth from a monocomponent system (melt) can be transferred to a multicomponent system. In that case the drive rates have to be reduced by a factor between 10^4 and 10^5 since the growth rate is essentially determined by the mass transfer according to Eq. (2) or by surface reaction kinetics.

In addition to the selection of the temperature, the choice of fluid composition is possible. Hence crystal growth can be carried out under isothermal conditions. Four principal isothermal procedures may be employed:

1. slow cooling (or heating) of the fluid-crystal system;
2. slow enrichment of the fluid with crystal species, e.g., by evaporation of the solvent or by injection of an additional fluid;
3. continuous supply of a supersaturated nutrient into the isothermal environment of the crystal (temperature difference technique); and
4. decomposition or heterogeneous chemical reaction in an isothermal environment.

1.5 Selection of Crystal-Growth Method

The choice of crystal-growth method depends on

1. external aspects such as desired size, shape, purity, doping, and perfection of the crystal;
2. economics;
3. available equipment; and
4. thermodynamic behavior.

The thermodynamic properties such as structural change below the melting point, peritectic reaction, decomposition, sublimation, reaction of the melt with the environment, and solubility in a suitable solvent exclude or indicate the applicability of a particular fluid and therefore determine the growth method.

In particular, the structural change limits heating to below the transition temperature so that solvents, reactive vapors, or eventually sublimation have to be employed. If an incongruent melting compound A_mB_n, e.g., forms solid A and a liquid at the peritectic temperature T_p, then crystals of A_mB_n can be grown below T_p from a solution which is obtained from a relevant mixture of B and A_mB_n. Materials which decompose into one or more (e.g., into two in the case of CdS) volatile components can be grown from the melt in closed containers or by encapsulation in a suitable liquid. Crystal growth through sublimation is possible at reasonable rates for vapor pressures higher than 100 Pa.

2. BASIC COMPONENTS OF CRYSTAL-GROWTH EQUIPMENT

2.1 Containers

In most crystal-growth methods containers are used to hold the fluid. Commercially available containers are often made of fused silica, sintered alumina, graphite, tungsten carbide, platinum, and iridium. It is necessary that the container material does not contaminate the fluid and does not react with the ambient atmosphere. In addition, when the crystal reacts with the crucible wall, the contact between crystal and wall must be prevented. For many substances these conditions of compatibility are hard to fulfill.

The principal designs of containers can be classified as regular crucibles, boats, tubular crucibles, and ampoules as shown schematically in Figs. 5(a), 5(b), 5(c), and 5(d), respectively.

2.2 Heating Apparatus

2.2.1 Resistance Heater. Resistance furnaces are employed for the heating of fluids; therefore, most heaters consist of tubes which closely surround the crucibles or ampoules. Ceramic tubes wound with NiCr or AlFe alloy wires and SiC resistor tubes are commonly

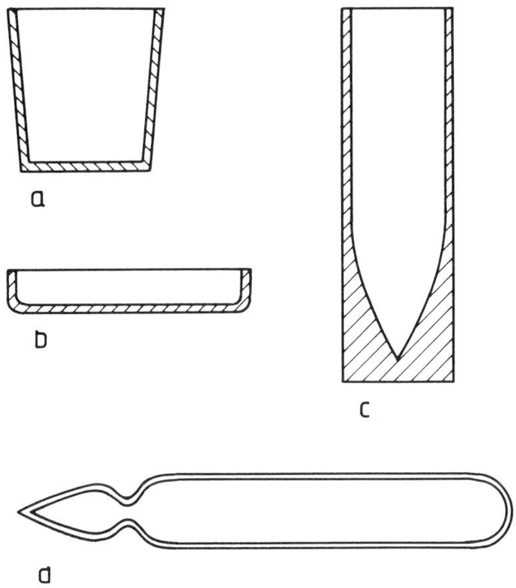

FIG. 5. Principal container designs for crystal growth. (a) Regular crucible, (b) boat, (c) tubular crucible, and (d) ampoule.

used in air. They are enclosed in an insulating housing to reduce the heat loss. Lengthwise-sliced cylinders of graphite or high-melting metals surrounded with metallic heat shields are preferred in oxygen-free equipment. Distinct temperature distributions are obtained through separate heating elements along the furnace axis. Chamber furnaces are mainly employed for growing crystals from high-temperature solutions through slow cooling. The available temperature of a resistance furnace depends on the corrosiveness of the ambient atmosphere. Temperatures up to 1800 °C can be reached in air with rhodium or $MoSi_2$ heaters. In vacuum and noble gases, temperatures up to 3000 °C are obtainable with tungsten, while in nitrogen, noble gases, and vacuum, graphite can be heated up to 2600 °C.

Advantages: Except those with noble-metal wiring, resistance furnaces have a low initial cost. Their temperature can be easily controlled. *Disadvantages:* In general, a desirable high temperature gradient cannot be achieved.

2.2.2 Induction Heater. An induction heater consists of a water-cooled copper coil which is supplied with an alternating voltage at a frequency between 100 kHz and 13.6 MHz. A body within the coil is heated by an induced current if the electrical conductivity of the body is sufficiently high and if the distance between coil and body is sufficiently small. Thus conductive fluids can be directly heated without using a container. Metal and graphite crucibles as well as any appropriately shaped conducting part (susceptor) can be inductively heated, while the surrounding wall can be kept cool with water [Fig. 6(a)]. Moreover, direct heating of a corrosive fluid is possible in a cold-wall boat [Fig. 6(b)], obtained by indenting a copper tube, or in a cold-wall crucible [Fig. 6(c)] described, e.g., by Aleksandrov *et al.* (1978). In both cases the melt is held by a shell which is formed between melt and crucible.

Advantages: Inductive heaters offer the possibility of directly heating the fluid and susceptors at higher temperatures than those which can be reached with resistance heaters. *Disadvantages:* High-frequency generators have a high initial cost, high consumption of cooling water, and difficulty in making an efficient power transfer.

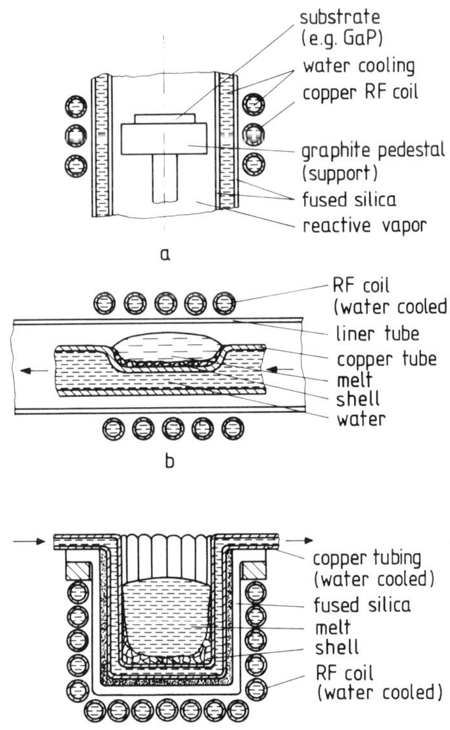

FIG. 6. Cold-wall heating designs. (a) Substrate heating enclosed in a cold-wall tube. (b) Cold-wall boat. (c) Cold-wall crucible.

2.2.3 Other Heat Sources. The temperature of flames is difficult to control. They are used for the Verneuil method only.

Very high temperatures at an optimal efficiency can be obtained by electron bombardment; its application is preferred for floating-zone crystallization of conductive materials with a high melting point and low vapor pressure, as reviewed by Shah and Wills (1975). In the simplest assembly the fluid acts as an anode which is surrounded by a negative tungsten glowing filament. The method requires a vapor pressure less than 10^{-2} Pa of the environment and material. Insulating materials can be bombarded using a water- cooled slitted-ring anode.

Pure optical radiation is suitable for formation of molten zones in nonmetallic rods. The highest temperatures can be obtained by irradiation with a CO_2 laser beam. The light from the filament of a halogen lamp or arc of a xenon-tungsten lamp can be transferred by imaging with elliptical reflectors.

Heat pipes consist of a closed double-wall cylinder. The space between the walls is partly filled with a liquid. Heat is steadily transferred by boiling the fluid at some points and condensing it at others by placing the tube in a resistance furnace. Heat pipes provide an extremely spatially constant temperature distribution. Detailed descriptions are given by Chisholm (1971) and Winter and Barsch (1971).

Fluid baths are preferred for heating or cooling of tanks used for crystal growth from low-temperature solutions. The baths are heated with immersed resistance elements.

Other heating methods such as gas discharge arcs and plasmas are of minor importance. They have been reviewed by Shah and Wills (1975).

2.3 Temperature Control

To grow high-quality crystals and films it is very important to provide fluid and crystal with a temperature distribution according to a prescribed function of time. This is achieved with commercially available closed-loop control systems (personal computer). In general, the input function of the control system is a voltage which results from the difference between prescribed (set point) and measured (actual) temperatures. Temperature sensors are usually thermocouples. Platinum resistors and thermistors are frequently used at low temperatures (<200 °C) and pyrometers at very high temperatures (>1500 °C). The output function of the controller is a voltage which drives the power of the heat source, e.g., the current of the resistance element through a thyristor or the power of a high-frequency generator through the gate of the oscillation tube.

2.4 Growth-Rate Control

Growth-rate control is required for optimal crystal perfection and dopant distribution. With respect to process parameters (input functions), which can be varied during processing, the growth systems can be classified into two groups: (1) systems which are driven with one input function and (2) those with several input functions.

1. When only one input function is needed, the control of growth rate is reduced to control of that function (open-loop system). For optimal operation the functional relation between input and crystal shape have to be known. In all cases the input rates have to stay below a critical amount to prevent degeneration of the crystal interface. Bridgman, liquid-zone, and Pizzarello techniques are examples where the input is the displacement velocity between the temperature distribution and the fluid. These systems are approximately linear when the crystal grows at a constant diameter and can be described by the governing equation

$$\beta d(t) = a(t) + \tau \frac{da(t)}{dt}, \tag{11}$$

where $\beta \approx 1$ is the transition factor describing the temperature change through the crystal, $d(t)$ the input function (ampoule displacement), $a(t)$ the response (crystal length), and τ the time constant. The relaxation time is a function of kinetics, mass, and heat transfer. Figure 7 shows solutions of Eq. (11) for a ramp input.

Other examples of linear behavior are solvent evaporation, where the input function is the evaporation rate, and solution cooling (with a linear relation between saturation and temperature) where the input is the cooling rate. Both systems are approximately linear when the rate of crys-

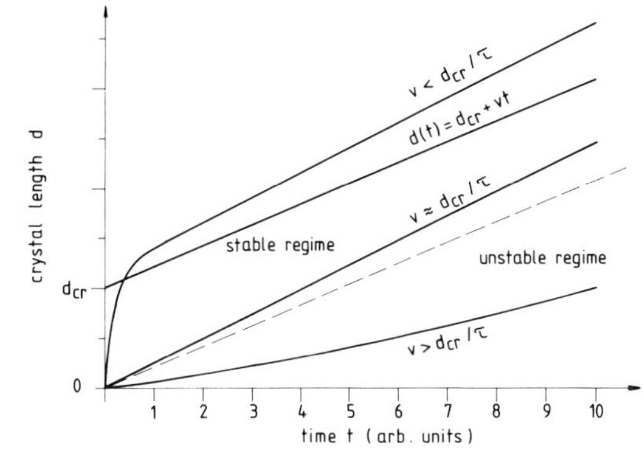

FIG. 7. Crystal lengths in response to different drive rates v as functions of time, according to Eq. (11), using the ramp function $d(t)=d_{cr}+vt$ as an input function. The drive length d_{cr} determines the critical undercooling of seed formation. Crystal growth degenerates for too high drive rates $v>d_{cr}/\tau$.

tallized mass is considered as the response. The growth rates are therefore monotonic and not constant for constant input rates.

2. For complex systems with more than one input possibility, closed-loop systems (or the skill of the crystal grower) have to be applied, where the shape of the crystal is measured and compared with a prescribed one. The Czochralski method is an example where the change of the melt temperature and pulling rate are two independent input functions. Two types of sensors, namely, a video camera for measuring the crystal diameter [described by Draper (1975)] and a balance for weighing the crucible [described by Bardsley *et al.* (1977) and Meeseki *et al.* (1987)], are preferably employed. The signals of the sensors are used to control the heating rate at a prescribed pulling rate.

2.5 Environment

The environment around the crystal and fluid may have different purposes:

1. to prevent contamination of the crystal, e.g., by oxidation,
2. to protect the laboratory from poisoning,
3. to reduce evaporation of the fluid, and
4. to stabilize decomposing compounds.

In most cases the crystal-fluid assembly is therefore enclosed in a modified housing which can be supplied with vacuum, static, or flowing high-purity inert gas. From the kind of heating design two types of housings can be classified: hot wall and cold wall. The hot-wall housings are heated by a resistance furnace from the outside. They are generally simple in design, i.e., consist, e.g., of sealed silica ampoules. Cold-wall housings are either heated inductively from the outside or enclose (in a more or less sophisticated manner) the heating elements.

The purest available environment is a vacuum. It can be applied to techniques of crystal growth from the melt if the vapor pressure of the melt is sufficiently low.

Slight evaporation of the fluid from open crucibles is reduced through heavy ambient inert gases (N_2, Ar, and Kr) at pressures of about 10^4 Pa. Strong evaporation of fluids at pressures $>10^4$ Pa can be suppressed by covering the fluid with an inert liquid and applying a sufficiently high ambient gas pressure. The liquid must be of a lower density than that of the fluid and must also adhere to the crucible and crystal.

The decomposition of a fluid A_mB_n into a liquid phase and vapor component B [e.g., $2Li_3N(l)=6Li(l)+N_2(g)$] is prevented through an excess vapor of B (e.g., N_2 atmosphere).

3. TECHNIQUES OF CRYSTAL GROWTH

3.1 Techniques of Crystal Growth from the Vapor Phase

3.1.1 Pizzarello Method. A fused silica ampoule is supported with source material either together with a transport agent or a high vacuum ($<10^{-2}$ Pa), then sealed off and hung in a vertical two-zone furnace. The growth method, reviewed by Schönherr (1980), is illustrated in Fig. 8. The upper end of the ampoule is distinctly shaped with a pointed tip and a neck [Fig. 9(a)] or capillary [Fig. 9(b)] for seeding. Starting with the upper seeding end in the middle of the upper heating zone (Fig. 8), the ampoule is slowly ($\approx 2\times 10^{-8}$ m/s) pulled upwards so that the upper end becomes cooler than the source. After a crystal is formed its growth velocity matches the drive rate after an initial high speed as shown in Fig. 7.

The mounting of a seed is possible in similar designs as illustrated in Fig. 9(c).

According to Eq. (10), small amounts of an impurity vapor reduce the mass transfer rate

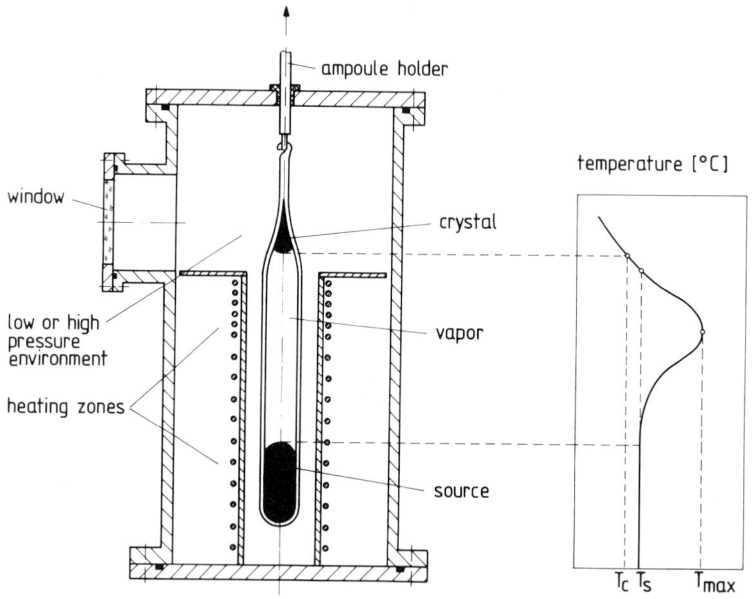

FIG. 8. Pizzarello method. (See text.)

dramatically. Foreign vapors are removed by using "almost-closed" crucibles with a small hole ($\approx 0.1 \times 10^{-3}$ m) [Fig. 10(a)], or with self- sealing ends [Fig. 10(b) and 10(c)] which are enclosed in an evacuated tube as described by Faktor and Garrett (1974). In particular, partial pressures of materials which dissociate into two vapor components [e.g., $2CdS(s) = 2Cd(g) + S_2(g)$] can be adjusted to the stoichiometric ratio through almost closed ampoules.

Advantages: The Pizzarello technique is achieved with low-cost equipment. The easy processing only requires a constant temperature distribution and a constant drive rate. High crystalline perfection and high chemical purity usually result. *Disadvantage:* The crystals grow at low rates of about 10^{-8} m/s.

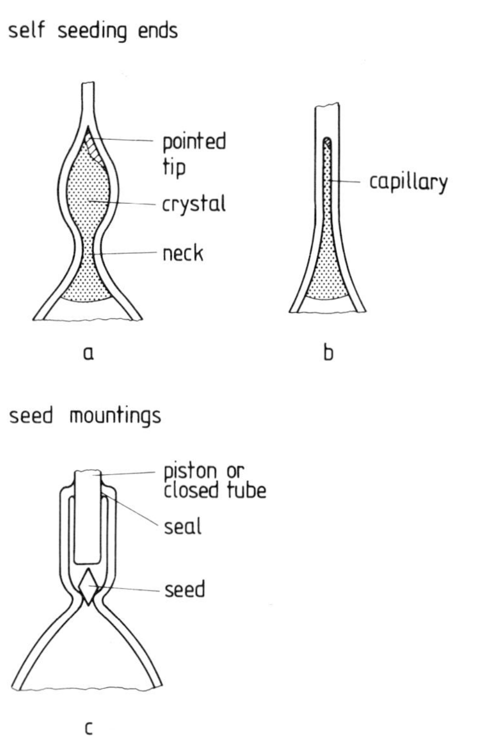

FIG. 9. Designs of ampoule ends for seeding. (a) Seeding tip with constriction and (b) capillary for seed selection. (c) Seed support in a closed ampoule.

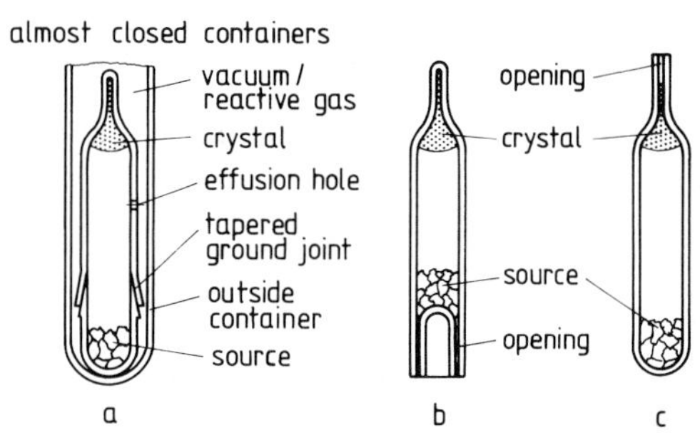

FIG. 10. Designs of almost closed containers. (a) Two-section ampoule with effusion hole. (b) Ampoule with self-sealing source end. (c) Ampoule with self-sealing seeding end.

3.1.2 Temperature Gradient Technique. This technique is used for growing large crystals from the vapor phase at low temperatures (≈120 °C), e.g., HgI_2 crystals as described by Kaldis *et al.* (1987). The principle of the method is shown in Fig. 11. A seed crystal is mounted on a coolable pedestal at temperature T_c in the center of a cylindrical, evacuated, sealed glass container. The source material initially deposited at the upper part of the container wall is held at a temperature T_s higher than T_c of the pedestal. A heating ring at the bottom at a temperature T_r higher than T_s of the source prevents parasitic nucleation on the container bottom and pedestal wall. Crystal growth is forced by lowering the bottom temperature T_r while the pedestal T_c and source temperature T_s are kept constant.

Advantages: The low-cost equipment offers the possibility of growing crystals with high crystalline perfection and high chemical purity. A seed crystal may be easily supported. Although low linear growth rates of 10^{-8} m/s result, a large total crystalline mass is condensed because of radial growth. *Disadvantage:* The processing is rather complicated since a variation of source and crystal temperature are required to control the crystal growth.

3.1.3 Temperature Difference Technique. For synthesis of new compounds through chemical vapor transport, an ampoule is filled with source material together with an appropriate reaction agent and sealed at sufficiently low pressure. Then the nutrient is distributed at one end of the capsule. After that procedure

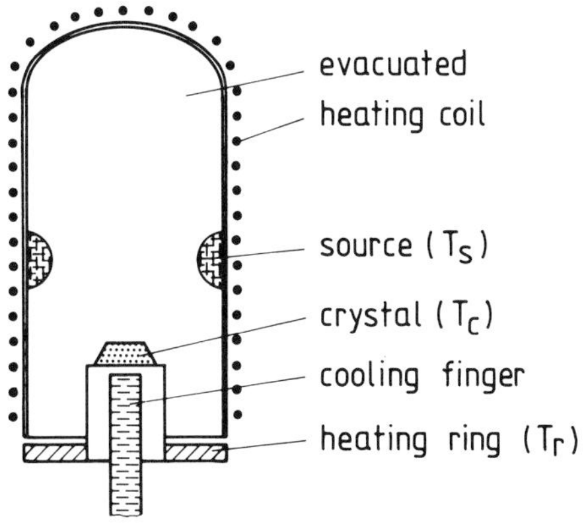

FIG. 11. Design of a closed vessel for crystal growth from vapor phase. (See text.)

the ampoule is placed in a horizontal two-zone furnace which heats the ampoule ends to different constant temperatures. The furnace is tilted slightly to increase the fluid velocity by gravitational convection. Numerous small crystals are formed opposite the source since no single distinctly supersaturated small area can be achieved.

Advantage: The technique needs very low-cost equipment only. *Disadvantage:* A single crystal cannot be achieved.

3.2 Techniques for Crystal Growth from the Liquid Phase

3.2.1 Czochralski Technique. The modern Czochralski technique (Fig. 12) results from the ideas of Nacken (1915), Czochralski (1918), and Kyropoulos (1926). The primary steps are

1. the melting of the source material in a crucible,
2. the preheating of a seed by holding it close above the melt,
3. touching the melt with the seed,
4. slight backmelting of the seed to achieve coherent growth,
5. pulling of the seed with decreasing diameter (necking) to reduce or eliminate structural faults (e.g., grain boundaries and dislocations) of the seed, and
6. pulling with increasing diameter until the desired radius is reached. The crystal shape is controlled by variation of the melt temperature and pulling rate. The crystal and crucible are rotated in opposite directions at 5–40 rpm to achieve a radially symmetric temperature distribution.

Quickly evaporating materials can be crystallized by the liquid-encapsulation technique where the melt is covered with a suitable liquid and the vessel filled with an inert gas of sufficiently high pressure. The Czochralski technique is also employed for growing crystals from solution, as described by Elwell and Scheel (1975).

Advantages: Fast crystal-growth rates of between 10^{-6} and 10^{-5} m/s are attainable. Very large crystals (e.g., silicon 2.2 m in length and 0.16 m in diameter) can be grown. Crystals of highest perfection (e.g., dislocation-free silicon) can be obtained. *Disadvantages:* Medium- to high-cost equipment is required.

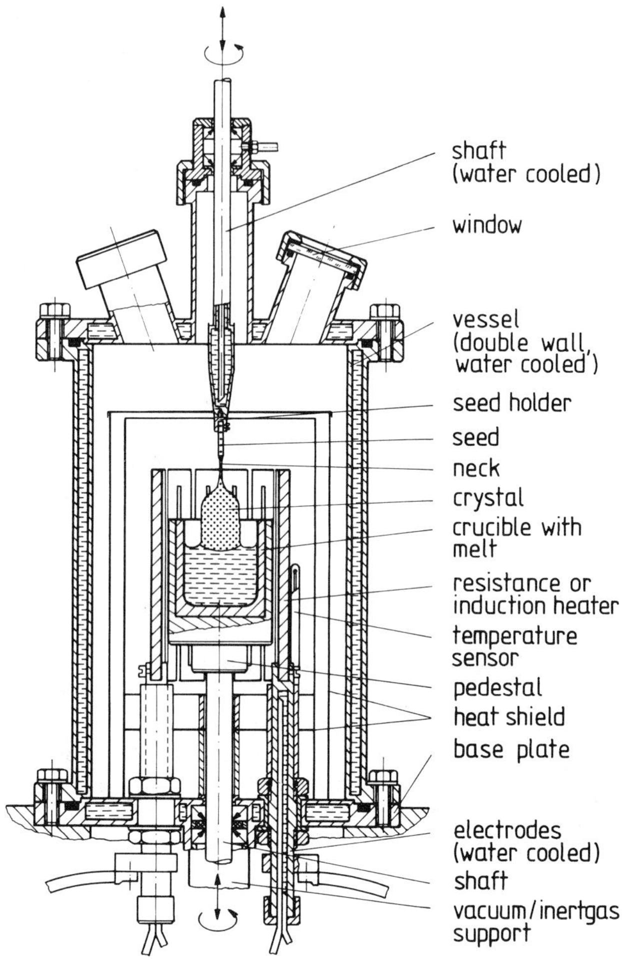

FIG. 12. Czochralski puller.

The crystal growth has to be continuously controlled since the heat-flow conditions change with crystal length.

3.2.2 Bridgman Technique. In the Bridgman method, crystallization of a melt is carried out in a boat [Fig. 5(b)], vertical tubular crucible [Fig. 5(c)], or ampoule [Fig. 5(d)]. Tubular containers are designed with a seeding end at the bottom similar to that shown in Figs. 9(a) and 9(b), while boats can have a seed installed at one end [Fig. 13(a)]. Figure 14 displays the principle of the vertical technique. The container is initially hung in the upper hot zone of a furnace for complete melting of the source material. Then the crucible is moved down into the cooler furnace zone, where an initial nucleus is formed in the seeding end. The growth of the nucleus is achieved by a sufficiently low drive rate. Alternatively, the heater can be moved while the container is held in a fixed position. Quickly evaporating materials can be processed in closed hot-wall containers (Fig. 14). Decomposition of a compound which forms a volatile constituent B can be prevented by an excess of B in a closed container at a sufficiently high wall temperature. Crystals of lower density than that of their melt (e.g., Ge) are processed in horizontal boats with a trapezoidal cross section where the upper side is greater than the bottom side.

Advantages: In general, low-cost equipment is sufficient. Medium growth rates between 10^{-7} and 10^{-6} m/s are reached. The processing is very simple, since it employs a constant temperature distribution and constant drive rate. The method offers the possibility of

1. crystal growth from solution, e.g., growth of peritectically melting compounds,

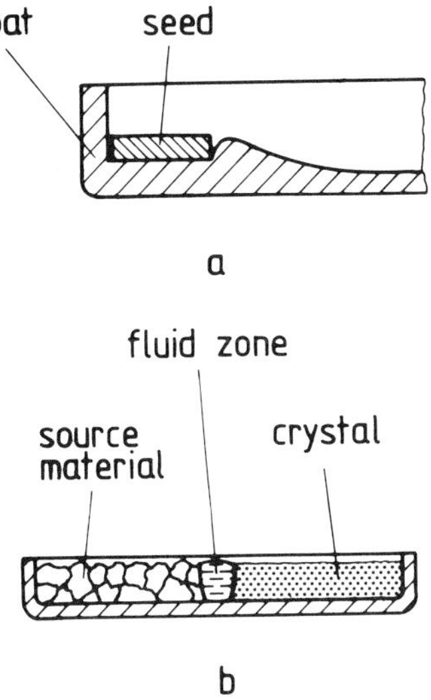

FIG. 13. Use of a boat for zone melting. (a) Seed mounting. (b) Crystal-growth method.

2. crystal growth of volatile or decomposing materials, and
3. doping with volatile elements.

Disadvantage: Dislocations may be formed if the thermal expansion of the container differs from that of the crystal.

3.2.3 Temperature Gradient Technique. The temperature gradient technique is, in principle, a modified vertical Bridgman method, where the crucible is always maintained at an initially fixed position in a particular furnace. The temperature of the heater locally decreases from the upper crucible part to the seeding end. Nucleation and crystal growth is forced by slow cooling of the furnace while the slope of the temperature distribution stays approximately constant.

Advantage: No movable mechanical parts are required. *Disadvantages:* The disadvantages are the same as those of the Bridgman techniques.

3.2.4 Liquid-Zone Techniques

3.2.4.1 Liquid-Zone Techniques with Crucible. The liquid-zone techniques are very similar to those of Bridgman with one difference: Only a small portion [Fig. 13(b)] of the source material is melted by heating with a short resistance oven or a short induction coil around the container. Crystal growth is achieved by driving the fluid zone from the seeding end to the opposite end by moving the container or heater. The method is widely applied for material refining through the use of a multi-heating-cooling zone assembly.

Advantage: A nearly constant (macroscopic) doping distribution may be obtained as shown in Fig. 22 (Sec. 5.2.2). *Disadvantages:* The disadvantages are the same as those of the Bridgman techniques.

3.2.4.2 Floating-Zone Methods. In the floating-zone techniques the melt hangs on a vertical feed rod (or crystal) and is supported by the vertical crystal (or feed rod) below without use of a container. The zone is achieved by heat sources, modified for particular materials as described in Sec. 2.2.3. The liquid is held in position by the surface tension σ which compensates for the weight. The resulting maximum crystal radius is given by

$$r_{max} = 0.92\sigma^{1/2}(\rho_f g)^{-1/2}, \tag{12}$$

where ρ_f is the density of the melt and g is the acceleration due to the gravity, as described by Brice (1965). The method is mainly applied to industrial production of silicon crystals of low oxygen content (Sec. 6.2).

Advantages: The method provides the fluid and crystal with the purest environment. The highest growth velocities (e.g., 10^{-4} m/s for silicon) can be achieved. *Disadvantage:* A

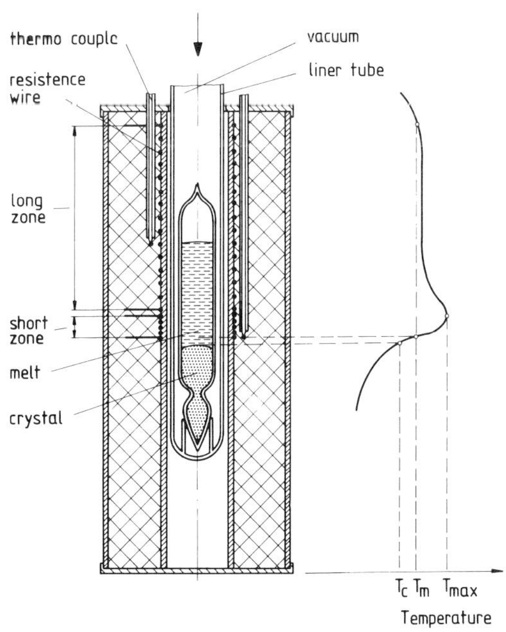

FIG. 14. Vertical Bridgman apparatus.

complex growth control is required by setting power, movement rate, and zone length. A suitable temperature distribution has to be applied for sufficient structural perfection.

3.2.4.3 Verneuil Technique. The Verneuil technique, described, e.g., by Falkenberg (1978), consists in passing powdered source material through a flame downwards to the top of a crystal on which a thin molten feeding zone is formed by the heat of the burner (Fig. 15). Crystal and flame are surrounded with an insulating ceramic tube to reduce heat loss. Oxygen and powder [(60–125)$\times 10^{-6}$ m in diameter] are conducted through the center tube and hydrogen is passed through the outer tube. The powder chamber is divided by a fine-mesh screen into an upper large part which contains the powder source and a lower tapered part through which oxygen is conducted. The powder is mixed with the oxygen flow through the screen by tapping the chamber at a frequency of about 45 taps/min. Crystal growth from the thin molten zone is achieved by lowering the crystal while preserving the distance to the burner. Higher temperatures than those of a oxygen-hydrogen burner are obtained with an induction-coupled plasma as reviewed by Brice (1973) and Laudise (1970).

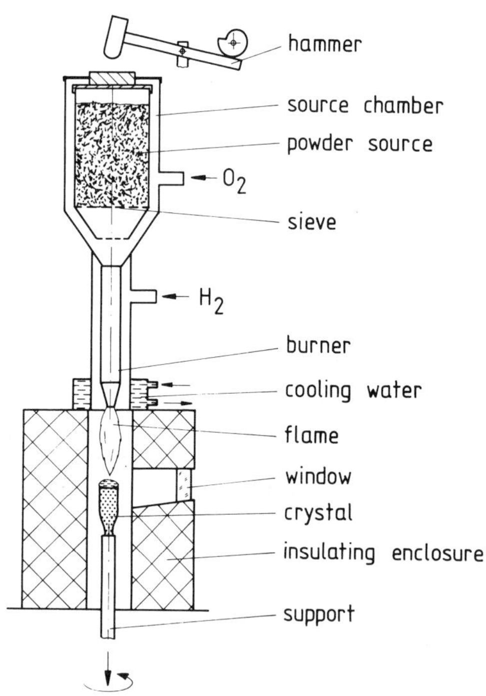

FIG. 15. Verneuil technique.

Advantages: Only low-cost equipment is needed for crystal growth of high melting oxides at temperatures below 2800 °C. *Disadvantages:* The heat control of the burner is difficult. The growth control is complicated, requiring variation of powder supply and crystal movement.

3.3 Techniques of Crystal Growth from a Solution

3.3.1 Solvent Evaporation. In general, a solute crystallizes from a solution at the container wall by slow evaporation of the solvent. A single crystal can be grown if one of the deposited crystals is selected and suspended in that (saturated) solution. The solvent evaporation is controlled by effusion through a hole in the closing top plate. The method requires a solvent whose vapor pressure is much greater than that of the solute. Therefore, the procedure is mainly restricted to low-temperature solvents, e.g., water. The technique is improved by temperature control, stirring, and feedback of the solution. Feedback is achieved by placing a sieve plate together with solute material between the top plate and the solution surface. The evaporating solvent condenses at the cooled top plate, flows back, and is saturated by solute material on the sieve.

Advantages: The method can be carried out with equipment of lowest cost. Large crystals can be grown by very simple processing. *Disadvantages:* The reachable growth rates are low (10^{-9}–10^{-7} m/s) and the optimum growth rate is difficult to control.

3.3.2 Temperature Difference Techniques

3.3.2.1 Normal Solution Conditions. In the temperature difference techniques a saturated aqueous solution, together with a great excess of solute material (nutrient), is maintained in a tank at the temperature T_s. The solution is continuously pumped through pipes to a second tank and back to the source tank. To grow a crystal the temperature T_c of the second tank is held lower than that of the source tank. The principle of the method is illustrated in Fig. 16. Here, in addition, the saturated solution first passes an overheated tank at the temperature T_u on the way to the

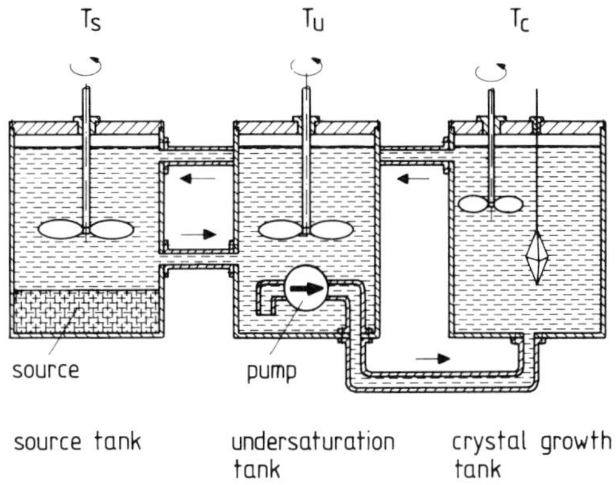

FIG. 16. Three-tank technique for growth of large ADP ($NH_4H_2PO_4$) crystals from aqueous solution, after Walker and Kohman (1948).

growth tank. Thereby residual floating nuclei of the source solution are dissolved.

Advantages: Isothermal crystal growth can be achieved at a constant supersaturation $C(T_s)$–$C(T_c)$. The growth rate is easily controlled by adjusting the temperatures. *Disadvantages:* The attainable growth rates are low (10^{-9}–10^{-7} m/s).

3.3.2.2 Hydrothermal Solution Conditions. Aqueous solutions under hydrothermal conditions (i.e., at temperatures higher than 100 °C and pressures greater than 10^5 Pa) are industrially applied to the growth of β-quartz crystals. The method requires a high-pressure autoclave as, e.g., described by Laudise (1970) and depicted in Fig. 17. The autoclave, of special steel (up to 1 m in diameter and 4 m in length), contains in the lower part a grit of natural quartz as source material (nutrient) and in the upper part suspended single-crystalline quartz seed plates. Nutrient and crystals are separated by a baffle plate with a small opening (5%) to reduce convective instabilities. For crystal growth the nutrient is held at 400 °C and the crystals at 380 °C (temperature difference technique). An 80% (at room temperature) filling with water results in a pressure of about 1.4×10^8 Pa. The solubility of quartz is increased by adding NaOH. A growth rate of 10^{-8} m/s is obtained with a 1.0 M NaOH solution. A review of hydrothermal synthesis of other materials is given by Rabenau (1985).

Advantages: Low-cost equipment is sufficient for synthesis of small crystals of new materials. *Disadvantages:* The costs for safe processing increase enormously with increasing autoclave volume.

3.3.3 Techniques with Gel Media. Most kinds of fine-crystalline precipitation in solutions at room temperature can be directed to single-crystal growth by precipitation in a gel as described by Henisch (1988) and Arora (1981). The possible chemical processes are, e.g., the formation of a reaction product, reduction of a compound, decomplexing of complex compounds, or outsalting. Water glass (sodium metasilicate solution), agar (carbohydrate polymer), and gelatin (protein) are the most commonly used gels which sufficiently reduce the nucleation frequency.

Figure 18 shows two principles of reaction-product precipitation in a silica hydrogel. The gel is prepared from Na_2SiO_3, H_2O, and a suitable acid at a *p*H value between 6 and 8. In Fig. 18(a) the gel contains the acid (here 2 M $C_4H_6O_6$) as one reactant. The other reactant (here 1 M $CaCl_2$) is dissolved in water which is placed above the set gel. Crystals of the solid reaction product (here $CaC_4H_4O_6$) are formed in the gel. In the U-tube method [Fig. 18(b)], the gel contains a minimal amount (here 0.2

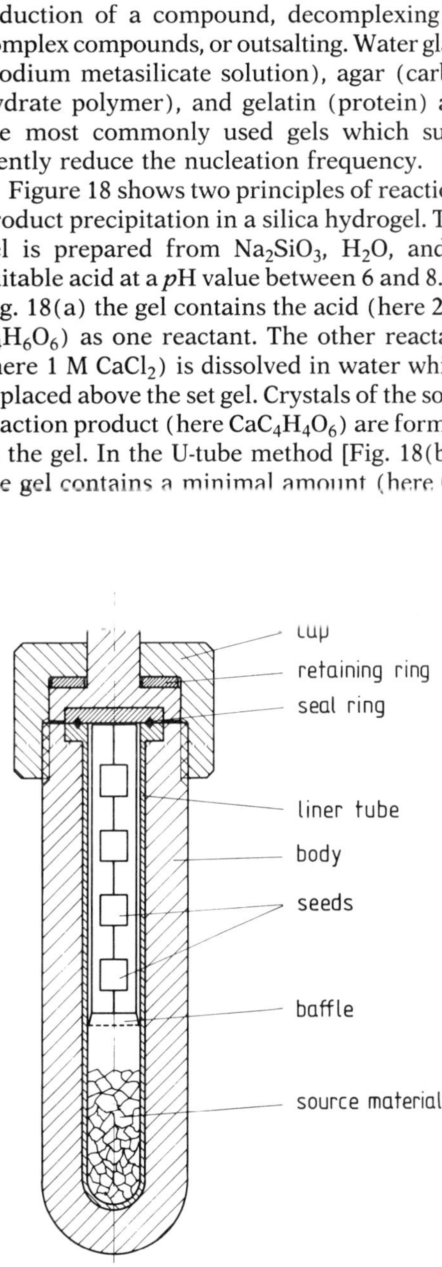

FIG. 17. Autoclave for quartz growth under hydrothermal conditions.

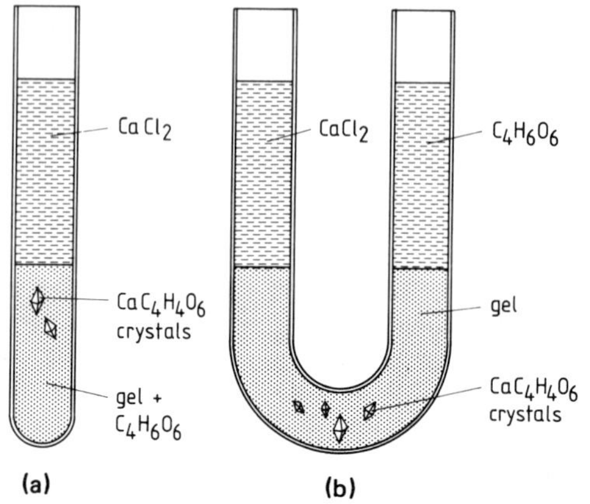

FIG. 18. Crystal growth in gel.

M $C_4H_6O_6$) of acid, which is sufficient forsetting. The reactants (here 1 M $CaCl_2$ and 2 M $C_4H_6O_6$) are placed on opposite sides of the set gel.

Advantages: Crystal growth in gels is one of the cheapest methods. *Disadvantages:* The attainable growth rates are very low (10^{-9}–10^{-8} m/s) and crystals above 10^{-2} m are hard to obtain.

3.3.4 Slow-Cooling Techniques. Growth of crystals by slow cooling is widely applied with aqueous, metal, and fluxed melt solutions. Tanks of glass enclosed in fluid baths and equipped with stirrers, and rotatable seed holders are used for aqueous solutions. Molten salt solvents with fluoride constituents are mainly employed for growth of oxide crystals. Detailed descriptions are given by Elwell and Scheel (1975) and Wanklyn (1975). High-temperature solutions are contained in crucibles of noble metals (platinium, iridium, etc.). They are sufficiently closed to release the pressure without noticeable solvent evaporation. The growth procedure is composed of several steps and illustrated in Fig. 19, where $LiAlSiO_4$ is used as a typical example.

1. The procedure is started by slowly heating (e.g., at 50 °C /h) the source materials in a chamber furnace to the soaking temperature T_{soak}. This temperature is preset 20–50 °C above the saturation temperature T_i.
2. The soaking temperature is held between 2 and 24 h for complete dissolution.
3. Then follows a fast cooling (e.g., at 100 °C /h) to the initial cooling temperature T_i, and

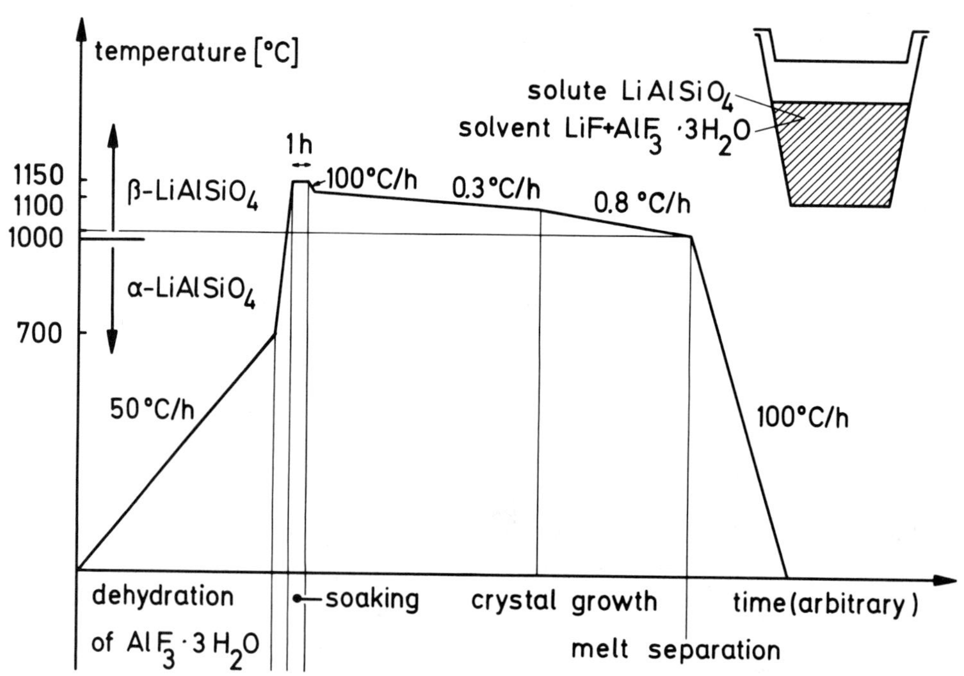

FIG. 19. Growth of eucryptite from fluoride flux as an example of the high-temperature solution technique (see text).

4. a slow cooling (e.g., at 0.3 °C/h) over several days. The slow cooling achieves a low number of nuclei and their undisturbed growth.
5. The slow cooling is continued at an increased rate (e.g., at 0.8 °C/h) since the normal growth rates decrease with the cube root (for linear solubility) of the cooling rate.
6. The crystal growth is terminated by decanting the solution from the crystals.
7. A slow cooling (e.g., at 20 °C) to room temperature prevents mechanical damage of the crystals.

A sophisticated example (growth of $Y_3Fe_5O_{12}$ crystals) with a preset seed is described by Tolksdorf and Welz (1978).

Advantages: The method is easily carried out and generally requires low-cost equipment. *Disadvantages:* The properties of fluid, conditions of surface reactions, and incorporation of the crystal species change with decreasing temperature.

3.4 Techniques of Crystal Growth from the Solid Phase (Recrystallization)

It is fundamentally difficult to grow a single crystal from a solid phase since the mobility of the crystal-forming species is strongly restricted. Precipitation of a dissolved phase or devitrification of glass leads to very small crystals in general. Only materials of high dislocation mobility, like metals, can in some cases be transformed by recrystallization into single crystals. The process consists of the release of Helmholtz free energy through migration of grain boundaries, i.e., through the growth of large grains by the consumption of adjacent small ones. The procedure is achieved in two steps: the sample is (a) loaded with high free energy by straining and (b) either annealed in a homogeneous temperature field or moved through a temperature distribution in the direction of increasing temperature (in contrast to the Bridgman technique). The methods are described in detail by Laudise (1970) and Burgers (1963).

Advantage: The method generally requires low-cost equipment. *Disadvantages:* The method is restricted to materials of high dislocation mobility. Large single crystals are hard to obtain.

4. TECHNIQUES OF THIN-FILM GROWTH

Thin ($<100\times10^{-6}$ m) crystalline layers of various materials are of great importance for modern electronic devices. Crystalline films are grown by deposition or precipitation of the film species on a particular plate, called the substrate. The procedure is called (homo)epitaxy if the film consists of the same material as the substrate, heteroepitaxy if not (see HETEROSTRUCTURES AND SUPERLATTICES, METALLIC; HETEROSTRUCTURES AND SUPERLATTICES, SEMICONDUCTOR; THIN FILMS, NONELECTRONIC PROPERTIES). The control of layer thickness requires sufficiently low growth velocity, which is obtained by using a low-density fluid. Therefore layers are grown from a solution, i.e., by the technique of liquid-phase epitaxy, and from various kinds of vapor phases for which the techniques of chemical vapor epitaxy [see CVD (CHEMICAL VAPOR DEPOSITION)] and molecular-beam epitaxy (*q.v.*) are most widely used. Methods with low growth rates, low processing temperatures, and fast exchangeable fluids are applied to produce technologically important multilayers which consist of a periodic sequence of very thin ($<0.1\times10^{-6}$ m), variably doped semiconducting films.

4.1 Liquid-Phase Epitaxy

Liquid-phase epitaxy is mainly applied to layer growth of III-V compounds (e.g., GaAs) using the metal of the compound as a solvent. The methods are characterized by the procedure of attachment between substrate and solution:

1. Tipping: The solution and substrate are separately placed in a modified (graphite) boat. The solution is tipped over the substrate by tilting the boat.
2. Dipping: Similar to the Czochralski technique, the substrate attached to a rod is immersed into the solution.
3. Sliding: The substrate is held in a tray of a horizontal (graphite) plate. A second plate on top of the support plate contains a sufficiently large bore filled with solution and is slid over the substrate. Multilayers can be grown using a support plate with several solution chambers.
4. Centrifuging (especially designed for grow-

ing multilayers) as described by Bauser (1987): A rotating vertical graphite cylinder holds the solution at low revolution rates in two opposite inner tanks around a rotation axis, and at high revolution rates in two opposite outer storage channels. The substrates are mounted in radially connected channels between inner and outer storage chambers. The solution comes in contact with the substrate when the initially low revolution rate is increased.

In each method the layer growth is achieved either

1. by cooling, i.e., a saturated solution is brought in contact with the substrate and is cooled by a linear or sudden temperature change, or
2. by isothermal growth, i.e., a supersaturated solution is brought into contact with the substrate on which the crystal then grows at constant temperature by the consumption of the supersaturation.

4.2 Chemical Vapor Epitaxy

In most cases epitaxial layer growth through a heterogeneous chemical reaction is achieved in an open transport system (reactor), i.e., in a fused silica tube or vessel which contains the substrate and eventually source materials. The reactor is fed by appropriate vapor species (precursor molecules) in an H_2 or He carrier gas through a pipework which controls the flow rate. The various reactor designs are directed to different purposes—e.g., a laminar gas flow over the substrate, small exchangeable gas volume for multilayers, and high mixing of the vapors for homogeneous growth. The simplest type consists of a horizontal chamber with an obliquely mounted substrate or a vertical tube with the substrate attached to a horizontal pedestal. Two different principal reactor types result from the means of heating: namely, the cold-wall reactor where only the substrate support is (inductively) heated [Fig. 6(a)], and the hot-wall reactor where the substrate is heated by a resistance furnace outside of the reactor.

Three different thermochemical processes —pyrolysis, pyrolytic synthesis, and chemical transport—are employed for epitaxial deposition.

1. Pyrolysis (thermal decomposition) is preferred for deposition of elements and is achieved in a cold-wall reactor. An example is silicon epitaxy from a $SiHCl_3$-H_2 feed which decomposes at 1100 °C, i.e.,

$$2SiHCl_3(g) \rightarrow 2Si(s) + H_2(g) + 3Cl_2(g).$$

2. Pyrolytic synthesis consists of the layer formation from a reaction of at least two introduced vapor compounds. An example is the MOVPE (metalorganic vapor-phase epitaxy) of III-V compounds from a metal alkyl and phosphine in a cold-wall reactor, e.g., indium phosphide according to the reaction

$$(C_2H_5)_3In(g) + PH_3(g) \rightarrow InP(s) + 3C_2H_6(g)$$

at about 550 °C.
3. Chemical transport is a reversible process. It additionally requires a heated source chamber. Hot-wall reactors are preferred. An example is the chloride procedure according to the (simplified) reaction

$$2GaP(s) + 2HCl(g) = 2GaCl(g) + P_2(g) + H_2(g)$$

at about 850 °C. The deposition rate is adjusted by the H_2 flow rate, chlorine/hydrogen concentration ratio, and temperature difference between source and substrate.

4.3 Molecular-Beam Epitaxy

A molecular-beam apparatus consists of a very sophisticated assembly (see MOLECULAR-BEAM EPITAXY). It achieves precisely controlled epitaxial layer growth by incident fluxes [molecular beams, 10^7–10^8 molecules/(cm^2 s)] of atoms and molecules in an ultrahigh vacuum ($<10^{-8}$ Pa). Compound layers, e.g., $Ga_{1-X}Al_XAs$, are synthesized by the appropriate fluxes of Ga, Al, and As_2 species. Each species is evaporated from a particular Knudsen effusion cell. Monolayers of monatomic height are obtained by closing and opening the cells with specific shutters. Several apparatus are integrated such as

1. a mass spectrometer to control the vapor phase,
2. an electron gun to analyze the surface structure by angle-resolved photoelectron emission spectroscopy and reflection high-energy electron diffraction, and

3. an ion gun to determine chemical composition by secondary-ion mass spectroscopy.

5. CHARACTERIZATION OF CRYSTALS

5.1 Morphology

Depending on the growth procedure, a single crystal can display rounded or flat as-grown faces. Primarily, the morphology of growing crystals, i.e., the macroscopic shape, is determined by the density of free bonds at the surface or alternatively by the number of periodic bond chains (PBC) in different directions parallel to the interface within a distinct slice. According to the PBC theory, which is discussed by Hartman (1973), three types of interfaces can be classified—namely, kinked, stepped, and flat faces. Kinked (*K*) faces contain no PBC and are therefore rough, hence fast growing and unlikely to appear. Stepped (*S*) faces contain one type of PBC and provide the surface with parallel growth steps. Flat (*F*) faces contain at least two types of PBC, need step sources, grow slowly, and are therefore most likely to appear. In general, *F* faces have the highest lateral atom density, lowest Miller indices, and lowest surface energy. Reaction of free bonds with impurities or formation of surface bonds from neighboring free bonds (reconstruction) can lead to other faces as the PBC analysis predicts.

The formation of rounded faces depends on the temperature, the form of the fluid isotherms, and fluid composition. Strong rounding is not observed in crystal growth far below the melting point. In the case of crystal growth from solution or reactive vapor (e.g., chemical transport), the final morphology is strongly polyhedral: composed of *F* faces and eventually *S* faces. When the crystal is dissolved the morphology becomes different from that of growth in general. While the growth morphology consists of slowly growing faces, the solution morphology is preferably composed of fast-growing faces.

In the case of growth from a nonisothermal melt or vapor the crystal form approaches the surrounding isotherms. Faceting is pronounced by the presence of impurities in the fluid. Figure 20 illustrates an example. The faceting and rounding of the crystal interface are represented by interface contours. The sample, an iodine crystal, was grown by the Pizzarello technique (Fig. 8) from iodine vapor which contained a small amount of krypton. The initial (F_i) and final (F_f) interfaces were obtained by holding the ampoule at fixed positions x_i and x_f, respectively. The sequential growth contours from F_i to F_f result from a sudden ampoule displacement from x_i to x_f. Only rounded faces are observed for backsublimation where the interface moved from F_f to F_i after the reverse displacement from x_f to x_i.

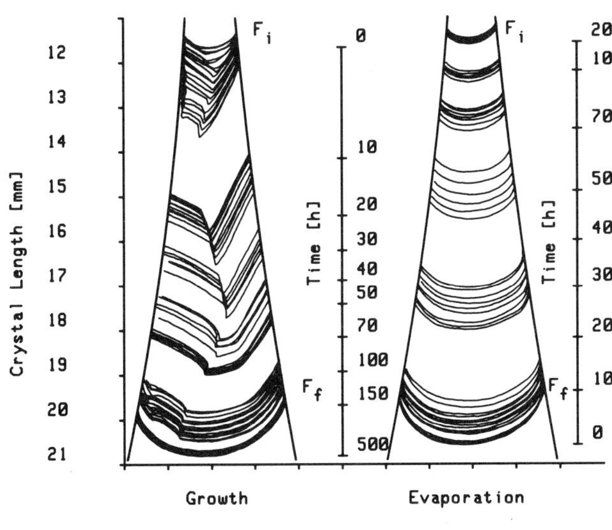

FIG. 20. Vapor-solid interface contours of iodine in the presence of krypton (400 Pa) at different times: growth at about 62 °C through sudden ampoule displacement; backsublimation (evaporation) after sudden reverse displacement.

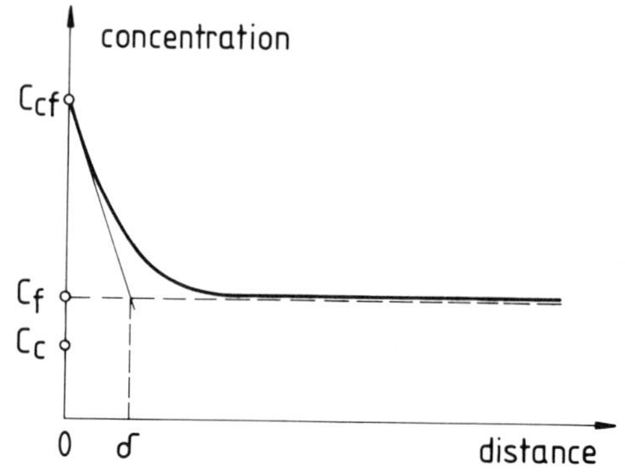

FIG. 21. Impurity distribution in front of a crystal growing from a melt. C_c is the concentration of the solute in the crystal, C_{cf} at the interface, and C_f in the melt at a large distance from the interface. δ is defined as the width of the boundary where transport of melt is achieved by diffusion.

5.2 Segregation of Impurities

5.2.1 Segregation Coefficients. In general, impurities dissolve in the coexisting melt and solid phases at different concentrations. The ratio $k_{eq}=C_{ce}/C_{fe}$ of the crystal impurity concentration (C_{ce}) to that (C_{fe}) of the melt is called the equilibrium segregation (or distribution) coefficient. The coefficient is obtained from the appropriate binary phase diagram. If $k_{eq}<1$ the impurity becomes enriched in front of the growing crystal interface as illustrated in Fig. 21 and discussed, e.g., by Pfann (1952, 1966), Carruthers (1975), and Brice (1965, 1973, and 1986). The impurity enrichment leads to surface concentration C_{cf}. The local (exponential) change of the impurity concentration in the fluid is approximated by a linear change which leads to the thickness δ of the so-called solute boundary layer. Two additional segregation coefficients are defined: the interfacial $k^*=C_c/C_{cf}$ and the effective $k_{eff}=C_c/C_f$ where C_c is the impurity concentration in the crystal near the interface and C_f the concentration in the bulk fluid. Under steady-state conditions the two coefficients are related by the crystal-growth rate V_c and diffusion coefficient D through

$$k_{eff}=k^*/[k^*+(1-k^*)\exp(-V_c\,\delta/D)]. \qquad (13)$$

The concentration distribution of the crystal is thus severely affected by fluctuations of the growth rate V_c and boundary-layer thickness δ. Fluctuations of both parameters are caused by fluctuations in the fluid which may arise from stirring or gravity-driven convection. If free-fluid surfaces are exposed to thermal gradients additional, irregular convection (Marangoni convection) appears driven by the difference in surface tension. The fluctuating impurity incorporation leads to changes of impurity densities perpendicular to the growth fronts, i.e., striations parallel to the interface, which can be revealed by a suitable etchant.

5.2.2 Impurity Distribution through Zone Refining. The difference in impurity concentrations in coexisting fluid and solid phases is used for purification through sublimation, distillation, and zone refining. For zone refining the principles of floating-zone techniques used for crystal growth are applied. A high efficiency is maintained by designs with several heating zones. Techniques are described by Pfann (1966) and Parr (1960). When a single melt zone of a width L is passed through a solid rod the concentration profile of the ingot assumes the form

$$C_c(x)=C_i[1-(1-k_{eff})\exp(-k_{eff}x/L)] \qquad (14)$$

[after Pfann (1966)], where C_i is the initial mean concentration of the ingot (molten zone) and k_{eff} an effective segregation coefficient. In the Bridgman and Czochralski processes the solute concentration of the crystal of a total length L_t is given by

$$C_c(x)=k_{eff}C_i(1-x/L_t)^{k_{eff}-1} \qquad (15)$$

for complete solidification of the melt and a constant crystal cross section. Concentration

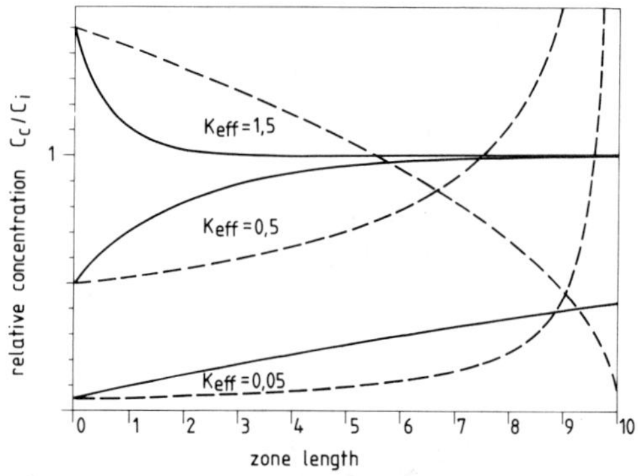

FIG. 22. Impurity distribution in a crystal for different segregation coefficients. Solid line for zoning. Dashed line for Bridgman procedure.

profiles according to Eqs. (14) and (15) are shown for different segregation coefficients in Fig. 22.

5.2.3 Precipitation. When the solubility of a solute decreases with decreasing temperature the solute may be precipitated in the crystal as small particles. Transparent crystals become opaque by this process. Preferably precipitation occurs along dislocations and grain boundaries.

5.3 Structural Defects

There are three classes of structural defects—

1. point defects,
2. line defects, and
3. two-dimensional defects—

that are or may be formed during crystal growth. Their formation is discussed by Vere (1987) and Chernov (1984). (See also IMPERFECTIONS IN CRYSTALS.)

5.3.1 Point Defects. The requirement of a minimum value of the Helmholtz free energy leads to some disorder of the structure above $T > 0$ K. There are two types of disorder:

1. vacant lattice sites, the so-called vacancies or Schottky defects, and
2. interstitials (Frenkel defects) which occupy sites other than those of the perfect structure.

Other point defects are formed by impurities which can occupy vacancies or interstitials. Point defects mainly influence the electrical and optical properties.

5.3.2 Line Defects. Line defects are called dislocations. They cannot terminate within the crystal, and so they either run between two surfaces, form a closed loop, or form a branch of another dislocation. A dislocation is characterized by the direction of its axis and Burgers vector. The vector results from comparison of the perfect structure with the actual structure around the dislocation. The defect is called a step dislocation when the Burgers vector runs perpendicular to the dislocation axis and a screw dislocation when the vector runs parallel to the axis. The screw dislocations serve as growth centers.

In Czochralski or floating-zone techniques, dislocations usually originate from the seed. If the dislocations are sufficiently strongly bonded by the lattice (e.g., silicon) they can be eliminated by growing a neck after seeding. During the crystal growth additional dislocations can be formed by shear strains which may result from

1. radial temperature gradients in the crystal,
2. precipitated impurities,
3. trapped impurity particles, or
4. sticking of the crystal to the container wall.

The formation of new dislocations is enhanced by multiplication of already existent dislocations as described by Hull (1975) and Friedel (1964).

5.3.3 Two-Dimensional Defects. Essentially there are three types of two-dimensional defects: (a) grain boundaries, (b) twins, and (c) stacking faults.

(a) Low-angle grain boundaries are mainly formed by multiplication of dislocations due to a straining stress field. Therefore, in the simplest case, low-angle boundaries consist of an array with parallel dislocations. For example, one family of step dislocations forms a pure tilt grain boundary where the dislocation axis is parallel to the tilt axis, while a family of screw dislocations forms a pure twist grain boundary where the dislocation axis is perpendicular to the boundary plane. The structure of high-angle boundaries is complex. They arise from intergrowth of differently oriented grains or from noncoherent nucleation on the growing surface (see GRAIN BOUNDARIES).

(b) Twin boundaries form a mirror plane where the structure is reflected. These particular faults originate either from a noncoherent nucleation or from mechanical deformation. Twinning is frequently observed for III-V and II-VI compounds which crystallize in the sphalerite structure. Here a twin is formed by a 180 °C rotated nucleus on a {111} plane. In contrast to grain boundaries, these twins are unmovable during heat treatment. Multiple twinning is caused by a too high supersaturation. Particular multitwin structures lead to dendritic growth as described, e.g., by Lindberg and Faust

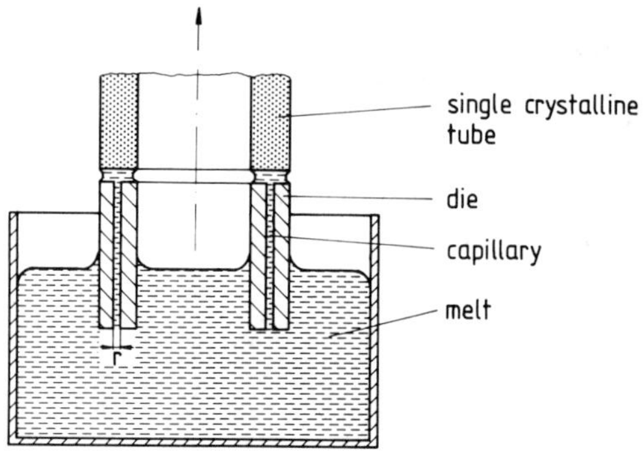

FIG. 23. Growth of a tube by the edge-defined film-fed method.

(1962).

(c) Stacking faults are mainly formed by precipitation of vacancies or interstitials so that either an atomic layer (intrinsic) is left out or an additional one (extrinsic) is inserted. Stacking faults are often observed on {111} planes of the sphalerite and diamond structures. Stacking faults extend within the crystal over a limited area, in contrast to twins, grain boundaries, and dislocations. More or less periodically different stacking (polytypism) of crystallographic planes occurs in some layered structures. The particular stacking faults are caused by dislocation movements. The faults are described by Trigunayat and Verma (1976) and Farkas-Jahnke (1983).

6. EXAMPLES OF APPLICATIONS

6.1 Czochralski Techniques

6.1.1 Edge-Defined Film-Fed Growth. Particular shapes of cross sections can be made through the Czochralski technique by placing a wettable die with pores between the melt and the crystal interface as illustrated in Fig. 23 and described by Ciszek (1981) and Wald (1981). The crystal is supplied with melt through a thin layer on top of the die. The layer is fed by the fluid which creeps through the pores if the die thickness is less than $2\sigma_f \cos(\phi)/\rho_f g r$, where σ_f is the melt surface tension, ϕ the wetting angle, ρ_f the melt density, g the acceleration due to the gravity, and r the radius of the capillary bores. Preferably, the method is applied to the growth of thin sheets (e.g., Si with a graphite die) and tubes (e.g., Al_2O_3 with a Mo die).

6.1.2 Dendritic Silicon Web Growth. Dendritic Si webs, candidates for solar cells, are grown by the Czochralski technique as indicated in Fig. 24. The sheets contain at least two coplanar twin planes parallel to the wide (111) sides; in addition, they are bounded by two [112] dendrites and grown in the [112] direction where the particular structure of two neighboring twin boundaries acts as a reentrant corner for the Si incorporation. The growth of the webs is started with a thin dendritic [112] seed (10^{-3} m diameter) from a melt of asymmetric temperature distribution. A review is given by Seidensticker (1982).

6.2 Float Zoning of Silicon

Figure 25 shows the principle of (a) normal and (b) pedestal zoning of silicon with inductive heating at a frequency between 3 and 5 MHz as reported by Keller and Mühlbauer (1981). A solid wedge-shaped coil (pancake coil with inner diameter of 35×10^{-3} m) with bores for water cooling (10^3-A current) superposes on the surface tension an electrodynamical force. In this way, a five times larger diameter (10^{-1} m) than that given in Eq. (12) can be achieved. The feed rod and crystal rod are oppositely rotated and displaced with respect to the fixed coil. The procedure of seeding and necking for disloca-

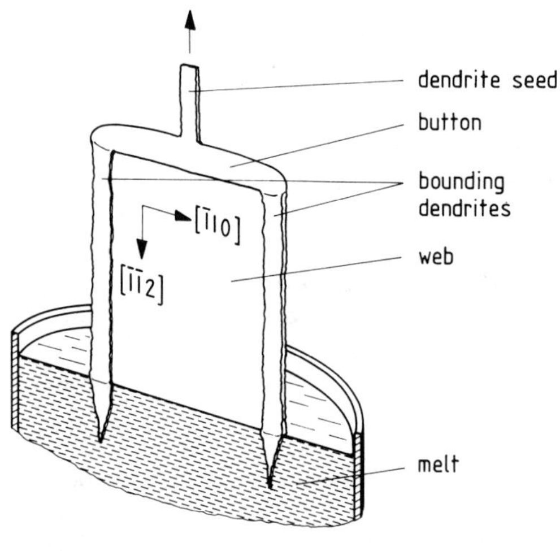

FIG. 24. Growth of a (111) silicon web.

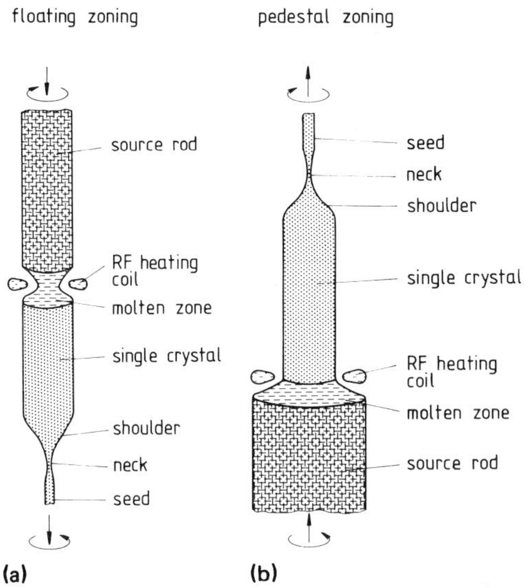

FIG. 25. Silicon crystal growth by float zone methods.

tion-free growth is, in principle, equal to the Czochralski method.

6.3 High-Pressure Technique of Diamond Growth

Diamonds of one carat (5×10^{-3} m in diameter) can be grown within a week from graphite dissolved in nickel at an external pressure of 6 GPa and a temperature of 1400 °C. The principle of this highly sophisticated method has been reviewed by Bundy *et al.* (1973), Wentorf (1974), and Nassau and Nassau (1979), and is schematically illustrated in Fig. 26. (See also DIAMONDS AND DIAMONDLIKE CARBON; HIGH-PRESSURE EFFECTS IN SOLIDS; HIGH-PRESSURE TECHNIQUES.) Two sections of a graphite heating tube are filled with a mixture of nickel and graphite powder. For better pressure control a diamond grit (synthetic or natural) is preferred instead of graphite powder. The two mixture compartments are lined with pyrophyllite [$Al_2Si_4O_{10}(OH)_2$]. The particular material is insulating and plastic at the employed high pressures and temperatures. A small seed plate of diamond is placed at the upper and lower pyrophyllite covering. The graphite tube is closed with nickel plates for current transfer and placed in a special belt press. The pestles and inner ring are made of cobalt-bonded tungsten carbide (carballoy: about 90% WC+10% Co). Both are surrounded by several belts of special steel. The free space between the tapered pistons and spheroid belts as well as that between the inner belt and graphite tube are insulated by pyrophyllite. By applying an electric voltage between the pistons the graphite tube is resistively heated slightly hotter at the middle than at the ends. Therefore the diamond grit dissolves in the middle and precipitates epitaxially at the seeds mounted at the tube ends (temperature difference technique).

Small diamond particles of about 0.25 $\times\ 10^{-3}$ m in diameter are crystallized within a short time (a few minutes) in a pyrophyllite

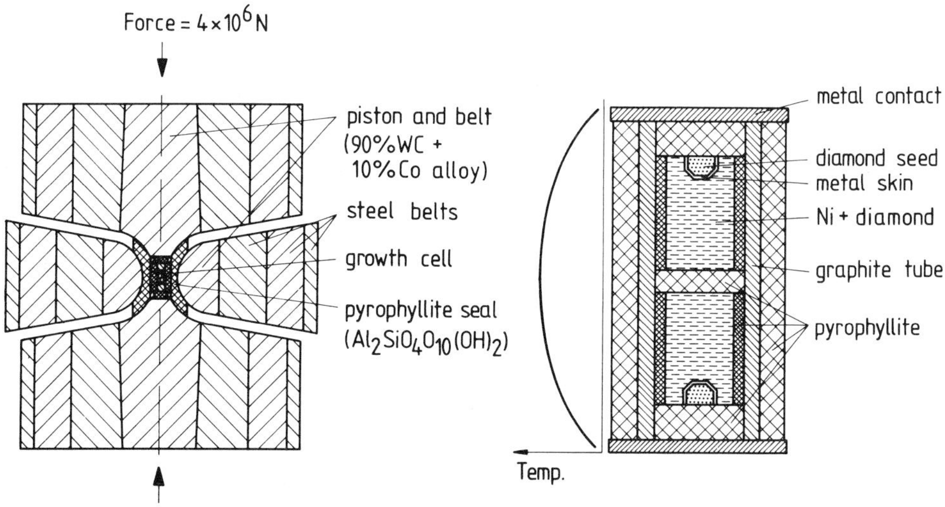

FIG. 26. Diamond growth. (See text.)

tube from a nickel graphite solution as shown in Fig. 26. The graphite is completely converted into diamond since the solubility of graphite is higher than that of diamond.

Advantages: Diamond grit can be fabricated at reasonable costs, and the grains are more suitable for tools than crushed natural diamond since they are free from cracks. *Disadvantage:* The fabrication of large crystals of high quality is more expensive than the cost of natural diamonds.

GLOSSARY

Associative Sublimation: Atoms of a solid join together to form molecules before entering the vapor phase.

Chemical Vapor Transport: Reversible transfer of a solid as a volatile compound through reaction with a gas.

Dendrite: Needlelike, often with branches, fast-growing crystal.

Devitrification: Transformation of glass into crystalline state.

Dissociative Sublimation: Decomposition of molecules of a solid when entering the vapor phase.

Dopant: A small defined quantity of substance added to the crystal.

Cluster: A group of like atoms or molecules forming an unstable condensate.

Nucleus: The smallest group of like atoms or molecules forming a stable condensate.

Peritectic Reaction: Formation of a solid phase β and liquid phase l from a solid phase α by heating of α, i.e., $\alpha \rightarrow \beta + l$.

Peritectic Temperature: The temperature at which the three phases α, β, and l coexist.

Segregation: Change of solute concentration during crystallization.

Stefan Problem: Problem of solving the heat-flow equation with a moving boundary.

Striations: Narrow bands of nearly constant solute concentration in crystals.

Substrate: A thin plate (in most cases a single crystal) onto which a thin solid layer is deposited from a vapor or liquid phase.

Supersaturation: State of vapor (or solute concentration) at pressures (concentrations) greater than that in equilibrium with the crystal.

Undercooling: Supercooling, state of melt at temperature below the melting point.

List of Works Cited

Aleksandrov, V. I., Osiko, V. V., Prokhorov, A. M., Tatarintsev, V. M. (1978), in: E. Kaldis (Ed.), *Current Topics in Materials Science*, Vol. 1. Amsterdam: North-Holland, pp. 421–480.

van Arkel, A. E. (1923), *Physica* **3**, 76–87.

Arora, S. K. (1981), *Prog. Cryst. Growth Charact.* **4**, 345–378.

Bardsley, W., Hurle, D. T. J., Joyce, G. C., Wilson, G. C., (1977), *J. Cryst. Growth* **40**, 21–28.

Bauser, E. (1987), in: R. F. C. Farrow, S. S. P. Parkin, P. J. Dobson, J. H. Neave, A. S. Arrott (Eds.), *Thin Film Growth Techniques for Low-Dimensional Structures*, New York: Plenum, pp. 171–194.

Bennema, P., Gilmer, G. H. (1973), in: P. Hartman (Ed.), *Crystal Growth: An Introduction*, Vol. 1, Amsterdam: North-Holland, pp. 263–327.

Bird, R. B., Stewart, W. E., Lightfood, E. N. (1960), *Transport Phenomena*, New York: Wiley.

Bockris, J. O'M., Razumney, G. A. (1967), *Fundamental Aspects of Electrocrystallization*, New York: Plenum.

Bohm, J. (1985), *Acta Phys. Hungar.* **57**, 161–178.Brice, J. C. (1965), in: E. P. Wohlfarth (Ed.), *Selected Topics in Solid State Physics*, Vol. 5, *The Growth of Crystals from the Melt*, Amsterdam: North-Holland.

Brice, J. C. (1973), in: E. P. Wohlfarth (Ed.), *Selected Topics in Solid State Physics*, Vol. 12, *The Growth of Crystals from Liquids*, Amsterdam: North-Holland.

Brice, J. C. (1986), *Crystal Growth Processes*, Glasgow: Blackie (Wiley).

Bridgman, P. W. (1925), *Proc. Amer. Acad. Arts Sci. (Boston)*, **60**, 303–383.

Buckley, H. E. (1951), *Crystal Growth*, New York: Wiley.

Bundy, F. P., Hall, H. T., Strong, H. M., Wentorf, Jun, R. H. (1955), *Nature (London)* **176**, 51–55.

Bundy, F. P., Strong, H. M., Wentorf, Jr., R. H. (1973), in: P. L. Walker, Jr., P. A. Thrower (Eds.), *Chemistry and Physics of Carbon*, Vol. 10. New York: Marcel Dekker, pp. 213–263.

Burgers, W. G. (1963), in: J. J. Gilman (Ed.), *The Art and Science of Growing Crystals*, New York: Wiley, pp. 416–450.

Burton, W. K., Cabrera, N., Frank, F. C. (1951), *Philos. Trans. R. Soc.* London, A **143**, 299–358.

Carruthers, J. R. (1975), in: N. B. Hannay (Ed.), *Treatise on Solid State Chemistry*, Vol. 5. New York: Plenum, pp. 325–406.

Chernov, A. A. (1984) in: B. K. Vainshtein, A. A. Chernov, I. A. Shuvalow (1984) (Eds.), *Modern Crystallography III. Crystal Growth*, Berlin: Springer, pp. 1–297.

Chisholm, D. (1971), in: G. Cook (Ed.), *M & B Technical Library, TL/ME/2, The Heat Pipe*, London: Mills & Boon.

Ciszek, T. F. (1981), in: J. Grabmaier (Ed.), *Crystals Growth, Properties, and Applications*, Vol. 5, *Silicon*, Berlin: Springer-Verlag, pp. 109–146.

Crank, J. (1988), *Free and Moving Boundary Problems*, Oxford: Clarendon.

Czochralski, J. (1918), Z. *Phys. Chem.* **92**, 219–221.

Draper, R. C. J. (1975), in: B. R. Pamplin (Ed.), *Crystal Growth*, Oxford: Pergamon, pp. 497–520.

Elwell, D., Scheel, H. J. (1975), *Crystal Growth from High-Temperature Solutions*, London: Academic.

Faktor, M. M., Garrett, I. (1974), *Growth of Crystals from the Vapour*, London: Chapman and Hall.

Falkenberg, R. (1978), in: C. H. L. Goodman (Ed.), *Crystal Growth Theory and Techniques*, Vol. 2. New York: Plenum, pp. 109–184.

Farkas-Jahnke, M. (1983), *Prog. Cryst. Growth Charact.* **7**, 163–210.

Friedel, J. (1964), *Dislocations*, Oxford: Pergamon.

Hartman, P. (1973), in: P. Hartman (Ed.), *Crystal Growth: An Introduction*, Vol. 1. Amsterdam: North-Holland, pp. 367–402.

Henisch, H. K. (1988), *Crystals in Gels and Liesegang Rings*, Cambridge: Cambridge University Press.

Hershey, D. (1973), *Transport Analysis*, New York: Plenum.

van Hook, A. (1961), *Crystallization, Theory and Praxis*, New York: Reinhold.

Hull, D. (1975), *Introduction to Dislocations*, Oxford: Pergamon.

Jackson, K. A. (1975), in: R. Ueda, J. B. Mullin (Eds.), *Crystal Growth and Characterization*, Amsterdam: North-Holland, pp. 21–32.

Jones, M. E., Shaw, D. W. (1975), in: N. B. Hannay (Ed.), *Treatise on Solid State Chemistry*, Vol. 5. New York: Plenum, pp. 283–323.

Kaldis, E., Cadoret, R., Schönherr, E. (1987), in: H. U. Walter (Ed.), *Fluid Sciences and Material Science in Space*, Berlin: Springer-Verlag, pp. 355–404.

Keck, P. H., Golay, M. J. E. (1953), *Phys. Rev.* **89**, 1297–1227.

Keller, W., Mühlbauer, A. (1981), in: W. R. Wilcox (Ed.), *Preparation and Properties of Solid State Materials*, Vol. 5, *Floating-Zone Silicon*, New York: Marcel Dekker.

Koref, F. (1922), *Z. Elektrochem.* **28**, 511–517.

Kossel, W. (1927), *Nachr. Ges. Wiss. Göttingen, Math.-Phys. Kl.*, pp. 135–143.

Krüger, F., Finke, W. (1910), Patentschrift Nr. 22 82 46, Klasse 12c, Gruppe 2, Berlin: Reichsdruckerei.

Kyropoulos, S. (1926), *Z. Anorg. Allg. Chemie* **154**, 308–313.

Laudise, R. A. (1970), *The Growth of Single Crystals*, Englewood: Prentice-Hall.

Lindberg, O., Faust J. W., Jr., (1962) in: R. K. Willardson, H. L. Goering (Eds.), *Compound Semiconductors*, Vol. 1. New York: Reinhold, pp. 294–310.

Mecseki, A., Mezö, B., Voszka, A., Galántai, A., Szabó, B. (1987), *Cryst. Res. Technol.* **22**, 677–683; **22**, 685–693.

Mullin, J. W. (1985), "Crystallisation," in: *Encyclopedia of Chemical Technology*, Vol. 7. New York: Wiley, pp. 243–285.

Mutaftschiev, B. (1980) in: F. R. N. Nabarro (Ed.), *Dislocations in Solids*, Vol. 5. Amsterdam: North-Holland, pp. 57–126.

Nacken, R. (1915), *Neues Jahrb. Mineral. Geol. (Abt. B.)* **2**, 133–164.

Nacken, R. (1950), *Chem. Ztg.* **74**, 745–749.

Nassau, K., Nassau, J. (1979), *J. Cryst. Growth* **46**, 157–172.

Obreimow, I., Schubnikow, L. (1924), *Z. Phys.* **25**, 31–36.

Parker, R. L. (1970), in: H. Ehrenreich, F. Seitz, D. Turnbull (Eds.), *Solid State Physics*, Vol. 25. New York, Academic, pp. 152–299.

Parr, N. L. (1960), *Zone Refining and Allied Techniques*, London: G. Newnes.

Peube, J.-L. (1977), in: R. Balian, J.-L. Peube (Eds.), *Fluid Dynamics*, London: Gordon and Breach.

Pfann, W. G. (1952), *Trans. AIME* **194**, 747–753.

Pfann, W. G. (1966), *Zone Melting*, New York: Wiley.

Pizzarello, F. (1954), *J. Appl. Phys.* **25**, 804–805.

Rabenau, A. (1985), *Angew. Chem. Int. Ed. Engl.* **24**, 1026–1041.

Ramachandrarao, P. (1978), *Z. Metallk.* **69**, 656–660.

Rosenberger, F. (1979), *Fundamentals of Crystal Growth I*, Berlin: Springer-Verlag.

Sauveur (1912), *Proc. Int. Assoc. Test. Mater.* **2** (11), No.II.6.

Schäfer, H. (1964), *Chemical Transport Reactions*, New York: Academic.

Schönherr, E. (1980), in: H. C. Freyhardt (Ed.), *Crystals Growth, Properties, and Application*, Vol. 2. Berlin: Springer-Verlag, pp. 51–118.

Seidensticker, R. S. (1982), in: H. C. Freyhardt (Ed.), *Crystals Growth, Properties, and Application*, Vol. 8, *Silicon, Chemical Etching*, Berlin: Springer-Verlag, pp. 145–172.

Shah, J. S., Wills, H. H. (1975), in: B. R. Pamplin (Ed.), *The Science of Solid State*, Vol. 6, *Crystal Growth*, Oxford: Pergamon, pp. 326–396.

Sigsbee, R. A. (1969), in: A. C. Zettlemoyer (Ed.), *Nucleation*, New York: Marcel Dekker, Chap. 4.

Scheffen-Lauenroth, Th., Klapper, H., Becker, R. A. (1981), *J. Cryst. Growth* **55**, 557–570.

Spezia, G. (1905), *Atti R. Accad. Sci. Torino* **40**, 254–262.

Stockbarger, D. C. (1936), *Rev. Sci. Instrum.* **7**, 133–136.

Stranski, I. N. (1928), *Z. Phys. Chem.* **136**, 259–278.

Strickland-Constable, R. F. (1968), *Kinetics and Mechanism of Crystallization*, London: Academic.

Tammann, G. (1914), *Lehrbuch der Metallographie*, Leipzig: Leopold Voss.

Tolksdorf, W., Welz, F. (1978), in: C. J. M. Rooijmans (Ed.), *Crystals Growth, Properties, and Applications*, Vol. 1. Berlin: Springer-Verlag, pp. 1–52.

Trigunayat, G. C., Verma, A. R. (1976) in: F. Lévy (Ed.), *Crystallography and Crystal Chemistry of Materials with Layered Structures*, Vol. 2. Dordrecht: D. Reidel, pp. 269–340.

Vere, A. W. (1987), *Crystal Growth Principles and Progress*, New York: Plenum.

Verneuil, M. A. (1902), *C. R. Acad. Sci. (Paris)* **135**, 791–794.

Voronkov, V. V. (1983), in: A. A. Chernov, H. Müller-Krumbhaar (Eds.), *Crystals Growth, Properties, and Applications*, Vol. 9. Berlin: Springer-Verlag, pp. 75–111.

Wald, F. V. (1981), in: J. Grabmaier (Ed.), *Crystals Growth, Properties, and Applications*, Vol. 5, *Silicon*, Berlin: Springer-Verlag, pp. 147–198.

Walker, A. C., Kohman, G. T. (1948), *Trans. Am. Inst. Electr. Eng.* **67**, 580–585.

Walton, A. G. (1969), in: A. C. Zettlemoyer (Ed.), *Nucleation*, New York: Marcel Dekker, Chap. 5.

Wanklyn, B. M. (1975), in: B. R. Pamplin (Ed.), *Crystal Growth*, Oxford: Pergamon, pp. 217–288.

Wentorf, Jr., R. H. (1974), in: R. H. Wentorf, Jr. (Ed), *Advances in High-Pressure Research*, Vol. 4. London: Academic, pp. 249–261.

Wilcox, W. L. (1971), in: R. A. Lefever (Ed.), *Preparation and Properties of Solid State Materials*, Vol. 1. New York: Marcel Dekker, Chap. 2.

Winter, E. R. F., Barsch, W. O. (1971), *Adv. Heat Transfer* **7**, 219–320.

Further Reading

Brice, J. C. (1973), in: E. P. Wohlfarth (Ed.), *Selected Topics in Solid State Physics*, Vol. 12, *The Growth of Crystals from Liquids*, Amsterdam: North-Holland.

Brice, J. C. (1986), *Crystal Growth Processes*, Glasgow: Blackie (Wiley).

Buckley, H. E. (1951), *Crystal Growth*, New York: Wiley.

Chalmers, B. (1964), *Principles of Solidification*, New York: Wiley.

Elwell, D., Scheel, H. J. (1975), *Crystal Growth from High-Temperature Solutions*, London: Academic.

Faktor, M. M., Garrett, I. (1974), *Growth of Crystals from the Vapour*, London: Chapman and Hall.

Gilman, J. J. (Ed.) (1963), *The Art and Science of Growing Crystals*, New York: Wiley.

Goodman, C. H. L. (Ed.) (1974), *Crystal Growth Theory and Techniques*, Vol. 1. New York: Plenum.

Goodman, C. H. L. (Ed.) (1978), *Crystal Growth Theory and Techniques*, Vol. 2. New York: Plenum.

Henisch, H. K. (1988), *Crystals in Gels and Liesegang Rings*, Cambridge: Cambridge University Press.

van Hook, A. (1961), *Crystallization, Theory and Praxis*, New York: Reinhold.

Laudise, R. A. (1970), *The Growth of Single Crystals*, Englewood: Prentice-Hall.

Lawson, W. D., Nielsen, S. (1958), *Preparation of Single Crystals*, London: Butterworths.

Mullin, J. W. (1972), *Crystallisation*, London: Butterworths.

Mullin, J. W. (1985), in: "Crystallization," in: *Encyclopedia of Chemical Technology*, Vol. 7. New York: Wiley.

Nassau, K. (1980), *Gems Made by Man*, Radnor: Chilton.

Pamplin B. R. (Ed.) (1975), *Crystal Growth*, Oxford: Pergamon.

Parr, N. L. (1960), *Zone Refining and Allied Techniques*, London: G. Newnes.

Pfann, W. G. (1966), *Zone Melting*, New York: Wiley.

Rosenberger, F. (1979), *Fundamentals of Crystal Growth I*, Berlin: Springer-Verlag.

Strickland-Constable, R. F. (1968), *Kinetics and Mechanism of Crystallization*, London: Academic.

Tarján, I., Mátrai, M. (1972), *Laboratory Manual on Crystal Growth*, Budapest: Akadémiai Kiadó.

Vainshtein, B. K., Chernov, A. A., Shuvalow, I. A. (Eds.), (1984), *Modern Crystallography III. Crystal Growth*, Berlin: Springer.

Vere, A. W. (1987), *Crystal Growth Principles and Progress*, New York: Plenum.

Verma, A. R. (1953), *Crystal Growth and Dislocations*, London: Butterworths.

Wilke, K. Th. (1988), *Kristallzüchtung*, Frankfurt/M: Verlag Harri Deutsch.

Series and Journals

Crystals Growth, Properties, and Applications, Berlin: Springer-Verlag (1978).

Crystal Res. Tech. (1981).

Current Topics in Materials Science, Amsterdam: North-Holland (1978).

Growth of Crystals (Translated from Russian), New York: Consultants Bureau (1959).

J. Cryst. Growth (1967).

North-Holland Series in Crystal Growth, Amsterdam: North-Holland (1973).

Preparation and Properties of Solid State Materials, New York: Marcel Dekker (1981).

Prog. Cryst. Growth Charact. (1977).

CRYSTALLINE STATE

A. I. GOLDMAN, *Ames Laboratory of the United States Department of Energy and Department of Physics, Iowa State University, Ames, Iowa, U.S.A.*

INTRODUCTION: CRYSTALLINE AND NONCRYSTALLINE SOLIDS

Historically, we have categorized solids as either crystalline or glassy (often referred to as amorphous solids) as illustrated in Fig. 1. A simple operational definition of crystallinity, which will be expanded upon in detail below, is the state of matter characterized by long-range positional order of its constituent atoms. That is, there is a set of well-defined rules that may be used to specify the positions of all atoms in the solid. In general, the task of describing the various crystalline structures is simplified by the presence of periodic translational order. Once the local arrangement of atoms in one region of the crystal is described, this basic structural unit may be translated through some set of well-defined vectors to construct the entire solid. For glassy materials, on the other hand, no such set of rules or relations exists beyond an average local structure enforced by the forces that bind atoms together. This means that although local short-range positional order may exist in glassy solids, there is no corresponding long-range positional order. A third class of structures which, though well-ordered, are not periodic crystals includes incommensurate structures and quasicrystals. These are solids which possess long-range positional order which is aperiodic. Although our primary purpose here

1-56081-063-7/92/$5.00+.50

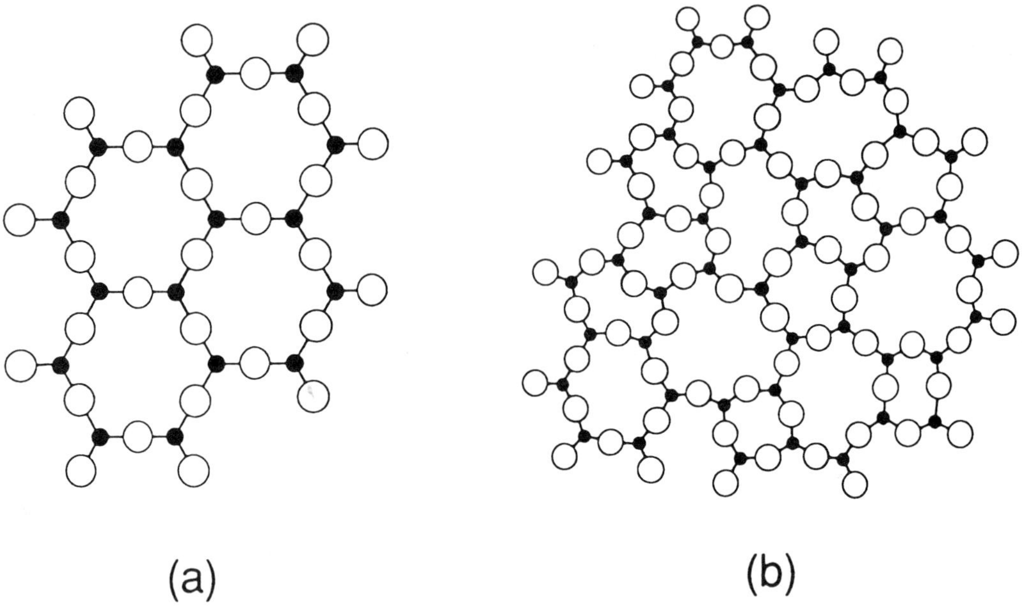

FIG. 1. Crystalline (a) and glassy (b) networks of atoms. (After Kittel, 1986.)

is the description of periodic crystalline solids, the latter two classes of structures will also be discussed where appropriate.

1. CRYSTAL STRUCTURES

This section deals with the definition and details of periodic crystal structures. In particular, we consider what restrictions periodicity places on the symmetries of crystal structures and the placement of atoms within these structures. We first develop the language of crystallography by considering two-dimensional crystal lattices and then move on to three-dimensional structures.

1.1 Two-Dimensional Crystal Lattices

1.1.1 Bravais Lattices and Translational Symmetry. Consider the two-dimensional array of points shown in Fig. 2. These points are said to belong to a *Bravais lattice* if the arrangement and orientation of lattice points, as viewed from one particular point, is identical to that viewed from all other points in the lattice. All points in Fig. 2 satisfy this condition. The hexagonal net of Fig. 3, however, clearly does not satisfy this requirement since points X and Y are not equivalent. A more formal definition of a two-dimensional Bra-

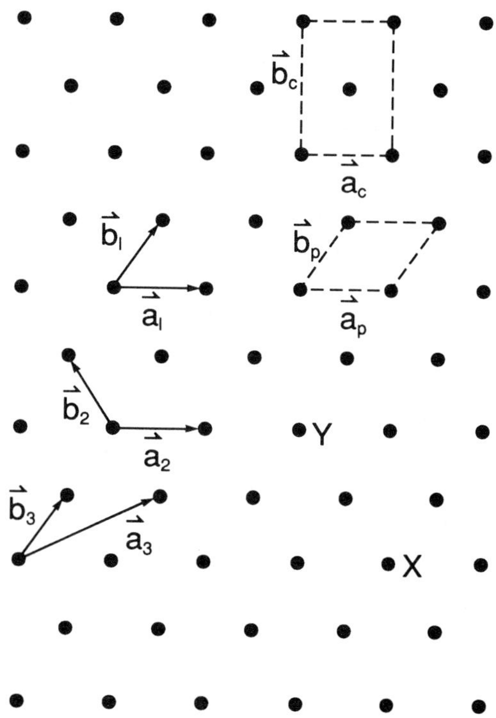

FIG. 2. The centered rectangular Bravais lattice. Primitive translation vectors and primitive and conventional unit cells are illustrated.

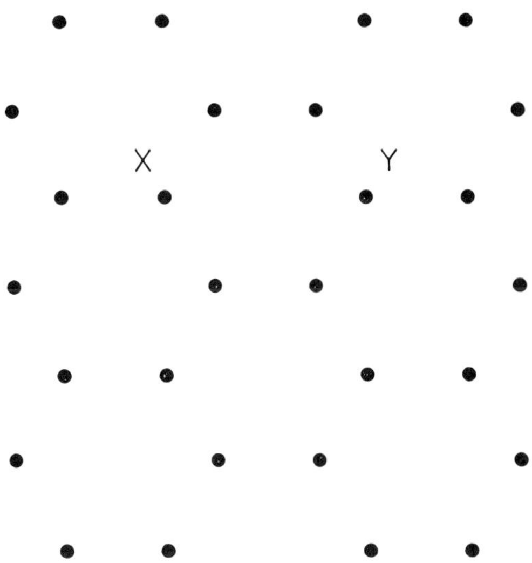

FIG. 3. Two-dimensional hexagonal net. Points X and Y are not equivalent.

vais lattice is that all points may be reached by a translation of the form

$$\mathbf{R}=n_1\mathbf{a}+n_2\mathbf{b},$$

where n_1 and n_2 are integers, and **a** and **b** are referred to as the *primitive translation vectors*. In three dimensions this definition must be modified to include translations out of the plane along a third primitive translation vector, **c**. A general translation in three dimensions is then described by the vector

$$\mathbf{R}=n_1\mathbf{a}+n_2\mathbf{b}+n_3\mathbf{c}.$$

For the two dimensional Bravais lattice in Fig. 2, we can see that there are many possible choices for **a** and **b**. In terms of the three sets of primitive vectors illustrated in Fig. 2, a translation from point X to point Y can be represented by $-2\mathbf{a}_1+2\mathbf{b}_1$ or $-2\mathbf{b}_2$ or $-2\mathbf{a}_3+4\mathbf{b}_3$.

In the same way that we chose primitive translation for a Bravais lattice, we can define a *primitive unit cell* as the minimum-volume cell that will completely tile the plane. Note that the particular choice of this cell is also nonunique. One example of a primitive unit cell for the Bravais lattice of Fig. 2 is illustrated by the parallelogram spanned by vectors $\mathbf{a}_p$ $(\mathbf{a}_1)$ and $\mathbf{b}_p$ $(\mathbf{b}_1)$. If we now take this parallelogram and translate it through all linear combinations of $n_1\mathbf{a}_p+n_2\mathbf{b}_p$, for integer n_1 and n_2, we can tile the entire plane without overlap or gaps. It is often more useful, however, to define a second type of unit cell, termed a *conventional unit cell*, as depicted by the rectangle in Fig. 2. The reason for this is that the rectangular unit cell more clearly illustrates the underlying rectangular symmetry of the lattice. Note that while the primitive unit cell contains one lattice point, the conventional unit cell in this figure contains two lattice points. The four lattice points at the corners are shared by four unit cells while the lattice point at the center is completely within the domain of a single cell. Again, if we translate the rectangle by integer combinations of $\mathbf{a}_c$ and $\mathbf{b}_c$, we can tile the entire plane without overlap or gaps.

Before going on to describe the symmetries of lattices, we complete the picture by pointing out that the crystal structure is then fully determined by associating a set of atoms with each lattice point or unit cell. Clearly, if we decorate each unit cell with atoms in an identical manner, the same translation vectors used to move between points on a Bravais lattice can be used to move between identical atoms in all unit cells. In other words, the atomic arrangement is also periodic.

1.1.2 Point Groups and Space Groups. Although there are an unlimited number of possible lattices in two dimensions, there are only five distinct two dimensional Bravais lattices, as shown in Fig. 4. The most general Bravais lattice is the oblique lattice since it places no restrictions on **a**, **b**, or the angle (γ) between them.

The symmetries of a crystal are defined by the kinds of operations that leave the structure invariant. In addition to the translations described above, there are also *point-group operations* to consider. The point group is the

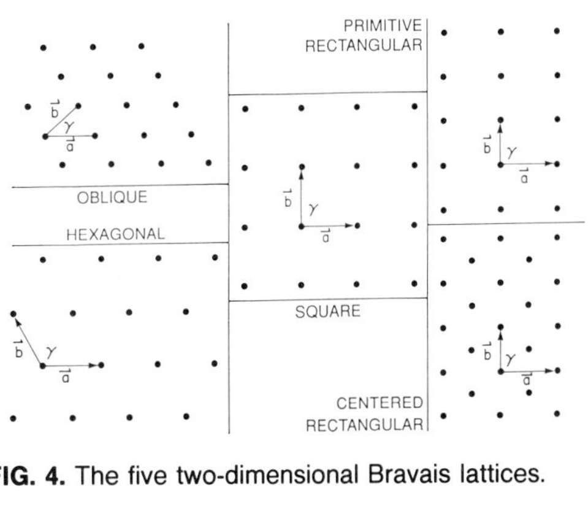

FIG. 4. The five two-dimensional Bravais lattices.

set of all operations which leave one point fixed. In two dimensions these operations include rotations about a fixed point and reflection across a line. The group of such actions which leave the lattice invariant (map the lattice points back onto themselves) defines the point symmetry of the lattice. The oblique lattice, for instance, may be left invariant only under rotations of π (180°) or 2π (360°). If we rotate the lattice points about one fixed point by these values, the rotated lattice is again coincident with the unrotated lattice. By further restricting the possible values of **a**, **b**, and γ, we obtain the special lattice types in Fig. 4. For instance, by restricting $a=b$ and $\gamma=90°$, we see (Fig. 5) that the square Bravais lattice is invariant under rotations of $\pi/4$, as well as reflections across lines x, y, x', and y'. Similarly, one can place constraints on **a**, **b**, and γ that lead to hexagonal Bravais lattices which are invariant under rotations of $2\pi/3$ and $2\pi/6$.

It should be pointed out here that the requirement of periodic translational order restricts the allowed rotational symmetries of two- and three-dimensional lattices to rotations by 2π, $2\pi/2$, $2\pi/3$, $2\pi/4$, and $2\pi/6$. In other words, in addition to a simple rotation by 2π (360°), periodic crystals may have twofold, threefold, fourfold, and sixfold rotation axes, but not five-fold or seven-fold or higher rotational symmetry. The reason for this is illustrated in Fig. 6 for the case of fivefold rotational symmetry. Consider two nearest-neighbor lattice points **A** and **B** separated by some minimum distance d. If we rotate **B** by an angle of $2\pi/4$ about **A** to get **B**$'$, and rotate **A** about **B** by the same angle to get **A**$'$, then both **A**$'$ and **B**$'$ are also lattice points of a square lattice with a minimum separation distance of d. If we perform the same set of operations on **A** and **B**, now specifying the rotation angle to be $2\pi/5$, the new lattice points **A**$'$ and **B**$'$ are now separated by less than d, contradicting the original assumption that d is the minimum distance. The conclusion from all of this is that periodic lattices may not have fivefold rotation axes.

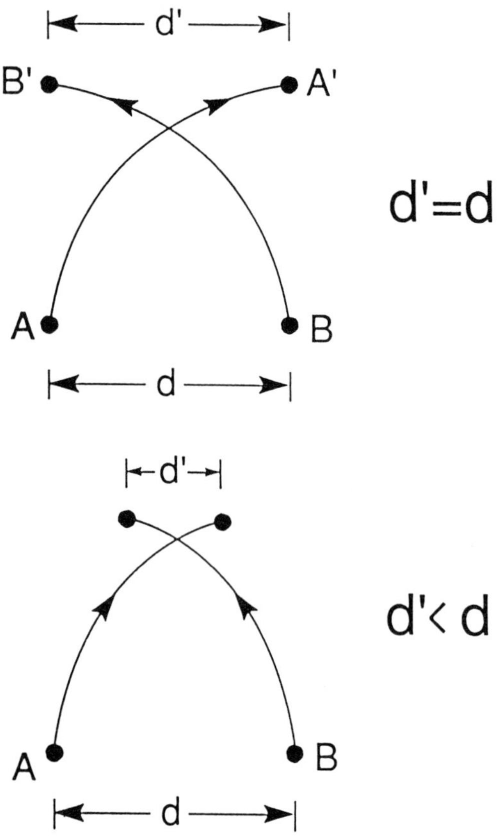

FIG. 6. An illustration of why fivefold rotation axes are not permitted for periodic structures.

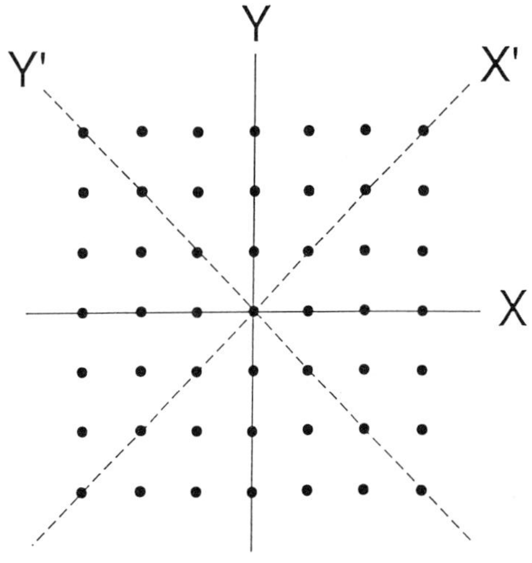

FIG. 5. Square Bravais lattice showing reflection symmetry across the lines x, y, x', and y'.

The combined actions of all point-group operations (rotations and reflections) result in a total of ten two-dimensional point groups which are illustrated in Fig. 7. The oblong, triangular, square, and hexagonal filled symbols represent twofold, threefold, fourfold, and sixfold rotation axes, respectively. The heavy lines denote *mirror lines*. The presence

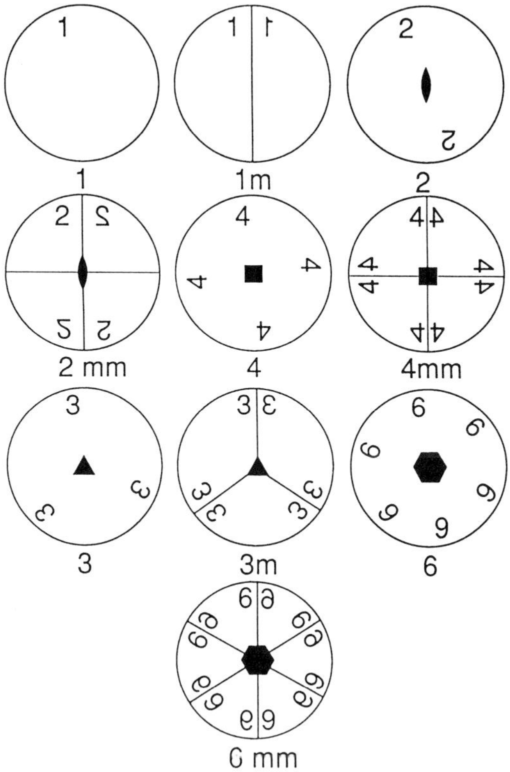

FIG. 7. The ten two-dimensional point group operations. The oblong, triangular, square, and hexagonal symbols denote twofold, threefold, fourfold, and sixfold rotation axes. The heavy lines represent mirror lines. (After the *International Tables for X-ray Crystallography.*)

of a mirror lines indicates that reflection across that line leaves the pattern invariant. It is interesting to note that the combination of certain point-symmetry elements lead to others. For example, for the group 2*mm* in Fig. 7, the combination of a twofold rotation axis and a mirror line results in a second mirror line perpendicular to the first.

New symmetries appear when we consider the effect of compounding point-group and translational symmetries. For example, the point group 4 translated through the vectors of a square lattice results in a set of twofold axes midway between the lattice points and fourfold axes at the center of each unit cell (Fig. 8). Another operation, termed a *glide,* is composed of a reflection across the glide line and a translation of one-half the repeat distance along that line (see Fig. 12). Taken together, the combinations of point-group operations with translations, and glides, yield 17

FIG. 8. The point group 4 translated through the vectors of a square lattice.

two-dimensional *space groups.* In other words, there are 17 ways of arranging a set of identical objects periodically in two dimensions. All two-dimensional periodic structures belong to one of these space groups.

The three two-dimensional space groups of fourfold (square) symmetry are shown in Fig. 9 along with the diagram of spatially distributed symmetry operations found in the *International Tables for X-ray Crystallography.*

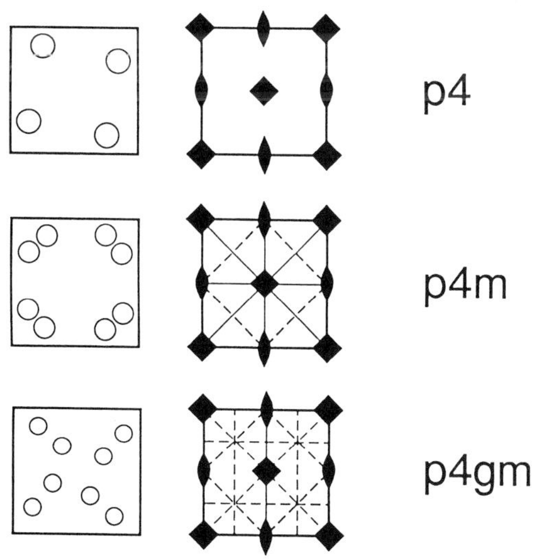

FIG. 9. The three two-dimensional space groups derived from the square space lattice and point-group operations. (After *The International Tables for X-ray Crystallography.*)

The left-hand figure illustrates an important point concerning the demands that various symmetry operations place upon the distribution of atoms in the unit cell. The space group label *p*4 indicates that the group is primitive (one lattice point per unit cell) and has fourfold rotational symmetry. If we place an atom at some *general position* (x,y) within the unit cell (where x and y are some fraction of the repeat distance a along the two orthogonal axes), three other atoms must be placed at $(-x,-y)$, $(y,-x)$, and $(-y,x)$ to satisfy the requirement of fourfold rotational symmetry. Likewise, the other space groups associated with the square lattice, which include additional mirror and glide symmetries, pose their own requirements for the symmetric placement of atoms. Note that there are also *special positions* such as (0,0), which only require one atom per unit cell.

1.2 Crystallography in Three Dimensions

All of the language that we developed in the last two sections carries over to the description of three-dimensional periodic crystalline lattices. In three dimensions there are 7 crystal systems (summarized in Table 1) and 14 Bravais lattices with conventional unit cells as shown in Fig. 10. Each Bravais lattice may also be described by a minimum-volume primitive unit cell as described in Sec. 1.1.1. For instance, the primitive unit cell of the face-centered-cubic (fcc) structure is a 60° rhombohedron defined by the vectors connecting a corner of the conventional unit cell with the lattice points at the centers of the three faces. The primitive unit cell of the fcc (cubic *F*-type) lattice contains one lattice point, while the conventional unit cell contains four (each of the corners is shared by eight unit cells, while each of the six faces connects two unit cells). As before, the conventional unit cell is preferred since it more closely reflects the symmetry of the underlying lattice.

Including the 10 two-dimensional point groups, which remain valid operations, there are 32 three-dimensional point groups which include new symmetry operations such as mirror planes perpendicular to the axis of rotation, rotary-inversion axes (where a rotation about an axis of $2\pi/n$ is accompanied by a reflection across a plane perpendicular to

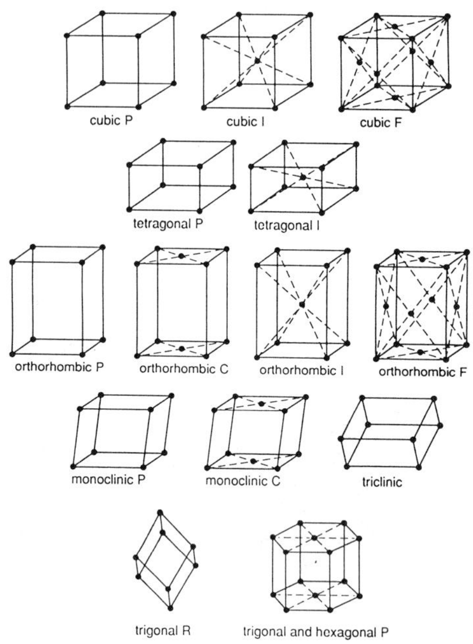

FIG. 10. Conventional unit cells of the 14 three-dimensional Bravais lattices. (After Kittel, 1986.)

TABLE 1. The seven crystal systems in three dimensions.

System	Conventional cell edge lengths	Restrictions on angles
Cubic	$a=b=c$	$\alpha=\beta=\gamma=90°$
Tetragonal	$a=b\neq c$	$\alpha=\beta=\gamma=90°$
Orthorhombic	$a\neq b\neq c$	$\alpha=\beta=\gamma=90°$
Hexagonal	$a=b\neq c$	$\alpha=\beta=90°$ $\gamma=120°$
Trigonal	$a=b=c$	$\alpha=\beta=\gamma\neq 90°$ but less than 120°
Monoclinic	$a\neq b\neq c$	$\alpha=\gamma=90°$ $\beta\neq 90°$
Triclinic	$a\neq b\neq c$	$\alpha\neq\beta\neq\gamma$

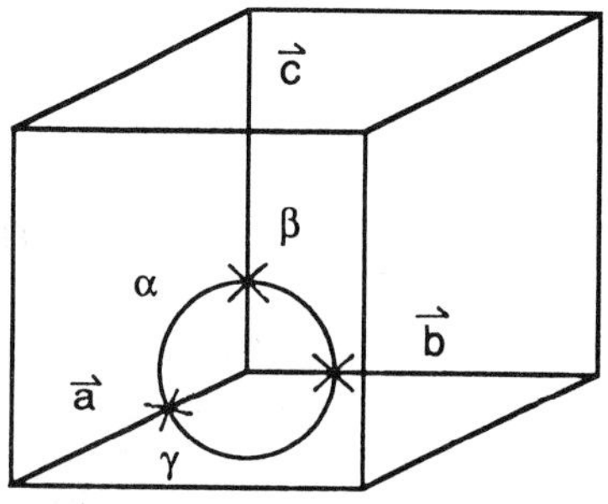

the line), and inversion through a point [where points $(x,y,z) \rightarrow (-x,-y,-z)$].

There are two additional operations which are required to enumerate the 230 three-dimensional space groups completely. First, a *screw axis* is defined by a rotation accompanied by a translation along the rotation axis by a vector which is some rational fraction of a Bravais lattice vector. Figure 11, for instance, shows examples of right-handed and left-handed screw tetrads. In this case, equivalent sites are related by a 90° rotation and a translation by $c/4$, where c is the repeat distance along the screw axis. One can similarly define screw diads, triads, and hexads. The second operation defines a *glide plane*, and consists of a translation through a vector which is not in the Bravais lattice and a reflection across a plane containing that vector. A glide plane consisting of a translation by $c/2$ is illustrated in Fig. 12.

For further details on space groups and point groups in crystallography, we refer the reader to the first volume of *The International Tables of X-ray Crystallography.*

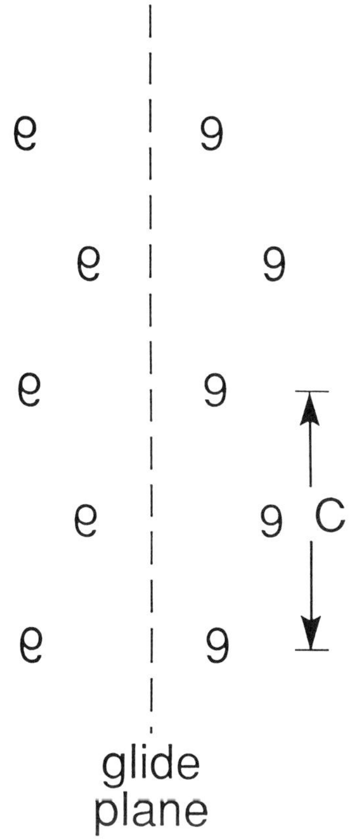

FIG. 12. A glide plane. The arrangement to the right of the plane is obtained by translating the assembly on the left by $c/2$ and reflecting across the plane.

1.3 Describing Planes and Directions

Quite often we are concerned with the arrangement of atoms in a particular plane of a three-dimensional crystal or the direction of the normal to that plane. Lattice planes in the cubic system are shown in Fig. 13. The plane itself is specified by a set of three *Miller indices* (h,k,l) which are defined as a set of integers which are inversely proportional to the intercepts of the plane along the crystal axes. For example, the (2,1,1) plane in Fig. 13 intercepts the a, b, and c axes at points $a/2$, b, and c. Similarly, the $(-1,1,1)$ or $(\bar{1},1,1)$ planes lies along the body diagonal of the cube and has intercepts of $-a$, b, and c. We may also group symmetry-related planes in a crystal into a family denoted by $\{h,k,l\}$. For instance, in the cubic system the planes (1,0,0), (0,1,0), (0,0,1), $(\bar{1},0,0)$, $(0,\bar{1},0)$, and $(0,0,\bar{1})$ all belong to the {1,0,0} family of planes. In contrast, for a crystal with tetragonal symmetry the (0,0,1) and $(0,0,\bar{1})$ planes do not belong to the {1,0,0} family. The direction of the plane is defined by the normal to the lattice plane of interest, and denoted by the same set of Miller indices enclosed in square brackets, $[h,k,l]$. The [2,1,1] direction in Fig. 13 is therefore normal to the (2,1,1) plane.

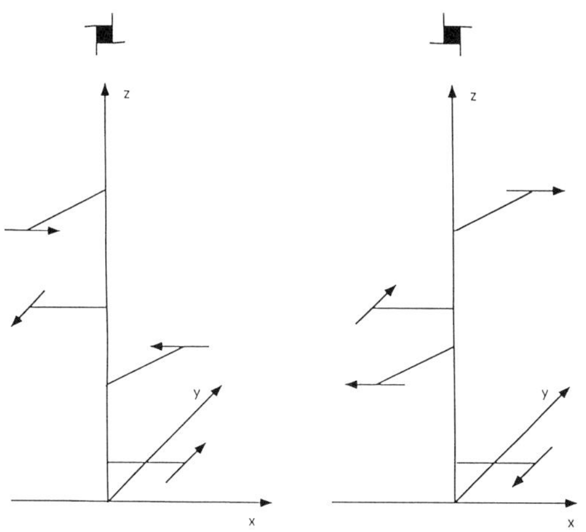

FIG. 11. Right-handed and left-handed screw tetrads. The arrangements in planes displaced by $c/4$ are related by a 90° rotation about the line.

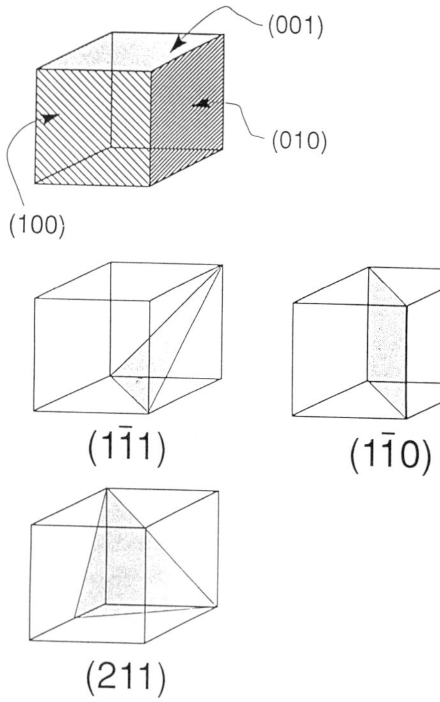

FIG. 13. Lattice planes in a cubic crystal. Directions are specified by the outward normals to these planes.

1.4 Examples of Crystal Structures

1.4.1 Close-Packed Structures. With few exceptions, metals crystallize into highly symmetric, densely packed structures of the cubic and hexagonal lattices shown in Fig. 14. As described below in Sec. 2.4, this arises from the nondirectionality of metallic bonding.

The *face-centered-cubic* (fcc) close-packed structure [Fig. 14(a)] may be described in terms of a cubic conventional unit cell with four atoms at positions (0,0,0), $(0,\frac{1}{2},\frac{1}{2})$, $(\frac{1}{2},0,\frac{1}{2})$, and $(\frac{1}{2},\frac{1}{2},0)$. Each atom is in contact with 12 nearest neighbors along the face diagonals of the unit cell at a center-to-center separation of $a/\sqrt{2}$, where a is the edge length of the unit cell. The density of the fcc structure may be calculated as follows: The volume of the cubic unit cell is a^3, and the total volume occupied by the four atoms per unit cell is $4\times 4\pi R^3/3$. If we take the radius of each atom as half the nearest- neighbor distance the packing fraction of the fcc close-packed structure is 74%.

The *hexagonal-close-packed* (hcp) structure itself is not a Bravais lattice, but may be viewed as two interpenetrating hexagonal lattices displaced from each other by $\mathbf{a}/3$, $2\mathbf{b}/3$, and $\mathbf{c}/2$. The atomic positions of the two atoms per hexagonal unit cell are then (0,0,0) and $(\frac{1}{3},\frac{2}{3},\frac{1}{2})$. Each atom is again coordinated by 12 nearest neighbors (six in the plane and two sets of three from the planes above and below). For the ideal close packing of hard spheres, the ratio of the c-axis lattice constant to the in-plane (a axis) lattice constants is $c/a = \sqrt{8/3}$.

There is a close similarity between the fcc close packing and hexagonal close-packing schemes. This is most clearly seen if we consider the packing of atoms in the planes perpendicular to the [111] directions of the fcc lattice or the [001] axis of the hcp structure. As illustrated in Fig. 15, the closest packing of atoms (equal sized spheres) in a plane is

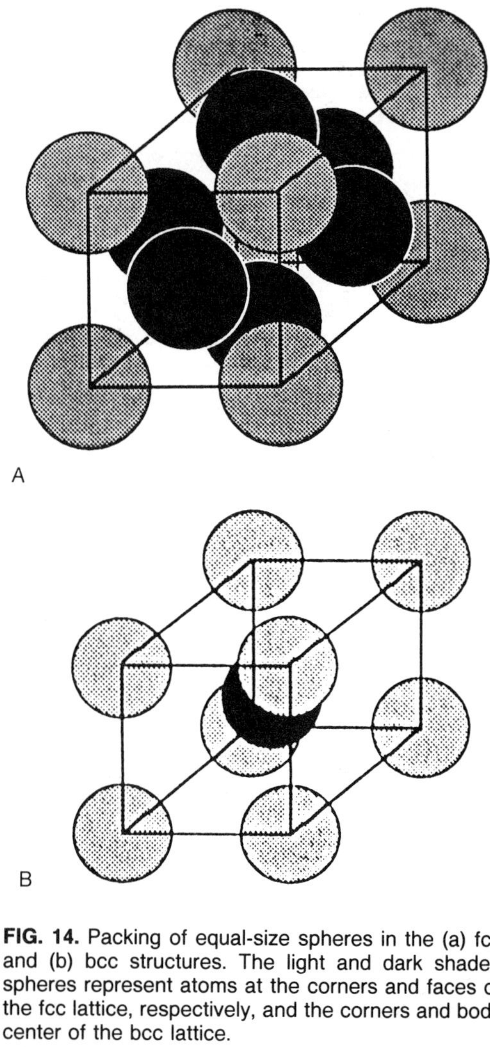

FIG. 14. Packing of equal-size spheres in the (a) fcc and (b) bcc structures. The light and dark shaded spheres represent atoms at the corners and faces of the fcc lattice, respectively, and the corners and body center of the bcc lattice.

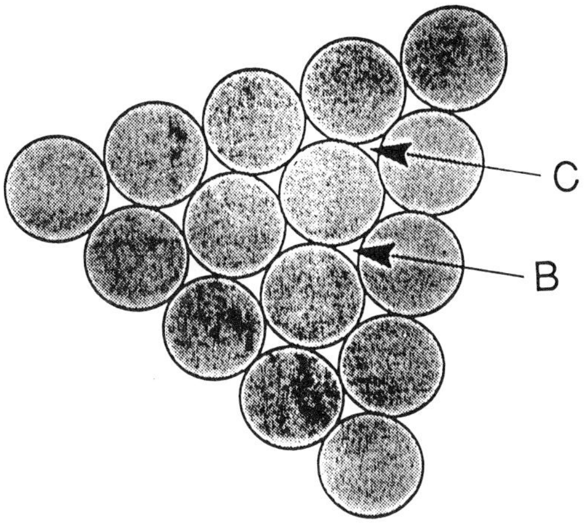

FIG. 15. Atomic packing in the (111) or (001) planes of the fcc or hcp structures. Interstices *B* and *C* denote sites available for atoms in the next layer.

accomplished by sixfold coordination of each atom. The pattern of atoms is precisely analogous to the two-dimensional hexagonal Bravais lattice that has threefold and sixfold symmetry. The voids between the atoms in this plane define the set of positions available for atoms in the next plane. Note that once one of the voids labeled *B* or *C* is chosen as the site for an atom in the next layer, all atoms in that layer sit in the *B* sites. If we continue up one more layer, there is a choice. The next layer of atoms may be situated either in the interstices labeled *C*, or directly over the atoms in layer *A*. If the packing proceeds in a sequence *ABCABCABCABC*,..., the fcc close-packed structure results, with the normals to these planes defining the [111] direction in the cubic lattice. On the other hand, if the stacking sequence is *ABABABABABAB*,..., the layers correspond to the planes perpendicular to the [001] direction of the hcp structure. A third stacking sequence, observed in some rare-earth metals, is *ABACABACABACABAC*,..., corresponding to the double hexagonal close-packed arrangement (dhcp).

The *body-centered-cubic* (bcc) close-packing arrangement [Fig. 14(b)] consists of two atoms per cubic unit cell at (0,0,0) and $(\frac{1}{2},\frac{1}{2},\frac{1}{2})$. Each atom is coordinated by eight nearest neighbors along the body-diagonal directions at a separation of $\sqrt{3}a/2$, and six atoms along the cube edge directions at a separation of a. The packing fraction of this structure is somewhat lower than the fcc structure (67%). The closest-packed planes are the (110) layers where the atoms are arranged in a square rather than the triangular network of the fcc (111) layers.

1.4.2 The Diamond, Cubic Zinc-blende, and Hexagonal Wurtzite Structures. The *diamond* structure is commonly found for covalently bonded semiconductors such as silicon and germanium. The structure (Fig. 16) is based upon a fcc space lattice with a two atom basis in the unit cell at (0,0,0) and $(\frac{1}{4},\frac{1}{4},\frac{1}{4})$. The unit cell, therefore, contains eight atoms. The resulting structure may be viewed as two equal-sized, interpenetrating fcc sublattices with a relative displacement of one-quarter of the distance along the body diagonal. Each atom is tetrahedrally coordinated by four nearest neighbors at a separation of $\sqrt{3}a/4$, and the packing fraction of these structures is quite low (34%). The *zinc-blende* structure (Fig. 17) of many other structures with covalent or partially covalent bonding is a simple modification of the diamond structure. In ZnS, e.g., the Zn atoms are situated on one of the fcc sublattices while the S atoms sit on the other. Each atom is tetrahedrally coordinated by atoms of the other species.

The relationship between the hexagonal *wurtzite* and zinc-blende structures is analogous to that between the fcc and hcp structures described in Sec. 1.4.1. Viewing along the [111] direction of the zinc-blende struc-

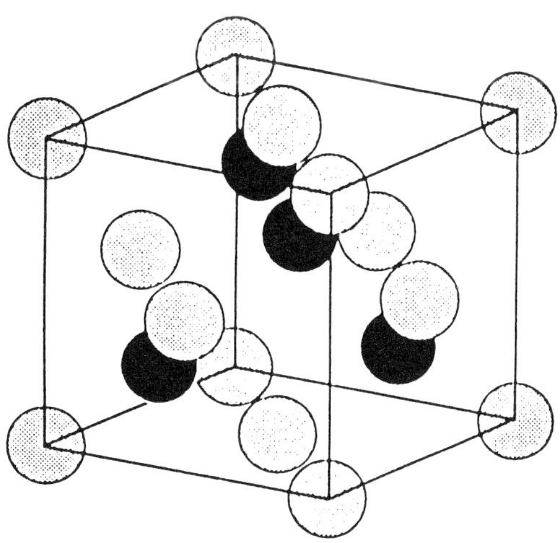

FIG. 16. Atomic packing in the diamond lattice. The light and dark shaded spheres represent atoms on the two fcc sublattices which are displaced along a body diagonal by $(\frac{1}{4},\frac{1}{4},\frac{1}{4})$.

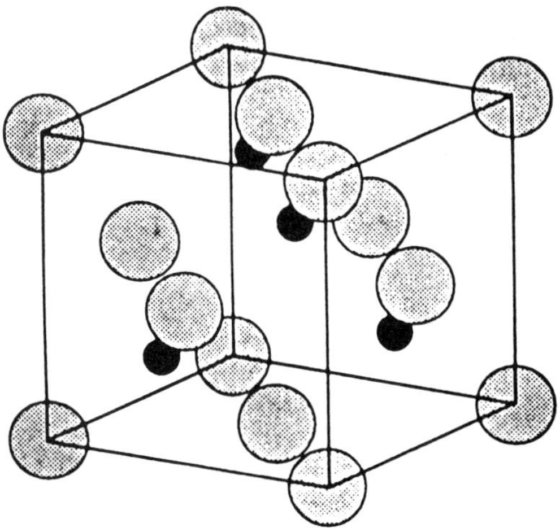

FIG. 17. Packing of different size atoms in the cubic ZnS lattice. The large spheres represent Zn while the small spheres represent the S atoms. Note the similarity to the diamond structure.

ture, we see planes of atoms of the same species stacked in the *ABCABCABC...* sequence. These same planes, stacked in the *ABABABAB...* sequence along the [001] direction, are the basal planes of the wurtzite structure. The space lattice of the wurtzite compounds is hexagonal with four atoms per unit cell. One species is situated at the position (0,0,0) and $(\frac{1}{3},\frac{2}{3},\frac{1}{2})$, while the other is found at $(0,0,u)$ and $(\frac{1}{3},\frac{2}{3},\frac{1}{2}+u)$. The offset u depends upon the compound but is close to $\frac{3}{8}$.

1.4.3 The Sodium Chloride and Cesium Chloride Structures. Many ionic solids crystallize into the NaCl or CsCl structures (Fig. 18). In the *sodium chloride* arrangement, each ion is coordinated by six oppositely charged ions in an octohedron. The space lattice is face-centered cubic with the halogen at positions (0,0,0) and the alkali atom at $(\frac{1}{2},\frac{1}{2},\frac{1}{2})$. The *cesium chloride* structure is simple cubic with the alkali ion at (0,0,0) and the halogen at $(\frac{1}{2},\frac{1}{2},\frac{1}{2})$. Each ion is coordinated by a cube of eight ions of opposite charge.

1.4.4 The Calcium Fluoride Structure. Many compounds of the form AB_2 crystallize in the CaF_2 structure (Fig. 19). The space lattice is fcc with a two-point basis and eight atoms per unit cell. The structure may be viewed as three interpenetrating fcc sublattices. Atoms of type A, and half of the B atoms, are found in the ZnS arrangement described

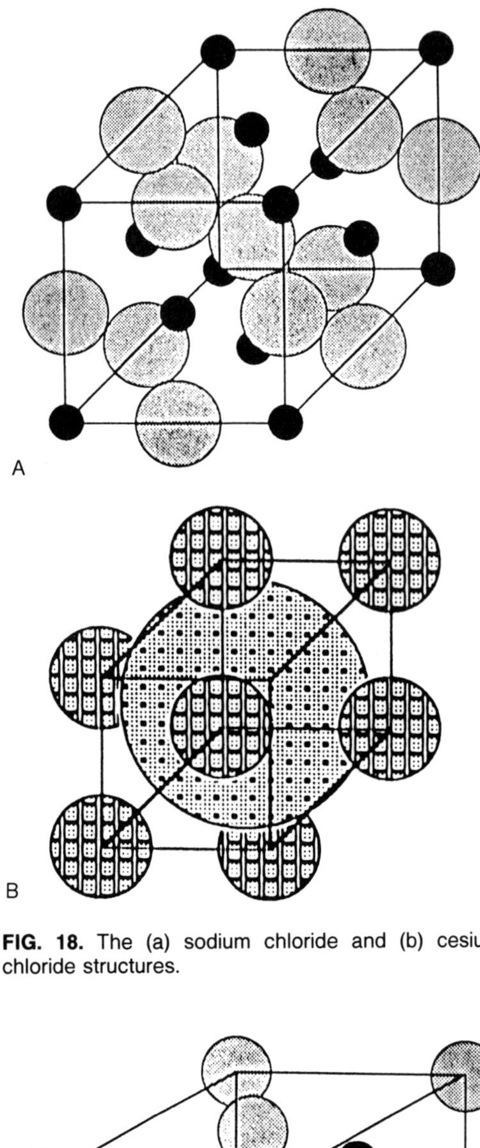

FIG. 18. The (a) sodium chloride and (b) cesium chloride structures.

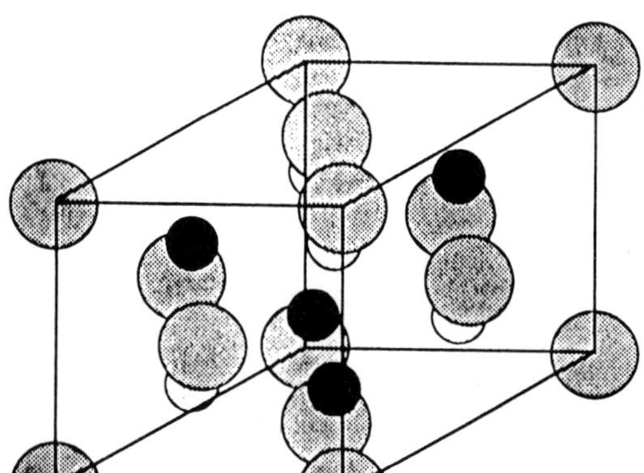

FIG. 19. The CaF_2 structure. Ca atoms are found at the corners of the cubic cell. Half of the F atoms (lightly shaded) are found at the *S* sites of the ZnS arrangement,and the remaining F (darkly shaded) are located on the third fcc sublattice as described in the text.

in Sec. 1.4.2. The remaining B atoms are found at positions $(-\frac{1}{4},-\frac{1}{4},-\frac{1}{4})$ plus face-centered translations. Each A atom is coordinated by a cube of eight B atoms, and each B atom is found within a tetrahedron of A atoms.

1.5 More Complicated Crystal Structures

All of the crystalline structures described in the last few sections are relatively simple, involving only a few atoms per unit cell. It is important to realize that this is not the general case. Unit cells with tens, hundreds, or even thousands of atoms are not uncommon and a description of the local environment of a particular component may be quite difficult. As an example, consider the structure of Al_6Li_3Cu illustrated in Fig. 20. The space lattice is bcc ($a = 13.9$ Å) with 160 atoms in the unit cell. The structure may be more simply described as a bcc packing of clusters of atoms centered at the corners and body center positions. Each cluster contains 105 atoms and has the shape of a truncated icosahedron.

1.6 Surface Structures

The surface of a crystal contains only a very small fraction of the total number of atoms in the solid, but is of importance since it represents the interface between the bulk crystal and its external environment. The structure of the surface layers may be essentially unchanged from that of the bulk crystal except for some modification in layer spacing due to the absence of atomic layers above the surface. On the other hand, there may be significant *reconstruction* of the surface, so that the top layer is out of registry with lower lying layers.

The structure of the surface layer may be cast in terms of one of the 17 space groups detailed in Sec. 1.1.2. Figure 21 shows three possible configurations, or *nets*, for the fcc (111) or hcp (001) surfaces. Here we imagine that the surface atoms are placed at the intersections of the net. The $p(1\times1)$ net is an unreconstructed surface. The $p(2\times2)$ net represents a reconstructed surface with primitive translation vectors twice as long as those of the unreconstructed $p(1\times1)$ net. The $(\sqrt{3}\times\sqrt{3})R30°$ net is also a reconstruction with translation vectors $\sqrt{3}$ times longer than the $p(1\times1)$ net, and rotated 30° with respect to the in-plane translation vectors of the bulk.

1.7 Incommensurate Crystals and Quasicrystals

While the crystal structures of the vast majority of elements, compounds, and alloys we encounter may be classified in terms of one of the 230 three-dimensional space groups, these do not exhaust all of the possibilities.

Incommensurate solids are materials which do not have periodic translational order but exhibit only the crystallography allowed rotational symmetries. As the term implies, incommensurability is derived from a periodic crystalline structure through the modulation of some property of that solid. The modulation itself often results from interactions which have a periodicity that is different from that of the basic lattice. In metals, e.g., instabilities can arise from interactions between the conduction electrons resulting in a modulation of the density of electrons in the solid (a charge density wave). The charged ion cores are also driven to readjust their position through their interaction with the electron gas. Therefore, as illustrated schematically in Fig. 22, the positions of like atoms in neighboring unit

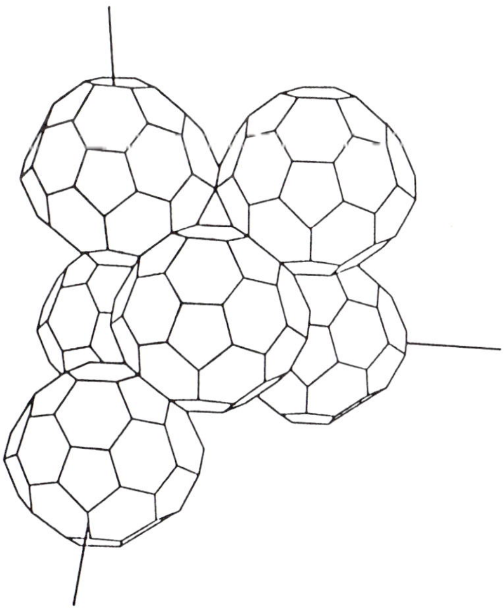

FIG. 20. The cubic phase of Al_6Li_3Cu may be viewed as a bcc packing of clusters of 105 atoms in the shape of a truncated icosahedron. Only the overall shape of the clusters is shown.

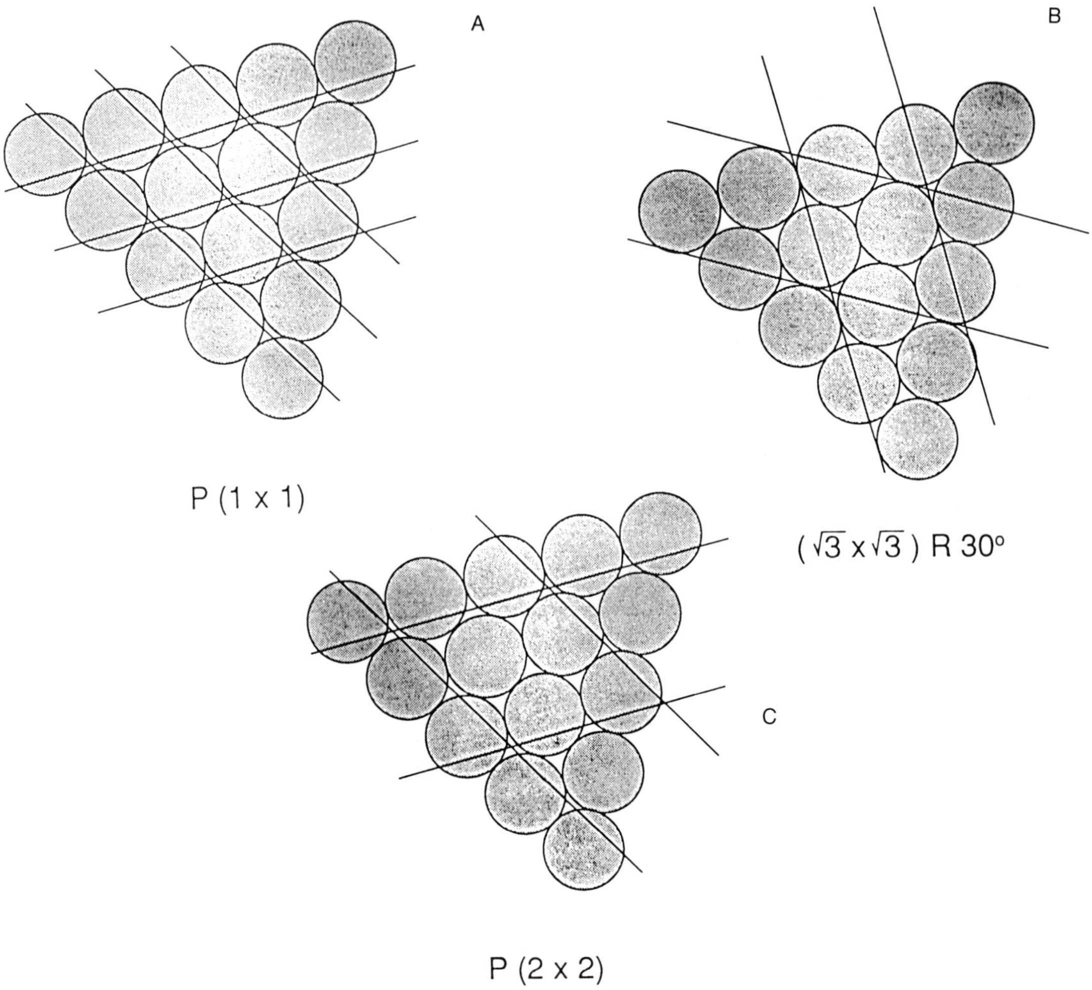

FIG. 21. Two-dimensional nets on the surface of the (111) planes of an fcc structure or (001) planes of an hcp structure.

cells are no longer connected by a simple translation vector, but by a translation plus a displacement from the average position. This displacement is also periodic, but may not have a period that is some rational multiple of the underlying lattice. In general, the incommensurability of the structure varies as a function of temperature and may exhibit "lock-in" transformations to a commensurate structure at particular temperatures.

Quasicrystalline solids not only lack periodic translational order, but exhibit rotational symmetries (e.g., five-fold rotation axes) which are crystallographically forbidden. The first quasiperiodic solid was discovered in 1984 in an alloy of aluminum and manganese with icosahedral symmetry (see Fig. 23). Since the icosahedral group admits rotations of $\pi/5$, the structure is inconsistent with any of the 230 periodic space groups discussed in Sec. 2.2. It is important to note, however, that the lack of periodic translational order does not imply that these materials are disordered glassy solids. For example, Fig. 24 shows a one-dimensional quasiperiodic Fibonacci sequence of atoms separated by two distances, $L(=\tau)$ and $S(=1)$. The sequence of atomic positions, R_n, from some origin may be defined by the rule

$$R_n = \frac{n}{1+1/\tau^2} - \frac{1}{\tau}\{n\tau\},$$

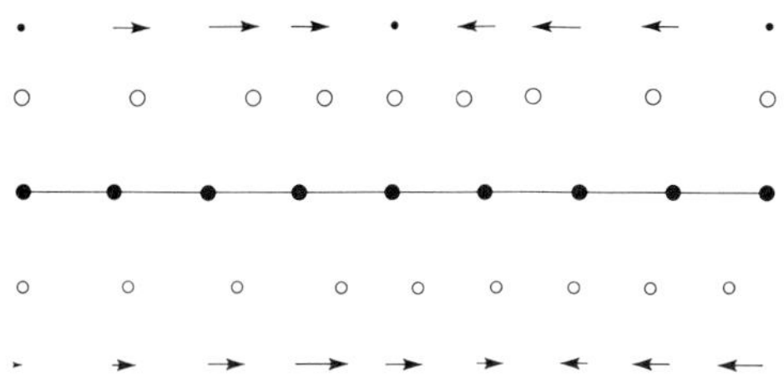

FIG. 22. One-dimensional modulated structures. The filled circles represent the atomic positions in the absence of a modulation, while the open circles denote the modulated structure. Top: The period of the modulation is a rational multiple of the underlying lattice. Bottom: The modulation is incommensurate with the underlying lattice.

where $\tau=\frac{1}{2}(\sqrt{5}+1)$ is known as the golden mean, n is an integer, and the curly braces denote the fractional part of the enclosed expression. The very existence of a rule specifying the atomic positions means that the structure is well-ordered, albeit aperiodic. Since our example here is one dimensional, the idea of forbidden rotational symmetries does not enter the discussion. Therefore, the Fibonacci sequence is perhaps more correctly described as an incommensurate structure. In two dimensions, however, one can construct a quasiperiodic tiling of the plane which is closely related to the one-dimensional Fibonacci sequence, using two rhombic "unit cells." The *Penrose tiling* shown in Fig. 25, or rather its three-dimensional cousin, is one of the candidate structures for the icosahedral alloys.

2. BINDING FORCES AND COHESIVE ENERGY

Having described and discussed the various classifications of crystalline solids, we now turn to a description of the interactions which bind atoms together into these structures. Binding forces act to lower the free energy of the solid with respect to an assembly of isolated atoms. The *cohesive energy* (*q.v.*) of a solid is, in fact, defined as the difference in total energy between the solid and its isolated constituents.

2.1 The Van der Waals–London Interaction

The weakest of the crystal binding forces is found in the inert-gas solids such as krypton, neon, xenon, and argon. These atoms have closed-shell configurations and are highly stable. When two atoms closely approach each other, however, an attractive interaction can result from induced dipole moments. This attractive interaction is a strongly decreasing function of interatomic separation (R), falling off as R^{-7}. There is also, of course, an even shorter-range repulsive interaction which results from electron overlap at very small separations. Taken together, the two terms combine to yield a total energy for the solid which is usually written as

$$U(r)=2\epsilon\left[A_{12}\left(\frac{\sigma}{R}\right)^{12}-A_{6}\left(\frac{\sigma}{R}\right)^{6}\right],$$

and is known as the Lennard-Jones 6-12 potential. Here σ and ϵ are parameters derived from the gaseous state data, and the constants A_{12} and A_6 depend upon the crystal structure. It should be pointed out that the $1/R^{12}$ dependence of the repulsive energy is introduced for calculational convenience and because it seem to work well for the description of binding energies in the inert-gas crystals. In fact, we only require that the repulsive interaction be much shorter-ranged than the attractive forces, and other forms for this term have been used.

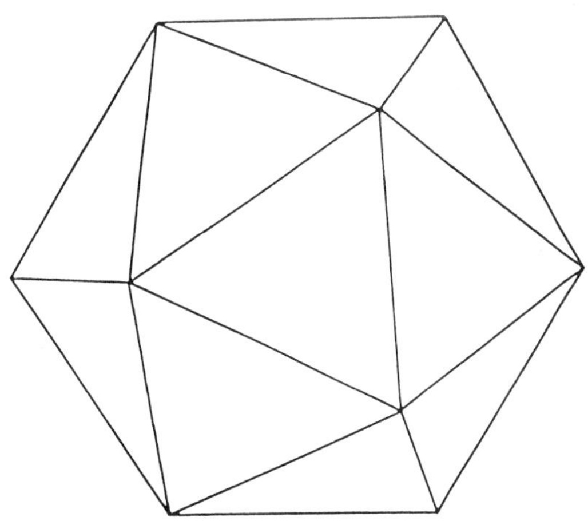

FIG. 23. An icosahedron is characterized by 12 vertices (fivefold axes), 20 triangular faces (threefold axes), and 30 edges (twofold axes).

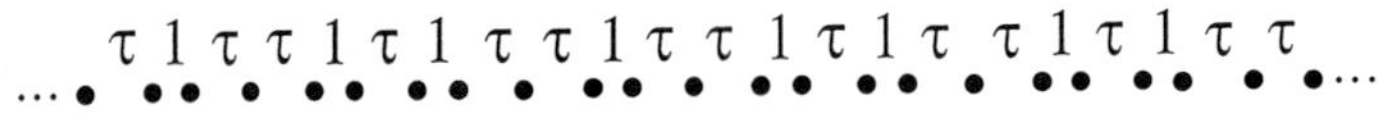

FIG. 24. A one-dimensional quasiperiodic Fibonnaci sequence. Atoms are separated by either a long (τ) or short (1) distance. The ratio of the number of L to S separations approaches τ as the length of the sequence increases.

All of the inert-gas solids crystallize into the fcc structure, and are characterized by binding energies on the order of 0.01 eV per atom. Since the binding is weak, solidification takes place only at low temperatures (< 10 K) and zero-point fluctuations play an important role in the stability of these structures. Since there are no free electrons in these solids, they are extremely good insulators.

2.2 Ionic Bonding

For ionic crystals, the dominant contribution to the binding energy is the interionic Coulomb attraction. NaCl is an example of a purely ionic solid; a Na $3s$ electron is transferred to fill the $3p$ shell of Cl, resulting in a positively charged Na^+ and negatively charged Cl^-. The stability of the structure is owed to the fact that the total energy decrease due to the interionic electrostatic attraction is greater than the energy required to transfer electrons between isolated neutral atoms. The electrostatic interaction is much stronger than the dipole interaction, and is longer ranged (the electrostatic force decreases like $1/R^2$). The total cohesive energy, including a term for the repulsive interaction between ion cores at small separation, can be written as

$$U(R) = C/R^n - \alpha q^2/R.$$

Here, C is a constant that can be determined from the known equilibrium separation of the ions and q is the net charge of the ions. The *Madelung constant* α essentially sums the contributions to the electrostatic potential between an ion and all of its neighbors (both near and distant).

In contrast to the weakly bound inert-gas crystals discussed above, ionic crystals are usually hard and have high melting points. Since the excess charge is highly localized around the negative ion, there are no free electrons to carry charge. Ionic crystals, in the absence of defects such as vacancies, are therefore poor conductors.

Alkali halide ionic solids crystallize into either the sodium chloride or cesium chloride cubic lattice (see Sec. 1.4.3). Although most crystallize into the NaCl structure, it is interesting that the electrostatic contribution to the cohesive energy is actually a bit larger (by about 1%) for the CsCl structure. The repulsive interaction between ions, however, increases with increasing coordination number and slightly favors the NaCl arrangement. The particular choice of structure is largely governed by the relative sizes of the ions.

2.3 Covalent Bonding

Covalent bonding in crystals is characterized by the sharing of electrons between constituent atoms. For instance, the $2p$ orbital of carbon, the $3p$ orbital of silicon, and the $4p$ orbital of germanium are all four electrons short of being filled. The energy of an assembly of carbon atoms, e.g., can be lowered by overlap between the unfilled orbitals, thereby sharing the valence electrons. Because of this sharing, the charge distribution is not completely localized around each ion as is found

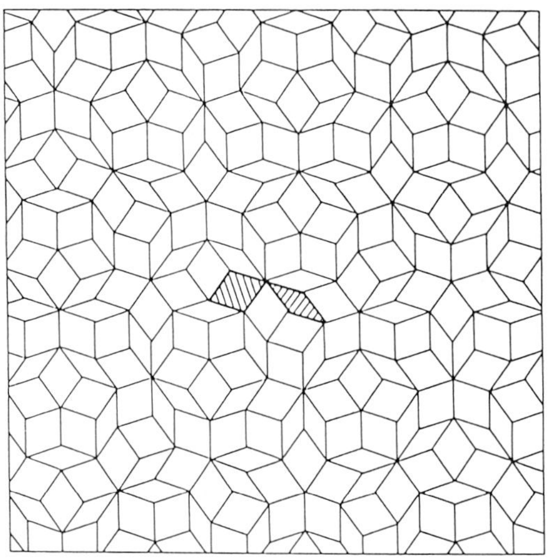

FIG. 25. A two-dimensional Penrose tiling of "fat" and "skinny" rhombuses. Note the presence of local regions with clear fivefold rotational symmetry. The three-dimensional analog of this tiling is one model for the icosahedral phase alloys.

for the ionic crystals. In particular, substantial electron density is found in the interstitial region between atoms. Covalent bonds are highly directional, forming tetrahedrally coordinated structures of the diamond lattice type (see Sec. 1.4.2), and very strong, making them hard and difficult to deform. Since the electron distribution is at least partially delocalized in covalent bonding, covalent crystals are not as good insulators as the ionic or inert-gas crystals. All semiconductors (i.e., Ge, Si, GaAs, etc.) are fully, or nearly, covalent.

For some other compounds, the cohesive forces may best be described as mixed ionic and covalent bonding. The bonding in systems such as ZnO, ZnTe, BeO, ZnSe, CdS, and CdTe, e.g., is characterized as $\approx$60%–70% ionic. The first three crystallize into the hexagonal wurtzite structure and the latter three are found in the cubic zinc-blende structure (see Sec. 1.4.2). At the other end, the bonds between atoms in semiconductors such as InAs, GaAs, and GaSb are described as mostly covalent with some ionic character.

2.4 Metallic Bonding

Metallic bonds differ from the covalent bonds of semiconductors in that the valence electrons are shared among all atoms in the crystal and are relatively free to wander through the crystal lattice. This accounts for the good conductivity of metals. The mechanisms which give rise to metallic bonding are complex and must account for the interaction between the ion cores and the electron gas, as well as the interaction between the electrons themselves. Since the electrons are not assigned to any particular ion core, or pair of ions, metallic bonds are nondirectional. Therefore, metals tend to crystallize into the dense, highly symmetric arrangements—the close-packed structures described in Sec. 1.4.1.

3. DEFECTS IN CRYSTALS

The Gibbs free energy of a system at some temperature, T, may be written as $G=H-TS$. At $T=0$ K an ideal crystal, free of defects, is the state of minimum free energy. Defects increase both the enthalpy H and the entropy S of a system. Consequently, the free energy can be a minimum for a crystal with defects at finite temperature. Even for cases in which G is not a minimum, however, defects in crystals may be stabilized for kinetic reasons since their removal usually requires the diffusion of atoms.

All real crystals contain defects. One must keep in mind, though, that in many cases the presence of defects is not undesirable. For instance, pure semiconducting crystals of Si require the intentional addition of certain impurities to become useful solid-state devices. The presence of dislocations at the surface of a crystallite generally enhances the rate of growth at the solid–liquid interface.

3.1 Vacancies, Interstitials, and Impurities

The simplest defect is a *vacancy*. As illustrated in Fig. 26, a vacancy corresponds to an empty site in the structure which is normally occupied in the ideal crystal. Another simple point defect corresponds to an atom situated at an *interstitial position*, not normally occupied in the ideal crystal.

The concentration of isolated defects in thermal equilibrium may be calculated if we know the energy E required for their formation. For N atoms in the crystal, the number n of defects of some particular type is given by

$$n=Ne^{-E/k_BT},$$

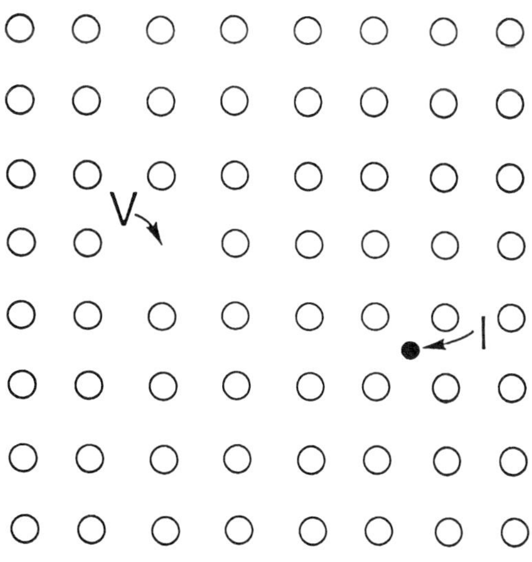

FIG. 26. A schematic illustration of simple defects in a monatomic Bravais lattice. The absence of a circle denotes a vacancy, while the filled circle represents an interstitial atom.

where k_B is the Boltzmann constant and T is the temperature. Similar calculations for clusters of defects are more complicated since the presence of one defect may influence the energy required to form another nearby. There may be other constraints placed on the number and type of simple defects in crystals. For instance, the relative number of positive-, or negative-ion vacancies of interstitials in ionic crystals is regulated by the requirement of charge neutrality.

Chemical impurities can play a crucial role in modifying the physical properties of materials, even in minute concentrations. The electrical properties of semiconductors, e.g., depend largely upon the number of electrons that can be excited across the narrow energy gap separating the valence band from the conduction band. Electrons can be excited across this gap thermally (so that the conductivity of semiconductor increases with temperature), or by an applied electric field. The intentional addition of certain chemical impurities can introduce energy levels within the gap close to either the valence or conduction band, thereby reducing the temperature or applied potential required to populate the conduction band, or alternatively, depopulate the valence band. Chemical impurities may be either *substitutional*, sitting at a site normally occupied by an atom of the pure crystal (e.g., an arsenic impurity at a germanium site in a germanium crystal), or interstitial.

3.2 Chemical Disorder in Alloy Systems

In *ordered alloy* systems, specific sites are assigned to the constituent elements (Fig. 27). For instance, at low temperature, the equiatomic alloy of CuZn (brass) is chemically ordered and may be described as simple cubic with the copper atoms situated at the corners of the cube and Zn atoms at the body center (the CsCl structure). As the temperature of this system is raised above some critical temperature, T_C, there is an *order–disorder transformation* (see ORDER–DISORDER TRANSITIONS) which results in a mixing of the Cu and Zn sites. That is, both the corner and body center sites are occupied by copper or zinc with equal probabilities. The structure is then more correctly described as body-centered cubic. If this alloy is cooled rapidly from some temperature above T_C, the high-temperature phase may be "frozen in," resulting in a metastable disordered phase with some degree of *site-exchange disorder*.

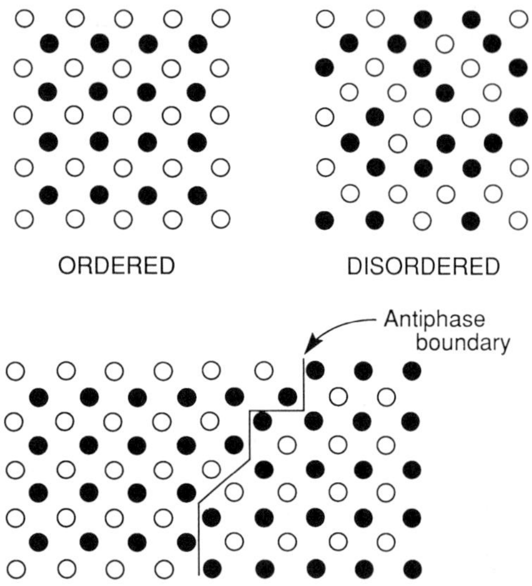

FIG. 27. Ordered (left) and disordered (right) alloys of A (open circles) and B (closed circles) atoms. The full line in the bottom panel represent an antiphase domain boundary between two ordered domains.

Antiphase domains may also occur in chemically ordered alloys. The sites of A and B atoms within any given domain are well-defined, but the A sites in one domain become the B sites in the adjacent domain (Fig. 27).

3.3 Stacking Faults and Line Defects

In contrast to the situation for point defects, extended defects in crystals are metastable since the energy required for their formation is not fully compensated by the increased entropic contribution to the free energy. Therefore, most extended defects may be removed by appropriate heat treatment or *annealing*.

A common defect found in fcc and hcp metals is the *stacking fault*. As described in Sec. 1.4.1, both structures may be viewed as a stacking of close-packed atomic planes. The layer sequence *ABCABCABC*... corresponds to the stacking of planes along the [111] direction of the fcc lattice, while the sequence *ABABABAB*... corresponds to layers stacked along the [001] direction of the hcp lattice. There are, in fact, an infinite number of such possible sequences which correspond to the intermixture of local fcc or hcp order. For instance, the sequence *ABCABABCABCABC*... corresponds to an fcc crystal with a stacking

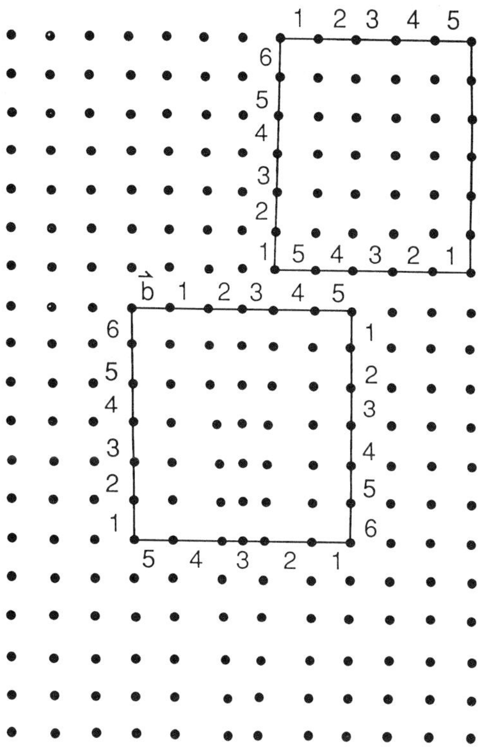

FIG. 28. An edge dislocation. The solid line at the upper right shows a Burgers loop around a defect-free region of the crystal. The solid line in the middle describes the same set of translations around the edge dislocation. The Burgers vector of the dislocation is indicated by the arrow which closes the loop.

fault (a missing *C* layer in the sixth position), which produces a local region of hexagonal order.

These sequences may be used to describe another type of planar defect commonly seen in crystals, termed a *twin*. Consider the sequence of layers *ABCABCACBACBA*..., where the ordering of the layers reverses after the seventh layer. The sequences on either side of this *twin boundary* are mirror images.

Line defects, also termed *dislocations*, have a profound effect upon the mechanical properties of crystals. An *edge dislocation*, depicted in Fig. 28, may be constructed by removing a plane of atoms from one-half of the crystal, leaving the other half unmolested. Alternatively, one may view an edge dislocation as arising from the insertion of a half-plane of atoms. The *dislocation line* runs parallel to the end of the half-plane, out of the plane of Fig. 28. Note that in order to maintain long-range order throughout the entire crystal, the two halves of the crystal must be rejoined so that, near the dislocation line, atoms are crowded together on one side and pushed apart on the other. The presence of edge dislocations greatly reduces the crystal's resistance to shear deformation since the mobility of this defect is quite high. Dislocations may be pinned, however, by impurities.

If we cut partway through a crystal, and then shear the two halves of the cut as shown in Fig. 29, the resulting defect is termed a *screw dislocation*, defined by the dashed line in the figure. At the surface of the crystal, a screw dislocation produces a ledge. Sites immediately adjacent to the ledge on the surface are particularly favorable for the addition of new atoms. Therefore, screw dislocations perform the important function of providing nucleation sites for the next layer of surface atoms during the growth of a crystal from the melt (see Sec. 4). The surface of the crystal grows in a spiral fashion around the dislocation line.

Other extended defects may be formed from combinations of edge and screw dislocations, and may be described by a *Burgers vector*. The Burgers vector is defined by constructing a closed path around the region of the dislocation. In a perfect crystal, a closed loop may be accomplished by a series of translations through the Bravais lattice vectors as shown in Fig. 28. If a dislocation is present within this region, the same sequence of translations will not yield a closed loop. The vector joining the end of the sequence to

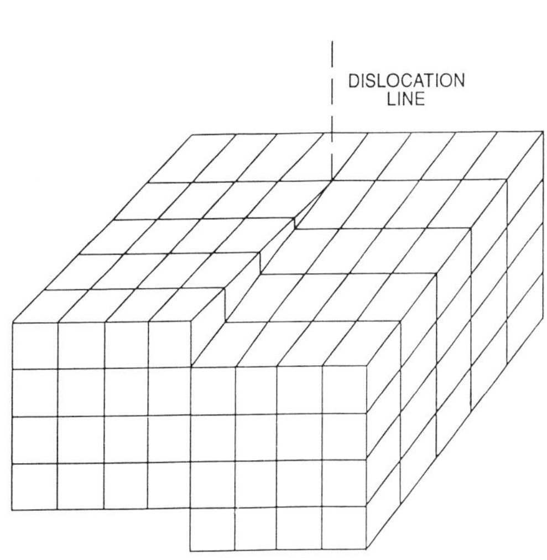

FIG. 29. A screw dislocation. The dislocation line is indicated by the dashed line.

the beginning is the Burgers vector of the dislocation. The Burgers vector is perpendicular (parallel) to the dislocation line characterizing an edge (screw) dislocation.

Low-angle *grain boundaries* are extended defects that separate crystallites, and may be constructed from arrays of edge dislocations. The relative orientation of the two crystallites depends upon the Burgers vector and spacing of the array of dislocations. The presence of grain boundaries often enhances reactions in solids since they provide easy pathways for the diffusion of atoms.

4. CRYSTALLIZATION

In this section, we are primarily concerned with the growth of crystals via a first-order transformation from a liquid phase (see PHASE TRANSITIONS). Below the melting point, the free energy of the well-ordered crystalline phase is lower than that of the liquid, favoring crystallization.

In the next sections, we consider only the case of transformations between a single-phase liquid and a single-phase solid at the same composition. More complex transformations, such as the solidification of eutectic alloys, will not be addressed here.

4.1 Nucleation

There are two distinct stages in the crystallization process: nucleation and growth. For the case of *homogeneous nucleation*, seeds of the new crystalline solid are dispersed uniformly throughout the volume of the melt. The formation of these nuclei is a dynamic process where microscopic regions of crystalline order form and then dissolve through thermal fluctuations. The growth of atomic clusters of the crystalline phase is favored by its lower bulk free energy, but inhibited by the unfavorable positive free energy associated with the surface of the cluster. Since the surface-to-volume ratio decreases with increasing cluster size, clusters larger than some critical radius are stable and can grow; they become the nuclei of the solid phase.

The nucleation process can be enhanced by the presence of interfaces or foreign particles in the liquid. *Heterogeneous nucleation* occurs when the seeds of the solid phase form about these defects, offsetting the positive surface energy of the cluster with the decrease in free energy associated with removing the defect. This process is analogous to "seeding" a cloud with foreign particles to nucleate rain drops.

4.2 Growth

Once nuclei are formed, their growth is controlled by a complex array of processes at the interface between the nuclei and liquid. These include the success of atoms in finding the proper sites on the surface of the crystallite; the probability that atoms will "stick" to this site; and the rate at which excess energy, released in the form of heat once the atom joins the cluster, is removed from the region of the interface.

For some period of time after nucleation, the crystallites grow independently at a rate controlled by the factors mentioned above. As the size of the crystallites increases, neighboring crystallites of different orientations will come in contact. At low enough temperatures (e.g., relatively fast cooling rates), polycrystalline material results. At a high temperature, however, the larger grains can grow at the expense of smaller grains yielding a finite number of single crystals. Certain techniques can be used to promote the growth of large single crystals.

The morphology of grain growth in a liquid medium depends largely upon the rate of cooling. For strong departures from equilibrium conditions, such as large undercooling, grains can be highly irregular or *dendritic*, consisting of tree-like projections into the melt (Fig. 30). These dendrites have a preferred orientation, with their axes along the direction of fastest growth (e.g., the [100] direction in fcc metals).

Under conditions closer to equilibrium, the grain often develops into a well-defined polyhedron closely related to the symmetry of the underlying crystal lattice. *Faceted interfaces* occur along directions of slow growth (Fig.

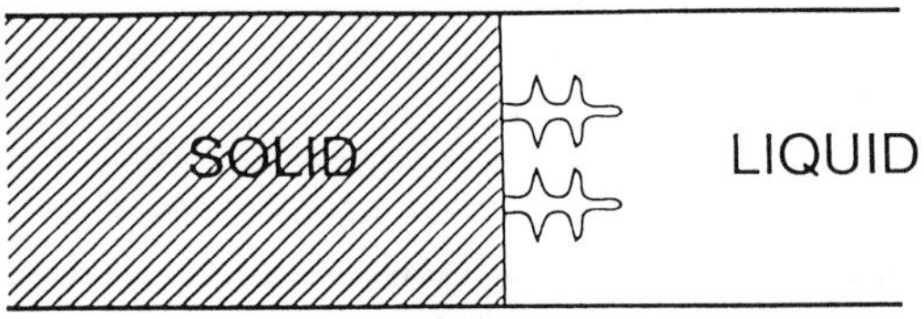

FIG. 30. Dendritic growth at a solid–liquid interface.

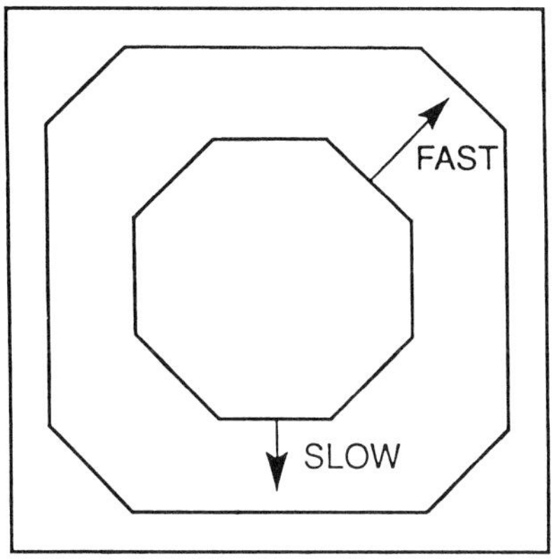

FIG. 31. The development of facets during the equilibrium growth of crystals. Note that the faces perpendicular to the slow growth directions dominate the shape of the crystal at later stages of growth.

31) which generally correspond to directions normal to the most dense planes of the structure. In an fcc metal, e.g., faceting is most likely to occur along the [111] directions.

4.3 The Recrystallization of Glasses

If the rate of undercooling is extremely large, the liquid can solidify into a glassy structure. The recrystallization of glasses often proceeds in a manner analogous to the nucleation and growth process described above. At temperatures far below the melting point, the free energy of the crystalline phase is significantly lower than that of the glass. Crystallization is inhibited, however, by slow kinetics (atoms have to move some distance to effect the transformation). As we heat the system above the glass transition temperature T_g, seeds of the crystalline phase nucleate and grow. Since the difference in free energy between the glass and crystal is so large, the rate of homogeneous nucleation in the glass phase is much larger than that found for liquids; the critical size for nuclei is much smaller. Recrystallized glasses, therefore, are generally much finer-grained than crystals grown from the melt.

5. SUMMARY

In this article, some of the basic features and language of crystalline solids including crystal structure, binding forces, defects, and growth have been introduced. All of these topics have been the subject of intense investigation over the past century which has led to the development of important new technologies such as semiconductor devices. Given the brevity of this article, many interesting and important topics could not be covered in any detail. The reader is referred to the list below for a more comprehensive treatment of the subject.

List of Works Cited

Henry, N. F. M., Lonsdale, K. (1952), *International Tables for X-ray Crystallography*, Birmingham, England: Kynoch Press, Vol. 1.

Kittel, C. (1986), *Introduction to Solid State Physics*, New York: Wiley.

Further Reading

Ashcroft, N. W., Mermin, N. D. (1976), *Solid State Physics*, New York: Holt, Rinehart and Winston.

Wyckoff, R. W. G. (1974), *Crystal Structures*, New York: Interscience.

Jaric, M. J. (Ed.) (1988), *Introduction to Quasicrystals*, New York: Academic.

Verhoeven, J. D. (1975), *Fundamentals of Physical Metallurgy*, New York: Wiley.

CRYSTALLOGRAPHY

W. B. PEARSON AND C. CHIEH, *Department of Chemistry, University of Waterloo, Ontario, Canada*

INTRODUCTION

Crystallography is the science of crystals and crystalline materials, which include most terrestrial and industrial solids. The science includes the study of crystal structure on an atomic scale and also the shapes, forms, defects, growth, transitions, and properties of crystals. Chemical analysis to determine the elemental composition of a substance and the determination of its crystal structure are the important experiments of the solid state. Principally, x-ray- or neutron-diffraction experiments are used to determine crystal structures, that is to say, the crystal symmetry, the relative positions, and the elemental nature and arrangement of the atoms in the structure. X-ray diffraction by a crystal was first demonstrated in 1912, following the discovery of x rays by Röntgen in 1896. Nevertheless, using earlier mathematical developments and measurements of the angles between the faces of crystals, all of the rudiments of three-dimensional crystal symmetry had been worked out by this time and were at hand to be applied to the new atomic-scale crystallography. Prominent among these early developments were observations on crystal form by René Haüy (1784), Wollaston's (1809) construction of the optical goniometer, Hessel's (1830) demonstration of the 32 symmetry classes, Bravais's (1848) demonstration of the 14 lattices, Sohncke's (1867) discovery of new symmetry elements, and finally, the derivation of the 230 space groups independently by Schoenflies (1891) and Fedorov (1891).

Thus, when in 1912 von Laue, Friedrich, and Knipping demonstrated the diffraction of x rays by crystals, the door was opened to the possibility of gaining exact information on the

1-56081-063-7/92/$5.00+.50

arrangement of atoms within a crystal. This discovery was rapidly followed by the Braggs' (1913) interpretations of their experiments on rocksalt and other crystals; however, the wavelengths of characteristic x radiations still had to be determined. So began a science of the most fundamental significance. The history of crystallography is complicated and enlightening, and a short description thereof can be found in *Fifty Years of X-Ray Diffraction* by Ewald (1962). Over the years the subject became so important that in 1947–48 the International Union of Crystallography (I.U.Cr.) was created to be concerned with all viable aspects of crystallography through its Commissions, its triennial General Assembly and convention, and its publishing activities.

Much of the realm of crystallography is covered in specific articles elsewhere in the *Encyclopedia*, and some of these articles are cross-referenced in the text. Therefore this article introduces crystallography to the general reader, providing an overview, yet without sacrificing physical and mathematical precision to simplicity. With very many known crystal structures at hand, we can only illustrate this article with information on a few simple types, upon the basic structural principles of which many more complex types are formed.

Finally, it is observed that the rate at which the current technological and medical revolutions are proceeding results in no small degree from the information provided by crystallographic studies.

1. THE CRYSTALLINE STATE

A crystal is characterized by a long-range order. That is to say, the atomic arrangement is periodic in a three-dimensional space throughout the crystal (see CRYSTALLINE STATE). Other solid material may be amorphous, so that there is only short-range order that extends merely for a few shells of atoms in the surroundings of any given atom (see AMORPHOUS MATERIALS, STRUCTURE OF).

Any solid that one views is likely to be made up of crystals and/or amorphous material, and of crystals of more than one type and chemical composition. A compound may exist in the form of gas, liquid, or several forms of solids. Each of these has been conventionally called a *phase*, which is a homogeneous portion separated by a boundary from other portions in a multicomponent system. Thus, the term *crystal* here is the solid of a *single phase*, and the term *crystal structure* refers to the geometry, symmetry, and the arrangement of atoms or molecules of the phase. The solid of a single phase generally is composed of many independent crystals, variously oriented with respect to their symmetry axes, the whole being joined solidly together. Such a state is referred to as *polycrystalline* (see GRAIN BOUNDARIES). In general, solids may also contain crystals of several compounds with quite different chemical compositions and crystal structures or several types of crystals of the same compound. When an element or chemical compound, e.g., SiO_2, exists in several different structural arrangements, this is known as *polymorphism* (see PHASE EQUILIBRIA and PHASE TRANSITIONS). A phase is stable at some temperatures and pressures, and a metastable phase stays indefinitely at temperatures and pressures in which it is energetically unfavorable. Phase transitions take place by changing either the temperature or the pressure or both (see HIGH-PRESSURE EFFECTS IN SOLIDS). For example, diamond, the metastable phase at room temperature and pressure, must be synthesized at high temperature and pressure.

When one sets out to determine the atomic arrangement of a crystal, one not only seeks a sample of the solid that is pure (i.e., that does not also contain crystals of other compounds), but one that is a single crystal, wherein one crystal orientation extends throughout the solid. This is a practical requirement (see CRYSTAL GROWTH), as the diffraction patterns from polycrystalline samples confuse the data analysis since crystal orientation is random, thus limiting the useful information obtainable.

A crystalline arrangement of atoms is described by the laws of symmetry, which include the translation operations. Theoretical consideration of crystal symmetry verified 230 *space groups* (see Sec. 1.4), to one of which the crystal structure must belong. Although the number of space groups is limited, the number of different crystalline arrangements is essentially unbounded. This is because these laws permit continuously variable atom positions to be assumed, although they are fixed in any particular crystal structure.

Furthermore, the positions can be occupied by any combinations of about 100 natural and manmade elements.

A periodic crystalline arrangement gives rise to well-developed crystal faces of single crystals that are independently grown. These faces display the symmetry elements inherent in the crystal type. Sometimes such crystals exhibit a phenomenon known as *twinning*, where, for example, two different orientations are related by a *mirror plane*, whereon they unite coherently.

A crystal may contain organized coherent defects such as twinning, or a given crystal structure may be composed, e.g., of layers of two simple structural types that join coherently and repeat in regular sequence (see IMPERFECTIONS IN CRYSTALS). It may also exhibit irregular or random defects; indeed a polycrystalline material exhibits such defects along the boundaries of the crystallites of different orientations. In other crystals where the atoms occupy ordered sites, the ordering may "get out of step," e.g., be displaced in one region of the crystal compared to another, either on a random or regular basis. Regular ordering sequences or regular atom-displacement sequences may cause a *superstructure*, where a new unit cell that recognizes the regular ordering or displacements relative to the basic unit cell (see Sec. 1.1) may require multiplication of the edges by 2, 3, etc., along the axes **a**, **b**, or **c** concerned. Thus the ordered tetragonal Al_3Ti is an F_{112} superstructure (multiplication $2\times\mathbf{c}$) of the cubic copper structure, F (see Sec. 1.2 and HETEROSTRUCTURES AND SUPERLATTICES, METALLIC; HETEROSTRUCTURES AND SUPERLATTICES, SEMICONDUCTOR).

Liquid crystals are ordered fluids. The ordering in one, two, and three directions gives rise to phases called *nematic, smectic,* and *cholesteric* liquid crystals, but they are not considered solid states (see LIQUID CRYSTALS). On the other hand, the recently discovered *quasicrystals* are solids, but their symmetry and geometry do not follow those observed for crystals. Nevertheless, their existence was confirmed by diffractions, which proclaimed the presence of noncrystallographic fivefold symmetry (see Sec. 1.3) axes in these solids (see QUASICRYSTALS).

1.1 Unit Cells and Crystal Systems

Since a single crystal is an aggregate of millions of atoms periodically arranged in space, systematic methods of description have been devised. The first thereof is the *unit cell*, which is (generally) the smallest parallelepiped that stacked repetitively in three dimensions generates the structure of the crystal, regardless of the atomic content within the parallelepiped. This unit cell is defined by three noncoplanar vectors of magnitudes a, b, and c and three angles between them, α, β, and γ (Fig. 1). Such unit cells must belong to one of seven *crystal classes* that are defined purely by symmetry, although they also have metrical constraints as indicated in Table 1. Thus even though a unit cell may have $a=b=c$ and $\alpha=\beta=\lambda=90°$, if the atomic arrangement does not exhibit at least the minimum symmetry of a cubic system (*point group* 23, see Sec. 1.3), the unit cell must belong to a class lower than cubic. Further remarks on the choice of unit cells are made in the next section.

1.2 Lattices

The concept of a *lattice* is essential to the study of crystal structure. A lattice is an infinite periodic array of points at the termini of the vectors $\mathbf{L}_{uvw}=u\mathbf{a}+v\mathbf{b}+w\mathbf{c}$, where **a**,**b**,**c** are the three noncoplanar vectors defining a primitive unit cell and u,v,w are integers. Thus the integers $[uvw]$ are used to define a lattice direction (note the use of [ ] to define directions, whereas ( ) are used to define lattice planes (see Sec. 1.5)).

There are seven primitive lattices (one lattice point per cell) corresponding to the seven crystal classes. However, because of

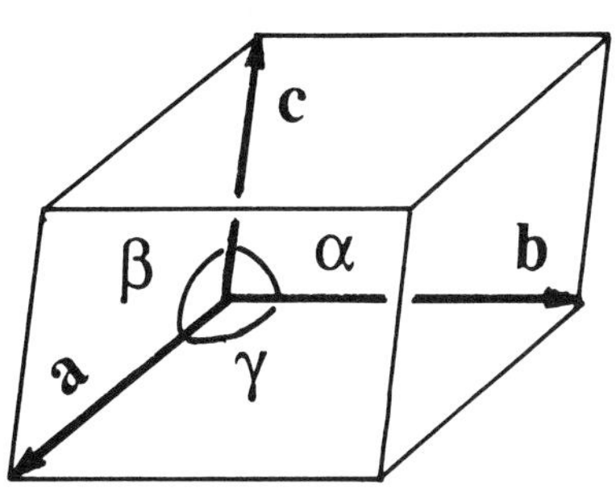

FIG. 1. Quantities a, b, c, α, β, and γ used in defining a unit cell. A right-handed coordinate system is always used.

Table 1. Axial and angular relationships, characteristic symmetry, and the 14 Bravais lattice types of the seven crystal systems.

System	Axial and angular relationships	Min. symmetry and (point group)[a]	Lattice type and symbol[b]
Triclinic	$a \neq b \neq c$; $\alpha \neq \beta \neq \gamma \neq 90°$	$1(\bar{1})$	*P*
Monoclinic	$a \neq b \neq c$; $\alpha = \gamma = 90° \neq \beta$	$2(2/m)$	*P,C* [c]
Orthorhombic	$a \neq b \neq c$; $\alpha = \beta = \gamma = 90°$	222 (mmm)	*P,I,F,C* [d]
Tetragonal	$a = b \neq c$; $\alpha = \beta = \gamma = 90°$	$4(4/mmm)$	*P,I*
Cubic	$a = b = c$; $\alpha = \beta = \gamma = 90°$	23[e] ($m3m$)	*P,I,F*
Trigonal	$a = b = c$; $\alpha = \beta = \gamma$	$3(\bar{3})$	*R*
Hexagonal	$a = b \neq c$; $\alpha = \beta = 90°$, $\gamma = 120°$	$6(6/mmm)$	*P*

[a]The point group is the maximum symmetry.
[b]Lattice symbols: *P*, primitive; *C*, *ab*-face centered; *I*, body centered; *F*, all-face centered; *R*, rhombohedral.
[c]Lattice type *C* is for *b*-axis unique or *A* for *c*-axis unique.
[d]Depending on the choice of axes, lattice type *A* (*bc*-face centered) or *B* (*ca*-face centered) may also be assigned.
[e]The twofold and threefold axes for cubic are along the face and body diagonals, respectively. The angle of intersection is 35.26°.

symmetries permitted in some of these lattices, it is more convenient for structural description to choose the face-centered lattices *A*, *B*, or *C*, the body-centered lattice *I* (all with two lattice points per cell), and the all-face-centered lattice *F* (four lattice points per cell), rather than the primitive lattices. This was already recognized by Bravais (1848), who saw that there were 14 such lattices; they now bear his name (Table 1 and Fig. 2). Although some crystal structures such as that of copper do have atoms only at lattice points, so that it is correct to refer to either the *Cu lattice* or the *Cu structure*, in general this is not so, and it is incorrect to use *crystal lattice* as a synonym for *crystal structure*.

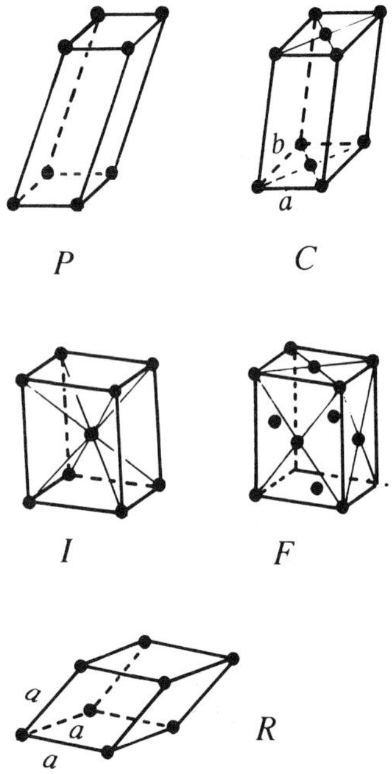

FIG. 2. Primitive *P*, *ab*-face-centered *C*, body-centered *I*, all-face-centered *F*, and rhombohedral *R* (trigonal) lattices (see Table 1).

1.3 Point-Group Symmetry

The manifestation of crystal symmetry is present both in the exterior forms and in the atomic arrangement within the crystals. This is how all of the elements of the crystal symmetry could be determined before any of the internal arrangements of atoms therein were discovered. Thus early morphological and mathematical studies led to the correct conclusion that there are 32 symmetry types that are called *point groups*. The morphology of crystals led to the recognition of *n*-fold rotation axes (where $n = 1, 2, 3, 4$, or 6, but not 5), mirror planes *m*, and a possible center of inversion *i* or $\bar{1}$. Combinations of these symmetry elements, which leave at least one point fixed, give rise to the 32 point groups. It is worth noting that some texts include *axes of rotary inversion* $\bar{n}$ or *improper rotation axes* as symmetry elements, but they result from combining rotation axes with a center of inversion or with a mirror plane.

Notations for describing point groups make use of some characteristic symmetry operation in the group, but they do not give all symmetry elements. Sometimes symmetry elements n or m in three directions are involved. When a mirror plane perpendicular to the n-fold axis is a major feature, it is represented by n/m. A twofold axis, if present, is often perpendicular to another major n-fold axis, and mirror planes can contain the major axis; these symmetry elements are placed side by side. For example, the symmetry point group of a cube with equal parts cut off from the corners is 432, representing the fourfold, threefold, and twofold rotation axes as major features. The pyramidal arrangement of H and N atoms in NH_3 gives rise to a threefold axis and mirror planes, its point group being $3m$. The planar benzene molecule C_6H_6 has a mirror perpendicular to the sixfold axis and two other types of mirror planes; its symmetry belongs to the point group $6/mmm$. Whereas these symbols of the Hermann–Mauguin notation are popular with crystallographers, the Schoenflies symbols are used more often by spectroscopists. The Schoenflies notations of the above point groups are $D_{6h}=6/mmm$, $C_{3v}=3m$, and $O=432$. Schoenflies symbols stress the nature of the group, being dihedral (D), cyclic (C), or octahedral (O). Although point groups were of interest to mineralogists at some time in the past, they do not play a key function in crystal structure descriptions today; however, they are still significant in crystallography as they provide an easy classification of crystals. A regular tetrahedron belongs to the point group 23 (Schoenflies symbol T), where the twofold axes go through the midpoints of opposite edges and the threefold axes go through the vertices and the geometric center. The relationship between a regular tetrahedron and a cube illustrates that 23 is a point group of the cubic crystal system.

1.4 Space Groups

Crystallographic lattices are generated from a point by translations, which are thus a symmetry operation. Hence it is apparent that a complete description of crystal symmetries must include translations as well as those present in point groups. Again the study is executed by the techniques of group theory. Combination of crystallographic point-group symmetry operations with translations gives new symmetry elements such as *screw axes* (2_1, 3_1, 4_1, and 6_1) and *glide planes* (axial glides a,b,c, diagonal glide n, and diamond glide d). The theory showed that there are 230, and only 230, distinct combinations of symmetry elements in three-dimensional space, which are known as the 230 *space groups*. These are each described in the Hermann–Mauguin notation to show essential symmetry elements, e.g., $P6_3/mmc$. The *International Tables for Crystallography* (Hahn, 1983) of the International Union of Crystallography, the *vade mecum* of crystallographers, give essential data for all space groups such as the possible sets of symmetry-equivalent positions therein, their multiplicity (number of equivalent positions), Wyckoff notation (see the Glossary), and point symmetry, and their coordinates. The space group to which a crystal belongs is an essential piece of information, without which the description cannot be completed from the published set of crystal data.

A space-group symbol consists of two parts: a letter depicting the lattice type and some essential symmetry elements. The triclinic system has two primitive space groups, $P1$ and $P\bar{1}$. Crystal structures belonging to the latter have a center of inversion, i.e., if there is an atom located at a point x,y,z, there is also one at $-x,-y,-z$; thus only half of the atomic coordinates in the unit cell needs to be given, as the other half is implicit or equivalent by symmetry. By way of illustration, the 13 space groups of the monoclinic system are listed in Table 2 with a remark about the symmetry features. Space groups with the same symmetry feature, but of different lattice type, are given on the same line. Table 2 also illustrates the relationship between space-group and point-group symmetry. The translation component of the former is absent in the latter.

Table 2. The 13 space groups of the monoclinic system.

Space group	Symmetry feature	Point group
$P2$, $C2$	Twofold axis	2
$P2_1$	2_1 screw axis	2
Pm, Cm	$m \perp$ the unique axis[a]	m
Pc, Cc	c glide	m
$P2/m$, $C2/m$	$m \perp 2$	$2/m$
$P2_1/m$	$m \perp 2_1$	$2/m$
$P2/c$, $C2/c$	c glide $\perp 2$	$2/m$
$P2_1/c$	c glide $\perp 2_1$	$2/m$

[a]⊥ means perpendicular to.

Therefore screw axes in space groups become simple n-fold rotation axes in point groups. Similarly, glide planes of space groups become mirror planes of point groups.

A symmetry element or operation will transform a set of coordinates x,y,z to a new location. For example, a twofold axis, if parallel to the b axis, gives $-x,y,-z$, and a 2_1 screw gives $-x,\frac{1}{2}+y,-z$. A c glide gives a new location at $x,-y,\frac{1}{2}+z$. Symmetry operations in each space group can be specified by a set of coordinates, which generate all atomic positions in a unit cell. For the space group of common occurrence, $P\,2_1/c$, the set of general coordinates (b axis unique) is: x,y,z; $-x,\frac{1}{2}+y,\frac{1}{2}-z$; $-x,-y,-z$; $x,\frac{1}{2}-y,\frac{1}{2}+z$. A diagram depicting these points is given in Fig. 3. However, if x, y, or z have values such as 0 or $\frac{1}{2}$, the set of four points degenerates into a set of two points only. These are called *special positions*. The special values give rise to several distinct sets, each with its own number of positions or multiplicity, and site symmetry. The sets are denoted by Wyckoff notations. A complete list of all Wyckoff sites for the space group $P\,2_1/c$ mentioned above is given in Table 3. Obviously, the origin for the coordinate system given in this example is at the center of inversion. The choice of origin and setting complicates matters for the comparison and classification of structure types.

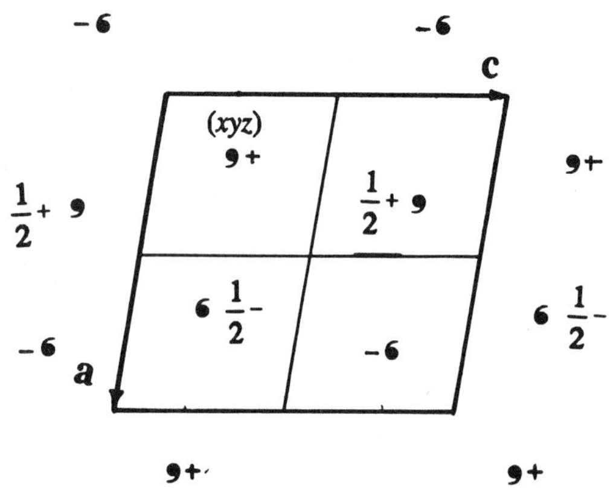

FIG. 3. Symmetry-equivalent images of a pattern represented by a comma at x,y,z (fractional coordinates with respect to a,b,c) in the space group $P2_1/c$. The + signs beside the comma indicate a point below the page. The coordinates $\frac{1}{2}\pm y$ of the points are represented by $\frac{1}{2}\pm$. Depending on the value for y, these points can be above or below the page. Inverted commas are due to mirror planes or centers of inversion.

1.5 Lattice Planes and Crystal Faces

For use in crystal diffraction and for morphological studies, it is necessary to consider a set of parallel planes so constructed that every lattice point lies on some member of it. These are called lattice planes, rational planes, or simply *planes*. When such planes cut an axis of the unit cell, they always divide it into an integer number of equal parts. If we start with some lattice point on a plane as the origin and proceed from it along the axes to the first member of the same set of planes, the equation for the first member plane can be expressed in terms of the intercepts. If the intercepts along the three axes are a/h, b/k, and c/l, respectively, the (h,k,l) are referred to as the Miller indices of the set of planes. The commas may be omitted if the integers or h, k, and l lead to no ambiguity. For example, referred to a lattice point as the origin, the first member of planes (321) intercepts the three axes at $a/3$, $b/2$, and c, respectively; see Fig. 4. The set of planes (100) is parallel to the bc plane. It should be noted, however, that the distance d_{100} between (100) planes is not a, except in orthorhombic, tetragonal, and cubic systems, which have orthogonal axes. The interplanar distance d_{hkl} for the general triclinic system is rather complicated, but for structures with orthogonal axes it is easily recognized.

Miller indices may be used in many ways: to designate a set of lattice planes, a particular member of the set, or a crystal face parallel to the set. The plane (100) intersects at unit length on the axis [100], which represents a lattice point and a direction (see Sec. 1.2). The direction [100] is not perpendicular to the plane (100) except in crystal systems with orthogonal axes, but [100]* referred to a direction in reciprocal space is perpendicular to (100).

The Miller indices are sometimes used to represent a *form* $\{hkl\}$ which represents a set of symmetry-equivalent planes. For example, {100} includes (100), $(\bar{1}00)$, (010), $(0\bar{1}0)$, (001), and $(00\bar{1})$, which generate a cube for a cubic crystal. For a triclinic crystal, {100} represents only (100) and its symmetry-equivalent plane $(\bar{1}00)$; {010} represents (010) and $(0\bar{1}0)$, and {001} represents (001) and $(00\bar{1})$. The faces or planes represented by the form $\{hkl\}$ depend on the crystal system.

Table 3. Multiplicity, Wyckoff notation, site symmetry, and coordinates of symmetry-equivalent sites in the space group $P\,2_1/c$.

Multiplicity	Wyckoff notation	Site symmetry	Coordinates of equivalent points
4	*e*	1	$x,y,z;\ \bar{x},\bar{y},\bar{z};\ \bar{x},\frac{1}{2}+y,\frac{1}{2}-z;\ x,\frac{1}{2}-y,\frac{1}{2}+z$
2	*d*	$\bar{1}$	$\frac{1}{2},0,\frac{1}{2};\ \frac{1}{2},\frac{1}{2},0$
2	*c*	$\bar{1}$	$0,0,\frac{1}{2};\ 0,\frac{1}{2},0$
2	*b*	$\bar{1}$	$\frac{1}{2},0,0;\ \frac{1}{2},\frac{1}{2},\frac{1}{2}$
2	*a*	$\bar{1}$	$0,0,0;\ 0,\frac{1}{2},\frac{1}{2}$

All ($hk0$) planes are parallel to **c** or [001], and all ($hk0$) faces have one direction [001] in common. If the crystal is elongated in [001], these faces make it appear as a prism. Thus these are called *zonal planes* or *zonal faces*. The *zones* formed by these faces are convenient in the description of morphology and diffraction patterns. Diffracted beams from zonal planes form a cone with the apex at the crystal.

1.6 The Reciprocal Lattice

A crystal is an identity in a three-dimensional space, and its repeating lattice is called a *direct lattice*, for which a *reciprocal lattice* can be defined either for the convenience of diffraction studies or for other applications. A

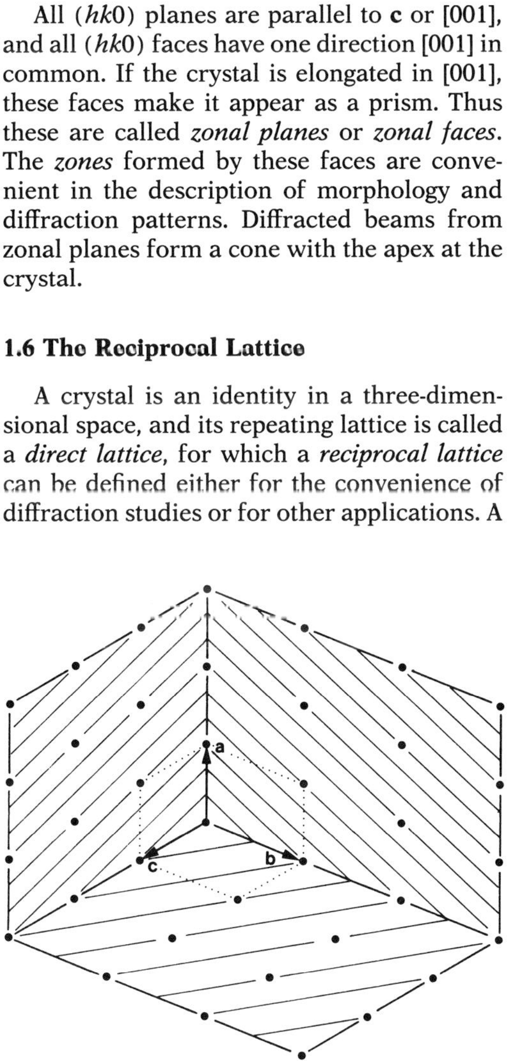

FIG. 4. A set of planes with Miller indices (321) showing the intercepts on axes *a*, *b*, and *c*. Every lattice point lies on some member of the set. The diagram includes three unit cells (one outlined by dashed lines) in each of the three directions.

reciprocal-lattice point is defined to be lying on the normal of a crystal plane (hkl) at a distance K/d_{hkl}, K being a constant. The three reciprocal-lattice vectors **a***, **b***, and **c*** are normals to the planes (100), (010), and (001) with magnitudes of K/d_{100}, K/d_{010}, and K/d_{001}, respectively. Note that since $2/d_{100} = 1/d_{200}$, the reciprocal point corresponding to (200) is twice the distance from the origin as that of (100) or **a***. By the same token, other points in the reciprocal space can be deduced, generating the reciprocal lattice using the basic set of vectors **a***, **b***, and **c***.

When applied to x-ray diffraction studies, the constant K can be taken as λ (wavelength of the x rays used); in this case **a*** ($=\lambda/d_{100}$), **b***, and **c*** are unitless pure vectors. The convenience for such a choice has a direct connection with the Bragg equation written in the form $2\sin\theta=\lambda/d_{hkl}$ (see Sec. 2.2). Alternatively, K may be taken as 1, and the reciprocal-lattice space has a unit of reciprocal length. Using vector notation, the relationships between reciprocal and direct lattice vectors are

$$\mathbf{a}^*=\frac{\mathbf{b}\times\mathbf{c}}{V},\quad \mathbf{b}^*=\frac{\mathbf{c}\times\mathbf{a}}{V},\quad \mathbf{c}^*=\frac{\mathbf{a}\times\mathbf{b}}{V},$$

where V is the volume of the unit cell in direct space, and it has the expression

$$V=abc(1+2\cos\alpha\cos\beta\cos\gamma-\cos^2\alpha-\cos^2\beta-\cos^2\gamma)^{1/2}.$$

The reciprocal-lattice angles α^*, β^*, and γ^* bear the following relationships:

$$\cos\alpha^*=\frac{\cos\beta\cos\gamma-\cos\alpha}{\sin\beta\sin\gamma},$$

$$\cos\beta^* = \frac{\cos\gamma \cos\alpha - \cos\beta}{\sin\gamma \sin\alpha},$$

$$\cos\gamma^* = \frac{\cos\alpha \cos\beta - \cos\gamma}{\sin\alpha \sin\beta}.$$

Of course, the above relationships are simple for orthogonal systems, in which $\alpha=\beta=\gamma=90°$.

2. CRYSTAL STRUCTURE DETERMINATION

Although determination of chemical composition and crystal structure are the first two steps in the characterization of a substance, they are but a component in the investigation of its chemical, physical, and other properties. Crystal structure determination provides a three-dimensional map of the atoms in the solid. Thus it gives information on the crystal symmetries involved in the atomic arrangement, on the dimensions and crystal class of the repeat volume or unit cell, and on what atoms occupy which crystal sites fully, partially, and in ordered or random fashion, as well as giving the values of the atomic coordinates. Thus information is obtained about the surroundings of each atom in the structure and the distances and angles between it and its neighboring atoms.

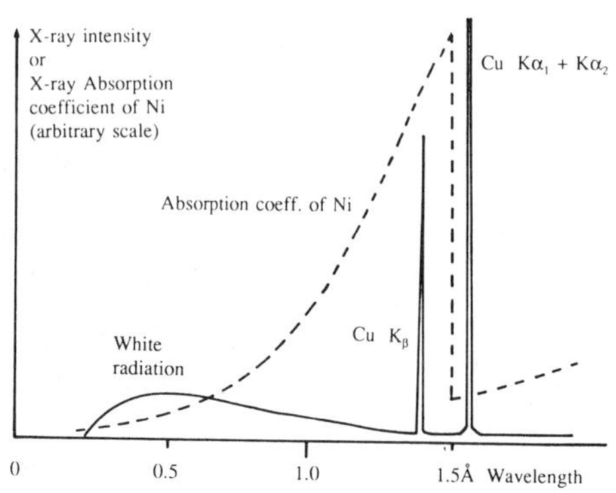

FIG. 5. A typical x-ray spectrum generated by bombarding a copper target with high-energy electrons; also shown are the absorption coefficients of nickel. The sharp drop in absorption coefficients at wavelength between $K\beta$ and $K\alpha$ is called the absorption edge.

2.1 Radiations for Crystallography

X-ray, neutron, and electron beams are the major radiations used for crystal structure determination. Each radiation has a unique characteristic and diffraction mechanism.

2.1.1 X-Ray Emission Spectra. Figure 5 shows a typical low-wavelength x-ray intensity spectrum, here generated by 50-kV electrons from a copper target. It is characterized by relatively low-intensity *white* or continuous radiation above a certain wavelength that is made use of in Laue photographs (see Sec. 2.2.1) and by the relatively intense $K\beta$ and $K\alpha_1$ and $K\alpha_2$ peaks at specific wavelengths, with intensities $K\alpha_1 > K\alpha_2 \gg K\beta$. Wavelength separation between $K\alpha_1$ and $K\alpha_2$ is small, and both are used for single-crystal and powder-diffraction photographs and diffractometer data. To simplify the data, the $K\beta$ radiation can be removed by placing a thin nickel foil in the path of the entrant beam, since it strongly absorbs $K\beta$ but not $K\alpha_1$ and $K\alpha_2$. The sharp difference in absorption coefficients within a short range of wavelength is called the ***absorption edge.*** The absorption edge of nickel falls between $K\beta$ and $K\alpha$ of copper x rays, i.e., nickel strongly absorbs $K\beta$ but absorbs little $K\alpha$ (Fig. 5). If the $K\alpha_1$ and $K\alpha_2$ frequencies have to be separated, it requires very careful setting of a *bent-crystal* monochromator, but generally this refinement is considered unnecessary. The principles outlined here are generally applied appropriately to x radiation from molybdenum or other metal targets. Of course, each target metal has characteristic radiations of some unique wavelengths with high intensities and corresponding filter material. The intensity of the x radiation is a function of the accelerating potential and intensity of the electron-generating beam. The potential and intensity are limited by the probability of "burning through" the water-cooled metal target. Stronger x-ray beams can be generated by a rotating-target generator, and very much more intense x-ray sources have now become available at specific *synchrotron radiation* (SR) locations. These are pulsed (intermittent) x-ray beams of continuous wavelength generated when electrons are accelerated to an energy of the order of 10^9 eV or 1 GeV (see SYNCHROTRON RADIATION).

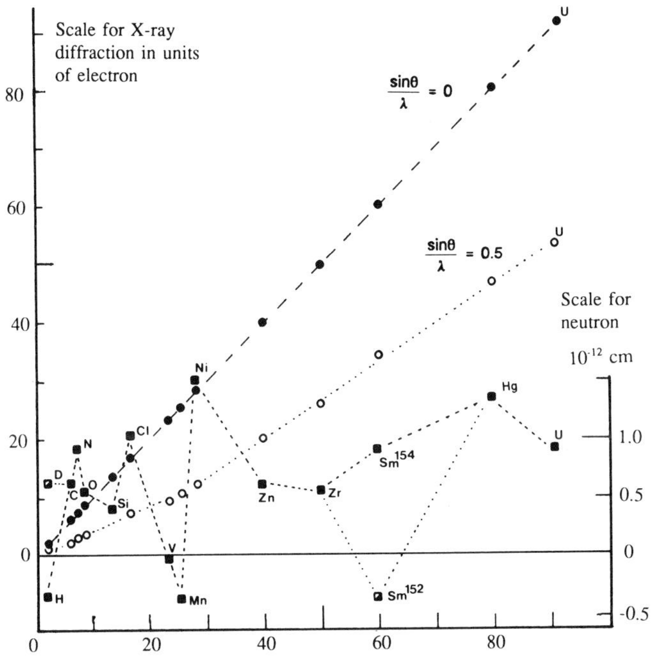

FIG. 6. Comparison of atomic-scattering factors for x rays and for neutrons. Two sets of values are given for x rays, filled circles for $\sin\theta/\lambda=0$ and open circles for $\sin\theta/\lambda=0.5$. Corresponding neutron data of the most abundant isotope are represented by filled squares. Isotopes of H and Sm are given for contrasting their difference. Units for the factors are electrons for x rays, and 10^{-14} m for neutrons.

2.1.2 X-Ray Atomic-Scattering Factors. X rays are scattered by the shells of electrons surrounding the atomic nuclei; the amplitudes of x rays scattered by an element, defined as the *atomic-scattering factors*, are positive and increase monotonically with the number of electrons or atomic number (Fig. 6). This has several important consequences in structural analysis.

1. *Location of atoms.* Since for hydrogen, particularly, the scattering factor is very small, very little electron density will appear in Fourier syntheses (see Sec. 2.2) at the locations of the hydrogen atoms, especially if the substance contains other atoms of much greater scattering power. This is why locations for hydrogen atoms cannot be determined with confidence.
2. *Assignments of elements to crystal sites.* When the atoms in a structure have scattering powers that differ little from the average value, it is difficult to assign elements to specific crystal sites. Furthermore, superstructure reflections that might be expected due to atomic ordering will be very weak and probably not detectable. In special cases this situation may be improved by judicious choice of x radiation and making use of the anomalous scattering factor, which is pronounced in the vicinity of an element's absorption edge.
3. *Phase-angle determination.* If a heavy atom is introduced in organic compounds that contain elements H, C, N, and O, its dominant atomic-scattering factor can be used to assist phase-angle determination (see Sec. 2.2).

Finally, since x rays are scattered by the electrons surrounding an atomic nucleus, interpretation of x-ray-diffraction experiments gives information not on the locations of the atomic nuclei but on the electron densities surrounding the nuclear sites.

2.1.3 Neutron Atomic-Scattering Factors. Neutrons, discovered in 1935, are particles emitted from the nuclei of atoms in nuclear reactions such as the fission of heavy atoms in reactors. The wave properties of neutrons make them suitable for diffraction studies. First, whereas x rays are scattered by electrons, neutrons are scattered by the atomic nuclei, whose radii are about $\frac{1}{100\,000}$ of those of atoms. Furthermore, the combined effect of nuclear scattering and magnetic scattering due to intrinsic spin and magnetic moment of the neutrons and of atomic nuclei gives rise to a phase-angle shift of the diffracted neutrons.

An out-of-phase (angle) shift results in a negative neutron-scattering factor. Therefore neutron atomic-scattering factors can have positive or negative values, or indeed have a value approximating zero (Fig. 6). They may even differ appreciably between atomic isotopes, as between hydrogen and deuterium. Indeed, with respect to other elements, the hydrogen and deuterium neutron-scattering factors are relatively much larger than their x-ray-scattering factors, which generally makes neutron-diffraction experiments more effective than x-ray experiments for locating the positions of these atoms in structures. Second, the particular values of neutron-scattering factors can be used with judicious advantage in structure determinations. For example, in the case of an ordered Ni-V alloy, neutron-diffraction experiments can well be combined with x-ray-diffraction experiments to locate the sites of the atoms, since the neutron atomic-scattering factor for vanadium is approximately 0. Finally, since neutrons have intrinsic spin and magnetic moment, diffraction experiments can be used to gain information on atomic moments and their orientations for magnetic atoms in structures by separating from the observed scattering that which is normally expected, thus revealing the *magnetic* scattering, which then may be interpreted.

2.2 X-Ray Crystal Structure Analysis with Single Crystals

W. L. Bragg's interpretation of the x-ray-diffraction patterns from NaCl, KCl, KBr, KI, and ZnS laid the foundation for x-ray structure analysis. A simple illustration of the principle of x-ray diffraction is the Bragg equation: $2d_{hkl}\sin\theta=\lambda$, where d_{hkl} is the interplanar spacing for (hkl), θ the angle of inclination of the x-ray beam to the crystal plane, and λ the wavelength of radiation used (Fig. 7). Thus a constructive beam results when both the wavelength and orientation of the crystal with respect to the beam are proper. Note that the diffracted x-ray beam or *reflection* [of plane (hkl)] deviates by an angle of 2θ from the incident beam (see X-RAY DIFFRACTION).

Bragg's law gives the condition for a reflection to develop, while the intensity I_{hkl} of the diffracted x-ray beam is related to the arrangement of atoms in the crystal. In short, I_{hkl} is proportional to the square of a quantity known as the *structure factor* F_{hkl}, which can be expressed as

$$F_{hkl}=\sum_{j=1}^{j=n} f_j \exp[2\pi i(hx_j+ky_j+lz_j)],$$

where $i=\sqrt{(-1)}$; x_j,y_j,z_j are the coordinates of the atom j expressed as fractions of the unit-cell lengths; f_j is the atomic scattering factor; and the summation is taken over all atoms (number n) in the unit cell. Generally, the higher the atomic number of the atom, the larger the f_j value, which decreases as θ increases. When a reflection develops, the path difference of x rays scattered from neighboring planes (hkl) is one wavelength, or a complete cycle of 2π (Fig. 7). The atom j with atomic scattering factor f_j located away from the origin at x_j,y_j,z_j will have a phase-angle difference of $2\pi(hx_j+ky_j+lz_j)$ relative to that from the origin. X-ray diffraction is the superposition of waves from all atoms, and thus the sum is the structure factor, which when squared is proportional to the intensity of the reflection. Thus structure determination involves the investigation of diffraction patterns, measurements of the angles 2θ and intensities of the reflections, and comparisons of the structure factors calculated for the proposed structural arrangement with those derived from experiment. However, to derive the structure from x-ray measurements is not

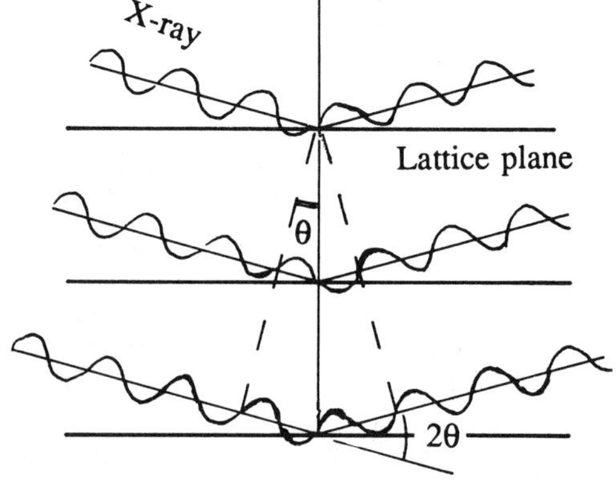

FIG. 7. A simple illustration of the principle of x-ray diffraction, showing the path-length difference of one wavelength for reflected x rays from adjacent lattice planes.

a trivial matter. If the structure factors are known, the electron density $\rho(x,y,z)$, the peaks of which correspond to the positions of atoms, can be calculated from the Fourier series (see FOURIER AND OTHER MATHEMATICAL TRANSFORMS)

$$\rho(x,y,z)=\frac{1}{V}\sum_{-h}^{h}\sum_{-k}^{k}\sum_{-l}^{l}F_{hkl}\exp[2\pi i(hx+ky+lz)],$$

where x,y,z is a point in the unit cell. Unfortunately, the diffraction experiments only provide the magnitudes $|F_{hkl}|$, but not the phase angles α_{hkl}, of $F_{hkl}\,[=|F_{hkl}|\exp(2\pi i\alpha_{hkl})]$. Thus, the diffraction pattern is analogous to only half of a holographic image. Therefore, solving a crystal structure is essentially a matter of solving the *phase-angle problem* or what crystallographers call the *phase problem*. Various methods have been devised for solving this problem, and the Patterson method invented in 1934 dominated until the 1980s (Glusker *et al.*, 1987). The Patterson function employs the square of the structure factors in a Fourier synthesis,

$$P(xyz)=\frac{1}{V^2}\sum_{-h}^{h}\sum_{-k}^{k}\sum_{-l}^{l}|F_{hkl}|^2\cos[2\pi(hx+ky+lz)],$$

where $P(xyz)$ is the Patterson density at the coordinates x,y,z, and its peak positions correspond to the interatomic vectors transferred to a common origin. Analysis of the Patterson map often reveals relative pairs of atoms. Since heavy atoms, such as chlorine or iron, say, in an organic substance, dominate the x-ray scattering, vectors between them in the crystal are commanding features of the Patterson map, from which locations of heavy atoms can easily be determined. Once the locations of the heavy atoms are found, structure factor calculations using them give phase angles that approximate those using all atoms. Thus borrowing the phase angles derived from the heavy atoms, but using the measured amplitudes for the Fourier synthesis, reveals the locations of the light atoms N, O, C, and H of most organic compounds. This is known as the *heavy-atom method* of Robertson (1953) for solving organic and protein crystal structures (see Sec. 2.3). However, the situation for metallic alloys is more difficult, since generally atomic-scattering factors of the different atoms may differ little from the average value [see Sec. 2.1.2, point **2**].

Hauptman and Karle (1953) applied statistics to the intensities of reflections, and their ideas led to wide success in solving the phase-angle problem during the 1970s. These are called *direct methods*, because the phase angles are derived directly from a mathematical treatment of the intensities themselves.

When phase angles for many reflections are derived from the locations of some atoms or by direct methods, these reflections can be used in the Fourier synthesis to reveal more atomic positions. Successive rounds of calculations generally lead to a satisfactory structure, and when all atoms are located with the help of chemical knowledge, better values of the atomic coordinates can be obtained by carrying out least-squares refinements. These minimize the factor R, which is the normalized difference between calculated and observed intensities of all reflections for a given structural arrangement:

$$R-\left(\frac{\Sigma^{hkl}(|F_{hkl}|_o-|F_{hkl}|)^2}{\Sigma^{hkl}|F_{hkl}|^2}\right)^{1/2}.$$

Early structure determinations from film methods gave R values of 10% or more, but modern data collected using diffractometers give R values between 9% and 2%.

Failure to obtain, after successive refinements, what would be considered a satisfactorily low R value should indicate possibly

1. a wrong choice of space group,
2. significant errors in phase-angle determination,
3. the incorrect positioning of some atoms in the structure,
4. incorrect assignment of elements to crystal sites or unrecognized random distribution of elements among the sites, or
5. any combination of points **1–4**.

A final check of the correctness of the structure can be obtained by calculation of a *difference Fourier synthesis*, $\Delta\rho$, which gives a three-dimensional map. Here the calculated electron-density map is subtracted from the observed one. Significant maxima or minima of electron density in the difference map indicate the possible presence of unplaced atoms, misplaced atoms, or incorrect assignment of atoms to crystal sites.

So far we have ignored two major factors that influence the observed intensities of x-ray reflections, compared to those calculated for a proposed structure:

1. *Thermal vibrations of the atoms.* This is taken into account by introducing into calculations a term known as the *temperature factor.*
2. *Absorption of x rays in the crystal studied.* To minimize the effects of x-ray absorption relative to crystal shape, ideally very small spherical single crystals should be examined.

When this is not possible, especially when the substance contains elements of high atomic number, estimates of the effects of absorption on the observed intensities of reflections relative to crystal shape, orientation, and dimensions must be made and taken into account.

Although modern crystal-structure analysis is now based on computer-controlled diffractometers and does not require much use of photographic techniques, many structures have been determined previously using various methods, so we give a general outline of these here.

2.2.1 The Laue Methods. The Laue methods (Amorós *et al.*, 1975) place a flat photographic film in a plane perpendicular to a beam of x rays of continuous wavelength to record the diffraction pattern from a stationary single crystal (Fig. 8). The film can be placed on either the same or the opposite side as the x-ray source, giving back-reflection or transmission Laue photographs, respectively. The Laue method is now used only to study crystal symmetry, perfection, or alignment for other experiments, although it is the easiest experiment to perform. Reflections from some zones appear on the film, and spots of each zone form ellipses on transmission and hyperbolas on back-reflection Laue patterns.

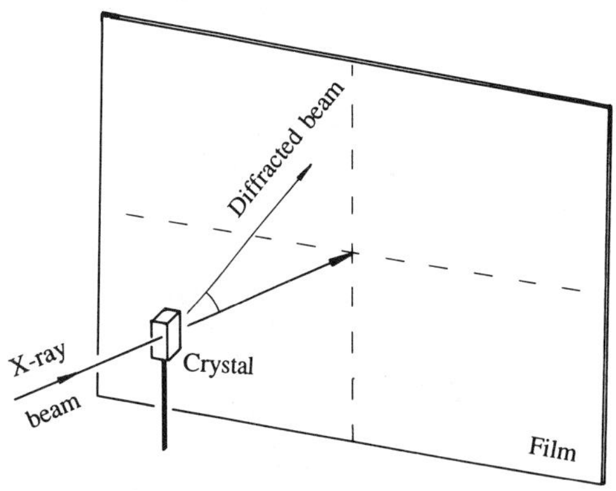

FIG. 8. A simple sketch showing the major arrangements in the Laue methods of x-ray diffraction.

2.2.2 Cylindrical-Film Methods. A flat film in the Laue method covers only a limited area for recording the reflections. More data can be collected when the film forms a cylinder coaxial with the rotation axis of the crystal. The axis is normal to the x-ray beam, giving rotation or oscillation photographs. Furthermore, if the film is allowed to translate while the crystal rotates, and if only one zone of reflection is allowed to be recorded at a time, a distorted reciprocal-lattice plane, e.g., ($hk2$), is recorded. This technique is called the Weissenberg method, which allows reflections from each set of planes (hkl) to be separated and their intensities measured. In the simple rotation methods, many reflections coincide in a single spot. Before 1970, the majority of intensity data for crystals was collected using the Weissenberg method, and the instrument, the Weissenberg camera, could also take cylindrical-film Laue, oscillation, and rotation photographs.

2.2.3 Precession Methods. In the precession method (Buerger, 1964), a simple direction, e.g., [100], of the crystal is aligned in the direction of the x-ray beam and is then tilted with respect to and caused to precess around it. This gives reflections of undistorted reciprocal-lattice planes for selected zones. Techniques used in aligning the crystal and selecting zones are useful in the study of crystal orientation as well as symmetry investigations. Careful study of such photographs is required to discover superstructures, especially if reflections due to the superstructure are weak. Precession cameras can also be used to take Laue photographs.

In summary, x-ray photographs reveal crystal symmetries and make possible the determination of cell dimensions, and generally of the space group. Estimation of the intensities of reflections from films was the major technique for crystal structure determination until the 1970s, when the use of automatic diffractometers that record intensities with scintillation detectors became common.

2.2.4 Diffractometers. A diffractometer is an instrument that rotates the crystal about three independent directions using three concentric circles. The angles on the circles are called ϕ, χ, and ω. The detector scans through the angle 2θ to measure the x-ray intensity. Thus the instruments are called four-circle diffractometers. The required orientations of the crystal for it to reflect the x-ray beam from (hkl) planes towards the scintillation detector used to measure the intensity are calculated from the unit-cell dimensions and orientation already determined. Computers are essential not only for these calculations, but also for controlling the angle driver and the detector. By the early 1980s, computer programs for solving the crystal structure had been integrated with those for diffractometer control to create complete operating packages.

2.3 Macromolecule Crystallography

Structure determination of proteins is a young science started in the 1930s, but its progress was slow at first, rapid growth only taking place since the 1970s. Compounds with large molecules include enzymes, redox proteins, carrier and storage proteins, hormones, antibodies, and viruses (see PROTEINS). Proteins are thousand-atom, chainlike, very large molecules, the building blocks of which are amino acids. The long chains of proteins are always folded into a globular shape (Blundell and Johnson, 1976).

Compared to small-molecule structural studies, the most difficult task in macromolecule crystallography has been the growth of suitable crystals and the preparation of heavy-atom derivatives. Nowadays, conditions for growing some protein crystals are well understood. Suitable methods have been found for growing an increasing number of proteins and enzymes. Growing crystals with heavy atoms inserted made the solution of the phase-angle problem easier.

Once the crystals are available, data collection, phase-angle determination, calculation of electron-density maps, interpretation of the maps, and least-squares refinements can follow, as in normal crystallography. However, because of the large molecular size, the unit cells of their crystals are very large, and many more reflections have to be collected, often using numerous crystals, since prolonged exposure to x rays may be destructive (see RADIATION DAMAGE OF CRYSTALS), and the calculation and interpretation of data are truly tedious. Many years of work used to be required for such a structure analysis; since faster data collection facilities and high-speed computing facilities have become available, this is no longer the case. When initially a small number of intensities is used, a low-resolution blurry image of the molecule may appear. When more reflections are added as time goes on, better resolution may reveal the backbone of the molecule and eventually even atomic sites. Thus the solution of protein and similar structures proceeds, as indeed do all structural solutions, by successive iterations whereby in the case of proteins, the resolution is successively increased, say from a scale of 10 to 5 to 2 Å, when individual major atom positions may appear in electron-density maps or sections. At 1 Å resolution, all atomic positions are accurately determined. Enzyme projects include the study of interactions between the protein and the *substrate* or between the protein and the *inhibitors.* These results lead to the understanding of biochemical reactions at the molecular level that are of great consequence in medicine and drug design. Furthermore, computer displays are used to model interactions between enzyme molecules, and these facilities enable biochemists and medical scientists to "see" biochemical reaction mechanisms and to facilitate the design of certain drugs.

Many countries have sites for macromolecular data collections, e.g., Cornell High Energy Synchrotron Source (CHESS), Stanford Synchrotron Radiation Laboratory (SSRL), National Synchrotron Light Source (NSLS) in the U.S.A.; Laboratoire d'Utilisation de Rayonnement Electromagnetique (LURE) in France; European Molecular Biology Laboratory (EMBL) in Germany; and Daresbury Laboratory in the United Kingdom.

Diffraction studies of synthetic and natural polymers usually give information about crystallinity and texture of the sample; low-angle scattering experiments are often employed. However, high-angle scattering experiments for polymeric molecules have lately been made possible by use of synchrotron radiations from which a suitable wavelength can be selected from a continuous spectrum of high intensity.

2.4 X-Ray Analysis of Polycrystalline Samples

Before 1960, many simple and even complex structures, particularly of metals and alloys, were solved by cylindrical-film powder methods, because single crystals could not be obtained. However, since crystal orientation is random in polycrystalline materials, reflections due to equivalent planes appear as arcs on the cylindrical film, and much information is lost compared to that obtainable from Weissenberg photographs of single crystals. More recently, computer-aided data analysis has increased the effectiveness of structure determination by powder methods when single crystals are unavailable.

Generally, powder-diffraction techniques are most useful for accurate determination of unit-cell dimensions and therefore for measurement of thermal expansion coefficients, since room-temperature films can be used to record reflections from samples cooled to as low as liquid-helium temperatures, or, enclosed in thin silica tubes, heated to high temperatures. However, diffractometer techniques are now used also for these purposes. Since most natural or industrial materials are polycrystalline, powder-diffraction methods are widely used for their identification, particularly in the metal and mineral realms. To this end, the Joint Powder Diffraction Commission (JPDC) maintains both printed and computerized indexes of powder-diffraction patterns of all known substances for their identification. Access to these can be had by various means but principally by the matching interplanar (d) spacings of the three most intense lines in the powder-diffraction pattern.

2.5 Other Diffraction Methods

X rays are electromagnetic waves, whereas neutrons and electrons are usually considered particles; however, a beam of each has the dual property that it can be treated either as a wave motion or as a beam of particles. In diffraction experiments, the former property is of interest. When just emitted from the nucleus of an atom during a nuclear reaction or fission, neutrons have an energy of 0.5 to 5 MeV, but are quickly slowed down when they collide with light atoms such as hydrogen in water, and become *thermal neutrons* of energy 0.025 eV at 293 K, with a de Broglie wavelength of $\lambda = h/mv = 18$ nm (1.8 Å). On the other hand, only high-energy electrons give a desirable wavelength for diffraction experiments.

2.5.1 Neutron Diffraction. The main thermal-neutron sources are nuclear reactors. Elaborate equipment is required to protect researchers and detectors from the harmful radiation; otherwise, the geometric requirements are similar to those for x-ray-diffraction experiments (see NEUTRON DIFFRACTION).

Since neutrons are scattered by the nuclei, rather than by the electrons, neutron-diffraction experiments give the locations of the nuclei in structures. Although neutron diffractions "see" time-averaged crystal structures, the inelastically scattered neutrons can also be measured by comparing the energies of neutrons before and after they interact with atoms in solids. Inelastic scattering experiments make possible the study of dynamic phenomena in the solid state. Thermal neutrons have velocities that are comparable to the thermal waves in solids; hence neutron-scattering experiments allow the observation of the Doppler shift of the scattered neutrons. This property makes possible the investigation of the dynamics (vibration spectrum) of atomic systems, by separating the inelastic and elastic scattering data (Bacon, 1975).

Compared to x-ray diffraction, neutron-diffraction experiments offer three distinct advantages:

1. Since neutron atomic-scattering factors can be either positive, negative, or approximately 0, judicious design of a neutron-diffraction experiment or comparison of the results of x-ray- and neutron-diffraction experiments can give specific information on the location of elemental species on crystal sites.
2. Since the atomic-scattering factor of hydrogen (and more so of its isotope deuterium) by neutrons is relatively much larger than that by x rays (Fig. 6), neutron-diffraction experiments are greatly favored over x-ray experiments for determining the location of hydrogen atoms in structures.
3. Since the neutron diffraction consists of "normal" and magnetic interactions, the two components of the scattering observed can be separated. Thus magnetic properties

associated with certain atoms and their specific sites in crystal structures can be studied (see MAGNETIC ORDERING IN SOLIDS).

2.5.2 Electron Diffraction. Electrons in a modern electron microscope operating at a potential of 50 000 to 10^6 V can penetrate thin materials of 0.1 μm thickness with a magnification power of 10^7 or more. For diffraction purposes, the effective de Broglie wavelengths range from 0.01 to 0.1 Å, so that the Bragg angles are about 1°. Nevertheless, diffraction patterns of very small crystals can be recorded photographically and used for structural investigation. However, because of the sensitivity of orientation, a small misalignment caused by crystal defects is observable in electron-diffraction patterns. The diffracted electron beam gives the image of the area irradiated by electrons; thus it is used for the study of crystal perfection so that the properties of the material can be correlated to sample treatment (see ELECTRON DIFFRACTION).

Because of the charge on the electron, an electron beam only penetrates slightly into the surface of a solid and can only pass through a very thin film of material. Therefore, the electron-diffraction pattern is essentially only two dimensional, although use of special techniques, such as slightly changing the orientation of the surface of the thin film or solid relative to the beam, may improve this situation when structure determination is necessary. Nevertheless, electron diffraction can well be used to identify small crystals settled in a thin-film metal matrix, and the shallow penetration is used with advantage to determine the arrangement of atoms on the surface of a crystal, a technique known as low-energy electron diffraction (LEED), although precise interpretation of the results of such experiments is subject to difficulties.

3. THE RESULTS OF CRYSTAL STRUCTURE DETERMINATIONS

Here we are able to provide only brief versions of atomic arrangements in the thousands of different crystal structures that have been determined. Broadly speaking, these fall into two classes:

1. structures in which single atoms can be considered bonded together, including metals and insulators/semiconductors, where the atoms either are covalently bonded together, or are charged ions;
2. molecular crystals in which groups of atoms forming molecules are covalently bonded together so that the peripheral atoms of the molecules exhibit filled valency subshells to those of neighboring molecules. Thus cohesion between the molecules in the crystals is very weak (van der Waals forces), and such crystals are generally soft, cleave easily if the geometry is appropriate, and have relatively low melting points.

3.1 Metals

The main characteristic of the crystal structures of metals and alloys is that the atoms generally, though not invariably, have high coordination numbers (CN) (see METALS AND ALLOYS, STRUCTURE OF). The "ideal" metallic arrangement is found in the close-packed face-centered cubic structure of copper in which each atom is surrounded by 12 neighbors in a cubo-octahedron (Fig. 9) or in the close-packed hexagonal structure of magnesium, where the surrounding is a twinned cubo-octahedron. This is because 12 is the maximum number of equal-sized spheres that can pack around a central sphere, all spheres being mutually in contact. This close-packed cubic arrangement is the basis of industrially used metals and alloys (see Sec. 4). When some atoms in an alloy have much greater diameters than others, coordination numbers up to 24, or even greater, may be found.

Another common metallic structure is the body-centered cubic type of wolfram (W), or the primitive β-CuZn type (two different components in ratio 1:1) in which each atom is considered to be surrounded by 8 atoms in a cube or 14 in a dodecahedron, depending on the relative sizes of the atoms (Fig. 10). Such alloys are of less use for industrial construction, but the basic arrangement may be of interest for magnetic and other properties. The close-packed and body-centered models are the basis of many more complex superstructural alloy structures.

Four convex coordination polyhedra that have triangulated surfaces are important in the structures of alloys. These are the CN12 icosahedron, the CN14 two-capped hexagonal

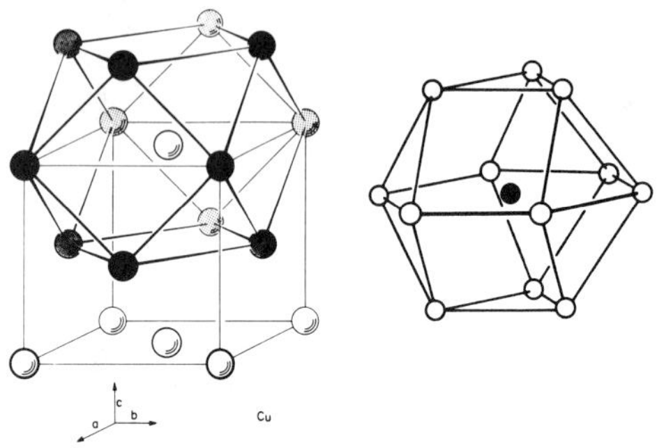

FIG. 9. The cubo-octahedron formed in all-face-centered cubic (fcc) crystals of copper (Cu), and the twinned cubo-octahedron in hexagonal-closest-packed (hcp) magnesium, which results when any one of the triangles is turned about its center by 60°.

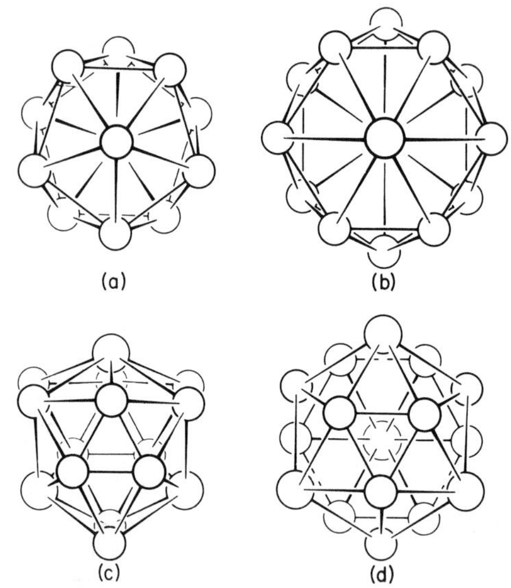

FIG. 11. Pictorial diagrams of CN12 (a) icosahedron, and (b)–(d) CN14, 15, and 16 coordination polyhedra of Frank and Kasper, central atom omitted (Frank and Kasper, 1958).

antiprism, the CN15 three-capped truncated trigonal prism, and the CN16 four-capped truncated tetrahedron (Fig. 11). Also the trigonal prism, that may be capped by one, two, or three atoms, occurs frequently in the structures of alloys (Fig. 12).

3.2 Insulators/Semiconductors

The tetrahedron and the octahedron are the most important CN polyhedra of the covalent and ionic insulators/semiconductors (see INSULATORS, STRUCTURE OF). Here the basic covalent structures are the cubic diamond and sphalerite (ZnS), or hexagonal wurtzite (ZnS) types of semiconductors, where each atom is surrounded by a tetrahedron of unlike neighbors so that adamantine arrays of covalent bonds run throughout the crystal in three dimensions (Fig. 13; see DIAMOND AND DIAMONDLIKE CARBON). Many more complex variants of these basic structural types are known. One of the basic ionic types is the rocksalt structure (NaCl), wherein each atom is surrounded by an octahedron of unlike ions

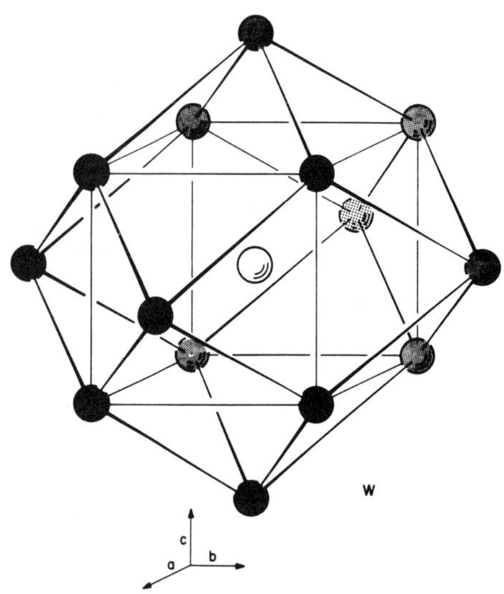

FIG. 10. The dodecahedron, CN14, in the body-centered-cubic (bcc) structure of tungsten (or wolfram W). The 14 neighbors are divided into sets of 8 (corners of cube) and 6 (body centers of neighboring cells).

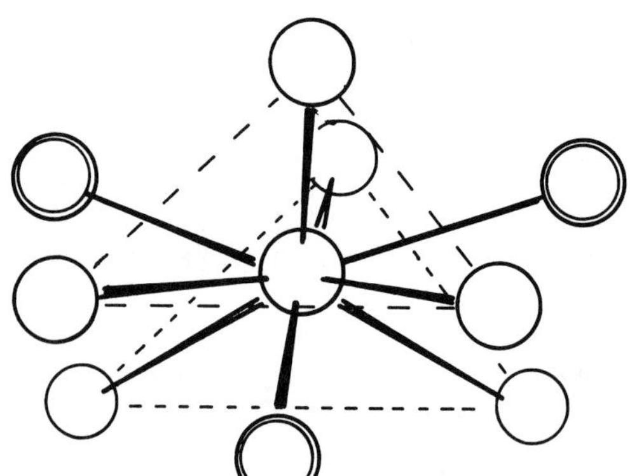

FIG. 12. A tricapped trigonal prism configuration.

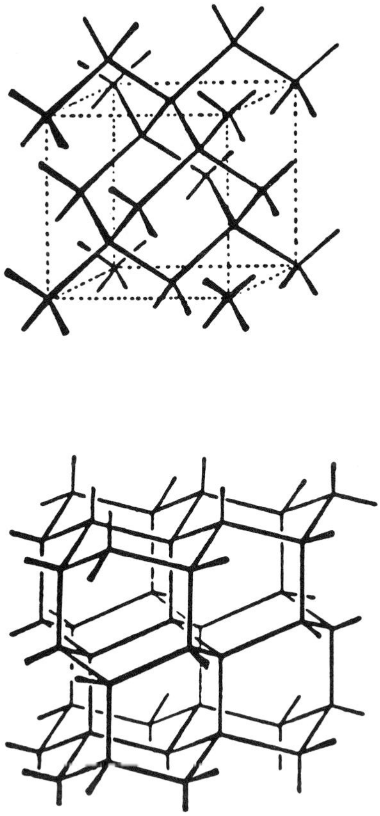

FIG. 13. The tetrahedral bonded framework of diamond or ZnS in the forms of cubic sphalerite (top) or hexagonal wurtzite type (bottom) (Wells, 1962).

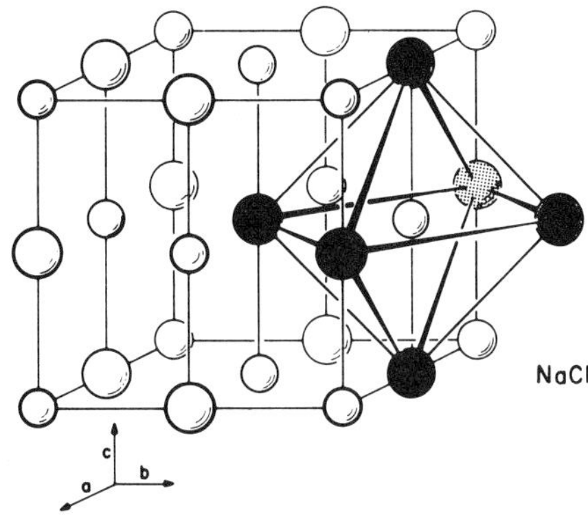

FIG. 14. The rocksalt structure as one of the basic structure types, with one octahedral coordination polyhedron outlined.

(Fig. 14). This arrangement satisfies the central forces of the predominantly ionic bonds. Many complex variants of this basic structure are also known, particularly in oxides, as are very many structures of inorganic compounds and minerals that embrace both tetrahedral and octahedral surroundings, such as spinels (Fig. 15), silicates, sulfosalts, etc. Another basic ionic structure of importance is the CsCl type (identical to the metallic β-CuZn mentioned above) in which each ion is surrounded by a cube of unlike ions. Finally, we draw attention to the frequent occurrence of trigonal prismatic coordination (Fig. 12) also in the structures of insulators/semiconductors.

3.3 Molecular Crystals

These mainly embrace organic materials (substances composed largely of C, H, O, and/or N), but they are also formed by covalent and ionic bonded inorganic materials. Thus the common elemental structures of Se and As form "infinite" molecules that run throughout the crystal in one and two dimensions, respectively (Figs. 16 and 17); the alloy GaTe forms an "infinite" two-dimensional molecule (Fig. 18). Many inorganic substances contain polyatomic anions or polyatomic cations in which a group of atoms is covalently bonded together to form a molecule that has an overall ionic charge, the charges being neutralized in the crystal by other ions of opposite charge: e.g., Na_2TaF_8 (Fig. 19) and $Mo[S_2CN(C_2H_5)_2]I_3$ (Fig. 20).

As well, metallic or insulator structures may contain uncentered clusters of atoms, the

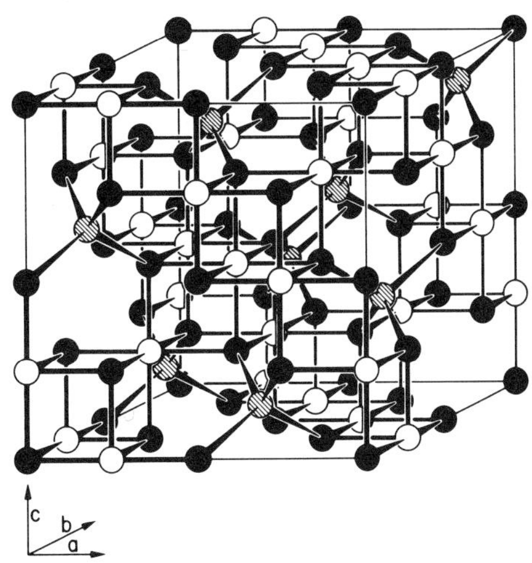

FIG. 15. The spinel crystal structure type.

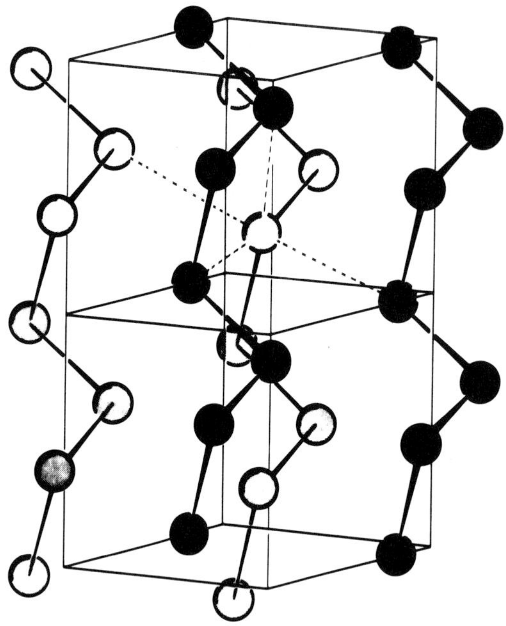

FIG. 16. The structure of Se and Te forming infinite chains in the crystal.

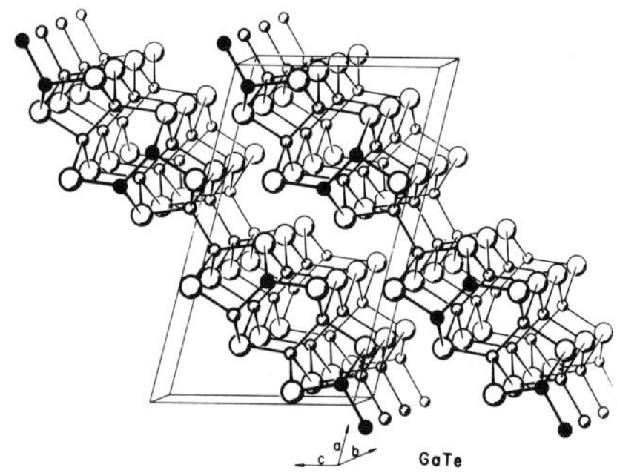

FIG. 18. The infinite-sheet structure of GaTe, with Ga and Te atoms represented by small and large spheres, respectively.

residual valency requirements of which are satisfied by forming occasional covalent bonds to neighboring clusters or intermediate atoms, or by the presence of charged ions in the spaces between the clusters—e.g., boron icosahedral clusters (Fig. 21) or Mo_3Se_4, CN14 (cube+octahedron clusters).

The crystallography of organic molecules is clearly a function of the most efficient packing together of the various molecular shapes as discussed by Kitaigorodski (1963) (see e.g., Fig. 22). However, ionic charges, polarity, and hydrogen bonding are some of the features that can greatly influence molecular packing and affect the cohesion between molecules. Atoms in some organic molecules called *zwitterions* are the sites of distinct positive or negative charges. Such molecules pack together, as far as shape permits, so that negative charges of one molecule attract positive charges of others. The separation of charges is never complete in a bond involving two different elements, and these bonds are said to be *polar*. Molecules usually pack in such a way that their polar branches are attracted to each other, whereas the nonpolar segments also tend to pack together (see Fig. 23). The —OH and —NH groups in a molecule tend to seek negative sites of polar bonds, and *hydrogen bonds* are formed via the positive

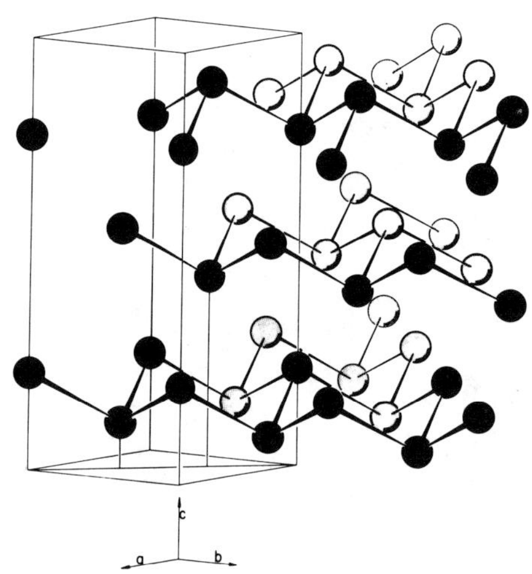

FIG. 17. The structure of sheets of the element arsenic, As.

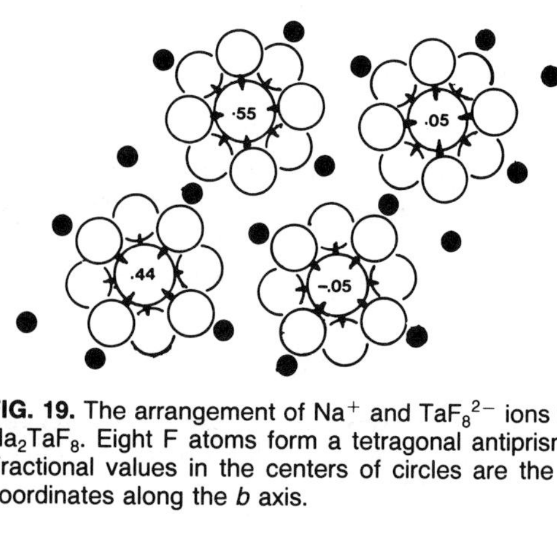

FIG. 19. The arrangement of Na^+ and TaF_8^{2-} ions in Na_2TaF_8. Eight F atoms form a tetragonal antiprism. Fractional values in the centers of circles are the *y* coordinates along the *b* axis.

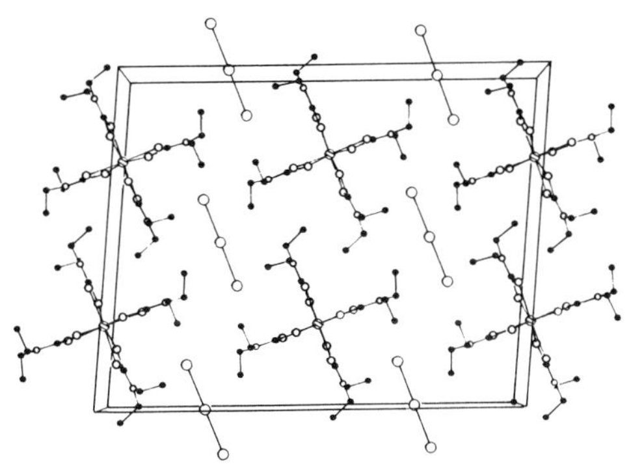

FIG. 20. Large cations $Mo[S_2CN(C_2H_5)_2]^+$ and linear I_3^- anions in the crystal of $Mo[S_2CN(C_2H_5)_2]I_3$.

nature of the H atoms in them. The electronegative atoms O and N in carboxyl —C(=O)OH, hydroxyl —OH, amine —NH and other groups or ions attract these hydrogen atoms. Since molecules may contain many of these groups, their packing together is arranged so as to permit the maximum number of intermolecular hydrogen bonds to form, thus considerably increasing the intermolecular cohesion (Fig. 24).

Finally, there are the structures of very large organic molecules such as proteins and viruses that can be (extracted and) grown in crystalline form for structural analysis. The crystals of these compounds fasten the molecules in rigid positions for x-ray studies. However, the features of their molecular structures are of prime interest. In Fig. 25, a picture of the structure of deoxyribonucleic acid (DNA) is shown.

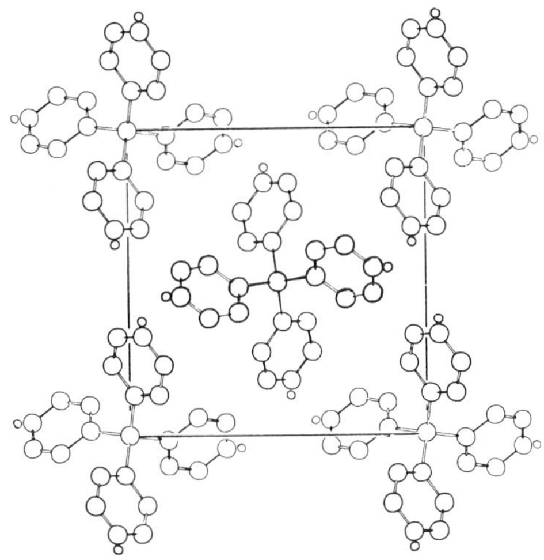

FIG. 22. The packing of tetraphenylmethane in a tetragonal system, a structure common to all tetraphenyl derivatives of the carbon group (C, Si, Ge, and Pb). The molecule in the center is located halfway up along **c**.

It should also be noted that the structures of inorganic polyionic molecules containing water of crystallization (see Fig. 24) can be greatly influenced by hydrogen-bond formation.

3.4 Assessment and Collation of Crystal Structure Data

With the number of determined crystal structures exceeding a hundred thousand, there is an obvious problem of recovering data when required, and perhaps more signifi-

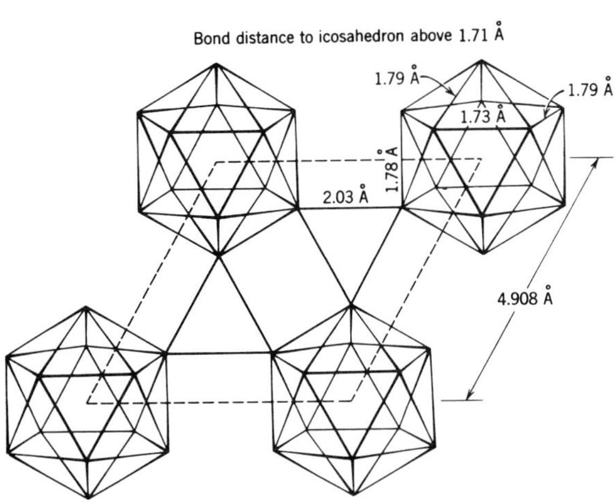

FIG. 21. Predominant icosahedral clusters of boron are the feature of the α-B structure (Decker and Kasper, 1959).

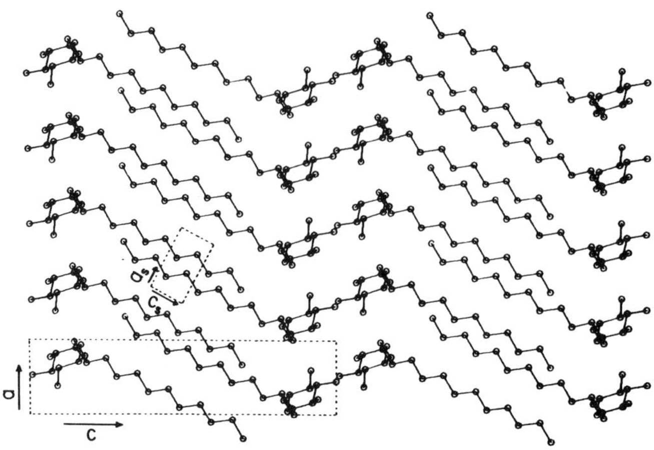

FIG. 23. Molecular packing of 1-decyl α-D-glucopyranoside, $C_{16}H_{32}O_6$, projected along *b*. The polar heads attract each other, leaving the nonpolar chains or tails staying together by themselves. The hydrocarbon subcell also appears in the projection (Moews and Knox, 1976).

cantly of usefully comparing and collating the data at hand. The first problem is now well served by computer data banks that permit retrieval of data under many headings. Of the many such private and public data banks, we shall mention only a few: Cambridge Structural Database (U.K.), organic; National Research Council Canada Metals Data Files (Canada); Gmelin/FIZ Inorganic Crystal Structure Database (Germany); Brookhaven Protein Databank (U.S.A.); and European Molecular Biology Laboratory Database (France).

The second problem of assessment and collation of data by the human mind is much more difficult, since even a mass of computer printouts is, by virtue of its very volume, likely to be intractable. However, in the metal (particularly) and inorganic realms, the concept of a *crystal structure type* developed over the years has proved useful. Although such a type, defined by the crystal's space group and the partially or fully occupied Wyckoff sites therein, may be unique to a single substance, generally it is adopted by several substances and occasionally by a hundred or even several hundred substances. Thus with this concept, the same structure type of many substances can be considered a single unit of information, and can be related to other structure types, and to the observed physical and other properties of its adherents. The advantage of this classification is that it groups together the structures of substances, regardless of their relative unit-cell sizes, relative ratios of the

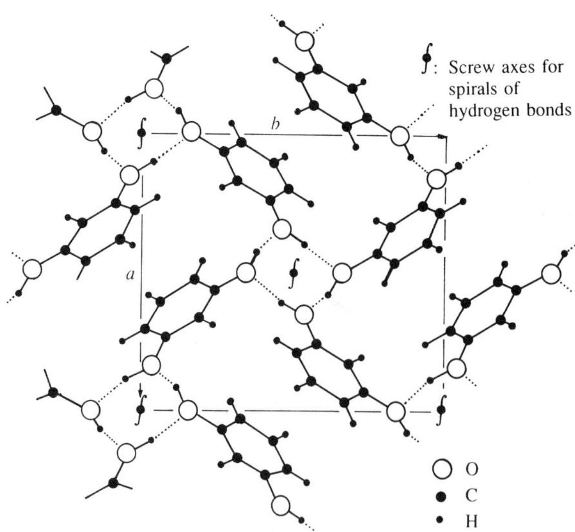

FIG. 24. A projection of the crystal structure of α-resorcinol, $C_6H_4(OH)_2$, showing two hydrogen bonds for each –OH group. The spirals of hydrogen bonds follow the crystallographic screw axes 2_1 which is perpendicular to the paper. The hydrogen positions have been determined by neutron diffraction.

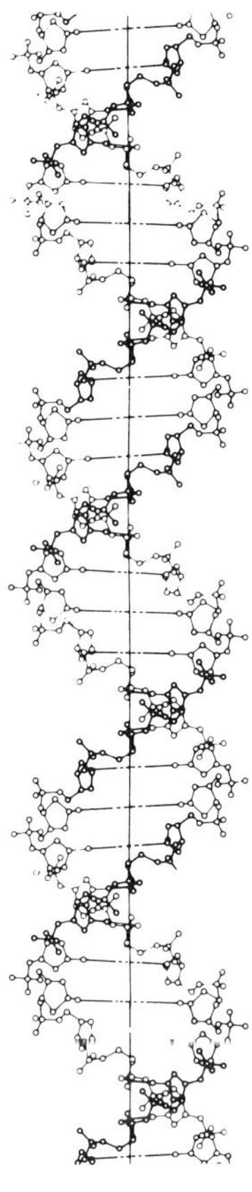

(*b*)

FIG. 25. The structure of DNA, where the hydrogen bonds between the pairs of bases are responsible for keeping the two helices together (Roller-Masser, 1973, where credit is given to M. H. F. Wilkins).

lengths of the unit-cell edges, unit-cell angles, and relative values of variable atomic coordinates x,y,z. Its disadvantage is that it may asssemble together structures in which the coordination of the atoms differs. Generally this consequence is not serious compared to its simplicity and advantages, and it can be taken into account by noting the combinations of structural parameters at which changes of coordination occur.

The advantages of being able to interrelate many different structure types include

1. comparison of structures with similar atomic coordinations that may be significant in relation to particular physical or other properties, and
2. showing that a particular complex structure with a large unit cell is but a superstructure of a much simpler type, the superstructure being caused, e.g., by the ordering requirements of the atoms in the complex structure, or by atomic displacements from regular crystal sites caused by the particular chemical bonding needs of the atoms in the complex structure.

This concept has been basic to many famous books that have been written over the years; some of these are given in the Further Reading list.

4. CRYSTALLOGRAPHY IN INDUSTRY AND MEDICINE

Crystallographic knowledge is basic to the industrial use of materials, which is now known as *materials science*. For example, close-packed cubic metals and alloys are used for construction and mechanical devices because they are isotropic and because the close packing of the equilateral triangular networks of atoms to which the structure gives rise readily permits six active slip systems that facilitate plastic deformation (forming) of a polycrystalline material. Another mechanism for plastic deformation is mechanical twinning, in which the atomic planes slip through a definite fraction of the crystallographic repeat distance, in such a way as to produce a new crystal in twinned orientation relative to the original. Hardening generally results from precipitation of very small crystals of other symmetry in the cubic matrix as a result of suitable adjustment of composition by addition of minor quantities of other elements and annealing at elevated temperatures, followed by rapid cooling (quenching). Information on such precipitates derives from crystallographic studies, as does information on "textures" (crystal orientation) following forming operations. Metals are also important as electrical conductors, a property that arises from the unsaturated metallic bonding or partly filled overlapping energy bands, but which is influenced by interatomic distance, high purity, and perfect (not disordered) occupation of crystallographic sites in intermetallic compounds. Useful magnetic properties of alloys depend on composition ("magnetic" metallic elements present) and particularly on the crystallographic properties of atomic arrangement, coordination, and interatomic distances.

Superconductivity does not have its immediate origins in crystallographic arrangement, but it is interesting to note that all ordinary superconductors ($T_c < 23$ K) with high transition temperatures have one of some half-dozen structure types. Among the new superconductors with much higher transition temperatures, not only is composition important but also structural arrangement and the degree of order or disorder of the components on the crystal sites.

The important electrical properties of inorganic materials (insulators/semiconductors) and their practical use in devices depends not only on composition but also on crystallographic knowledge and crystal-growth techniques for the preparation of highly pure crystals and the creation of graded junctions, etc. Superionic conductivity is directly dependent on crystallographic knowledge, since the structures of such conductors must have linear channels of diameter sufficient for single conductor ions to move through unimpeded on application of an electric field. Magnetic properties of these materials (such as oxides) are, or were, exceedingly important to the recording industry, and structural information has played its part in the development of industrial magnets.

Crystallographic information has been a component in the pharmacological development of "drugs" that have so greatly influenced the modern practice of medicine. However, the greatest triumph of crystallography for modern medicine was Watson and Crick's (1953) determination of the structure of DNA, which resulted also from the concomitant work of many other able crystallographers. This, followed by the determination of the structures of proteins, has lead to the unravelling of the genetic code, the possible recognition of the causes of bodily malfunction that result from defects therein, etc. In addition, diffraction experiments have been used in examining the function of cells and cell membranes.

5. EPILOGUE

This article would not be complete without positive evidence of the importance attached to crystallography by the scientific community, and what better way is there than by listing the commanding number of Nobel Laureates who have been recognized for their achievements related to crystallography. Despite the inexhaustiveness, such a list marks the major milestones since the inauguration of the prize.

Physics:

- 1901 Wilhelm Röntgen, discovery of x rays
- 1914 Max von Laue, discovery of diffraction of x rays by crystals
- 1915 William Bragg and Lawrence Bragg, analysis of crystal structure by means of x rays
- 1929 Prince Louis de Broglie, discovery of the wave nature of electrons
- 1935 James Chadwick, discovery of the neutron
- 1937 Clinton Davisson and George Paget Thomson, discovery of electron diffraction by crystals.

Chemistry:

- 1936 Peter Debye, studies of dipole moments and the diffraction of x rays and electrons in gases
- 1946 James Sumner, discovery of enzyme crystallization
- 1954 Linus Pauling, study of the nature of the chemical bond
- 1962 John C. Kendrew and Max F. Perutz, determination of the structure of hemoproteins
- 1964 Dorothy M. C. Hodgkin, determination of the structure of vitamin B_{12}
- 1976 William N. Lipscomb, revelation of the structure of boranes
- 1980 Paul Berg, Walter Gilbert, and Frederick Sanger, studies of structure and biochemical function of DNA
- 1982 Aaron Klug, determination of the structures of viruses
- 1985 Herbert Aaron Hauptman and Jerome Karle, development of the procedures for the solution of the phase-angle problem in x-ray diffraction
- 1988 Johann Deisenhofer, Robert Huber, and Hartmut Michel, structure determination of a large (more than 10 000 atoms) bacterial protein complex responsible for photosynthesis.

Physiology and medicine:

1962 Francis H. C. Crick, James D. Watson, and Maurice H. F. Wilkins, elucidation of the structure of DNA.

GLOSSARY

Absorption Edge: The sharp difference in absorption coefficients of x rays within a short range of wavelength. The absorption edges are unique for each chemical element.

Anomalous Scattering: The change of atomic scattering amplitude due to a phase-angle shift for scattered x rays from a chemical element in a crystal, particularly noticeable when the element has an absorption edge near the wavelength of the radiation used for the diffraction study.

Crystal Structure: The arrangement of ions, atoms, and molecules in a crystal. The symmetry and geometry exhibited by their arrangement are part of the structural properties.

Crystal Symmetry: The balanced proportions present in cubes, hexagonal or rectangular prisms, and other polyhedra; see **Point Group**.

Diffraction: The interference phenomenon of waves when they travel through a grating in their path. In x-ray diffraction, the crystal is the grating.

Lattice: An infinite periodic array of points at the termini of vectors $\mathbf{L}_{uvw}=u\mathbf{a}+v\mathbf{b}+w\mathbf{c}$, where **a,b,c** are three noncoplanar vectors.

Miller Indices: Three integers enclosed in parentheses (h,k,l) used to represent a set of (parallel) crystal planes. When enclosed in brackets $[h,k,l]$, they represent a direction normal to the plane; and $\{h,k,l\}$ is a form representing a set of symmetry-equivalent planes.

Organic Compounds: Compounds whose molecules are made up mainly of carbon, hydrogen, oxygen, and/or nitrogen atoms. However, in terms of crystal structure classification, these compounds may also contain other elements such as metals, even though they are then called organometallic compounds by chemists.

Phase: A crystalline, liquid, or gaseous state of matter, as used in Gibbs's phase rule or phase transitions.

Phase Angle: A quantity used in the analysis of interferences of waves, for which sinusoidal models are used. However, most crystallographers simply call it a phase. This term is used here to avoid confusion between it and the Gibbs phase.

Point Group: A group consisting of symmetry elements: n-fold rotation axes, mirror planes m, and/or center of inversion i. The property that makes a crystal indistinguishable from its images after being turned (along an axis) by an angle of $2\pi/n$ is called an n-fold rotation axis. When two halves of a crystal are mirror images of each other, the plane separating the halves is a mirror plane m, and a point from which two opposite directions are equivalent is called a center of inversion i.

Space Group: A group consisting of symmetry elements: those of point groups (see above), screw axes, glide planes, and lattice translations.

Symmetry Element: A process that brings the crystal or object back to itself. For example, a threefold rotation is a rotation of the crystal by 120° $(2\pi/3)$. For crystal structure description, symmetry elements includes n-fold rotations, mirror planes m, center of inversion i, glide planes, screw axes, and lattice translations.

Symmetry Group: A point group or space group consisting of symmetry elements. A group consists of a set of elements with a defined group pairing operation, and it satisfies four basic axioms: closure, identity, inverse, and associative; see **Point Group** and **Space Group**.

Unit Cell: The smallest parallelepiped which when stacked repetitively in three dimensions generates the entire crystal structure.

Wyckoff Notation: A letter representing a set of symmetry-equivalent positions in a space group. A general point in a unit cell of a crystal structure belonging to any space group has coordinates x,y,z. However, if one or more of the coordinates has a special value, the number of symmetry-equivalent positions is reduced. Most space groups have several sets of such points, and they are represented by letters a,b,c, etc., which are called Wyckoff notations.

List of Works Cited

Amorós, J. L., Buerger, M. J., de Amorós, M. C. (1975), *The Laue Method*, New York: Academic.

Bacon, G. E. (1975), *Neutron Diffraction*, 3rd ed., Oxford: Clarendon.

Blundell, T. L., Johnson, L. N. (1976), *Protein Crystallography*, New York: Academic.

Buerger, M. J. (1964), *The Precession Method*, New York: Wiley.

Bunn, C. W. (1958), *Chemical Crystallography*, Oxford: Oxford University Press.

Decker, B. F., Kasper, J. S. (1959), *Acta Crystallogr.* **12**, 503–506.

Evans, R. C. (1939), *An Introduction to Crystal Chemistry*, Cambridge, England: Cambridge University Press.

Ewald, P. P. (Ed.) (1962), *Fifty Years of X-Ray Diffraction*, Utrecht: For I.U.Cr., N.A.V. Oosthoek's Uitgeversmaatschappij.

Frank, F. C., Kasper, J. S. (1958), *Acta Crystallogr.* **11**, 184–190.

Glusker, J. P., Patterson, B. K., Rossi, M. (1987), *Patterson and Pattersons*, New York: Oxford Science Publications and International Union of Crystallography.

Kitaigorodski, A. I. (1961), *Organic Chemical Crystallography*, New York: Consultants Bureau.

Hahn, T. (Ed.) (1983), *International Tables for Crystallography*, Vol. A. Boston: International Union of Crystallography and D. Reidel.

Hauptman, H. A., Karle, J. (1953), *Solution to the Phase Problem: I. The Centrosymmetric Crystal*, Ann Arbor: American Crystallographic Association.

Moews, P. C., Knox, J. R. (1976), *J. Am. Chem. Soc.* **98**, 6628–6633.

Reller-Massar, Ann (1973), *Discovering the Basis of Life: An Introduction to Molecular Biology*, New York: McGraw-Hill.

Robertson, J. M. (1953), *Organic Crystals and Molecules*, Ithaca: Cornell University Press.

Watson, J. D., Crick, F. H. C. (1953), *Nature (London)* **171**, 737–738.

Wells, A. F. (1962), *Structural Inorganic Chemistry*, 3rd ed., Oxford: Oxford University Press.

Further Reading

Barrett, C. S., Massalski, T. B. (1966), *Structure of Metals*, New York: McGraw-Hill.

Brice, J. C. (1986), *Crystal Growth Process*, Glasgow: Blackie.

International Tables for X-Ray Crystallography, 3rd ed., Vols. I (1969), II (1972), and III (1968), Birmingham, England: Kynoch.

Jaric, M. V. (Ed.) (1988), *Introduction to Quasicrystals*, Boston: Academic.

Kelker, H., Hatz, R. (1980), *Handbook of Liquid Crystals*, Weinheim: Verlag Chemie.

Pauling, L. (1960), *The Nature of the Chemical Bond*, 3rd ed., Ithaca: Cornell University Press.

Pearson, W. B. (1972), *The Crystal Chemistry and Physics of Metals and Alloys*, New York: Wiley-Interscience.

Villars, P., Calvert, L. D. (1985), *Handbook of Crystallographic Data for Intermetallic Phases*, Metals Park, OH: American Society for Metals.

CVD (CHEMICAL VAPOR DEPOSITION)

BERNARD S. MEYERSON, *IBM Research Division, IBM T. J. Watson Research Center, Yorktown Heights, New York, U.S.A.*

INTRODUCTION

Chemical vapor deposition (CVD) is the process wherein a gaseous source species is decomposed to form a film upon a substrate. This process requires some form of energetic input, whereby the source species is dissociated in the course of film growth. This is distinct from a condensation process, where a vapor is condensed to form a solid, with the source species undergoing only a phase transition. In its most common form, CVD relies upon elevated temperatures to pyrolyze the gaseous precursor for film growth. However, in instances where high growth temperatures are incompatible with the substrate materials, or where metastable materials are to be grown, a wide variety of energetic inputs have been employed to facilitate film growth at reduced temperatures. Examples are plasma excitation (dc, rf, and microwave), photoexcitation (lamp and laser driven), and remote thermal excitation (direct heating of the gaseous source).

CVD is at present the method of choice for the deposition of both mechanical and electronic coatings in the industrial setting. This reflects the numerous desirable characteristics associated with the growth of a material from a gaseous source, key features being excellent film conformality, film uniformity, high crystalline perfection, tight compositional control, high process throughput, and relatively low cost. The intent of this article is to provide an overview of the state of the art in CVD, and to serve as a resource to those who require information on CVD methods for film deposition within a particular material system.

1. CLASSIFICATION BY EXCITATION

CVD processes, independent of the film to be grown, are commonly named employing two distinct strategies. One grouping is done in terms of the excitation source employed, thus the nomenclature plasma-enhanced CVD, photolytic CVD, photothermal CVD, etc. A second grouping is performed in terms of the characteristics of gaseous transport within the

1-56081-063-7/92/$5.00+.50

reaction vessel. First to be considered will be the grouping of CVD processes by excitation source.

1.1 Thermal CVD

The earliest forms of CVD relied upon the simple pyrolysis of a gaseous source to form film, and this remains the most commonly employed method to date. A wide variety of heating methods are employed, those being the use of an isothermal furnace, lamp-heated or rf-heated susceptors (energy-absorbing substrate holders), photothermal methods using lamps or laser radiation, or the Ohmic heating of a substrate itself. Variants include the heating of a gaseous source so as to prepyrolyze it, forming highly reactive radicals such that films may be grown on substrates held at temperatures lower than would otherwise have been required.

1.2 Energetically Enhanced CVD Processes

The second broad category encompasses those methods where electronic excitation is employed to activate film growth chemistry directly. The primary reason for the development of this methodology was the desire to go to ever lower processing temperatures, thus achieving the growth of nonequilibrium layers, and allowing the use of a wider variety of substrate materials. Two subcategories exist, those being techniques that are based upon electronic versus optical excitation.

When electronic excitation is employed, rf, dc, or microwave energy is imparted to a gaseous precursor. This can be a direct process, where the discharge is within the gaseous source itself, or may be remote, exciting another gaseous species which then equilibrates with the growth source above the substrate to be coated. The excited precursor molecules then decompose to form the layers desired. The use of remote excitation is particularly desirable when employing an unstable growth precursor, or depositing upon a substrate sensitive to the high-energy emissions (e.g., UV radiation) nominally present in the region of plasma excitation.

In the instance of photoexcitation, a light source is employed to impart energy to the gaseous precursor. Direct excitation requires that the growth precursor have an adequately strong absorbance at wavelengths readily accessible to conventional light sources. This coincidence is frequently absent, and many photolytic methods must therefore rely upon the use of a chemically inert sensitizing agent. As an example, minute quantities of mercury vapor may be entrained in a gaseous source, which can then be excited employing a high-power mercury light source. Photolytic CVD must be distinguished from simple photothermal methods, where a substrate may be locally heated with a focused IR source or the like. This is simply a thermally driven growth process, modified with the intent to localize film growth to a desired area.

For completeness, it must be noted that there will be as many energetically enhanced growth methods as there are ways to couple other than thermal energy into a gaseous source. An electron beam may also be employed to impart energy to an adsorbed species such that it decomposes to form film, but it and all other excitation methods share the common trait of decoupling the temperature/energy of the gaseous precursor from that of the substrate to be deposited upon.

2. CLASSIFICATION BY REACTANT TRANSPORT

Without regard to the method of excitation employed, a second commonality among the numerous CVD methods known is the mechanism of gaseous transport within the region of film growth. Reactant transport through a CVD apparatus is a key determinant in how one must approach the application and analysis of a given CVD method. The characteristic of gaseous flow through a reactor—viscous flow at the high-pressure limit, molecular flow in the low-pressure limit—determines such critical reactor parameters as reactant residence time, the homogeneity of reactant distribution, reactor memory of past gaseous content, and the rate at which the source species comes to chemical and thermal equilibrium with its environment. For historical reasons, reactant transport considerations have been included implicitly in the categorization of CVD processes by a nomenclature based around commonly employed operating pressures. This scheme is reasonable as gaseous transport and pressure are coupled for a given geometry. As such, the commonly employed terminology for classifying CVD processes is adhered to below. A cautionary note is that the true determinant of how gaseous species are transported within a given reactor geometry is the ratio between the mean free path (λ) and some typical system dimension (D). For example, where $\lambda/D \gg 1$, molecular flow, transport involving only molecule–surface interactions, dominates. It is possible in principle to achieve this at any given pressure, 1 atm for example, so long as D is appropriately small in value. This would be the case for the vapor deposition of a metallic catalyst on the interior of a pore in a catalyst pellet. This is an extreme case, but it is worth noting that this is a limitation of classifying CVD by any one system parameter.

2.1 Atmospheric-Pressure CVD

Atmospheric-pressure CVD (APCVD) processes are the simplest example of growth from a gaseous source. A gaseous source is flowed through a reactor configured as shown in Fig. 1, where a highly dilute gaseous source is made to flow at high velocity through the cooled quartz vessel. Cooling of the vessel is employed to prevent the desorption of contaminants from the vessel walls, to prevent the homogeneous nucleation of particulates due to gas-phase pyrolysis of the precursor, and to suppress the growth of film on the reactor vessel itself (to minimize particle generation). Gas velocity is maintained at a level adequate to prevent the back-diffusion of contaminating species into the reactor at the necked- down end of the apparatus. The simplicity of this apparatus belies the complexities of its chemical and hydrodynamic behavior.

APCVD processes can display complex flow geometries. Flow in an APCVD reactor, shown in Fig. 1, is a strong function of the boundary conditions of the reactor (e.g., wall temperature, gas velocity, entrance and/or exit geometries, susceptor position and temperatures, etc.). Employing the injection of smoke particles to trace flow contours in a horizontal APCVD reactor, early workers (Eversteijn *et al.*, 1970) demonstrated the existence of roll cells, where two symmetric vortices are formed along the reactor's flow axis. As seen schematically in view A of Fig. 1, roll cells spiral such that the source gas rises above the heated substrate, and returns to the bottom of the reactor cross section along the cooled walls. This greatly complicates film growth because of the associated variations of reactant thermal history. Contaminating species, dopant as an example, evaporated from the leading edge of a substrate and carried off in the spiral gas flow, may reincorporate at a point one full rotation of the cell further along the reactor axis, modifying the chemical composition of the film from that which was intended. An additional issue is that of heterogeneous (surface) versus homogeneous (gas-phase) reactor chemistry. In that conditions of viscous flow exist, the fluid stream (growth source) experiences drag due to nearby surfaces, and a slowly flowing layer of gas is found in the region ($\leqslant$1 cm) immediately above the heated susceptor. As the gaseous source in this slowly flowing region comes to thermal equilibrium with the hot susceptor and substrate, significant gas-phase pyrolysis of the source species may occur, forming a more or less reactive sequence of by-products. This leads to spatial variations of gas composition within the reactor, and complicates the attainment of layers of uniform composition. Thus, the proper design of reactor hydrodynamics is a key issue if uniform, reproducible film growth is to be achieved. To this end, extensive modeling of reactor hydrodynamics has been performed.

Gaseous transport within a reactor operating at or near atmospheric pressure has been studied by a number of workers (Coltrin *et al.*, 1984; Jensen *et al.*, 1989). Complex behaviors such as those in Fig. 1 are reproduced in

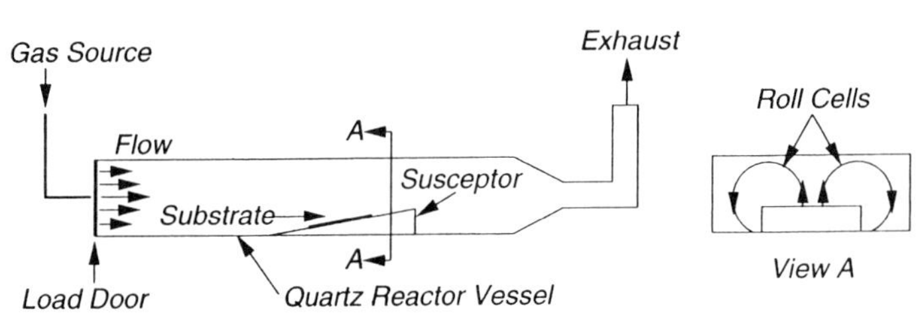

FIG. 1. Atmospheric-pressure CVD system. The tilted susceptor serves to enhance film uniformity. Shown in view "A" are two roll cells, induced by the heating of the gaseous source above the susceptor.

models, but such behavior cannot be generalized in that every reactor geometry produces a unique flow structure. The only generalization made is that of the mathematical formalism employed in the modeling of such complex systems. Hitchman and Jensen (Hitchman *et al.*, 1991) have recently reviewed the state of the art in the area of reactor modeling for a wide variety of reactor configurations.

2.2 Subatmospheric-Pressure CVD

The reduction of operating pressures has been a major factor in the advancement of CVD from a means to grow substrate-like blanket films to a method for the preparation of highly complex layered structures. A liability of many atmospheric-pressure growth processes is, as noted above, the complex hydrodynamic behavior of the system. Reactant residence times in such reactors can be long, in that recirculating flows act as storage nodes for reactants. This leads to an intermixing of layers deposited sequentially, where more abrupt compositional transitions would have been desired. In the instance where a substrate contains volatile species, such as the instance of dopants in silicon, these may be evolved into the gas phase, and reincorporate into film being deposited downstream. To reduce or eliminate such effects, a majority of CVD methods in use today operate in the subatmospheric-pressure regime.

For a fixed flow, an order of magnitude reduction in operating pressure results in an equal reduction of gas residence times within the reactor, facilitating the growth of films with controlled depth-varying compositions. Further, operating at reduced pressures has other fundamental advantages. Low-pressure operation eliminates deleterious flow effects such as the roll cells seen in APCVD. Also, by reduction of gas residence times, undesirable homogeneous chemistry is suppressed, eliminating one mechanism of reactant depletion and/or particle nucleation. When pressures reach the range of 1 to 5 Torr, difficulties encountered with vortices and other "dead spaces" seen in APCVD are absent. The resulting enhancements in the CVD process are remarkable. As an example, one may pack wafers at ever increasing density when gas-phase reactant transport is both rapid and uniform, as is the case at reduced operating pressures. Using an example from silicon technology, consider a process known generically as low-pressure chemical vapor deposition (LPCVD).

LPCVD is a hot-wall CVD growth method, where wafers are packed at high density in a quartz furnace tube. It is common to have 50 or more 125-mm silicon wafers packed coaxially at a spacing of less than 1 cm apart. Shown in Fig. 2, this high packing density makes such systems very efficient. In spite of the high packing density, film thickness across a wafer is uniform to 5% or better by this method for numerous growth processes. Unlike atmospheric pressure CVD, modeling of hydrodynamics in this system is relatively intractable, in that flow in this pressure regime is in transition from laminar, viscous flow to molecular transport (e.g., flow dominated by gas/surface interactions).

A very recent subset of subatmospheric-pressure CVD processes are those operating under conditions where gaseous transport is by molecular flow. In such instances, the assumptions of viscous transport break down, as molecule–molecule collisions become infrequent as compared to molecule–surface interactions. The onset of this condition cannot be specified by a pressure, but rather by the ratio of the molecular mean free path (λ)

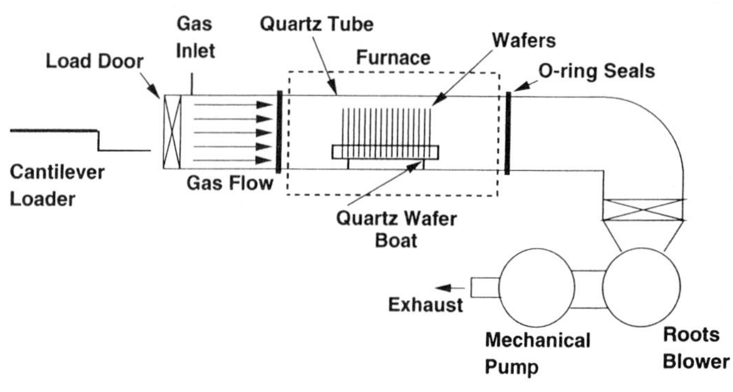

FIG. 2. A schematic of low-pressure CVD as applied to batch processing in silicon technology. High-density wafer packing is possible because of uniform gas distribution at low pressure, although other wafer carrier geometries may be employed.

to a typical system dimension (D), where $\lambda > D$. Using an interwafer dimension of 1 cm, such as in the LPCVD geometry, molecular flow will dominate for $P < 10^{-2}$ Torr. A growth method employing this approach in silicon technology is UHV/CVD (Meyerson, 1986). Similarly, molecular beam epitaxy (MBE) apparatus are being adapted to this approach, where beams of gaseous molecular growth precursors are made to impinge upon heated substrates to form film. This is in contrast to the conventional mode of MBE operation, where a beam of atoms from an evaporative source is employed (see MOLECULAR BEAM EPITAXY). CVD operated at this extreme of pressure creates significant opportunities, in that layer thickness is controllable to better than atomic monolayer dimensions, and applications exploiting this new capability will be discussed further on.

3. CHEMICAL VAPOR DEPOSITION

What has been elaborated thus far is a framework within which various methods of CVD will be discussed. However, it is not within the scope of this article to discuss at great detail any one method of CVD. Rather, this entry is to serve as a resource for workers requiring access to information about a specific growth technique or materials system. A single growth technique, plasma-enhanced CVD (PECVD), may be employed to prepare films of widely varying chemical and structural makeup. Similarly, a single materials system may be deposited employing widely varying deposition processes. Consequently, presented in Tables 1 and 2 are overviews of CVD. Table 1 is broadly organized by CVD growth technique, listing the ranges of operating parameters, and common applications for each method. The exceptionally wide range of operating conditions listed comes about from the large scope of applications for which each method is employed. Application-specific information is found in Table 2, covering silicon-technology-based processes, metal CVD, and compound semiconductor growth by CVD. These entries are organized by material system. It is common for one CVD growth technique to appear under several material systems, it being adapted to the differing requirements by modification of growth chemistry. Noted earlier was the proliferation of acronyms for various deposition processes. Such acronyms frequently attempt to give the reader a hint at the approach to film growth that is being employed. However, a tabulation by either material system or growth method acronym cannot reveal the strategy underlying the development of a given CVD process. For completeness, what follows is a series of entry-level summaries of representative CVD processes in current use.

3.1 Atmospheric-Pressure CVD

Atmospheric-pressure CVD (APCVD) is commonly employed for commercial, high-temperature silicon epitaxy, as well as to prepare materials in many compound semiconductor systems. Producing layers of extraordinary perfection ($\leqslant 10$ defects/cm^2), this method is appealing for its simplicity. High temperatures are employed for substrate preparation, desorbing contaminants, and stabilizing the substrate surface prior to deposition. Examples of this surface preparation chemistry for silicon materials would be the desorption of silicon monoxide at 1000 °C into a hydrogen ambient, while GaAs would be stabilized by a more moderate bake at 800 °C in an arsine-rich hydrogen carrier gas. Growth precursor, diluted in an inert ambient, is then flowed through the cold-walled reactor and pyrolyzed on a heated substrate. High deposition rates are readily obtained, of order several thousands of angstroms per minute. APCVD is primarily employed for the growth of thick, compositionally uniform layers, although work at the research level is under way to lower overall processing temperatures with the goal of producing highly controlled abrupt compositional transitions in film alloy and dopant content. The primary approach to the reduction of the thermal budget for APCVD is to employ rapid thermal processing methods, where samples may be cleaned at high temperatures, followed by rapid temperature reduction to the desired growth temperature. In parallel with this, efforts are under way to operate such reactors in the methodology of UHV, controlling chamber contamination levels by using load-locked introduction chambers and high-purity gaseous sources. Although total pressure remains atmospheric, the partial pressure of contaminating species in such reactors can be reduced to $\simeq 10^{-6}$

Table 1. Operating parameters of commonly employed chemical methods.

Growth technique	Deposition Pressure (Torr)	Deposition Temperature (°C)	Excitation	Film growth applications	Figure	Reference
Atmospheric-pressure CVD (APCVD)	760	150–1200	Thermal	Silicon epitaxy, silicon carbide, III-V and II–VI MOCVD, atomic-layer epitaxy of III–V compounds, tungsten silicide	1	a
Reduced-pressure CVD (RPCVD)	10–80	150–1050	Thermal	Silicon epitaxy, silicon dioxide, silicon nitride, III–V MOCVD	3	b
Low–pressure CVD (LPCVD)	0.2–2.0	150–650	Thermal	Polycrystalline silicon, amorphous silicon, silicon dioxide, silicon nitride, tungsten, copper, aluminum	2	c
CVD in molecular flow	$\leqslant 10^{-3}$	400–800	Thermal	Low-temperature silicon epitaxy, gas-source MBE of silicon and III–V alloys, atomic-layer epitaxy of III–V compounds	4	d
Plasma–enhanced CVD (PECVD)	0.001–0.5	25–800	rf, microwave, ECR	Amorphous silicon, silicon dioxide, silicon nitride, amorphous carbon, diamond, silicon carbide, low-temperature silicon epitaxy	5	e
Photolytic CVD (Photo-CVD)	80	150–600	Photolytic	Silicon dioxide, silicon nitride, low-temperature silicon epitaxy, metal (Cu, Au, W,..,)	6	f

[a]Hammond, 1988; Mayer and Shea, 1964; Eversteijn, 1971; Bloem and Classen, 1980; Beers and Bloem, 1982; Fujino *et al.*, 1990; Giunta *et al.*, 1990; Dobkin *et al.*, 1990.
[b]Ogirima *et al.*, 1977; Reif, 1982; Hammond, 1988.
[c]Hitchman, 1990; Brekel and Bollen, 1981; Meyerson and Olbricht, 1984; Gorowitz *et al.*, 1987; Rojas *et al.*, 1990; Beach *et. al.*, 1990.
[d]Meyerson, 1986; Meyerson *et al.*, 1987; Meyerson *et al.*, 1988b; Donahue and Reif, 1985; Yew *et al.*, 1987.
[e]Donahue and Reif, 1985; Gorowitz *et al.*, 1987; Kamins and Chiang, 1982; Emesh *et al.*, 1989; Wong and Saraswat, 1988; Kim *et al.*, 1987; Batey and Tierney, 1986; Lucovsky *et al.*, 1986; Yew *et al.*, 1987; Manabe and Mitsuyu, 1989; Pai *et al.*, 1990; Oehr and Suhr, 1988.
[f]Yamazaki *et al.*, 1990; Yamazaki *et al.*, 1984; Suzuki *et al.*, 1989; Jansinski *et al.*, 1987; Markwalder *et al.*, 1989.

Table 2. Materials commonly prepared by chemical vapor deposition.

Material deposited	Growth technique	Deposition pressure (Torr)	Deposition temperature (°C)	Precursor	Excitation	Comments	Reference
Silicon (cr)	Atmospheric-pressure CVD (APCVD)	760	800–1100	$SiCl_{4-x}H_x$ ($x \leqslant 4$), B_2H_6, PH_3, AsH_3	Thermal	Commercial silicon epitaxy	a
Silicon (cr)	Reduced-pressure CVD (RPCVD)	10–80	800–1150	$SiCl_{4-x}H_x$ ($x \leqslant 4$), B_2H_6, PH_3, AsH_3	Thermal	Commercial silicon epitaxy	b
Silicon (cr)	Plasma-enhanced CVD (PECVD)	0.001–0.1	750–800	SiH_4, Si_2H_6, B_2H_6, AsH_3	Plasma preclean, thermal growth	Low-temperature epitaxy	c
Silicon (cr) and silicon:germanium (cr)	Rapid thermal CVD (RTCVD)/(LRP)	1.0–760.0	600–1200	$SiCl_{4-x}H_x$ ($x \leqslant 4$), B_2H_6, AsH_3, GeH_4	Thermal	Low-temperature epitaxy 600–750 °C, high temperature preclean ≃1000 °C	d
Silicon (cr) and silicon:germanium (cr)	Ultrahigh vacuum/ CVD (UHV/CVD)	10^{-3}	425–800	SiH_4 PH_3, B_2H_6, GeH_4	Thermal	Low-temperature epitaxy	e
Silicon (cr) and silicon:germanium (cr)	Photolytic CVD (Photo-CVD)	80	250–600	SiH_4 $Si_2H_6PH_3$, B_2F_6, GeH_4	Photolytic	Low-temperature epitaxy	f
Si (polycrystalline)	Low-pressure CVD (LPCVD)	0.3–1.0	500– 700	SiH_4, Si_2H_6, B_2H_6, PH_3, AsH_3	Thermal	Commercial process for poly-Si	g
Silicon (amorphous)	PECVD	10^{-3}–10^{-1}	25–425	SiH_4, Si_2H_6, SiF_4, PH_3, B_2H_6, AsH_3	Plasma (rf, dc, microwave)	Typical process for thin-film transistor fabrication	h
Silicon dioxide / silicon nitride	LPCVD	0.5–2.0	400–700	$Si(C_2OH_5)_4$, SiH_4, O_2, NO_2, NH_3	Thermal	Commercial CVD oxide/nitride growth processes	i
Silicon dioxide/ silicon nitride	PECVD	0.1–1.0	80–400	$Si(C_2OH_5)_4$, SiH_4, O_2, NO_2, NH_3	Plasma (rf, dc, microwave)	Low-temperature nitride/oxide CVD processes	j
Silicon dioxide/ silicon nitride	LPCVD	0.5–2.0	400–700	$Si(C_2OH_5)_4$, SiH_4, O_2, NO_2 NH_3	Thermal	Commercial CVD oxide/nitride growth processes	i
GaAs, GaAlAs, InP, ZnSe, CdS, CdTe	Metal-organic CVD (MOCVD/OMCVD)	10–760	400–800	$Ga(CH_3)_3$, $In(CH_3)_3$, $Al(CH_3)_3$, $Ga(C_2H_5)_3$, $In(C_2H_5)_3$, $Al(C_2H_5)_3$, PH_3, AsH_3, $Zn(CH_3)_2$...	Thermal	Commercial compound semiconductor device and substrate fabrication	k

(Continued).

Table 2. Materials commonly prepared by chemical vapor deposition. (Continued.)

Material deposited	Growth technique	Deposition pressure (Torr)	Deposition temperature (°C)	Precursor	Excitation	Comments	Reference
GaAs, GaAlAs, InP, ZnSe, CdS, CdTe	Atomic-layer epitaxy (ALE)	10^{-6}–760	200–700	$Ga(CH_3)_3$, $In(CH_3)_3$, $Al(CH_3)_3$, $Ga(C_2H_5)_3$, $In(C_2H_5)_3$, $Al(C_2H_5)_3$, PH_3, AsH_3, $GaCl_3$	Thermal	Self-terminating monolayer by monolayer growth	l
Aluminum, copper, gold	Metal-organic CVD (MOCVD)	1–760	150–500	$Al(C_4H_9)_3$, AlH_3-N-$(CH_3)_3$, $Cu(C_6H_7O_2)_2$, $Cu(C_5F_6HO_2)_2$, $Au(C_5F_6HO_2)(CH_3)_2$, $Cu(C_5H_5)P(C_2H_5)_3$	Thermal or photolytic	Low-temperature metal deposition	m
Tungsten, tungsten silicide	LPCVD, APCVD	1–760	200–400	WF_6, $W(CO)_6$, SiH_4, Si_2H_6	Thermal	Commercial deposition of tungsten for microelectronic applications.	n

[a] Hammond, 1988; Mayer and Shea, 1964; Eversteijn, 1971; Bloem and Classen, 1980; Beers and Bloem, 1982.
[b] Ogirima *et al.*, 1977; Hammond, 1988.
[c] Yew *et al.*, 1987; Donahue and Reif, 1985.
[d] Gibbons *et al.*, 1985; Garone *et al.*, 1990.
[e] Meyerson *et al.*, 1988a; Meyerson, 1986; Greve and Racanelli, 1990.
[f] Yamazaki *et al.*, 1984; Yamazaki *et al.*, 1990.
[g] Hitchman, 1990; Brekel and Bollen, 1981; Meyerson and Olbricht, 1984.
[h] Kanoh *et al.*, 1990; Hamaya *et al.*, 1990; Hishikawa *et al.*, 1991; Kushner, 1988.
[i] Rojas *et al.*, 1990; Hitchman, 1990; Matsumura, 1989; Kanoh *et al.*, 1990.
[j] Kamins and Chiang, 1982; Emesh *et al.*, 1989; Kiermasz and Beekman, 1990; Pai *et al.*, 1990; Suzuki *et al.*, 1989; Manabe and Mitsuyu, 1989.
[k] Chen *et al.*, 1989; Bose *et al.*, 1990; Bhat *et al.*, 1990; Mullins *et al.*, 1990; Kuech *et al.*, 1984.
[l] Watanabe *et al.*, 1990; Ohta *et al.*, 1990; Hayafuji *et al.*,; Chiu *et al.*, 1990; Jin *et al.*, 1990; Bedair, 1990.
[m] Saxena and Luehrs, 1990; Beach *et al.*, 1990; Markwalder *et al.*, 1989; Oehr and Suhr, 1988; Kodas *et al.*, 1988; Baum and Jones, 1985.
[n] Gorowitz *et al.*, 1987; Kobayishi *et al.*, 1991; Dobkin *et al.*, 1990.

Torr, with a remarkable improvement (Sedgewick *et al.*, 1991) in film quality at a given growth temperature.

3.2 Reduced-Pressure CVD (RPCVD)

Reduced-pressure CVD (RPCVD) is a term reserved primarily for the subatmospheric-pressure deposition of epitaxial silicon, as shown schematically in Fig. 3. RPCVD is identical to APCVD, other than in the utilization of moderate vacuum (10–80 Torr) to sharpen dopant transition widths in films deposited upon doped substrates. The use of high temperatures and resultant high growth rates are common to both techniques. In the MOCVD (metal-organic CVD) adaptation of this method, it has been shown that the reduction of operating pressures allows the fabrication of layered III–V superlattices, as well as lasers, which have overall quality approaching that found by earlier beam techniques such as MBE.

3.3 Low-Pressure CVD (LPCVD)

Low-pressure CVD (LPCVD) is one of the most versatile film growth methods. Widely used throughout silicon technology, this method operates at significantly lower pressure ($0.2 \leqslant P \leqslant 2.0$ Torr) and temperature ranges ($150 \leqslant T \leqslant 650$ °C) than APCVD or RPCVD.

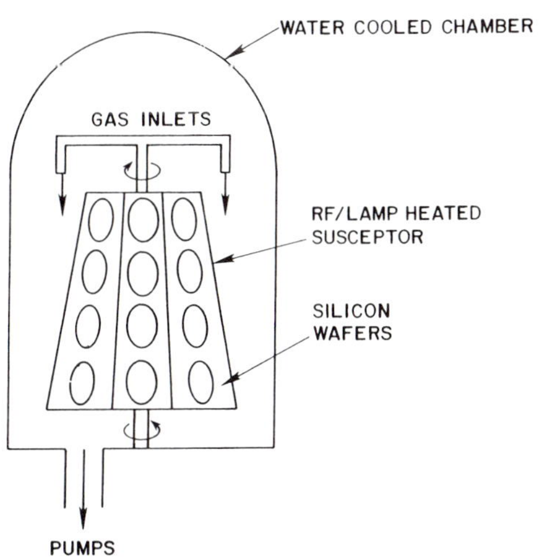

FIG. 3. A typical reduced-pressure CVD reactor as employed in silicon epitaxy or GaAs growth. The rotation of the growth susceptor, gas inlet, and in some cases the individual wafers themselves aids in attaining uniform film thickness and chemical composition.

LPCVD is most commonly utilized to deposit noncrystalline layers such as silicon dioxide, silicon nitride, metals, polycrystalline silicon, etc. This method is favored for these applications by virtue of the coatings being both uniform and conformal when deposited. The level of vacuum employed here is moderate, being of order 1.0 Torr, requiring in general a combination Roots blower and mechanical pump to handle the gas loads ($\simeq$100 std. cm^3/min) at these pressures.

3.4 CVD in Molecular Flow

There are now several CVD growth methods which operate in this low-pressure regime which accords them numerous desirable characteristics, and such processes constitute the most recent entry into the field of CVD. Gas transport for these processes is by molecular flow. The elimination of molecular collisions in the gas phase leads to reduced reactant depletion by homogeneous chemistry, and, where gas–surface interactions are inefficient, can result in films of extraordinary thickness and compositional uniformity. The appropriate view of a reaction chamber operating in this regime is that of a bath, where reactant distribution is completely isotropic. Transition widths between sequentially grown layers, no longer limited by gas-phase memory effects, become abrupt on the atomic scale, making feasible the large-area production of quantum-scale devices. This method has been employed in the III–V material system during chemical beam epitaxy, where a beam of metal-organic precursor molecules is made to impinge upon a heated substrate in a UHV chamber. UHV/CVD, a batch growth process in the silicon system, also employs reactant transport in this flow regime, as seen in Fig. 4.

Atomic-layer epitaxy (ALE) may also be performed under these flow conditions, where a beam is used to create a self-terminating monolayer of material A on a surface, at which time material B is deposited over A to form another self-limiting monolayer of material.

Self-limiting growth processes are most frequently encountered in polar systems, such as GaAs. Under commonly employed growth conditions, an upper bound of one complete monolayer of arsenic may be adsorbed on a free gallium surface. Upon completion of the monolayer, arsenic no longer accumulates, as

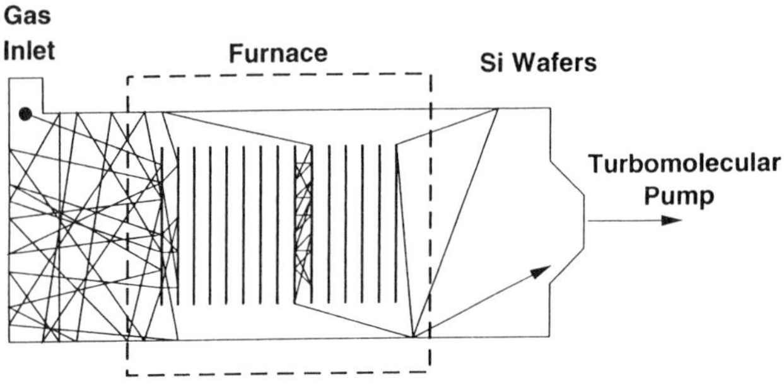

FIG. 4. Molecular transport in a batch-process reactor such as UHV/CVD. In that gaseous transport is to a good estimate similar to that of a particle in a box, concepts such as "upstream" and "downstream" are not applicable.

arsenic clusters are not stable at the temperatures employed in ALE. The gaseous source may then be exchanged for one bearing gallium, again allowing time for saturation coverage to be reached. In a similar way, other III–V species may be introduced with precision, so long as saturation coverage of the growth surface is achieved before the next gaseous source is introduced. This serves to produce artificially structured, highly ordered materials, and is applied most often to III-V systems. In such apparatus, the preferred pumping method is the use of a turbomolecular pump–mechanical pump combination. In processing at such low pressures, the choice of pumping strategy is critical for the following reason.

In the millitorr pressure regime, concepts such as upstream and downstream in a reactor fail. Any contaminant reaching the reaction vessel is uniformly distributed. There is no longer a laminar flow of gas, as in APCVD, which serves to prevent the backstreaming of contaminants. Thus, the maintenance of an appropriately clean reactor environment is determined solely by the nature of the vacuum and gas-handling equipment employed. Turbomolecular pumps, ideally suited to this application, are opaque to the backstreaming of contaminants such as hydrocarbon pump oil, and are unreactive, as pumping is purely by mechanical rather than chemical means as in an ion pump. The use of cryopumping with nominally toxic and explosive growth precursors is hazardous in that condensing out such species accumulates quantities of order many grams over time, which are then released rapidly during regeneration cycles. Thus, great care must be taken when working with CVD processes in this operating regime.

3.5 Plasma-Enhanced CVD

Plasma-enhanced CVD (PECVD) (see Fig. 5) is widely employed where a decoupling of substrate temperature from reactant temperature or energy is required. So that device dimensions may be reduced, semiconductor technologies require that the thermal budget of any given device fabrication scheme be minimized, and thus the trend toward the use of excitation schemes that allow a reduction of substrate temperatures. Plasma excitation for film growth has found application in virtually all aspects of semiconductor growth, insulator growth, metal deposition, and hard-coating deposition. Films are not generally as conformal or uniform as in thermally driven CVD processes, but engineering of discharge geometries has brought these issues to ade-

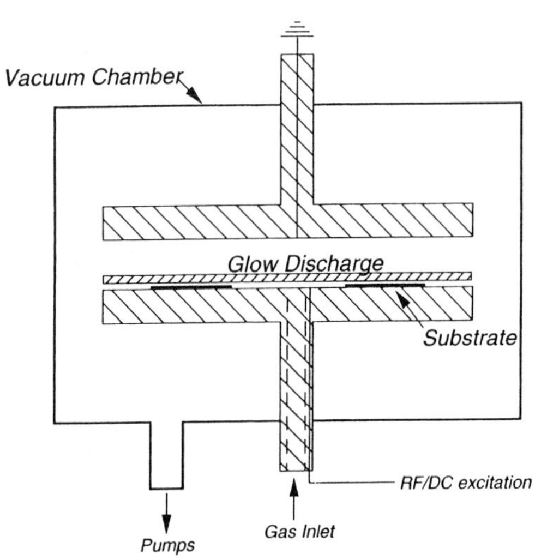

FIG. 5. Parallel-plate rf/dc plasma-enhanced CVD apparatus. Note that although wafers are shown on the powered electrode, this is not universally the case. Further, a dc bias voltage is frequently applied in addition to the rf excitation signal.

quate levels of perfection. A liability of plasma-enhanced methods is that the energies employed are adequate both to fragment the growth precursor, and to sputter impurities from chamber surfaces. Various means have been developed to minimize this effect, one being the confinement of a plasma via electron cyclotron resonance methods, or the formation of a plasma in a clean, remote chamber, from which a beam of excited species is extracted. In spite of some difficulties, PECVD is prevalent throughout the research environment, and in numerous commercial applications as well.

3.6 Photolytic CVD

Photolytic CVD (see Fig. 6) is an example of an alternative strategy for decoupling precursor temperature or energy from the temperature of the substrate. In contrast to PECVD,this method enables one to localize film growth, a trait utilized in most of the applications of this method. Direct photolysis of a source species, such as a metal-organic precursor, is frequently employed in industry to perform local repairs of circuit wiring or lithography masks. Although not well suited to large-area film growth, the prospect of directly writing a circuit by this method maintains a significant level of interest in this area. There are several methods of photo-CVD which employ large- area, high-fluence light sources to perform blanket deposition on wafer surfaces, but the control of film uniformity, and difficulties in eliminating undesirable film growth on chamber optical ports, have slowed the widespread implementation of this method.

4. MODELS—EQUILIBRIUM THERMODYNAMICS VERSUS KINETICS

Modeling of CVD is an active field, where general, accurate, predictive models are noticeably absent. The inability to generalize modeling of CVD comes about through the enormous complexity of the chemistries employed, coupled with the lack of fundamental data required to complete any given model. Moreover, before one even attempts to model a given CVD process, a basic choice must be made as to the approach to be taken. In the limiting case of a closed system, at low growth rates and high temperatures, it is frequently assumed that such a system is in a near-equilibrium state and may be characterized using thermodynamic equilibrium considerations. In the opposed limit of low-temperature, high-flow systems, it is often assumed that reaction kinetics are dominant (rate limiting). Both strategies have been employed in characterizing the CVD process. Hess *et al.* (1985) have recently reviewed this issue.

4.1 Equilibrium Thermodynamics

Equilibrium thermodynamics is frequently employed in the analysis, at least qualitatively, of a CVD process. It is attractive for its simplicity, as details of the exact growth mechanism drop out of the calculation. Rather, one must identify the number n_i of each atomic and molecular species present, as well as the chemical potential μ_i of each species. The Gibbs free energy of the system at a single operating point with well-defined temperature and pressure is simply

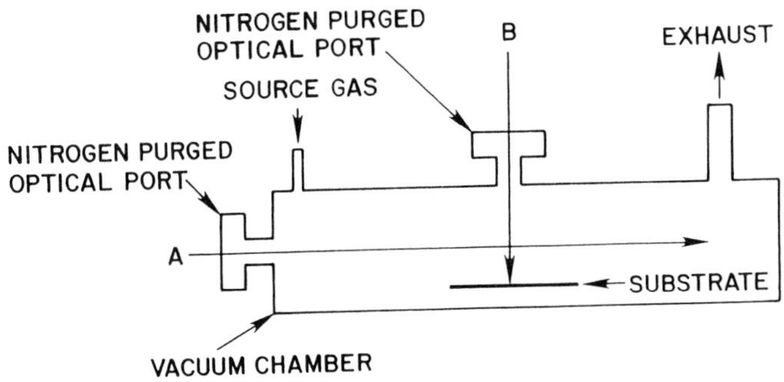

FIG. 6. Photo-CVD reactor shown in two configurations. Optical input via "A" is employed to produce gas-phase reactive intermediates for film growth over large areas. Input "B" is employed for local photothermal or photolytic CVD, allowing direct writing of features upon the substrate.

$$G = \sum_{i=1}^{i=N} n_i \mu_i .$$

Minimization of G for a given set of operating conditions, constrained by the initial values for species present, allows the calculation of the final distribution of species after equilibration with a new environment. The ability to compute the equilibrium distribution of chemical species in the reactor environment allows one to manipulate film growth in the following manner.

In the instance of silicon epitaxy, such a calculation has been performed by Bloem and Giling (1985), for the case of deposition in the silicon:hydrogen:chloride system. Figure 7 displays the calculated distribution of gaseous species present in equilibrium with solid Si over a span of temperatures, holding the Cl/H ratio fixed. From this data, one can compute the total solubility of Si in this gaseous mix when in contact with a silicon surface, shown in Fig. 8. The solubility of Si in the gas phase is limited by the atom count of silicon that may be bound in the computed distribution of hydride and halide species. As an example, for a fixed input of chlorine, conditions favoring

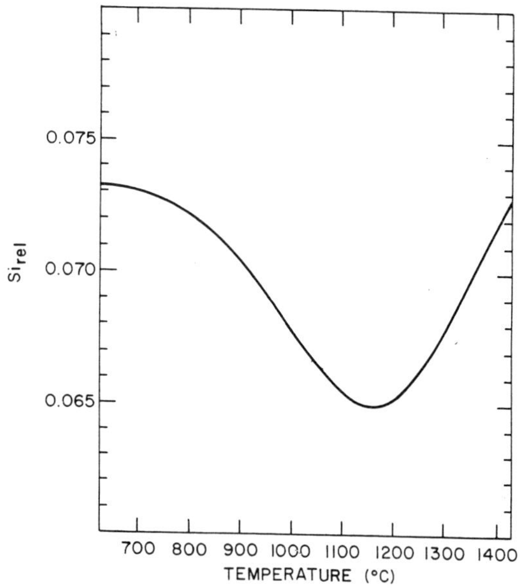

FIG. 8. Relative Si solubility in the gas phase as a function of temperature (see the text). After Bloem and Giling (1985).

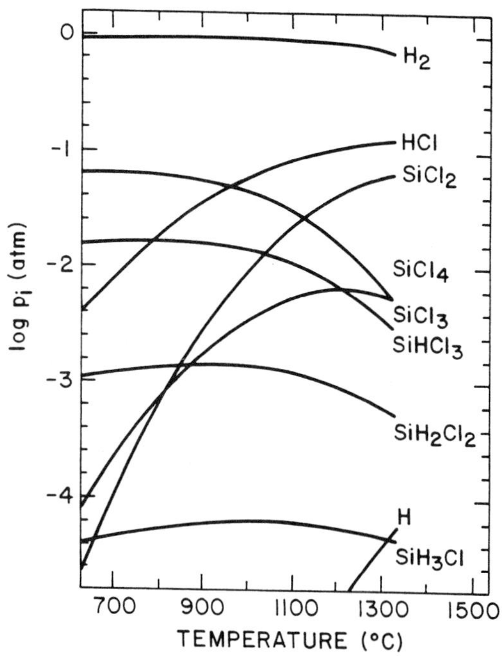

FIG. 7. Calculated equilibrium distribution of gaseous species during APCVD of epitaxial silicon using silicon chloride source species. Data after Bloem and Giling (1985).

the formation of hexachlorodisilane (Si_2Cl_6), at a resultant Si:Cl ratio of 1:3, permit the gaseous source to support a higher silicon fraction than conditions favoring the formation of silicon tetrachloride ($SiCl_4$), at a Si:Cl ratio of 1:4. Thus, using temperature to alter the distribution of silicon-bearing halide species, one may control film growth under near-equilibrium conditions. For example, when the Si-H-Cl chemical system is at equilibrium at 1300 °C, and then flowed into a reactor region held at 1100 °C, the Si content of the gaseous mix must decrease (see Fig. 8) to remain in thermodynamic equilibrium. The system therefore returns to equilibrium by reducing total silicon content by the deposition of Si onto any preexisting Si surfaces. The degree of Si supersaturation that is induced by the rapid temperature drop determines the resulting film growth rate. If this gas mixture were now allowed to equilibrate at 1100 °C, a subsequent reduction in temperature from 1100 to 800 °C would result in the etching of a silicon substrate placed in this flow. This is seen in Fig. 8, where the solubility of Si rises again at 800 °C. This ability to create controllably a supersaturation of Si in the gas phase has been widely employed in conducting selective Si epitaxy. Surfaces other than silicon, such as silicon dioxide, require a finite, non-

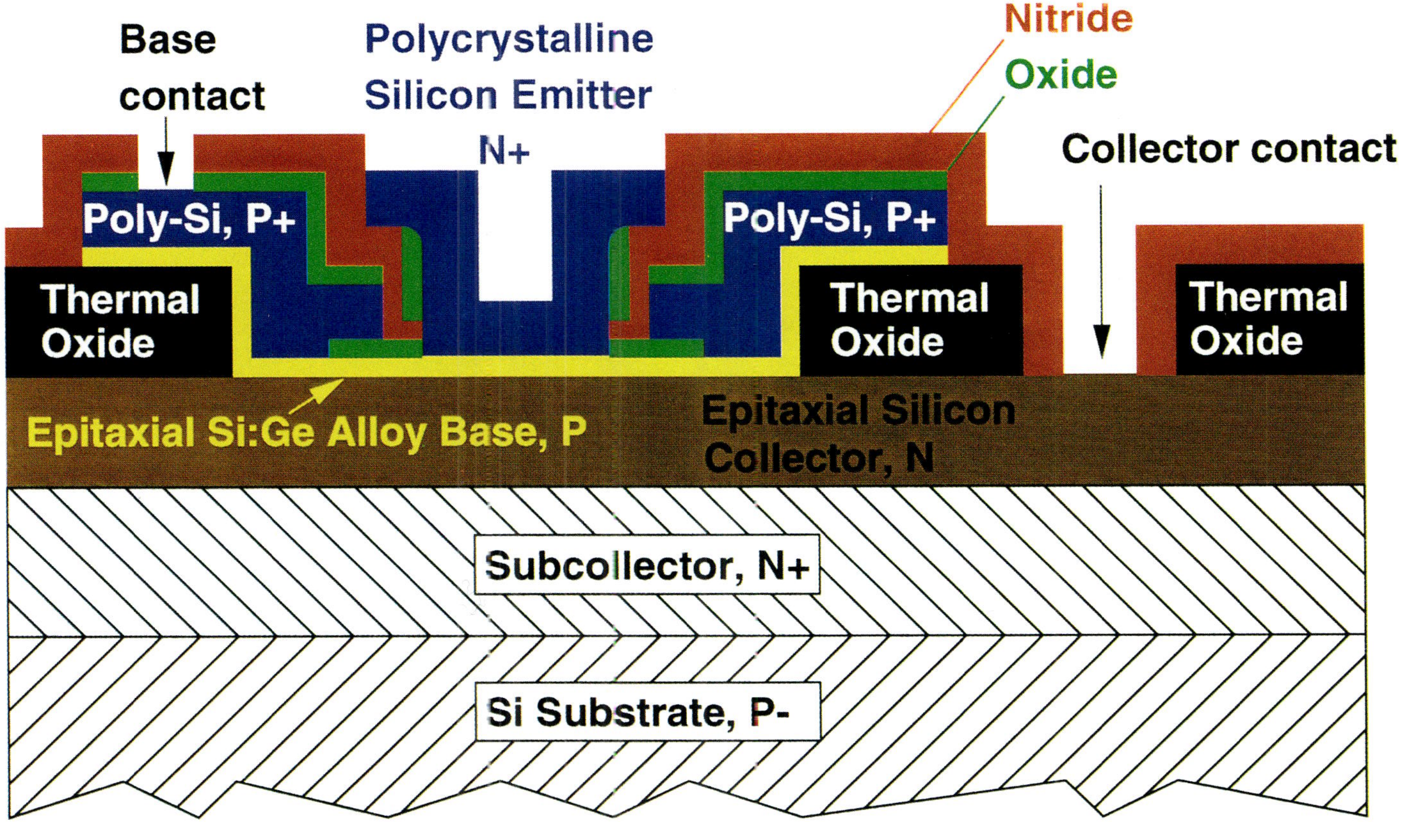

PLATE 1. A cross-sectional view of a high-performance epitaxial base heterojunction bipolar transistor. All features other than those in black and white are formed using various CVD processes.

zero supersaturation of silicon such that film nucleation and growth proceeds. By contrast, silicon growth on silicon has no minimum requirement on supersaturation for growth to proceed. It is therefore possible to induce selective growth of silicon on silicon for small supersaturations, while silicon dioxide present in the same environment remains free of film. The trends predicted by equilibrium thermodynamic considerations as given in this example are indeed observed, but quantitative agreement between experiment and theory is rare. Further, a cautionary note is due here.

Another system where one may attempt to apply considerations as above would be the growth of III–V materials—e.g., GaAs. One of the difficulties in the growth of GaAs by MOCVD is the unintended incorporation of carbon into GaAs from the Ga precursor molecule, $Ga(CH_3)_3$. If one replaces trimethylgallium with triethylgallium, $Ga(C_2H_5)_3$, the ratio of carbon to gallium increases, and, to first order, one would expect an increase in film carbon content. The opposite occurs, as the use of triethylgallium greatly reduces the formation of surface-active methyl radicals, CH_3, and thus eliminates what was found to be an efficient source for carbon in GaAs. Using only equilibrium thermodynamic considerations to model this chemistry, even qualitatively, would predict the wrong result. This points to the power of having available a full chemical description of a reactive system, so that one may explicitly look at the dynamics of growth chemistry, rather than the equilibrated state, which may not in fact be attained in the time that a mixture is present within the reactor environment.

4.2 Kinetics-Based Reactor Modeling

In contrast to equilibrium-based models, models based upon the detailed chemistry of growth are by definition quantitative, and as perfect as one's choice of chemical model and one's knowledge of all relevant reaction rates. For precisely this reason, models based upon kinetic considerations are for the most part qualitative, and are not suited to general, predictive use. This seeming contradiction results from our lack of knowledge at present as to the exact reaction pathways and the relevant reaction rates for most CVD systems of interest.

The methodology of kinetic modeling requires that one set forth a chemical pathway by which a film is to be deposited. This must include all those reactions postulated to be relevant to film growth, both heterogeneous and homogeneous. The form such a chemical system would take for homogeneous chemistry during Si epitaxy from silane (SiH_4) could be

$$SiH_4 \rightarrow SiH_2 + H_2, \tag{1}$$

$$SiH_4 + SiH_2 \rightarrow Si_2H_6, \tag{2}$$

$$SiH_2 + Si_xH_{2(x+1)} \rightarrow \rightarrow \rightarrow Si_{x+1}H_{2(x+2)} (x \geqslant 2) \tag{3}$$

.

.

.

The heterogeneous component would be described in detail as well, with adsorption (4), heterogeneous decomposition (5), and product evolution (6) represented as follows, where $*$ denotes an active adsite:

$$SiH_4(\text{gas}) + * \rightarrow SiH_4(\text{ads.}), \tag{4}$$

$$SiH_4(\text{ads.}) \rightarrow SiH_x(\text{ads.}) + (4-x)H(\text{ads.}), \tag{5}$$

$$SiH_x(\text{ads}) + (4-x)H(\text{ads}) \rightarrow Si(\text{cr}) + 2H_2(\text{gas}) + * \; (x \leqslant 3). \tag{6}$$

The six reactions enumerated here are a small fraction of the several hundred that need to be and have indeed been considered as part of such a model. It is found, via sensitivity analysis, that many reactions are superfluous to the growth process, and can be eliminated. However, this process relies upon one's making "reasonable" guesses as to the rate constants, forward and reverse, for each of the reactions enumerated above. Thus, though chemical kinetic modeling is potentially an exact science, it suffers from the lack of an adequate database for chemistries of general interest. In that the silane system is among the most intensely studied, Coltrin *et al.* (1984) and Jensen *et al.* (1989) have carried out the full finite-element treatment of silane-based CVD, solving a coupled hydrodynamic and chemical code from the system of reactions as above.

A more common application of kinetic modeling to CVD has been the study of rate-limiting phenomena during film growth. In this simple application of kinetic modeling, one prepares a skeletal series of growth-related chemical reactions, similar to the framework above, but assuming foreknowledge of the active growth species (e.g., the parent molecule, a homogeneously generated radical species, etc). Using what is assumed to be an accurate chemical model, one then derives the functional dependence of film growth rate upon a series of system parameters, such as reactant partial pressure, temperature, and total pressure. This approach will highlight those reactions that are rate-limiting bottlenecks in the film growth process, and has the longest history of application (Bloem and Claasen, 1980) in the analysis of CVD. Upon completion of a model, validation is by experimental verification of the projected dependence of growth rate upon process parameters. However, it must be understood that this does not prove the model to be correct, in that there is no assurance that the model assumed is unique. In spite of this considerable flaw, such early forms of kinetic modeling were frequently useful as qualitative guides as to the optimization of silicon-hydride- and -halide-based CVD.

5. APPLICATIONS

The applications of CVD are as varied as all those applications in which thin-film deposition is required. The current view of where CVD is the method of choice for film deposition is for applications that are mature, where large-area or high-throughput thin-film fabrication is required for commercial products. This is in contrast to beam techniques, notably MBE, where the emphasis is on research and early-stage development of materials and structures. Though this broad assessment has been true in the past, numerous recent applications of the CVD method have shown that the earlier designation of CVD as more of an industrial than a research tool is changing rapidly.

5.1 Passive Circuit Elements and Structures

In technology, CVD is employed to the exclusion of other growth methods in those instances where conformal coatings are required. An extreme example of the conformal nature of CVD growth is shown in Fig. 9. Shown is an advanced trench capacitor, the heart of a standard computer memory chip, and among the most common electronic components in the semiconductor industry. The aspect ratio of the trench shown is 16:1 depth to width. Fabrication consists of the use of anisotropic etching to form a periodic array of trenches in a silicon substrate. The walls of the trench are first thermally oxidized, then further insulated by the deposition of a thin layer of CVD silicon nitride. The trench is then filled conformally by the LPCVD growth of polycrystalline silicon, resulting in the trench capacitor structure shown here. The utilization of the inherent conformality of the CVD process is pervasive. Other applications are the CVD of hard coatings on tool bits employed in high-speed machining, scratch-resistant or antireflection coatings on complex optics, and the use of chemical vapor infiltration (filling of voids in woven carbon fibers in similar fashion to trench filling) to produce lightweight carbon composite parts such as disk-brake rotors.

5.2 Active Circuit Elements and Device Structures

CVD is most widely employed for the conformal growth of passive (e.g., dielectric) device and/or structure components as above. However, CVD is making rapid advances in the deposition of critical electrically active layers. MOCVD is now making inroads into the fabrication of III–V devices, and is of increasing importance at present as III–V (GaAs/GaAlAs) devices and integrated circuits are being commercialized. A more common application of MOCVD is for the fabrication of optically active devices, primarily LEDs and lasers, again in the commercial environment.

CVD, as applied in silicon device technology, has recently come into its own as a method by which active device regions themselves are formed. In existing silicon technology, CVD has long been employed as a means to prepare highly uniform substrate-like material upon which active device regions were defined by either diffusion or ion implantation. For the growth of substrate-like blanket epitaxial layers, straightforward high-temperature epitaxial growth methods were employed—e.g., APCVD and RPCVD. In the past five years, there has been a virtual revo-

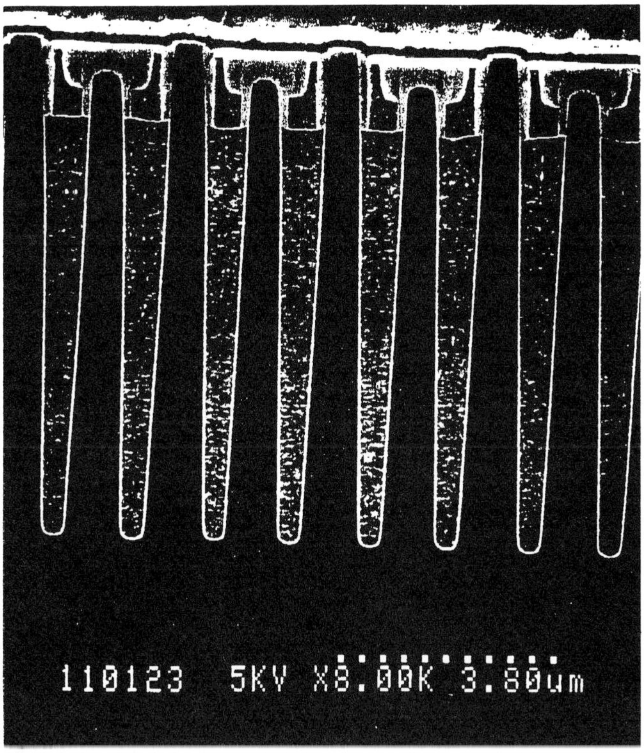

FIG. 9. The conformal nature of the CVD process was employed to fill the high-aspect-ratio features of this deep trench capacitor, the storage node in a modern dynamic random-access memory (DRAM) cell. These trenches are lined with a CVD silicon oxide/silicon nitride dual dielectric, and then filled by the CVD of polycrystalline silicon. The center-to-center spacing between neighboring pairs of trenches is 2.8 μm.

lution in silicon-based devices as a consequence of the application of CVD to this field.

With the advent of several methods of low-temperature silicon epitaxy as enumerated in Tables 1 and 2, new device designs have been implemented, employing *in situ* dopant incorporation, and utilizing as-deposited layers as the active regions of various bipolar and FET structures. The freedom of device design, and the precise control made possible, have rejuvenated the field of Si-based device design. Plate 1 is an example of a Si:Ge-based heterojunction bipolar transistor, and most noteworthy is that all colored regions of the plate represent layers prepared by CVD. Virtually all active and passive device components above the substrate are prepared by CVD, primarily because of the precision possible and the low thermal budget associated with those processes. By working at temperatures below those associated with dopant diffusion, $T \leqslant 700$ °C, the integrity of deposited dopant and alloy profiles is maintained throughout device fabrication.

As a result of advances in device design and materials deposition techniques, presumed limitations on the performance of Si-based devices have been overcome through the application of many of the band-structure tailoring methods applied earlier in the III–V system. One example of this trend was the recent fabrication, by UHV/CVD, of high-performance Si:Ge heterojunction bipolar transistors, having a unity-gain cutoff frequency of 75 GHz at room temperature (Patton *et al.*, 1990), and 94 GHz at 77 K (Crabbe *et al.*, 1990). In these devices, the nominally implanted base region of a bipolar transistor was replaced by a boron-doped, tailored Ge profile in Si. The Ge profile was tailored so as to produce a conduction band grading to accelerate electrons across the base region of the device, resulting in a doubling of previously observed Si device speed, and exceeding the projected limits of Si-based device technology.

5.3 Fundamental Materials Studies

At the forefront of the applications of CVD are those applications where film preparation is being performed (or attempted) for the creation of artificially structured materials.

The intent here is to generate, via atomic layering, materials having unique electronic and optical properties. In the least demanding (but hardly trivial) implementation, quantum-effect devices are fabricated, resonant tunneling diodes being an example (Chang *et al.*, 1983). Dimensional control in such structures must be on the order of a few atomic lengths along one dimension, film thickness in general. This field, heretofore dominated by beam methods (MBE), is now being broadened by the application of CVD methods, showing equal or better dimensional and compositional control in some instances. In the III–V system, atomic-layer epitaxy is employed to prepare structures with atomic-layer control along the growth axis, allowing the fabrication of artificially structured materials. In the case of silicon, recent work reporting the first electron-based double-barrier resonant tunneling diodes fabricated in the silicon system was CVD based, and room-temperature operation of these quantum-effect devices was demonstrated (Ismail *et al.*, 1991).

In the most aggressive application of materials fabrication, one attempts to alter the fundamental band structure of a given material system by appropriate interleaving of constituent atomic layers. This is going a step beyond quantum electronics, in that the operation of a given quantum-effect-based device is governed by the local electronic structure of the host material. Thus, one attempts to create a unique function in known device structures by alteration of the fundamental nature of the material system employed. An example is ongoing work where strain-symmetrized Si and Ge layers are interleaved in an effort to produce "zone folding," the artificial alignment of the valence-band maxima with the conduction-band minima, such that the material becomes a direct-gap semiconductor. The modification of this group-IV alloy system to produce structures with efficient optical activity would open the way for Si-based integrated optoelectronics, in the manner of the far more successful III–V-based systems. In that the outcome of this effort is far from clear, new growth techniques, CVD among them, are being brought to bear to facilitate the preparation of such demanding structures.

6. SUMMARY

What has been discussed in this article is an overview of a field that is evolving at an extremely rapid pace. Although CVD has long been employed in a variety of technical applications, it is now being widely employed within the scientific community as well. Recent advances in the field have allowed atomic-length scale control of materials properties with both abruptness and dynamic range heretofore unavailable, resulting in materials, structures, and devices having unique physical and electronic structure. There are individual articles in the *Encyclopedia* which focus on specific CVD methods discussed in broad terms above, and it is recommended that this text serve as an entry point to the field of chemical vapor deposition.

GLOSSARY

Adsite: Adsorption site at which an impinging species may bind to a surface.

Band Structure: The fundamental electronic structure of a given material, determining the transport properties of carriers in that material.

HBT: Heterojunction bipolar transistor, a bipolar transistor utilizing dissimilar semiconductor materials at its emitter-base or base-collector junctions.

Heterogeneous Chemistry: Reactions occurring at a gas/surface interface.

Homogeneous Chemistry: Reactions occurring in a single reactant phase, that being the gas phase as used in this article.

LED: Light-emitting diode.

Photolysis: Decomposition induced by the electronic energy imparted to a species by incident light.

Radical: A molecular fragment not fully coordinated, which is therefore highly reactive.

RTD: Resonant tunneling diode.

Susceptor: Energy-absorbing substrate holder employed inside a growth chamber to couple to an external energy source.

UHV: Ultrahigh vacuum, generally defined as pressure less than or equal to 10^{-9} Torr.

List of Works Cited

Batey, J., Tierney, E. (1986), *J. Appl. Phys.* **60**, 3136–3146.

Baum, T. H., Jones, C. R. (1985), *Appl. Phys. Lett.* **47**, 538–540.

Beach, D. B., LeGoues, F. K., Hu, T.J. (1990), *Chem. Mater.* **2**, 216–2I9.

Bedair, S. M., (1990), *Acta Polytech. Scand. Chem. Technol. Metall. Ser.* **Ch195**, 17–37.

Beers, A. M., Bloem, J. (1982), *Appl. Phys. Lett.* **41**, 153.

Bhat, R., Kapon, E., Werner, J., Hwang, D. M., Stoffel, N. G., Koza, M. A. (1990), *Appl. Phys. Lett.* **56**, 863–865.

Bloem, J., Claasen, W. A. (1980), *J. Cryst. Growth* **49**, 435–444.

Bloem, J., Giling, L. J. (1985), *VLSI Electron. Microstruct. Sci.* **12**, 89–139.

Bose, S. S., Szafranek, I., Kim, M. H., Stillman, G. E. (1990), *J. Appl. Phys.* **56**, 752–754.

Brekel, C. H., Bollen, L. J. (1981), *J. Cryst. Growth* **54**, 310–322.

Chang, L. L., Esaki, L., Tsu, R. (1983), *Appl. Phys. Lett.* **43**, 588.

Chen, W. K., Chen, J. C., Anthony, L., Liu, P. L. (1989), *Appl. Phys. Lett.* **55**, 987–989.

Chiu, T. H., Cunningham, J. E., Robertson., Jr., A., Malm, D. L. (1990), *J. Cryst. Growth* **105**, 155–161.

Coltrin, M. E., Kee, R. J., Miller, J. A. (1984), *J. Electrochem. Soc.* **131**, 425–434.

Crabbe, E. F., Patton, G. L., Stork, J., Meyerson, B. S., Sun, J. (1990), *1990 IEDM Tech. Dig.* **17**.

Dobkin, D., Bartholomew, L., McDaniel, G., DeDontney, J. (1990), *J. Electrochem. Soc.* **137**, 1623–1626.

Donahue, T. J., Reif, R. (1985), *J. Appl. Phys.* **57**, 2757.

Emesh, I. T., Dasti, G., Mercier, J. S., Leung, P. (1989), *J. Electrochem. Soc.* **136**, 3404–3408.

Eversteijn, C. (1971), *Phillips Res. Repts.* **26**, R757.

Eversteijn, F. C., Severin, P. J., Brekel, C. H., Peek, H. L. (1970), *J. Electrochem. Soc.* **117**, 925.

Fujino, K., Nishimoto, Y., Tokumasu N., Maeda, K. (1990), *J. Electrochem. Soc.* **137**, 2883–2887.

Garone, P. M., Sturm, J. C., Schwartz, P. V., Schwarz, S. A., Wilkens, B. J. (1990), *Appl. Phys. Lett.* **56**, 1275–1277.

Gibbons, J., Gronet, C., Williams, K. (1985), *Appl. Phys. Lett.* **47**, 721–723.

Giunta, C. J., Chapple-Sokol, J. D., Gordon, R. G. (1990), *J. Electrochem. Soc.* **137**, 3237–3253.

Gorowitz, B., Wilson, R. H., Gorczyca, T. B. (1987), *Solid State Technol.* **30**, 97–103.

Greve, D. W., Racanelli, M. (1990), *J. Vac. Sci. Technol. B* **8**, 511–515.

Hamaya, H., Murai, H., Kobayashi, K. (1990), *J. Appl. Phys.* **67**, 1356–1360.

Hammond, M. L. (1988), *Solid State Technol.* **31**, 159–164.

Hayafuji, N., Miyashita, M., Kumabe, H., Murotani, T. (1990), *J. Cryst. Growth* **106**, 421–425.

Hess, D. W., Jensen, K. F., Anderson, T. J. (1985), *Rev. Chem. Eng.* **3**, 99–186.

Hishikawa, Y., Tsuge, S., Nakamura, N., Tsuda, S., Nakano, S., Kuwano, Y. (1991), *J. Appl. Phys.* **69**, 508–510.

Hitchman, M. L. (1990), *Vacuum* **41**, 880–884.

Hitchman, M. L., Jensen, K. F. (1991) *Chemical Vapor Deposition*, New York: Academic.

Ismail, K., Meyerson, B. S., Wang, P. J. (1991), *Appl. Phys. Lett.* **59**, 973–975.

Jasinski, J. M., Meyerson, B. S., Nguyen, T. N. (1987), *J. Appl. Phys.* **61**, 431–433.

Jensen, K. F., Fotiadis, D. I., Mountziaris, T. J. (1989), *Proc. Mater. Res. Soc.* **145**, 107.

Jin, Y., Kobayashi, R., Fujii, K., Hasegawa, F. (1990), *Jpn. J. Appl. Phys.* **29**, L1350–1352.

Kamins, T. I., Chiang, K. L. (1982), *J. Electro-Chem. Soc.* **129**, 2331–2335.

Kanoh, H., Sugiura, O., Breddels, P. A., Matsumura, M. (1990), *IEEE Electron. Dev. Lett.* **11**, 258–260.

Kiermasz, A., Beekman, K. (1990), *Semiconduc. Int.* **13**, 108–111.

Kim, S. S., Tsu, D. V., Lucovski, G. (1987), *J. Vac. Sci. Technol. A* **6**, 1740–1744.

Kobayashi, N., Goto, H., Suzuki, M. (1991), *J. Appl. Phys.* **69**, 1013–1019.

Kodas, T. T., Baum, T. H., Comita, P. B. (1988), *J. Cryst. Growth* **87**, 378–382.

Kuech, T. F., Meyerson, B. S., Veuhoff, E. (1984), *Appl. Phys. Lett.* **44**, 986–988.

Kushner, M. J. (1988), *J. Appl. Phys.* **63**, 2532–2551.

Lucovsky, G., Richard, P. D., Tsu, D. V., Lin, S. Y., Markunas, R. J. (1986), *J. Vac. Sci. Technol. A* **4**, 681–688.

Manabe, Y., Mitsuyu, T. (1989), *J. Appl. Phys.* **66**, 2475–2480.

Markwalder, B., Widmer, M., Braichotte, D. van-den-Bergh, H. (1989), *J. Appl. Phys.* **65**, 2470–2474.

Matsumura, H. (1989), *J. Appl. Phys.* **66**, 3612–3617.

Mayer, E., Shea, D. E. (1991), *J. Electrochem. Soc.* **111**, 550.

Meyerson, B. S. (1986), *Appl. Phys. Lett.* **48**, 797.

Meyerson, B., Uram, K., LeGoues, F. (1988), *Appl. Phys. Lett.* **53**, 2555–2557.

Meyerson, B. S., Olbricht, W. L. (1984), *J. Electrochem. Soc.* **131**, 2361.

Meyerson, B. S., LeGoues, F. K., Nguyen, T. N., Harame, D. L. (1987), *Appl. Phys. Lett.* **50**, 113.

Meyerson, B. S., Uram, K. J., LeGoues, F. K. (1988), *Appl. Phys. Lett.* **53**, 2555.

Mullins, J. T., Clifton, P. A., Brown, P. D., Brinkman, A. W., Woods, J. (1990), *J. Cryst. Growth* **101**, 100–104.

Oehr, C., Suhr, H. (1988), *Appl. Phys. A* **45**, 151–154.

Ogirima, M., Saida, H., Suzuki, M., Maki, M. (1977), *J. Electrochem. Soc.* **124**, 903.

Ohta, S. I., Kobayashi, S., Kaneko, F., Kashiro, K. I. (1990), *J. Cryst. Growth* **106**, 166–174.

Pai, C. S., Chang, C. P., Baiocchi, F. A., Swiderski, J. (1990), *J. Appl. Phys.* **68**, 2442–2449.

Patton, G. L., Comfort, J. H., Meyerson, B. S., Crabbe, E. F., Scilla, G. J., DeFresart, E., Stork, J. M., Sun, J. Y., Harame, D. L., Burghartz, J. (1990), *Electron. Dev. Lett.* **11**, 171.

Reif, R. (1982), *J. Electrochem. Soc.* **129**, 1122–1128.

Rojas, S., Modelli, A., Wu, A., Borghesi, A., Pivac, B. (I990), *J. Vac. Sci. Technol. B* **8**, 1177–1184.

Saxena, A. N., Luehrs, F. (1990), *J. Electron. Mater.* **19**, 271–276.

Sedgewick, T. O., Agnello, P. D., Ngoc, D. N., Kuan, T. S., Scilla, G. (1991), *Appl. Phys. Lett.* **58**, 1896–1898.

Suzuki, N., Yoshikawa, T., Masu, K., Tsubouchi, K., Mikoshiba, N. (1989), *Jpn. J. Appl. Phys. 2 Lett.* **28**, L2316–2319.

Watanabe, A., Kamijoh, T., Hata, M., Isu, T., Katamaya, Y. (1990), *Vacuum* **41**, 965–967.

Wong, M., Saraswat, K. C. (1988), *IEEE Electron. Dev. Lett.* **9**, 582–584.

Yamazaki, T., Ito, T., Ishikawa, H. (1984), *1984 VLSI Symposium, Digest of Technical Papers,* Tokyo: Japan Society of Applied Physics, pp. 56–57.

Yamazaki, T., Watanabe, S., Ito, T. (1990), *J. Electrochem. Soc.* **137**, 313–318.

Yew, T. R., Comfort, J. H., Garverick, L. M., Burger, W. R., Reif, R. (1987), *Mater. Res. Soc. Symp. Proc.* **75**, 705–712.

CYBERNETICS

See INFORMATION SCIENCE

CYCLOTRONS

G. H. MACKENZIE, P. W. SCHMOR, AND H. R. SCHNEIDER *TRIUMF, Vancouver, British Columbia, Canada*

INTRODUCTION

The cyclotron is the founding member of a class of cyclic particle accelerators that accelerate charged particles to high energies by repeated passage of the beam through one or more gaps across which an rf voltage is maintained. It was conceived in 1929 by E. O. Lawrence, and the version built by Lawrence and M. S. Livingston at the University of California, Berkeley, campus in 1931 (Lawrence and Livingston, 1932) was the first machine to accelerate protons to an energy greater than 1 MeV. In a cyclotron the repeated traversal of the accelerating gap is achieved by use of a dc magnetic field to deflect the particles in circular arcs whose radius increases as their energy increases. Operation of the Lawrence or classic cyclotron is based on the fact that if relativistic effects can be ignored, i.e., when particle velocities are much less than the speed of light, the angular frequency of particle rotation is independent of energy and determined only by the particle charge-to-mass ratio and the magnetic induction. So long as both are constant, the particle orbit frequency is constant. Acceleration then is possible if the oscillation frequency of the gap voltage is equal to this or in some cases a harmonic of the

1-56081-063-7/92/$5.00+.50

orbit frequency.

At the time of its invention, the cyclotron significantly extended the range of particle energies available in the laboratory, and for many years this classic cyclotron, accelerating light atomic and molecular ions, was the mainstay of a number of nuclear research facilities. The restriction, however, to energies for which relativistic effects are negligible meant, for example, that proton energies were limited to about 15 MeV.

Over the years a number of improvements have been made to the original concept. Development of the synchrocyclotron in 1946 to overcome the nonrelativistic energy limitation of the classic cyclotron followed the enunciation of the principle of phase stability by Veksler in 1944 and independently by McMillan in 1945. This showed that stable bunches of particles could be accelerated to high energies by appropriate reduction of the frequency of the accelerating voltage in accordance with the reduction in orbit frequency due to the relativistic mass increase with increasing particle energy. Later in the 1950s an idea originally proposed by Thomas in 1938, to employ a periodic variation of the magnetic field strength to provide a transverse focusing force on the beam, was revived. Such an azimuthally varying field (AVF) eliminated the need for a small negative radial gradient in the magnetic field, required for focusing in the classic cyclotron. This then permitted the magnetic field to increase with radius to maintain a constant orbit frequency (isochronism), and cw operation, to the relativistic energies achievable with synchrocyclotrons, but with much greater beam intensities.

In 1962 M. E. Rickey and R. S. Smythe successfully accelerated H^- ions, i.e., hydrogen atoms with an extra electron, loosely bound, and extracted a beam of protons by stripping the associated electrons in a thin foil. This simple technique allowed a single cyclotron to deliver, simultaneously, several extracted beams with independently variable energy and inspired the construction of machines designed specifically for H^- acceleration.

The higher magnetic inductions achievable with superconducting magnets found an application in cyclotron design with the development of the superconducting heavy-ion cyclotron in the 1970s.

Today there are about 150 cyclotrons operating around the world, and this number is increasing at the rate of about five per year. The use of cyclotrons in nuclear research is less widespread than before; however, other fields of research and application have been developed. Many cyclotron laboratories offer customers an ion beam service. The beams are being used, for example, to label items with radioactive elements, to detect contaminants present in trace amounts, and to modify the properties of materials. Cyclotron beams can generate high fluxes of secondary particles, such as neutrons; several hospitals operate cyclotron-based neutron therapy facilities, and other cyclotrons are used in an industrial setting for neutron radiography. The majority of new cyclotrons, however, are intended to produce isotopes for incorporation into radiopharmaceuticals. These, and other, applications are described in Sec. 4.

1. TYPES OF CYCLOTRON

1.1 Classic Cyclotron

A charged particle of mass m and charge qe, and moving with a velocity $\mathbf{v}$ in a magnetic field of induction $\mathbf{B}$, is subject to a force $qe\mathbf{v}\times\mathbf{B}$. For the important special case of a uniform magnetic field with $\mathbf{v}$ perpendicular to $\mathbf{B}$ this results in a circular particle orbit in the plane containing $\mathbf{v}$ and normal to $\mathbf{B}$. Now a particle with an orbit radius r experiences a centripetal acceleration v^2/r, so by Newton's second law of motion

$$qevB=mv^2/r. \tag{1}$$

From this it follows that the particle orbit radius is directly proportional to the particle momentum mv, and also, as Lawrence recognized, that the particle orbit frequency f_{ion}, or angular velocity $\omega=2\pi f$ given by

$$\omega_{\text{ion}}=qeB/m, \tag{2}$$

is constant, independent of particle energy, so long as relativistic effects are negligible and m may be considered constant.

These equations are the basis of the classic cyclotron illustrated in Fig. 1. An essentially uniform magnetic field is generated between the circular poles of an electromagnet, and a hollow cylindrical electrode, split along a

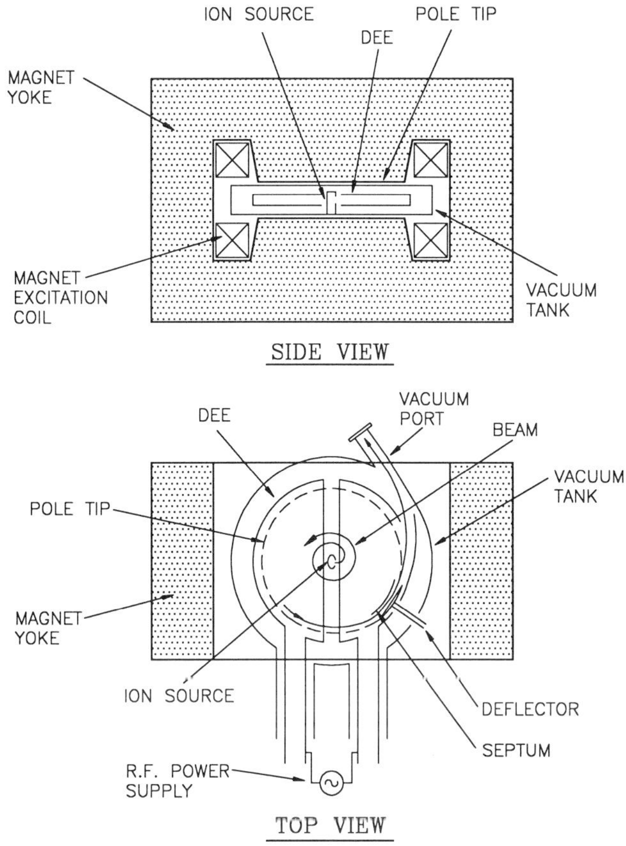

FIG. 1. A classic cyclotron, shown in section to reveal the major components.

diameter, is installed in this gap. In plan views the two sections resemble the letter *D* and are therefore referred to as "dees." By maintaining a high-frequency alternating voltage between the dees, an alternating electric field is established in the gap between them. During half the dee oscillation period the electric field direction is appropriate for extraction of ions from a plasma created by an arc discharge in an ion source located between the dees near the magnet center. If the oscillation period of the applied dee voltage is chosen in accordance with Eq. (2), then particles extracted when the electric field is at or near its maximum travel in a circular path, shielded from the electric field while inside the dee, to return to the dee gap half a period later when the field is again at a maximum and in the direction of the particle motion. The beam, therefore, consists of a series of bunches of particles separated by the rf period. At each successive gap crossing the particle energy is increased with a concomitant increase in orbit radius until the maximum energy and orbit radius, as determined by the magnet pole size, is reached. At this point, in early cyclotrons, the particle beam hit an internal target and initiated nuclear reactions. It is preferable, however, to extract the beam from the cyclotron and lead it to a separate experimental area. This reduces the activation of cyclotron components and permits a much wider range of applications.

The beam is extracted by means of an electrostatic deflector. This consists of a pair of curved electrodes, one a thin grounded metal plate, called a septum, that is interposed between the penultimate and final turns of the accelerated beam, and the other a matching electrode located at a slightly larger radius and carrying a high dc voltage. The transverse electric field between the two electrodes provides sufficient radial deflection to enable the beam to be clear of the magnetic field in less than one turn.

Ions that collide with gas molecules while being accelerated may be lost from the beam. To reduce this loss the volume between the magnet poles, including the dees and other components, is enclosed in a chamber and evacuated.

To produce useful beam intensities in a cyclotron, it is essential that particles having orbits that deviate from the ideal trajectory should also be accelerated and be extractable. An elementary discussion of the beam stability required to achieve this may ignore the effects of acceleration and consider only the motion of particles of selected energies about their corresponding closed or equilibrium orbits. For the classic cyclotron of Fig. 1 these are circles in the midplane, concentric with the magnet center. The magnetic field in this case is azimuthally uniform but may vary slowly with radius, resulting in curved field lines and nonzero radial field components $B_r(r,z)$ off the midplane, as illustrated in Fig. 2. To obtain the conditions for focusing in the axial and radial directions we begin by describing the local midplane field variation by an index $k(r)$ such that

$$B_z(r)=B_z(r_0)(r/r_0)^{k(r_0)}. \quad (3)$$

Here B_z is the midplane magnetic induction, r_0 the equilibrium orbit radius, r a nearby radius, z the axial direction perpendicular to the orbit plane, and

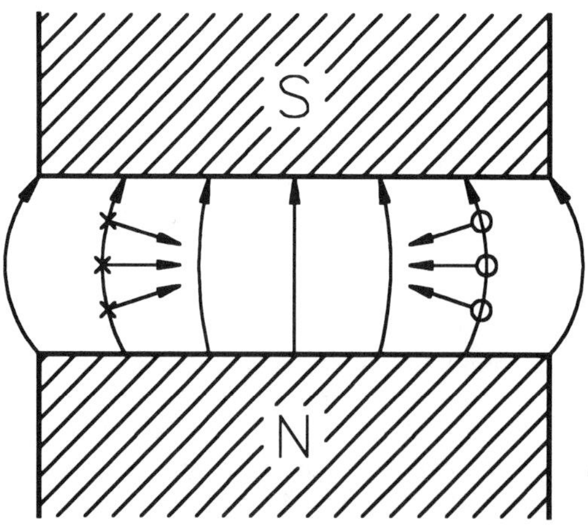

FIG. 2. The vertical focusing effect of a radially decreasing magnetic field. Shown are lines of magnetic induction and the forces experienced by protons traveling in clockwise orbits when seen from above.

$$k=(\Delta B/B)/(\Delta r/r)=(r/B_z)(dB_z/dr). \quad (4)$$

A particle displaced axially experiences a radial field component $B_r(r,z)$ (Fig. 2), and an axial force F_z. By retaining only the linear term in the Taylor expansion for B_r about the midplane, $z=0$, and using the fact that curl $\mathbf{B}=\mathbf{0}$ in the magnet gap, we find that

$$B_r(z)=z(\partial B_r/\partial z)_{z=0}+\cdots=z(\partial B_z/\partial r)_{z=0}+\cdots \quad (5)$$

and the axial force is

$$F_z=qev_\theta B_r=m\ddot{z}=qev_\theta z(k/r)B_z+\cdots . \quad (6)$$

Particles with small displacements from the magnet midplane experience a linear restoring force provided k, or more specifically the gradient $\partial B_z/\partial r$, is negative. Using Eqs. (1) and (6), the equation of motion may be written

$$\ddot{z}-\omega^2 kz=0, \quad (7)$$

which, when k is negative, describes simple harmonic motion about the midplane with frequency $\omega_z=\sqrt{(-\omega^2 k)}$. The oscillation frequency is more commonly expressed in terms of the ion rotation frequency,

$$Q_z=\omega_z/\omega_{\text{ion}}=\sqrt{-k}, \quad (8)$$

and Q_z termed the axial betatron frequency, since the theory was first developed for the betatron accelerator (see BETATRONS). Should k be positive, Eq. (7) describes a steadily increasing axial displacement and particles would eventually collide with the dee or other structures.

A somewhat more complex derivation shows that particles displaced radially from their equilibrium orbits are focused provided that $k>-1$, and that they execute linear betatron oscillations with $Q_r=\omega_r/\omega_{\text{ion}}=\sqrt{(1+k)}$. (The origin of the restoring force may be comprehended by noting that the centripetal force required for circular orbits decreases as $1/r$ [Eq. (1)]. Should the magnetic field decrease more slowly than this, then a particle at a radius larger than its equilibrium orbit will be bent more and focused back toward the equilibrium orbit.) Values of $k<-1$ occur in the edge field of the magnet, the orbit unwinding in a strongly defocusing (radially) field.

For both radial and axial motion to be stable the midplane field must decrease with radius such that

$$-1<(r/B_z)(dB_z/dr)<0. \quad (9)$$

A radially decreasing field means that the resonance condition, i.e., the equality between the orbit frequency, Eq. (2), and a fixed dee frequency, cannot be satisfied everywhere. If $\omega_{rf}=\omega_{ion}$ at the cyclotron center, then the revolution time of particles with higher energy will exceed $2\pi/\omega_{rf}$ and the particles arrive at the dee gap at progressively later times with respect to the peak of the accelerating voltage. If we write the energy gained each turn in the two-dee cyclotron of Fig. 1 as

$$E_G=2qe2\hat{V}\cos\phi, \quad (10)$$

where $\hat{V}$ is the peak voltage on one dee and ϕ the voltage phase when the particle crosses the dee gap center, then progressively later arrival times correspond to a progressive increase in ϕ until, eventually, the accumulated phase slip exceeds $\pi/2$, at which point acceleration ceases and particles begin to decelerate back toward the machine center where they are lost. The lack of isochronism and its consequent phase slip is exacerbated by the relativistic increase in mass, $m=\gamma m_0=(1+T/E_0)m_0$, which causes the rotation time to increase as the kinetic energy T increases. To postpone the inevitable deceleration to the highest possible energy, classic cyclotrons operated with the highest sustainable dee voltage. However, the practical limit corresponded to 100 or 200 turns and $\gamma_{final}\approx 1.02$.

The U-150 machines built in the Soviet Union are examples of late-period classic cyclotrons. Protons were accelerated to 12 MeV, deuterons to 20 MeV, and alphas (^{4}He) to 40 MeV in a 330-ton magnet with a pole face diameter of 1.5 m. The d voltage was 85 kV at 10 MHz and the electrostatic deflector held 70 kV. Beam currents up to 250 μA (1.5×10^{15} proton/s) were obtained.

1.2 Synchrocyclotron or FM Cyclotron

Two methods have been devised to deal with the problem of phase slip and thus permit acceleration to high energies. Frequency modulation, described below, was implemented first; an alternative method, due to Thomas, will be discussed in Sec. 1.3.

In the frequency-modulated (FM) cyclotron, or synchrocyclotron, the frequency of the accelerating voltage is lowered progressively so that it matches the rotation frequency of a hypothetical reference ion, termed the synchronous particle, as it is accelerated. A cluster of bunches captured at about the same time can thus be accelerated to energies far beyond the synchronism limit of the classic cyclotron. Following extraction of the cluster, the rf oscillator is returned to its starting frequency, another batch of ions captured, and the process repeated.

If it were essential that the ion orbit frequency and the accelerating frequency match exactly, then the phase acceptance and capture time interval would be small and the beam current low; it would also be difficult to program the rf. Fortunately Veksler (1944) and McMillan (1945) showed, independently, that exact matching is not necessary because, within certain limits, ions whose energy or phase deviate somewhat from the synchronous ion will oscillate, in energy and phase, about that particle and gain energy at the same average rate. This behavior, termed the principle of phase stability, makes synchrocyclotrons, and also synchrotrons and linear accelerators, practical machines.

The principle was first demonstrated experimentally by Richardson *et al.* (1946). The first synchrocyclotron to surpass the classic cyclotron limit operated at Berkeley, also in 1946, and accelerated deuterons and α particles to 95 MeV/amu. About 20 synchrocyclotrons have been built. The repetition rates for frequency modulation range from 50 to 4000 Hz, while ions are captured into phase-stable orbits for, typically, 0.01 to 0.1 ms; consequently the duty factor and beam intensity are about 1% of that from a cw cyclotron. The long acceleration time means that ions make many turns and dee voltages are low, typically 10 to 50 kV.

The highest proton energies, 1000 MeV, are achieved in the synchrocyclotron at the St. Petersburg Nuclear Physics Institute, Russia. The machine at the European Centre for Nuclear Research (CERN), Geneva, Switzerland, has accelerated 7 μA of protons to 600 MeV and $\sim$0.2 particle μA of carbon, oxygen, and neon to $\sim$80 MeV/amu. The 3000-ton magnet has a 5-m pole-tip diameter and a field falling from 1.94 to 1.81 T. A single dee is used with a voltage 20 kV. A frequency swing between 30.4 and 16.6 MHz for protons and, for example, 7.6 to 6.6 MHz for $^{20}Ne^{5+}$ is provided by a rotating capacitor at 360 Hz.

Synchrocyclotrons pioneered research in meson physics, studies of nucleon-nucleon forces, and particle therapy for cancer and other diseases. These roles are being taken over by AVF machines, although there are proposals to build new synchrocyclotrons for therapy since currents of a few nA are adequate and simplicity of construction and operation may give some competitive advantage.

1.3 The Sector-Focused, Isochronous, or AVF Cyclotron

In order to maintain an ion rotation time that is independent of energy (isochronism), the midplane magnetic induction B_z of the classic cyclotron should increase with radius as

$$\begin{aligned} B_{\text{isoch}}(r) &= \gamma(r)B_0 \\ &= B_0(1-\beta^2)^{-1/2} \\ &= B_0[1-(r\omega/c)^2]^{-1/2}, \end{aligned} \tag{11}$$

where γ and β are the usual relativistic quantities. Such a field is axially defocusing. To overcome the defocusing force and provide adequate axial focusing, Thomas (1938) proposed that an azimuthal variation in B_z be introduced by installing three or more wedge-shaped iron sectors (often called hills) on opposing magnet pole faces, Fig. 3. The particle path has a smaller radius of curvature in the hill and a larger one in the intervening valley. The closed, or equilibrium, orbit therefore has now a radial modulation, termed scalloping, about its average value. Particles have a radial velocity component v_r at the sector edges. Those particles displaced from the midplane experience an azimuthal component of induction B_θ, resulting in an axial force $qev_rB_\theta(r,\theta,z)$ that is focusing at both edges of a sector.

Additional focusing can be realized by having the sectors follow a spiral rather than a radial locus. The curved hill edges then produce a radial component of B off the midplane that results in an axial force $qev_\theta B_r(r,\theta,z)$. This force is focusing at the convex hill edges and defocusing at the concave edges. Following a well-known principle of classical optics for a series of focusing and defocusing lenses of equal strength (the strong-focusing principle), the net effect is focusing. In fact the focusing is even stronger because, as detailed calculations for cyclotron beam optics show, the focusing edge is stronger than the defocusing one.

A Fourier series can be used to describe the midplane induction and the orbit scalloping:

$$B_z(r,\theta) = \overline{B}_z(r)\left[1+\sum_p (H_p(r)\cos p\theta + G_p(r)\sin p\theta)\right], \tag{12}$$

$$r = \overline{r}\left[1+\frac{1}{p^2}\sum_p (H_p \cos p\theta + G_p \sin p\theta)\right]. \tag{13}$$

The magnitude of the azimuthal variation of midplane induction is termed the flutter $F(r)$, where

$$\begin{aligned} F &= \frac{1}{2}\sum_p (H_p^2+G_p^2) \\ &\approx (B_{\text{hill}}-\overline{B})(\overline{B}-B_{\text{valley}})/\overline{B}^2. \end{aligned} \tag{14}$$

The betatron frequencies cannot be given in a closed form for the AVF cyclotron; however, several authors, e.g., Hagedorn and Verster (1962), have obtained expressions for ν_z and ν_r in the form of infinite series the leading terms of which are useful approximations, viz.

$$Q_z^2 = -\overline{k} + [N^2/(N^2-1)]F(1+2\tan^2\xi) + \cdots \tag{15a}$$

$$Q_r^2 = 1+\overline{k}+\cdots, \tag{15b}$$

where $\overline{k}(r)$ is now the average field index, N the number of sectors, and $\xi(r)$ the spiral angle [Fig. 3(a)].

For perfectly isochronous machines $\overline{k}$ may be replaced by γ^2-1. From Eq. (15a) it can then be seen that the axial focusing strength will tend to decrease with increasing particle energy. To counteract this, especially in high-energy cyclotrons, designers increase the spiral angle with radius, up to a practical limit ~70°.

Isochronous cyclotrons can both accelerate a wide variety of ion species and vary the energy of the beam extracted by altering the dee frequency and making the appropriate adjustments to field strength and shape. The relation between the kinetic energy T, the field, and radius may be obtained by writing Eq. (1) in its relativistic form,

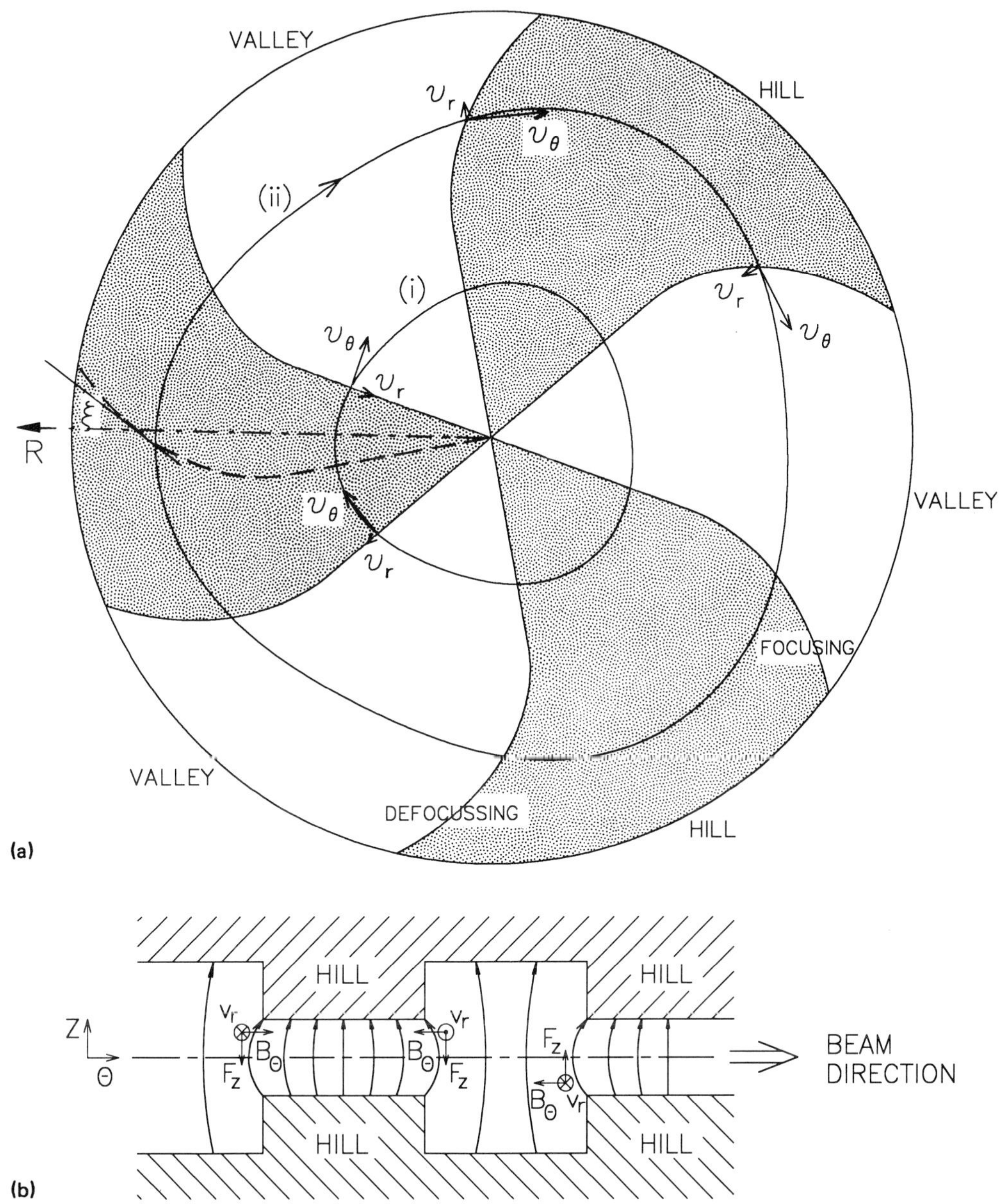

FIG. 3. (a) Equilibrium orbits in a sector-focused (AVF) cyclotron. The inner orbit (i) experiences Thomas focusing from radial sectors; orbit (ii) also experiences strong focusing from spiral sectors. (b) Vertical forces seen by orbit (i) at sector boundaries.

$\overline{Br}=\gamma m_0 \beta c/qe$, and recalling that $m_0 c^2 \gamma = E_0 + T$, where E_0 is the rest energy of the ion. The result

$$\overline{Br}=(T^2+2TE_0)^{1/2}/qec \qquad (16)$$

is sometimes more conveniently expressed in terms of energy per nucleon T/A, where A is the ion mass in amu. This relation is plotted in Fig. 4 for several charge/mass ratios. The diagram can be made more useful for defining the operating range of a cyclotron by adding a scale corresponding to fr, where f is the ion orbit frequency. The frequency range demanded from the rf can be reduced by operating, for some T/A, at a harmonic of the rotation frequency, i.e., $f_{\mathrm{rf}}=hf_{\mathrm{ion}}$. Two isochronous cyclotrons, of the type discussed above, which have accelerated many different ion

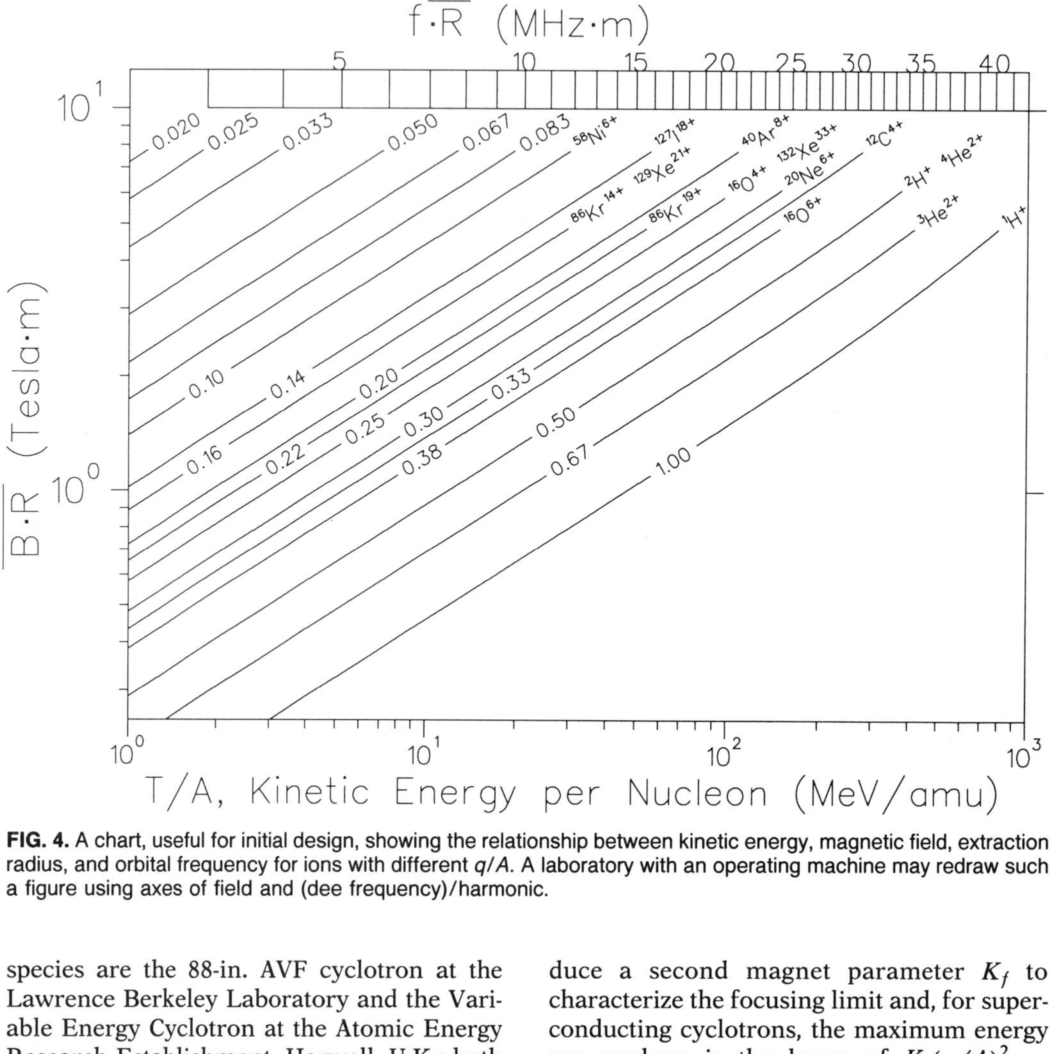

FIG. 4. A chart, useful for initial design, showing the relationship between kinetic energy, magnetic field, extraction radius, and orbital frequency for ions with different q/A. A laboratory with an operating machine may redraw such a figure using axes of field and (dee frequency)/harmonic.

species are the 88-in. AVF cyclotron at the Lawrence Berkeley Laboratory and the Variable Energy Cyclotron at the Atomic Energy Research Establishment, Harwell, U.K.; both were built in the early 1960s.

The maximum energy attainable by an ion in a given magnet may be set either by its ability to bend the trajectory into a closed orbit [Eq. (16)] or by its ability to provide axial focusing [Eq. (15a)]. The bending power is sometimes characterized by K_b, where, in the nonrelativistic approximation, $T/A \leqslant K_b(q/A)^2$. Saturated iron gives fields in the region of 2 T; however, superconducting excitation coils can contribute an extra 4 T and $K_b \approx 1000$ T m can be achieved in compact, single-pole machines with extraction radius 1 m. The iron section in these superconducting cyclotrons are saturated at all working fields; the field modulation is constant and therefore the flutter is inversely proportional to $\overline{B}^2$. This behavior led Blosser and Johnson (1974) to introduce a second magnet parameter K_f to characterize the focusing limit and, for superconducting cyclotrons, the maximum energy per nucleon is the lesser of $K_b(q/A)^2$ or $K_f(q/A)$. The situation is different for cyclotrons using unsaturated iron; for these the flutter is almost independent of excitation. Superconducting machines to accelerate heavy ions began operating at the National Superconducting Cyclotron Laboratory (NSCL-MSU), East Lansing, Michigan, and at Chalk River National Laboratories, Canada, in the 1980s.

Prior to this, during the 1960s, the concept of the separated-sector cyclotron (SSC) had been developed, first at the Paul Scherrer Institute (PSI), Villigen, Switzerland (Willax, 1963), then at the Indiana University Cyclotron Facility (IUCF), U.S. In these machines individual sector magnets are arranged in a pinwheel fashion (Fig. 5), and powered in series. The absence of iron in the valleys

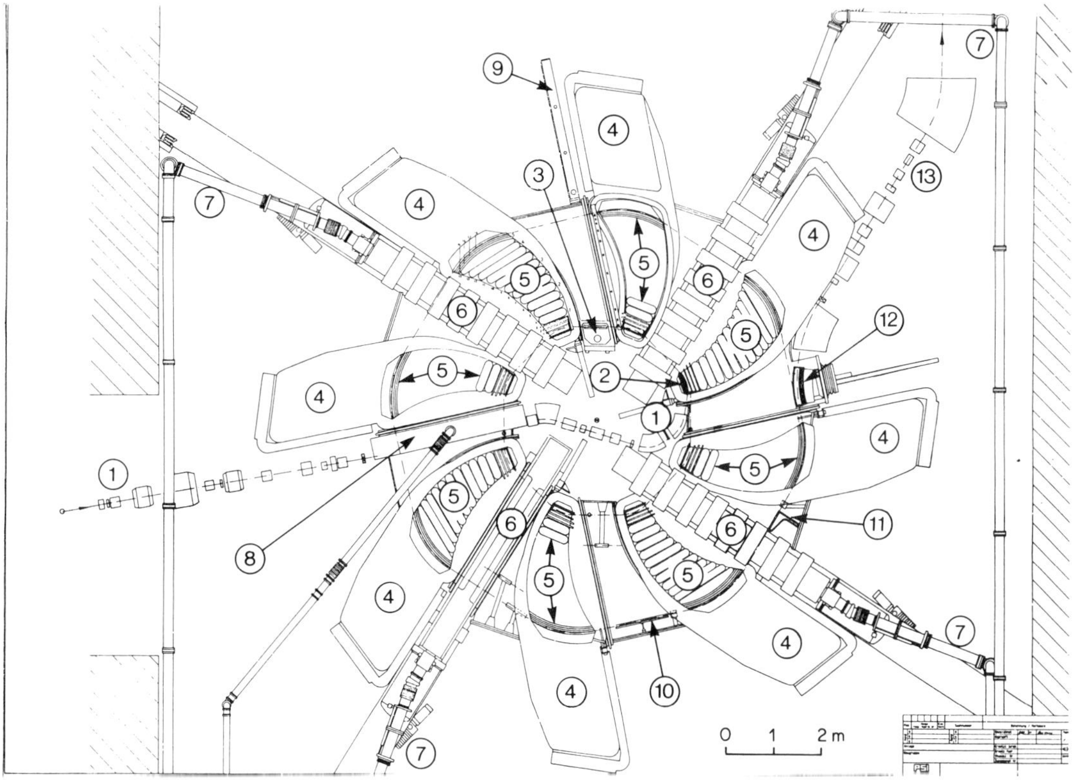

FIG. 5. The PSI separated-sector cyclotron. This machine accelerates protons from 72 to 590 MeV. (1) Beam line from 72-MeV injector cyclotron, (2) magnetic inflector channel, (3) electrostatic inflector, (4) one of eight sector magnets, (5) set of correcting (trim and harmonic) coils wound on a magnet pole, (6) one of four 50-MHz accelerating cavities, (7) coaxial rf power transmission line, (8) 150-MHz "flat-top" cavity, (9) beam diagnostic probe (radial drive), (10) electrostatic deflector, (11) magnet to focus the deflected beam, (12) septum magnet, (13) extracted beam line (590 MeV). (Courtesy of the Paul Scherrer Institute, Villigen, Switzerland.)

provides easier access and space for other equipment, especially for efficient, high-Q, rf cavities operating at high voltage that, in turn, lead to more efficient beam extraction. The valley field is small and the flutter F approaches unity. The lower average field, however, means a larger, more costly, cyclotron, and the absence of a conventional center region means that ions must be preaccelerated to an energy corresponding to the first turn.

Sector-focussed cyclotrons are examples of fixed-field alternating-gradient (FFAG) accelerators. This term, however, is usually reserved for machines with magnets of SSC style and with a field shaped such that the betatron frequencies are independent of radius (energy). This would allow ions to be accelerated to high energies, GeV/amu, while avoiding resonances in betatron motion, but, since the field is no longer isochronous, the accelerating frequency must be modulated as in a synchrocyclotron. Several proton FFAG machines of this latter type have been proposed but only models, accelerating electrons, have been built.

More than 100 AVF machines have been built since the first isochronous cyclotron, equipped with radial sectors and accelerating protons to 12 MeV, operated at Delft, The Netherlands, in 1958. They range from small machines accelerating protons to a few MeV to machines several meters in diameter some of which accelerate intense beams of protons to hundreds of MeV, while others are capable of accelerating virtually any element in the periodic table to energies of ~100 MeV per nucleon. The distribution of K values for machines currently operating is given in Fig. 6. Virtually all new cyclotrons are of the AVF

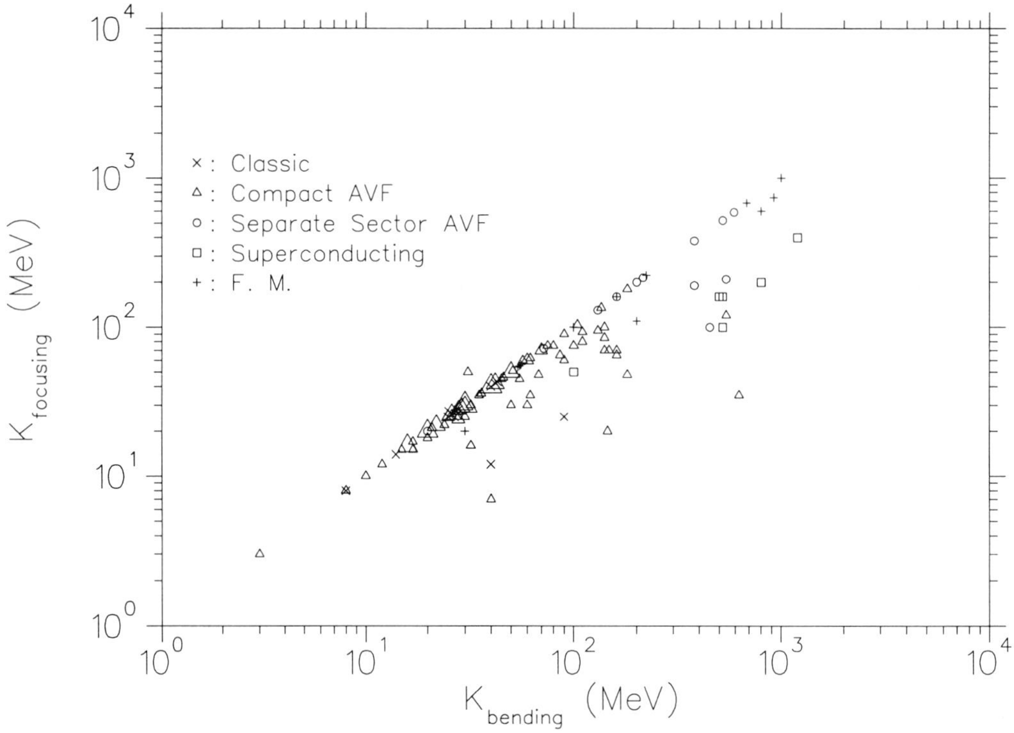

FIG. 6. Parameters describing the maximum energy/nucleon attainable in 143 cyclotrons of different type. The symbol size is proportional to the number of machines with the same parameters; however, cyclotrons manufactured commercially are under-represented. $(T/A)_{max}$ in MeV/amu is the lesser of $K_b(q/A)^2$ or $K_f(q/A)$. (From the compendium attached to the proceedings of the 11th International Cyclotron Conference.)

type, and some older machines have been converted; for example, the synchrocyclotrons at the Joint Institute of Research, Dubna, Russia, and the University of Uppsala, Sweden, now have sectors added to their magnet poles to improve performance. Many new cyclotrons accelerate negatively charged hydrogen ions, H^-, in order to exploit the simple and versatile stripping process for extraction. They are intended for nontraditional fields of research and commercial applications. One such machine is illustrated in Fig. 7, and the parameters of another are given in the Appendix.

Perhaps the most novel cyclotron under construction is a separated-orbit cyclotron, called TRITRON, at the Technical University, Munich, Germany. It employs 12 superconducting sector magnets and 6 superconducting rf cavities, each capable of 0.5 MV. Ions are injected from a potential-drop accelerator and their energy increased five-fold in only 20 turns. Orbits are separated by ~4 cm; conductors can be placed around each turn in each sector and the magnetic field and gradient, and hence steering and focusing, adjusted at each step.

1.4 The Microtron (Electron Cyclotron)

Microtrons are constant-field, constant-rf, cw electron accelerators. The low electron mass (m_0c^2=0.511 MeV) means that the particles are relativistic at very low energies and that the orbit frequency is (28 GHz/T)B/γ. Electrons would rapidly lose synchronism in a conventional cyclotron; however, Veksler showed in 1944 that a suitable energy gain E_G could be found for which the phase slip for each turn corresponded to an integral number of rf cycles and, moreover, that the principle of phase stability applied. The orbits are not concentric but arranged to have a common tangent at which point an accelerating cavity is placed. This cavity may even be a small linear accelerator. The relative energy gain per turn is high, orbits are well separated, and extraction is straightforward. The highest-

FIG. 7. A modern compact cyclotron accelerating H^- ions to 30 MeV. (Courtesy of I.B.A., Belgium.)

energy microtron is the 850-MeV stage at Mainz, Germany, which is intended for nuclear research. Most microtrons operate at tens of MeV and are used as injectors for other electron accelerators or as compact sources for radiation therapy. Currents are of the order of 100 μA. Although occasionally termed electron cyclotrons, microtrons are usually considered to be a separate field and the reader is referred to the general references below.

2. DESIGN CONSIDERATIONS AND SUBSYSTEMS FOR AN AVF CYCLOTRON

The design of a cyclotron is an iterative process during which many parameters, which interact with one another, are brought to an overall optimum. Some design considerations are introduced below, but the most complete sources of information are the published proceedings of conferences on accelerators and related topics referred to for further reading.

2.1 Beam Physics

2.1.1 Resonances. Ions accelerated in an AVF machine undergo many transverse oscillations that can exhibit the same phenomena, e.g., amplitude growth at resonance or coupled motion, seen in other periodic systems. For example, the inclusion of higher-order terms in the field expansion [Eq. (5)] transforms Eqs. (6) and (7) from a description of free to one of forced oscillations. For a field harmonic p [Eq. (12)], the general resonance condition is

$$mQ_r+nQ_z=p, \quad m,\ n,\ p \text{ integer.} \tag{17}$$

Energy may, in principle, be transferred between transverse and longitudinal motion; however, while the phase in an AVF cyclotron may deviate from an average value it does not perform periodic oscillations. This is not the case for synchrocyclotrons, and a term involving the frequency of phase oscillations would be included in Eq. (17).

Resonances in betatron motion influence the choice of Q_z, the number of magnet sectors and accelerating gaps, and the tolerances of manufacture and assembly.

In an isochronous cyclotron $Q_r \approx \gamma$ and so may start near 1 and increase with energy; also Q_z frequently rises from a low value near the center of the machine where the field modulation F is small and changes again on approaching the focusing limit or the fringe field where k deviates from the isochronous requirement. In superconducting cyclotrons different magnet excitations have a marked effect on F and Q_z. Consequently, one or more resonances may be crossed. The Mathieu–Hill differential equation describes particle motion with periodic focusing; analysis shows that the phase shift of a betatron oscillation is $2\pi Q_{r,z}/N$ per sector and that motion becomes unstable when this phase shift is π. Consequently, stable radial motion is not possible in a two-sector cyclotron; at least three sectors are required at low energy. For each N there is a corresponding limiting value for Q_r. Should $N=3$, for example, then $Q_r \leqslant \frac{3}{2}$, which, in the case of protons in an isochronous cyclotron, corresponds to 469 MeV. In practice the limit is somewhat lower.

Low-order—m (or n)$=\pm 1$ or ± 2—resonances may be crossed provided the pth amplitude or gradient is small; small coils are often placed in the resonance region and their currents adjusted empirically to cancel any magnetic field errors remaining after assembly. Higher-order resonances, $m=\pm 3, \pm 4 \cdots$, correspond to nonlinear terms and may be traversed provided the betatron amplitudes are small (Hagedorn and Verster, 1962).

Cyclotrons intended for nuclear physics research may be required to preserve polarization of the beam during acceleration. The angular momenta, or spins, of atomic nuclei are usually randomly distributed. Polarized-ion sources select and populate one spin orientation. These ions are injected with spins aligned with the magnetic field and precess about this axis at the Larmor frequency. Those particles whose motion causes them to experience, in the rest frame, a quasiconstant transverse field component will begin to precess about the resultant, nonvertical, axis. Conditions for depolarization exist whenever

$$\gamma G + mQ_r + nQ_z = p, \tag{18}$$

where G is 1.79 and -0.143 for protons and deuterons, respectively, and -3.79 and -1.86 for H^- ions and D^- ions. Driving terms arising from field imperfections can, again, be compensated by coils. Depolarization has not been observed in positive-ion cyclotrons, but the large negative G values for H^- and D^- make resonances with the sector symmetry more likely, and H^- polarization changes of 7% have been observed at the six-sector TRIUMF cyclotron, Vancouver, Canada, and complete depolarization at the University of Manitoba, Canada, four-sector machine.

2.1.2 Electric Forces. The forces exerted on an ion by a magnetic field are proportional to both charge and velocity, while those from an electric field depend on the charge alone. The two are equivalent when $E=(300\ \text{MV}\ \text{m}^{-1}\,\text{T}^{-1})\beta B$. Electric fields, therefore, are relatively more important at low energies.

The electric field at an accelerating gap is illustrated in Fig. 8. An axially focusing kick is experienced by particles entering the gap, a defocusing kick on leaving it. At low energies the transit time may be a substantial fraction of the rf period leading to a net focusing effect for particles in the gap when the voltage is falling (positive phases). Flutter vanishes at the magnet center, and electric focusing can so dominate the first few turns that the defocusing of ions with negative phases can lead to their removal from the accelerated beam. Another defocusing force arises from the mutual repulsion of the ions. This space-charge force, which is proportional to the charge density, is the major factor limiting the beam current at low energies. At any energy the distribution of accelerating impulses around a turn may have a pth harmonic satisfying Eq. (17); gap-crossing resonances may also be driven by the beat frequency arising from the combination of dee and sector symmetries.

The forces from the accelerating structure depend on the ion phase; so, usually, do the space-charge forces.

2.1.3 Charge-Changing Phenomena. An ion that picks up an electron, or has one stripped from it, will alter its charge/mass

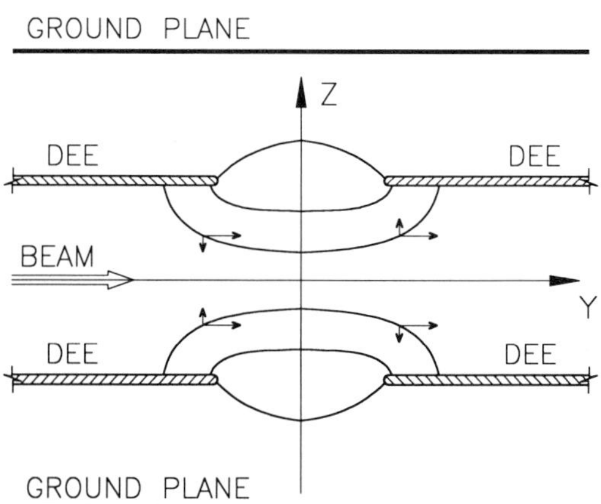

FIG. 8. Electric field lines and their components in the acceleration gap.

ratio, drop out of synchronism, slip in phase, and eventually be decelerated and lost. Either may follow from a collision with a residual gas molecule, however, and except for highly ionized heavy ions, stripping dominates once acceleration has begun. The cross section per atom $\sigma(T)$ for stripping H^- to H^0 in air is usefully approximated by $7\times10^{-19}/\beta^2$ cm^2 and the fraction lost by $2.36\times10^4\,P(E_0/E_G)R_e$ where P (bar) is the operating pressure and R_e (m) the extraction radius for a cyclotron with $\gamma\approx1$. The cross section for stripping in hydrogen is eight times smaller. A vacuum of 10^{-10} bar seems adequate for compact heavy-ion or H^- cyclotrons, 7×10^{-11} bar for larger machines.

At high energies the Lorentz electric field $\mathbf{E}=\gamma\mathbf{v}\times\mathbf{B}$ lowers the potential barrier of weakly bound negative ions sufficiently to allow quantum-mechanical tunneling to occur and an electron to escape. Measurements and theory are in good agreement, the lifetime for an H^- ion under such conditions being

$$\tau=(a/E)\exp(b/E), \tag{19}$$

where $a=2.47\times10^{-6}$ V s/m and $b=4.49\times10^9$ V/m. Figure 9 gives, for H^- ions of several energies, the fraction dissociated per meter path length as a function of the magnetic field strength. The steepness of Eq. (19) means that this electromagnetic stripping is localized to the hills and the highest energies of a given cyclotron.

2.1.4 Computer Codes. Analytic methods are complemented by the extensive use of computer codes. These are especially important in calculating trajectories of ions in electric and magnetic fields and thence the betatron frequencies and deviation from isochronism (Gordon, 1984). The fields themselves, when they cannot be measured, may be calculated using two- or three-dimension finite-element codes.

2.2 Magnet

The maximum magnetic rigidity ($\overline{Br}$) required can be obtained by use of Eq. (16) or Fig. 4. The peak hill field is usually at or near magnetic saturation. The field increase required for isochronism and the required flutter F then determine the maximum central field value B_0. This is generally used in order to minimize magnet size and cost. Exceptions are the higher-energy H^- cyclotrons where

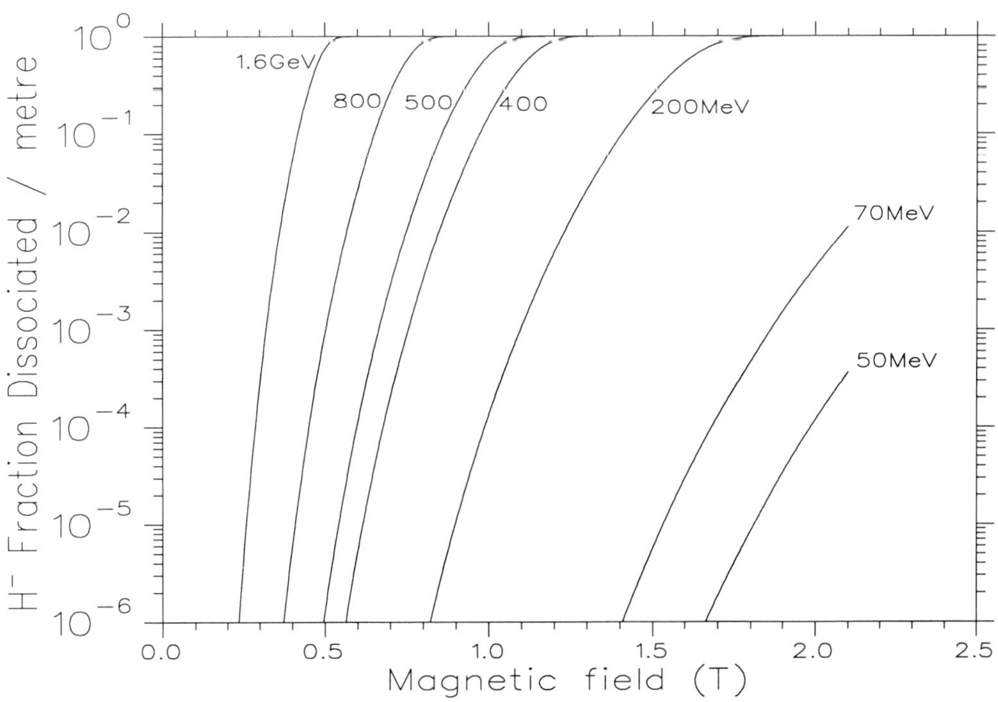

FIG. 9. Dissociation of H^- ions as a function of field strength and energy (pathlength = 1 m).

ion dissociation (Fig. 9) restricts the hill field. The gap field of a magnet falls rapidly approaching a pole edge; consequently the maximum isochronous orbit radius is smaller than the pole radius by at least half the hill gap width.

The yoke serves as a low-reluctance flux-return path. Early cyclotrons employed the H-frame design illustrated in Fig. 1. Beam ports, dee stems, and other services were inserted through the open sides and modifications were easily made. Superconducting machines and modern compact cyclotrons have a more efficient flux-return structure that surrounds the pole and has midplane and vertical penetrations for extracted beam and other services. This provides, also, some shielding from induced radioactivity (Fig. 7 and Appendix). Access to the interior of the vacuum vessel is obtained by raising the upper yoke, pole, and vessel lid.

Both the poles and yoke are fabricated either from low-carbon mild steel in the form of forgings or rolled plate, or from soft iron castings. The magnetomotive force is provided by coils fitted around the poles, generally wound from hollow, water-cooled, copper conductor. Power consumption ranges from 30 kW for a minicyclotron to almost 1 MW for separated-sector machines with large K_b.

Two- or three-dimensional finite-element codes can be used to determine flux distributions in the pole and yoke, the ampere-turns required for excitation, and to some degree the average flux distribution in the gap. With their aid, magnets can be designed that generate fields within 1% of a specified design objective. This is generally not precise enough for cyclotron operation, and an iterative process of field mapping followed by shimming, i.e., adding or removing small amounts of iron at various locations in or near the gap, must be included as part of magnet fabrication and commissioning. This procedure reduces deviations from the ideal field to a few mT, which may be adequate for low-energy cyclotrons. For others with tighter tolerances auxiliary coils are included for empirical adjustments of the isochronous field, or correction of field harmonics, based on measurements of beam properties. These are termed trim coils and are wound on the pole face or vacuum vessel (e.g., Fig. 5). Similar coils, of larger power, are used to produce the changes in $B(r)$ required for isochronous operation with different γ_{final}.

The superconducting cyclotron employs superconducting windings in a magnet with iron yoke and poles at room temperature. The high current density, typically 3000 A/cm^2, achievable in superconductors makes possible magnets with large (6-T) gap inductions. This results in a considerable economy when designing a cyclotron to accelerate, for example, partially ionized heavy ions with low q/A and high magnetic rigidity. All coils built to date use copper-stabilized multifilament NbTi superconductor, and, because of the large stored energies involved, all are cryostatically stable. A high copper-to-superconductor ratio, typically 20:1, is used to make a winding with many internal channels, allowing good thermal contact between the liquid helium bath coolant and the conductor. The iron of the hills and pole is completely saturated and can be treated as uniformly magnetized volumes for design purposes. The coils contribute directly to the gap induction and are generally divided into a pair of coils close to the midplane and a more distant pair; each pair is energized independently to provide some control over the radial variation of $\overline{B}(r)$. Design currents range from 700 to 2300 A, stored power from 18 to 60 MJ.

At the Chalk River superconducting cyclotron laboratory an array of 13 cylindrical steel rods with axes perpendicular to the midplane penetrate the hills. Axial adjustment of a rod produces a local change in the hill fields to provide both radial variation of the field and harmonic adjustment.

2.3 Accelerating Systems

Isochronous cyclotrons do not necessarily require a large dee voltage to reach high energies, and in some cases the second dee of Fig. 1 is replaced by a narrow grounded electrode termed a dummy dee. A large energy gain per turn E_G, however, relaxes magnetic field tolerances, opens the center region geometry, assists in conventional extraction, and reduces stripping losses. The dees of an AVF cyclotron are often triangular or spiral shaped, designed to fit within the valleys of the magnet poles. The increased number of gaps increases E_G. The smaller hill gap improves magnet efficiency.

The dees are made part of an rf resonant circuit, usually as the capacitive termination of a shorted quarter-wave line resonator. This generates higher dee voltages and also facilitates impedance matching to the power source. The resonator can also provide the mechanical support for the dees, obviating the need for failure-prone insulator supports. The shorting plate is usually movable for coarse frequency tuning adjustment and to accommodate different operating modes. The high thermal and electrical conductivity of copper makes it the preferred material. The center conductor of the resonator is usually water cooled; other surfaces are also often water cooled to minimized dimensional changes due to thermal expansion and maintain a stable resonant frequency.

The accelerating fields for many separated-sector cyclotrons are developed in more efficient resonant cavities installed in several of the spaces between magnet sectors (Fig. 5).

Rf power is coupled to the dees or cavities by either terminating the drive line from the rf source with a plate that is located in a high rf electric field region near a dee or by terminating the drive line with a loop near a magnetic field maximum, e.g., the short. Plate or loop dimensions and orientation are adjusted to match source and load impedances.

Cyclotrons tend to operate between 5 and 75 MHz and rf power requirements range from about 15 to 1 MW. To generate these a variety of gridded tubes manufactured for use in short-wave broadcasting or telemetry are available. They are used in either of two basic circuits. The earliest and simplest circuit is the self-excited oscillator, illustrated schematically in a grounded-grid configuration in Fig. 10. It uses positive feedback to the power-tube cathode, and the dee and line resonator then determine the resonant frequency. The second type of circuit is the master oscillator–power amplifier. In this case the feedback line is replaced by a signal obtained from a drive chain consisting of a crystal oscillator and several stages of amplification. In either case an automatic motor-driven capacitor is usually installed near the dee to maintain the desired resonant frequency by fine tuning.

For a synchrocyclotron, rf power must be supplied as the frequency sweeps over its operating range. A self-excited oscillator system is often used with the transmission-line short replaced with a variable capacitor, either a motor-driven rotating capacitor or a vibrating vane, and the high-power rf is swept in frequency in accordance with the varying resonator capacitance.

2.4 Ion Sources for Cyclotrons

Cyclotron ion sources must operate continuously, unlike those used with pulsed machines. Until the 1960s most ion sources were variants of a type developed at the Oak Ridge National Laboratory, Tennessee based on the PIG (Penning ionization gauge). In these

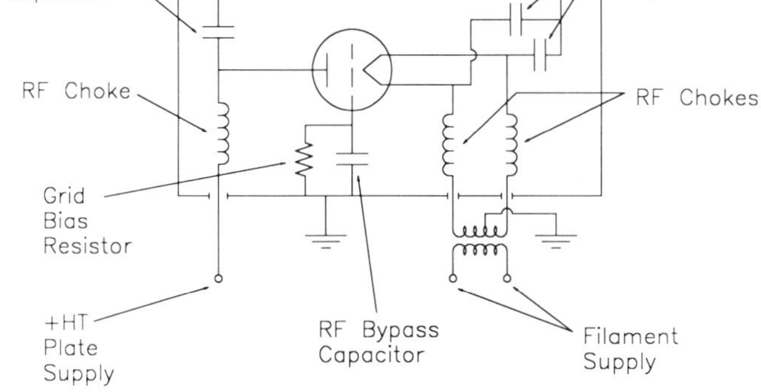

FIG. 10. A simplified grounded-grid self-excited oscillator circuit for cyclotron dee excitation.

sources (Fig. 11), an arc discharge is maintained inside a hollow chimney (anode), aligned parallel to the magnetic field, and passing through the midplane near the cyclotron center. A cathode is placed at one end of the chimney and biased by a few hundred volts with respect to the anode. Electrons drawn from the cathode are constrained by the magnetic field to pass through the arc chamber toward an anticathode placed at the other end of the chimney that reflects them back toward the midplane. The cathode may be a filament or a block heated indirectly by electron bombardment. Each electron can oscillate through the arc chamber many times before collision, thus increasing the probability of ionization. A current of 0.1 to 10 A at 100 to 1000 V maintains the discharge. The flow of gas (or vapor) to be ionized is carefully regulated through a small line to the arc chamber. The electric field of the dee penetrates a small slot in the chimney to extract ions from the plasma. The system is otherwise closed and a relatively small gas flow maintains an adequate pressure. These sources provide both light and heavy positive ions as well as negative hydrogen and deuterium ions. Higher electron energies produce multiply charged ions; however, the cathode lifetime decreases because of ion bombardment. The chimney diameter must be small so as not to interrupt any of the first turn; the relatively small source volume limits the maximum input arc power and consequently the maximum charge state.

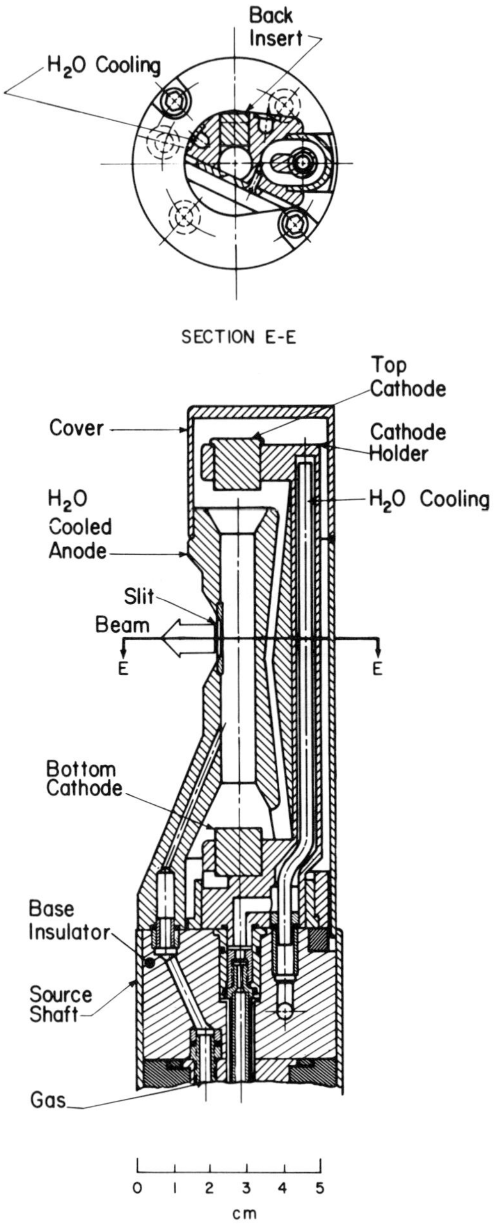

FIG. 11. Schematic cross section of the internal PIG source used until 1985 for high–charge-state ions at the 88-in. cyclotron. (Courtesy D. J. Clark, Lawrence Berkeley Laboratory.)

An ion source can produce several ion species, with different q/A, at the same time. The requirement that the rf be equal to, or a harmonic of, the orbit frequency means that only certain species will be accelerated. The selection of species can be refined further by placing posts between early turns.

High-charge states, more current, or polarized beams can be provided by larger sources placed outside the cyclotron. These inject their ions either through a valley (Fig. 5) or down the magnet axis. Axial injection was first demonstrated at Birmingham University, U.K., in 1964. A hole is bored along the axis of the magnet yoke and ions, with tens of keV/amu, led through an evacuated tube to a point just prior to the magnet midplane (Fig. 12). They are bent into the plane of acceleration by means of either an electrostatic mirror (a grounded grid placed above a biased electrode, both at an angle of about 45°) or, preferably, by an inflector consisting of a pair of curved electrodes carrying voltages of opposite sign. Some electrostatic or magnetic focusing elements may be required in the tube. Detailed field calculations are necessary

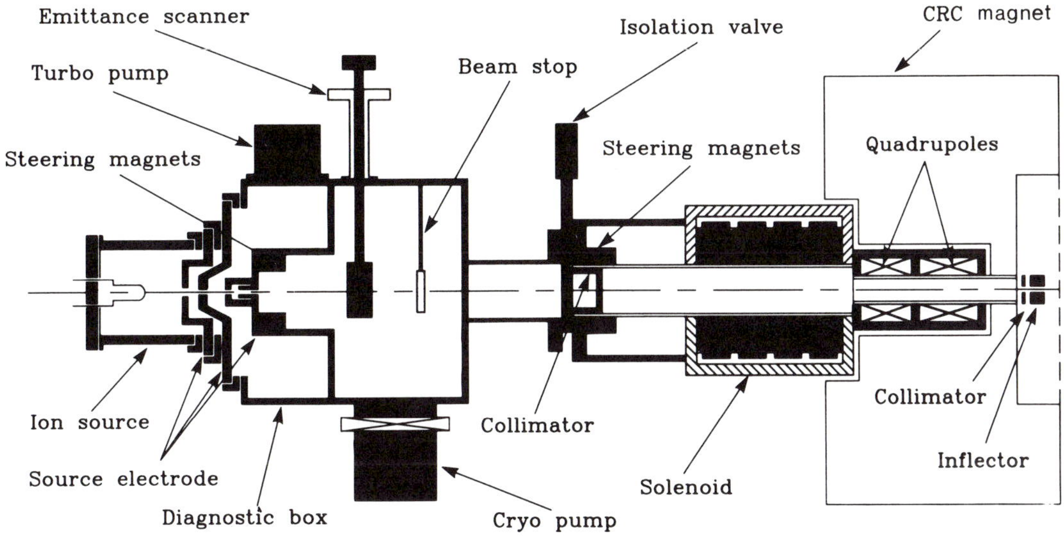

FIG. 12. Axial injection from a cusp source providing H^- ions into a 1-MeV test-bed cyclotron (CRC) at TRIUMF. The plane of acceleration is vertical in this cyclotron.

to ensure a smooth match between cyclotron and injection system optics.

Electron cyclotron resonance ion sources (ECRIS) were developed at the Centre d'Etudes Nucléaires, Grenoble, France, in the 1970s and are used by most new heavy-ion cyclotrons. Microwaves at a fixed frequency, somewhere between 2 and 20 GHz, are fed into a cavity located within a solenoid. The solenoid is designed to provide at least two surfaces where field and rf satisfy the resonance condition [Eq. (2)] for electrons and where power can be transferred from microwaves to electron motion. In the absence of collisions the electrons can reach energies of several MeV. When gas or vapor is introduced these high-energy electrons collide with atoms or molecules yielding ions and more electrons for the microwaves to accelerate. Multipole (sextupolar or octupolar) magnetic fields confine the plasma radially, which increases the ion lifetime and hence the probability of multiple collisions and consequently the maximum charge state. Several charge states are extracted and the desired species is selected by a bending magnet and slits prior to injection. The source contains neither filament nor cathode, nor is an arc required to obtain a plasma. Elements that do not easily vaporize may be ionized by placing solid pieces in the resonance zone.

Plasma can be confined within a cylindrical chamber by rows of permanent dipole magnets; a row with north poles facing the plasma alternates with a row with south poles. The shape of the magnetic field flux lines led to the term cusp or multicusp ion source. The chamber is the anode, a heated filament the cathode. The source chamber may be divided into two regions by a small transverse magnetic field allowing independent optimization of two-step processes. An arc of 25 A at 100 V provides very intense continuous beams of negative hydrogen and deuterium, particularly suited for isotope-producing cyclotrons. The source shown in Fig. 12 delivers 7 mA of H^-. The Paul Scherrer Institute has developed a proton version capable of 25 mA.

Nuclearly polarized ions of ^1H, ^2H, ^3H, and ^3He are, in the main, produced by the three different types of source discussed below, namely atomic (ground state), Lamb shift (metastable), and optically pumped.

An atomic source uses an rf discharge to dissociate $^1H_2$ or $^2H_2$ molecules. The atomic beam is passed through a tapered magnetic sextupole that deflects out of the beam one-half of the hyperfine states that have a particular electron spin direction (Stern–Gerlach effect). The remainder then pass through an rf field that induces transitions from states containing unwanted nuclear spin alignments to the just depopulated hyperfine states. A pass through a second tapered sextupole removes those atoms that have undergone rf transitions. The result is a beam of nuclearly polarized atoms that may now be ionized to $^1H^+$ or $^2H^+$ in helium or to negative ions formed by charge exchange with alkali atoms. Proton currents are about 500 μA, H^- currents are of the order of 10 μA, and polarization is usually about 90%.

Lamb-shift type polarized-ion sources begin by directing a 550-eV proton beam (or equivalent velocity ^2H, ^3H, or ^3He beam) through a cesium-vapor charge-exchange target. About one-third of the atoms emerging from the cesium are in the $2S_{1/2}$ metastable state. This atomic beam then passes through a magnetic solenoid containing an rf electric field. The magnetic field and rf frequency are chosen such that metastable states with an unwanted nuclear spin become degenerate with the $2P_{1/2}$ state and quickly decay to the ground state. Finally, the atomic beam is passed through an ionizer (iodine for positive ions, cesium for negative) that selectively ionizes the metastable atoms rather than the ground-state atoms. Polarizations of about 80–85% are obtained at currents of 1–2 μA.

An optically pumped polarized-ion source is used at TRIUMF for experiments requiring more intense proton beams. Circularly polarized light from dye lasers tuned to the rubidium D_1 line polarizes the outer electrons of the atoms in a rubidium-vapor charge-exchange target immersed in an axial magnetic field. Protons at ~3 keV have a large probability of picking up a polarized electron in this target to produce an electronically polarized beam of hydrogen atoms. The atomic beam leaves the first solenoid to enter a second solenoid whose magnetic field is in the opposite direction and for which the field gradients on axis have been designed to induce a Sona type transition (i.e., only those atoms with the hyperfine states having opposed electron and nuclear spins change their nuclear spin direction). This results in a nuclearly polarized beam of hydrogen atoms that may then be ionized to positive or negative ions. 10 μA of H^- with 78% polarization has been achieved to date; positive-ion currents should be ten times larger.

2.5 Extraction

The two principal methods for extracting the beam from a cyclotron are electrostatic and magnetic deflection and extraction by stripping. In the latter process the beam passes through a charge-changing medium, usually a very thin foil, located so that particles with altered q/A leave the machine between magnet yokes. A simple change in the radial position of the foil changes the energy extracted; several narrow foils, or ones partially inserted into the beam, allow the simultaneous extraction of several beams with different energies. This method is now chiefly used for negative ions of hydrogen but, in the past, has been used for molecular ions H_2^+, HD^+, etc.

The stripped electrons dissipate their energy in the foil; passage of ions both heats the foil and causes radiation damage. Refractory materials, such as carbon or aluminum oxide, are commonly used. Amorphous graphite foils, 200 $\mu g/cm^2$, extract up to 200 μA from a Cyclotron Corporation CP-42 cyclotron, operated by Nordion International, Canada, at energies between 20 and 40 MeV. Fifteen foils, each lasting about 10 mA h, are mounted in a cartridge.

Deflecting systems are used for more highly ionized ions or when the charge state must be preserved. The electrostatic deflector of Fig. 1 may be followed by one or more magnetic channels, which produce a local field reduction to augment the deflection. These channels may be coils powered to oppose the main field or iron bars placed either side of the deflected beam to divert magnetic flux. The fringe field from such channels often perturbs the orbits prior to extraction, and additional coils are incorporated to cancel such effects.

Any beam incident on the septum of an electrostatic deflector causes local heating and makes both the septum and neighboring components radioactive. Secondary particles affect voltage holding. Unfortunately the turn-to-turn separation arising from acceleration falls rapidly with energy, as $(B_0\beta\gamma^3)^{-1}$, and turns normally overlap at extraction radius. The turn width depends on both radial emittance and the energy spread associated with the phase band being accelerated. Some machines eliminate extraction losses by eliminating overlap between turns at the deflector entrance. This may be achieved by a combination of

1. slits near the machine center to restrict emittance and phase width at the cost of intensity,
2. a large energy gain per turn, and/or

3. the excitation of a coherent radial betatron oscillation.

The latter alternately adds to and subtracts from the turn spacing resulting from acceleration.

The relative alignment of beam and deflectors is quite critical and is usually adjusted empirically. Once set up this system demands a better machine stability than that required just for acceleration to the final energy or that required for stripping extraction.

The magnet fringe field is a powerful lens, and optics calculations should be performed to ensure that the beam matches smoothly into the external beam transport system.

2.6 Vacuum

The tightest constraints for vacuum in cyclotrons arise from either the inability of electrodes to support high voltage or unwanted charge-changing collisions between ions and residual gas molecules. Acceptable operating pressure ranges from 10^{-8} bar for a small proton machine to 10^{-11} bar for H^- or high–charge-state heavy ions. The initial outgassing rate for clean stainless steel, or nickel-plated mild steel, is about 10^{-8} bar L s^{-1} m^{-2}, and is chiefly water vapor. Over several days the evolution of water declines to match the steady-state evolution of hydrogen of 10^{-10} bar L s^{-1} m^{-2}. Provided that little organic material is incorporated, there are no leaks, and there is no internal ion source, these gases usually constitute the continuing gas load.

The speeds of pumps commonly used lie between 10^3 and 10^5 L/s; to achieve this pumps should be close to, or preferably inside, the cyclotron and be unaffected by magnetic field. Cryopanels cooled to 20 K with 80 K baffles are preferred to the older oil diffusion pumps. Additional hydrogen pumping may be provided by charcoal or zeolite at 20 K or by condensate cryopumps. Initial pumpdown is provided by mechanical pumps located in regions free from magnetic field. The magnetic field may also affect the calibration of ion gauges or the operation of mass spectrometers used to analyze the residual gas. Pumpdown time can be shortened by venting with dry nitrogen and warming the vacuum tank while roughing. New equipment should be cleaned with solvent or detergent and baked before installation to remove volatile organics and adsorbed water from the surface.

2.7 Beam Instrumentation

The beam parameters measured, in a rough order of frequency of measurement, are

1. both the amount of charge delivered and that lost,
2. the transverse position of the bunch center, its energy, and its phase with respect to rf,
3. the beam size and distribution as a function of these coordinates,
4. polarization, and
5. emittance and correlations between parameters.

In general these quantities may be measured directly in a beam line, whereas in a cyclotron they must sometimes be inferred by observing the dependence of a related parameter on a machine variable. Most cyclotrons have at least one movable probe giving beam intensity, phase, and density as functions of radius. However, a cyclotron whose operating tolerances are comfortably within the precision of manufacture may monitor only current and the position and size of the beam spot near the production target. A research facility will have the most extensive array of instrumentation.

The sensing instrument may intercept all or part of the ion beam or detect the electromagnetic field associated with a moving bunch of ions. Scintillation screens glow when hit by a beam with density $\gtrsim 1$ nA/cm^2. This glow, when observed through a window by a TV camera, gives an inexpensive display in a control room of beam position, size, and transverse correlations. The system is, however, nonlinear and subject to radiation damage. The beam can be stopped in an electrically insulated conducting target and the incident charge taken via a vacuum feedthrough to a measuring instrument that may be a meter, amplifier, oscilloscope, etc. Information on transverse distributions may be obtained from a subdivided target or from a wire target swept through the beam. At higher energies the thickness to stop the beam may be inconveniently large, and secondary particles, e.g., electrons or x rays, escaping from the surface of a thin target may be detected. Electrons in a cyclotron, including those stripped from H^- ions, spiral along magnetic field lines until collected. In most environments an incident current >1 nA may be measured. Ion chambers or electron multipliers that boost incident charge have been used for lower currents. Plates or loops placed near the beam detect the electromagnetic field to give information on intensity, position, and phase without perturbing the ions. Various signal-processing techniques enhance the signal-to-noise ratio, but, in general, they are less sensitive than intercepting methods. At IUCF a monitor that capacitively senses the beam provides a 0.3-mm resolution at 50 nA current.

The cyclotron operating conditions for an ion species sparsely produced by an ion source may be found by using a more intense species with very similar q/A (analog beam).

It is important to be able to bring signals generated by beam users for their own purposes into the central control room. They are concerned that beam properties match their requirements and may be able to generate information beyond the resources of the cyclotron crew.

2.8 Computer Control and Monitoring

This is a rapidly changing field. Simple cyclotrons are operated using commercial process controllers. Control systems in research laboratories, however, have to service many more set points and need more frequent revisions to incorporate new equipment or modes of operation; on-line data-reduction algorithms are common, also. These systems, therefore, should be compatible with off-line software development. The performance of some devices should be logged regularly and automatically for fault diagnosis and failure prediction. Closed-loop regulation of subsystems frequently includes microcomputers that exchange data with the main control system; however, closed-loop operation of the entire cyclotron is still rare.

3. CYCLOTRONS AS INJECTORS OR POSTACCELERATORS

Several laboratories generate relativistic beams using two or more accelerator stages. A single-stage cyclotron that provides a large energy increase and has a peak field chosen to satisfy the requirements of isochronism and turn spacing at extraction and to avoid saturation will also have a low central field B_0. Relatively few turns take place in the center, and magnets and accelerating structures can be used more economically by a two-stage process that separates the optimization of injection and initial acceleration from the extraction of a high-energy beam. The beam transport line between two machines provides an opportunity to trim off unwanted particles or to increase the charge state of partially ionized atoms by stripping.

The momentum of the beam from an energetic preaccelerator is too high for axial injection into a cyclotron, and injection schemes tend to mirror the extraction schemes of Sec. 2.5. Fully stripped ions may be directed roughly radially along the valley of a separated-sector cyclotron (Figs. 5 and 13) and an electrostatic inflector used to steer them onto the accelerated equilibrium orbit. The radius gain per turn should be such that the next turn clears this inflector. Partially stripped ions may be directed across the face of a solid-pole magnet onto a foil placed at the common tangent of their path and the orbit of stripped particles of the same energy.

The rf of a second cyclotron should be the same as, or a low multiple of, the frequency of the injector cyclotron and the phase width of the injected beam much less than the half-period of the new machine. An rf "bunching" cavity is often installed between machines. This buncher improves beam capture by decelerating the beam during one half-cycle and accelerating it during the other to produce a "focus in time" at the first accelerating gap of the next machine.

While cyclotrons are natural partners, with 11 pairs in operation worldwide, other machines have also been used as pre- or postaccelerators. Four laboratories use potential-drop machines as injectors; they have the advantage, for negative ions, of an external ion source and the opportunity to strip off electrons in the terminal, thus increasing q/A. Two laboratories use potential-drop postaccelerators to provide simple and precise energy control over a wide range. A 70-MeV/amu variable-frequency linear accelerator provides heavy ions for the $K_b=540$ separated-sector cyclotron operated by the Institute of Physics and Chemistry (RIKEN), Saitama, Japan.

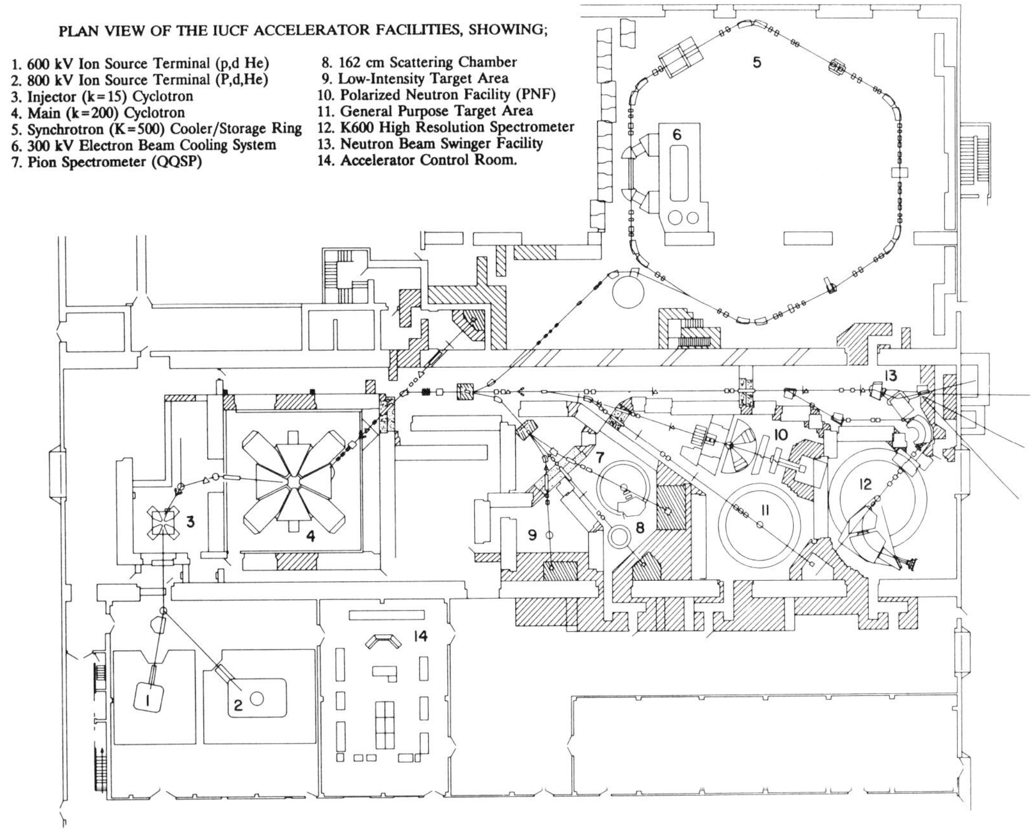

FIG. 13. Plan view of accelerator facilities at the Indiana University Cyclotron Facility. Shown are four stages of acceleration from the ion-source terminals, which are potential-drop accelerators, through two cyclotron stages to a cooler/synchrotron. (Courtesy of D. Friesel, Indiana University)

Cyclotron beams are also injected into storage rings where they may be accumulated over many turns and "cooled," i.e., have betatron and synchrotron oscillation amplitudes reduced either by applying a series of small transverse or longitudinal kicks to the beam or by transferring oscillation energy to an electron beam that shares part of the ring circumference. The experimenter's target is also placed within the ring. The recirculation of the beam through the target increases the interaction rate while the cooling system removes the "heating," or beam degradation, caused by the target so as to maintain the beam size and narrow energy spread. Most rings include an rf cavity so that, after accumulation, the magnet and rf may be ramped, as in a synchrotron, and the beam accelerated to higher energy. Five cooler/synchrotrons have been built, or are under construction, the first (Fig. 13) being at Indiana University with $K_b=560$. The first cooling of a beam from a cyclotron occurred in 1984 at the 8-MeV/amu test accumulation ring TARN I at the Institute of Nuclear Study, Tokyo.

It is proposed, at TRIUMF, to produce 100 μA of 30-GeV protons using the existing cw cyclotron as an injector for a chain of rapid-cycling synchrotrons. Since synchrotrons, like synchrocyclotrons, capture particles over a relatively short injection period, an interjacent storage ring would be necessary in order to utilize most of the cyclotron beam. H^- ions would be extracted from the cyclotron at $\sim$450 MeV and injected into the ring by stripping to protons. The protons would be accumulated, at 450 MeV, over $\sim$20 ms ($\sim 10^4$ turns) and extracted in one turn for transfer to a synchrotron, whereupon accumulation would start again. The synchrotron would have a 50-Hz repetition frequency.

In general a postaccelerator demands better beam quality and stability of operation from an injector than most other applications.

4. APPLICATIONS OF CYCLOTRONS

The development of early cyclotrons was driven by the needs of basic research in physics and chemistry; however, workers in other fields were soon experimenting with the use of isotopes as tracers and particle beams in medicine. Construction of cyclotron-based facilities intended primarily for use in fields other than nuclear physics began in the 1950s. A number were intended specifically for isotope production [e.g., Amersham International (U.K.)], other for general science and engineering (e.g., Harwell), and others for neutron therapy [e.g., Hammersmith Hospital, London (U.K.)]. The characteristics of cyclotron beams overlap those of potential-drop and linear accelerators, and the machines share some common applications.

In the past, cyclotrons have been designed and sold by large engineering companies, e.g., Philips (The Netherlands), A.E.G. (Germany), Thomson–CSF (France), and Japan Steel Works. They have been joined, in the past 15 years, by smaller specialized firms such as Scanditronix (Sweden), IBA (Belgium), RDS–Siemens (U.S.), and Ebco Technologies (Canada).

4.1 Interaction of Particle Beams with Matter

Most cyclotron applications depend on the way their beams interact with matter. These processes are summarized below and are described more completely in the articles on atomic and nuclear physics. Charged particles with energies up to a few MeV interact chiefly with the Coulomb field of electrons and nuclei. Sufficient momentum may be imparted to atoms to lead to lattice dislocations. Electrons may be detached from atoms (ionization) or excited to more energetic states with subsequent emission of characteristic x rays. Molecules may be dissociated (radiolysis), and the resulting ions or radicals may be very reactive chemically. Detached electrons or displaced atoms may be sufficiently energetic to cause further ionization or dislocation.

At energies above the Coulomb potential barrier (4 MeV for protons on oxygen), charged particles may enter the nucleus and increase nucleon orbital energy with subsequent emission of characteristic γ rays, or initiate a reaction transforming the atom to a different isotope or element. The deuteron binding energy is only 2.2 MeV; at energies above this the nuclear Coulomb field may polarize, then dissociate, an incoming deuteron. The component neutron may enter a nucleus, since neutrons are not inhibited by the Coulomb barrier, and initiate reactions.

Particles with tens of MeV may eject several nucleons or light nuclear fragments from a nucleus. Figure 14 shows the probability of multiparticle ejection and the creation of different isotopes as the bombarding energy increases. At higher energies copious numbers of energetic secondary particles are emitted and cascade reactions ensue. Meson production begins around their rest energy, 106 MeV for muons and 140 MeV for pions.

Atomic interactions are many orders of magnitude more likely than nuclear interactions and ions lose their energy in many small steps. The energy deposited along the track, averaged over many samples, is shown in Fig. 15. Ions, in contrast with neutrons, photons, and electrons, have a specific range at the end of which there is a peak, the "Bragg peak," in differential energy loss. There is a statistical variation in range caused by absorption reactions and angular scatter. Very low-energy π^- are captured in a nucleus that disintegrates into short-range fragments and neutrons enhancing the Bragg peak. Figure 16 shows the energy dependence of the differential energy loss, or stopping power, and the mean range for several particles in carbon and lead.

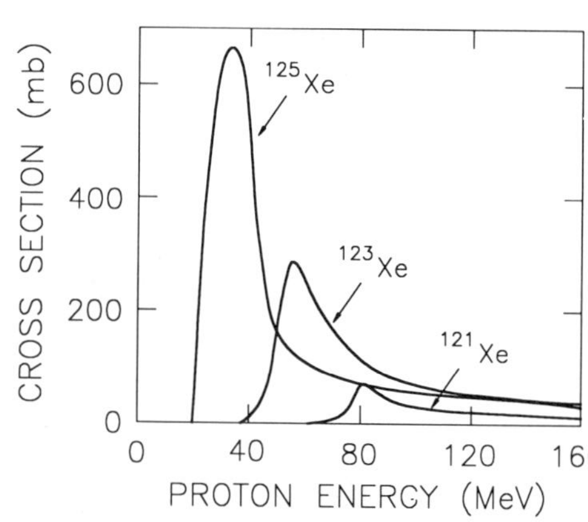

FIG. 14. Production of xenon isotopes as a function of bombarding energy; reaction $^{127}I(p,xn)Xe$. (After Syme *et al.*, 1978.)

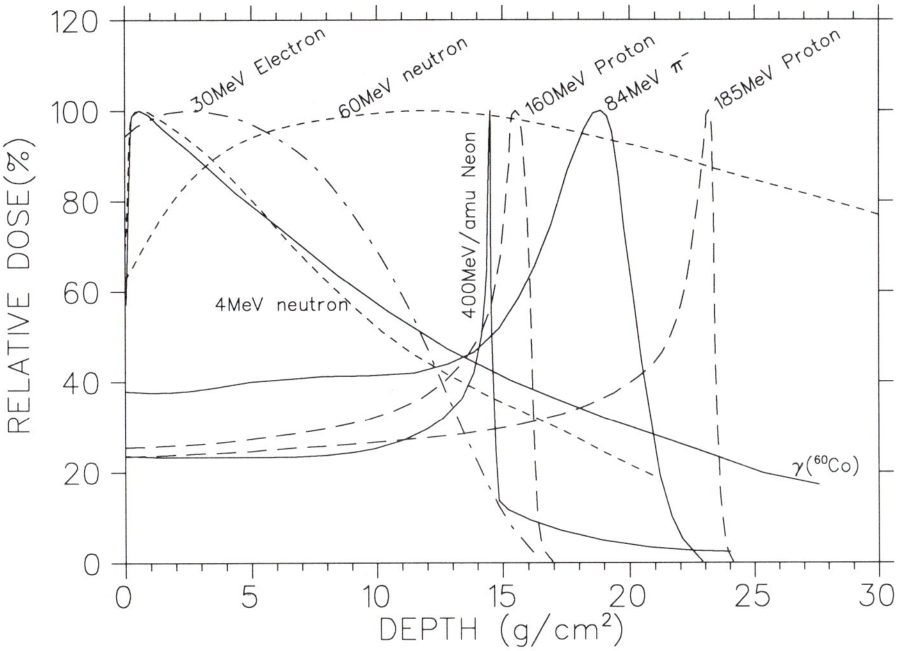

FIG. 15. Dose (related to local ionization) as a function of penetration depth in water or tissue for several particle beams and gamma rays (1.17 and 1.33 MeV) from ^{60}Co.

4.2 Basic Research in Nuclear Science

The separate articles NUCLEAR STRUCTURE, NUCLEAR REACTIONS, and PHYSICS PARTICLE give a general treatment of this subject.

The first generation of AVF cyclotrons, with $K \lesssim 100$ and accelerating light ions ($A \lesssim 7$), contributed beams that were more intense and higher in energy than were available from potential-drop machines. The energy resolution could be made comparable by the use of slits to reduce phase-space volume and by magnetic analysis of the extracted beam coupled with magnetic analysis of reaction products. The beam bunch length could be made small enough that reaction products could be identified from their differences in flight time from target to detector. On the other hand, the quasi dc macroduty factor, when compared with that of a pulsed linear accelerator, gave better rejection of uncorrelated events for those experiments that depend on the coincident detection of several correlated reaction products. Some examples of fields investigated using these cyclotrons are quantitative descriptions of nuclear potentials, identification of nuclear states excited at tens of MeV, and reactions with several emergent nucleons.

Nuclear structure and reaction mechanism studies continue at the higher-energy separated-sector cyclotrons and meson factories where, e.g., the nucleon-nucleon potential could, be explored using spin-polarized beams and the charge distribution within nuclei obtained by comparing π^+ and π^- scattering. The new, cyclotron-injected, cooler-synchrotrons offer nuclear physicists superior energy resolution ($\Delta E/E \approx 10^{-5}$) at high energies, from tens of MeV/amu to 2 GeV. The high flux from meson factories facilitated particle physics experiments, in particular the search for those rare decay modes of mesons that may violate the presently accepted model of strong, weak, and electromagnetic forces or hint at conservation laws not yet included. At bombarding energies of hundreds of MeV many different isotopes are produced, many of which are short lived. An electromagnetic analysis system that can isolate individual species on line is used by hundreds of experimenters at the CERN synchrocyclotron.

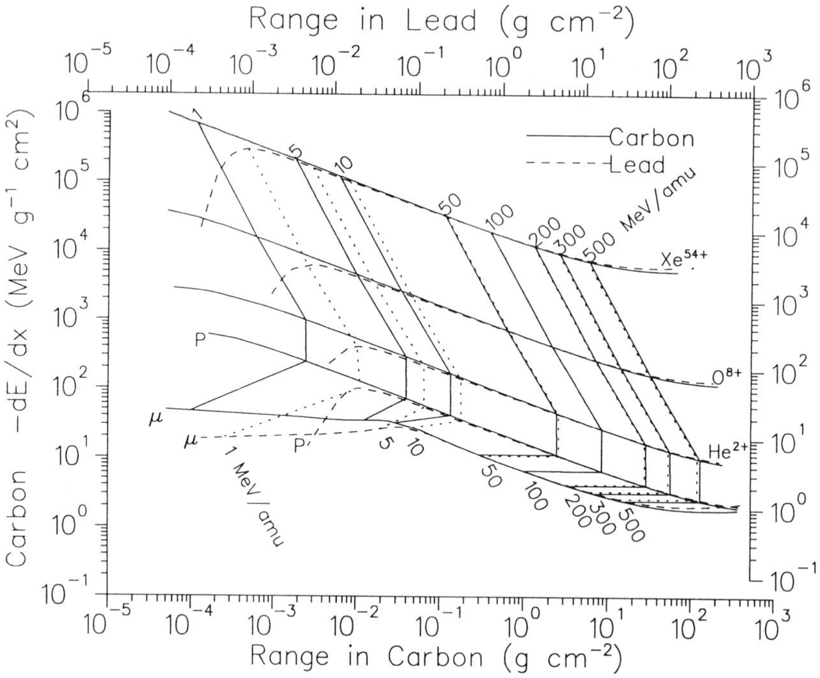

FIG. 16. Range and stopping power of muons, protons, alphas, oxygen, and xenon ions in carbon and lead (vertical lines join particles with the same energy/amu and are not intended for interpolation).

Beams from heavy-ion cyclotrons may have less energy/nucleon than from meson factories, but their higher total energy make them suitable for study of the bulk properties of condensed nuclear matter in a background relatively free from the products of competing, e.g., nucleon-nucleon, reactions. Heavy projectiles can generate excited nuclear states with high rotational energy or extreme modes of vibration. A large energy input creates compound nuclei with high thermal energy; subsequent evaporation of nuclear matter produces transuranic elements and nuclei far from the line of stable elements on the *A-Z* diagram.

Astrophysicists and nuclear chemists are interested in reactions initiated by unstable isotopes and have proposed facilities to produce, collect, ionize, and accelerate radioactive particles. The first such facility in operation is at the Université de Louvain, Louvain-la-Neuve, Belgium, where ^{13}N, with a 10-min half-life, is generated using 30-MeV protons from one cyclotron, ionized in a radiation-hard ECRIS placed next to the production target, and then accelerated in the adjacent $K=120$ variable-energy cyclotron CYCLONE.

4.3 Basic Research in Other Physical Science

Neutrons have long been used in condensed matter research; see e.g., NEUTRON DIFFRACTION. Accelerator-based neutron sources now complement the traditional reactors and are more easily licensed at present. Storage-ring/synchrotron sources produce pulses of neutrons with a higher instantaneous flux and offer good resolution through time-of-flight techniques. The integrated thermal-neutron flux from the beam dump of a cw-cyclotron meson factory approaches that from research reactors ($2\times10^{14}/\text{cm}^2\cdot\text{s}$); the cold-neutron flux is superior. An ambitious project under construction at PSI will direct a 1.4-mA beam from their 600-MeV proton cyclotron vertically into a lead target surrounded by horizontal neutron guide tubes. Liquid D_2 or H_2 moderators are placed close to this almost point source. The neutron flux at the experiments, per mA of protons, is 10^{10} $n/(\text{cm}^2\text{ s nm})$ at 0.1 nm to 10^9 $n/(\text{cm}^2\text{ s nm})$ near 1-nm wavelength.

The application of muons, a more recent and rapidly evolving field, is discussed in the article MUON SPIN RESONANCE. The muon, μ^- or μ^+, has a mass 206 times that of an electron, or one-ninth that of a proton. It experiences electromagnetic and weak forces but not the strong (nuclear) force. The μ^- mimics the response of a heavy electron to such forces, but the μ^+ acts more like a light proton, often forming the "element" muonium ($\mu^+ e^-$) that has almost the same size and binding energy as the chemically reactive hydrogen atom but only one-ninth the mass and three times the thermal velocity. The proton radius is 10^{-13} cm, whereas the μ^+ is pointlike. The muon half-life, 2.2 μs, vastly exceeds the thermalization time (of order 1 ns) when stopping in matter and is long with respect to many chemical and physical processes. Their range may be estimated from Fig. 16.

The muon beams originate from pion decay and are spin polarized along, or opposite to, their direction of motion. In a local magnetic field, B, a muon will precess at a Larmor frequency of (135.5 MHz/T)B. The precession of the triplet state of muonium at low fields is (13.9 GHz/T)B. Muons themselves decay into a neutrino, an antineutrino, and an electron or positron; the latter are emitted preferentially along the muon spin axis with energies up to 52.8 MeV and are easily detected. In a typical muon spin-rotation experiment, a sample is placed in a homogeneous magnetic field and the time interval between a μ^+ entering the sample and its decay positron entering a detector nearby is recorded for millions of events. In the example shown in Fig. 17 a fast precessional signal is seen to be superimposed on the slower muon decay. The precession frequency depends on the microscopic magnetic environment in which the muon finds itself, e.g., a certain crystal lattice site or chemical species. The amplitude depends on the number of muons finding such sites, which may be related to their availability. The term μSR also includes longitudinal spin relaxation experiments carried out in zero field or with a bias field (anti)parallel to the spin direction and muon spin resonance where transitions between states are detected by the absorption of microwave power. Data acquisition may be triggered by a step change in the sample environment, e.g., illumination from a pulsed laser. Negative muons are more likely to depolarize during capture by atoms and to undergo nuclear capture. When not captured they probe magnetism at sites different from μ^+SR.

Currently μSR is used as a probe of structure (e.g., measurement of magnetic penetration depth in superconductors, or the study of defect states in semiconductors) and dynamics (e.g., quantum diffusion in solids, or following μ^+-substituted radicals in chemical reactions).

Muons can bind certain nuclei sufficiently closely, e.g. 700 fm for $^2H\mu^-\,^3H$, that nuclear fusion may occur. The μ^- is released to participate in about 150 similar events before decay. Muon-catalyzed fusion may be a distance source of energy, and the process is being studied in detail at several laboratories.

The present meson-factory beam lines deliver μ^+ fluxes from 10^5 to 10^7/cm^2·s. Upgraded facilities at PSI will increase this severalfold, while the proposed kaon factories would provide a further increase.

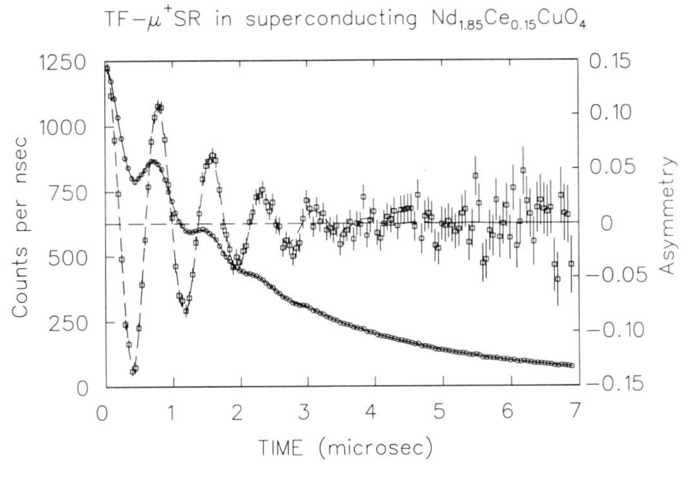

FIG. 17. Solid line: Positron counting rate from μ^+ stopped in a superconducting compound immersed in a 0.01-T transverse magnetic field. Dashed line: The same data corrected for background and muon decay. (Courtesy of J. Brewer, University of British Columbia.)

The local structure of materials is also examined using heavy ions, e.g., at the Hahn-Meitner Institute, Berlin, Germany. Hyperfine structure depends on both nuclear and atomic parameters and the latter are influenced by the solid-state environment. Nuclear probes, which may be constituent or impurity ions, are implanted and the hyperfine levels measured by nuclear techniques such as the Mössbauer effect or magnetic resonance.

4.4 Isotope Production

Radioactivity is a well-established field, discussed in a separate article, and radioisotopes have been used for decades, primarily as tracers and for nuclear medicine. Most commercial products incorporate isotopes from reactors. Neutron-deficient isotopes, however, require accelerators, and the accelerator of choice is a high-current cyclotron with variable-energy beam. The international trade in cyclotron products amounts to $100 million annually, mostly radiopharmaceuticals, and is increasing. Some targets are used as irradiated—e.g., in wear studies of machine parts—but usually the radioisotope has to be separated from the original target material and other reaction products and synthesized into a chemical compound. For example, ^{18}F-labelled 2-fluoro-2-deoxy-D-glucose is used to study glucose metabolism of the brain; ^{13}N- and ^{15}N-labelled compounds are used to study the fast and slow uptake of nitrates by plants. The investment in hot cells and radiochemical equipment can be comparable to the cost of a cyclotron. The target, bombarding particle, and energy (Fig. 14) are usually chosen to optimize the desired cross section and to reduce contamination from competing reactions. Table 1 lists several commercial isotopes and their production reactions.

Short-lived isotopes give a stronger signal with less environmental dose; the use of cyclotron-produced ^{123}I in thyroid studies reduced patient dose fiftyfold compared with the longer-lived, reactor-produced ^{125}I or ^{131}I. Production, however, must be sited near a user or near excellent transportation facilities. The isotopes ^{11}C, ^{13}N, ^{15}O, ^{18}F, used in positron emission tomography, have half-lives of minutes. Fortunately they can be produced by low-energy (<20 MeV proton) cyclotrons that can be installed in a hospital. Production facilities at TRIUMF are 2.5 km from the university hospital; however, radiopharmaceuticals are transferred within 2 min by means of an underground pneumatic tube (rabbit).

In addition to research laboratories and hospitals with radioisotope production facilities, there are also several companies that make, process, and distribute isotopes and isotope-labelled compounds. These include Amersham International (U.K.), DuPont-Merck and Mallinckrodt (U.S.), Nihon Mediphysics (Japan), and Nordion International (Canada). Most use proton beams below 50 MeV and have two or more cyclotrons to guarantee production. Isotopes from spalla-

Table 1. Examples of popular cyclotron-produced medical isotopes.

Isotope	Production reaction	Beam	Use[a]
Thallium-201	^{203}Tl$(p,3n)$ ^{201}Pb ^{201}Pb $\rightarrow$ ^{201}Tl	29-MeV protons	Heart studies (SPECT)
Gallium-67	^{68}Zn$(p,2n)$ ^{67}Ga	26-MeV protons	Soft tumor imaging (SPECT)
Iodine-123	^{124}Xe$(p,2n)$ ^{123}Cs ^{123}Cs $\rightarrow$ ^{123}Xe $\rightarrow$ ^{123}I and ^{124}Xe(p,pn) ^{123}Xe $\rightarrow$ ^{123}I	30-MeV protons	Thyroid imaging and screening of acute stroke patients
Xenon-127	^{133}Cs$(p,$spallation$)$	High-energy, e.g. 500-MeV protons	Lung agent
Carbon-11	^{16}N(p,α) ^{11}C	12–14 MeV protons	Brain function (PET)
Fluorine-18	^{18}O(p,n) ^{18}F	~11-MeV protons	Brain function (PET)
Rubidium-82	^{85}Rb$(p,4n)$ ^{82}Sr ^{87}Rb$(p,6n)$ ^{82}Sr $^{82}Sr \xrightarrow[25d]{} {}^{82}Rb \xrightarrow[1.2m]{\beta^+} {}^{82}Kr$	70-MeV protons 85-MeV protons	Heart function (PET) and brain metastases

[a]SPECT: single-photon-emission computed tomography. PET: positron-emission tomography.

tion reactions, e.g., ^{82}Sr and ^{127}Xe, require much higher energies; however, they are sometimes made parasitically, e.g., in targets placed just before the beam dump of a line with primary targets, devoted to other work, upstream.

A modern low- or medium-energy cyclotron can produce 15 kW beam power, which is more than most production targets can withstand. A multifoil H^- or D^- cyclotron can simultaneously share this beam between several targets at the same or different energies.

4.5 Particle Beams in Medicine

Cancerous tumors and other structures inaccessible to conventional surgery have been eradicated by using penetrating radiation to ionize and destroy malignant cells; see BIOMEDICAL USES OF RADIATION. Most treatments employ x rays or γ rays. Some tens of thousands of patients, however, have been treated with neutron or charged-particle beams that, although less tractable, can offer improved localization of dose and interact differently with tissue. Proton and α treatments were developed, starting in the mid-1950s, at the Berkeley, Uppsala, and Harvard synchrocyclotrons and at the ITEP (Moscow) synchrotron. Neutron studies, begun at Berkeley, were revived at the Hammersmith cyclotron in the 1960s. Heavy-ion (HI) and π^- therapy began in the 1970s.

To reduce the dose to intervening tissue a patient is irradiated from two or more sides with the beam energy adjusted to place the Bragg peak (Fig. 15), where applicable, at the treatment site. The energy may be altered and the beam steered transversely to distribute the dose over an irregular volume. The choice of particle is influenced by

1. radiobiological effect [HI, π^-, and neutrons have high linear energy transfer (LET) at the treatment site],
2. the ratio of end-of-range to transit dose (favors π^-, protons and HI),
3. the absence of dose beyond a sharp range end (favors protons and α), and
4. dose delineation (primary beams are smaller than secondary).

A therapeutic dose is 10–50 Gy, delivered in several fractions. Moderate energies are required—250-MeV protons reach the deepest tumors—and since 1 nA/cm^2 delivers 1 Gy/min, synchrocyclotrons or synchrotons are adequate for primary beams. Secondary beams, π^- or neutrons, require cw cyclotrons or linacs.

There are 20 neutron treatment centers worldwide, 13 with cyclotrons. Attention has focused on neutron radiobiology, especially at hypoxic sites. A typical installation would accelerate 100 μA of deuterium to 50 MeV, and bombard a beryllium target to obtain 20–40-MeV neutrons that are then collimated into a fixed beam for treatment. A gantry-mounted, 50-MeV deuteron, superconducting cyclotron, however, designed by H. Blosser and installed at the Grace Harper Hospital, Detroit, is able to rotate completely around the axis of a patient (Fig. 18).

About a dozen laboratories operate adjunct medical facilities performing radiosurgery by means of charged-particle beams. Charged-particle treatments have been especially advantageous where dose localization is important, e.g., in the pelvic region, for tumors of the pituitary (adjacent to the spinal chord and brain stem), in choroidal melanoma (adjacent to the optic nerve), and in destruction of arteriovenous malformations in the brain. During the last 6 years 1000 patients with ocular melanoma have been treated using 70-MeV protons from the PSI injector cyclotron. The success rate, 95%, is the same as for the traditional treatment, which involves removal of the eye, but proton treatment also preserves vision in most cases. Recent improvements in medical imaging (CAT, MRI, etc.) make it possible to exploit the 1-mm dose delineation possible with proton beams. Ten new charged-particle treatment facilities are under construction, half with dedicated medical accelerators, six associated with proton cyclotrons.

4.6 Elemental Analysis of Materials

Many physical and chemical methods are used to determine the elemental constituents of a material. Those associated with accelerators include

1. the identification of atoms from their characteristic x rays emitted under bombardment (PIXE),
2. the identification of nuclei from their decay products following irradiation,
3. the identification of nuclei from elastic-

FIG. 18. A superconducting, 50-MeV deuteron, cyclotron being mounted on a cylindrical rotating gantry at the Harper Hospital (Detroit). Neutrons are directed toward a patient located near the cylinder axis. The patient position is adjustable and irradiation can take place from all sides. (Courtesy of H. G. Blosser, NSCL-Michigan State University.)

scattering cross sections and angular distributions, and

4. mass analysis following ionization and acceleration of a small sample, a technique termed accelerator mass spectrometry (AMS).

These techniques are frequently nondestructive and may be considered when only a small sample (milligrams) is available, isotopes, and not just the elements, must be identified, or the spatial distribution within a matrix is required. Absolute resolutions are of the order 1 ppm or less, i.e., subnanogram quantities; AMS isotopic ratios, e.g., $^{14}C/^{12}C$, have a resolution $\sim 10^{-14}$. In the main they are the domain of low-energy potential-drop machines; however, cyclotrons have advantages in certain cases and, worldwide, more than 10 cyclotron laboratories offer a service to external users and some derive a substantial fraction of their income in this way.

The proton microprobe uses a collimated incident beam focused into a spot ~ 10 μm diameter, and x-ray analysis takes place as the beam is slowly scanned over the sample. The cross section for *K*-shell x-ray production is greater at cyclotron energies, but so is the background; *L*-shell x rays are also produced, however, which may resolve ambiguous cases. A cyclotron beam penetrates further into a sample; however, the spatial resolution is somewhat worse than the standard potential-drop laboratory arrangement. Cyclotron laboratories offering PIXE services include the University of Hamburg and the University of California, Davis Campus. Applications include analysis of thousands of specimens of airborne pollutants per day and the distribution of aluminum in biological tissue.

Proton-induced radioactivity is suited to the detection of trace amounts of light and medium-mass nuclei, whereas neutron activation offers better resolution for heavy elements. Some medium-energy cyclotron laboratories offer both techniques and also irradiation by other ions. Resolutions range from 1 ppb to 10 ppm depending on the element of interest and the matrix. Commercial applications include the determination of oxygen in metals and impurities or dopant in semiconductors. Many laboratories engage in this work; the more experienced include the Radiation Studies Centre (CERI), Orleans, France; University of California, Davis (U.S.), and RIKEN (Japan).

Cyclotrons have been used as mass spectrometers since, at any one time, only ions whose q/A lies within a narrow band can be accelerated and extracted. However, the measurement of a ratio, say ^{14}C:^{12}C, for dating purposes would require changes in frequency or field, and potential-drop machines are preferred by AMS facilities that process many samples. The mass resolution of a conventional cyclotron, $\Delta m/m \sim 10^{-3}$, can be enhanced by further analysis of the extracted beam or by designing a machine with very tight synchronism tolerances. The Institute of Nuclear Research, Shanghai, is constructing such a compact and inexpensive cyclotron able to distinguish between ^{14}C, ^{13}CH, and $^{12}CH_2$.

The resolution of isobars, nuclei with the same mass, is effected by measuring another physical property, e.g., stopping power. Such measurements are easier at the higher energy available from cyclotrons.

4.7 Transmission Radiography

The absorption and scatter coefficient for x rays increases smoothly with atomic number, whereas the coefficient for neutrons is scattered over 5 order of magnitude and varies considerably from element to element, even from isotope to isotope. This selectivity enables neutron radiography to reveal the distribution of those elements with high cross section, e.g., hydrogenous or nitrogenous explosives encased in materials of high Z. The isotopic sensitivity makes it possible to observe the distribution of enriched uranium in fuel rods. In other materials neutrons have high penetrability and large castings can be inspected for voids and aircraft structures for stress cracks and corrosion. The object is backed by material to convert neutrons to electromagnetic radiation, and radiographs are made with x-ray film or pictures recorded in real time using a TV camera. The neutron flux from an easy-to-operate 50-μA, 20-MeV compact cyclotron approaches that from a 5-MW reactor. Japan Steel Works and Sumitomo Heavy Industries have both installed cyclotron-based radiography systems. Oxford Instruments (U.K.) manufactures a transportable, superconducting, 200-μA, 17-MeV H^- cyclotron. The machine consumes only 20 W, weighs 2 tonnes, and may be mounted on a "cherry picker" device. The flux of thermal neutrons is estimated to be 2×10^{11} $n/cm^2\cdot s$.

Proton radiography is used less extensively; it requires energies above 100 MeV for a useful range. Marginal-range radiography depends on the rapid change in differential energy loss near the end of the Bragg peak. The proton energy is adjusted to match the thickness of the object and resolutions of 0.1% for objects a few mm thick and 1% for 50 mm are obtained, which is almost ten times better than for x rays. In scatter radiography a broad, parallel beam illuminates the object. Protons passing through the object near the edge will be scattered into the unperturbed stream enhancing the delineation of the edge. This subject has been developed extensively at the Harwell 160-MeV synchrocyclotron.

These techniques leave little residual activity and are thus nondestructive. Since neither protons nor neutrons are common products of radioactive decay, the methods above may be used, with appropriate detectors, on radioactive objects.

4.8 Wear and Corrosion

The rate of wear caused by operation of equipment can be measured by irradiating part of the surface subject to wear, then reassembling and operating the equipment while monitoring the radioactivity. In some cases abraded material may be carried by lubricants, or other fluids, past a radiation detector to give a dynamic measure of wear as a function of machine operating conditions. In cases where this is not possible, e.g., locomotive wheels, the activity remaining is measured periodically while logging the amount of use. The amount of activity required is small; the irradiated area may be large or as small as 1 mm and located to 0.1 mm. Thickness changes may be measured to 1 nm. Different radionuclides may be created on different components of the same machine and relative wear observed in a single experiment. The Kernforschungzentrum Karlsruhe (Germany) cyclotron has treated 500 parts/year for many years and for different customers. A recent development in this field is the implantation of radioactive ^{7}Be and ^{22}Na carried out at NSCL-MSU.

4.9 Materials Modification and Processing

The dislocations, ionization, and reaction products from ion bombardment can alter the physical properties, e.g., conductivity, of a target and also lead to chemical change. Normal chemical incompatibilities may be overcome and unusual alloys and compounds may be formed at low temperature and in the solid state. The bombarding energy may be chosen to concentrate the changes at the surface or in a layer buried beneath. Chemical change may be minimized by using one of the target elements as the ion beam; conversely, physical dislocations can often be removed by annealing. The $K \approx 100$ cyclotrons at the Oak Ridge National Laboratory, Harwell, KfK Karlsruhe, and the Joint Institute of Research, Dubna, have exploited such processes for decades. The newer, high-K, machines, in particular the chain of cyclotrons at GANIL, Caen, France, extend the range of ion species and energies, and consequently the depth of dose and LET, available. Irradiations may take place under vacuum, but the majority are carried out in air, the beam emerging through a thin metal window at the end of an evacuated beam pipe.

Light- and heavy-ion beams are used to render stainless steel surfaces more resistant to corrosion, friction, and wear. Heavy-ion and alpha beams are used to simulate the damage caused by neutrons in fission or fusion reactors. The neutrons cause dislocations in structural materials and helium, from (n,α) reactions, accumulates in the voids. Ion bombardment produces similar effects, but at much higher rates, and also allows mechanical tests to be performed during irradiation.

Ionizing particles may cause intermittent malfunctions in electronic circuits; heavy exposure may lead to failure. Ion beams are used to measure the radiation resistance of military or space-bound electronics in the laboratory. The French agency MATRA ESPACE simulates space radiation effects by using 5- to 50-MeV/amu carbon to xenon ions from the GANIL cyclotron to obtain penetration depths from 5 to 1300 μm and LET between 0.2 and 68 MeV/(mg cm)^{-2}.

Micropore filters are used to remove contaminants from fluids (e.g., the water used in printed circuit manufacture), to separate biological cells, and as semipermeable membranes to preserve the freshness of fruit and vegetables. Superior filters (Fig. 19) are manufactured at Dubna and also by the company Biopore using beams from GANIL. Rollers feed polycarbonate film, 10–50 μm thick, through 10^{12}-s^{-1} beams of 5- to 10-MeV/amu xenon or krypton and then into etching and washing solutions. These preferentially dissolve material around the ion path to produce cylindrical holes with a high degree of uniformity and a diameter determined by choice of ion species, energy, and etching process. Diameters range from 0.05 to hundreds of microns. Filters may have 10^9 holes/cm^2, whereas membranes used for cell counting have a single hole.

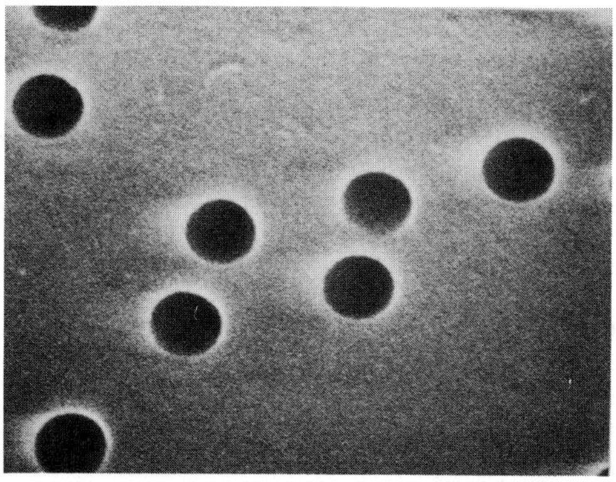

FIG. 19. Holes with a 0.2-μm diameter produced in polycarbonate film by heavy ion bombardment followed by etching. [Courtesy of C. Bieth (GANIL) and Biopore S.A., France.]

Embryonic commercial processes include the generation of color centers in crystals, and the manufacture of field-emission surfaces and of nonreflective glass and plastic. Multi-MeV beams can bury oxide or nitride layers deep in a semiconductor, thus permitting the development of three-dimensional integrated circuits, optical circuits in semiconductor sandwiches, and the micromachining (via etching) of mechanical devices in silicon.

5. FACILITY DESIGN AND SHIELDING

Cyclotron laboratories employ conventional construction methods for the most part, but have the additional requirements of shielding personnel and experimental equipment from primary and secondary particles and of monitoring and minimizing radioactivity in the workplace, at points of egress, and in air and water exhausts.

For beam energies above 10 MeV/amu a shielding wall thick enough to attenuate the neutrons produced to a safe level will also attenuate γ rays and stop charged particles. The number of fast neutrons rises rapidly with beam energy [Fig. 20(a)]. They are slowed, initially, by nuclear elastic or inelastic scattering in the shield and some nuclear reactions may occur. They reach thermal equilibrium at some distance from the point of origin and are ultimately captured in a nucleus,

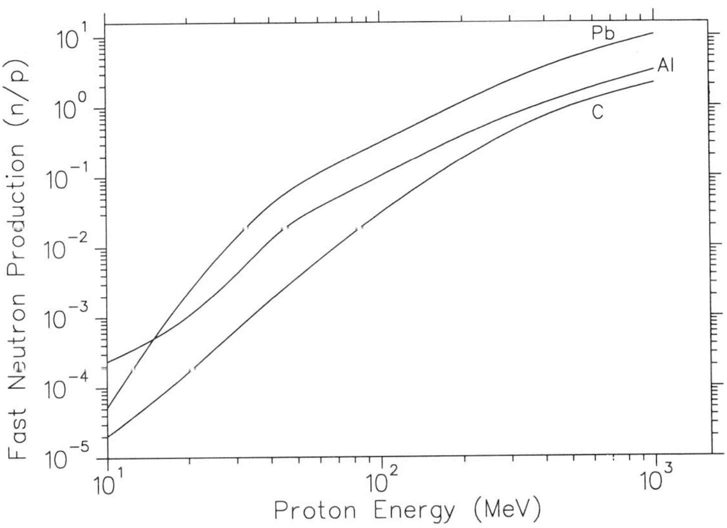

(a)

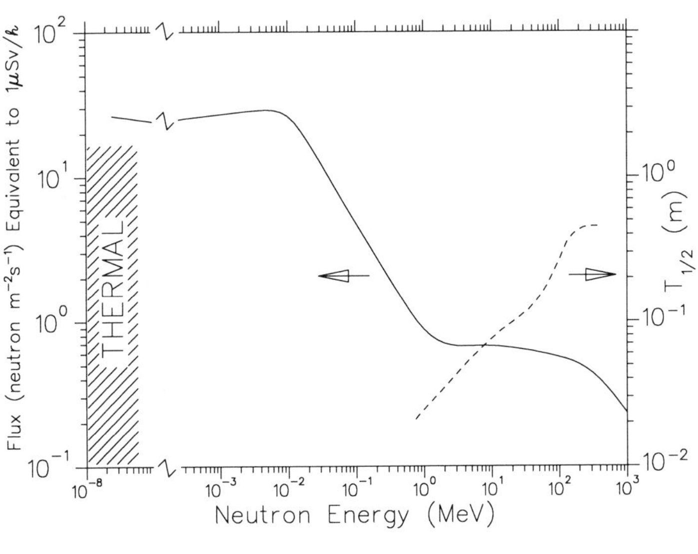

(b)

FIG. 20. (a) Yield of fast neutrons produced by protons stopping in thick targets. (b) (Solid line) Neutron flux giving unit dose and (dashed line) the thickness of concrete (density 2.4 g/cm^3) required to halve the neutron flux as a function of neutron energy.

usually with the release of a γ ray that may necessitate an additional layer of shielding.

The maximum permissible dose for radiation workers is currently 0.05 Sv/year. This may be transformed to neutron flux and an approximate estimate of shielding thickness made with the aid of Fig. 20(b). While hydrogenous materials have high neutron-scattering cross sections, materials with high atomic number are better attenuators of γ rays. Concrete is the usual low-cost compromise. Shielding calculations can be quite complicated since both particle production and absorption are multibranched energy-dependent processes, and Monte Carlo simulations are often used.

In practice the greatest dose to humans comes from induced radioactivity and from activities such as processing targets and from cyclotron or beam line maintenance. Some beam loss is inevitable; good design minimizes the amount and concentrates its deposition in absorbers such as carbon, which minimize residual activity, and where the loss can be monitored by cyclotron operators. Particles likely to be lost, i.e., those with large betatron amplitude or extreme phase, may be trimmed at low energy. The absorbers may be removed quickly and safely during maintenance or themselves shielded by a portable lead curtain. Lead 10 cm thick attenuates 10-MeV gammas 100-fold and stops betas and alphas. Impurities in construction materials can cause significant residual radiation problems; two important examples are sodium in concrete (which may be a few percent unless controlled) yielding ^{24}Na, and trace amounts of cobalt in steel producing ^{60}Co. The vault and high-intensity beam lines should be sealed and volatile radiation products and airborne radioactivity pulled through a monitored exhaust. The cooling water for components in areas of high loss should recirculate through ion exchangers to remove products of radiolysis. Personnel exposure should be considered at the equipment design stage. For example, screws may be superseded by quick-release clamps, quick disconnects used for electrical and fluid services, and remote removal and handling features incorporated.

Most organic materials and electronic equipment begin to deteriorate at an accumulated dose of 10^4 Sv. Some are more resistant, e.g., ethylene/propylene O rings (10^5–10^6 Sv), fiberglass, and certain epoxies (5×10^6 Sv). At higher levels inorganic insulators and metal vacuum seals are used.

Ancillary equipment, such as production targets or particle detector arrangements, are reconfigured quite frequently and may not be commissioned to the same level of reliability as the cyclotron. Some targets may be more radioactive than the cyclotron itself. These facilities are best placed outside the cyclotron vault in their own shielded rooms that can provide low-background areas for experiments requiring them, easier access for maintenance of both cyclotron and equipment, and, when magnets can switch the beam between two or more such areas, a more productive use of cyclotron time. The switching magnets, and any initial collimation or common beam-focusing elements, are best placed in the cyclotron vault. Site plans for all types of facilities may be found in the proceedings of conferences; Figs. 13 and 21 are examples. While poured-in-place concrete is cheaper and services may be conveniently attached, movable blocks are commonly used to subdivide areas and for some outer walls in order to facilitate future rearrangement and expansion.

APPENDIX. DESCRIPTION OF A TYPICAL ISOTOPE-PRODUCTION CYCLOTRON

The TR30 500-μA, 30-MeV H^- cyclotron (Fig. 22), built by Ebco Technologies, Vancouver, Canada, is located at the TRIUMF site (Fig. 21), and operated by Nordion for isotope production.

The compact design has four radial sectors. The magnet is approximately square in shape, 2.3 m from side to side and 1.26 m high, and weighs approximately 46 tonnes. It is split at the midplane, allowing four hydraulic jacks located in the magnet supports to elevate the upper part for access to the cyclotron interior. Two 37 000-A-turn coils mounted on the upper and lower poles provide the magnet excitation. No trim coils are used. Tolerances for isochronism, vertical focusing, and avoidance of resonances were met by a slight modulation of hill gap and by shims attached to the hill sides; the latter received their final machining following magnetic measurements.

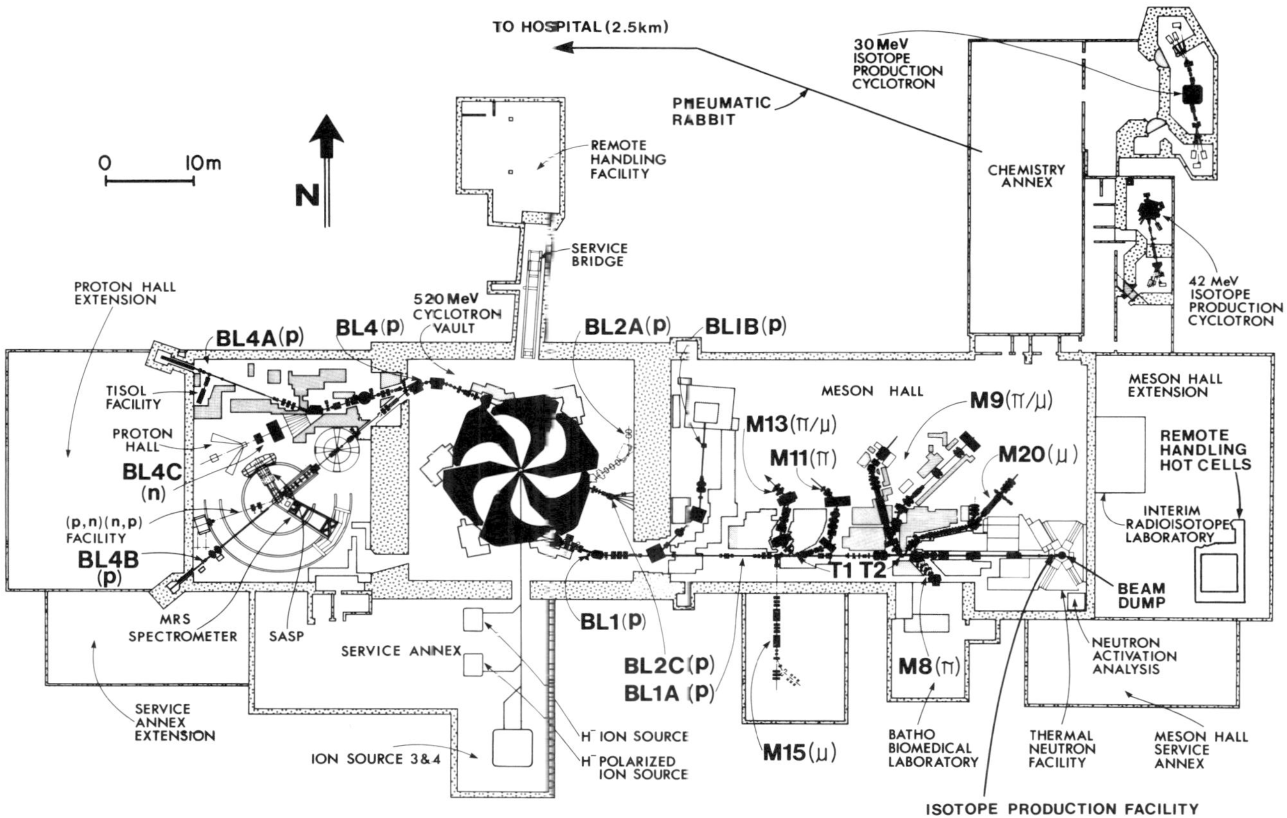

FIG. 21. TRIUMF site plan showing the 520-MeV, 42-MeV, and 30-MeV H^- cyclotrons. The meson production beam line, BL1A, services secondary particle channels M8 (π^- therapy), M20, M15, and M13 (μSR), M9 (particle physics), and M11 and M13 (π-nuclear physics). Spallation isotope production and neutron applications take place at the beam dump. The dump window limits proton currents to 200 μA. Beams of energy 65–100 MeV are used in BL2C for isotope production. BL4A, 4B, and 1B accept beams from 180 to 520 MeV with maximum currents 10 μA, 0.2 μA, and 0.01 μA, respectively. Polarized H_2 and $^2H_2$ targets in 4A are used for nucleon-nucleon physics; a liquid $^2H_2$ target produces polarized neutrons for BL4C. The magnetic spectrometers MRS and SASP are used for nuclear physics studies. Large equipment is assembled and checked out in the E and W extensions; the whole area is serviced by a pair of 50-ton cranes. Power supplies, data acquisition and control equipment, control rooms, and offices are located at several levels in the service annexes and the north side of the meson hall. Hot cells for isotope separation and manufacture of radioisotopes are located in the chemistry annex. Speckle indicates concrete poured in place; the unshaded shielding is movable concrete blocks.

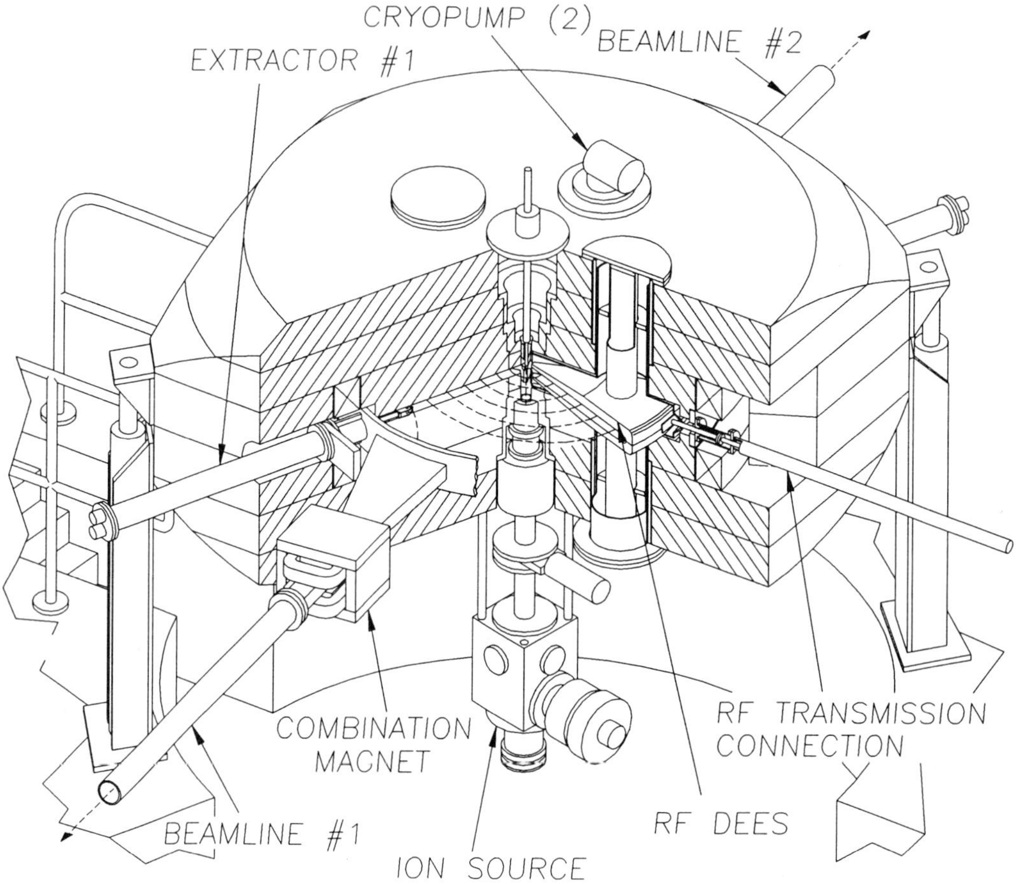

FIG. 22. Major components of a 30-MeV H^- cyclotron intended for isotope production. (Courtesy of Ebco Technologies, Canada.)

The cyclotron is mounted over a pit to allow installation of the external H^- ion source below and thus minimize cyclotron-vault head-room requirements. The H^- beam is injected vertically upward toward an electrostatic spiral inflector that bends it into the median plane. Two 45° dees located in opposite valleys operate at 73 MHz, the fourth harmonic of the orbit frequency, and provide acceleration at four gap crossings per orbit. The dee voltage is 50 kV. RF power is delivered to the dees through a capacitive coupling to a 50-Ω transmission line that passes through a port in the vacuum-tank wall. For ease of maintenance the entire 35-kW rf amplifier system is located outside the cyclotron vault.

Four large holes through the yoke in the dee valleys accommodate the coaxial dee resonator stubs. To maintain fourfold magnetic symmetry there are four additional holes in the unoccupied valleys. Two of these are used as vacuum pump ports in which two 8-in. cryopumps are installed. The vacuum enclosure is defined by the nickel-plated upper and lower pole surfaces and a cylindrical aluminum wall that is sealed to the poles by a double O-ring gasket. The operating vacuum is 7×10^{-10} bar.

Thin graphite stripping foils mounted on two extraction probes traveling in opposite hill gaps deliver two independent, variable-energy, external beams. The basic cyclotron parameters are given in Table 2.

ACKNOWLEDGMENTS

The authors gratefully acknowledge the assistance of those colleagues who provided information for this article and helped with proofreading.

Table 2. Principal parameters for the TR30 cyclotron.

Magnet	Average induction	1.20 T
	Hill induction	1.90 T
	Valley induction	0.55 T
	Hill gap	4 cm
	Valley gap	18 cm
	Pole radius	76 cm
RF	Frequency	73 MHz
	Harmonic	4
	Dee voltage	50 kV
	Power (for 500 μA)	32 kW
Ion source	Type	H^- cusp
	Output current	5 mA
	Biased at	25 kV
Injection	Energy	25 keV
	Method	axial
	Inflector field	20 kV/cm
Extraction	Energy	15–30 MeV
	Method	stripping
	Number of extracted beams	2

GLOSSARY

amu: Atomic mass unit, currently $\frac{1}{12}$ of the mass of the carbon isotope ^{12}C.

AVF Cyclotron: An isochronous cyclotron with an aziumthally varying field to provide focusing and thus permit cw operation at relativistic energies.

Betatron Motion/Frequency: Oscillations in the transverse [axial (z), radial (r)] planes executed by the particles constituting a focused beam. The frequency of such oscillations expressed in terms of the particle rotation frequency ($Q_{z,r}=\omega_{z,r}/\omega_{\text{rot}}$).

Bunch: A collection of particles accelerated within one rf cycle.

cw Cyclotron: One with a constant accelerating frequency, and hence a continuous pulse train and macro duty factor of 1.

Dee: A hollow electrode, semicircular in early cyclotrons, carrying the rf accelerating voltage. Part of the beam path lies within the dee, acceleration occurring at entrance and exit.

Emittance/Acceptance: A quantitative description of the quality of a beam. The transverse emittance is a combination of the beam size and divergence; the longitudinal emittance, a combination of momentum spread and bunch length. The acceptance of an accelerator corresponds to the emittance of the beam that would just fill it.

Equilibrium (closed) Orbit: A conceptual unaccelerated particle trajectory closing smoothly on itself after one turn. Betatron oscillations can be imagined to take place about this orbit.

FM Cyclotron or Synchrocyclotron or Phasotron: A cyclic particle accelerator in which the accelerating frequency is varied during acceleration to match the particle rotation frequency. The beam is produced in a series of pulses with macro duty factor 0.1–10% at repetition rates from 0.02 to 2 kHz.

Isochronism: Equality between particle rotation frequency and accelerating frequency.

K-Factors: Parameters associated with a cyclotron magnet. The maximum energy of an ion, charge qe, mass A, which can be contained in a closed orbit is $(T/A)=K_b(q/A)^2$ MeV/amu; the maximum energy which can be focused axially is $(T/A)=K_f(q/A)$ MeV/amu.

Macro Duty Factor: The ratio of pulse length to repetition time.

Meson Factory: An accelerator laboratory where intense beams of particles are produced whose energy is well above the threshold for meson production.

Particle Microamp: Cyclotron beam intensities are usually expressed in microamperes, 1 μA being 6.24×10^{12} electronic charges/s. For an ion of charge qe the current may be expressed in particle microamperes where 1 pμA is 6.24×10^{12} particles/s, the electrical current being q times larger.

Pulse (Train): A sequence of bunches.

Stripping: Removal of some, or all, of the atomic electrons from an accelerated ion.

Synchrotron Oscillations: Oscillations in energy and phase about those of an ideal, synchronous particle.

List of Works Cited

Blosser, H. G., Johnson, D. A. (1974), *Nucl. Instrum. Methods* **121**, 301–306.

Gordon, M. M. (1984), *Particle Accelerators* **16**, 39–62.

Hagedoorn, H. L., Verster, N. F. (1962), *Nucl. Instrum. Methods* **18,19**, 201–228.

Lawrence, E. O., Livingston, M. S. (1932), *Phys. Rev.* **40**, 19–35.

McMillan, E. M. (1945), *Phys. Rev.* **68**, 143–144.

Richardson, J. R., MacKenzie, K. R., Lofgren, E. J., Wright, B. T. (1946), *Phys. Rev.* **69**, 669–670.

Rickey, M. E., Smythe, R. (1962), *Nucl. Instrum. Methods* **18,19**, 66–69.

Syme, D. B., Wood, E., Blair, I. M., Kew, S., Perry, M., Cooper, P. (1978), *Int. J. Appl. Radiat. Isot.* **29**, 29–38.

Thomas, L. H. (1938), *Phys. Rev.* **54**, 580–590.

Veksler, V. (1944), *C. R. Acad. Sci. USSR* **43**, 329–331; **44**, 365–368 (1945); *J. Physics (USSR)* **9**, 153–158.

Willax, H. A. (1963), *Proceedings of the International Conference on Sector Focused Cyclotrons, 1963,* edited by F. T. Howard and N. Vogt Nilsen, CERN Publ. No. 63-19, 386–397.

Further Reading

Humphries, S., Jr. (1986), *Principles of Charged Particle Acceleration*, New York: Wiley.

Kapitza, S. P., Melekhin, V. N. (1978), *The Microtron*, New York: Harwood Academic.

Livingston, M. S., Blewett, J. P. (1962), *Particle Accelerators,* New York: McGraw-Hill.

Scharf, W. (1986) *Particle Accelerators and their Uses,* New York: Harwood Academic.

The best sources for detailed information are the proceedings of international accelerator conferences. The most recent in each series are the following:

Bennett, F., Kopta, J. (Eds.) (1989), *Proceedings of IEEE Particle Accelerator Conference,* New York: IEEE Publishing Services.

Duggan, J. L., Morgan, I. L. (Eds.) (1991), Proceedings of the 11th International Conference on the Application of Accelerators in Research and Industry, *Nucl. Instrum. Methods* **B56/57**.

Marin, P., Mandrillon, P. (Eds.) (1990), *Proceedings of the Second European Accelerator Conference,* Gif-sur-Yvette: Editions Frontières.

Martin, B., Ziegler, K. (Eds.) (1991), *Proceedings of the Twelfth International Conference on Cyclotrons and their Applications,* Singapore: World Scientific.

DATA-ACQUISITION SYSTEMS

WALTER N. GEKELMAN, *Department of Physics, University of California, Los Angeles, California, U.S.A.*

INTRODUCTION

For the most part, the world we live in is complex, and most events stem from situations which have many degrees of freedom. One approach we use to deal with this is to average over quantities; for example, thermodynamics is used to attempt to describe a system in terms of a few variables such as temperature, pressure, and flow velocities. Even in systems such as these, variables of interest may vary rapidly in space and time. If these variables are recorded in any detail, a data set for a given case can easily contain tens of millions of numbers. As an example, consider an experiment in plasma physics done by the author (Gekelman *et al.*, 1987) in which a fully three-dimensional magnetic field changes its topology rapidly (within 100 ms). The vector field is measured at 4000 locations and sampled at thousands of time steps at each position. Magnetic-field data sets in excess of 10^9 numbers are commonplace. If the experiment is fully characterized by measurement of the electric field, $\mathbf{E}(\mathbf{r},t)$, and particle distribution functions, $f_\alpha(\mathbf{r},\mathbf{v},t)$, gigabyte (GB) data sets are collected for each set of initial conditions. This enormous influx of data is becoming the rule in experimental physics as elaborate detectors capable of detailed measurement are designed. This trend in measurement is most welcome since with it comes the opportunity for detailed comparison with theory, but it is no accident that the theories and

1-56081-063-7/92/$5.00+.50

experimental techniques have evolved in parallel with data-acquisition technology.

The list of environments which rely heavily on data-acquisition systems is nearly endless, but the following are a few examples: plasma physics and fusion, high-energy physics supercolliders, wind tunnels, laser physics and optics, process monitoring and control in manufacturing, etc. Many devices, such as commercial aircraft and automobiles, have *intelligence* built into them. This is nothing other than one or more sensors attached to a microprocessor which uses the data collected to feed back to the machine operation and report malfunctions. As these data-acquisition systems become cheaper, more and more ordinary devices will contain them. One day we may have "smart" toasters.

The data-acquisition systems (DAS's) that have evolved to meet these demands can range from relatively simple turnkey commercial packages to incredibly complex ones. The former are used in small laboratory experiments, and the latter are used at facilities such as CERN, where high-energy physics detectors with several hundred thousand channels deliver 500 MB of raw data per second (Rüden, 1989). This article will introduce the components within a DAS and attempt to explain how they interact with the experiment and with each other, and finally how data might be stored and reduced. There are, of course, a variety of approaches to this which depend on the speed required, the number of data channels, and on-line processing requirements. Several of these will be reviewed. Examples of DAS's of varying complexity will be presented, and in doing so we will discuss several important elements, such as digitizers, pulse counters, clocks, and control devices. Finally, we will speculate on what will happen in this arena in the near future.

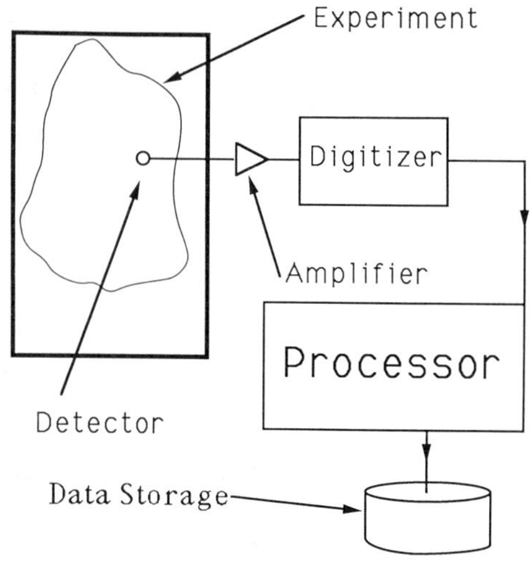

FIG. 1. The basic elements in data acquisition. A detector generates a signal which is usually conditioned by an amplifier and then input to a digitizer. The analog signal is converted to a binary number and then sent via a processor to storage media such as a computer disk or tape.

1. DEFINITION AND CRITICAL COMPONENTS OF A DATA-ACQUISITION SYSTEM

A data-acquisition system is anything which accepts a signal from a detector and stores it so that it may be analyzed at a later time. Its basic elements are shown in Fig. 1. Here, the signal source is shown to occur in an experiment, but this definition is too narrow. Anything worth understanding or diagnosing serves as the experiment. This could be the chemicals pouring from a smokestack, the temperature and gas mixture in an automobile engine, or the vital signs of a patient on an operating table. The detector is a device which responds to something within the experiment (fluid flow, temperature, presence of an electric charge, etc.), and usually presents a voltage at its output which is related to that thing. Sometimes the detector output is easily interpreted. The output voltage of a thermocouple is directly proportional to the temperature of the thermocouple. In other cases, the detector output bears no simple relationship to the measured quantity and has to be analyzed on a computer. This is the case for Thomson laser scattering diagnostics in fusion experiments, where the final results are the plasma electron temperature and density. Since detectors vary enormously from case to case, little more will be said about them here. A detector may produce an output signal which is a time-dependent voltage, charge, or current or optical signal. In plasma physics research, for example, sensors typically generate output voltage wave forms. Data acquisition consists of digitizing and recording these rapidly varying analog signals, with a voltage-sensitive device, after a start trigger, or before a stop trigger is issued. The gain of any signal-conditioning amplifier is a dimensionless number since the sensor output and digitizer input are both voltages.

In high-energy physics experiments, photomultipliers or wire detectors are used to record the passage of particles. These events produce pulses of charge at the detector output, although the signal generated by each varies and must be separately conditioned. In this case analog-to-digital converters with charge-sensitive input amplifiers are used to encode the signal. The charge present at the amplifier input is integrated over a short period of time ($50 \text{ ns} \leqslant t_{gate} \leqslant 2\ \mu\text{s}$) determined by an input gate pulse. The output of the digitizer is proportional to the number of picocoulombs (pC) of input charge. This is very different from a digitized wave form which may represent the time history of a magnetic field. The gain of charge-sensitive amplifiers is expressed in units of V/pC; that is, the peak output voltage is proportional to the charge received. A third type of conditioning amplifier, the transimpedance amplifier, takes current as the input signal and produces an output voltage proportional to it. The gain of this device has units of V/A or resistance. These amplifiers may have to operate in an electrically noisy environment, and elaborate shielding, or the transformation of the electrical to an optical signal for the purpose of transmission, may be necessary at this stage.

The next operation is signal storage. Sometimes the signal is processed in real time before storage to save storage space. There are a number of ways to accomplish this, and some will be discussed in the next section. In this article, we will concentrate on digital data storage. This involves the transformation of the analog voltage to a time series of digital numbers sampled at various positions on the input wave form. Although, as will be discussed later, information is always lost during sampling, the digitized signal is perfect for use by a computer. In most cases, digitizers have their own local memory and the data must be subsequently transferred to a computer where it may be analyzed mathematically, plotted, or stored on a disk or tape after acquisition.

2. THE EVOLUTION OF DATA-ACQUISITION SYSTEMS

2.1 NIM

Until the 1960s, commercial products such as digitizers did not exist, and the only way to record nonrepetitive, rapidly varying data was by taking pictures from an oscilloscope screen. The budget of a research laboratory may have included tens of thousands of dollars per year for fast Polaroid film. Data analysis involved laboriously measuring the amplitude of the recorded signals from the scope traces with a ruler and entering them by hand into computers when necessary. This meant that experimental researchers generally eschewed the use of computers, and data were sparse. Most experimental results were easily plotted as line drawings. Needless to say, detailed comparison between experiment and theory was not possible.

There was, however, a need for uniformity in the measurement equipment utilized in large high-energy physics experiments. A power supply for devices such as counters and amplifiers with a fixed instrument input voltage and specifications for signal input and output levels was agreed upon by the Nuclear Instrumentation Methods (NIM) committee in the United States. The power supply was cratelike in appearance and allowed for the easy insertion and removal of a variety of modules. The NIM crates had no provision for data readout and any device control had to be done via inputs on the devices themselves.

2.2 CAMAC

The biggest breakthrough in data acquisition occurred with the advent of large-scale integrated (LSI) circuit chips and the introduction of microcomputers which relied on them. This new technology allowed for the control and readout of devices in crates and an interface standard for CAMAC (computer automated measurement and control) was developed in 1969 by the European Standards on Nuclear Electronics Committee and then with further collaboration from the NIM committee in the U.S. (Horelick and Larsen, 1976; U.S. NIM Committee, 1981). The CAMAC standards were updated several times to allow for multiple crate systems that could be connected in parallel and series, and to allow for

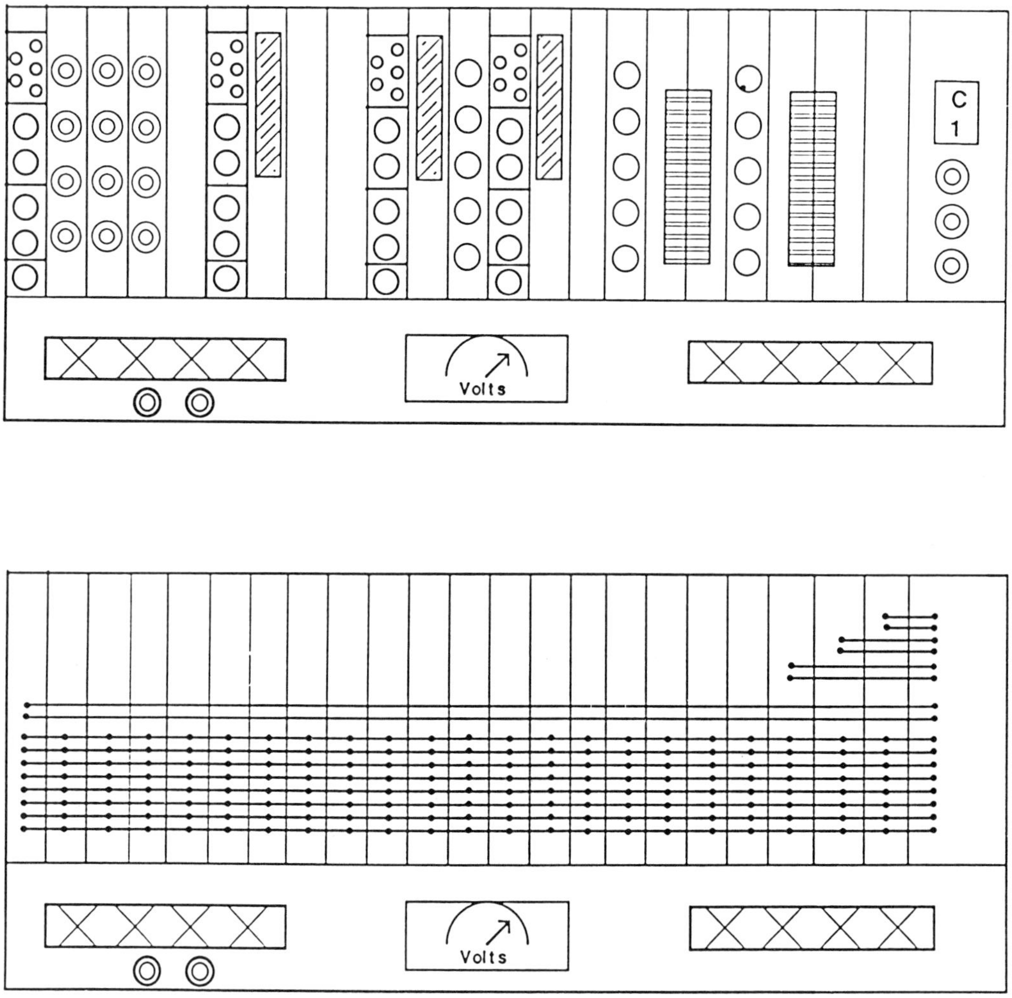

FIG. 2. (a) The front view of a CAMAC crate. There are a number (usually 25) of slots into which various devices such as analog-to-digital converters, memory modules, or clocks (see Table 1) may be placed. The leftmost slots are reserved for the crate controller module(s). (b) A view of the bus arrangement for a CAMAC crate. The top shows four of the point-to-point wires. These go from the controller to each slot and are used for individual addressing. One of the wires is the *N* line, used for station addressing, while the second is the *L* line, used for interrupts. The bussed wires, shown on the bottom, connect all the modules and are used for data transmission.

distributed processor intelligence. These techniques will be discussed in this article. The latest standard was issued by the Institute of Electrical and Electronic Engineers (IEEE, 1982).

CAMAC crates differ from NIM crates in that the former have a built-in computer bus called a CAMAC dataway. The crate has as many as 25 slots, each a standard size, with power and signal contacts at its rear specified by the IEEE convention. The slots are made to house data-acquisition or experimental control devices [Fig. 2(a)]. The variety of devices that can be inserted into CAMAC crates is large and ever growing. An example, which is not exhaustive, is provided in Table 1.

2.2.1 CAMAC Bus. The first slot, or the first and second slots, in a CAMAC crate is usually occupied by a *crate controller*, a microprocessor enabling it to communicate with external computers or other crates. On the back of the slot is an 86-pin connector to the dataway [Fig. 2(b)]. The dataway has two point-to-point wires for each station within it, as well as 24 read and 24 write lines which are shared by all the modules. These lines permit 24-bit parallel data transfers. The point-to-point lines are used to send commands to an individual station or to allow that station to send a signal, such as "now I have data, look at me," a LAM. Other bus lines concern themselves with clock signals and status information about data transfers. The clock signals, or strobes, are generated by the crate controller, which is also the source for all commands to the devices in the crate as well as a funnel through which the device's data move to an external computer. All the CAMAC devices are controlled by issuing instructions to the crate controller, which in turn reissues them to the entire crate or an individual device within it. The commands have been given standard names but always result in words with specific bit patterns being sent along the dataway.

2.3 GPIB and VME

At some point, the crate must interact with a computer to get instructions and send data for storage. The simplest interface is an RS-232 serial port. For this interface, one must buy an RS-232 module for the crate and plug the output from this module into the com-

Table 1. A sample list of devices used in CAMAC and FASTBUS crates.

Device	Purpose	Properties
Analog-to-digital converters	Analog-to-digital signal	16 bit 30 kHz to 8 bit 1 GHz
Programmable amplifier	Computerized signal gain	125-MHz gain, offset, impedance
Digital voltmeter	Digitize many (64) input channels	Slow, 1 mV to 10 V autoranging
Programmable clock	Dynamically control digitization rates	Multiple outputs, 5 ns to 1 ms/tick
Memory modules	Storage of data from digitizers	May be partitioned up to 1 Mbyte each
Stepping motor controllers	Digital control of motors	To move detectors in experiments
Digital-to-analog converters	To operate relays, analog device control	Multiple output channels, 16 bits in
Programmable trigger generator	Generates reliable recording and stop triggers	Variety of triggers, multiple outputs can be combined
Multichannel input gates	Determines electrical state of contacts	32 or more channels; 0=closed contact
Spincoder	Measures shaft rotation	Can feed back to a stepping motor
Gated integrators	Integrates charge	Multichannels (32) for phototube output
Multiplexer	Combines up to 16 input channels	600-ns switch time sequence by computer
Contact closure	Closes physical relays	12 contacts controlled by computer command
Histogrammer	Creates histogram from memory data	10^6 inputs @ 4096 counts/input

puter. The computer then regards the crate as a terminal. The transfer rate is extremely slow (maximum of 19.2 kilobits or 800 24-bit words/s). The next step up is the general purpose interface bus (GPIB). This was first developed by Hewlett Packard in 1965 (and called the HP-IB), and, because of its popularity, it was accepted as an IEEE standard interface. One may now purchase GPIB interface modules for CAMAC crates, available from several vendors. Although quite slow, they are a widely used interface to IBM personal computers and their clones, as well as a number of instruments such as digital oscilloscopes, spectrum analyzers, etc. A faster way for crate/processor communication is to connect the crates directly to the bus of the computer, as shown in Fig. 3(a). There are a variety of standard busses (see BUS SYSTEMS AND COMPUTER INTERFACING), e.g., the Unibus, Qbus, and BI bus for DEC computers, and the VME bus for 32-bit workstations such as SUN computers. Several companies market CAMAC interfaces for these busses. The interface consists of a board set (one for the crate and one for the computer) and software programs called drivers that permit data-acquisition software to allow command and data transfers to occur.

2.4 Microprocessor Control

The introduction of cheap, powerful microprocessors, namely the PDP-11 series by the Digital Equipment Corporation (DEC), enabled a computer to be installed directly in the crate, as shown in Fig. 3(d). This made it possible to deal with situations requiring rapid response, if data had to be preprocessed or moved out of the digitizer's local memory in time to handle the next event. A number of startup companies bought processor chips from DEC and packaged them to fit in the crates along with a crate controller. These intelligent crate controllers had serial ports so that terminals and printers could be connected. This arrangement became an immediate success and is still used in many laboratories. However, new technologies drive the insatiable need for more (in this case, channels of data), and it became necessary to configure systems containing many crates.

2.4.1 CAMAC Processor Connections. A number of CAMAC crates may be connected

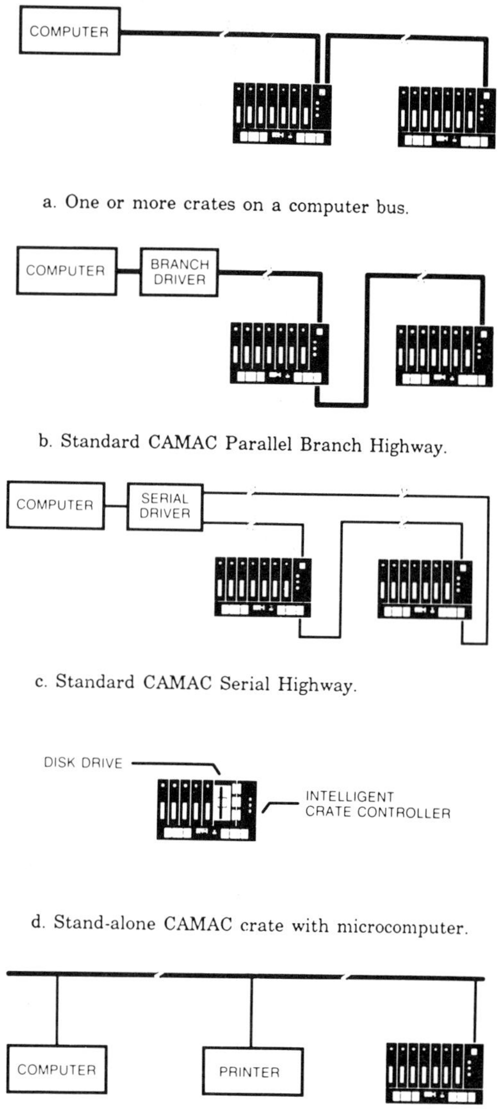

a. One or more crates on a computer bus.

b. Standard CAMAC Parallel Branch Highway.

c. Standard CAMAC Serial Highway.

d. Stand-alone CAMAC crate with microcomputer.

e. LAN-interfaced CAMAC crate.

FIG. 3. Several approaches to CAMAC computer interfacing.

in series (serial highways) or in parallel (parallel branch highways). A CAMAC serial branch highway is shown in Fig. 3(c). A serial driver board is installed in the host computer and then connected to the first of up to 64 crates. Each crate has a controller in it which examines the command from the host computer. If it is to one of its modules the command is carried out, if not the command is passed on to the next crate. One of the lines from the host is a clock line to set the timing for all of this. The clock rate can be as high as five MHz, and block transfer rates as high as 2.8×10^5 24-bit words/s have been measured (Cleary, 1985).

It is always faster to interact with many devices in a parallel mode and so parallel branch highways for CAMAC came into being. To handle this the crate interconnections are done by use of a 66-pair ribbon cable as shown in Fig. 3(b), and a special driver board in the host computer. Only seven crates may be connected to the interface (as opposed to 66 in the serial mode), and there is a 50-m length limitation of the cabling, but the data-transfer rate is an order of magnitude higher in this mode.

2.4.2 CAMAC LAN Connection. It is possible to hook a CAMAC crate directly to a local-area network or LAN, with the appropriate crate controller [Fig. 3(e)]. This has the advantage of having the crate accessible by more than one computer on the network. The LAN can be fiber-optics based which makes it immune from electromagnetic noise. This is a very important consideration in many environments. A LAN based on Ethernet hardware can run at up to 10 megabits/s and crate controllers for this are available now. The next technology for LANs is FDDI (fiber distributed data interface) and will be about ten times faster. It is expected that crate controllers for this will be available by mid-1992.

2.5 FASTBUS

The high data rates from some modern experiments have strained serial and parallel CAMAC highways. A response from the industry has been the introduction of FASTBUS which is an IEEE standard (standard No. 960-1986) which allows for the fastest, highest-density data acquisition and transfer today. In addition, researchers in the past few years have taken advantage of the introduction of powerful and inexpensive 32-bit workstations to the computing scene. Since the CAMAC dataway is 24 bits wide, it is incompatible with these processors; 32-bit transfers would take more than one CAMAC cycle. The industry responded with a new type of crate for FASTBUS. Like CAMAC it is a modular data bus but it is 32 bits wide; the crate has additional lines for timing and control status. The FASTBUS crates can be interconnected in a variety of ways which will be outlined in a section to follow. The FASTBUS backplane achieves a bandwidth of 40 MB/s as compared to CAMAC which saturates at about 5 MB/s. For example, a single crate might have analog-to-digital converters with 5 MB of memory to be transferred after the analog-to-digital converters (ADCs) are filled and generate a LAM. At a maximum transfer rate of 1×10^6 24-bit words/s (Cleary, 1985) it would take at least 5 s to move the data to a computer. The actual number depends upon the clock rate of the computer and the software employed, and this is a best-case scenario. One must also consider that a large experiment may have more than one crate delivering data to the host computer. FASTBUS can transfer 32-bit words at about 8 times the above speed. This difference can be critical in experiments with a high repetition rate. There is, however, a cost tradeoff; FASTBUS crates are more expensive than those for CAMAC, and there are far fewer modules available for them at this time. Research groups wishing to use them often are faced with designing their own FASTBUS ADCs.

2.6 Speed

The revolution in 32-bit microcomputers has led to the standardization of several computer bus architectures. Some computer manufacturers have opened up the details of their architectures so that others could build peripheral devices to connect to their machines, thus enhancing the popularity of their product. Two such busses are Multibus and VME. The latter has been made popular by the success of SUN Microsystems, which makes the largest selling workstations in the United States. Crates containing both these busses in their backplanes have been manufactured. The VME crate supports 32-bit transfers just as FASTBUS does. The difference here is that now the crate is an integral part of the host computer. No intermediary crate controller is required. Since VME has been widely adopted there are a large number of cheap memory and processor cards and other modules. It is possible, therefore, to buy your own processor and build a powerful computer directly into the crate. The raw speed of the VME bus is quite high, up to 100 MB/s. This system can stand alone for real-time data processing. A VME interface to a disk can allow for data storage. Unfortunately, there are no fast ADCs which are commercially available for VME busses. Those who require them have had to build their own, which is quite a task. Just as in the case of CAMAC the VME crates can have local processing power, and, in addition, a direct memory access (DMA) interface to the memory of a larger computer which can do further data processing and graphics. We note here that data can be transferred between the memories of two autonomous devices in one of two ways. The first, and much slower, method involves one device placing a single number in a memory register which is mapped into a single memory register of the second device. A flag is raised by device 1 advising the second that the number is there; device 2 copies the number into its memory and sends acknowledgment to the first. The second and much faster way is a DMA data exchange. Device 1 requests the use of the device-2 bus and some of its memory. When the request is granted (usually after device 2 is finished with its latest chore) device 1 fills the allocated memory of device 2 with a stream of data. In this method the handshaking that goes with the transfer of every number is eliminated.

Multiple crates can be connected using a network such as Ethernet (which runs at 10 Mb/s). One such system employing multiple VME crates with 96 8-bit, 100-MHz flash ADCs per crate has been built for a nuclear experiment at CERN, in Geneva (D'Isep *et al.*, 1989). The network can be fiber-optics based and, therefore, immune to electrical noise. In addition, networks even faster than Ethernet, such as FDDI, have developed in parallel with these systems. Interfaces to VME for FDDI already exist. Experimental networks with Gb/s rates are under development, and although there are as yet no standards for them, one company (ULTRANET) has a commercially available product. The trend in the high end of data-acquisition systems is toward systems with networked distributed processing.

Just as important as the leading-edge technology is the availability of very inexpensive and easy-to-use data-acquisition systems for personal computers (PCs). Some inexpensive modules can be installed directly on the backplane of MS-DOS machines and Apple computers. The latest generation of NeXT computers, as well as a new product from Silicon Graphics, have both voice and data digitizer hardware as well as the software needed to use them easily. Menu-driven software packages for DOS computers, that allow the user immediately to acquire data and plot it, are available from a number of third-party vendors. There also exist interfaces and CAMAC crate controllers for PCs, along with the appropriate software. These systems cannot support high-speed data transfers, and are limited by the speed of the PC, but they are perfect for small experiments or student teaching laboratories. In the following sections we present, in greater detail, the elements of some of these systems and give specific examples.

3. TRIGGERS

A simple trigger is an electrical pulse which is issued to stop or start data acquisition at a critical moment during an experiment. Large-scale experiments may involve many levels of triggering. This is the case in high-energy physics or fusion physics where thousands of detectors may be involved in a single experiment. These detectors may generate a signal pattern which, if of interest, may be stored or used to trigger other groups of detectors. In high-energy physics a great number of uninteresting events occur, and these data must be rejected. Multilevel triggers with successively complex discrimination logic are employed in this task. This is interwoven with the need for accurate timing and triggering at each stage. The results from one set of detectors could be compared to a data mask; if a valid event has occurred a secondary trigger is generated and delivered to other parts of the data-acquisition system which are working in parallel. Each trigger level provides a filter for the next level of acquisition.

3.1 Specialized Instrumentation

Generally, these decisions have to be made on microsecond time scales. A great deal of specialized equipment has been developed for this (LeCroy, 1990). Fast-pulse instrumentation often uses discriminators which take input pulses with varying wave forms and produce output pulses of standard amplitude and duration. Discriminators must deal with the varying duration and rise time of the input pulses as well as bursts of pulses which can exceed their own input bandwidth. Logic units are also employed in triggering decisions. These generate an output pulse whenever a number of input pulses occur within a preselected time interval. Logic circuits may produce an output which is either reinforced or inhibited by associating the input pulses to them with preprogrammed logic diagrams. Time intervals between events are measured by time-to-digital converters, which generate binary output. The temporal resolution is a function of the number of digitization bits available and the total time interval involved. Accuracy of tens of picoseconds, which is necessary in time-of-flight measurement, is possible. Internal buffering in time-to-digital converters allows for counting to be resumed before previous data are read out.

4. DIGITIZERS

A great variety of modules are available for CAMAC-based data-acquisition systems. A sample of these is presented in Table 1. The elements listed in the table can be roughly grouped into two categories. The first comprises modules used for timing and control of the experiment. The second, and the key component in data acquisition, includes the digitizers or analog-to-digital converters. The signal to be recorded from the experiment and later processed is assumed to be a current or voltage wave form. Since all digitizers require an input voltage (usually in the range of 0.0–0.5 V) the signal must be properly conditioned before applying it to the digitizer. This conditioning process can be as simple as placing the signal across a resistive voltage divider, if it is too large, or may involve the use of differential amplifiers with variable gain if there is an impedance mismatch between the source and digitizer. Signal conditioning is an enormous topic and is best left to a reference (Horowitz and Hill, 1989).

4.1 Resolution

There are three main features to any digitizer which warrant discussion here. They are the number of bits of resolution, the resident memory, and the digitization rate. In general the more of each feature that a unit possesses, the better it will perform, and the more expensive it will be. Digitizers work by mapping the value of an analog signal (i.e., a voltage) measured at a particular time into an n-bit binary number between 0 and 2^n-1. The number of times per second that a signal is measured and translated into binary is called the digitization rate. Since the measured signal is assigned a discrete value there will always be a quantization error associated with the process. In general, this error is one part in 2^n for an n-bit converter. For example, consider an 8-bit converter which accepts a signal of $0 \leqslant V \leqslant 0.5$ V. The number is stored in the converter as an integer between 0 and 255. One bit, which in this case is roughly 2 mV, is about 0.4% of the maximum signal value. If the signal is close to the maximum that the digitizer can handle, the stored signal will closely resemble the input signal as shown in Fig. 4(a). However, if the signal (or a portion of it) is small, it will appear as a staircase of discrete steps and a considerable error is introduced [Fig. 4(b)]. If the signal exceeds the input range of the converter, it will be assigned the maximum value and information will be lost as shown in Fig. 4(c). There are two ways to avoid large digitization error. The first is to ensure that the input signal is close to the full operating range of the digitizer. This may be accomplished with the use of a preamplifier or, in the case of a signal which may change slowly in time, a computer-controlled programmable preamplifier. The second way to attack the problem is to use analog-to-digital converters of 12 and 16 bits. A 16-bit converter has only a 0.000 15% error. For low digitization rate devices (< 50 kHz) these converters are inexpensive, but their price escalates with frequency. At the time of this writing, there are no digitizers with greater than 8-bit accuracy with sampling rates above 200 MHz.

4.2 Sampling Rate

The second criteria for digitizers is sampling rate. Many inexpensive PC-based data-acquisition systems come with digitizers that work at speeds below 100 kHz (10^5 samples/s). This speed is more than adequate for digitizing voice signals (the human ear generally cannot hear above 20 kHz) and sampling signals of many everyday phenomena. A car

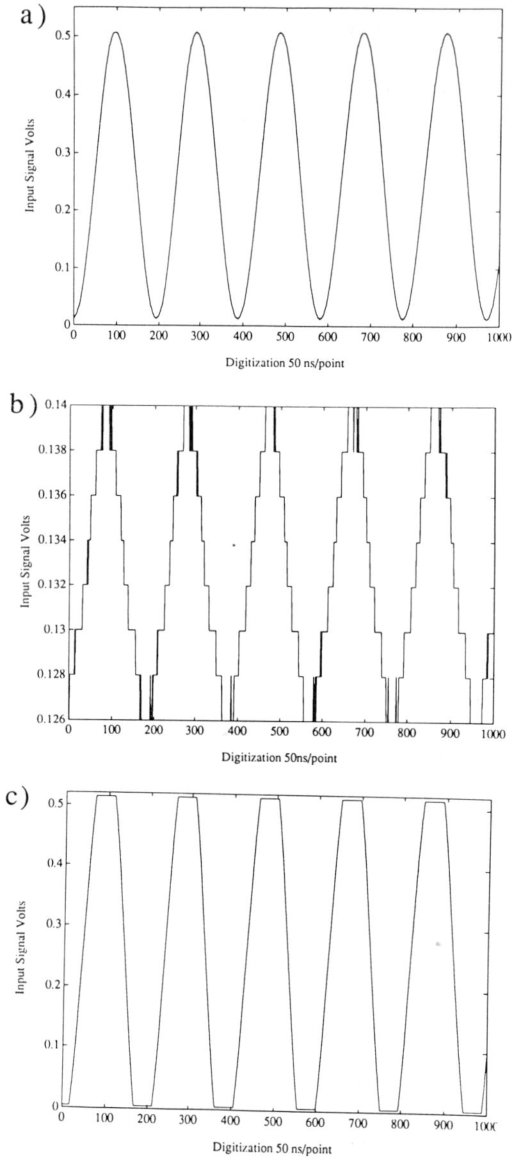

FIG. 4. (a) A digitized 100-kHz sine wave. The digitizer has 8-bit accuracy and can accept a signal of $0 \leqslant V \leqslant 0.5$ V. Since the digitizer is used over its full working range, signal reproduction is excellent. The 1000 samples correspond to a time interval of 50 μs. (b) The same digitizer is used on a signal of very small amplitude. The signal is less than 3% of the 0.5-V working range so that bit noise is apparent. (c) A signal in excess of the 0.5-V limit on the digitizer results in clipping.

moving at 60 mph will move 2.7 mm in one sample time. However, there are a great number of other phenomena such as the oscillations of electrical circuits, or the change of parameters in a plasma, that occur on much briefer time scales. The fastest digitizers today work at about 1 GHz (10^9 samples/s) and are about 100 times too slow for some of these applications. One must, therefore, consider what happens if there are elements present in a wave form which are above the maximum frequency that the digitizer can handle. Figure 5(a) shows an example of a 110-kHz signal digitized at a rate of 1 MHz (1 μs/sample). There are 11 samples on each cycle of the wave form. The theoretical limit for a particular sampling rate is called the Nyquist limit (Bendat and Piersol, 1971). At the Nyquist limit there are only two samples on every cycle, and the wave form, as shown in Fig. 5(b), is not discernable to the unaided eye. Finally, if one introduces a wave form of higher frequency than the digitizer can handle, a curious phenomenon occurs. As shown in Fig. 5(c), a lower-frequency component, which is nonexistent in the original data, appears. The occurrence of this spurious signal is explained in Fig. 6. In the upper graph the original high-frequency signal is shown with dots on it. The dots correspond to the sampling times of the slow digitizer. Suppose it is known that the input signal is a pure sine wave. If the sampled signal points (the dots) are used to reconstruct a sine wave, as shown in Fig. 6(b), a lower-frequency, incorrect wave form results. It is therefore important to exclude all signals of bandwidth greater than that of the digitizer. This can be accomplished with the use of a low-pass filter. Failure to filter the input signal properly can result in a corrupted stored signal, as seen in Fig. 7. The first part of the figure shows the power spectrum of the correctly digitized 10-kHz signal of the type shown in Fig. 5(a). The power spectrum is a measure of how the input signal power varies as a function of frequency, and is found by Fourier transforming the original time-series signal and multiplying by its complex conjugate. In Fig. 7(a) there is a clear peak at 10 kHz corresponding to the input signal. (There is always a peak at zero frequency which is suppressed in these plots.) When an improper digitization procedure is used, such as 1-MHz sampling of a 20-MHz signal [as in Fig. 5(c)], a signal not present in the original wave form appears, as shown in Fig. 7(b). No signal should have been detected at $f < 0.5$ MHz, but the undersampling gives a nonexistent peak at lower frequency.

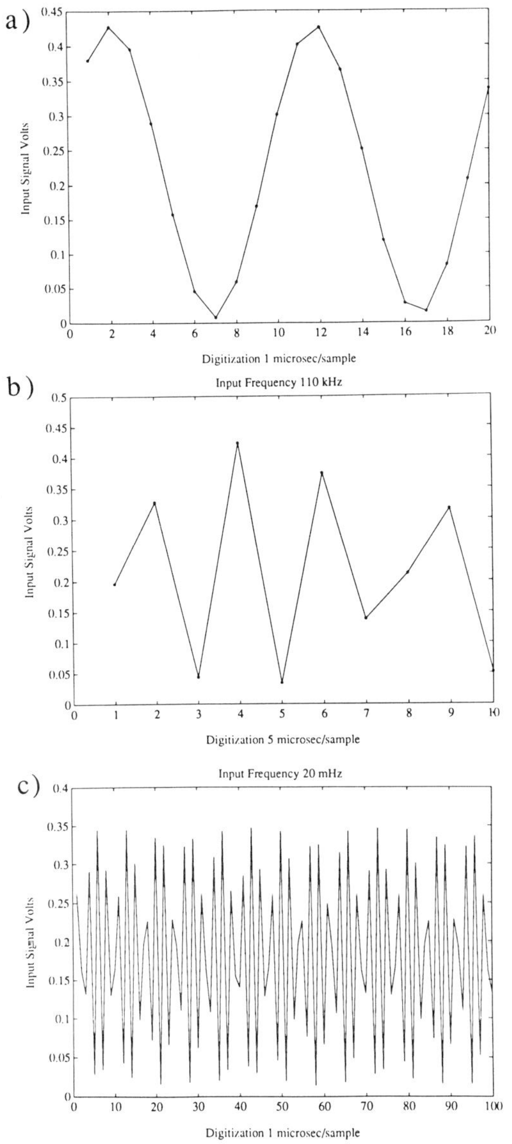

FIG. 5. (a) A 110-kHz signal digitized at 1 μs/sample. The dots on the curve are the digitized data and are connected by straight lines for clarity. The sine wave is choppy [unlike that of Fig. 4(a)] but easily recognizable. (b) The same signal digitized at 5 μs/sample, just above the Nyquist limit. This is truly a limit for the sine wave in (a) is not recognizable. The record shown is 50 μs in length. (c) The effect of undersampling. A 20-mHz signal is digitized at 1-μs intervals; a factor of 20 too slow. The time interval shown is 100 ms but a wave form of five oscillations is not visible.

4.3 Memory

Every analog-to-digital converter must be coupled to a memory. The digitizer performs a conversion of the input signal into a binary number which must then be read into a memory before the next conversion is attempted. In many cases this memory is closely coupled to the converter and is not the same as the memory of the computer which controls the process. When the digitizer's memory is full, its contents must be transferred to the host computer. The cost of an analog-to-digital converter depends upon how much resident memory it has, and how fast that memory must be addressed. Clearly, a digitizer which does one conversion per nanosecond must be matched to a memory fast enough to accept numbers at that rate.

5. HOW A DATA-ACQUISITION SYSTEM WORKS

As outlined above, there are many hardware variations within a DAS, but the process of acquiring data shares many common procedures. As an example of these common elements, we describe the features a software package should have for controlling the interaction between a digitizer (ADC) and a CAMAC-based DAS. When the DAS control software is initiated, the user decides which devices in the crates will participate in the experiment. The software generates a table containing which crate and slot the device occupies. If an experiment is to be repeated, these tables may be read from a file which defines the experiment.

5.1 Software

The software then issues an instruction to the crate controller, which in turn initiates the ADC. Before starting, the ADC must receive a LAM (look at me) enable instruction from the crate controller. In some crate controllers, an instruction to enable an interrupt of the processor when its data acquisition is complete must be issued also. Next, the crate receives a clear signal from the processor controlling it, all the device registers are cleared, and digitization begins. The ADC continually reads and converts data from its input signal. The digitized data are transferred into its local memory. When the memory is filled, the digitizer overwrites the start of memory and continues until a stop trigger is received from the control software. The trigger forces the ADC

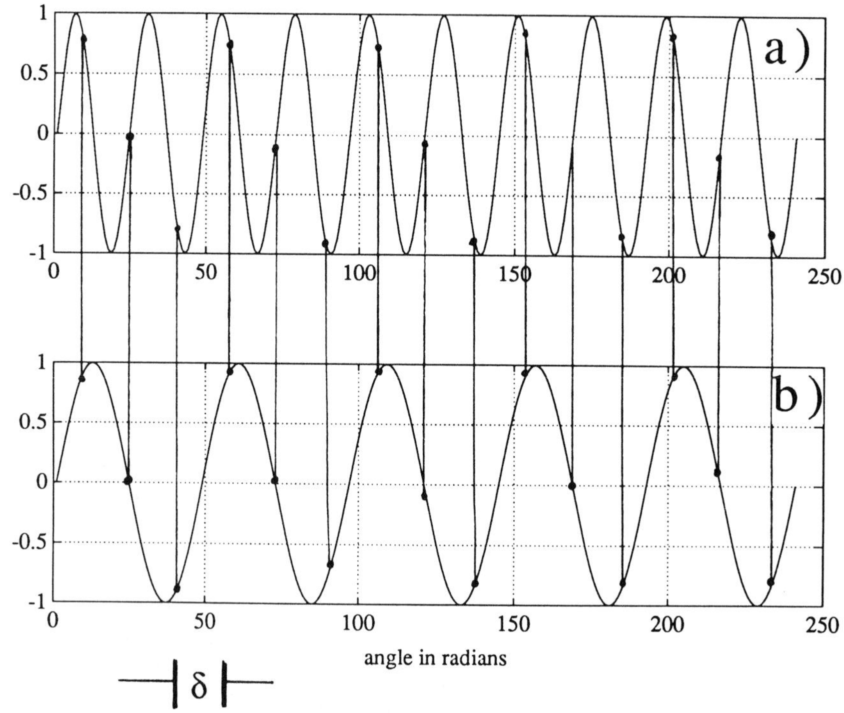

FIG. 6. Effect of undersampling a high-frequency wave form. A sine wave shown in (a) is sampled at an interval δ. Each sample time is denoted by a dot. If it is known that a sinusoidal wave form was input to the digitizer, an incorrectly reconstructed wave form, shown in (b), may result. Undersampling a high-frequency signal can therefore result in low-frequency *ghost* modes not present in the original one.

immediately to stop acquiring data and raise a LAM. When the crate processor detects a LAM, it immediately stops what it was doing and jumps to a location in its memory where it expects to find the starting address of an executable program. Such a program is called an interrupt handler.

5.1.1 Interrupt Handlers. These programs were formerly laboriously written in assembler language by data-acquisition programmers. If there was a programming error and the computer jumped to the wrong address, or there was nothing at the right address, the computer simply crashed. Now, however, companies which sell crate controllers also sell higher language subroutines (usually written in FORTRAN, PASCAL, or C) which activate handlers they have written. The handler is usually quite simple. For example, it might check to see that one of the devices authorized to participate in the experiment generated the interrupt. If an illegal device interrupted, that device could be ignored or the software could branch to an error-handling routine. If a legal device interrupted, the handler disables the LAM and branches off to another program which may be a user-written routine to read data. The interrupt handlers provide an interface between the software control package and internal crate operations.

There is another way to service the device without generating an interrupt. All the above steps are performed, but the device is not authorized to issue a processor interrupt. It does, however, issue a LAM. The software control package must enter a loop in which it looks at all the devices that could raise a LAM and waits for one.

5.2 Sample Program

A generic routine (in FORTRAN) might look like this:

```
        time = secnds(0.0)                        ! set clock to zero
        do i = 1,no__crates                       ! loop over crates
          if(status(crate(i),crate__stat).eq.0)   ! check for problem
            call error__message()                 ! something wrong, print
          endif                                   ! an error message
10      if(crate__stat. eq. LAM__mask)then        ! we got a LAM
        {
                    .......do things like read data and turn off the LAM
                    }
        endif
        else                                      ! if we didn't get a LAM
          if (secnds(time).lt.timeout)then        ! continue to check
          goto 10
        else                                      ! we waited long enough
          goto 100                                ! leave the loop
        endif
        enddo                                     ! finish looping over valid modules
100     continue                                  ! jump out of search for LAMs
```

Here, the crate controller has a word called crate__stat which it passes to the microprocessor. If the word is blank, there is a problem, and an error message is printed. Further action could be taken here as well. If the word returned from the crate controller equals the bit pattern (the correct sequence of zeros and ones) in the word LAM__mask, the *i*th device has raised a LAM and is then serviced. Finally, a timing routine must be included so that, if nothing happens for a while (timeout seconds), the computer can go on to do other things. Otherwise, it will be caught in an infinite loop!

5.3 Processor–Crate Communication

How do the crate controller (which is a microprocessor) and the master computer actually communicate with each other? This is done by sharing memory locations. Furthermore, the dataway lines previously discussed are directly mapped into these locations. This mapping is done by the hardware in the crate controller. Before the data-acquisition control software is run, the memory locations in the host computer permanently allocated for sharing with the crate controller must be assigned. For some small computers directly housed in the crate, this assignment can be done by setting dip switches in the back of the computer. For larger systems, a software *driver* must be loaded into memory when the computer is booted or turned on. This driver is a short program which permanently resides in memory with the computer's operating system and provides all these memory addresses. When the crate controller interrupts the computer, the driver takes over, finds the correct servicing program, and calls it. When you want to do something such as clear a LAM, the driver actually issues the instruction, which might be something like:

```
MOVB #10., @#164040.
```

This moves a byte, or 16-bit word, which contains the decimal number 10 (or 0 000 000 000 001 010), into memory location 164040, shared by both the controller and host computer. The crate controller then puts this number out onto the dataway, which has the effect of putting positive pulses on two of the signal lines, which in turn forces the hardware to turn off the LAM. Presently, the programmer who writes control software no longer needs to contend with these low-level instructions. Software that handles these processes may be purchased from the company that sells the crate controller.

Once the data is stored in the ADCs, the next step is transferring it to the host computer. In early crate controllers, this process was slow, since data could only be transferred word by word. A CAMAC command initiated the transfer and also passed the address of the data array in the host computer. Each time a word was transferred, the controller also set standard CAMAC bits (the Q and X bit) indicating that all went well. The user could check these bits with the main data-acquisition program. Later models of crate controllers allow for block transfers of data. Blocks of up to 32K (32 767) 24-bit words can be transferred all at once. This procedure saves time as there is considerable overhead in getting the host computer and crate controller to start the transfer (a memory region must be allocated and other parameters initialized). Actual transfer rates are hardware specific, but for a MicroVax (μVAX) running the VAXELN operating system (which is a Digital Equipment Corp. real-time operating system), the first word transfer takes about 700 μs and each additional word 2.3 μs (Antone *et al.*, 1989).

6. SPEED AND RESPONSE OF DATA-ACQUISITION SYSTEMS

6.1 Transfer Rate to Host Computer

The response required of a DAS is entirely dictated by the experimental situation. The relevant parameters are the repetition rate of the experiment (if indeed it is repetitive; a nuclear explosion is not!), the number of data channels and digitizers, and the amount of memory each digitizer must eventually transfer to another computer. The TFTR (Tokomak fusion test reactor) at Princeton is an experiment that lasts for several seconds and has a repetition rate of about 8 min. It can produce 24 MB of data in each shot (Randerson *et al.*, 1986), or about 2.4 GB per day. The experiment utilizes 280 CAMAC crates which contain over 5000 modules (Rauch, 1986). Because there is a fairly long time between shots in the TFTR experiment, the main problem is getting all the data out of the digitizers into storage. Real-time processing is not a great issue. The crates are connected to computers through eight separate serial highways. The CAMAC services a number of Gould SEL computers which gather the data. Data are then transferred to several VAX computers where it is written to disk. Some of it is immediately plotted to give the scientists an idea of what happened on the shot. After a day or two, the data are transferred to magnetic tape where it is archived. One obstacle the TFTR DAS overcame was that no fast digitizer has enough memory to store several seconds worth of data. The problem was solved by designing the system to generate preprogrammed timing pulses to start and stop digitizing. Fortunately, there are modules for CAMAC that can do this.

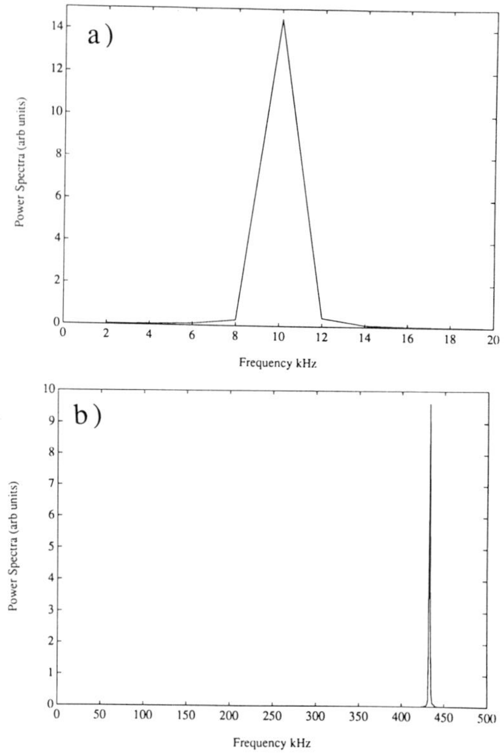

FIG. 7. (a) The power spectrum of a properly digitized 10-kHz signal. There is a peak at the appropriate frequency. The zero-frequency peak has been suppressed. (b) The power spectrum of an improperly digitized wave form. Here a 20-MHz signal is digitized at a 1-MHz sampling rate. Because of the Nyquist limit on the 1-MHz sampling the spectrum extends from 0 to 0.5 MHz. A peak at 440 kHz not present in the original signal is evident.

6.2 Real-Time Analysis

There are other cases where collection of all the data from the modules and storage for subsequent processing is not possible. This occurs in the LEP electron-positron collider at CERN in Geneva (Rüden, 1989). The detector there has several hundred thousand channels and produces 500 MB of data per second. It must be processed in real time and compressed to about 100 kB per second. Each channel, however, may or may not generate

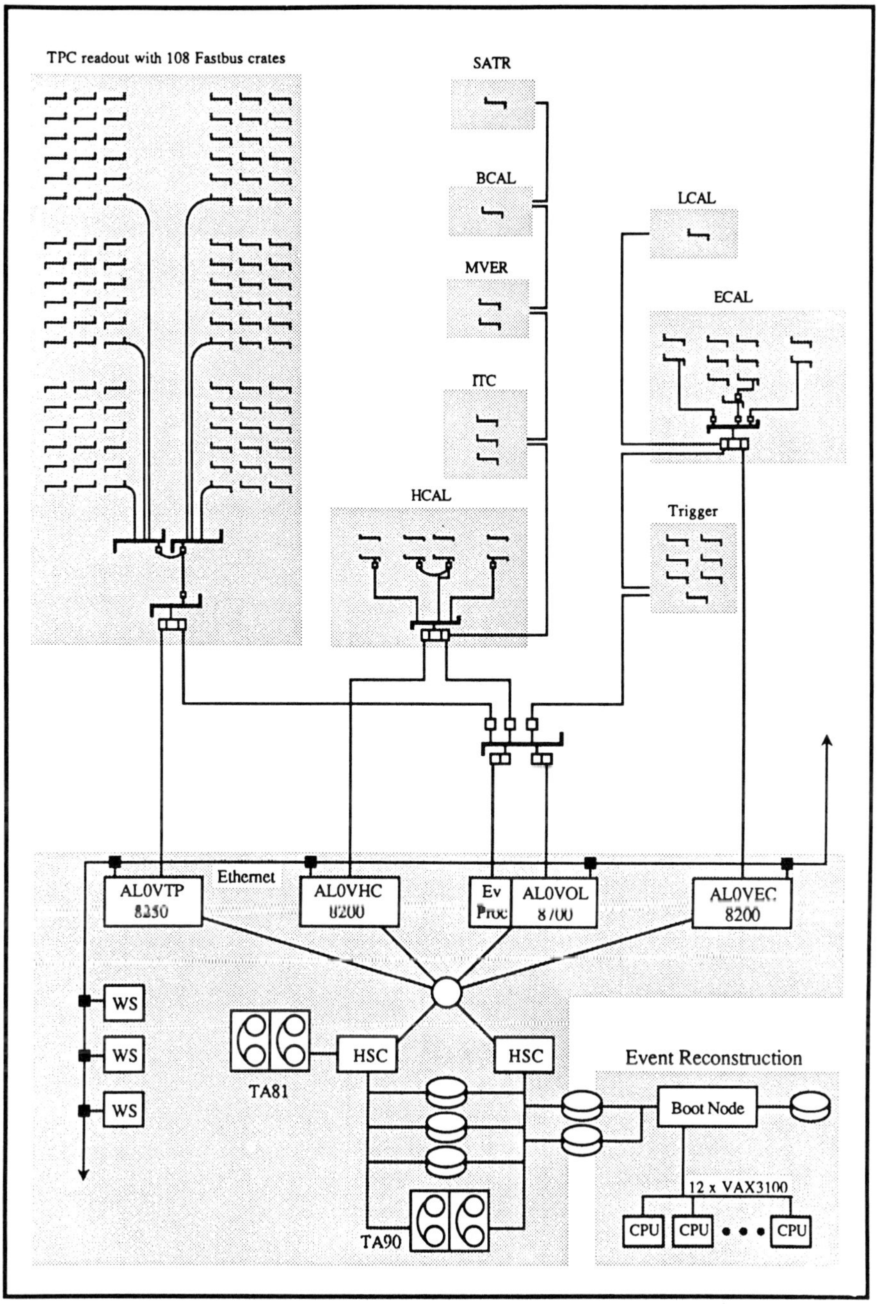

FIG. 8. A large data-acquisition system used in a high-energy physics experiment at CERN. The symbols denote FASTBUS crates each with its own processor. Various abbreviations are as follows: TPC, time projection chamber; SATR, BCAL, HCAL, LCAL, various calorimeters; ITC, inner track chamber; MVER, minivertex detector; ALOV**, VAX computers; HSC, high-speed channel, WS, workstation; TA90, high-speed magnetic tape drive. The system works in parallel to discard an enormous amount of uninteresting data from the 500 MB/s which are continuously acquired.

data depending upon whether a particle passed its detector. In fact, a great many of the channels are empty at any given time. A schematic of the setup is shown in Fig. 8. In this schematic, each shaded area on the top of the page refers to data acquisition associated with a different detector (such as TPC, time projection chamber detector) or monitoring of the machine operation itself. The lower shaded area encloses the group of computers used to acquire, store, and analyze the data. The larger VAX computers are named AL0V** and are networked to each other, the tape drives, and the workstations. Data reduction is accomplished in several stages by use of distributed processing. In the first stage of processing, analog hardware compares signal levels against a threshold and rejects signals that are too small. Meaningful data, properly scaled by the 32-bit FASTBUS crate computers, is next transferred from the crates to a series of DEC KA800 computers through a network. These computers, which are not shown in this diagram, are placed directly on the bus of a larger DEC computer (a VAX 8700 denoted AL0VOL) and rapidly process the event and make a final decision as to whether it is worth saving. If the answer is yes, it is transferred to the memory of the VAX 8700. Final processing of the data is done on a cluster of MicroVAX workstations (Jost, 1989).

6.3 Choice of Operating System

Another issue regarding performance is the choice of computer operating systems. These can be broadly classified into two types, real-time and multitasking. A real-time operating system is stripped down for action. It is geared to serving one process only, the data-acquisition program. These operating systems were first available on single-user microcomputers and occupied very little memory. Real-time operating systems are designed to handle device interrupts quickly. When an interrupt occurs, they can stop the current process and initiate the interrupt handler within several microseconds. These small microprocessors are still extremely popular in DAS's. A multiuser operating system, such as UNIX or VMS, has a large built-in overhead since these systems are continually switching between many processes. A response to an interrupt can take several milliseconds to service, so they cannot work in strictly real time. It is possible to optimize performance by assigning a high priority to the DAS program and carefully designing the software (Appalaraju and Cleary, 1989; Atac *et al.*, 1989). It is also possible to buy an operating system which is tailored for real-time applications, such as VAXELN.

7. AN EXAMPLE OF A DATA-ACQUISITION SYSTEM

Let us consider a specific example that has some of the requirements cited above, i.e., the need for speed and distributed processing, but is not overly complex. The experiment involves the study of magnetic fluctuations in a plasma (a partially ionized gas) embedded in a magnetic field (Gekelman and Stenzel, 1984). When an electrical current is passed through an initially quiescent plasma, large fluctuations in the magnetic field are observed. The strength of these fluctuations is a function of space and time. The experimental goal is to find the source of these fluctuations. They are thought to be due to the random generation of waves in the plasma, but which waves are responsible? Since the waves are chaotic, it is necessary to perform correlation measurements at a number of frequencies to determine the cross spectral function (CSF) of the magnetic-field fluctuations (Bendat and Piersol, 1971). The data-acquisition procedure involves digitizing and storing three signals (B_x,B_y,B_z) that measure the three components of the fluctuating vector magnetic field. The signals from each of two probes are measured at 1024 time steps for an ensemble of 80 shots at 300 spatial positions on several planes. There is insufficient disk space to store all this data, so the CSF is evaluated at each position after the ensemble of data is acquired. The DAS used is shown in Fig. 9. The ADC's reside in CAMAC crates. The crates have a controller as well as an LSI 11/23 computer in them (Gekelman and Xu, 1986). The LSI runs a real-time operating system (RT-11) which rapidly responds to LAMs in the crates and services them. All of the software controlling the devices in the crates and reading data from them resides in the LSI. The LSI in turn is networked to a VAX computer with Ethernet hardware. The VAX has an array processor, which is a very fast calculating engine

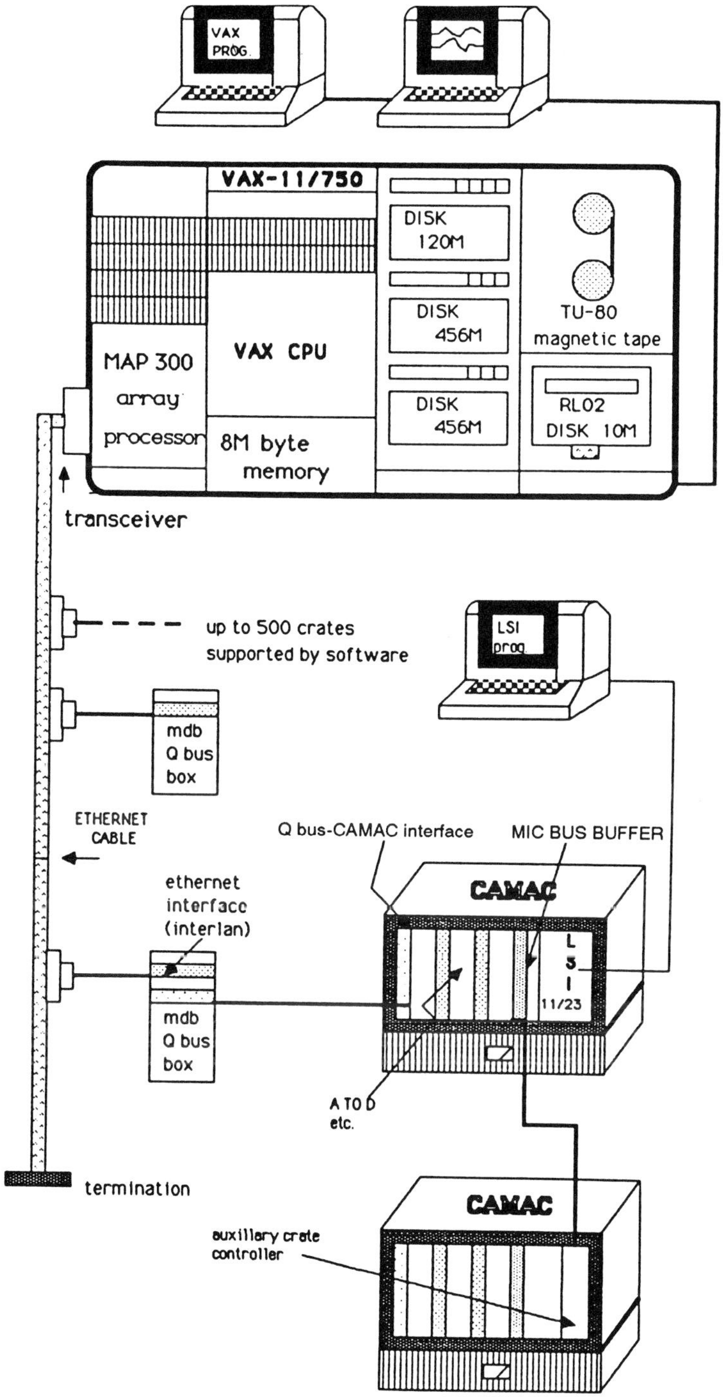

FIG. 9. A small data-acquisition system in a university physics laboratory. A series of CAMAC crates are interfaced via a microcomputer and network to a minicomputer. The array processor in the minicomputer allows for real-time data analysis such as digital filtering. This combines elements found in Figs. 3(a), 3(d), and 3(e).

capable of directly accessing its memory through a DMA interface. The main data-acquisition program runs on the VAX. The system is configured to make the LSI microcomputer appear as a disk to the VAX. Whenever a command is sent to the LSI, a write statement to a certain device number is issued which activates a device driver setting up a dialog through the network to the LSI. Whenever the LSI sends data to the VAX, the reverse happens. The pseudodisks the two computers read from and write to are divided into logical areas. One area is reserved for instructions and, when data appear, those data are interpreted as a command. Another much larger area is reserved for the actual data. The system can throughput about 250 kB of data per second between the CAMAC crate and the VAX. As soon as data appear in the VAX, a DMA transfer is initiated to the array processor. After 80 shots, the array processor takes the data and calculates four components of the cross spectral tensor. This calculation takes a few seconds, and then the results are sent back to the VAX for storage. In this mode of operation, only 4K real (32-bit) numbers need to be stored for each probe position instead of 480K.

Technology is changing so rapidly that, as far as computing is concerned, systems can become obsolete a year after purchase. In the four years since the above system was designed, extremely powerful workstations have become available, some of them with vector processors. They make attached array processors obsolete. Fast crate controllers networked to these workstations allow for a direct interface to the host computer, making the separate LSI crate computers unnecessary. However, in some other applications, such as the system at CERN, distributed computing is still necessary. This progress is welcome since programming is simpler because it can generally be done entirely in a higher-level language.

8. DATA MANAGEMENT AND VISUALIZATION

The growth of powerful data-acquisition systems has enabled vast amounts of data to be collected in relatively short periods of time. Experimentalists are then faced with trying to analyze enormous data sets. In a situation with many investigators, there may be many separate channels of data, with each scientist chiefly interested in his or her own data. The requirement here is to organize and archive the data to make it simple to retrieve for subsequent analysis. This may involve data compression and migration of data to secondary computers or tapes as it ages. One way to deal with large data sets is to use commercially available database-management software. This has been done on a fusion mirror machine experiment (Stewart *et al.*, 1986), the TEXT Tokomak experiment (Wiley *et al.*, 1989), the Doublet experiment at General Atomics (McHarg, 1989), and others. The key to understanding complex multidimensional data sets is the ability to visualize them. Until recently, this was a difficult process. Data were generally displayed as a series of contour maps or surface plots for separate planes in whatever space (position, momentum, etc.) was of interest. Trying to understand volumetric data involved comparing many of these plots; information was inevitably lost or obscured in the procedure. Recently, a number of vendors have come out with extremely powerful three-dimensional graphics workstations. It is possible to transfer data to one of these and make a full color rendering of volume data as a set of nested isosurfaces of differing transparency, or as voxels, which are color pixel representations. By using appropriate display filters, such as changing the way colors represent key features in the data, many subtle features can be brought out. In addition, the data may be magnified, translated, or rotated in real time. Interfaces to video equipment and computer-controlled slide- and print-making equipment are also available and becoming reasonably priced. Finally, there have been some startling advances in software such that expertise in graphics programming is not required. There are several new commercially available software packages which are menu driven and do all of the above as well as instantly creating vector fields, streamlines, and cut slices through data sets with any number of dimensions. Presently, they run only on powerful graphics computers which are rather costly. However, it is anticipated that the price of these workstations will plummet, and someday they will be as affordable as personal computers. This will herald a new age of discovery of subtle and complex phenomena which now are buried in huge data sets.

9. FUTURE TRENDS

9.1 Communication

From the tone of the preceding text, it is not hard to predict the evolution of DAS's for the next five years. The emphasis on the high end will be speed. This will entail the use of multiprocessors, as mentioned previously, but the old way of connecting them via a bus may change radically. The basic limitation in signal transfer is the speed of light which will not allow an arbitrary number of information packets, each addressed to a different processor, to share one data path effectively. One possible solution is the CSI, coherent scalable interface (Gustavson, 1989), which is basically a point-to-point datapath between processors. It will be designed for GB/s processor-to-processor communication. The CSI is now in a standardization process phase before the IEEE.

9.2 Computer Farms

There is a great deal of experimentation with high-speed DAS's at the large accelerator laboratories. A team at the Fermi National Accelerator Laboratory (Atac *et al.*, 1989) has designed and built a system which links devices from both FASTBUS and VME crates to a farm of computers using a very fast crossbar switch manufactured by Texas Instruments. The switch can make eight simultaneous connections out of a possible 28, each connection having a 20-MB/s bandwidth for a total speed of 160 MB/s. Each connection can be between a crate and a group of parallel processors which receive the data and preprocess it before sending it to storage. As soon as a group of processors have data to work with, the switch changes position and connects more data channels to other free processors. Another approach was used in a nuclear physics experiment in Heidelberg (Männer and Stucky, 1990) in which a multiple 32-bit bus system is used instead of a switch to connect FASTBUS crates to an array of 30 processors. In this experiment, data flows in at a rate of 10^6 events per second, with up to 10^3 parameters per event. The processor farm filters this down to 10^5 numbers per second which actually must be stored, a data reduction of 4 orders of magnitude. Efforts such as the two described here are large because they involve the construction of the hardware and writing of the software to support it. They are, however, the precursors of what may be commercially available systems utilizing very rapid communication technologies between data channels and multiple processors.

9.3 Parallel Processors

One of the recent advances in high-end computation is the introduction of highly parallel processors (see PARALLEL COMPUTING). These computers contain from 64 to 16 384 individual processors. There are various schemes for interprocessor communication and use of memory, but the basic idea is that, if a problem can be broken up into many parallel segments which run independently, significant increase in speed may be attained using processors running in parallel. Although several companies manufacture these systems, they are still largely in the experimental stage since compilers which can generate efficient code for parallel CPUs have yet to be developed. In the future, one could consider the development of a DAS in which one CPU is dedicated to each input data channel. Interprocessor communication would make it possible to compare data from neighboring probes instantly to determine what should be stored, or to do mathematical operations such as the instantaneous evaluation of the curl of vector fields. This will be some time in coming, however, since these multiprocessor machines do not easily interface to any of the standard data-acquisition gear such as FASTBUS or CAMAC. Thus far, the only parallelizing is in the type of system at CERN described earlier, where many FASTBUS crates, each with its own processor, are brought to bear on the problem.

9.4 Inexpensive, Simple Systems

On the other end of the DAS spectrum, we see a trend in development of easy-to-use, off-the-shelf DAS's for small laboratories. A number of digital instruments, such as oscilloscopes and wave-form generators, have GPIB outputs and can be controlled by software running on a PC. Setting them up to acquire, store, and plot data on digital *X-Y* recorders can take several hours. The most expensive piece of equipment is usually the measurement instrument, not the DAS. As time goes on, these systems will continue to drop in price and become as easy to use as digital voltmeters.

10. SUMMARY

Fifteen years ago, data acquisition was in its infancy. Data were taken in an analog fashion using Polaroid photographs of oscilloscope traces or read off of plots on *X-Y* recorders. No detailed experiment could be performed on a complex multiparameter system. The introduction of data-acquisition standards and microcomputers has radically changed this. Experiments now routinely collect and process billions of numbers from a large variety of detectors. As the price of computers and high-speed instrumentation tumbles, and the ease of implementing them improves, this technology will become available to any research group that needs it.

ACKNOWLEDGMENTS

Writing this made me recall the many hours I spent writing data-acquisition software with Tim Mc Fadden and Lingyu Xu. I have enjoyed working with them and am indebted for all they have taught me. I would like to thank Dr. R.T. Cleary for contributing Fig. 3 and Dr. Wolfgang von Rüden for Fig. 8. I am also indebted to Dr. James Maggs and Marilyn Pagnusat for proofreading.

GLOSSARY

Byte: 8 bits, where a bit is a zero or a one.

CAMAC: Computer automated measurement and control; standardized (IEEE 583) modular data handling system, 1-MHz 24-bit wide data path.

CERN: Center for European Nuclear Research.

CSI: Coherent scalable interface.

DAS: Data-acquisition system.

DMA: Direct memory access; a device which can take over a computer bus and access its memory.

Ethernet: A local-area network with hardware which can sense and correct for collisions between data from different computers; limited to 10 megabits per second.

FASTBUS: Fastest high-density data-acquisition standard available; 32-bit, 40 MB/s (IEEE standard 960-1986).

FDDI: Fiber distributed data interface; fiber-optics–based high-speed network with a token ring topology.

Gigabyte: One billion bytes or 8 billion bits.

GPIB: General purpose instrument bus; most popular instrument bus today; serial transfers limit it to 400 kB/s.

LAM: Look at me; signal generated by device wishing to be serviced by a processor.

LAN: Local-area network; a data communications system for moderately high-speed traffic, usually limited to a building or group of buildings.

LSI: Large-scale integrated circuit.

Megabyte: One million bytes or 8 million bits.

MIPS: Millions of instructions per second.

VOXELS: Representation of volumetric data as colored spots.

List of Works Cited

Antone, I. D., Mandrioli, G., Matteuzzi, P., Sanzani, G., Bloise, C., Grillo, A. F., Marini, A., Ronga, F., Baldini, A., Mancarella, G., Palamara, O., Surdo, A., Lamanna, A., Petrera, S. (1989), *IEEE Trans. Nucl. Sci.* **36**, 1602–1607.

Appalaraju, R., Cleary, R. T. (1989), *IEEE Trans. Nucl. Sci.* **36**, 606–610.

Atac, R., Cook, A., Deppe, J., Fischler, M., Gaines, I., Husby, D., Nash, T., Pham, T., Zmuda, T., Eichten, E., Hockney, G., Mackenzie, P., Thacker, H. B., Toussaint, D. (1989), *IEEE Trans. Nucl. Sci.* **36**, 726–730.

Bendat, J. S., Piersol, A. G. (1971), *Random Data: Analysis and Measurement Procedures*, New York: Interscience.

Cleary, R. T. (1985), *Enhanced Serial Highway System*, Kinetic Systems Corporation Technical Note TN-102, Lockport, IL: Kinetic Systems.

D'Isep, D., Maggiora, A., Minetti, B., Maron, G., Morando, M., Balestra, F., Masoni, A. (1989), *IEEE Trans. Nucl. Sci.* **36**, 713–717.

Gekelman, W., Urrutia, J. M., Stenzel, R. L. (1987), *Magnetotail Physics*, Baltimore, MD: Johns Hopkins University, pp. 261–274.

Gekelman, W., Xu, L. (1986), *Rev. Sci. Instrum.* **57**, 1851–1855.

Gekelman, W., Stenzel, R. L. (1984), *J. Geophys. Res.* **89**, 2715–2733.

Gustavson, D. B. (1989), *IEEE Trans. Nucl. Sci.* **36**, 811–812.

Horelick, D., Larsen, R. S. (1976), *IEEE Spectrum* **13**, 50–55.

Horowitz, P., Hill, W. (1989), *The Art of Electronics*, Cambridge, UK: Cambridge University Press.

Jost, B. (1989), *IEEE Trans. Nucl. Sci.* **36**, 1580–1584 (1989).

LeCroy (1990), *Instrumentation Research Catalog*, Chestnut Ridge, NY, LeCroy Corporation.

Männer, R., Stucky, O. (1990), *Comput. Phys.* **4**, 267–274.

McHarg, B. B. (1989), Proceedings of 13th Symposium on Fusion Engineering, Knoxville, Tennessee (unpublished).

IEEE (1982), *Modular Instrumentation and Digital Interface System (CAMAC)*, ANSI/IEEE Std. 583, New York: IEEE.

Randerson, L., Chu, J., Ludescher, C., Malsbury, J., Stark, W. (1986), *Rev. Sci. Instrum.* **57**, 1892–1894.

Rauch, W. A. (1986), *Rev. Sci. Instrum.* **57**, 1898–1900.

Rüden, W. (1989), *IEEE Trans. Nucl. Sci.* **36**, 1444–1448.

Stewart, M. E., Carter, M. R., Casper, T. A., Meyer, W. H., Perkins, D. E., Whitney, D. M. (1986), *Rev. Sci. Instrum.* **57**, 1929–1931.

U. S. NIM Committee (1981), *Definitions of CAMAC Terms*, U. S. Department of Energy, Office of Energy Research, Washington, DC 20545.

Wiley, J. C., Miner, Jr., W. H., Bravenec, R. V., Patterson, D. M., McCool, S. C. (1989), *Rev. Sci. Instrum.* **60**, 1286–1296.

Further Reading

Carr, J. J. (1978), *Digital Interfacing with an Analog World*, Blue Ridge Summit, PA: Tab Books.

Malcolme-Lawes, D. J. (1988), *Microcomputers and Laboratory Instrumentation*, 2nd ed. New York: Plenum.

DATA ANALYSIS

See STATISTICAL DATA AND ERROR ANALYSIS

DEFORMATION OF MATERIALS

See MATERIALS DEFORMATION

DENRITES

See WHISKERS AND DENDRITES

DETECTORS, PARTICLE, CALORIMETRIC

CHRISTIAN W. FABJAN, *CERN, Geneva, Switzerland*

INTRODUCTION

The Task of Particle Detectors

Several areas of experimental physics (solid-state, nuclear, and particle physics) demand the measurement of kinematical properties of particles or quanta of radiation. The determination of the four-momentum vector $(\mathbf{p},E)$ calls for the measurement of the momentum $\mathbf{p}$ and the energy E, or equivalently the particle's velocity v and its mass m [$\mathbf{p}=m\beta\gamma$, $E=m\gamma c^2$, $\beta=\mathbf{v}/c$, and $\gamma=(1-\beta^2)^{-1/2}$], or momentum $\mathbf{p}$ and velocity $\mathbf{v}$ (Fernow, 1986).

A range of different particle sensors or particle detectors has been developed, frequently with very specialized functions. The momentum of charged particles is determined in magnetic spectrometers, using the dispersive power of magnetic fields: particles with momentum p (GeV/c) in a field B (T) describe a circular trajectory with radius r (m), such that $p=0.2998Br$. Position detectors are required to trace these trajectories. For the velocity measurement, the methods used depend on the range of $\beta=v/c$ to be covered. The determination of the flight time of particles traversing a known distance ["time-of-flight" (TOF) method] requires sensors reacting to transit time; accuracies of better than 100 ps (10^{-10} s) can be achieved quite routinely with a number of different detectors. This method is therefore applicable to the identification (i.e., mass determination) of particles with velocities up to $\beta\leqslant0.98$, when combined with a momentum determination. For particle velocities approaching $\beta=1$, one relies on the velocity dependence of the electromagnetic interaction between the charged particle and the medium. If the particle velocity is larger than the velocity of light in the medium c/n, n being the index of refraction, the medium radiates photons, at a characteristic angle $\theta=\arccos(1/\beta n)$. This is called Cherenkov radiation, after P. Cherenkov, who first observed it in 1937. In modern Cherenkov detectors, the angle of the emitted photons is measured with a typical accuracy of $\sigma\leqslant10^{-3}$ rad, providing useful velocity deter-

1-56081-063-7/92/$5.00+.50

mination up to $\beta \leqslant 0.9999$ and hence particle identification up to ≈ 100 GeV/c.

These momentum and velocity measurements have found numerous and diverse applications despite two major shortcomings: one is fundamental, since the method is limited to charged particles only; the other, a practical limitation, is the range of particle momenta that can be analyzed magnetically. Given that the relative accuracy scales as $\Delta p/p \propto p$, huge engineering efforts are required, such as very large superconducting magnets and position detectors comprising typically 10^5 position sensors, to provide adequate information for momenta above 100 GeV/c. For the measurement of neutral and very energetic particles, an additional technique is required: calorimetry.

Energy Measurement with Calorimeters

The measurement of the particle's energy E relies on "total absorption" in the detector and takes place within a few to a few tens of nanoseconds. The kinetic energy of the particle is transformed into excitation or ionization energy of the absorber atoms, which can be used for a measurable signal related to the particle's energy. After approximately 10 μs, all energy appears in thermal form, hence the name "calorimeter."

The concept of total-absorption energy measurement has been used extensively in nuclear and intermediate-energy physics for many decades. One of the first applications of calorimetry was the measurement of the average energy of β particles emitted by a radioactive nucleus, ^{210}Bi. It was found to be considerably smaller than the maximum available energy, a result which led to the formulation of the neutrino hypothesis. In nuclear physics, the energy of particles (protons, nuclear fragments, and electrons) or gamma quanta is typically in the MeV range, and relatively small volumes of absorber (a few tens of cm^3) are sufficient (Sec. 1).

In particle or high-energy physics (Sec. 2), energies range from 100 ($\equiv$0.1 GeV) to 10^6 MeV ($\equiv$1 TeV). While the same principle of total absorption is used, the energies are such that different types of interactions (nuclear or "strong" interactions, high-energy electromagnetic processes, or even weak interactions) are involved. There are several reasons why calorimeters are widely used in experimental particle physics:

1. They are sensitive to both charged (e.g., protons, pions, kaons, and electrons) and neutral (e.g., neutrons, neutral pions, and photons) particles;
2. the absorption process is statistical, so that in well-designed instruments the relative accuracy of energy measurements improves with the energy E, in principle as $\sigma(E)/E \propto E^{-1/2}$;
3. the total depth D necessary to fully absorb (or "contain") the particle increases only slowly with energy as $D \propto \ln E$, an important practical advantage; and
4. the fundamental constituents of matter (called quarks) manifest themselves as tightly collimated groups of charged and neutral particles, whose total energy can only be measured with calorimeters.

The most original and diverse developments are for astro-particle physics (Sec. 3). It requires detection methods for the widest spectrum of particles—known and unknown—as well as the largest energy range, from meV to PeV (10^{15} eV) or higher.

The layout of the principal detectors in a generic particle physics experiment is shown in Fig. 1. The momentum of charged particles is measured from the curvature of the track in a magnetic field. The required position measurements barely alter the energy. The energy of particles, charged or neutral, is measured in calorimeters. Different types are used, depending on the interaction between the particle and matter. Photons, electrons, and positrons require relatively little mass to be absorbed through a cascade of increasingly lower-energy particles. Hadrons, interacting predominantly through the strong interaction, require considerably more material for total absorption, again through cascades of secondary-particle generation. Calorimeters are instrumented blocks of mostly dense absorbers that have sensors (shown in the figure as triangles) embedded to detect the strength of the particle cascade and hence derive a signal related to the energy of the absorbed particle. The destructive nature of the calorimetric energy measurement imposes the spatial sequence of the measurement tasks and hence fixes the layout of experiments.

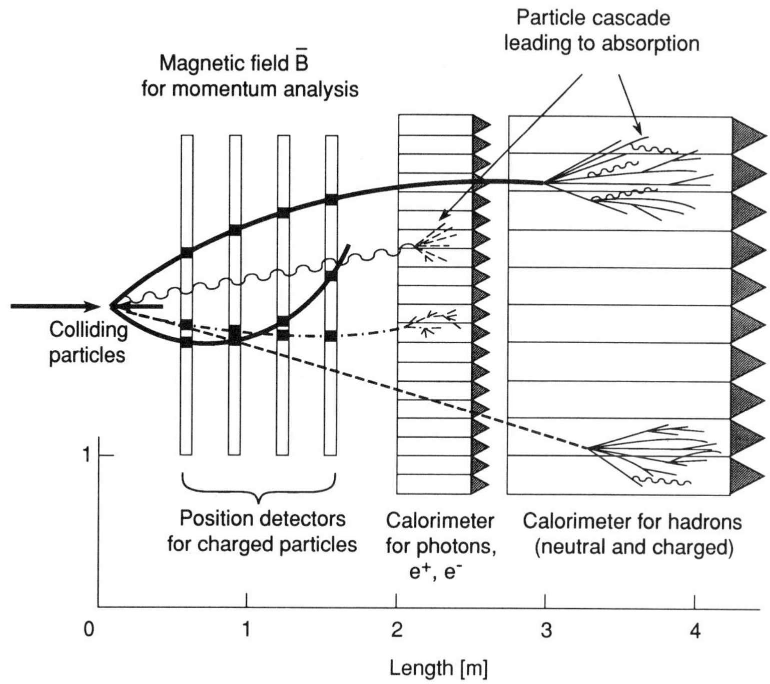

FIG. 1. Conceptual layout of detectors in a particle physics experiment. In the collision of particles (solid arrows) a great number of different types of particles at different energies may be produced: charged hadrons (pions, kaons, and protons) (solid lines), neutral hadrons (neutrons) (dashed lines), photons (wavy lines), and electrons and positrons (dash-dotted lines). See the text for explanations.

1. CALORIMETRY METHODS IN NUCLEAR PHYSICS

1.1 Total Absorption and Range of Particles

In nuclear physics, the energy measurement of charged particles (protons, alphas, nuclei and fission fragments, electrons, and positrons) and neutral particles (neutral pions and, in particular, neutrons) is frequently required. Precision energy measurement of photons emitted from excited nuclear states is also important. The energies are typically in the range of 100 keV to 100 MeV.

First, we discuss the absorption of charged particles, which, in the energy range considered here, proceeds exclusively through the electromagnetic (e.m.) interaction.

The mean energy loss per unit path length x, called the "stopping power," is given to a good approximation by the Bethe–Bloch equation:

$$\left[\frac{dE}{dx}\right]_{\text{inc}} = \frac{CZ_{\text{med}}\,\rho_{\text{med}}}{A_{\text{med}}}\left[\frac{Z_{\text{inc}}}{\beta}\right]^2 \times\left[\ln\left(\frac{2m_e\gamma^2\beta^2c^2}{I}\right)-\beta^2 + \text{correction terms}\right],$$

where $C=0.307$ MeV cm^2 g^{-1}, $Z_{\text{med,inc}}$ are the charges of the absorbing medium and incident particle, respectively, m_e is the electron mass, and I is the effective ionization potential of the absorber. The global dependence $1/\beta^2$ on the particle velocity has an important consequence: most of the energy loss occurs near the end of the path, where the velocity is smallest (Ahlen, 1980). For a unit we have chosen, instead of a geometrical length l [cm], the more natural quantity length$\times$density [g cm^{-2}], which expresses the number of atomic electrons, i.e., the number of scattering centers, encountered.

The depth at which half the initial particles are stopped or "ranged out" is called the "mean range" $R(E)$. It is related to the energy loss by

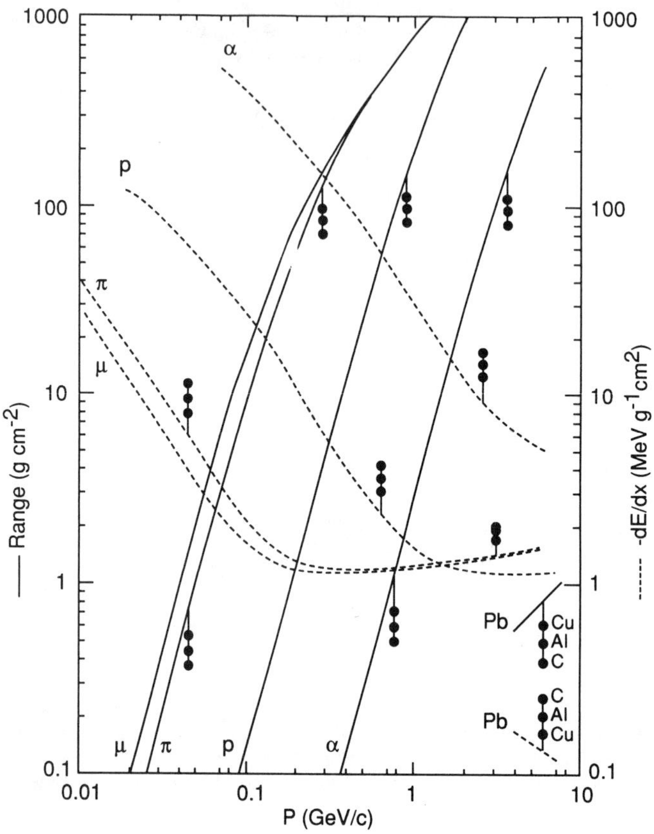

FIG. 2. Mean range (left ordinate, solid lines) and energy loss (right ordinate, dashed lines) due to ionization for the indicated particles in Pb, with scaling to Cu, Al, and C indicated, using the Bethe–Bloch equation, including corrections (Particle Data Group, 1988).

$$R(E) = \int_E^0 -dE/(dE/dx).$$

Typical values of the mean energy loss and the range of various particles (Particle Data Group, 1988) are given in Fig. 2 and can be used to estimate the calorimeter depth required to absorb or range out charged particles in this relatively low-energy range. As an example, consider a pion with a momentum of 100 MeV/c. According to Fig. 2, it has a range of approximately 10 g cm^{-2} in lead, i.e., 10 cm/(density in lead=11.4 g cm^{-3})=0.9 cm: a layer of lead somewhat thicker than 1 cm will be adequate to absorb or range out such a particle.

Next we consider the absorption of neutral particles. Neutrons with energies in the MeV range are most commonly detected through elastic scattering (via the nuclear or strong force) off the light nuclei of the detector. The scattering process transfers some portion of the neutron kinetic energy to the target nucleus, usually hydrogen, deuterium, or helium, resulting in a recoil nucleus. The recoil nuclei lose their energy in turn through ionization of the detector medium. This process is repeated until essentially all of the kinetic energy of the neutron has been transferred to the detector nuclei.

The atomic absorption processes for gamma quanta are shown in Fig. 3. Low-energy quanta in the eV to keV range will ionize the atoms of the detector medium and produce detectable electrons (atomic photoeffect). At higher energies, pair production, the conversion into electron–positron pairs in the e.m. field of the nucleus, dominates. Once the photoelectron or the e^+e^- pair has been produced, energy loss by ionization will degrade the energy as described above.

1.2 Principal Techniques: Scintillators and Solid-State Detectors

In the previous section, we described absorption processes relevant for particles in the MeV region. In this section, we will describe the two principal techniques with which mea-

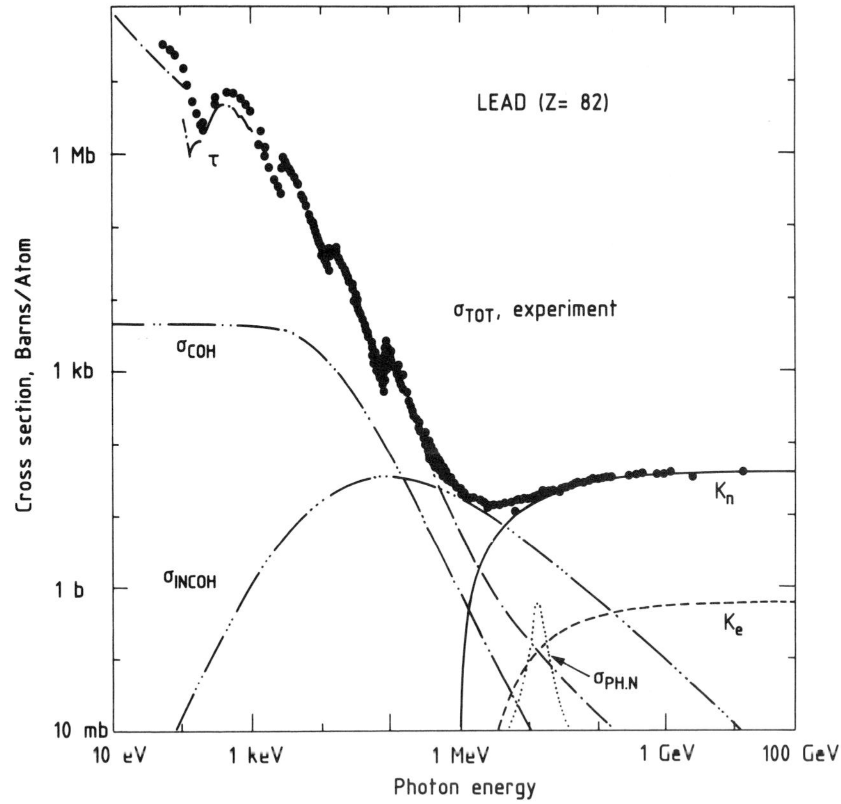

FIG. 3. Photon total cross sections as a function of energy in lead, showing the contributions of different processes (Particle Data Group, 1988): τ, atomic photoeffect (electron ejection and photon absorption); σ_{COH}, coherent scattering (Rayleigh scattering—atom neither ionized nor excited); σ_{INCOH}, incoherent scattering (Compton scattering off an electron); K_n, pair production, nuclear field; K_e, pair production, electron field; and $\sigma_{PH.N.}$, photonuclear absorption (nuclear absorption, usually followed by emission of a neutron or other particle)

surable signals can be obtained from this energy deposit.

In scintillators, the energy deposited by a charged particle produces a flash of light with typical time constants of 10^{-9} to 10^{-6} s. The scintillation process is closely related to fluorescence, which is the emission of a light quantum subsequent to photon absorption (see DETECTORS, SCINTILLATION).

Both inorganic (crystals) and organic scintillators are being used. The inorganic scintillators are usually grown with an admixture of dopants, which act as recombination centers for ionization charge. Recombination results in excitation followed by light emission. Certain inorganic scintillation crystals are compounds of heavy elements and hence have a large absorption coefficient and an excellent energy resolution for gamma rays (see Sec. 2). More recently, the scintillating properties of noble liquids, in particular liquid xenon, have been considered. These liquids combine the high gamma absorption cross section with a very fast time response.

There exists a great variety of organic scintillators, in crystal, liquid, or amorphous (plastic) form. The scintillation mechanism depends on the molecular structure of the medium: the passage of a charged particle will result in the excitation of the many degrees of freedom of the medium (rotational or vibrational levels, lattice vibrations).

In general, only a relatively small fraction of the absorbed energy, typically a few percent, is converted into visible light that can be detected subsequently with photon detectors (e.g., photomultipliers). Nevertheless, the light output is quite sufficient; in "good" scintillators there may be as many as 10^4 photons produced for 1 MeV of absorbed energy. This method works because the fraction of energy converted into light is rather constant, so that the light output is proportional to the energy deposited.

The second important category of detectors, also pioneered by nuclear physicists, uses the measurement of ionization in semiconductors (predominantly silicon and germanium) (see DETECTORS, SEMICONDUCTORS). In these materials, the energy ϵ required to create the basic measurable excitations, electron-hole pairs, is approximately $\epsilon \approx 3$ eV, compared to approximately 100 eV needed on average per photon in a scintillator. Therefore, for a given energy deposit, the statistical signal fluctuations σ are smaller and the energy resolution can be better. Experimentally, one finds $\sigma^2(E) = (FE/\epsilon)\epsilon^2$, and a "Fano" factor $F \approx 0.1$ for Si and Ge; the limit of energy resolution is therefore $\sigma \sim 500$ eV for 1-MeV energy deposit.

Semiconductor detectors work as "ionization chambers," in which the charge created by the energy deposit is collected under the influence of an externally applied electric field. Registration of this ionization charge is, however, only possible if it is not masked by free charges, thermally generated or due to the presence of impurities. Adequate suppression is possible through purification of the material. For germanium, impurity levels as low as 10^9 atoms per cm^3 have been achieved! Such ultrapure crystals are fabricated in sizes of hundreds of cubic centimeters, making them the detector of choice for γ-ray spectroscopy in the energy range of 10 keV to a few MeV. Silicon cannot yet be purified to this level, limiting the useful depth of Si detectors to a few millimeters. Such detectors are still well suited for the energy measurement of heavy charged particles, such as protons, alphas, and fission fragments.

2. CALORIMETRY METHODS IN PARTICLE PHYSICS: PHOTONS, LEPTONS, AND HADRONS

2.1 Absorption Mechanisms

In the "high-energy" domain ($E \geqslant 100$ MeV) of calorimetry the incident particle is sufficiently energetic to initiate particle production. This provides a further mechanism for absorption in addition to those previously discussed (ionization, excitation, etc.). With increasing energy of the incident particle, the energy of the secondary particles may be sufficient for subsequent, i.e., tertiary production, and so forth. The image of an "avalanche" or shower of particles appropriately describes this process of energy degradation. The particle multiplication continues until the energy of the last generation drops below the energy threshold for further production; at this point, the shower has reached the maximum number of particles—the shower maximum—after which further degradation continues via the low-energy processes described in Sec. 1.

There are two distinct types of particle showers. The first type is initiated by energetic electrons (positrons) and photons and propagates through the electromagnetic interaction with the atoms of the absorber. The second type of cascade is produced by particles strongly interacting with the absorber, i.e., hadrons, such as pions, kaons, and protons. It is characteristic of both types of cascades that the relevant collision cross sections reach values which are rather energy independent for particle energies $\geqslant 1$ GeV. In other words, the cascade dimensions are described by the mean free path λ between collisions. Each absorber is characterized by its specific value of λ (which in a given material is generally different for the electromagnetic and hadronic cascades). This argument suggests that the shower size in different absorbers, when expressed in units of the mean-free path, should follow a scaling law describing its geometrical dimensions.

The principle of the calorimetric energy measurement of energetic particles may be formulated as follows: particle cascading (through the electromagnetic or strong interactions) degrades the energy over rather small geometrical distances. The energy of the charged component in the cascade is determined through its ionizing or excitation effect on the medium, which provides a signal approximately proportional to the energy of the incident particle (Fabjan, 1982).

2.2 Calorimetry for Energetic Electrons (Positrons) and Photons

Calorimeters specifically designed for the measurement of electrons, positrons, and photons with energies above ~ 100 MeV represent an important class of instruments. Although

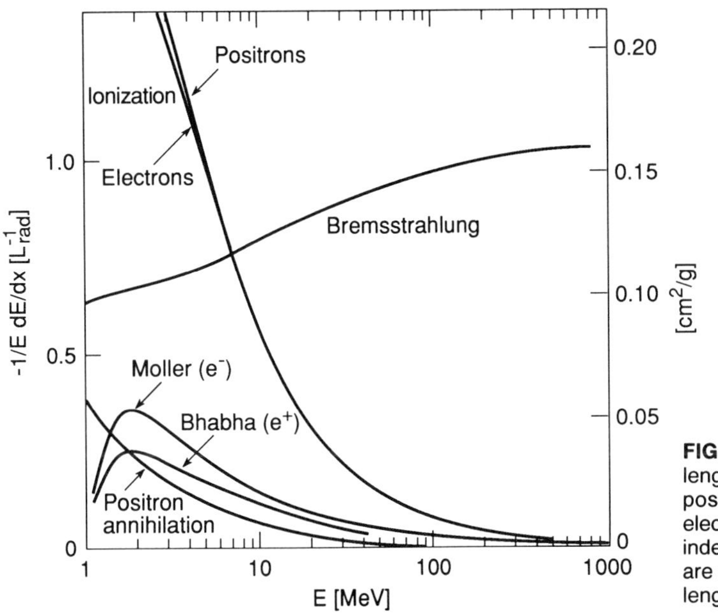

FIG. 4. Fractional energy loss per radiation length in lead as a function of electron or positron energy. The development of electron–photon cascades is approximately independent of absorber when the results are expressed in terms of inverse radiation length (i.e., scale on left of plot).

they have been used for several decades already, these instruments are continually being refined and important developments are still taking place.

These "electromagnetic calorimeters" (EMCs) and the physics constraints on their performance are well understood, since the energy-loss processes are described within the frame of quantum electrodynamics (Tsai, 1974). The energy-loss mechanisms for electrons and positrons (for energies above ~100 MeV) are dominated by bremsstrahlung of photons. The fractional energy loss incurred over the characteristic length (mean free path in the nomenclature of Sec. 2.1), the "radiation length" X_0, is shown in Fig. 4. The radiation length is defined through the relation

$$(\Delta E/\Delta x \text{ by bremsstrahlung}) = -E/X_0.$$

It is the distance over which the energy E_0 is reduced to $E_0 e^{-1}$ by the radiation process. The radiation length is a characteristic constant of materials and to a good approximation energy independent for $E \geqslant 1$ GeV. For the chemical elements, it can be numerically approximated by X_0 [g cm^{-2}]$\approx 180A/Z^2$ (for $Z \geqslant 13$).

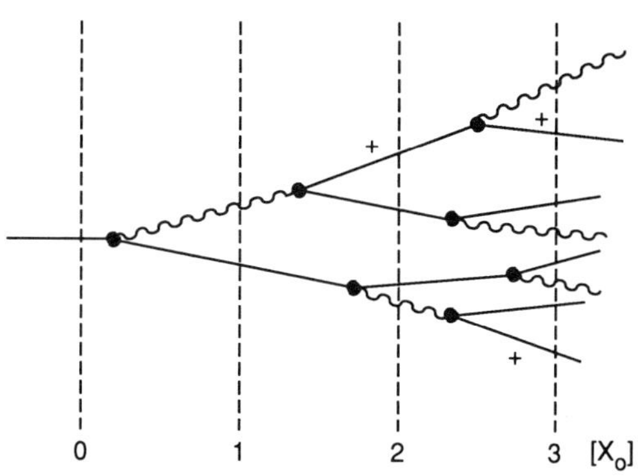

FIG. 5. Simplified model of an electromagnetic cascade, qualitatively describing the longitudinal shower profile.

A simple model gives the correct qualitative description of the e.m. shower and is sketched in Fig. 5. An incident electron with energy E travels one radiation length and transfers Ee^{-1} to a bremsstrahlung photon. The photon travels one X_0 and undergoes pair production, with each particle created receiving half the energy of the photon. This process continues until the energy of electrons (positrons) is so low that radiation losses become unimportant. The number of charged particles rises rapidly to the shower maximum, followed by a reduction in particle density, as shown in Fig. 6. Also shown is the scaling of the longitudinal particle production, or shower profile, in different materials, when the shower depth is expressed in the specific radiation length of the material. The total longitudinal depth required for (essentially) total absorption is an important parameter: an electron of 1-GeV energy is absorbed in approximately

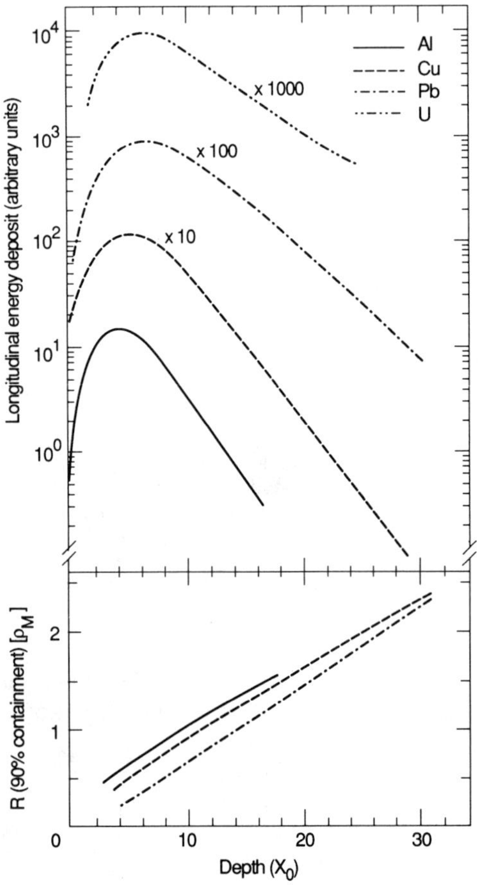

FIG. 6. (Top) Longitudinal shower development of 6-GeV/c electrons in four very different materials, showing the approximate scaling in units of radiation lengths X_0. (Bottom) The shower radius for 90% containment of the shower as a function of the shower depth. In the later development of the cascade, the radial shower dimensions scale with the Molière radius ρ_M [Fabjan and Ludlam, 1982; reproduced with permission of Annual Review of Nuclear and Particle Science.]

$8X_0$, whereas $\sim 25X_0$ are needed for electrons in the $\sim$100-GeV range.

The transverse shower dimensions are determined by the angular distribution θ of bremsstrahlung which is very collimated ($\theta \approx 1/\gamma$). However, scattering processes (multiple scattering of the electrons and positrons in the Coulomb field of the absorber atoms) contribute to a widening of the transverse distribution; this effect is stronger for low-energy electrons, i.e., for the deeper part of the shower. The theory of multiple scattering describes this shower widening; the characteristic transverse dimension is given by the "Molière radius" $\rho_M \approx 7A/Z$ [g cm^{-2}]. As an indication, a cylinder with radius $r = 3\rho_M$ contains approximately 90% of the shower (i.e., the energy of the incident particle), whereas a cylinder of $r \approx 5\rho_M$ is required for 99% energy containment (Fig. 6).

2.3 Performance of Electromagnetic Calorimeters

Armed with this knowledge on shower development and dimensions, the physics constraints which determine the accuracy of the energy measurement ("energy resolution") can be analyzed, i.e., effects leading to shower-by-shower fluctuations about the mean signal. First, we evaluate fluctuations due to the properties of shower development, as measured in a real instrument. These "intrinsic" fluctuations are caused by effective thresholds in detectable energy of the shower particles (or nonlinearities, such as saturation in detector response). In well-designed instruments, this contribution is rather small compared to other sources, and typically $\sigma_{\text{intrinsic}}/E \approx 0.001/\sqrt{E_{\text{inc}}}$ to $0.01/\sqrt{E_{\text{inc}}}$ (E_{inc} given in GeV). In addition to this incomplete signal observation, there are several other effects with larger contributions to the energy resolution. Incomplete longitudinal containment, for instance, is particularly harmful. As a rule of thumb, one may estimate that a 2% longitudinal noncontainment will contribute an average of 1% to the energy resolution.

Frequently, calorimeters are constructed in such a way as *not* to be uniformly sensitive to the signal (ionization or excitation), but with signal sensors interspersed in the absorber (see also Sec. 2.6 and Fig. 10). The shower profile is therefore sampled, hence the name for these devices: sampling calorimeters. The resulting disadvantage is incomplete signal information, which causes an additional source of fluctuations, sampling fluctuations, σ_{sampling}. The simple shower model alluded to earlier allows an order-of-magnitude estimate of σ_{sampling} to be made assuming the major contribution to be due to the Poisson fluctuation of the number of signal sensors activated. This gives

$$\sigma_{\text{sampling}}/E \approx 0.032[\Delta E(\text{MeV})/E(\text{GeV})]^{1/2},$$

where ΔE denotes the average energy loss in one cell, absorber layer plus sensor.

As an example, consider an electromagnetic calorimeter with a relatively fine sampling of 2-mm lead plates alternated with 2-mm scintillator sheets in the longitudinal direction. For this device, $\Delta E \approx 3$ MeV, and the sampling contribution $\sigma_{\text{sampling}}/E \geqslant 0.055/\sqrt{E}$ (GeV) will be considerably larger than the intrinsic fluctuations, previously estimated to be $<0.01/\sqrt{E}$ (GeV). Besides good energy resolution, a linear dependence of the signal on the deposited energy is another important consideration. In EMCs, a linear response is easily achieved because of the simple nature of an e.m. shower. Only instruments using sensors which exhibit severe saturation as a function of ionization density have shown a nonlinear behavior.

Besides an energy measurement, a calorimeter may also help determine the impact point, i.e., the position of incidence on the face of the instrument. This is achieved by probing the transverse shower distribution and by taking the particle position as its center of gravity. Given adequate transverse instrumentation—e.g., sensors of $\sim(0.5\rho_M)^2$—a position resolution of $\sigma_{\text{position}} \leqslant \rho_M/\sqrt{E}$(GeV) may be obtained. For practical instruments, this translates into position resolutions of better than 10 mm for 1-GeV particles and submillimeter accuracy for 100-GeV electrons.

An incident photon can be measured in the EMCs, as it transforms into an electron–positron pair in the Coulomb field of the absorber's nuclei (Fig. 3), after which the showers develop as an electron- and positron-initiated cascade.

We may summarize the preceding discussion on the energy measurement of e.m. showers:

1. It is fundamentally well understood, as it is based on the well-tested and reliable theory of e.m. interactions (QED);
2. it can therefore be well modeled using computer codes;
3. it can reach an accuracy of $\sigma(E)/E \approx 0.01/\sqrt{E}$(GeV), provided the absorber is homogeneously sensitive to the shower particles; and
4. it has a worse resolution in sampling calorimeters, for which $\sigma(E)/E \leqslant 0.1/\sqrt{E}$(GeV) is typically achieved.

2.4 Calorimetry for Energetic Hadrons

Conceptually, the energy measurement of hadrons is analogous to that of e.m. showers. The detailed understanding is, however, more difficult because of the much greater variety and complexity of the hadron-initiated processes that propagate the cascade. Typical for hadronic interactions is multiple particle production. The secondaries are mostly pions (charged and neutral) and nucleons, and their multiplicity increases with energy E as $\ln E$.

As in the case of e.m. cascades, a characteristic scaling length describes the longitudinal shower shape. This is the interaction length λ, defined as

$$\lambda = A/N_{\text{Av}}\sigma_i \ [\text{g cm}^{-2}],$$

where A is the mass number of the material, N_{Av} is the Avogadro number, and σ_i is the inelastic cross section. It describes the absorption of N_0 hadrons in matter through the equation

$$dN/dx = -N/\lambda.$$

Typical longitudinal shower profiles in different absorbers are shown in Fig. 7. The useful shower features can be parametrized approximately as follows:

1. shower maximum $S_{\max}\ [\lambda] \approx 0.2 \ln E$ (GeV) $+0.7$;
2. longitudinal depth for 95% containment $L(0.95)\ [\lambda] \approx S_{\max} + 2.5E^{0.13}$ (GeV); and
3. transverse radius for 95% containment $R[\lambda] \approx 1[\lambda]$.

This parametrization and the data shown in Fig. 7 indicate the first major difference compared with EMCs: in the same absorber material hadronic showers are considerably larger in size.

The achievable energy resolution in hadronic calorimeters (HCs) (again, in contrast to EMCs) is dominated by intrinsic fluctuations due to the complex nature of hadronic cascades. Two specific processes have been identified as the major culprits:

1. A sizable fraction of the shower particles are π^0's, which will decay on the scale of 10^{-16} to 10^{-15} s into two photons, which initiate e.m. showers.
2. The hadronic component will be partially degraded by excitation or breakup of the

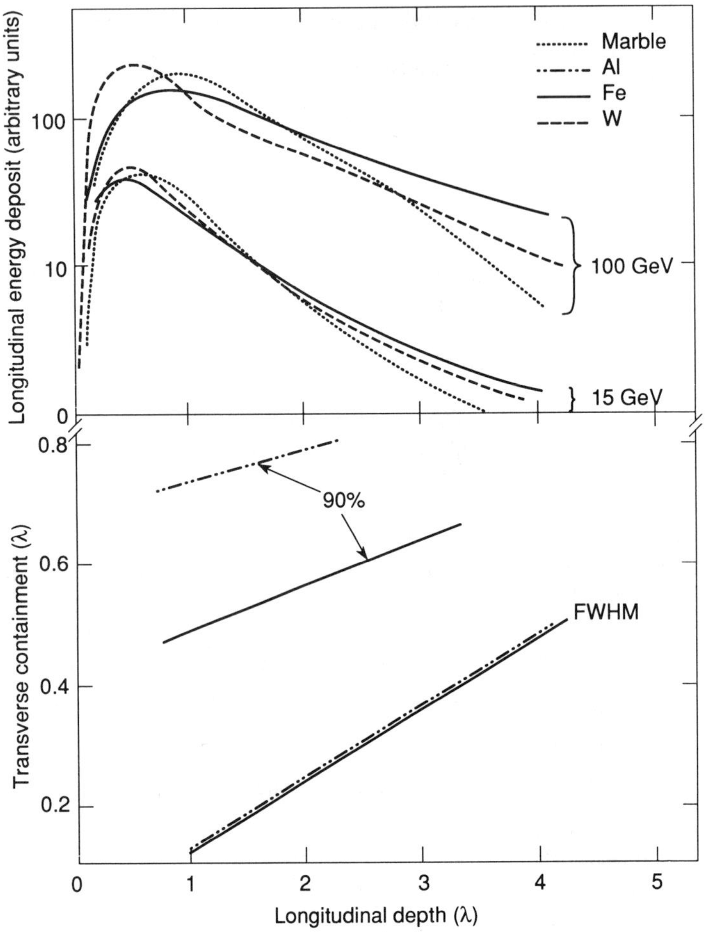

FIG. 7. (Top) Longitudinal shower development induced by hadrons in different materials, showing approximate scaling in interaction length λ. The shower distributions are measured with respect to the face of the calorimeter. (Bottom) For the transverse distributions as a function of shower depth, scaling in λ is found for the narrow core (FWHM) of the showers. The radius of the cylinder for 90% lateral containment is much larger and does not scale in λ [Fabjan and Ludlam, 1982; reproduced with permission of Annual Review of Nuclear and Particle Science.]

absorber nuclei. In this process, a sizable fraction of the particle's energy is spent to overcome the binding energy; in general, this energy is "lost" as it will not contribute to a measurable signal. The level of π^0 production is anticorrelated with the number of nuclear breakups and therefore will lead to strongly varying shower compositions.

In summary, the essential features of the hadronic cascade are

1. the creation of fast pions, protons, and kaons propagating through hadronic collisions;
2. the production of neutral pions propagating electromagnetically; and
3. nuclear reactions producing slow neutrons, protons, low-energy photons, and nuclear fragments.

These very different components usually provoke different responses in the sensors of the calorimeter which are no longer proportional to the energy of the cascade particles. The event-by-event fluctuations will determine the resolution $\sigma(E)$, and be at the origin of a non-Gaussian response function $f(E)$.

2.5 Compensation and Performance of Hadron Calorimeters

Based on our knowledge of the varied contributions to a hadronic cascade (Sec. 2.4), we expect that the energy resolution of HCs should improve to the extent that the response of the instrument to these different components can be equalized. A useful indicator of the degree of equality of response is e/h, the ratio of the signals from electrons and hadrons measured in the same device (Fig. 8). During the last decade, sampling calorimeters were developed to achieve this very desirable equalization or compensation in response, $e/h \approx 1$, by a careful consider-

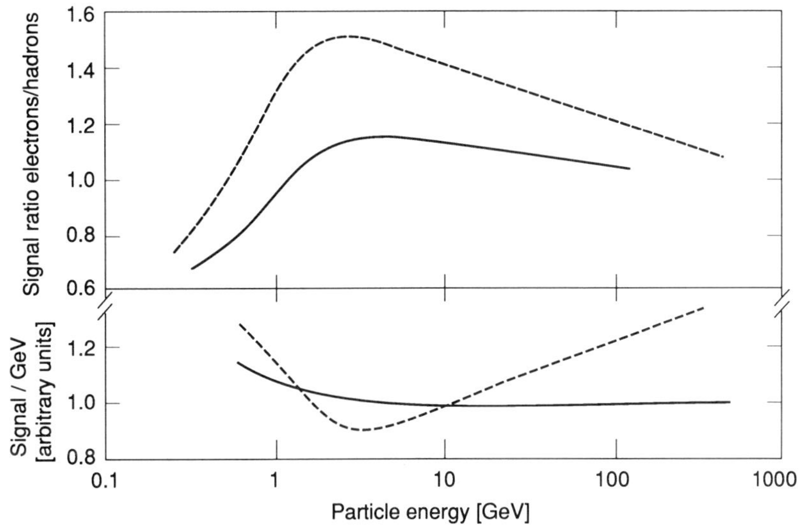

FIG. 8. The ratio e/h of response of a calorimeter to electrons and hadrons at a given energy deposition (upper graph). Most hadron calorimeters have $e/h \sim 1.3$–1.5 (dashed line). In certain cases the response towards hadrons may be improved, resulting in $e/h \sim 1$ (compensated calorimeter, solid line). As the average e.m. contribution increases with hadron energy, e/h approaches 1. This energy dependence gives rise to a nonlinear behavior vs energy (lower graph), which is more pronounced for noncompensated calorimeters (dashed line) than for compensated ones (solid line).

ation of the various shower components and their interactions with the readout medium (Fabjan, 1989). The key to this equalization is a calorimeter design in which the sensitivity to the e.m. component relative to the hadronic-nuclear component is controllable. So far, it has been most practical to select slow, MeV neutrons originating from nuclear excitations or breakup as a measure of the hadronic component. A suitable choice of thickness ratio between the passive absorber (controlling the signal of fast particles) and the active sensors (such as scintillators, which preferentially respond to neutrons) makes it possible to "tune" this compensation (Wigmans, 1991).

Today we know that the intrinsic energy resolution in these compensated calorimeters is $\sigma_{\text{intrinsic}}/E \leqslant 0.2/\sqrt{E(\text{GeV})}$, considerably worse than the intrinsic resolution in EMCs. However, in noncompensated calorimeters ($e/h \approx 1.4$ typically) this intrinsic resolution is much worse: $\sigma_{\text{intrinsic}}/E \approx 0.4/\sqrt{E(\text{GeV})}$.

Besides poor energy resolution, noncompensated calorimeters show another disease. The level of the hadronic component decreases as a function of energy, because an increasing fraction of the hadronic cascade is converted into π^0's, i.e., into an electromagnetic component. For calorimeters with $e/h \sim 1.4$, this results in a nonlinear behavior, as indicated in Fig. 8. Only for those calorimeters that achieve nearly equal response for electrons and hadrons can we expect an approximately linear behavior. The effect of noncompensation on the calorimeter response function is shown conceptually in Fig. 9.

Hadronic calorimeters are almost exclusively made in the form of sampling calorimeters. The fundamental reason for this can be understood from the previous discussion: only the careful choice of passive absorber combined with the active sensors permits the desirable tailoring of compensation, which is not achievable in homogeneously sensitive devices. The practical necessity to have reasonably compact devices leads to the use of rather dense (iron, lead, and uranium) absorbers. Sampling fluctuations are associated with this type of construction and its incomplete collection of shower information, and the following parametrization is found empirically:

$$\sigma_{\text{sampling}}^{\text{hadronic}}/E \approx 0.09\sqrt{\Delta E(\text{MeV})/E(\text{GeV})},$$

with ΔE being defined as previously for the e.m. sampling fluctuations. As an example, we note that uranium (2-mm-thick plates) alternated with scintillator plates (2 to 3 mm thick) or lead (10 mm)/scintillator (2 mm)

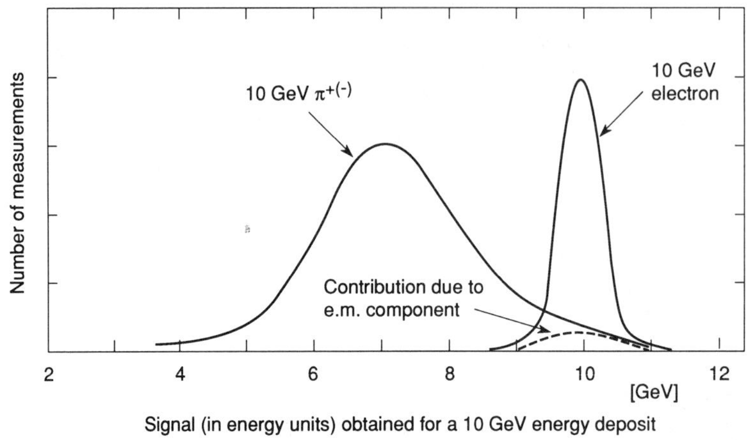

FIG. 9. Response function of calorimeters for electrons and hadrons. The curves are for an EMC with $\sigma/E=0.1/\sqrt{E}$ and for a HC with $\sigma/E=0.5\sqrt{E}$ and $e/h=1.4$. A fraction of the hadron-induced cascade propagates almost completely electromagnetically, broadening the response and producing non-Gaussian tails in the response function.

achieve close to perfect compensation. The corresponding sampling fluctuations will be approximately

$$\sigma_{\text{sampling}}/E{\sim}0.2/\sqrt{E}$$

for the uranium calorimeter,

$$\sigma_{\text{sampling}}/E{\sim}0.3/\sqrt{E}$$

for the lead design.

In summary, during the last 15 years, hadronic calorimetry has developed into a well-understood and mature measurement technology. The accuracy of the hadronic energy measurement cannot be made as good as that for electrons. Physics processes limit the intrinsic accuracy to approximately $\sigma/E\leqslant 0.2/\sqrt{E}$, resulting nevertheless in instruments that measure ~100-GeV particles or jets of particles with a few percent accuracy.

2.6 Calorimeter Systems

Many calorimeter design techniques have been developed to optimize one or several of the following performance parameters:

1. a linear response as a function of energy, covering a range from 100 MeV to ~1000 GeV at present-day facilities;
2. low instrumental noise of the sensors to avoid a degradation of the energy resolution;
3. a high degree of segmentation of the readout, which is required for position resolution and for discrimination between electrons (photons) and hadrons based on the differences in shower shape; and
4. insensitivity to magnetic fields, temperature variations, or ionizing radiation on the sensors.

Fundamentally, a distinction is made between homogeneously sensitive and sampling calorimeters. The former technique is essentially reserved for EMCs and results in a considerably improved performance. The latter is applied to both EMCs and HCs. The interest in high-performance, homogenous EMCs has motivated an active research and development effort during the last decade. Several novel crystals were developed with outstanding performance features (high density, fast scintillation) (Fabjan, 1987). One of these new crystals has become a detector of choice in positron emission tomography.

For sampling calorimeters, several sensor techniques are shown schematically in Fig. 10. Plastic scintillators are made of a matrix of polymethyl methacrylate, polystyrene, or polyvinyl toluene in which the scintillating agent is embedded; they can be mass produced in large sheets. The low-level light is directed towards light sensors by light guides (same principle as optical communication fibers) [Fig. 10(a)] or, alternatively, collected by "wavelength shifter" (WLS) plates [Fig. 10(b)]. These WLS plates are quite thin and are inserted perpendicularly to the absorber/scintillator arrangement. Scintillation pho-

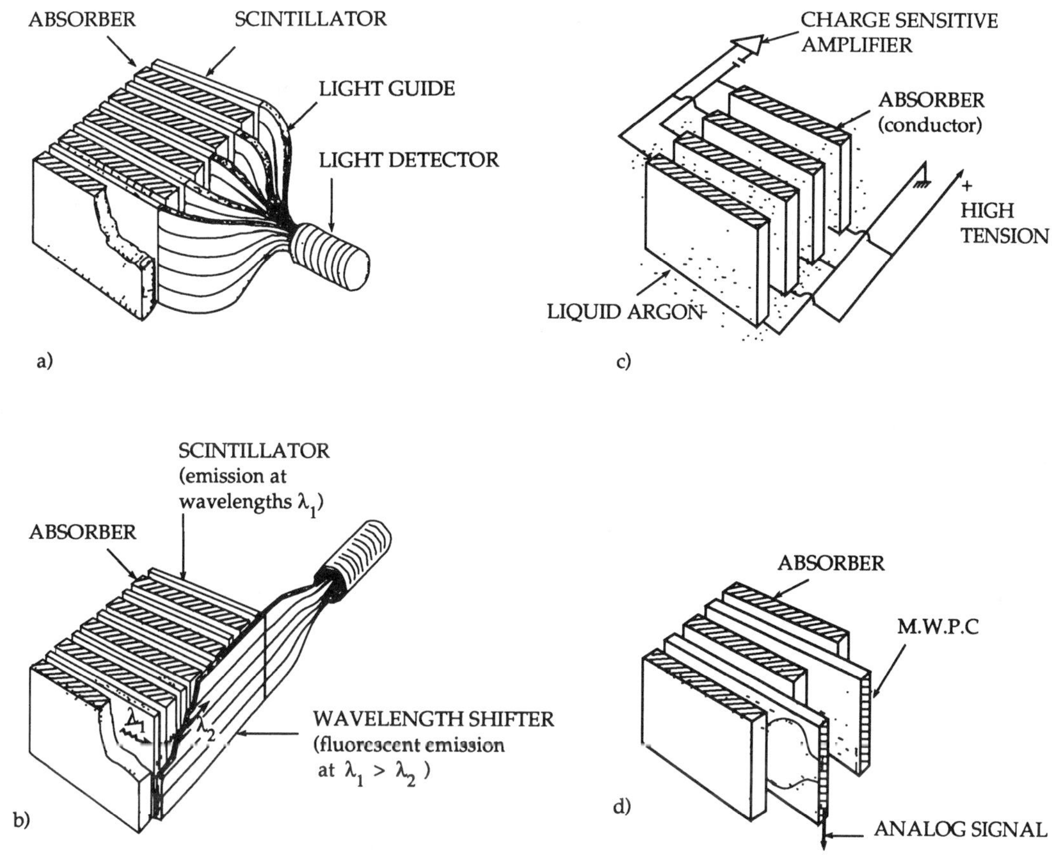

FIG. 10. Schematic representation of frequently used calorimeter readout techniques. (a) Plates of scintillator optically coupled individually to a photomultiplier. (b) Plates of scintillator read out by photon absorption and conversion in a wavelength-shifter plate. (c) Charge produced in an electron-transporting medium (e.g., liquefied or high-pressure argon) collected at electrodes, that may also function as the passive absorber plates. (d) Charge produced in a proportional gas and amplified internally on suitable readout wires.

tons emerge from the scintillator and penetrate into and are absorbed by the WLS; absorption of a photon results in reemission at longer wavelengths, to which the WLS is transparent and which can be recorded in an optical detector (photomultipliers, photodiodes, etc.). This method is widely used, as it allows reasonable segmentation and achieves acceptable uniformity in energy response.

The second class of readout senses the ionization charge. One of the most widely used sensors is liquid argon maintained at 80 K [Fig. 10(c)]. The absorber plates serve the dual function of particle absorber and charge-collecting electrodes. The method works well, provided attention is given to the electronic signal processing, which is similar to that used in nuclear spectroscopy applications with germanium counters (Sec. 1). In place of liquid argon, ionization in certain gases may provide a suitable measurement. Multiwire proportional chambers (MWPCs) or their derivatives have been used successfully for calorimetric applications [Fig. 10(d)].

The charge readout method is considered to offer the best control of systematic instrumental effects (e.g., uniformity and energy scale calibration). The light-collection methods are perhaps somewhat easier to realize and may offer a faster signal response.

Modern calorimeter facilities have a very large number of signal-processing channels, typically $> 10^5$. This has become possible only because of extensive electronics developments. In the future, this trend will continue with emphasis on large dynamic signal response, very high granularity, and very high rate capability (Donaldson and Gilchriese, 1990).

3. CALORIMETRIC MEASUREMENTS IN ASTRO-PARTICLE PHYSICS

During the past decade, astrophysics, cosmology, and particle physics have coalesced into a branch of physics—astro-particle physics—that synthesizes results from these different disciplines to study cosmology and to address problems in particle physics not accessible with experimentation at accelerators. Examples are the study of the very high-energy component of cosmic rays, the search for galactic objects emitting intense beams of particles (neutrinos and gammas), or the quest for relics of the "big bang," possibly too heavy to be produced by terrestrial machines. These studies require novel instrumentation mainly based on the absorption of particles (or radiation), i.e., application of the calorimetric detection principle.

One group of instruments is used to observe rare events with a very low interaction rate, e.g., the detection of galactic or extragalactic neutrino radiation. It requires very massive detectors with an instrumented absorber mass of 1000 to more than 10 000 tonnes. Such detectors need to be carefully shielded from the "mundane" cosmic particle flux and hence are usually buried in a mine below thousands of meters of rock. Several such facilities have already been built, resembling gigantic hadron sampling calorimeters of the type discussed in Sec. 2. However, certain reactions produce particles of rather low energies, in the GeV region, which can be more sensitively measured with homogeneous calorimeters; large underground "swimming pools" filled with ultrapure water have been constructed. The energy of the reaction products is inferred by the Cherenkov light (see Introduction) that these particles emit while they are slowed down and absorbed in the water. Perhaps the most significant result obtained to date from these facilities was the detection of the neutrino bursts emitted by the very recent Supernova SN1987A in the Magellanic Cloud. This observation provided an important confirmation of the theory of type II supernovae and the supernova synthesis of heavy elements (iron, nickel, etc.). The enormous mass of these calorimeters, the very low level of background signals, and the very sensitive electronic equipment of these instruments permit another very fundamental observation: the stability of nucleons. In certain fashionable theories explaining the strong and electroweak forces as different manifestations of an underlying "Grand Unified Theory," a unified force is mediated by a particle, 10^{15} times heavier than a proton. Such particles, which cannot be produced by terrestrial means, would manifest themselves through the decay of nucleons with lifetimes of approximately 10^{32} years. Searches to that level of sensitivity have been performed, but convincing evidence for nucleon decay has not yet been found. A similar technique, hadronic energy measurement via associated Cherenkov light emission, is being used for the ultimate calorimeter in size—using more than 1000 km^3 of the Earth's atmosphere as the absorber. Very high-energy cosmic-ray particles will interact in the upper layer of the atmosphere and initiate particle cascades. An array of almost 1000 ground-based light detectors records the light flashes. This facility has already permitted the observation of the cosmic-ray spectrum to energies beyond 10^{19} eV!

While high-energy astrophysical applications use known calorimetric principles in innovative ways, the need to detect very low-energy deposits in the meV to eV range has led to studies of totally new detection principles. Fundamental physics questions motivate these attempts, such as

1. improved lower limits on the rest mass of neutrinos;
2. search for weakly interacting massive particles (WIMPs), which may be responsible for the component of dark (invisible and not yet accounted for) matter in the Universe;
3. search for magnetic monopoles, which may be relics of the big bang; and
4. search for relic neutrinos from the big bang, which might have an energy of ~ 1 MeV.

The basic detection principle, registration of the interaction of the particle with the detector during the absorption process, is again calorimetry. However, the energies involved are below the atomic excitation levels, and different detection processes have to be invoked.

These calorimeters are characterized by operation at cryogenic temperatures, in the few-mK to ~1-K range. They exploit either some specific properties of the superconducting phase of matter or the reduction in the thermal noise of the detecting medium. Given the very low thresholds of the basic detection process, the number of elementary excitations can be considerably higher, and these detectors therefore hold the promise of much improved energy resolution (see Sec. 1.2).

While a great variety of detectors have been considered, all of them are based on one of the following properties of matter at low temperature:

1. Elementary excitations have very low energy. For example, Cooper pairs in superconductors have binding energies in the 10^{-6} to 10^{-3}-eV range and may be broken into quasiparticles by phonon (lattice vibration) absorption.
2. The specific heat c for dielectric crystals decreases as $\sim\exp(-1/kT)$, resulting in a measurable, relatively fast temperature rise for very low-energy absorption.
3. Energy absorption may induce a change in magnetization, latent heat release, or quasiparticle multiplication.

"Bolometric" detectors are an example of an application of concept **2**, the true calorimeters, in which the temperature rise due to the interaction of particles in an insulating crystal maintained at very low temperatures may be measured with a resistive thermometer. The energy resolution (neglecting noise in the readout electronics) is evaluated to be $\sigma(E)\propto T^{5/2}M^{1/2}$ (M being the mass of the bolometer); x rays have been measured with 0.1% energy resolution (at 6 keV), superior to that obtained with conventional semiconductor detectors. Development of detectors with 100–1000-g mass would permit, for instance, sensitive dark-matter searches. Instruments employing superheated superconducting granules (SSG) are based on concept **3**: small granules (10–100-μm diameter), made from a type-I superconductor, can exhibit a metastable superconducting state above the critical magnetic field. An energy deposit in a granule will heat it, drive it into the normal state, and change the magnetic flux which is sensed by macroscopic pickup coils. These detectors are being developed for dark-matter searches and solar or galactic neutrino observation. Their excellent sensitivity to very low-energy deposits suggests attractive applications outside particle physics (Pretzl *et al.*, 1987).

4. CALORIMETRIC INSTRUMENTATION IN APPLIED SCIENCES

The measurement techniques described in the previous sections were developed for the esoteric pursuit of fundamental research in nuclear, particle, and astro-particle physics. As a final comment we wish to sketch applications in areas such as nuclear medicine, geology, and materials testing.

Many radiation-detection problems occur in nuclear medicine. Medical diagnosis based on low-range x-ray imaging—using photographic recording, image intensifiers, or gaseous chamber techniques—is all pervasive. A modern, very powerful tool is positron emission tomography (PET), which measures the distribution of positron emitters in the human body through the characteristic signature of $e^+e^-\rightarrow\gamma\gamma$ with 511-keV energy each. Detection of these gammas is most frequently achieved with crystals (Sec. 2), such as bismuth germanate. The interest in higher spatial resolution and improved temporal response has motivated an active research and development program in several particle physics centers. While the search for the perfect crystal continues, particle physicists may have found the perfect liquid for high-quality x-ray detection. Liquid xenon is a very fast scintillator, with a light output comparable to that of sodium iodide (NaI), showing no radiation damage. Readout methods which would permit very high spatial localization of the gamma quanta are being evaluated.

For many years, x-ray detectors have been used by the oil industry to probe rock composition through the detection of characteristic radiation from certain isotopes. Typically, NaI crystals or semiconductor detectors are used for this type of measurement.

In Sec. 3, we mentioned the most recent methods aimed at very good x-ray energy resolution. When these devices become routinely available, they should have applications whenever energy resolution, or detection sensitivity for monochromatic x rays, is a premium: use in very low-level counting, such as archeological dating methods based on isotope abundance, or ultrasensitive chemical analysis based on x-ray fluorescence, will be attractive.

On a larger scale, calorimeters have been proposed to provide a "neutrino-ray" picture of the ore deposits in the Earth, sending a focused ν beam in various directions through the Earth's core and recording its absorption!

Calorimeters require signal-processing and pattern-recognition algorithms to reconstruct the particle's energy. Particle physics is the driving motor behind many state-of-the-art computer applications which arguably should become useful to the applied sciences within a decade.

Calorimetry is an essential detection technique in particle physics. Recent developments suggest that it will become an important technique in the other sciences, both fundamental and applied.

GLOSSARY

Calorimetry: In its most general definition, a technique to measure the energy of a reaction or a process. Sometimes, the energy measurement is related to a temperature measurement, hence the name "calorimetry." In particle physics, it is the name given to techniques that measure the energy of particles or quanta.

Gamma Quanta: Energetic quanta of electromagnetic radiation. They may be emitted from the deexcitation of nuclei or created in very energetic interactions of elementary particles with matter.

Hadrons: Generic name for those elementary particles which are carriers of the "strong interaction" (e.g., protons, neutrons, and pions).

Leptons: Generic name for those elementary particles which do not participate in the strong interaction (e.g., electrons, muons, and their corresponding neutrinos).

Nucleon: Generic name for neutrons and protons, the constituents of the atomic nucleus.

Particle Detectors: Instruments to detect or measure properties of particles or radiation quanta. Here, particles designate the microscopic constituents of matter or elementary particles created in energetic collisions of matter. Typically, the trajectory, velocity, energy, or mass of such particles is detected.

Particle Physics: Branch of physics devoted to the study of elementary particles and forces between them. Frequently very energetic accelerators are used to create or study such particles; this field is therefore often also called high-energy physics.

Scintillator: Any of several materials which are widely used for particle detection; it is excited by the passage of charged particles and emits photons (in the UV to visible range) in the deexcitation process. It has important applications outside particle physics.

Solid-State Detectors: Common class of particle detectors. Most frequently used are detectors based on very pure silicon, germanium, or gallium arsenide. The passage of a charged particle creates free charge carriers ("electron-hole" pairs), which may be collected on suitably arranged electrodes. They are used for the precision measurement of the energy loss or the position of charged particles.

List of Works Cited

Ahlen, S. P. (1980), *Rev. Mod. Phys.* **52**, 121.

Donaldson, R., Gilchriese, M. G. (1990), *Workshop on Calorimetry for the Super Collider, Tuscaloosa 1989*, Singapore: World Scientific.

Fabjan, C. W. (1987), in: Ferbel, T. (Ed.), *Experimental Techniques in High-Energy Physics*, Reading, MA:Addison-Wesley, p. 257.

Fabjan, C. W., Ludlam, R. (1982), *Annu. Rev. Nucl. Part. Sci.* **32**, 335.

Fabjan, C. W., Wigmans, R. (1989), *Rep. Progr. Phys.* **52**, 1519.

Fernow, R. C. (1986), *Introduction to Experimental Particle Physics*, Cambridge: Cambridge University Press.

Particle Data Group (1988), *Phys. Lett.* **240**, 1.

Pretzl, K. *et al.* (Eds.) (1987), *Low Temperature Detectors for Neutrinos and Dark Matter*, Berlin: Springer.

Tsai, Y. (1974), *Rev. Mod. Phys.* **46**, 815.

Wigmans, R. (1991), *Advances in Hadron Calorimetry*, CERN-PPE/91-39; to be published in Annu. Rev. Nucl. Part. Sci.

Further Reading

T. Ferbel (Ed.) (1987), *Experimental Tech-*

niques in High-Energy Physics, Reading, MA: Addison-Wesley.

Pretzl, K., *et al.* (Eds.) (1987), *Low Temperature Detectors for Neutrinos and Dark Matter*, New York: Springer.

Radeka, V. (1988), *Annu. Rev. Nucl. Part. Sci.* **38**, 217.

Williams, H. H. (1986), *Annu. Rev. Nucl. Part. Sci.* **36**, 361.

DETECTORS, PARTICLE, TRACKING

See PARTICLE DETECTORS, TRACKING

DETECTORS, RADIATION

See RADILATION DETECTORS

DETECTORS, SCINTILLATION

ERIC B. NORMAN, *Nuclear Science Division, Lawrence Berkeley Laboratory, Berkeley, California, U.S.A.*

INTRODUCTION

In many areas of basic scientific research, as well as in a variety of applied fields, one wishes to detect the passage of radiation through matter. The fact that some materials emit light when struck by radiation has been known for many years, and this principle is the basis for the scintillation detector. For example, in 1911, Ernest Rutherford observed scintillation light produced by the interactions of α particles with a zinc sulfide screen in his pioneering scattering experiments which established the existence of the atomic nucleus. Since then, a large number of different scintillating materials have been developed for specific applications in the fields of physics, chemistry, and biology. This article describes the scintillation process and some of the uses of scintillation detectors in the labo ratory and in industry.

Before describing the scintillation mechanism, it is useful to review briefly the processes by which radiation is absorbed by matter. For x rays and γ rays (high-energy photons) three processes are important: the photoelectric effect, Compton scattering, and pair production. In the photoelectric effect, the full energy of the photon is absorbed by a bound atomic electron. This ejects the electron from the atom with a kinetic energy equal to the photon energy minus the electron's original binding energy. In Compton scattering, the photon also interacts with an atomic electron, but only a fraction of its energy is transferred to the electron. The remaining energy is carried off by a secondary photon in an amount that depends on the angle of its emission relative to the direction of the primary photon. For photons with energies greater than twice the rest-mass energy of the electron ($2m_ec^2 = 1022$ keV) it is possible to convert into an electron-proton pair. The kinetic energy of the e^+e^- pair is then equal to the original photon energy minus 1022 keV.

When β particles (electrons) traverse matter, they produce atomic and molecular excitations and ionization through their collisions with atomic electrons. Heavy charged particles, such as protons, α particles (^{4}He nuclei), and nuclei of heavier atoms, also make inelastic collisions with atomic electrons that lead to atomic and molecular excitations as well as ionization. For neutrons, however, the situation is somewhat different. Since neutrons are electrically neutral, the primary way in which they lose energy in matter is through their nuclear interactions with atomic nuclei. They can be scattered off the nucleus or absorbed by it. In hydrogenous materials such as paraffin or polyethylene, neutrons scattering off hydrogen nuclei produce recoiling protons which then interact as described above. Some scintillation materials are doped with ^{6}Li, which has a large cross section for the ^{6}Li$(n,{}^3$H$)^4$He reaction. The recoiling triton and α particle then interact as described above.

1. PHYSICS OF SCINTILLATION DETECTORS

The mechanism by which energy deposited by radiation is converted into light within a scintillator is a fairly involved process and is

1-56081-063-7/92/$5.00+.50

somewhat different in inorganic and organic scintillators. In both types, however, the amount of light produced is proportional to the amount of energy deposited in the scintillator. Inorganic scintillators are made of crystals of materials such as NaI or CsI that are usually activated with an impurity such as Tl (thallium). It has been found that with an impurity concentration on the order of 1 part in 10^3 by number, there is a very efficient transfer of energy from the major constituent to the impurity. In NaI(Tl), for example, the result is that the capture of electrons and/or holes by the thallium impurity leads to excited states of the Tl^+ ion that decay by emitting the light which is observed from this scintillator. Through this process, approximately 12% of the incident energy is converted into Tl^+ light. Because the impurity concentration is so low, this type of scintillator is very transparent to its own light.

Liquid, plastic, and some kinds of crystal scintillators are composed of organic compounds known as aromatic hydrocarbons. These materials consist of molecules containing a benzene ring structure. Within these rings, not all of the electrons that bind the atoms together are strongly localized to individual atoms. These so-called valence electrons occupy the π molecular orbitals which extend above and below the plane of the molecule. Absorption of radiation by these electrons can produce excited states which decay to the ground state or other excited states by emitting light. If the decay proceeds to an excited state, then the photon that is emitted has too little energy to excite another molecule from its ground state to the appropriate excited state. Thus, such transitions produce light to which the scintillator is quite transparent.

Over the years, many different types of scintillators have been developed to meet a variety of radiation-detection needs. The physical properties of a number of different scintillation materials are listed in Table 1. Physical properties such as density, scintillation efficiency (i.e., a measure of the fraction of deposited energy that is converted into light), the wavelength at which the scintillation light appears, and the scintillation decay constant (i.e., the time interval over which light is emitted) vary widely among the different scintillators. These differences are used to decide which scintillator is used for a particular application. High-density materials containing elements with large atomic numbers have relatively high efficiencies for stopping γ rays. Thus, compounds such as NaI(Tl), BaF_2, or bismuth germanate (BGO) are used in such applications. Plastic scintillators have very short decay constants and thus are used in applications where there are needs for fast timing. Scintillators such as ^{6}LiI(Eu) contain nuclei that have high cross sections for neutron-induced reactions and thus are useful in neutron studies.

Table 1. Physical properties of various scintillator materials.

Material	Physical form	Wavelength of maximum emission (nm)	Decay constant (ns)	Density (g/cm^3)	Scintillation efficiency[a] (%)
NaI(Tl)	Crystal	415	230	3.67	100
CaF_2(Eu)	Crystal	435	940	3.18	50
CsI(Na)	Crystal	420	630	4.51	85
CsI(Tl)	Crystal	565	1000	4.51	45
BaF_2	Crystal	325	630	4.88	20
$Bi_4Ge_3O_{12}$	Crystal	480	300	7.13	12
^{6}LiI(Eu)	Crystal	480	1400	4.08	35
Anthracene	Crystal	447	30	1.25	43
NE102A	Plastic	423	2.4	1.032	28
NE115	Plastic	395	320	1.032	10
Pilot U	Plastic	425	1.36	1.032	29
NE213	Liquid	425	3.7	0.874	34

[a]Relative to NaI(Tl).

The process by which a complete scintillation detector system converts the energy deposited by the incident radiation into a usable signal involves several steps. Such a system is shown schematically in Fig. 1. As described above, the incident radiation ionizes and excites the scintillation material. Within the scintillator, a fraction of this energy is converted into light. This light travels through the scintillator and is then collected on a photosensitive surface. The resulting photoelectrons are multiplied and collected, thus producing a usable signal. Usually these functions are accomplished by coupling a scintillation material to a light pipe which guides the light to a photomultiplier tube. When photons strike the photocathode, electrons are ejected via the photoelectric effect. These electrons are then accelerated and focused onto a series of dynodes, where through collisions the number of electrons is greatly multiplied. By the time the signal reaches the collection anode, it may have been amplified by more than a factor of 10^6. This signal is usually then sent to an amplifier where further shaping and amplification of the pulse is done before it reaches the data-acquisition system. The size of this final signal provides a measure of the amount of energy deposited by the radiation in the scintillator.

An important characteristic of any type of energy-measuring device is its resolution. This is usually expressed as ΔE=the full width at half maximum (FWHM) of the peak observed from a source of monoenergetic particles. In the scintillation-detector system described above, the measure of the energy of the incident radiation is given by the amount of charge that comes out of the photomultiplier. However, as discussed by many authors (e.g., Rolfs and Rodney, 1988), at each step in the scintillation-detection process there are losses which contribute to the overall detector resolution. As discussed above, in NaI(Tl) detectors the efficiency, ϵ_1, of converting the incident energy into light is about 12%. The mean energy w of these photons is about 3 eV. Of these photons, only a fraction ϵ_2, of order 0.4, actually reaches the photocathode. Only another fraction ϵ_3, of order 0.2, causes the emission of a photoelectron. The fractional energy resolution $\Delta E/E$ of such detectors is limited by the statistical fluctuations in each of these processes and is given by the expression

$$\Delta E/E = 2.35(w/E\epsilon_1\epsilon_2\epsilon_3)^{1/2}.$$

This leads to an expected value of $\Delta E/E$ of approximately 4% for a 1.33-MeV γ ray. As will be shown below, in practice one usually observes slightly worse resolution than this because of imperfections in the detector crystal and/or electronics.

2. MANUFACTURE OF SCINTILLATION DETECTORS

Inorganic scintillators such as NaI(Tl) and CsI(Tl) can be produced as single-crystal ingots up to approximately 36 in. in diameter. To produce crystals of this size, grains or small chunks of the desired chemical are put into a large crucible. The crucible is heated to high enough temperature to melt the material. The crucible is then allowed to cool at a rate which is faster on the bottom than at the top.

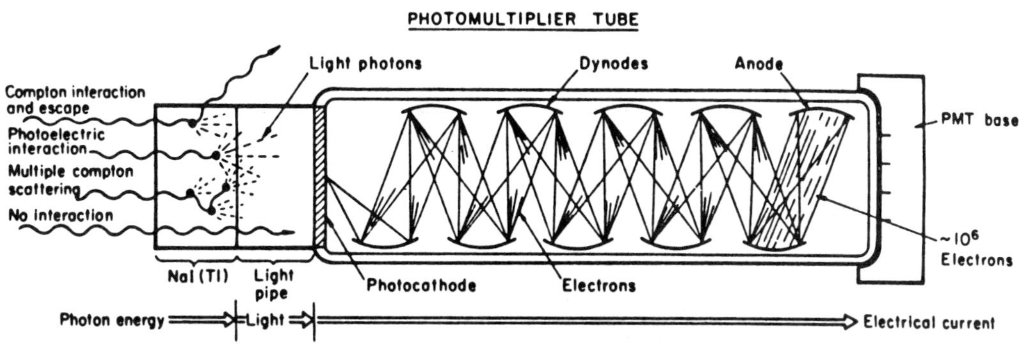

FIG. 1. Schematic view of a scintillation detector system. The interactions of photons in a NaI(Tl) scintillator are shown along with the transmission of the resulting optical photons through a light pipe to a photomultiplier where light is converted into an electrical signal. (Courtesy of S. E. Derenzo.)

Thus crystal growth begins at the bottom and proceeds upwards. In this way large single crystals can be grown. An example of this is shown in Fig. 2, where a large crystal of CsI has just been removed from the crucible in which it was grown. For more information on the production of crystals, see CRYSTAL GROWTH.

Plastic scintillators are usually manufactured in the form of cast sheets. However, these materials are also available in rods, fibers, or beads. They are produced by mixing together a suitable fluorescent organic compound with a polymer and allowing the mixture to solidify. Commonly used mixtures include *p*-terphenyl dissolved in either polystyrene or polyvinyltoluene. A typical plastic scintillator paddle with a light pipe attached is shown in Fig. 3.

Following the machining or forming process, the surfaces of a solid scintillator are polished. A highly reflective coating is usually applied to these surfaces to return the light to the crystal and thus aid in the transport of the scintillation light to the photomultiplier. Crystalline scintillators are normally packaged inside a metal canister so as to protect the scintillator from mechanical shocks. In the case of NaI such protection is also necessary because of the hygroscopic (i.e., moisture sensitive) nature of the material. Liquid scintillators are also often packaged in a metal container. In both cases, the can has a transparent window on one side on which the light pipe and then photomultiplier tube are mounted.

3. APPLICATIONS OF SCINTILLATION DETECTORS

In the areas of nuclear and particle physics research, scintillation detectors are used in a wide variety of experiments. In γ-ray-counting experiments, NaI(Tl) detectors are used in applications where high detection efficiency, good energy resolution, and reasonably good timing properties are needed. An example of low-energy γ rays observed in a small NaI(Tl) detector is shown in Fig. 4. A radioactive source of ^{60}Co, which emits a pair of coinci-

FIG. 2. An ingot of CsI as it emerged from the crucible in which it was grown. (Courtesy of Bicron Corp.)

FIG. 3. A plastic scintillator paddle with a light pipe attached to it. (Courtesy of Bicron Corp.)

dent γ rays with energies of 1173 and 1332 keV, was counted with a 7.6-cm-diam $\times$ 7.6-cm-long NaI(Tl) detector. The full-energy (photo)peaks of these lines are clearly seen along with a well-defined "sum peak" which is produced when the full energies of both γ rays are absorbed at the same time by the detector. The counts in the region below the photopeaks and in the region between the photopeaks and the sum peak are due to events in which one or both of the γ rays underwent Compton scattering in the crystal or in its surroundings. The energy resolution for the photopeaks is just over 5% FWHM. In contrast, solid-state germanium detectors (described in the article DETECTORS, SEMICONDUCTOR) typically achieve resolutions of approximately 0.15% at this energy.

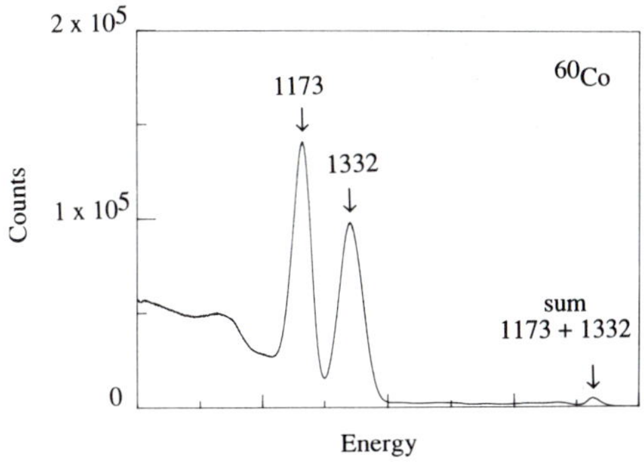

FIG. 4. γ-ray spectrum of ^{60}Co observed in a 7.6-cm-diam $\times$ 7.6-cm-long NaI(Tl) detector.

Another use of NaI(Tl) detectors is in nuclear reaction studies. In radiative capture reactions, the projectile fuses with the target nucleus and the resulting excitation energy is carried off by relatively high-energy ($\gtrsim 10$ MeV) γ rays. While solid-state germanium detectors would provide better energy resolution, their detection efficiency for such high-energy γ rays is too low for them to be useful. Therefore, in this type of research, NaI(Tl) detectors are often used. An example of a spectrum observed in the capture of protons by a target of ^{30}Si is shown in Fig. 5. The NaI(Tl) crystal used in this experiment is 25 cm in diameter and 28 cm thick. The proton bombarding energy is 25.5 MeV. The small peak near 32-MeV γ-ray energy corresponds to the case of leaving the resulting ^{31}P nucleus in its ground state; other peaks at lower γ-ray energy correspond to leaving the ^{31}P in one of its excited states.

Another application of scintillation detectors is in the area of Compton suppression. In a typical germanium detector, if a 1-MeV γ ray enters the detector, only about 20% of the recorded events will be in the full-energy peak (photopeak) and 80% will be in a continuous spectrum from zero up to the full γ-ray energy as a result of Compton scattering. These Comp-

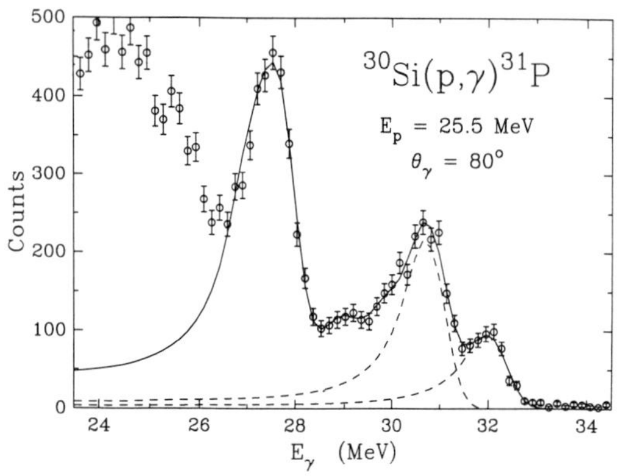

FIG. 5. γ-ray spectrum observed in a 25-cm-diam × 28-cm-long NaI(Tl) detector from the ^{30}Si(p,γ)^{31}P reaction at a proton bombarding energy of 25.5 MeV. (Courtesy of G. Feldman.)

ton events provide no useful information as to the original energy of the incident γ ray. Furthermore, they produce a background which can make it difficult to observe the often small full-energy peaks of interest. By surrounding a germanium detector with a suitable scintillator, one can look for events where both detectors fired simultaneously. Such events are produced by Compton scattering in the germanium detector into the scintillator and can be vetoed. An example of an annular NaI(Tl) scintillator acting as a Compton suppressor for a germanium detector is shown in Fig. 6. γ-ray spectra observed in the germanium detector in this system are shown in Fig. 7. A ^{54}Mn source, which emits monoenergetic γ rays at 835 keV, was counted with a 110-cm^3 germanium detector. The upper spectrum shows the response of the germanium detector without regard to what happened at the same time in the surrounding NaI(Tl) detector. One sees the 835-keV full-energy peak along with a Compton continuum. The lower spectrum is what is observed in the germanium detector when the NaI(Tl) did not observe an event in coincidence with the germanium detector. One can clearly see a reduction in the Compton continuum by a factor as large as 10 in some parts of the

FIG. 6. A 30-cm-diameter × 30-cm-long annular NaI(Tl) detector system surrounding a 110-cm^3 germanium detector. The NaI(Tl) crystal is optically divided into two halves, each viewed by seven photomultiplier tubes. In this system, the NaI(Tl) detector is used to veto events in which a γ ray undergoes Compton scattering in the germanium detector and the scattered photon goes into the NaI(Tl) detector.

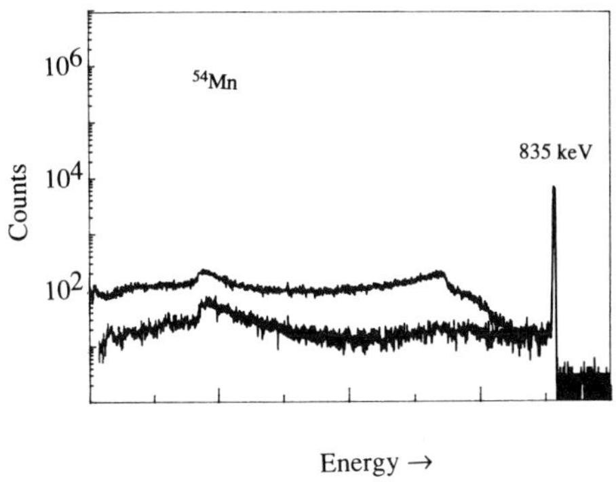

FIG. 7. γ-ray spectra observed in a 110-cm^3 germanium detector placed inside the NaI(Tl) annulus shown in Fig. 6. Data with (lower spectrum) and without (upper spectrum) the NaI(Tl) anticoincidence requirement are shown to illustrate the reduction of the continuum produced by Compton scattering.

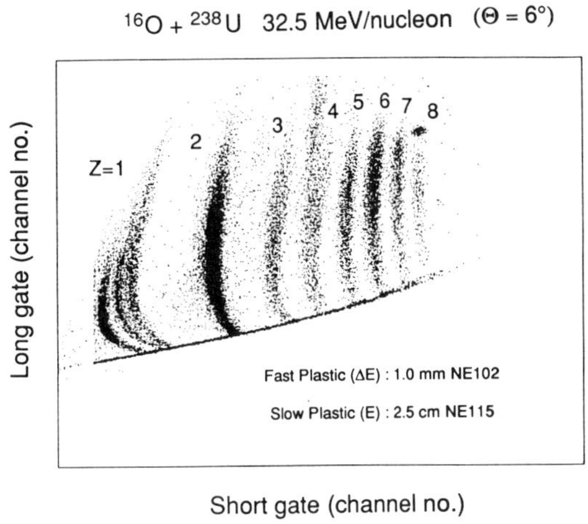

FIG. 9. An example of data obtained from an array of plastic phoswich detectors. By performing a fast integration of the photomultiplier signal, one obtains a measure of the light output from the thin front scintillator. A longer integration provides a measure of the energy deposited in the thicker and slower back scintillator. (Courtesy of Y. D. Chan.)

spectrum. In a number of multielement germanium detector arrays, such as the High Energy Resolution Array (HERA) at Lawrence Berkeley Laboratory, TESSA at Daresbury Laboratory, and the soon-to-be-built GAMMASPHERE and EUROGAM arrays, BGO anti-Compton shields are used instead of NaI(Tl). The reason for this is that the large atomic number of Bi and the high density of BGO mean that a much smaller-sized scintillator is needed to provide the same amount of Compton suppression. This permits the germanium detectors to be packed together closely, and thus provides high efficiency for detecting the full-energy γ ray in a germanium detector.

As can be seen in Table 1, the time dependence of the light output of scintillators varies widely. This variation has been put to practical use in what is called a phoswich detector. This consists of two different scintillators stacked one behind the other, both of which are viewed by one photomultiplier tube. An example of this type of detector is shown in Fig. 8. This detector consists of a thin crystal of CaF_2 mounted on the end of a long plastic scintillator. A charged particle which enters the CaF_2 and comes to rest in the plastic scintillator will produce scintillation light in both materials. Because of the difference in the decay time of the light output in these two materials, the photomultiplier output signal will have two components. There will be a fast component associated with the plastic scintillator and a slower one associated with the CaF_2. By appropriately integrating these signals, one can extract the individual signals from these two scintillators and obtain a measurement of both the energy lost in the front scintillator, ΔE, and the total energy of the charged particle, E. An example of the data resulting from a phoswich detector system is shown in Fig. 9. In this case the products of the fragmentation of an ^{16}O beam by a ^{238}U target were observed by an array of phoswich detectors consisting of two different types of plastic scintillator—one with a short decay time and one with a long decay time. From the "short" and "long" gates set on the photomultiplier output signals, one obtains

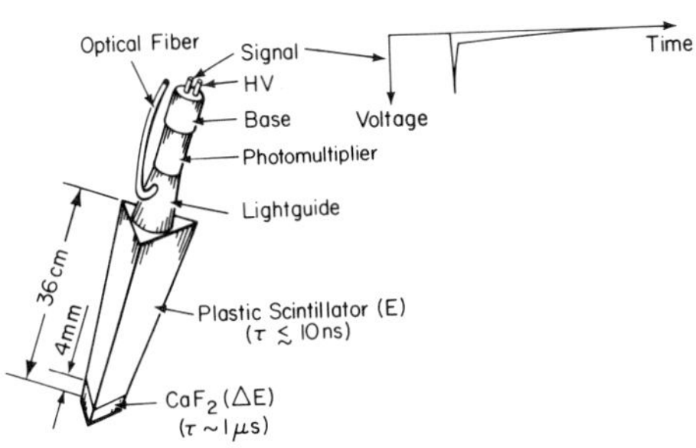

FIG. 8. Schematic view of a phoswich detector. Two scintillators with different decay constants are stacked one behind the other and are both viewed by a single photomultiplier tube. The resulting light signal then has two components which can be separated with appropriate electronics. (Baden *et al.*, 1982.)

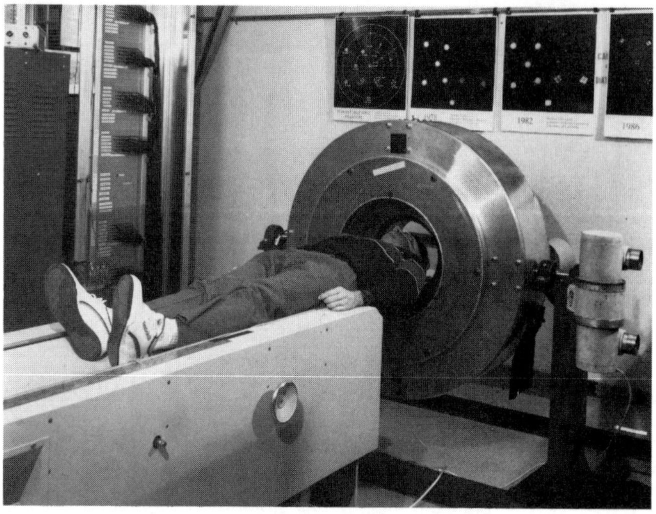

FIG. 10. A patient being examined with a positron-emission tomography system that consists of 280 NaI(Tl) crystals. (Courtesy of S. E. Derenzo.)

the ΔE and E parameters, respectively. The fact that, for a given kinetic energy, nuclei with different atomic numbers (Z) lose differing amounts of energy in the thin front scintillator gives rise to the various bands seen in this two-dimensional plot.

Scintillation detectors have also played a major role in astrophysical studies performed in space. For example, NaI(Tl) detectors on the Solar Maximum Mission detected nuclear γ rays at 847 and 1238 keV produced by the radioactive decay of ^{56}Co (half-life=77 days) coming from Supernova 1987A. This observation helped to confirm that the evolution of massive stars produces a core of iron which collapses and initiates the supernova explosion. This process had been suggested long ago, but the observation of these characteristic γ rays is one of the strongest pieces of experimental evidence that this type of nucleosynthesis actually occurs in supernovae.

In the field of medical physics, scintillation detectors have played major roles in a number of areas. Whole-body counters, which are used to determine the identities and quantities of radioisotopes contained in a person, often consist of large NaI(Tl) crystals. When housed in a low-background environment, such systems are capable of detecting very small (~nanocurie) amounts of γ-ray emitting radioisotopes.

Another area in which scintillators are now being widely used is positron-emission tomography, or PET scanning. In this procedure, a positron-emitting radioisotope, such as ^{18}F, is attached to a compound such as fluorodeoxyglucose and is then introduced into the subject's body. It is concentrated in an organ of

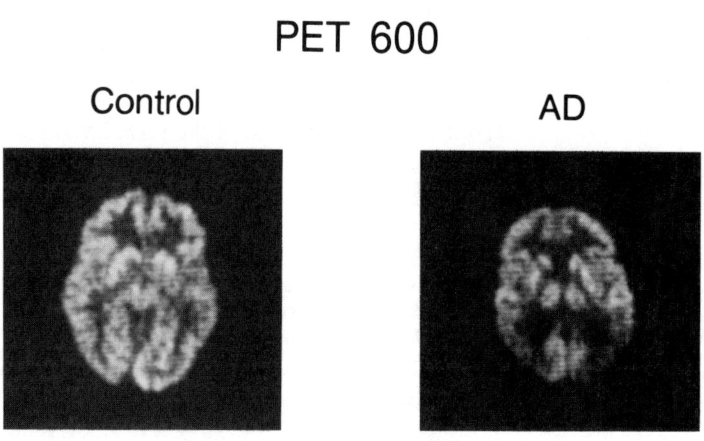

FIG. 11. An example of positron-emission tomography data taken with a PET system consisting of 600 scintillation detectors. Activity of the human brain is indicated by the white areas in the tomograph. A control patient is shown along with a patient suffering from Alzheimer's disease. (Valk *et al.*, 1990.)

interest where the positrons annihilate with electrons. The annihilation process produces two γ rays, each of 511 keV, which are emitted back-to-back. That is, they travel off in opposite directions from the point of annihilation. These annihilation photons are detected in coincidence in two diametrically opposed crystals. By surrounding the subject with a large number of detectors, and by finding the intersection point of all of the lines, one can determine the size and activity of the positron-emitting activity. By observing the time variations in such activity one can observe various body functions in progress. A subject being counted with a PET system of 280 bismuth germanate crystals is shown in Fig. 10. PET data showing activity of the human brain, taken with a system containing 600 individual crystals, are shown in Fig. 11.

In the area of geological exploration, scintillation detectors also play a major role. Measurement of natural radioactivity to locate deposits of uranium and thorium in the earth's crust is an activity in which scintillation detectors are widely used. Low-flying aircraft carrying large NaI(Tl) detectors make surveys of γ-ray activity searching for the characteristic γ rays emitted in the decay chains of ^{232}Th (2.6 MeV) and ^{238}U (1.76 MeV). Borehole measurements are performed in several different ways. Sometimes a scintillation detector is simply lowered down a hole and measurements are made of the local radioactivity. In other applications, sources of either neutrons or γ rays are also placed underground and the scintillation detector is used to measure the amount of scattered or induced activity.

In the area of nuclear safeguards, large plastic scintillators are often used in vehicle or personnel monitoring to look for unauthorized movements of radioactive materials. In such an installation, a vehicle or person drives or walks past the scintillation detector and the characteristic γ rays of ^{235}U or ^{239}Pu are sought for. Walk-through systems are now in place which can detect 0.02 g of ^{239}Pu or 1 g of ^{235}U.

In a number of industrial situations, one is interested in measuring the wearing away of material from an object as a result of its being used. This is the subject of tribology, and scintillation detectors are often used in this field. For example, in the automobile industry, engine wear can be determined by implanting a γ-ray-emitting radioisotope in the part of the engine of interest. A NaI(Tl) or BGO detector is then placed near this part and the γ-ray counting rate is monitored as a function of the engine use. As the part is worn away, one observes a decrease in the γ-ray activity. By taking an identically implanted part and measuring its γ-ray counting rate as one quantitatively removes material from the surface, one can calibrate the technique. Through such tests, it has been found that this method is sensitive to engine part wear as small as 0.00005 in.

Compton scattering of γ rays from an object can be used to determine the thickness and/or distribution of material within the object. This technique is often used in the area of quality control. For example, artillery shells coming off a production line have been irradiated with 662-keV γ rays from a ^{137}Cs source. By measuring the number of scattered γ rays and by comparing the result to that obtained from a standard, one can reject shells that contain voids or other imperfections. This method offers several advantages over more conventional schemes in that it is nondestructive and can be done very quickly.

4. CONCLUSIONS

The conversion of the energy deposited by radiation into light is the basis of the scintillation detector. Many different types of scintillators have been developed to meet a variety of needs in the field of radiation detection. Today scintillation detectors are used extensively in physics, chemistry, medicine, astrophysics, and in a number of industrial applications. With the ongoing development of new and improved scintillating materials, scintillation detectors will continue to play major roles in these fields for the foreseeable future.

List of Works Cited

Baden, A., Gutbrod, H. H., Lohner, H., Maier, M. R., Poskanzer, A. M., Renner, T., Riedesel, H., Ritter, H. G., Spieler, H., Warwick, A., Weik, F., Wieman, H. (1982), *Nucl. Instrum. Methods* **203**, 198–211.

Rolfs, C. E., Rodney, W. S. (1988), *Cauldrons in the Cosmos*, Chicago: Univ. of Chicago Press.

Valk, P. E., Jagust, W. J., Derenzo, S. E., Huesman, R. H., Geyer, A. B., Budinger, T. F. (1990), *Radiology* **176**, 783–790.

Further Reading

Birks, J. B. (1964), *The Theory and Practice of Scintillation Counting*, Oxford: Pergamon.

Brooks, F. D. (1979), *Nucl. Instrum. Methods* **162**, 477–505.

Heath, R. L., Hofstadter, R., and Hughes, E. B. (1979), *Nucl. Instrum. Methods* **162**, 431–476.

Knoll, G. F. (1989), *Radiation Detection and Measurement*, 2nd ed., New York: Wiley.

Leo, W. R. (1987), *Techniques for Nuclear and Particle Physics Experiments*, Berlin: Springer-Verlag.

Neiler, J. H., Bell, P. R. (1965), in: K. Siegbahn (Ed.), *α, β, γ Spectroscopy*, Vol. 1, Amsterdam: North-Holland, p. 245.

DETECTORS, SEMICONDUCTOR

JOHN WALTER, *IntraSpec Inc., Oak Ridge, Tennessee, U.S.A.*

INTRODUCTION

This presentation deals with semiconductor devices used for the detection and energy analysis of single photons or particles by measuring the quantity of free electrical charge generated in the semiconductor by the radiation. Devices that depend on integration over more than one incident radiation event or on optical transitions are not included. Also, devices that depend on ambipolar diffusion to produce charge collection are not included. Emphasis has been placed on Si and Ge devices since they encompass the majority of currently used detectors. Although reference is made to preamplifiers, shaping amplifiers, and multichannel analyzers, detailed discussion of these ancillary instruments is beyond the scope of this presentation. A detailed discussion of the basics of semiconductor physics is also beyond the scope of this article. It is assumed that the reader has a familiarity with the concepts of band gaps, electrons, holes, mobilities, phonons, etc. (Sze, 1981).

The detectors discussed here are really a special form of conduction counter where the leakage current is limited by the use of rectifying contacts. Generally, both signs of carrier are mobile. In principal, a large–band-gap semiconductor with its small number of intrinsic charge carriers would not need such a blocking contact. However, all of the detectors in common use today do require rectifying contacts. The original work on conduction counters was done by Van Heerden (1945), who performed extensive evaluations of diamond and a number of ionic crystals. None of this work produced really useful detectors, largely because imperfections in these materials caused extensive charge trapping and electrical polarization. In 1949, McKay demonstrated the first working model of a semiconductor detector, which he later used to measure the average energy ϵ required to create a free electron-hole pair (McKay, 1951). In his device, the sensitive region was only

1-56081-063-7/92/$5.00+.50

about 1 μm deep and could be reached by charged-particle radiation only from the narrow region where the edge of the junction intersected the surface of the crystal. By assuming that the largest observed pulses resulted from events where all of the α-particle energy was deposited close enough to the junction for the resulting free charge to be efficiently collected, McKay was able to estimate ϵ. His result is in surprisingly good agreement with the many precise measurements that were subsequently made with high-resolution detector systems. However, the shallow sensitive depth and narrow region of access made McKay's configuration not useful for practical applications.

Mayer and Gossick (1957) provided the next major advance by building a planar surface-barrier junction detector on Ge. Their devices had useful active areas of several mm^2 and a sensitive depth that was normal to this active area. However, the depletion depth was only a few microns, requiring charge carriers to be collected by slow ambipolar diffusion to the depletion region. This limited the useful sensitive depth to about 20 μm, enough for naturally occurring α particles, but not for higher-energy charged particles.

The first semiconductor detector that operated on the principal of depositing all of the useful energy within the depletion region of a junction was built by Walter *et al.* (Walter, 1958; Walter *et al.*, 1960). Their devices, which were planar surface-barrier junctions on high-resistivity Ge, had sensitive areas of several cm^2 and sensitive depths of up to 1 mm. This work was quickly followed by similar devices made with high-resistivity Si. Essentially all of currently used semiconductor radiation spectrometers depend on having a sensitive region whose depth exceeds that required to provide collection of all of the charge directly from the depletion region. The excellent energy resolution, fast response time, and large useful sensitive areas and volumes provided by this configuration resulted in dramatic changes in the field of charged-particle and photon spectroscopy.

A sensitive depth of 1 mm of Ge is not enough to provide good detection efficiencies for high-energy γ rays. The application of the Li-drift technique to compensate impurities, which was pioneered by Pell (1960), allowed sensitive depths of over 1 cm together with large active areas. This provided a basis for Ge γ-ray spectrometers. The subsequent development of very high-purity Ge by Hall (1969) allowed development of γ-ray spectrometers that do not depend on the Li-drift technology. Ge detectors are now the instrument of choice for γ-ray spectroscopy. They are widely used for environmental assay and industrial analysis of nuclear contamination, as well as for scientific research.

Elad (1965) demonstrated the feasibility of a Si spectrometer for characteristic elemental x rays by showing that combining a cooled, low-capacitance (deep depletion region) detector with a cooled field-effect transistor (FET) preamplifier could produce resolutions good enough to resolve the characteristic fluorescent x-ray lines of adjacent elements. This discovery was followed by rapid commercial development of fluorescent x-ray systems.

The major advantages of semiconductor detectors are their superior energy resolution, fast response time, small size, and insensitivity to magnetic fields. Their major disadvantages are sensitivity to radiation damage, cost, low efficiency for high-energy photons, and inability to operate at high temperatures.

Ge is the material of choice for high-energy γ-ray detectors because its higher Z makes it a much more efficient absorber than Si. However, Ge has a smaller forbidden band gap than Si, which results in the need to cool high-resolution Ge detectors to cryogenic temperatures in order to reduce the thermal noise.

Si is the usual material of choice for x-ray–energy spectrometers because Si detectors generally have lower leakage currents and thinner entrance windows. However, Ge is sometimes chosen for applications involving high-energy x rays.

There has been a considerable effort to develop alternatives to Si and Ge, most of it directed at materials with higher band gaps or higher Z. However, despite years of effort, there has been little real success, largely because of the difficulty of producing crystal quality comparable to what is available in Ge and Si.

1. PHYSICAL AND TECHNICAL PRINCIPLES

1.1 Detection of Photons and Ionizing Particles: Overview

When photons or ionizing particles traverse a semiconductor crystal, most of their lost energy is converted to positive and negative charges that can move within the crystal lattice.

In a highly perfect crystal of Ge or Si with suitable electrodes, the application of a bias voltage will establish a region depleted of mobile charge. Within this region, the mobile charges freed by the radiation separate and travel along the field lines towards the electrodes. Outside the depletion region, the electrons or holes recombine or diffuse to the surfaces; charge generated here contributes little or nothing useful for spectroscopy.

If the incident particle or photon is stopped within the depletion region, the free charge generated is proportional to the incident energy (less any energy lost in passing through any dead layers on the surface of the detector). With ancillary electronics, the charge is converted to a voltage pulse having an amplitude directly proportional to the energy of the initiating event. Thus the system becomes useful as a radiation spectrometer.

If not all of the energy of a charged particle is lost in the depletion region, the proportionality is lost. However, by making the detector thin so that only a fraction of the incident energy is lost, it becomes a transmission or dE/dx detector. Such devices are useful as time markers, particle mass identifiers, and sampling "calorimeters."

Figure 1 illustrates a planar semiconductor device for detection of charged particles. The contacts can be surface-barrier, ion-implant, diffused-junction, etc. However, the front contact must be a good rectifying junction to provide the charge-depletion region that forms the sensitive region of the device. The base material in Fig. 1 could just as easily have been *p* type, in which case the depletion would have started from an n^+ contact.

When a reverse bias voltage is applied to a rectifying contact, it expands the charge-depletion region that forms the sensitive region of the detector. The rectifying contact itself should be thin since it represents an absorbing dead layer that will degrade the performance of the device. If the depletion region approaches the rear contact, this contact must be noninjecting (Ohmic).

When a charged particle or photon loses energy in the depletion region by creating free electron-hole pairs, these charges are rapidly swept apart by the electric field, and the resulting displacement current is detected in the ancillary electronics illustrated in Fig. 2. The resulting voltage signal that is induced on the electrodes is given by Q/C. However, it is important to recognize that all of the charge does not need to reach the electrode before the signal is detected. What is observed externally is the effect of a displacement current that terminates when the last bit of charge is collected.

The preamplifier illustrated in Fig. 2 is the charge-sensitive type, which measures the integrated displacement current by using $R_f - C_f$ to hold the voltage step at the transistor input at zero. The charge-sensitive configura-

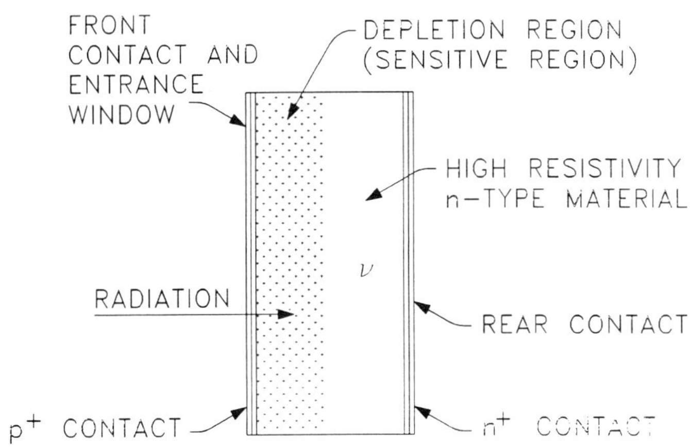

FIG. 1. Planar semiconductor radiation detector.

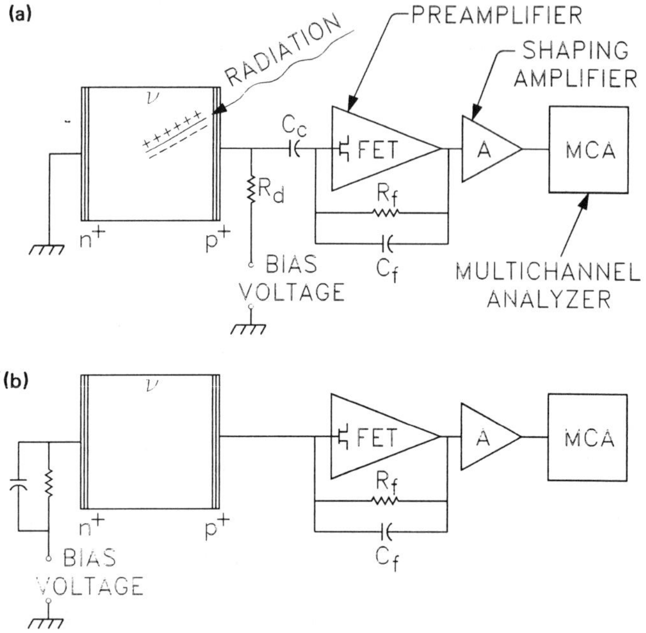

FIG. 2. Semiconductor radiation detector and signal-conditioning electronics.

tion is used in most preamplifiers since it makes the pulse height independent of changes in the detector capacitance. Figure 2(a) shows an ac-coupled configuration used for detectors with large leakage currents that cannot be compensated with a reasonable value of R_f. The dc-coupled configuration shown in Fig. 2(b) is preferable if the detector leakage current is small.

Leakage currents can be generated in the depleted bulk of the material by thermal generation of charge carriers and at the contacts and edges of the device. Most devices that operate at room temperature have their edges protected by an oxide coating or plastic encapsulation. Historically, the majority of charged-particle detectors were made by epoxy encapsulation techniques. However, that process has been largely replaced by oxide passivation. The shaping amplifier shown in Fig. 2 is used to optimize the signal-to-noise ratio. The multichannel analyzer is used to sort the events by energy (pulse height).

Figure 3 illustrates the geometry of a typical Ge γ-ray detector. The coaxial geometry provides a larger depletion volume at a given bias voltage than the planar geometry shown in Fig. 1. The limits on Ge detector volume, and therefore γ-ray detection efficiency, are set by the size and purity of available germanium crystals. All high-quality Ge coaxial detectors are totally depleted to prevent loss of energy resolution due to poor charge collection. Ge detector efficiencies are usually expressed as a percentage efficiency relative to 3×3-in. NaI crystal. Detectors with relative full-energy peak efficiencies in excess of 100% are becoming commonplace with no obvious limit in sight except cost.

1.2 Creation of Free Charge Carriers by Radiation

Radiation can interact with matter to lose energy by a number of mechanisms, including ionization to create free electrons, excitation of electrons followed by characteristic x-ray or light emission, nuclear inelastic scattering, and nuclear interactions that result in the creation of new species.

When a charged particle interacts with a semiconductor, the most important interaction is usually Coulomb scattering with the electrons in the solid to create free electrons and holes. These hot electrons usually have enough kinetic energy to create secondary, tertiary, etc., ionizations, with the result being a forward-peaked cascade of free electrons and holes. This process continues until the electron energy drops below the threshold for creation of a new electron-hole pair. The threshold energy depends on the width of the forbidden band gap, which explains why ϵ (the energy, in eV, required to create one free

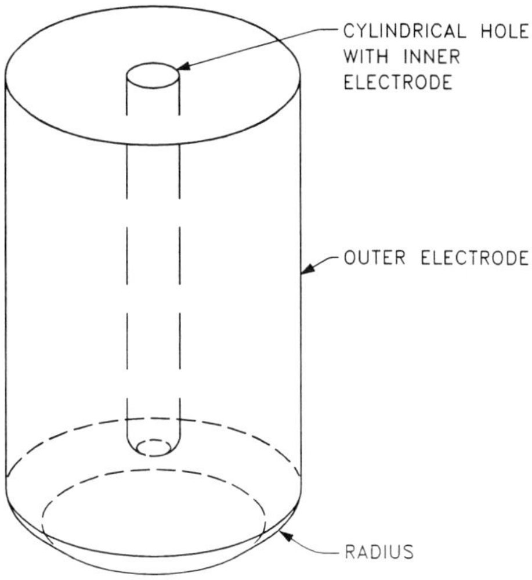

FIG. 3. Coaxial germanium detector, closed-end type.

electron-hole pair) is 2.8 in Ge compared to 3.6 in Si. The resulting cloud of free electrons and holes contains the basic information required for detecting and quantizing the energy of the particle.

After they fall below the threshold for creation of new pairs, the electrons are thermalized primarily by scattering with phonons. When a heavy charged particle nears the end of its track, it spends part of its time as a neutral particle and becomes involved in inelastic nuclear collisions. These nuclear interactions are responsible for the bulk of the radiation damage done by the particle and can also act as a competing source of energy loss that does not show up as free electron-hole pairs. Neither of these competing energy-loss mechanisms contributes to the desired electrical signal. Therefore they are a source of statistical variation in the number of free charge carriers created.

γ rays, x rays, and other photons lose energy in a semiconductor detector by three primary mechanisms, depending upon the energy of the photon. At low energies, the primary mechanism is photoelectric absorption, where the photon's energy is transferred to create a primary free electron; this can result in the creation of a cascade of secondary electrons as in the case of charged particles. The photoelectric cross section is of the form $\sigma \propto Z^5 E^{-3}$, which explains the desire for high-Z materials in γ-ray spectroscopy. The E^{-3} dependance makes the photoelectric effect dominant only for energies below about 60 keV in Si and 150 keV in Ge. This range includes most of the characteristic elemental x rays of interest for semiconductor detectors. Above about 200 keV in Ge, Compton scattering becomes the dominant mode of primary energy loss. Since Compton scattering does not exchange all of the incident energy, γ-ray events that result in only a single Compton interaction end up in the Compton background and not in the full-energy peak of the pulse-height spectrum.

At energies above about 7 MeV, pair production becomes a dominant mode for energy loss; however, this is not a major factor in most γ-ray spectroscopy applications.

Neutrons and other neutral particles lose energy in solids primarily by nuclear interactions, which are mostly of importance to the radiation damage problem. In cases where semiconductor detectors are used as neutron detectors or spectrometers, they depend on using a convertor target to create charged particles, which are then detected in the semiconductor.

1.3 Rectifying Contacts and the Properties of Space-Charge Regions in Semiconductors

All semiconductors in present use as radiation detectors require rectifying contacts to limit noise from leakage current associated with the normally present free charge carriers. Even for very-high-purity material with no ionizable impurities, the intrinsic free-carrier concentration is significant and is given by

$$n_i^2 \propto \exp(-E_G/kT), \tag{1}$$

where E_G is the forbidden band gap and T is the absolute temperature. At 300 K, E_G is 1.1 eV for Si and 0.7 eV for Ge. This results in room- temperature intrinsic resistivities of 2.3×10^5 Ω cm in Si and 60 Ω cm in Ge. In practical materials, there are always ionizable donor and acceptor centers that contribute additional carriers and lower the resistivity.

Figure 4(a) illustrates a simple energy-band diagram for a typical semiconductor such as Si or Ge. E_d is the energy separation between the donor defect level and the edge of

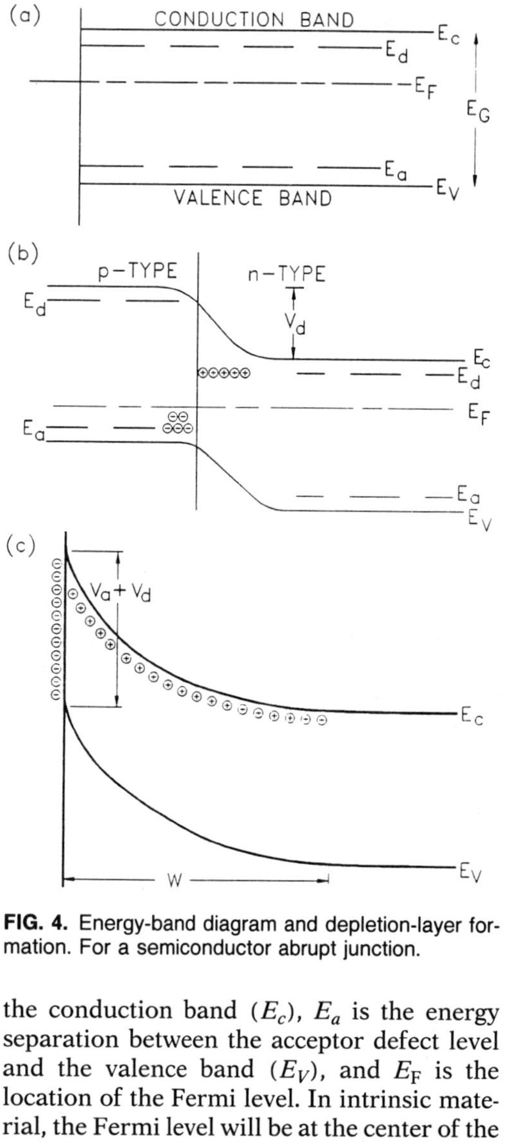

FIG. 4. Energy-band diagram and depletion-layer formation. For a semiconductor abrupt junction.

the conduction band (E_c), E_a is the energy separation between the acceptor defect level and the valence band (E_V), and E_F is the location of the Fermi level. In intrinsic material, the Fermi level will be at the center of the band gap, in *n*-type material (excess electrons) it will be above the center, and for *p*-type material (excess holes) it will be below the center. If the conductivity is dominated by the donors, the Fermi level will be near E_d, and if it is dominated by acceptors, the Fermi level will be near E_a. In either case, the charge-neutrality requirement is satisfied by the presence of immobile charged donor or acceptor sites to balance the free charge-carrier density.

If we create an abrupt transition from low-resistivity *p*-type material to high-resistivity *n*-type material, the band structure changes as shown in Fig. 4(b). Since the system is in thermodynamic equilibrium, the Fermi level must remain flat. This results in the creation of an extended charge dipole layer and warping of the band structure. In this simple model there are no free charge carriers in the dipole layer, only fixed charges corresponding to the ionized donors on the *n*-type side and an equal number of ionized acceptors on the *p*-type side. The resulting mobile-carrier–free region is variously referred to as the space-charge region, depletion region, or sensitive region of the detector. If the *p*-type region becomes vanishingly thin, all of the sensitive region exists in the *n*-type material. The built-in potential, V_d, is referred to as the diffusion potential or barrier height and must be less than $E_G - E_F -$ (a few kT), since at that point an essentially infinite supply of holes is available from the valence band to satisfy the charge-neutrality requirement. If an external reverse bias (V_a) is applied, the space-charge region grows as illustrated in Fig. 4(c). For the case of a very low-resistivity, very thin *p*-type layer, the Poisson one-dimensional equation for the space-charge region can be solved to obtain

$$W^2 = 2K\epsilon_0(V_a + V_d)q^{-1}|N_d - N_a|^{-1}, \qquad (2)$$

where W is the thickness of the space-charge region, q is the charge on the electron, K is the dielectric constant of the semiconductor (12 for Si and 16 for Ge), ϵ_0 is the permitivity of vacuum, N_d is the density of donor states, and N_a is the density of acceptor states. If the roles of the *n*- and *p*-type materials are reversed, Eq. (2) remains correct and Figs. 4(a) and 4(b) are simply inverted. Equation (2) correctly implies that the diode should work as a shallow detector even in the absence of external bias.

The junction where the depletion starts is customarily referred to as the front face or front contact. As increasing bias is applied, the space-charge region may reach the rear contact, and the device becomes totally depleted. If the rear contact on a totally depleted detector is thin, the device is referred to as a transmission detector. If the rear contact is not noninjecting, excess leakage current will result when the depleted region approaches within a diffusion length of the contact. Usually a $p^+\nu n^+$ or $n^+\pi p^+$ structure is preferable, where ν refers to high-resistivity *n*-type, π to high-resistivity *p*-type, p^+ to heavily doped *p*-type, and n^+ to heavily doped *n*-type material. The contact regions can be produced by a number of techniques including diffusion, ion implantation, and surface-barrier (SB) contacts. SB contacts owe their rectifying properties to a combination of surface states at the interface and in the thin interfacial oxide layer, which act as charge traps, and to the work-function differences between the metal contacts and the semiconductor. Equation (2) applies equally well to diffused junctions, ion-implant junctions, and surface-barrier and Schottky-barrier devices as long as the junction transition is abrupt.

The electric field distributions in partially depleted, fully depleted, and overbiased planar diodes are illustrated in Fig. 5, where the depletion voltage V_t is given by

$$V_t = qNt^2(2K\epsilon_0)^{-1}, \tag{3}$$

where t is the crystal thickness, $N \equiv |N_d - N_a|$, and $V_t \gg V_d$.

The resistivity of an extrinsic semiconductor is given by

$$\rho = (q\mu N)^{-1}, \tag{4}$$

where μ is the mobility of the majority carrier. Equation (2) is often simplified to

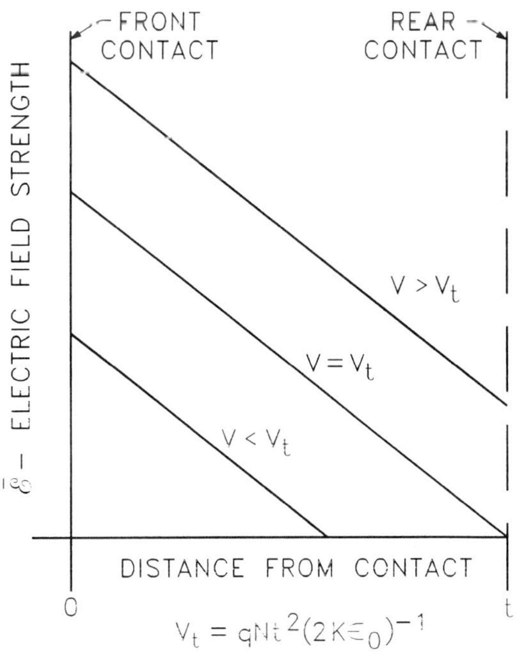

FIG. 5. Electric field distribution in a planar semiconductor radiation detector.

$$W = (2\mu K\epsilon_0)^{1/2}(\rho V)^{1/2}, \tag{5}$$

where $V = V_a + V_d$. Equation (5) can be restated for n-type (W_n) and p-type (W_p) silicon as

$$W_n \approx 0.54(\rho_n V)^{1/2} \quad \text{and} \quad W_p \approx 0.32(\rho_p V)^{1/2}, \tag{6}$$

where W is in μm, ρ is in Ω cm, and V is in V.

The distribution of electric field in a coaxial detector is quite different and is given by (Llacer, 1972)

$$|\mathscr{E}(r)| = qNr(2K\epsilon_0)^{-1} + [V - (eN/4K\epsilon_0) \times (r_o^2 - r_i^2)][r \ln(r_o/r_i)]^{-1}, \tag{7}$$

where r_o is the outer radius of the coax and r_i is the radius of the inner contact.

The capacitance of the detector is also an important variable since it affects both the signal-to-noise ratio (SNR) in the detection system and the response time of the detector. For a partially depleted planar detector with no edge effects, the capacitance per unit area is given by

$$C^{-2} = (2/qK\epsilon_0)VN^{-1}. \tag{8}$$

An experimental plot of C^{-2} vs V_a can be very informative (provided N is not too small). The slope of the curve yields the value of N, the extrapolated intercept at low voltage yields a value for V_d, and the break in the curve at high voltage gives a good way to estimate V_t.

The capacitance per unit length of a totally depleted coaxial detector with no end effects is given by

$$C = 2\pi K\epsilon_0[\ln(r_o/r_i)]^{-1}. \tag{9}$$

The need to keep N small to maximize sensitive depth is not easy to meet. If a 3-mm-deep planar Si detector at 500 V operating bias is required, the net uncompensated impurity density (N) must be $<8\times10^{10}$ cm^{-3} or <0.002 parts in 10^9 (atomic). For 2 cm of Ge at 3000 V, $N < 1.3\times10^{10}$ cm^{-3} is required. Obtaining such perfect materials explains in large part why large-volume semiconductor detectors are so expensive.

There are two basic approaches to reducing N, which often are used together: The first is to reduce the absolute values of N_d and N_a by material purification techniques, a task that becomes increasingly difficult as N gets small. The second is to compensate the material by adding impurities, N_d or N_a, to reduce the net difference (N) between the donors and acceptors. A serious limitation on the second approach is the inhomogeneity of the spatial distribution of the impurities, which produces obvious problems if you are dealing with small differences in large numbers. One method of compensating impurities is the lithium-drift technique. Lithium is a donor in Ge and Si. It is mobile and can be drifted through the crystal under the influence of an electric field. Since it is mobile in its charged state, it automatically compensates for inhomogeneity in the distribution of other impurities, i.e., any gradient in N will result in a field gradient that will automatically be compensated. This would seem to be an ideal solution to the problem, and for many years it was the major answer to the need for deeply depleted devices. However, Li compensation in Ge is unstable at room temperature and only quasistable in Si. This means that if a Li-compensated Ge device accidently warms up to room temperature, it is essentially destroyed. Ge(Li) detectors have been made obsolete by ultrahigh-purity Ge crystals. The Li in Si(Li) detectors also creates some serious handicaps. In addition to being only quasistable at room temperature, it prevents the use of high-temperature oxide passivation processes and the use of contact configurations that require high-temperature processing. In the future, there is the possibility that Si(Li) detectors will be replaced with ultrahigh-purity Si.

1.4 Signal Generation by the Motion of Charge Carriers

The high mobility of the free electrons and holes in Si and Ge makes semiconductor diodes among the fastest of the commonly used detectors. Pulse rise times can range from subnanoseconds to microseconds depending on the radiation, the device configuration, and the applied voltage. Normally both the electrons and holes contribute to the signal, but their relative importance depends on where the energy is deposited within the device. For example, if a short-range α particle deposits all of its energy within 20 μm of the front contact of a 2-mm-deep n-type device, the electrons generate almost all of the signal. For a minimum-ionizing particle that leaves a uniform track density all the way through the detector, the electrons and holes will make equal contributions in the absence of preferential charge trapping.

If the charge carrier velocities are not saturated, the relative contribution of any individual charge carrier will be directly proportional to the fraction of the total applied bias voltage that it transits. A quick reference to Fig. 5 and Eq. (7) will show that this is not at all the same as assuming that the contribution is directly proportional to the distance traversed. This distinction is significant in the calculation of pulse rise-time shapes and the analysis of the pulse-height distortion caused by trapping or weak-field regions.

The commonly used terms *charge collection* and *charge-collection time* should be treated with caution since it is not necessary for a charge to be actually collected at a contact in order to contribute to the signal. What is observed and measured in the external circuit is the behavior of the displacement current corresponding to the moving charge. Consequently, the signal results from the motion of the carrier and not the final act of its arrival at the electrode. Charge carriers that make only part of the trip, e.g., are stopped by a charge trap, make only part of their potential contribution. When the terms *charge collection, charge-collection efficiency,* etc., are used herein, they imply that for complete "charge collection" the charge moves from its point of origin to its final desired destination during the time interval of interest.

At moderate electric fields, the carrier velocity ν is proportional to the electric field ($\mathscr{E}$) and is given by

$$\nu=\mu\mathscr{E}. \qquad (10)$$

The mobilities (μ) of electrons and holes, respectively, in Si at 300 K are 1300 and 480 $cm^2\ (V\ s)^{-1}$ and improve to 21 000 and 11 000 at 77 K. In Ge at 77 K, the electron and hole mobilities are about equal at 40 000 $cm^2\ (V\ s)^{-1}$. At high fields, the velocities saturate at about 10^7 cm/s. If the fields become too high, avalanche multiplication can occur.

In addition to the variation in mobilities, electric field strengths, and original distribution of free charge carriers, there are several other factors that can affect the shape of the observed pulse and thereby the timing and energy resolutions of the detector. In the case of heavy ions, which leave densely charged tracks, the electric field cannot penetrate the dense plasma, and normal charge transit is perturbed until the plasma breaks up. The characteristic delay associated with this process is referred to as the plasma decay time and ranges from >10 ns for heavy fission fragments to several nanoseconds for α particles. For photons, β particles, and minimum-ionizing particles, this factor is negligible.

The shape of the rising pulse predicted from the charge transit characteristics is not necessarily what will be observed at the input of the preamplifier since the *RC* time constants of the equivalent circuit for the system will be superimposed on it. The simplest complete equivalent circuit for a semiconductor diode detector is shown in Fig. 6, where $V_A(t)$ is the time-dependent voltage signal at the preamplifier input, C_D is the capacitance of the depletion region, C_b is the capacitance of the undepleted bulk material, C_A is the effective input capacitance of the preamplifier plus any stray capacitance, R_b is the resistance of the undepleted bulk material, R_s is any series resistance in the leads or contacts, R_D represents the shunt leakage of the diode, and R_V is the resistor that decouples the bias supply. R_D and R_f affect only the decay time of the pulse but are included for completeness. The equivalent circuit in Fig. 6 has been analyzed in detail (Dabbs and Walter, 1961), but the complete results are beyond the scope of this presentation. However, there are several factors that deserve mention. The time constant R_bC_b is equal to $\rho K\epsilon_0$, which is the classical dielectric relaxation time of the material. It can be a significant factor for very high-resistivity silicon, which is another good reason to use fully depleted detectors. The term R_sC can be a very important factor and is a good reason for minimizing the contact resistance and electrode sheet resistance of the device. In fact, position-sensitive detectors that depend on signal rise-time differences use a distributed R_s to obtain the position information.

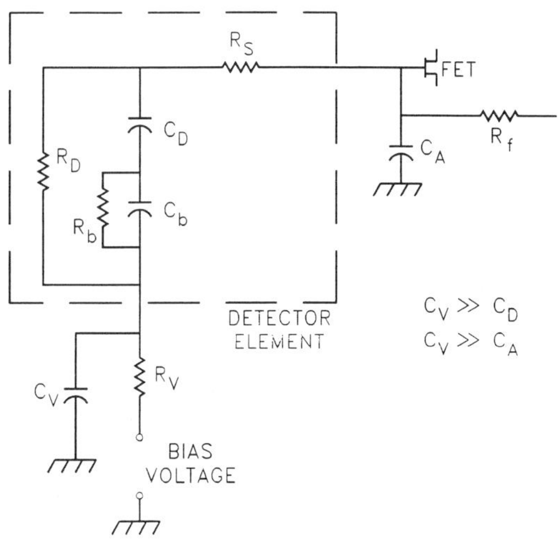

FIG. 6. Semiconductor detector equivalent circuit.

If no preamplifier were connected to the detector, the amplitude of the voltage signal would be closely approximated by Q/C_D, but the rise time would not be affected by C_D. However, connecting the preamplifier alters conditions at the interface because of the presence of R_s and C_A. If the preamplifier is the charge-sensitive configuration, C_A is much larger than C_D. The effect is to add a rise-time component of $\approx 2.2R_sC_D$.

2. SOURCES OF SPECTRAL DISTORTION

The spectral peak from a monoenergetic source will be broadened and distorted by a number of factors. The resulting peak width ΔE_S observed in the spectral energy resolution measurement can be expressed as

$$\Delta E_S^2 - \Delta E_A^2 + \Delta E_D^2 + \Delta E_{SL}^2 + \Delta E_o^2, \tag{11}$$

where ΔE_A and ΔE_D are the broadening contributions from amplifier noise and detector noise, respectively, ΔE_{SL} is the contribution from the statistical limit, and ΔE_o is the combined contribution from all other sources of broadening. ΔE is usually expressed as the full width at half maximum (FWHM) of the peak, in terms of energy. Adding terms in quadrature implies that they are independent and that their contributions are Gaussian in shape. The latter assumption is not always justified for ΔE_o since charge-collection problems, dead layers, etc., can cause non-Gaussian peak distortion. For this reason, Ge γ-ray spectrometers are additionally characterized by other specifications such as full width at one-tenth maximum, full width at one-fiftieth maximum, and peak-to-Compton ratio (ANSI/IEEE 1986). X-ray spectrometers are frequently characterized by a peak-to-tail ratio (Walter, 1970).

In order to minimize the contributions of ΔE_A and ΔE_D, it is necessary to limit the bandwidth over which the signal is collected by using a signal-shaping amplifier. The resulting limitation on charge-collection time changes the contribution of ΔE_o. Therefore selection of the best amplifier shaping time constant often involves a tradeoff between electronic noise and charge-collection problems.

2.1 Charge-Generation Statistics

The average energy required to produce a free electron-hole pair as previously defined implies that if an incident ionizing particle loses all of its incident energy, the resulting number of free charge pairs is Q/q. Since the early days of gas detectors, it has been conventional to describe the statistical standard deviation of Q/q as $\surd(Q/q)$. This implies Poisson statistics. In fact, there is no physical basis for assuming Poisson statistics since they apply only to random events (Walter, 1969). The cascade formation of free charge pairs is a highly correlated process, and if all of the energy went into the creation of free charge pairs, the variance in Q/q would vanish. Actually, the statistical variance in Q/q is dominated by the final interactions of the cascade in which the electrons and the primary particle slow down enough to allow competing energy-loss processes to become important. Consequently, the standard deviation of Q/q should be less than $\surd(Q/q)$ and is usually characterized by $\surd(FQ/q)$, where F is the Fano factor. For this case the statistical limit on resolution is given by

$$\Delta(Q/q)_{SL}=2.35(FQ/q)^{1/2}, \quad (12)$$

where $\Delta(Q/q)_{SL}$ is the statistical contribution to the full width at half maximum of the peak. If there were no variance in the charge collected, then Q/q and $\Delta Q/q$ could be simply replaced by E/ϵ and $\Delta(E/\epsilon)$. However, since this is never the case, a better approximation is given by

$$\Delta(E/\epsilon)_{SL}=2.35(FE/\epsilon\eta)^{1/2}, \quad (13)$$

where η is a factor introduced to account for the variance introduced by the still unresolved charge-collection efficiency problems (Walter, 1969). In many of the publications dealing with the subject of F, η has been ignored despite the fact that it is always F/η that is actually measured. This explains why the published measured values of F decreased monotonically through the years as detectors and their ancillary electronics improved. Measurement of F/η is not a trivial problem since it must be deduced by carefully unfolding all of the other known sources of resolution broadening. This usually involves a small difference between large numbers where the remainder is F/η. A number of attempts have been made to calculate F on a theoretical basis (van Roosbroeck, 1965). These theoretical estimates have also tended to become smaller as the experimental value of F/η decreased. The best published values for F (actually still F/η since they still include contributions from charge loss and ballistic deficit) are $F_{Si}=0.11$ and $F_{Ge}=0.08$ (Phel and Goulding, 1970). One theoretical estimate for F is 0.05 (Zulliger and Aitken, 1970).

One issue not covered by the preceding discussion is the case in which a particle passes through a transmission detector while losing only a small portion of its total energy. This case is governed by the Landau effect (Landau, 1944; Maccabee *et al.*, 1966); the resolution spread is non-Gaussian and much larger than predicted by $\delta E^{-1/2}$.

2.2 Windows; Entrance and Exit Dead Layers

Radiation must enter the device through some surface, usually one of the contacts. Energy lost in nonsensitive regions at the surface will affect the observed pulse height, its spread (resolution width), and, in the case of photon detectors, the efficiency and low-energy cutoff in the detector response. The effective window thickness on good Si detectors is <0.2 μm of equivalent Si. This corresponds to <40-keV energy loss for a 5-MeV α particle. Although it is possible to measure accurately the thickness of a diffused contact or a metal electrode, predicting the effective thickness of the dead layer and its effect on resolution broadening is not as simple a matter as it might first appear for several reasons: Deposited Au layers tend to be filamentary. The cascade of hot electrons originating in the dead layer is forward peaked and therefore directed out of the dead layer into the device. Hot electrons and radiation can be scattered back out of the active region into the dead layer. Charge carriers can diffuse over the barrier out of the device into the contact. The resolution-broadening effects of the dead layer are frequently lumped into F/η, although they more properly belong in ΔE_o.

In photon detectors, the effective dead layer is usually characterized by measuring the drop in detection efficiency as the energy of the incident photons is decreased. In semiconductor x-ray spectrometers and low-energy γ-ray spectrometers, dead layers and escape problems also produce non-Gaussian peak distortion in the form of a low-energy tail on the peak. This is frequently quantified by specifying the peak-to-tail ratio for low-energy characteristic x rays (Walter, 1970).

Detectors that must be cooled have another dead layer (window) associated with the container that provides the vacuum encapsulation. Low-energy photon detectors generally are provided with a thin cryostat entrance window made from a low-Z material such as Be. This window may be the limiting factor for the low-energy detection efficiency.

2.3 Radiation Escape, Scattering, and Peak-to-Compton Ratio

One major source of spectral distortion in γ-ray spectrometers arises from the Compton background. If a Compton interaction in the detector is followed by escape of the γ ray, the resulting signal ends up in the Compton distribution instead of in the full-energy peak. Although the Compton edge (the maximum possible signal for a Compton interaction) for this event is well below the full-energy peak, it produces a large background for lower-energy peaks. The peak-to-Compton ratio, which is defined as the ratio of the height of the full-energy peak to the height of the Compton distribution (just below the Compton edge) from the same incident γ-ray energy, is an important characteristic of the detector. It is customarily specified for the 1.3-MeV γ-ray from ^{60}Co (ANSI/IEEE, 1986). The peak-to-Compton ratio varies with a number of factors, the most important of which are energy resolution, detector geometry, detector full-energy peak efficiency, and the number of pulses lost from the full-energy peak because of poor charge collection.

2.4 Charge Collection Variations

Trapping of charge carriers and loss of carriers to surfaces or dead layers before the carriers make their full contribution to the signal are a major source of resolution broadening. Unfortunately, this cannot be described by a simple statistical treatment of charge-trapping probability for the following reasons: The charge traps may not be uniformly distributed. The probability of trapping is time dependent and therefore depends on electric field strength, which is not constant throughout the device. A charge may be trapped for only a short time, then be reemitted and partially contribute to the signal. The predominant traps may be hole or electron traps; the effect of the trapping will thus depend on which carrier should make the most contribution to the observed signal and therefore on where the carriers are generated with respect to the electrodes. Resolution deterioration by charge trapping can be reduced by higher electric fields and longer charge-collection times at the expense of diminished counting-rate capability and possibly increased electrical noise. High-resolution semiconductor detectors are so sensitive to trapping problems that they are the most sensitive means available to detect traps in Si and Ge. Crystal defect densities as low as 10^8 cm^{-3} can cause problems. Trapping centers may be introduced or removed during the high-temperature steps used in diode fabrication. Trapping centers are also introduced by radiation damage.

A related resolution broadening problem, referred to as ballistic deficit, occurs when slowly rising portions of the pulse are prematurely clipped by the low-frequency rolloff of the shaping amplifier. Because of their large volumes, large variations of electric field, and the extended nature of the original charge distribution, Ge coaxial detectors for γ rays are particularly susceptible to this problem. Special electronic circuits have been developed to minimize spectral distortion from ballistic deficits.

The very heavy ionization density in the charge track of a heavy ion, e.g., fission fragment, results in a loss of pulse height (and an associated deterioration in resolution) known as the pulse-height defect (Finch, 1987). This problem is related to charge recombination in the plasma mentioned previously and to the fact that charge loss in the dead layer is enhanced by the high density of free charge near the front electrode. An additional effect that may be observed with heavy ions is pulse-height multiplication. In this case, the high density of charge near the front electrode results in the injection of free electrons from the front contact that add to the pulse height (Walter, 1964).

Pulse-height multiplication by avalanches in high electric fields is also possible. This effect is used in some specialized devices to increase the signal size prior to the signal-conditioning electronics. Unfortunately, no one has yet succeeded in making a device with multiplication that is sufficiently uniform to preserve the spectroscopic qualities of the detector.

2.5 Electronic Noise

The detector, preamplifier, and other ancillary components can all contribute electronic noise to the system. The various noise sources can be categorized as series noise, parallel noise, microphonic noise, and a loosely defined category sometimes referred to as $1/f$ noise or flicker noise. Capacitance across the input of the preamplifier—which includes the detector capacitance—does not generate noise, but it degrades the SNR by attenuating the signal.

Resistors in the signal path between the detector and preamplifier (R_s and R_b in Fig. 6) appear as "series" noise sources. The input transistor in the preamplifier contributes a component of series noise in the form of an equivalent series resistor. The SNR due to the series noise component is degraded by the input capacitance because it attenuates the signal but not the noise. The amplitude of series noise broadening varies with the amplifier shaping time constant $t_{1/2}$ and the capacitive load on the preamplifier and is of the form

$$\Delta E_{\text{ser}} \propto C(t_{1/2})^{-1/2}, \qquad (14)$$

where the shaping time constant $t_{1/2}$ is defined (IEEE, 1988) as the FWHM of the shaped pulse and C is the total capacitance load on the input of the preamplifier including the detector capacitance, the effective input capacitance of the input element of the preamplifier, the effective value of C_f, and any stray capacitance associated with cables, interconnections, etc.

Bulk leakage current in the detector and resistors connected between the input of the preamplifier and signal ground (R_d and R_f in Fig. 2; R_D in Fig. 6) appear as "parallel" noise generators. The SNR of this component is not affected by the input capacitance because the capacitance attenuates the signal and noise equally. The amplitude of the parallel noise contribution ΔE_p varies as $(t_{1/2})^{1/2}$. For example, the noise from bulk leakage current in the detector is of the form

$$\Delta E_D \propto (It_{1/2})^{1/2}. \qquad (15)$$

The forms of Eqs. (14) and (15) indicate that there is an optimum noise corner at which the system should operate to minimize noise. This effect is illustrated in Fig 7. The noise corner is not necessarily the optimum time constant for best resolution. Slow charge collection and ballistic deficit considerations may dictate a longer shaping time constant. Microphonic noise or pulse-pileup problems at high counting rates may dictate the need for a shorter pulse-shaping time constant.

Microphonic noise is caused by relative motion between the elements connected to the input stage of the preamplifier (FET gate) and some other component, which is at a different potential; this is analogous to a capacitance microphone. The amplitude of the observed signal distortion is quite sensitive to the frequency of the exciting vibrations and whether they are airborne or mechanically coupled, because the relative motions are amplified by resonance effects. Detectors that require cryogenic cooling are particularly prone to microphonic noise because of their

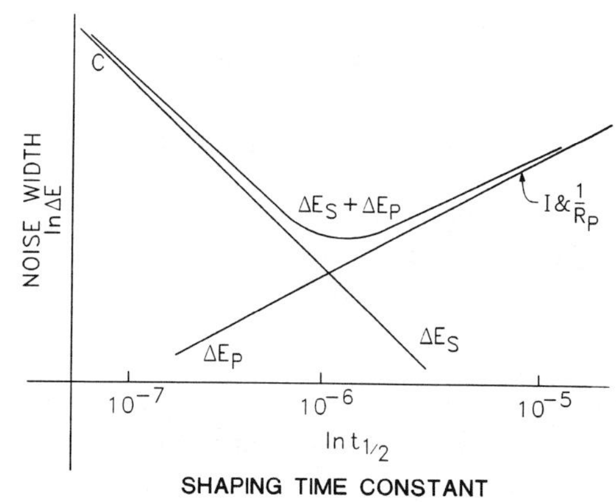

FIG. 7. Typical noise width in semiconductor detectors.

complex mechanical structure and low noise. Microphonic noise distortion can be quantified as equivalent noise at the input in much the same manner as other noise is quantified (IEEE, to be published).

2.6 Counting-Rate Effects

High counting rates can cause spectral distortion for a number of reasons. When a second event occurs before the preceding pulse has time to return to the baseline, the events are said to be piled up. The resulting combined event, as seen at the output of the shaping amplifier, will have an amplitude somewhere between that of the larger event and the sum of the two events depending on how badly the events are overlapped. If the counting rate gets too high, it can drive the amplifier out of its linear range and in extreme cases paralyze the amplifier or the multichannel analyzer (MCA).

If there is any ac coupling that generates a pole in the response function of the signal-processing chain that is not compensated with a corresponding zero, there will be an undershoot when the pulse returns to the baseline (see CIRCUITS, ANALOG). This undershoot can distort the next event if it follows too closely. Any capacitance in the processing chain, e.g., C_f in Fig. 2, can introduce a simple pole; distributed capacitance, e.g., in R_f, can introduce a complex pole that is difficult to cancel. Most modern amplifier systems contain PZ (pole-zero) cancellation adjustments, but seldom are all the poles completely canceled.

Shifts of the baseline in the signal-processing chain under the influence of high counting rates will also distort the spectrum. Most quality signal conditioning also has provisions for baseline restoration.

2.7 Environmental Effects

Compared to most other types, semiconductor detectors are relatively insensitive to environmental effects. However, some environmental effects deserve discussion.

The bulk generated leakage current in semiconductor detectors varies approximately as $\exp(E_G/2kT)$ which can make the noise level quite temperature sensitive in small–band-gap semiconductors such as Ge. Even in Si, it is necessary to cool the detector and the first stage of the preamplifier to cryogenic temperatures to obtain a satisfactory noise level for x-ray spectrometry.

Detectors with transparent entrance windows are also very efficient optical detectors. Si charged-particle detectors with thicker non-transparent windows are commercially available for situations for which the application itself, e.g., the scattering chamber, does not provide adequate optical shielding. High performance cryogenic detector systems contain internal shields to minimize infrared absorption.

Semiconductor detectors whose edges or electrodes are not adequately passivated are sensitive to the presence of contaminants that might induce excess leakage current and noise. In most cases of charged-particle spectrometry, the detector must be in a vacuum. However, backstreamed pump oil that accumulates on the detector surface can cause an increase in window thickness. Cracked pump oil that contaminates the detector may result in increased noise. Si detectors that are coated with a protective layer to protect them from harsh and corrosive environments are commercially available at the expense of increased window thickness.

Compared to most other detector types, semiconductor detectors are insensitive to magnetic fields.

The signal and noise levels in semiconductor detectors are low enough to require good shielding and good grounding practice in the part of the system near the detector and preamplifier.

High-purity Si and Ge are remarkably free of radioactive background, but the material used to mount and house the detector may be a source of undesirable background.

2.8 Radiation Damage

Compared to other types of detectors, one of the major disadvantages of the semiconductor variety is their sensitivity to radiation damage. In Si, usually the first symptom to appear is increased leakage current, which is caused by damage-induced defects that act as generation–recombination centers. Leakage current produces noise and bias shift. Another type of damage is the creation of defects that trap free carriers for times comparable to or longer than the amplifier shaping time, causing a degradation of energy resolution and pulse height. Finally, radiation damage can change the value of N by converting active donors such as phosphorus to an inactive complex and by creating new defects or defect complexes that act as acceptors or donors.

Ge γ-ray spectrometers are also sensitive to radiation damage from the fast neutrons encountered around accelerators and experimental reactor facilities. Because they are cooled to 78 K, the resulting charge-trapping problems are more serious than leakage-current problems. Coaxial detectors made from n-type material are much less sensitive than p-type detectors, not primarily because of the material difference but because the electrode configuration is reversed.

3. DETECTOR EFFICIENCY

The efficiency of Si charged-particle detectors is essentially 100% for particles that enter the sensitive window area with enough energy to penetrate the dead layer and deposit sufficient energy in the active region to be detected above the noise. However, unless a collimator is used, there may be some uncertainty about the exact active area because of edge effects.

The efficiency of semiconductor x-ray spectrometers is rolled off in the low-energy region by the dead layer on the detector and the window on the cryostat (ANSI/IEEE, 1984). The details of this rolloff vary from one detector to another. Si detectors capable of seeing the K x rays of C are possible. The dead layer on the detector not only decreases the low-energy efficiency, but also causes low-energy tailing of the peaks (ANSI/IEEE, 1984). These problems are much more pronounced for Ge detectors. At the high-energy end of the characteristic x-ray spectrum, the efficiency rolls off because of the finite thickness of the detector (ANSI/IEEE, 1984). The efficiency curves for both Ge and Si exhibit notches adjacent to the K absorption edge; this effect is very pronounced in Ge.

The low-energy efficiency of Ge γ-ray spectrometers is subject to the same considerations discussed above for x-ray detectors. A thin-window planar configuration or an n-type coax housed in a cryostat with a thin Be window is desirable for low-energy work.

The high-energy efficiency of Ge detectors is more complex since it varies nonlinearly with active volume and detector geometry. The usual way for specifying the efficiency of a Ge detector is to compare its full-energy peak efficiency to that of a 3×3-in. NaI crystal (ANSI/IEEE, 1986). Thus, the relative full-energy peak efficiency of a Ge detector can exceed 100%.

The peak-to-Compton ratio of a Ge detector improves with increased efficiency and with improved energy resolution. In most applications, the minimum detectable limit of a weak line in the presence of stronger lines is dominated by the peak-to-Compton ratio.

4. DETECTOR TYPES AND APPLICATIONS

The type of detector can be defined by the semiconductor involved, e.g., Si or Ge, its geometrical configuration or shape, the method used to create the contacts, and the method used to passivate the sensitive surfaces. If the depletion region reaches the rear contact, the device is referred to as a totally depleted detector. Totally depleted devices require bias voltage above that required for depletion to provide good charge collection for energy lost near the rear contact. If the rear contact on a totally depleted device has a reasonably thin window, the device is referred to as a transmission detector. Transmission detectors are frequently stacked into collinear arrays for δE and $\delta E+E$ measurements. The combination of δE and $\delta E+E$ can be used to determine the mass of a charged particle as well as its energy.

The desired geometric shape of the detector is application specific. The usual shape for a Ge γ-ray spectrometer is a coaxial cylinder with electrodes in the form of concentric cylinders. This geometry provides the maximum sensitive volume for a given bias voltage. Charged-particle and x-ray spectrometers usually are a planar configuration, i.e., in the form of a parallel-plate capacitor. Different types of electrodes and passivation materials are used to optimize entrance and exit dead layers, bias voltage, leakage current noise, stability, and cost.

Because of the many diverse applications of these detectors, they exist in a bewildering variety of types and sizes. In the following, some of the more popular applications of the three major classes of detectors are summarized. Space does not permit detailed discussion of all the applications or the many variations of detectors that have been built to satisfy them. Discussion will be limited to a brief description of a few examples in each class.

4.1 Ge γ-Ray Spectrometers

Ge detectors are used for nuclear structure physics research, reactor effluent and fuel process monitoring, neutron-activation analysis, environmental monitoring and analysis, space physics, γ-ray astronomy, medical and industrial isotope analysis, medical imaging, mineral surveys, nuclear safeguards, etc.

Except in special cases, Ge detectors are supplied by a commercial manufacturer complete with cryostat/vacuum enclosure and integral preamplifier. There are several reasons for this: The first FET in the preamplifier must be cooled for best low-noise performance and should be close to the detector to minimize stray capacitance and microphony. Most of the cost is for the Ge crystal, which does not have passivation to protect its sensitive surfaces from environmental hazards. Extreme cleanliness and careful techniques are required in handling the crystal and preparing the vacuum enclosure. These operations, which require special equipment and facilities, are as much art as science and the cost of failure can be high.

Most detectors use an internal, cryogenic sorption pump to maintain good vacuum in the enclosure. Instruments are available cooled with liquid nitrogen (LN) or with an external-compressor, expansion-type refrigerator.

For low-energy applications, the cryostat should have a thin entrance window, usually Be, and the detector element should have a thin dead layer as its entrance window. Since electronic noise tends to dominate the resolution at low energies, a small low-capacitance detector may be more desirable.

Although there is no simple answer to selecting the optimum detector for moderate- to high-energy applications, a useful rule of thumb is to use the most efficient detector that can be afforded without sacrificing too much in energy resolution and peak-to-Compton ratio.

The most exciting new use of Ge detectors is in large arrays of detectors for nuclear physics research. Arrays of this type were envisioned many years ago but are only now becoming economically feasible.

4.2 Semiconductor X-Ray Spectroscopy

Si and Ge low-energy photon detectors are used for materials analysis by analyzing the characteristic x rays produced when the material is excited by a fluorescing source. The exciting source may be an electron beam, radioactive source, x rays from an x-ray tube, a charged-particle beam, etc. These detectors are also used for research in nuclear physics, plasma physics, medicine, and space physics, and for surface analysis, forensics, and portable mineral assay-exploration.

The most common use of semiconductor x-ray detectors is for analysis of the characteristic elemental x rays emitted when an atom is excited with charged particles or other x rays. These characteristic x rays are excellent "fingerprints" of what elements are present and in what amounts. Thus Si detectors are powerful tools for materials analysis and have replaced the old cumbersome x-ray diffraction technique in many applications. The x-ray diffraction approach is basically a single-channel technique that scans one line at a time, while the semiconductor energy analysis is a multichannel technique that looks at all energies simultaneously. Semiconductor detectors are thus much faster; they are also smaller. However, they do not have the ultimate energy-resolution capability of the diffraction approach. The use of semiconductor x-ray energy spectrometers is sometimes referred to as EDS (energy-dispersive spectroscopy). This usage, which was probably derived from the older WDS (wavelength-dispersive spectroscopy), is an unfortunate choice of terms since it implies an energy dispersion process, which is technically incorrect. Better acronyms are XES (x-ray energy spectroscopy) or SXES (semiconductor x-ray energy spectroscopy).

Most semiconductor x-ray energy spectrometers use Si because of its thinner dead layer. The detectors are usually rather small planars (several millimeters in diameter) to reduce the capacitance and associated SNR deterioration. The typical sensitive depth is 2 to 3 mm. Because of their relatively large surface-to-volume ratio, the energy resolution is particularly sensitive to degradation by charge loss to the surfaces (edges). Careful collimation of the incident x rays is required to minimize this problem.

In most cases, the actual detector element comes as part of a complete package as in the case of Ge γ-ray detectors, and for some of the same reasons. A large number of cryostat configurations are commercially available to interface with the many brands of scanning electron microscopes, electron and ion microprobes, x-ray tube equipment, etc. Systems with proprietary thin windows on the cryostat and with "windowless cryostats" that open directly to the vacuum inside the source of x rays extend the useful range of the instrument to low-Z materials.

Portable detectors with lightweight cryostats are available for field assay and mineral surveys. Many of these portable systems use a collimated radioactive source to excite the characteristic x rays in the sample.

4.3 Silicon Charged-Particle Detectors

Si charged-particle detectors are used for research in nuclear-structure physics, space physics, plasma analysis, and medicine. They are also used for environmental and radiation safety monitoring, low-background counting of β and α particles, particle-beam–excited surface analysis, and thickness and surface gauging, and as tracking detectors and pixel arrays for sampling calorimeters in high- energy physics accelerators. Special configurations are used as neutron detectors, fast-neutron spectrometers, and neutron imaging arrays.

Si semiconductor detectors got their start in the nuclear-structure physics field, followed closely by space-physics applications. Their superior performance, small size, and versatility quickly made them the instrument of choice. The many different experimental requirements spawned a wide variety of detector configurations. Small Si transmission detectors as thin as a few microns and with thickness and flatness uniformities of a few tenths of a micron have been built. Si detectors that are not compensated with Li have been supplied with sensitive thicknesses as large as 3–5 mm, with availability dependent on the limited supply of suitable very high-resistivity Si crystals. Detectors with a hole in the center, which allows the primary accelerator beam to pass through while detecting other particles at small scattering angles, are standard commercial products. Detectors that simultaneously measure both energy and position of incidence have been built using the principle of voltage division across a resistive contact to supply the position information.

Si detectors were not rapidly accepted in the health safety and environmental monitoring fields, partially because of their higher cost and susceptibility to damage by hostile environments, but also because of the low priority and funding available for those applications. With decreases in cost, improvements in ruggedness, and improved funding, Si devices are now making a significant contribution to these areas.

The fastest moving and largest current application for Si detectors is in the high-energy particle-physics field where they are of interest for position sensors (tracking detectors) and pixel arrays for electromagnetic and hadron sampling calorimeters. Large hodoscopic arrays of this type were envisioned at the outset of the semiconductor detector era (Dabbs and Walter, 1961). The detector fabrication technology and funding have only now progressed to the point where such large systems are under serious study.

5. Further Reading

A large fraction of the published work on semiconductor detectors is in the IEEE Transactions on Nuclear Science and the Proceedings of the IEEE-NPSS Nuclear Science Symposia. The Nuclear Instruments and Detectors Committee of the IEEE Nuclear and Plasma Sciences Society has written a number of ANSI/IEEE Standards on nuclear instrumentation. In addition to techniques for measuring the performance of detectors (ANSI/IEEE, 1982, 1984, 1986) and their ancillary electronics (IEEE, 1988), these standards contain good tutorials and glossaries.

Radiation Detection and Measurement (Knoll, 1989) is a good comprehensive text that contains extensive references to semiconductor detectors and their ancillary electronics.

GLOSSARY*

Background (Associated with a Spectral Peak from a Semiconductor Detector): Nonideal spectral response that results from radiation that is not a part of the monoenergetic line of interest.

Ballistic Deficit: In an amplifier, the loss in pulse height that occurs when the rise time of the input signal is not negligible compared with the width of the shaped amplifier pulse.

Band Gap: The difference in energy between the energy level of the bottom of the conduction band and the energy level of the top of the valence band.

Baseline (at Pulse Peak): The instantaneous value that the voltage would have had at the time of the pulse peak in the absence of that pulse.

Baseline Restorer: A circuit that rapidly restores the baseline following an amplifier's output pulse (or series of pulses) to the level existing prior to it.

Bias (of a Semiconductor Radiation Detector): The voltage applied to a detector to produce the electric field to sweep out the signal charge.

Bias Resistor (of a Semiconductor Radiation Detector): The resistor through which bias voltage is applied to a detector.

Blocking Contact (of a Semiconductor Radiation Detector): The contact from which depletion proceeds into the semiconductor material under conditions of reverse bias.

Capacitance (of a Semiconductor Radiation Detector. The small-signal capacitance measured between terminals of the detector under specified conditions of bias and frequency.

Charge Carrier (of a Semiconductor): A mobile conduction electron or mobile hole.

*Reproduced from: IEEE Std. 301-1988, IEEE Standard Test Procedures for Amplifiers and Preamplifiers used with Detectors of Ionizing Radiation. Copyright © 1988 by the Institute of Electrical and Electronics Engineers, Inc., with the permission of the IEEE. IEEE Std. 759-1984, IEEE Std. Test Procedures for Semiconductor Charged Particle Detectors. Copyright © 1984 by the Institute of Electrical and Electronics Engeineers, Inc. with the permission of the IEEE. IEEE Std. 300-1982, IEEE Std. Test Procedures for Semiconductor Charged Particle Detectors. Copyright © 1982 by the Institute of Electrical and Electronics Engineers, Inc. with the permission of the IEEE.

Charge-Sensitive Preamplifier: A preamplifier in which the output signal is proportional to input charge, independent of input capacitance.

Dead-Layer Thickness (of a Semiconductor Radiation Detector): The thickness of an inactive region (in the form of a layer) through which incident radiation must pass to reach the sensitive volume.

Depletion Region (in a Semiconductor): A region in which the charge-carrier charge density is insufficient to neutralize the net fixed charge density of ionized donors and acceptors. In a diode-type semiconductor radiation detector the depletion region is the sensitive region of the device.

Depletion Voltage (of a Semiconductor Radiation Detector): The voltage at which a junction detector becomes fully depleted.

Diffused-Junction Detector: A semiconductor detector in which the *p-n* or *n-p* junction is produced by diffusion of donor or acceptor impurities.

Efficiency (of a Semiconductor Radiation Detector for a Monoenergetic Radiation Source): The ratio of the number of events in the spectral distribution to the total number of photons incident on the active detector volume during the same time interval.

Energy Resolution (FWHM) (of a Semiconductor Radiation Detector): The detector's contribution (including detector leakage current noise), expressed in units of energy, to the FWHM of a pulse-height distribution corresponding to an energy spectrum.

Full Energy Peak Efficiency (of a Semiconductor Radiation Detector): The ratio of the number of events in the full-energy peak of the spectral distribution to the total number of photons incident on the active detector volume during the same time interval.

Full Width at Fiftieth Maximum: Same as FWHM except that measurement is made at one fiftieth of the maximum ordinate rather than one half.

Full Width at Half Maximum: The full width of a distribution measured at half the maximum ordinate. For a normal distribution, it is equal to $2(2 \ln 2)^{1/2}$ times the standard deviation σ.

Full Width at Tenth Maximum: The full width of a distribution measured at one tenth of the maximum ordinate.

FWFM: Full width at fiftieth maximum.

FWHM: Full width at half maximum.

FWTM: Full width at tenth maximum.

Ion Implantation: A process in which a beam of energetic ions incident upon a solid results in the implantation of those ions into the material.

Intrinsic Semiconductor: A semiconductor containing an equal number of free holes and electrons throughout its volume.

Leakage Current (of a Semiconductor Radiation Detector): The total detector current flowing at the operating bias in the absence of radiation.

Lithium-Drifted Detector: A detector made by the lithium compensation process.

Multichannel Analyzer (MCA): An instrument that digitizes analog amplitude signal pulses and stores them in a memory as a function of their analog amplitude.

Noise Linewidth: The contribution of noise to the width of a spectral peak.

Noninjecting Contact (of a Semiconductor Radiation Detector): A contact at which the carrier density in the adjacent semiconductor material is not changed from its equilibrium value.

Position-Sensitive Detector: A detector that measures the impact position, in one or more dimensions, of ionizing radiation upon the detector surface.

Resolution (Energy): Unless otherwise specified, the full width at half maximum (FWHM) of a spectrum peak expressed as an increment of energy ΔE or as a percentage of the energy of the peak, $100\Delta E/E$.

Schottky-Barrier Contact: A metal–semiconductor contact structure whose rectification properties are heavily influenced by the difference in material work functions. These contacts frequently utilize an interfacial metal/semiconductor compound (for example, a silicide).

Semiconductor, Compensated: A semiconductor in which one type of impurity or imperfection (for example, donor) partially cancels the electric effects of the other type of impurity or imperfection (for example, acceptor).

Surface-Barrier Contact: A metal–insulator–semiconductor (MIS) contact structure whose rectification properties are dominated or heavily influenced by charge trapped at the interfaces and in the insulator.

Surface-Barrier Radiation Detector: A radiation detector for which the blocking contact is a surface-barrier contact.

Totally Depleted Detector: A detector in which the thickness of the depletion region is essentially equal to the thickness of the semiconductor material.

Transmission Detector: A totally depleted detector whose thickness including its entrance and exit windows is sufficiently small to permit radiation to pass completely through the detector.

List of Works Cited

ANSI/IEEE Standard 300-1982(1982), "IEEE Standard Test Procedures for Semiconductor Charged Particle Detectors," New York: IEEE.

ANSI/IEEE Standard 759-1984(1984), "IEEE Standard Test Procedures for Semiconductor X-Ray Energy Spectrometers," New York: IEEE.

ANSI/IEEE Standard 325-1986(1986), "IEEE Standard Test Procedures for Germanium Gamma-Ray Detectors," New York: IEEE.

Dabbs, J. W. T., Walter, F. J. (Eds.) (1961), *Semiconductor Nuclear Particle Detectors,* Publication No. 871, Washington, DC: NAS-NRC, pp. 237–239.

Elad, E. (1965), *Nucl. Instrum. Methods,* **37**, 327–329.

Finch, E. C. (1987), *Nucl. Instrum. Methods Phys. Res. A* **257**, 381–383.

Hall, R. N. (1969), in: W. L. Brown *et al.* (Eds), "Prospects for High-Purity Germanium," *Semiconductor Nuclear-Particle Detectors and Circuits,* Publication 1593, Washington, DC: National Academy of Sciences, pp. 171–184.

IEEE Standard 301-1988(1988), "IEEE Standard Test Procedures for Amplifiers and Preamplifiers used with Detectors of Ionizing Radiation," New York: IEEE.

IEEE Standard Test Procedure for Microphonics in Detector System (to be published).

Knoll, G. F. (1989), *Radiation Detection and Measurement,* 2nd ed., New York: Wiley.

Landau, J. (1944), *J. Phys. USSR* **8**, 201–205.

Llacer, J. (1972), *Nucl. Instrum. Methods* **98**, 259–268.

Maccabee, H. D., Raju, M. R., Tobias, C. A. (1966), *IEEE Trans. Nucl. Sci.* **NS-13**, 175–184.

Mayer, J., Gossick, B. (1957), "Use of Au-Ge Broad Area Barrier as Alpha-Particle Spectrometer," *Rev. Sci. Instrum.* **27**, 407–408.

McKay, K. G. (1951), "Electron-Hole Production in Germanium by Alpha Particles," *Phys. Rev.* **84**, 829–832.

Pell, E. M. (1960), *J. Appl. Phys.* **31**, 291–302 and 1675–1679.

Phel, R. H., Goulding, F. S. (1970), *Nucl. Instrum. Methods* **81**, 329–330.

Sze, S. M. (1981), *Physics of Semiconductor*

Devices, New York: Wiley.

Van Heerden, P. J. (1945), *The Crystal Counter, A New Instrument in Nuclear Physics,* Amsterdam: N. V. Noord Hollandsche Uitgevers Maatchappij.

van Roosbroeck, W. (1965), *Phys. Rev.* **139**, A1702–A1715.

Walter, F. J. (1958), "A Study of Germanium Surface Barrier Counters," University of Tennessee Thesis, ORNL Central Files No. 58-11-99.

Walter, F. J. (1964), *IEEE Trans. Nucl. Sci.* **NS-11**, 232–237.

Walter, F. J. (1969), "Comments on Measurements of Epsilon and the Fano Factor," in: W. L. Brown *et al.* (Eds.), *Semiconductor Nuclear-Particle Detectors and Circuits,* Publication 1593, Washington, DC: National Academy of Sciences, p. 63.

Walter, F. J. (1970), *IEEE Trans. Nucl. Sci.* **NS-17**, 196–214.

Walter, F. J., Dabbs, J. W. T., Roberts, L. D. (1960), *Rev. Sci. Instrum.* **31**, 756–762.

Zulliger, H. R. Aitken, D. W. (1970), *IEEE Trans. Nucl. Sci.* **NS-17**, 187–195.

DIALYSIS

ROBERT H. BARTH, M. D., *Department of Veterans' Affairs Medical Center, Brooklyn, New York, U.S.A. and State University of New York, Health Science Center at Brooklyn, Brooklyn, New York, U.S.A.*

INTRODUCTION

Generalities

Dialysis is a process whereby solutes of differing molecular weights are separated according to their varying abilities to diffuse through a membrane. The driving force is the chemical gradient across the membrane for each solute, with no convective exchange or movement of solvent through the membrane. Diffusion is spontaneous and irreversible, leading to an increase in the entropy of the system, and requires no external source of thermal or chemical energy. Dialysis is used in the laboratory, in medicine, and in industry as a separation or concentration process.

The word was coined and the process described during the 1850s by the Scottish chemist Thomas Graham. Graham recognized that certain substances, which he termed "colloids"—probably largely proteins—diffused much more slowly across a parchment membrane than other smaller molecular weight solutes that he called "crystalloids." Using a simple device with a "dialytic septum" of albumin-coated parchment paper, he was able to separate the components of urine, recovering a crystalloid in the diffusate that proved to be primarily urea.

There is little standardization of terminology. The solution on the feed side, with retained macromolecules, may be termed retentate, while the product solution containing diffused solutes is called permeate or diffusate. The term dialysate has been used by different authors for solution on both sides of the membrane. In hemodialysis, however, which is the most widespread application of the process, the feed solution is blood and the

1-56081-063-7/92/$5.00+.50

solution containing diffused solutes is always called dialysate.

The direct descendant of Graham's method is *equilibrium dialysis,* a laboratory procedure used primarily for purification of proteins, illustrated in Fig. 1. In this process, an aqueous solution containing both small molecules and macromolecules is placed in a collodion or cellophane bag, and the sealed bag placed in a large reservoir of buffer solution. The small molecules diffuse out of the bag along a concentration gradient, while the macromolecules are retained by the bag membrane. If the volume of buffer is large with respect to that of the bag—with, for example, continuous flux of fresh buffer—a relatively complete separation can be effected.

Applications

Industrial applications of dialysis include the recovery of caustic from hemicellulose in the production of rayon, salt removal in pharmaceuticals manufacture, and desalination of water, but the most extensive commercial use is for the medical treatment of end-stage renal disease. In patients with this condition, dialysis substitutes for the failed excretory function of the kidney, removing from the circulation the small [<1000 Da; 1 Da (dalton) $=1.007\,94$ u] and middle (~ 1000–5000 Da) molecular-weight metabolic end products that cause toxic symptoms if allowed to accumulate, while leaving behind proteins and other macromolecules. In 1988, over 110 000 patients received regular dialysis therapy in the United States alone, at a total cost of more than five billion dollars (United States Renal Data System, 1990).

Ion-Exchange Dialysis Methods

The retention of large solutes is ordinarily based on steric effects. The use of ion-selective membranes, which allow preferential passage of positively charged cations or negatively charged anions, has allowed the development of several specialized techniques. In *ion-exchange dialysis,* shown in Fig. 2(a), an anion-exchange membrane allows passage of anions, but excludes transfer of all cations except hydrogen ions. Thus if a solution of an acid and its metal salt is dialyzed against water, the anion diffuses along a concentration gradient and is accompanied by hydrogen ions to preserve electroneutrality. The net effect is separation of the acid from the metal salt, and the process is used for precisely this purpose in the recovery of metal ions in the electroplating industry.

Donnan dialysis, shown in Fig. 2(b), uses a cation-exchange membrane to concentrate a positive ion such as the cuprous ion shown in the figure. Hydrogen ions diffuse down a concentration gradient from the highly concentrated acid solution on the right side of the membrane. Since the membrane prevents diffusion of negative ions, the only way to maintain electroneutrality is to transfer the positive cuprous ion out of the left compartment against a concentration gradient, with resultant increase in concentration of the valuable metal ion by as much as one-hundredfold.

In *electrodialysis,* dialysis takes place in the presence of an electric field. The driving force for solute movement is the electrical attraction of anions to the positive anode and cations to the negative cathode. An array of alternating ion-selective membranes produces concentration and dilution of the ions in alternate cells of the dialyzer [Fig. 2(c)], and the process may be used to remove salts effectively from brackish water. Electrodialy-

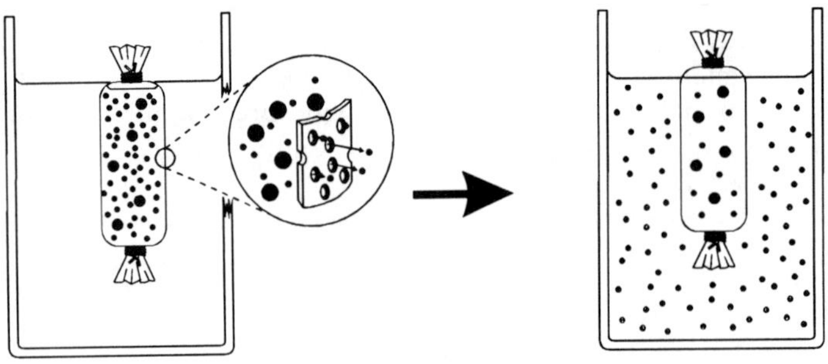

FIG. 1. Equilibrium dialysis. In this laboratory procedure, small and large molecules are separated by their differing ability to diffuse across a membrane.

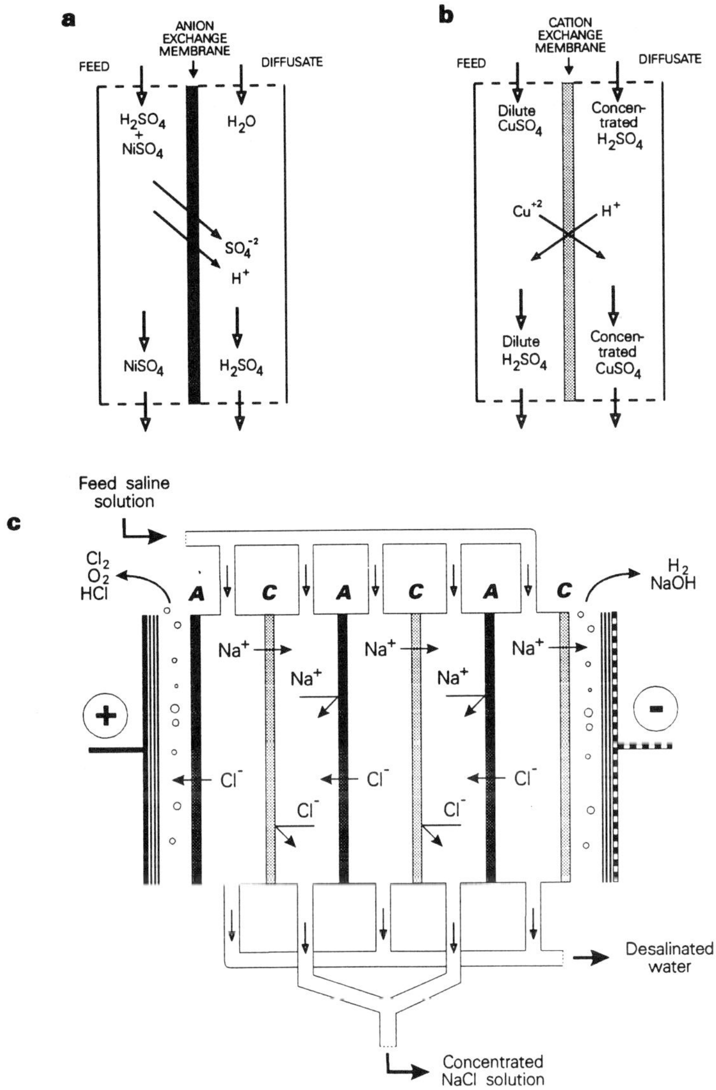

FIG. 2. Ion-exchange dialysis techniques. (a) Ion-exchange dialysis. (b) Donnan dialysis. (c) Electrodialysis; ***A***=anion exchange membranes, ***C***=cation exchange membranes. See text for description.

sis is widely used in the preparation of potable water from saline sources.

Hemodialysis as a Paradigm

Therapeutic hemodialysis represents the most highly developed and by far the commonest and most economically important use of the procedure, and has been subjected to the most extensive mathematical analysis. Because of this, and since much of that analysis may be generalized to any dialysis procedure, the remainder of this article will use hemodialysis as a paradigm. After a discussion of the functions of the kidney that hemodialysis must replace, the mathematics of mass transfer and apparatus and techniques will be examined in depth.

1. FUNCTIONS OF THE KIDNEY: THE GOALS OF DIALYSIS THERAPY

Therapeutic dialysis is a "renal replacement therapy," intended to substitute for the normal functions of the kidney that are lost in end-stage renal failure. The dialyzer is thus often referred to as an "artificial kidney," although its simulation of kidney function is incomplete and relatively inefficient.

The physiological role of the kidney includes regulation of body water and electrolytes, excretion of metabolic wastes, and regulation of metabolism through the production of a number of hormones. The first two of these are separation processes, although not the functions of a passive filter. The anatomical unit that is responsible for the formation of urine and adjustment of its solute concentrations is the *nephron,* shown schematically in Fig. 3. Each human kidney contains $(0.8\text{–}1.2)\times 10^6$ of these structures, which consist of a tuft of blood capillaries, the *glomerulus,* in intimate contact with a long duct, the *tubule,* which is lined with epithelial cells. Ultrafiltration takes place across the glomerular basement membrane, which divides the capillary lumen from that of the tubule. This membrane is quite permeable to water and to solutes of molecular weight less than about 10,000 Da, but provides an effective barrier to passage of macromolecules over 50,000 Da, especially those that carry a net negative charge at the physiologic *p*H of 7.40.

The rate of glomerular ultrafiltration is determined by the hydrostatic pressure difference between the glomerular capillary and tubular lumina, $P_{GC}-P_T$, and by the osmotic gradient produced by the retained macromolecules, $\pi_{GC}-\pi_T$, where π is the colloid osmotic pressure. This relationship can be expressed as

$$\mathrm{SNGFR}=K_f[(P_{GC}-P_T)-(\pi_{GC}-\pi_T)], \qquad (1)$$

where SNGFR is the ultrafiltration rate of a single nephron and K_f is the glomerular ultrafiltration coefficient, in turn the product of membrane hydraulic permeability and surface area. Using data obtained from rat glomeruli, Dworkin and Brenner (1985) have calculated K_f to be 42.1 nL/(s·cm^2·mm Hg) (1 mm Hg=133 Pa), a value that is one to two orders of magnitude greater than that of other capillaries. Glomerular filtration thus proceeds rapidly despite a relatively low hydro-

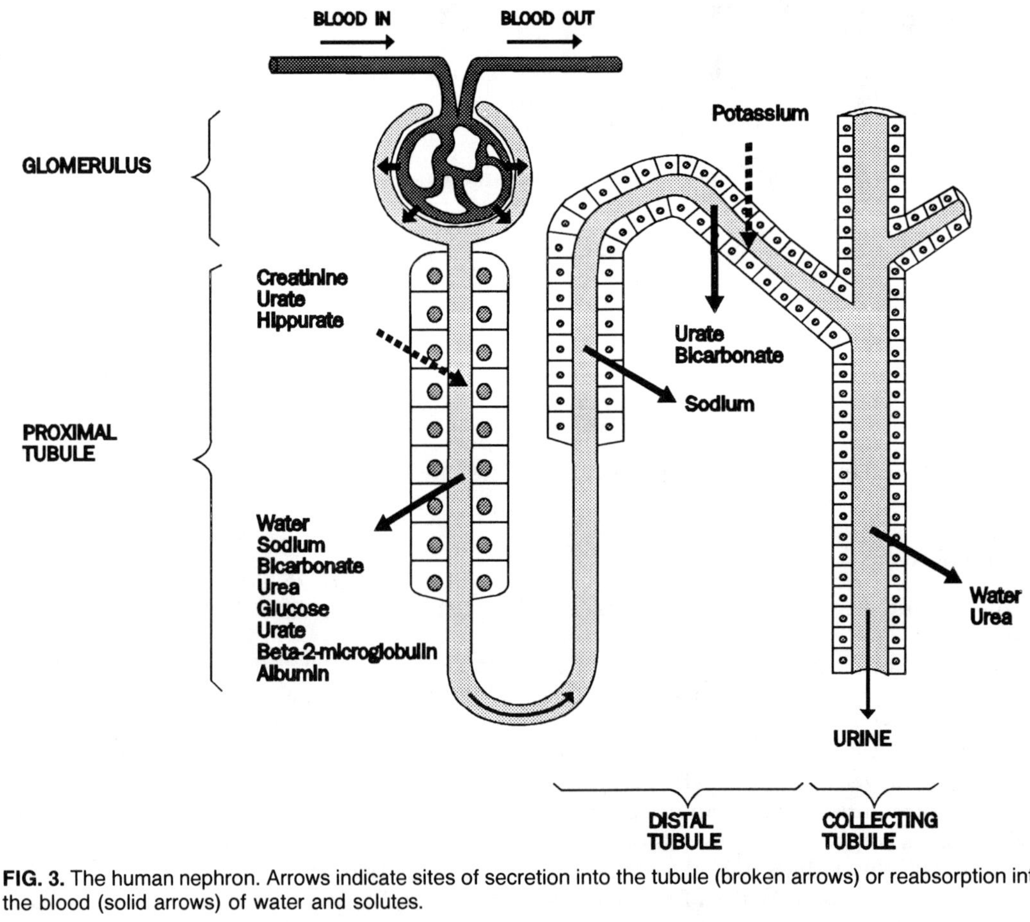

FIG. 3. The human nephron. Arrows indicate sites of secretion into the tubule (broken arrows) or reabsorption into the blood (solid arrows) of water and solutes.

static driving force in the range of 5–10 mm Hg, reaching levels of 120–130 mL/min, or about 180 L/d.

The ultrafiltrate formed by the glomerulus is practically free of protein, but contains small–molecular-weight solutes in the same concentrations found in plasma. As this solution travels down the tubule, about 99% of the filtered water is reabsorbed into the circulation. Some filtered solutes, such as inulin, a 5200-Da polysaccharide used in the measurement of renal function, remain entirely within the tubule and become progressively more concentrated as water is removed. Other solutes, including sodium, potassium, phosphate, bicarbonate, small proteins, and essential nutrients such as glucose and amino acids, are themselves reabsorbed, so that less than 1% of the solute filtered by the glomerulus is ultimately excreted in the urine (that is, the *fractional solute excretion* is less than 0.01). Finally, potentially toxic organic acids and bases such as hippurate and creatinine are secreted into the tubular fluid, so that their fractional excretion may exceed 1.0. Sites along the nephron where typical solutes are reabsorbed or secreted are indicated in Fig. 3.

By varying the rates of reabsorption and secretion, the tubule adjusts solute excretion to the current needs of the organism. At extremely low levels of salt intake, for example, or in states of dehydration, when there is need to minimize sodium loss, the urine sodium concentration may be virtually unmeasurable. During heavy salt ingestion, however, as much as 4 or 5 g of sodium per day may appear in the urine.

The kidney's ability to excrete a solute x may be represented by the *clearance* of x (C_x), defined as the mass removal rate divided by the plasma concentration. Clearance may also be conceptualized as the volume of plasma required to supply the amount of x appearing in the urine in a given time, usually expressed in mL/min,

$$C_x = U_x \dot{V} / P_x, \tag{2}$$

where U_x and P_x are the concentrations of x in the urine and plasma and $\dot{V}$ is the urine flow rate. Since the rate of excretion of x is equal to $U_x \dot{V}$, and rate of filtration of x is equal to the product of P_x and the glomerular filtration rate (GFR), the fractional excretion of x (FE_x) is

$$FE_x = U_x \dot{V} / (P_x \times GFR) \tag{3}$$

and is equal to the ratio of C_x to GFR. Thus, if x is freely filtered at the glomerulus and is neither secreted nor reabsorbed, then its rate of filtration is equal to its rate of excretion, $FE_x = 1.0$, and C_x is equal to GFR. In clinical practice, the clearance of *creatinine*, a product of muscle metabolism, is generally used as an estimate of GFR, despite the fact that some creatinine is secreted into the tubule and $FE_{creatinine}$ is slightly greater than 1.0.

2. HISTORY

In 1913, a little more than fifty years after Graham's first description of dialysis, *in vivo* dialysis was first performed at Johns Hopkins University, and the term "artificial kidney" entered the language. Abel, Rowntree, and Turner passed the blood of nephrectomized dogs through collodion tubes and demonstrated the removal of nonprotein nitrogen. In 1924 Georg Haas, in Germany, performed the first human dialysis, which lasted 15 min and demonstrated primarily that the procedure could be done without harm to the patient.

Two developments of the next decade had profound influence on the nascent hemodialysis. In 1928 the first practical anticoagulant—heparin—became available, making prevention of the coagulation of blood passing through the dialysis apparatus much more manageable. Secondly, cellophane membranes were developed and manufactured commercially, largely for sausage casings. This regenerated cellulose product was far more practical than the fragile collodion for the construction of dialyzers, and in slightly modified forms remains today the mainstay of hemodialyzer manufacture.

Clinical dialysis became a reality in the small town of Kampen, during the Second World War German occupation of the Netherlands. Using a large rotating drum wound with 2.5-in. cellophane sausage tubing and immersed in a 70-L tank of dialysate, Willem Kolff began in 1943 to dialyze patients suffering from uremia, and in 1945 succeeded in keeping a woman alive until her recovery from acute renal failure. After the war Kolff donated machines to clinics in Europe and the United States, and hemodialysis began slowly to grow. By the late 1950's it was an established therapy for acute renal failure, but long-term dialysis was not possible because of the extreme difficulty of repeated cannulation of a patient's arteries and veins.

The breakthrough came in 1960 with the development of the Quinton-Scribner shunt—a Teflon, and later Silastic, cannula that was implanted in an artery and vein in the arm and afforded permanent vascular access. This device was still fraught with problems of infection and coagulation of blood within its tubing, but it eliminated the need for repeated surgical cutdowns, and made chronic hemodialysis a possibility for the first time. Six years later Brescia, Cimino, and their colleagues in the Bronx surgically created an arteriovenous fistula in the wrist, and the resultant high flow and pressure in the draining cephalic vein caused it to dilate and its wall to become thickened. This dilated vein could then be accessed by venipuncture with a large needle, providing sufficient blood flow for effective dialysis. Routine intermittent access to the circulation became a reality, and the modern era of chronic dialysis was established.

Over the scant three decades of its existence as a practical treatment for kidney failure, dialysis has been a remarkable success. In the United States alone, well over 100 000 people receive regular dialysis treatment. The 1-yr mortality rate in industrialized countries is high, ranging from 7% to 23%, but many patients with renal failure, once doomed to a rapid demise from their disease, now continue to live independent and active lives for as much as 20 yr or more.

3. THE KINETICS OF DIALYSIS

Therapeutic dialysis replaces the filtration functions of the kidney—specifically, the establishment and maintenance of physiological solute concentrations and normal body water content. The basic elements, as illustrated in Fig. 4, are simple, with solution (blood) flowing on one side of a semipermeable membrane, and dialysate (an osmotically balanced electrolyte solution) flowing in a countercurrent direction on the other. Solutes diffuse down concentration gradients, and water removal is effected by ultrafiltration, driven by the hydrostatic pressure difference between solution (blood) and dialysate. A mathematical analysis of the process can be divided into several parts: mass transfer characteristics of the artificial kidney, or dialyzer; ultrafiltration; and the kinetics of solute removal from the body. The last of these will not be treated in this article.

3.1 Dialyzer Mass Transfer

Solute transfer across dialysis membranes occurs by diffusion, and is governed by Fick's law

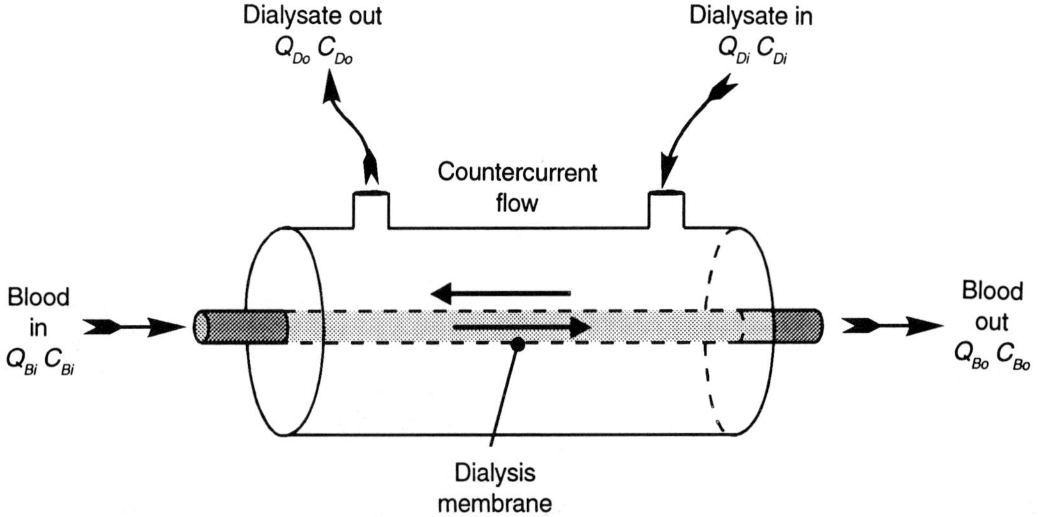

FIG. 4. Schematic representation of an artificial kidney. Blood and dialysate flow in opposite, or countercurrent, direction. Q=flow, C=concentration, Bi=blood inlet, Bo=blood outlet, Di=dialysate inlet, Do=dialysate outlet.

$$J_x = -D_x\, A \frac{dc}{dx} \quad (4)$$

or in integrated form

$$J_x = -(D_x\, A/L)(c_B - c_D), \quad (5)$$

which states that flux of solute x—J_x (mol/cm^2/s)—proceeds over the distance L at a rate proportional to the concentration difference from solution (blood) to dialysate, $c_B - c_D$, the surface area of the diffusing front, A, and the diffusivity, D_x, which is a constant property of the solute-solvent system at a given temperature. In a dialyzer D_x is constant for a given temperature and solute, and L is determined by membrane thickness and dialyzer geometry and is also constant, so that for a specified solute D_x and L may be combined in the overall mass transfer coefficient, k_0:

$$J_x = -k_0 A (c_B - c_D)_{\text{avg}}. \quad (6)$$

k_0 is a complex coefficient with components that depend on transport in solution (blood), membrane, and dialysate. Each of these phases has an associated mass transfer coefficient and reciprocal mass transfer resistance, R, related by

$$1/k_0 = R_0 = R_B + R_M + R_D, \quad (7)$$

$$1/k_0 = 1/k_B + 1/P_M + 1/k_D, \quad (8)$$

where the subscripts B, M, and D refer to blood, membrane, and dialysate, and P_M is membrane permeability. In some early hemodialyzers, thick and poorly distributed blood and dialysate boundary layers contributed as much as 75–80% of the total resistance to solute transfer. With modern dialyzer design techniques, this figure has fallen to less than 25% (Colton and Lowne, 1981).

Strictly speaking, k_0 is a local property of a point on the dialysis membrane, since k_B and k_D may vary as concentrations change with position along the dialyzer. In practice another value, K_0, is used, defined as the length-averaged value of k_0, and considered constant throughout the dialyzer. The product of K_0 and the membrane surface area, K_0A, is a fixed property of a dialyzer for a given solute (assuming aqueous solution and a fixed temperature, usually 37 °C), and expresses the dialyzer's intrinsic ability to filter that solute.

3.2 Dialysance

Potential dialyzer performance may be expressed by K_0, but in practice it is necessary to quantify the actual amount of filtration of specific solutes taking place under clinical conditions, that is, at known blood and dialysate flow rates. Diffusive dialysance D, the rate constant used to characterize dialyzer solute removal, was defined by Wolff (1951) as "the minute rate of net exchange of a substance between blood and bath fluid per unit blood-bath concentration gradient," or

$$D = \frac{\text{Rate of solute transfer}}{\text{Unit concentration driving force}}. \quad (9)$$

The concentration driving force appears in Eq. (6) as $(c_B - c_D)_{\text{avg}}$. The correct average concentration difference is the logarithmic mean of the inlet and outlet concentrations:

$$(c_B - c_D)_{lm} = \frac{(c_B - c_D)_1 - (c_B - c_D)_2}{\ln[(c_B - c_D)_1/(c_B - c_D)_2]}, \quad (10)$$

where 1 and 2 are the two ends of the dialyzer. Nevertheless, because of the inconvenience of measuring inlet and outlet concentrations in both dialysate and blood, in the clinical setting the driving force is approximated by the maximum concentration difference, $c_{Bi} - c_{Di}$, using the terminology of Fig. 4.

The mass transfer rate N, in moles per unit time, is given by the difference between incoming and outgoing solute mass (either in blood or in dialysate):

$$N = Q_{Bi}c_{Bi} - Q_{Bo}c_{Bo} = Q_{Do}c_{Do} - Q_{Di}c_{Di}. \quad (11)$$

Since in the absence of ultrafiltration $Q_{Bi} = Q_{Bo}$ and $Q_{Di} = Q_{Do}$, dialysance may be written as

$$D_B = \frac{Q_B(c_{Bi} - c_{Bo})}{c_{Bi} - c_{Di}} = \frac{Q_D(c_{Do} - c_{Di})}{c_{Bi} - c_{Di}}. \quad (12)$$

In modern hemodialyzers, blood and dialysate flows are always countercurrent, and for solutes like urea, creatinine, and phosphate, concentrations in the entering dialysate are 0. In this case, the denominator of Eq. (12) simplifies further to c_{Bi}, and the equation becomes the equivalent of Eq. (2), the definition of clearance for the natural kidney. By this analogy, the rate constant for transfer of solutes not present in entering dialysate is known as dialyzer clearance (K):

$$K = \frac{Q_B(c_{Bi} - c_{Bo})}{c_{Bi}} = \frac{Q_D(c_{Do} - c_{Di})}{c_{Bi}}. \quad (13)$$

Both dialysance and clearance may be calculated using either blood- or dialysate-side flow rates and concentrations, although in clinical practice the former is more common.

Dialysance is dependent on momentary conditions, that is, blood and dialysate flow rates, while K_0A is a fixed property of a dialyzer for each solute examined. The relationship between K_0A and dialysance is thus an important and useful one, necessary for prediction of dialyzer performance under changing conditions of flow. Equating J as defined in Eqs. (6) and (11), assuming countercurrent blood-dialysate flow and no net ultrafiltration, and substituting with Eqs. (10) and (12), yields

$$D = \frac{e^{-z} - 1}{e^{-z}/Q_B - 1/Q_D}, \quad (14)$$

where

$$z = \frac{K_0A(1 - Q_B/Q_D)}{Q_B}.$$

Equation (14) is plotted in Fig. 5 for a range of K_0A that varies from 50—a typical value for a large-molecular-weight solute such as vitamin B_{12}—to 900 mL/min, which is about the highest value of K_0A for the small molecule urea obtainable with modern high-performance dialyzers. Examination of this graph—which is calculated for a fixed dialysate flow of 800 mL/min—reveals the boundary conditions for dialysance. When K_0A/Q_B is large, as for small solutes at low blood flow rates, then $D \approx Q_B$. Conversely, when K_0A/Q_B is small, either because the solute is large and poorly diffusible or when blood flows are very high, then the limiting value is $D \approx K_0A$. Clinically attainable blood flows range from 200 to 600 mL/min; in this range small-molecular-weight solute dialysance is flow dependent with most dialyzers. For molecules larger than about 500 Da, increasing blood flow beyond 200 mL/min has little effect on dialysance in conventional cellulose dialyzers. Thus, to increase removal from the circulation of high-molecular-weight solutes, it is necessary to enlarge the surface area of the dialyzer or lengthen the treatment. The situation may, however, be different when newer, more highly permeable membranes, which have higher K_0A's for larger molecules, are used.

3.3 Ultrafiltration

Besides removal of waste solutes, therapeutic dialysis also has the task of removing excess water accumulated by ingestion and by

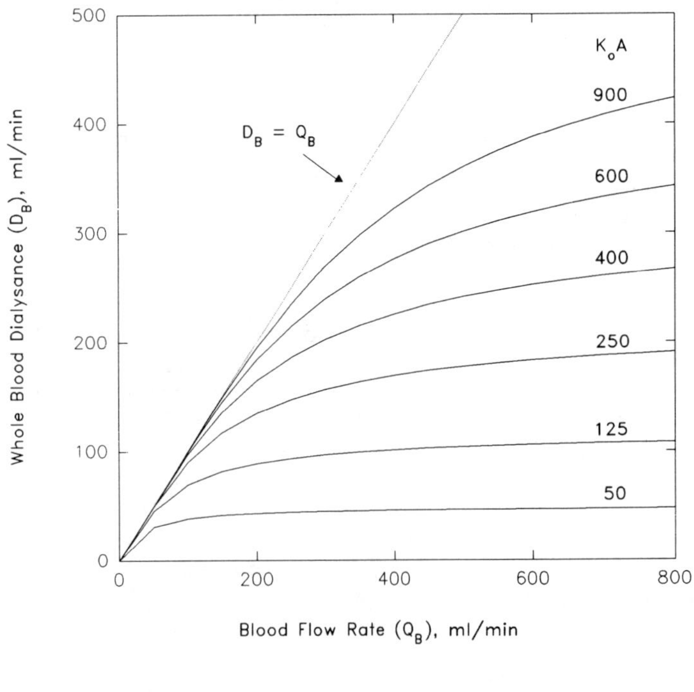

FIG. 5. Dependence of dialysance on blood flow rate, plotted for different values of K_0A.

metabolic formation, often as much as 1–2 L/d. This is accomplished by ultrafiltration across the dialysis membrane, a process described by

$$J_F = \frac{Q_F}{A} = \frac{L_p(\Delta P - \Delta\pi)}{\mu}, \qquad (16)$$

where J_F is the water flux (mL/cm^2/min), Q_F is the ultrafiltrate flow rate, A the membrane surface area, L_p the membrane hydraulic permeability coefficient, and μ the filtrate viscosity. The driving force is the sum of the hydrostatic and osmotic pressure gradients. In hemodialysis the transmembrane hydrostatic pressure (TMP) from blood to dialysate varies from 100 to 500 mm Hg, and almost invariably markedly exceeds the osmotic pressure generated by large molecules on the blood side. In clinical practice, an empirically defined ultrafiltration coefficient, K_{UF} [mL/(mm Hg)h], is used to describe dialyzer performance:

$$K_{UF} = \frac{\text{Water flux (mL/h)}}{\text{Transmembrane pressure (mm Hg)}} = Q_F/(P_B - P_D). \qquad (17)$$

Table 1 includes approximate ultrafiltration coefficients for a number of membrane materials, with kidney glomerular basement membrane added for comparison.

3.3.1 Convective Clearance. Frictional interaction between water and solute molecules—so-called solvent drag—results in transport of solute across the dialysis membrane with the ultrafiltrate. The restraining action of the membrane to this convective solute movement is expressed as the sieving coefficient S of a membrane for a given solute:

$$S = c_F/c_B. \qquad (18)$$

When $S=1$, there is no resistance to solute movement, and the concentration in the filtrate will be unchanged from that of blood. With cuprophane membranes, there is a rapid falloff in permeability with increasing molecular weight, but more "open" membranes such as sulfonated polyacrylonitrile (AN69) have sieving coefficients close to unity for molecules as large as 1500 Da (Table 2) (Green *et al.*, 1976). Since diffusive clearance of these larger molecules is low, convection plays a relatively important part in the transfer of these substances across open membranes.

When ultrafiltration is occurring,

$$Q_{Bo} = Q_{Bi} - Q_F \qquad (19)$$

Table 1. Hemodialysis hollow-fiber membrane materials, with glomerular basement membrane for comparison.

Membrane	Synonyms	Membrane thickness (μm)	K_{UF}[a]	Complement activation (%)[b]
Cellulosic				
Cuprammonium cellulose	Cuprophane	8–16	3–5	100
Saponified cellulose ester	SCE	25	3–4	100
Modified Cellulose				
Cellulose acetate	CA	40	4–10	48–67
Diethylaminoethyl cellulose	Hemophan	8	4	60
Cellulose triacetate	CT	85	20	
Synthetic hydrophilic				
Polycarbonate-polyether	PC-PE, Gambrane	15	5	40
Poly(ethylene-co-vinyl alcohol)	EVAL	32	8	
Synthetic hydrophobic				
Polyacrylonitrile	PAN	55	570	
Polymethylmethacrylate	PMMA	25	12	
Polysulfone	PS	40	40	24
Sulfonated polyacrylonitrile	AN69	45	25	5
Kidney glomerular basement membrane		0.2–0.3	1500	0

[a]Ultrafiltration coefficient, mL/(mm Hg) h m^2.
[b]Relative to cuprophan. [Data from Henderson and Chenoweth (1987).]

Table 2. Sieving coefficients for cuprophane and AN69 membranes.

Solute	Molecular weight (Da)	Sieving coefficient Cuprophane	AN69
Urea	60	1.000	1.000
Creatinine	149	0.891	0.985
Sucrose	342	0.791	0.982
Vitamin B_{12}	1355	0.629	0.940
Inulin	5200	0.309	0.776
Egg albumin	44000	0.002	0.044

Data from Green *et al.* (1976).

and

$$Q_{Do}=Q_{Di}+Q_F, \tag{20}$$

and Eq. (11) is thus no longer an accurate expression of solute transfer. Substitution of Eqs. (19) and (20) yields a new expression for dialysance:

$$D=\frac{Q_{Bi}(c_{Bi}-c_{Bo})}{c_{Bi}-c_{Di}}+\frac{Q_F\,c_{Bo}}{c_{Bi}-c_{Di}} \tag{21}$$

$$=\frac{Q_{Di}(c_{Do}-c_{Di})}{c_{Bi}-c_{Di}}+\frac{Q_F\,c_{Do}}{c_{Bi}-c_{Di}}, \tag{22}$$

which now consists of a diffusive term and a term containing Q_F, which represents the convective portion of dialysance. The equation for clearance can be similarly corrected, yielding the convective clearance term $Q_F\,c_{Bo}/c_{Bi}$, which cannot exceed the value of Q_F. Ultrafiltration rates in hemodialysis are rarely greater than 30 mL/min, a value that would range between 10% and 20% of the diffusive clearance for urea, but might account for the majority of removal of solutes with molecular weights over 1500.

3.3.2 Backfiltration. When no net ultrafiltration is occurring ($Q_{Do}=Q_{Di}$), the total net TMP is 0; nevertheless, at single points along the membrane this may not be the case. In fact, dialysate-side pressure is maximal at the blood outlet end of the dialyzer, where blood-side pressure is at a minimum. In the example shown in Fig. 6, net driving pressure on the blood side of the membrane varies from 42 mm Hg at the blood inlet to −16 mm Hg at the outlet. As a result, there is movement of ultrafiltrate out of blood at one end of the dialyzer, and an equivalent movement back from dialysate to blood at the other end, with total net ultrafiltration being 0.

With cellulose membranes, long the mainstay of hemodialysis, ultrafiltration coefficients range from 2 to about 4 mL/(mm Hg)h. Thus the maximum backfiltration rate at any point on the membrane at the pressures in Fig. 6 would be about 1.1 mL/min, probably not a clinically significant effect, especially since negative pressure applied to the dialysate side to produce net ultrafiltration further reduces the potential for backfiltration. With newer membranes, however, K_{UF} may be as high as 80–90 mL/(mm Hg)h, leading to a potential rate of back filtration as high as 24 mL/min at the blood outlet end if net Q_F is 0. At high backfiltration rates, contaminants in the dialysate, which is usually not sterilized, might be convectively transported into the blood by the back-filtrate. Endotoxin, a polymeric product of gram-negative bacteria, or its fragments, is a particularly feared contaminant, since exposure to even trace amounts could cause serious anaphylactic reactions.

4. COMPONENTS OF THE DIALYSIS SYSTEM

The dialysis system consists of a number of discrete components, each with characteristics that influence the process and that may vary considerably. The actual solute exchange takes place within one of several kinds of "artificial kidney," or dialyzer, among which membrane materials, geometry, and flow patterns may differ. Blood has chemical and rheological properties that strongly affect diffusion. The solutes contained in the dialysate determine concentration driving forces for diffusion of numerous small molecules, and the choice of dialysate buffer used to adjust *p*H may have profound effects on the patient. Finally, there is a variety of apparatus used to deliver blood and dialysate to the dialyzer and to control and monitor temperature, pressure, flow rates, and dialysate composition.

4.1 The Dialyzer

Early dialysis was performed with multiple-use devices that generally consisted of a large–surface-area blood-filled cellophane compartment suspended in an open dialysate bath. The ensuing three decades have seen the

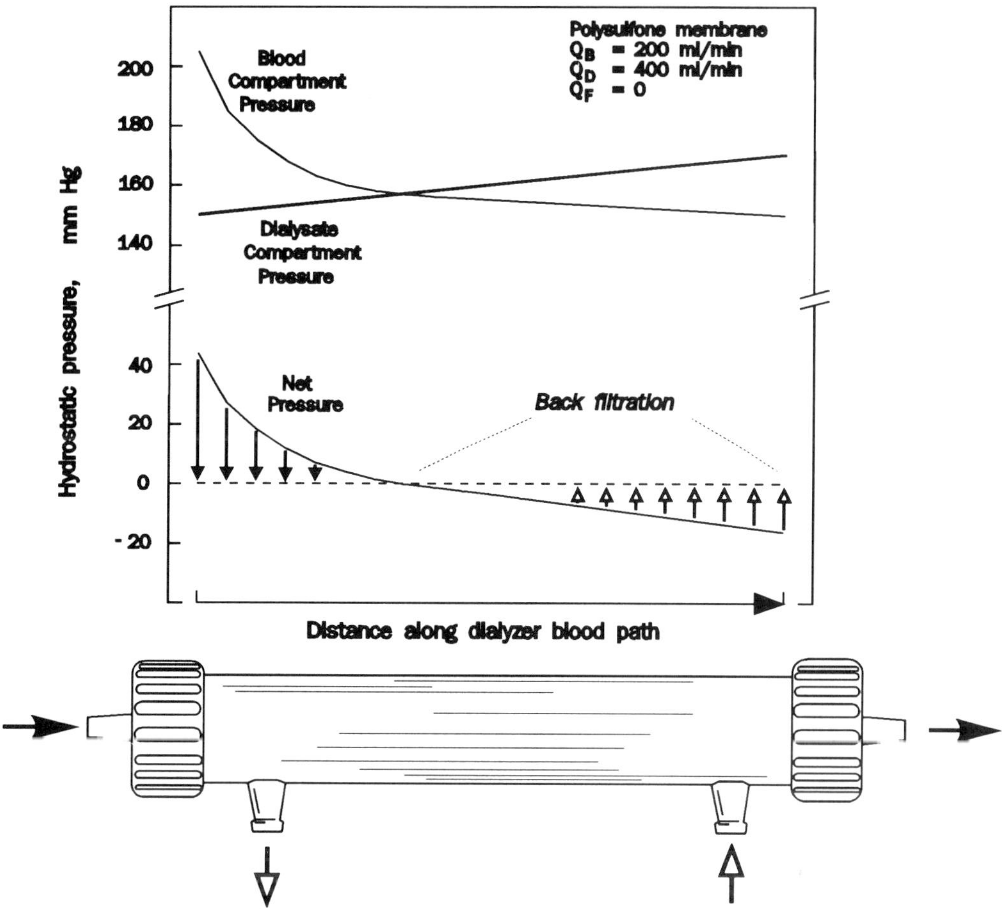

FIG. 6. Backfiltration. With no net ultrafiltration, blood-side hydrostatic pressure exceeds that on the dialysate side over about the first 40% of the blood path, and there is movement of ultrafiltrate out of blood. Over the last 60% of the blood path, the pressure relationships reverse and an equivalent amount of fluid moves from dialysate to blood. Filled arrows, blood compartment; open arrows, dialysate compartment. [Data from Streicher and Schneider (1985).]

development of disposable hemodialyzers, in which all transport processes take place within a self-contained small-volume unit. Cellulosic membranes remain the mainstay of therapeutic dialysis, but a number of new synthetic membrane materials have appeared that have distinctly different properties with respect to mass transfer and to biocompatibility.

4.1.1 Dialyzer Geometry. Hemodialyzers have been produced in several different dialyzer designs, with varying geometrical relationships between blood, membrane, and dialysate. The coil dialyzer dominated the 1950s and early 1960s, when it was succeeded by the parallel plate design. During the 1980s hollow-fiber dialyzers almost completely replaced the earlier designs in clinical use.

4.1.1.1 Coil Dialyzers. The first clinically viable dialyzer constructed consisted of a long cellophane sausage tubing wrapped around a cylindrical drum support. The tube was perfused with the patient's blood, and the entire apparatus was rotated in a tank of dialysis fluid. From this design descended the "twin coil" artificial kidney, introduced in 1956, which was small, efficient (capable of a $K_{D\,\text{urea}}$ of 140 mL/min), inexpensive, and disposable. The dialyzer was placed on end in a canister in an open 100-L tank, and dialysate was pumped crosswise through the supporting mesh. The twin coil was widely used throughout the 1950s and 1960s, and smaller, improved coils were used in many hospitals well into the 1980s, but by 1985 only 1–2% of dialyses in Europe, where the most complete data are available, were performed with this type of dialyzer (Hoenich *et al.*, 1989).

4.1.1.2 The Parallel Plate. The coil dialyzer has now been almost completely replaced by two other designs: parallel-plate and hollow-fiber dialyzers. Parallel-plate devices have evolved from the two-layer 40×14×3-in. Kiil dialyzer of the early 1960s, which was reassembled and fitted with new cellophane membranes before each dialysis. Current versions are small, sterile, disposable devices like the one shown in Fig. 7(a). Blood flows in flat chambers between sheets of membrane with dialysate moving in a countercurrent direction, as illustrated in Fig. 7(b). Modern flat-plate design innovations include precision membrane assembly that allows as many as 17 blood and 34 dialysate layers, and the formation of wavelike microchannels [Fig. 7(c)] in the blood and dialysate chambers to promote turbulent flow. The turbulence en-

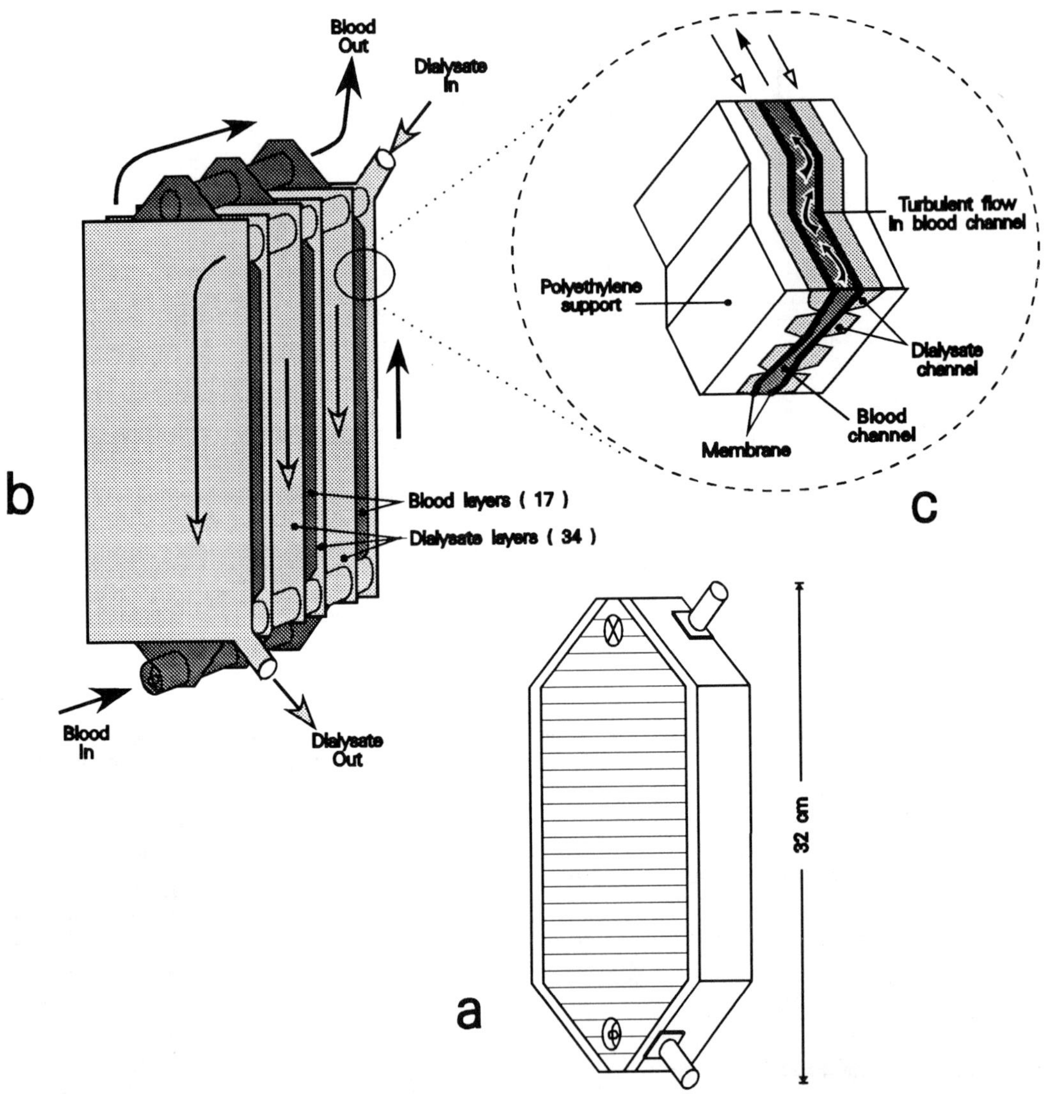

FIG. 7. Modern parallel-plate hemodialyzer. (a) "Flat-plate" dialyzers, used in about 20% of hemodialyses worldwide. (b) The blood and dialysate paths, flowing countercurrent in parallel flat chambers on opposite sides of a membrane sheet. (c) An enlarged view showing wavelike microchannels within the blood chamber that create turbulent flow and thus better mixing, flow distribution, and transport. Open arrows: dialysate path. Solid arrows: blood path.

sures active mixing, impeding the formation of large stagnant boundary layers, and thereby minimizes mass transfer resistance within the blood and dialysate phases. Parallel plates are probably used in about 20% of dialyses worldwide.

4.1.1.3 Hollow-Fiber Dialyzers. The most widely used design today is the capillary or hollow-fiber dialyzer (Fig. 8), first introduced by Cordis-Dow in the early 1970s. A typical hollow-fiber design consists of a bundle of 6000 to 20 000 capillary tubes, each with an internal diameter of approximately 200 μm and a wall thickness of 8–15 μm for cellulose and up to 40–50 μm for synthetic hydrophobic membranes. The fiber bundle is supported at each end by a "tube sheet" of potting compound, usually polyurethane [Fig. 8(c)]. Hollow-fiber dialyzers have several marked advantages over parallel-plate designs. The ratio of membrane surface area to device volume is greater, leading to a lighter, more compact dialyzer. The fibers are self-supporting, so that effective surface area is not lost through contact with an external rigid framework. Finally, the fibers are less compliant than the parallel-plate blood chambers, so that there is less expansion of the blood compartment during dialysis treatment and consequent decreased pooling of blood outside the body. This is important in minimizing the tendency toward falls in blood pressure in the patient undergoing treatment.

A disadvantage of the hollow-fiber design is its greater vulnerability to plugging of the fibers by particulate matter, making these dialyzers more susceptible to the effects of clot formation in the blood compartment. Severe anaphylactic reactions occasionally occur at the onset of dialysis, which are thought to be largely caused by allergy to residual ethylene oxide gas contained in the dialyzer (Daugirdas and Ing, 1988). This gas, used as a sterilant, is absorbed and slowly released by the polyurethane potting compound, which is not used in parallel-plate devices (Lee *et al.*, 1985); reactions occur over 10 times more commonly with hollow-fiber devices (Villarroel and Ciarkowski, 1985). Nevertheless, hollow-fiber dialyzers now account for up to 80% of dialysis treatments.

4.1.2 Membranes. Dialysis is, after all, a membrane separation process, and the characteristics of the membrane itself strongly affect the nature and efficiency of the dialysis treatment. A number of membranes are in use today, which differ in their structures, their transport properties, and their tendency to interact with blood components.

4.1.2.1 Classification. Dialysis membranes may be conveniently categorized in three groups: cellulose, modified cellulose, and synthetics, with the synthetic membranes further classified as hydrophilic or hydrophobic (Lysaght, 1988). All are natural or synthetic polymers, and are listed along with several properties in Table 1.

Cellulose Membranes. For the first 20 yr of chronic dialysis, there was really only one membrane material—regenerated cellulose. Reconstituted by the cuprammonium process or by saponification of heat-spun cellulose acetate, cellulose is prepared as flat-sheet or hollow-fiber membranes, and still is used in the large majority of dialysis treatments, principally because of its low cost. The structure of cellulose confers particular properties on a hemodialysis membrane. The density of hydroxyl groups on the cellobiose dimer (see Fig. 9) leads to a high polarity, high degree of intramolecular hydrogen bonding, and large affinity for water. The membranes swell markedly when wet, and form a *hydrogel*, a homogeneous single phase containing both water and polymer. Finally, the free hydroxyl groups readily participate in chemical reactions, and probably account for the high interactivity of cellulose membranes with blood cells and proteins.

Modified Cellulose Membranes. To reduce the chemical reactivity of cellulose, blocking groups may be used to replace some or all of the hydroxyl hydrogen atoms (Fig. 9). The most common membrane material of this type is cellulose acetate, in which 60–80% of the available hydrogen has been replaced by acetate radicals. The resultant membrane is less bioactive (see below) and somewhat less hydrophilic, forming a more "open" hydrogel with higher water permeability. The newer cellulose triacetate, with all hydroxyl groups bound to acetate, is so altered that its behavior is more similar to the synthetic hydrophobic materials than to native cellulose. Hemophan is cellulose with about 5% of hydroxyl hydrogen replaced by the diethylaminoethyl moiety, and has properties similar to cellulose acetate.

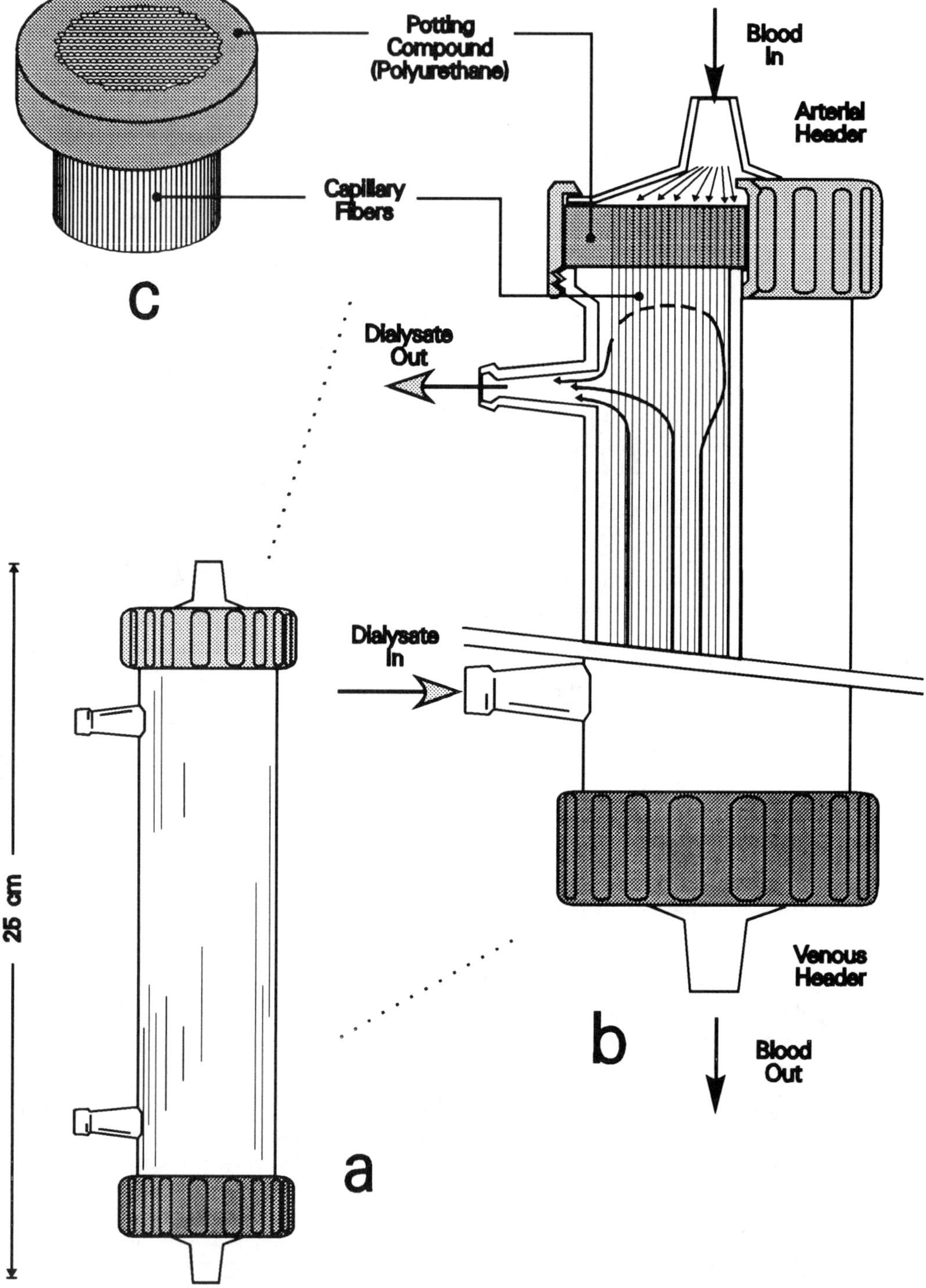

FIG. 8. Hollow-fiber hemodialyzer. In a typical hollow-fiber dialyzer (a) blood flows through small (~200-μm internal diameter) fibers that are bathed on the outside by dialysate (b). The fiber ends are supported by a disk of a potting compound such as polyurethane (c).

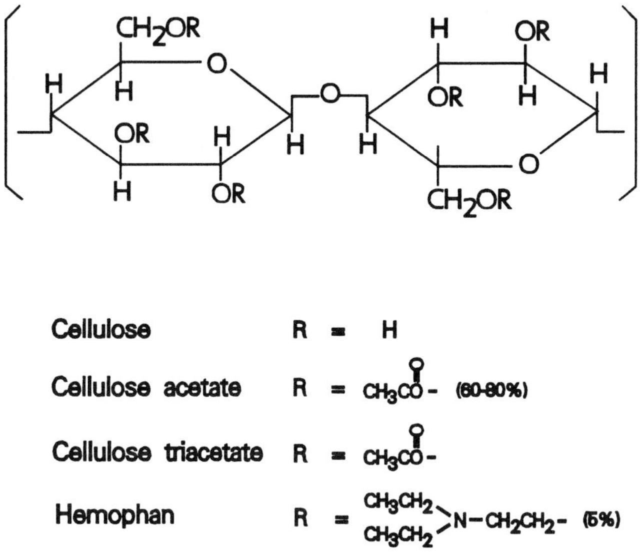

FIG. 9. Structure of cellulose and its derivatives.

Hydrophilic Synthetic Polymers. Polycarbonate-polyether (PC-PE) and poly-ethylvinyl alcohol (EVAL) are synthetic membranes with chemical and physical similarities to cellulose. PC-PE has many polar carbonyl groups, and EVAL has unsubstituted hydroxyls; thus both membranes are quite hydrophilic, and form hydrogels. Unlike cellulose, however, PC-PE does not swell when wet.

Hydrophobic Synthetic Polymers. These are nonpolar, resonance-stabilized polymers with low energy of interaction with water, listed in Table 1. Membranes are produced from these materials by a phase inversion process, in which the polymer is dissolved in a solvent and pumped through a spinning jet. As the polymer solution emerges from the jet, it is immersed in a nonsolvent precipitant (often water) that replaces the solvent and converts the solution into a two-phase system, the solid phase forming the rigid membrane structure and the liquid, polymer-poor phase forming pores. Membranes formed in this way have chemical and physical properties quite different from cuprophane and other hydrophilic materials. Their overall structure is that of an anisotropic foam. The inner, blood-contacting surface of a hollow fiber made from these materials is a thin skin, usually about 1 μm in thickness. This relatively dense skin determines the transport characteristics of the membrane. The surrounding open foam-like layer has very little flow resistance and functions mainly as mechanical support. The precise morphology of the membrane may be varied by changes in the polymer concentration, solvent, precipitant, or temperature of the phase inversion.

4.1.2.2 Transport Characteristics. The structural differences among these membranes lead to differences in transport capabilities. In general, clearances of small–molecular-weight solutes such as urea are similar for similar conditions of blood and dialysate flow rate, although individual dialyzer design and varying membrane thickness account for about a 40% difference between the highest and lowest urea clearance in the group CA > PS > PMMA > PAN > Cuprophane (Petersen *et al.*, 1987). Examination of Table 1, however, reveals a marked difference between hydrophilic and hydrophobic membranes in water permeability: K_{UF} for the hydrophobic synthetics is as much as 10 times higher than that of cuprophane. As a result, convective clearance is also higher with the hydrophobic group, and larger–molecular-weight solute removal is considerably more efficient. This marked advantage for control of uremic toxicity is partially offset by the problem of potentially excessive water loss, which necessitates the use of special volumetric ultrafiltration systems, by the increased risk of backfiltration (see Sec. 3.3.2), and by the generally much higher cost of the newer synthetic membranes.

4.1.2.3 Biocompatibility. Besides transport characteristics, there is another important property that determines the suitability of a membrane material for hemodialysis, and that is its propensity to interact, usually in deleterious ways, with components of the body. The term "biocompatibility" has come to be used in this context, defined in negative terms as the *absence* of (1) thrombogenic, toxic, allergic or inflammatory reaction, (2) destruction of formed elements of blood, (3) changes in plasma proteins and enzymes, (4) immunological reaction, (5) carcinogenic effect, and (6) deterioration of adjacent tissues (Klinkmann *et al.*, 1989). It is perhaps more accurate to speak of the occurrence of any of these reactions as "bioincompatibility."

Coagulation. Bioincompatibility of hemodialysis membranes occurs in several ways, primarily through interaction with the blood coagulation system, with the circulating inflammatory mediator proteins of the complement system, and with white blood cells. As occurs with almost any artificial surface, exposure of dialysis membranes to the circulation leads to rapid adsorption of a layer of plasma proteins and adhesion of platelets, the small formed elements in blood that initiate the formation of a "hemostatic plug." This is frequently followed by activation of the blood coagulation system and formation of a thrombus, necessitating the use of anticoagulants, usually heparin, during the dialysis treatment. Membranes differ in thrombogenicity, and although data are scarce, cellulose seems the most reactive, with a decrease in clotting tendency when derivatized to cellulose acetate or hemophan. The least reactivity is shown by the hydrophobic synthetic membranes.

Complement activation. The complement system is a group of more than 20 plasma proteins, normally inactive, which in response to appropriate stimuli may be enzymatically cleaved to yield a number of bioactive proteins and peptides that mediate host-defense and acute inflammatory reactions. Dialysis membranes may serve as the stimuli, and activation of complement to varying degrees is a well-established phenomenon in hemodialysis and a common measure of bioincompatibility. The effects of activated products include agglutination of white blood cells in the lung, leading to transient neutropenia and contributing to the hypoxemia associated with the first hour of dialysis. New, nonreprocessed cellulose membranes, with their exposed hydroxyl groups, are by far the strongest activators of complement, while the hydrophobic synthetics are the most biocompatible. Relative activities of various membranes are shown in Table 1.

Synthetic hydrophobic membranes seem to represent a step forward in hemodialysis technology, with improved transport properties and enhanced biocompatibility with respect to cellulosic membranes. There is also evidence that cellulose may induce the release into the circulation of β-2-microglobulin, a ubiquitous 11 800-Da cell-surface protein that accumulates in dialysis patients and may be deposited in bones and joints, leading to a painful and crippling arthropathy, while the synthetic membranes clear the molecule from the blood with varying effectiveness. Nevertheless, technical difficulties associated with high water permeability, and especially high cost, have greatly hindered acceptance of the synthetics, and in the United States cellulosic membranes are still used for 70–80% of hemodialysis treatments.

4.2 Blood

Much of the preceding analysis of diffusion phenomena in hemodialysis treats blood as a simple aqueous solution. The reality is a good deal more complex. Transport phenomena in blood are influenced by solute interactions with electrolytes, with suspended proteins, and with cellular elements—especially red blood cells—as well as by the particular rheological properties of blood.

4.2.1 Protein. The protein content of blood affects transport in several ways. First, small solutes may bind noncovalently to protein macromolecules, thereby becoming markedly less diffusible. In the extreme case, a small molecule may be almost completely unaffected by dialysis. The trivalent aluminum ion, for example, with a molecular weight of only 27, not only is not removed from the circulation by dialysis, but can actually be transported into the blood from contaminated water against an apparent concentration gradient. In fact, about 80% of blood-borne aluminum is bound to protein, and is therefore not available for diffusion, remaining entrapped in the intravascular space. A similar phenomenon occurs with some drugs, such as digoxin and glutethimide, which are highly protein-bound and not significantly removed by dialysis.

Hemodialysis membranes may themselves interact with plasma proteins, both specifically, as outlined in Sec. 4.1.2.3, and nonspecifically. The high hydraulic fluxes that are often attained with high permeability membranes carry along macromolecules, which accumulate in the immediate area of the membrane. This "concentration polarization," shown in Fig. 10, can lead to formation of a protein gel that may interfere with transport. In fact, perfusion of a polysulfone hollow fiber dialyzer with plasma may lead to a 60% reduction in K_{UF} compared to values obtained with saline perfusion (Bosch *et al.*, 1985).

4.2.2 Blood Cells. Blood is not, of course, a homogeneous liquid, but a suspension in which the dispersed phase—the blood cells—consists of liquid droplets each enclosed in a solid elastic envelope. Its resultant non-Newtonian rheological properties are complex, but probably not of major consequence to dialysis transport phenomena.

The distribution of solutes in the plasma and cell compartments, and the rate of transfer between them, however, have significant impact on the efficiency of dialysis. More urea is removed from aqueous solutions than from equimolar solutions in whole blood, partly because of the decreased volume fraction of free water in plasma and red cells (92% and 72%), and partly because of increased resistance to diffusion caused by plasma protein in the boundary layer. Equilibrium distribution coefficients ($c_{\text{red cells}}/c_{\text{plasma}}$) for various solutes are shown in Table 3. The average concentration in whole blood for most of these solutes, given by

$$\bar{c}=\phi c_{\text{red cells}}+(1-\phi)c_{\text{plasma}}, \tag{23}$$

where ϕ is the hematocrit, or volume fraction of cells, will be less than the concentration measured in plasma. Thus for most solutes the quantity passing through the dialyzer decreases as the hematocrit increases, and solute removal per unit time will correspondingly decrease.

Urea crosses cell membranes with great facility. During the 20 s or so that it takes for an erythrocyte to pass through the dialyzer, the urea concentration in its immediate environment falls by as much as 70–80%. Despite this drastic change in the cell's milieu, the

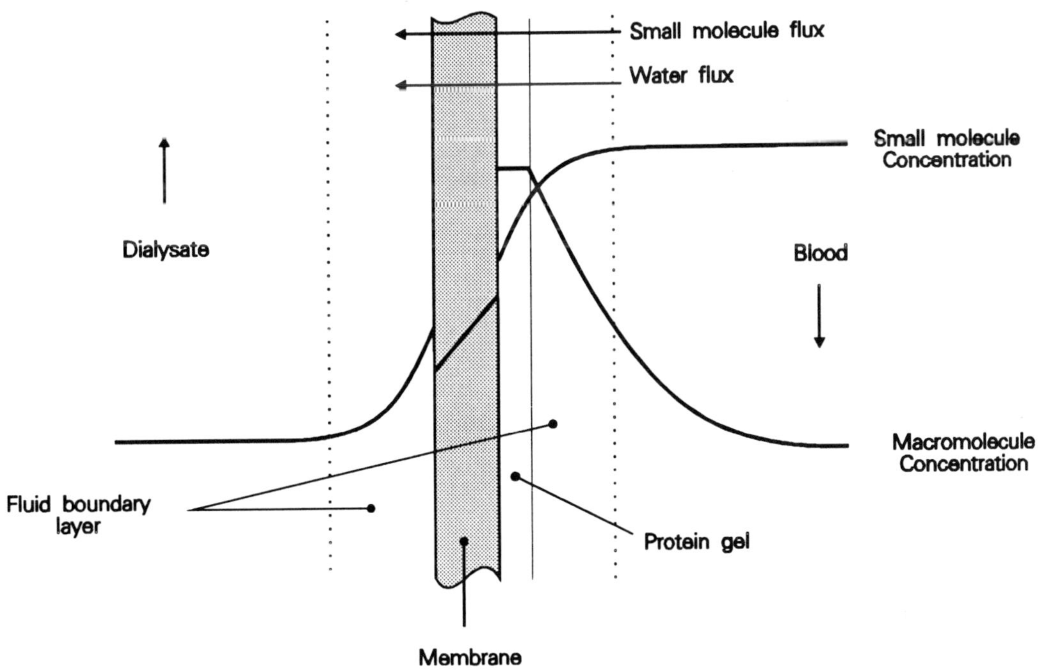

FIG. 10. Concentration polarization. Shown are concentration profiles for small molecules and macromolecules (protein) in the vicinity of a hemodialysis membrane. Water flux from blood to dialysate carries protein molecules to the membrane surface, where a gel may be formed that can interfere with transport of smaller solutes and water. Increasing blood flow rate increases shear forces which tend to limit gel formation.

Table 3. Equilibrium distribution coefficients, $c_{red\ cells}/c_{plasma}$.[a]

Urea	0.86
Creatinine	0.73
Uric acid	0.54
Amino acid nitrogen	1.54
Glucose	0.95

[a]Colton (1981).

ratio between intracellular and extracellular urea is virtually the same at the dialyzer blood inlet and outlet. Other substances, however, such as potassium and phosphate, diffuse more slowly across biological membranes, are retained within the red blood cell, and are relatively less available for removal by dialysis. Late equilibration between the intracellular and extracellular pools of these solutes after the dialysis treatment is finished contributes to a rebound rise in plasma concentration.

4.3 Dialysis Fluid

The driving force for transport of small molecules in dialysis is the transmembrane concentration gradient, which may be manipulated by altering the composition of the dialysis fluid.

4.3.1 Composition. Table 4 lists the composition of typical dialysate solution, with comparative values for serum. The presence of glucose in dialysate may be something of an anachronism. Its originally intended functions—stabilization of serum osmolality and osmotic extraction of water—have been rendered obsolete by high-sodium dialysate and by pressure- and volume-mediated ultrafiltration.

The need for an alkaline buffer to correct the accumulation of hydrogen ions in the patient with renal failure led to the early use of bicarbonate in dialysis solutions. Preparation of dialysate from concentrates, however, made the use of bicarbonate problematic. High concentrations of HCO_3^- are unstable in solution, as the high partial pressure of carbon dioxide leads to loss of the gas into the surrounding atmosphere and fall in HCO_3^- concentration. Further, the presence of calcium and magnesium ions in the dialysate causes precipitation of carbonates with these cations. During the 1960s, the problem was resolved by the use of acetate as the buffer anion. Once in the circulation, acetate is converted quickly and quantitatively to bicarbonate by normal metabolic pathways, and presents no problems of solubility or stability in dialysate solution.

As dialysis became more rapid and efficient, however, and transfer rates of acetate exceeded the rate of metabolic conversion, other problems arose. Acetate in the concentrations achieved in blood produced significant clinical toxicity, including hypotension, headaches, and vomiting. Patients differ in their sensitivity, but by now it is clear that there is a common and marked increase in

Table 4. Composition of serum and dialysate.

	Serum	Dialysate
Electrolytes (mEq/L):		
Sodium	140.0	140.0–145.0
Potassium	4.0	2.0
Chloride	100–110	95–109
Calcium (ionized)	1.8	2.5–3.5
Magnesium	1.6	1.0
Glucose (mg/dL):	120	0–200
Buffer (mEq/L):		
Acetate	0.2	35–40[a] 4[b]
Bicarbonate	25	0[a] 35–40[b]

[a]Acetate dialysate.
[b]Bicarbonate dialysate.

symptoms and hemodynamic instability associated with acetate use. The result has been a return to bicarbonate, but with more sophisticated dialysis delivery systems that allow mixing of two separate concentrates, one containing the buffer and the other the divalent cations Mg^{2+} and Ca^{2+}.

4.3.2 Water Preparation. In most modern dialysis systems the dialysate is prepared continuously by a proportioning system from concentrate and line water. During three 4-h hemodialysis treatments with 500-mL/min dialysate flow, a patient is exposed to 360 L of municipal water. This is roughly 25 times the average exposure by ingestion, and the transport of trace elements and toxins into the blood is far more direct. Moreover, dialysis patients are unable to clear toxins continuously by urinary excretion. Water that is suitable for drinking, therefore, requires further preparation for use in hemodialysis.

Most hemodialysis units use a serial water purification system with several components. After initial screening of particulate matter by a sediment filter, a water softener utilizes a cationic resin to exchange sodium ions for divalent cations, principally magnesium and calcium. An activated carbon filter is used to adsorb chlorine, chloramines, and dissolved organic material. Chlorine and chloramines are used extensively for bactericidal treatment of municipal water supplies, and in high dialysate concentrations can cause serious toxicity. A second sediment filter removes the small carbon particles released from the carbon filter.

For most water supplies, reverse osmosis (RO) is the method of choice for removal of solutes. Within the RO module, feedwater is in contact with a spirally wound or hollow-fiber semipermeable (often cellulose acetate) membrane, and a hydrostatic pressure is applied that exceeds the osmotic pressure generated by the dissolved solutes. As a result, water flows against the osmotic gradient, and a purified product is produced, even small ionic solutes having been removed by molecular sieving. Effectiveness of the device is monitored by measuring resistivity of feedwater and product water and calculating the percentage rejection of solutes, which ranges from 90% to 98% of monovalent and 95% to 99% of divalent ions. RO also effectively removes organic contaminants and bacteria, as well as particle matter; the combination of RO and pretreatment with water softener and carbon filter virtually always produces water pure enough for safe hemodialysis.

4.4 Dialysis Monitoring and Delivery Systems

Hemodialysis has evolved in 30 yr from a difficult and dangerous procedure, requiring constant supervision by a physician, to a routine treatment carried out with a minimum of technical intervention, even practicable in a patient's home. The dialysis apparatus has become more self-regulating, with the latest equipment capable of on-line automatic blood pressure recording and minute-to-minute variability in dialysate sodium concentration.

4.4.1 Functions. The delivery system, commonly referred to as the "dialysis machine," has four principal functions. First, it maintains flow through the blood chamber of the dialyzer, via the extracorporeal blood circuit. Second, it continuously prepares dialysate from line water and concentrate and keeps it flowing through the dialyzer. Third, it regulates ultrafiltration, usually by adjusting volume or hydrostatic pressure in the dialysate circuit. Fourth, it assures patient safety by monitoring and regulating temperature, pressures, and dialysate concentration, and detecting emergency situations like dialysis membrane ruptures or the presence of air in the blood returning to the circulation. A dialysis machine of the conventional "single-pass" type—in which dialysate flows in an uninterrupted path from line water to drain—is shown schematically in Fig. 11.

4.4.2 The Blood Circuit. In modern dialysis, blood is both withdrawn from and returned to the venous circulation; hence there is no pressure gradient to drive blood flow. The driving force for extracorporeal flow is therefore provided by a peristaltic roller pump in the blood circuit. This in turn necessitates a system of air traps and alarms to prevent air, which may be drawn into the system by the negative pressure created by the pump, from being returned to the patient's circulation.

An intermittent or continuous infusion of heparin is necessary to prevent coagulation of blood in the dialyzer or lines. A continuous pressure monitor on the returning ("venous")

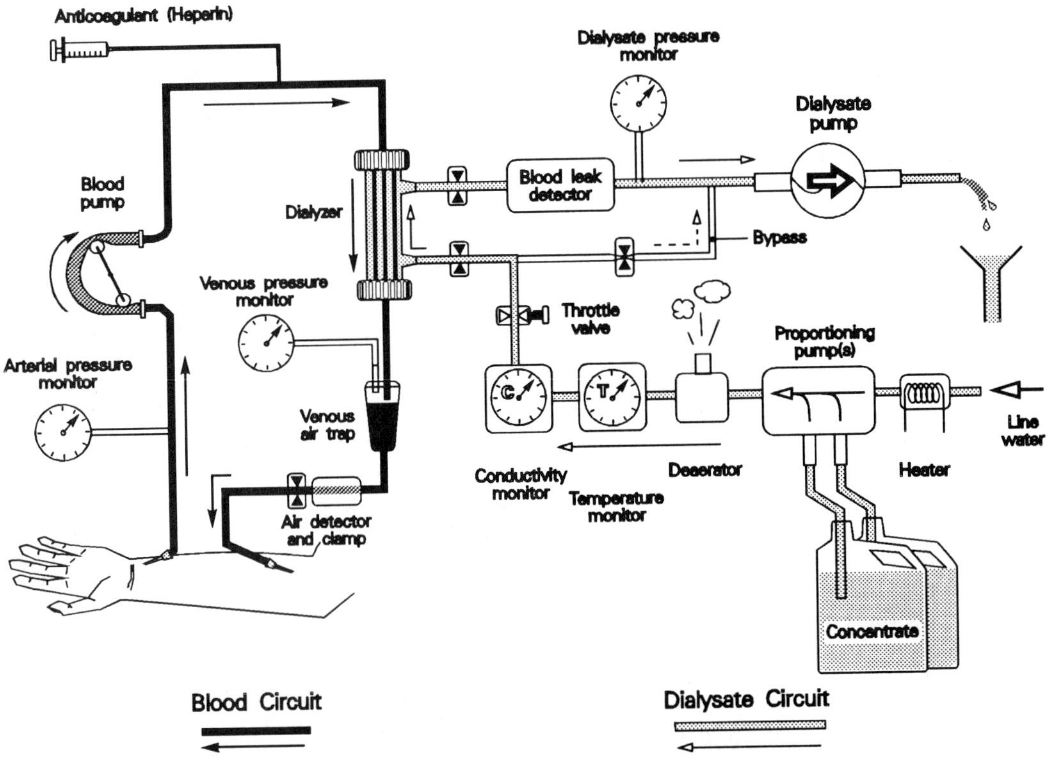

FIG. 11. Single-pass dialysis apparatus. This is still the most widely used configuration for dialysis delivery systems. Ultrafiltration is regulated by change in the tension on the throttle valve, thus varying relative negative pressure in the dialysate circuit.

blood line stops the blood pump if pressure exceeds or falls below preset limits. The dialyzer and all of the tubing that contacts blood are disposable. For economic reasons dialyzers are often reprocessed for repeated use by the same patient; the tubing is always discarded after one use.

4.4.3 The Dialysate Circuit. The dialysate is continuously prepared by a proportioning system from heated line water and bottled concentrate, usually mixing the two in a ratio of about 35 to 1. The solution is deaerated by subjection to negative pressure to prevent later spontaneous degassing, bubble formation, and "vapor lock" of blood channels in the dialyzer. The final composition is monitored by measuring the electrical conductivity of the solution; when the conductivity varies by more than 10% from the target value, an alarm condition results, and bypass valves are tripped that isolate the dialyzer from the dialysate path. A similar alarm system monitors temperature, and an optical sensor stops the blood pump if blood is detected in dialysate exiting from the dialyzer.

4.4.4 Ultrafiltration Systems. Control of the amount of fluid removed during a dialysis treatment requires special apparatus to regulate the hydrostatic pressure across the membrane. The earliest ultrafiltration control system, shown in Fig. 12(a), used with large open tanks of dialysate, consisted of a screw clamp (3) on the venous blood line. Increasing blood outflow resistance with the clamp raised the hydrostatic pressure in the dialyzer blood chamber (shown in figure as + signs) and increased the rate of ultrafiltration. The development of continuously proportioned single-pass delivery systems allowed regulation of the pressure on the dialysate side, and the system shown in Fig. 12(b) is now the most common method of ultrafiltration control. Raising inflow resistance with the throttle valve (4) causes the dialysate pump (5) to develop increasing negative pressure within

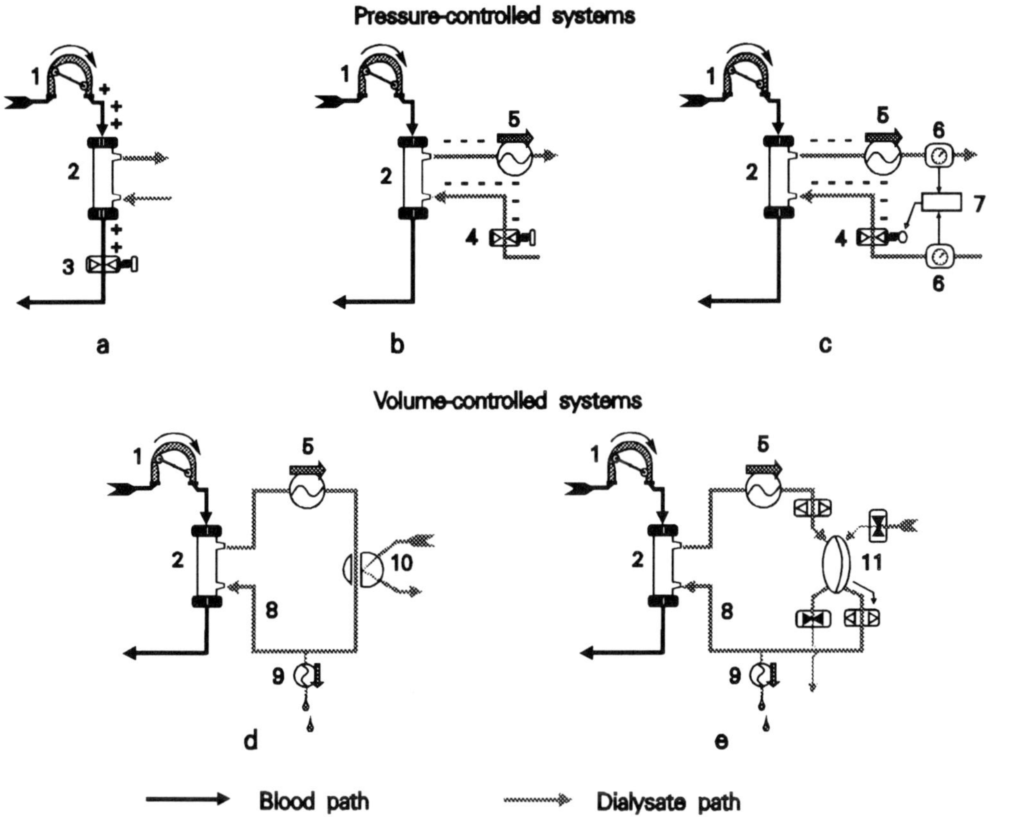

FIG. 12. Systems for ultrafiltration control. (1) blood pump; (2) dialyzer; (3) screw clamp on venous blood line; (4) dialysate line throttle valve; (5) dialysate pump; (6) flow sensor; (7) digital analyzer; (8) closed dialysate circuit; (9) ultrafiltration pump; (10) three-way valve for draining and refilling of dialysate circuit; (11) balancing chamber and valves. See text for explanation.

the dialysate circuit, which in turn increases water flux from blood to dialysate.

This system is vulnerable to a number of inaccuracies. Pressure measurement is not precise and varies with arm movements, fiber clotting, etc., and accurate clinical prediction of ultrafiltration requires exact knowledge of the dialyzer K_{UF} and blood oncotic pressure. A refinement of the system that seeks to overcome these problems [Fig. 12(c)] uses flow sensors (6) on the incoming and outgoing dialysate lines to monitor ultrafiltration more precisely, and to exert feedback control via a digital analyzer (7) that adjusts the throttle valve (4) to maintain the desired rate of fluid removal.

Pressure-based systems serve adequately for dialyzers with cellulose membranes, but use of the newer synthetics, with ultrafiltration coefficients greater than 10 mL/(mm Hg)h, has introduced another problem. The minimum ultrafiltration rate achievable with pressure-controlled systems is that rate generated by the minimum transmembrane hydrostatic pressure. This minimum pressure is the mean blood-chamber pressure, since the system does not generate positive pressure on the dialysate side. Flow resistances in the dialyzer and return lines will almost always produce a pressure of greater than 100 mm Hg; thus, for a dialyzer with a K_{UF} of 30 mL/(mm Hg)h (see Table 1), the minimum ultrafiltration rate would be 3000 mL/h, and could easily reach 5000–6000 mL/h. Removal of 10–20 L of fluid from a patient's circulation over a 3-h period could obviously lead to serious complications, and so in order to use high-UF membranes safely, new ultrafiltration control systems have been devised.

These "volumetric" systems [Fig. 12(d) and 12(e)] regulate ultrafiltration by removing fluid from a closed, rigid dialysate circuit (8) at a measured rate via an ultrafiltration pump (9). Since the dialysate loop is noncompliant, ultrafiltrate crosses the dialysis membrane at a rate that exactly matches the rate of removal by the ultrafiltration pump. Several variations exist, one of which [Fig. 12(d)] uses a closed loop that is drained and refilled at 2-min intervals via a three-way valve (10), and another [Fig. 12(e)] that uses balancing chambers separated by a diaphragm (11) to ensure the equality of dialysate inflow and outflow volume, with exiting dialysate displacing an exactly equal quantity of fresh dialysate into the circuit. In practice, two parallel chambers operating 180° out of phase are used to maintain more constant flow.

5. THE FUTURE

Dialysis is still in many ways a "half-way technology," which prolongs life but does not completely restore patients to a healthy, physiologic state. The definition of adequate dialysis remains elusive, despite our recent enchantment with blood urea. There is much yet to understand about the interaction of the dialysis apparatus with the immune system, and the influence of end-stage renal disease on immunity itself. Microminiaturization of dialysis components for the development of portable or wearable artificial kidneys, and the development of new, efficient membranes with transport profiles that more closely mimic the human kidney are priorities for technical development.

Despite its shortcomings, however, regular dialysis treatment has progressed in 30 yr from a relatively primitive, controversial, and short-term therapy to a highly technological and largely automated procedure backed by a complex and sophisticated body of pathophysiological knowledge, which keeps hundreds of thousands of patients alive for as much as 20 yr and more. If progress continues at that rate during the next few decades, concepts which now seem remote fantasy, such as implantable bioartificial kidneys, may well be realized.

GLOSSARY

Anion: A negatively charged ion.

Biocompatibility: Tendency of a foreign substance not to react with components of the body.

Cation: A positively charged ion.

Clearance: Solute removal capacity of a kidney or dialyzer, defined by total solute removal per unit time divided by plasma concentration.

Collodion: Precipitated nitrocellulose; Used as a dialysis membrane.

Colloid: Large–molecular-weight solutes that diffuse slowly or not at all across a semipermeable membrane.

Complement: A system of blood proteins that participate in host-defense and inflammatory reactions.

Convective Transport: Transfer of solute across a membrane by ultrafiltered solvent; transfer by "solvent drag."

Crystalloid: Small–molecular-weight solutes that diffuse readily across a semipermeable membrane.

Dalton: The unit of molecular weight; the mass of the nucleus of the hydrogen atom; equals 1.00794 u.

Dialysance: Solute removal capacity of a hemodialyzer, defined by total solute removal per unit time divided by plasma-dialysate concentration gradient.

Dialysate: In hemodialysis, the electrolyte solution on the diffusate side of the membrane.

Diffusate: The solution on the side of the dialysis membrane containing diffused crystalloids.

Fractional Excretion: The ratio of rate of urinary excretion to rate of glomerular filtration of a solute.

GFR: Glomerular filtration rate, the rate of ultrafiltration of plasma water at the glomerulus.

Glomerulus: The capillary tuft in the kidney where filtration takes place.

Hemodialysis: Dialysis of blood across an artificial membrane for treatment of kidney failure.

Middle Molecules: Molecules with molecular weight 1500–5000 Da, felt to be responsible for toxicity in kidney failure.

Nephron: The structural unit of renal function, consisting of a glomerulus and tubule.

Permeate: Diffusate.

Retentate: The solution on the side of the dialysis membrane containing retained colloid.

Reverse Osmosis: Transfer of water across a semipermeable membrane against an osmotic gradient, driven by hydrostatic pressure.

Uremia: The complex of symptoms of kidney failure.

List of Works Cited

Bosch, T., Schmidt, B., Samtleben, W., Gurland, H. J. (1985), in: E. Streicher and G. Seyffart (Eds.), *Highly Permeable Membranes (Contributions to Nephrology, Vol. 45)* Basel: Karger, pp. 14–22.

Colton, C. K., Lowrie, E. G. (1981), in: B. M. Brenner and F. C. Rector (Eds.), *The Kidney,* Philadelphia: Saunders, Chap. 47.

Daugirdas, J. T., Ing, T. S. (1988), *Kidney Inter.* **33** (Suppl. 24), S37–S43.

Dworkin, L. D., Brenner, B. M. (1985), in: D. W. Seldin and G. Geibisch (Eds.), *The Kidney: Physiology and Pathophysiology,* New York: Raven, Chap. 19.

Green, D. M., Antwiler, G. D., Moncrief, J. W., Decherd, J. F., Popovich, R. P. (1976), *Trans. Am. Soc. Artif. Internal Organs* **22**, 627–636.

Henderson, L. W., Chenoweth, D. (1987), *Blood Purif.* **5**, 100–111.

Hoenich, N. A., Woffindin, C., Ward, M. K. (1989); in: J. Maher (Ed.), *Replacement of Renal Function by Dialysis,* Dordrecht: Kluwer, Chap. 5.

Klinkmann, H., Falkenhagen, D., Stefoni, S., Bonomini, V. (1989), in: V. Bonomini, M. P. Scolari, S. Stefoni, and A. Vangelista (Eds.), *Biotechnology in Renal Replacement Therapy (Contributions to Nephrology, Vol. 70)*, Basel: Karger, pp. 213–226.

Lee, F. F, Durning, C. J., Leonard, E. F. (1985), *Trans. Am. Soc. Artif. Internal Organs* **31**, 526–533.

Lysaght, M. J. (1988), in: G. D'Amico and G. Colasanti (Eds.), *Nephrology and Dialysis Updated (Contributions to Nephrology, Vol. 61)*, Basel: Karger, pp. 1–17.

Petersen, J., Hyver, S., Yeh, I. (1987), *Trans. Am. Soc. Artif. Internal Organs* **33**, 265–268.

Streicher, E., Schneider, H. (1985), in: E. Streicher and G. Seyffart (Eds.), *Highly Permeable Membranes (Contributions to Nephrology, Vol. 45)*, Basel: Karger, pp. 1–13.

United States Renal Data System (1990), *1990 Annual Data Report,* Bethesda: National Institutes of Health.

Villarroel, F., Ciarkowski, A. A. (1985), *Artif. Organs* **9**, 231–238.

Wolff, A. V., Remp, D. G., Kiley, J. E., Currie, G. D. (1951), *J. Clin. Invest.* **30**, 1062–1070.

Further Reading

Brenner, B. M., Rector, F. C., Jr. (Eds.) (1991), *The Kidney,* Philadelphia: Harcourt, Brace, Jovanovich. (More than 2400 pages of extremely authoritative information on kidney physiology, kidney diseases, and their management.)

Daugirdas, J. T., Ing, T. S. (Eds.) (1988), *Handbook of Dialysis,* Boston: Little, Brown. (A concise, readable, introductory-level text on medical management that is also quite complete.)

Maher, J. (Ed.) (1989), *Replacement of Renal Function by Dialysis,* Dordrecht: Kluwer. (The most exhaustive and authoritative text on dialysis.)

Tuwiner, S. B., Miller, L. P., Brown, W. E. (1962), *Diffusion and Membrane Technology,* New York: Reinhold. (An older but classic work on industrial dialysis applications.)

United States Renal Data System (1990, 1991), *1990 and 1991 Annual Data Reports,* Bethesda: National Institutes of Health. (A statistical gold mine on the epidemiology and economics of dialysis and transplantation in the United States.)

Wankat, P. C. (1990), *Rate-Controlled Separations,* London: Elsevier. (An engineering approach to dialysis and related processes.)

DIAMAGNETISM

LEOPOLDO M. FALICOV, *Department of Physics, University of California, Berkeley, California, U.S.A., and Materials and Chemical Sciences Division, Lawrence Berkeley Laboratory, Berkeley, California, U.S.A.*

INTRODUCTION

The word diamagnetic was introduced—originally as a noun, later as an adjective—by Michael Faraday in 1846. In the *Philosophical Transactions* (Faraday, 1846) one reads, "By a diamagnetic I mean a body through which lines of magnetic force are passing, and which by their action does not assume the usual magnetic state of iron or loadstone." It was constructed from the Greek prefix $\delta\iota\alpha$ (which means by, through, across) and the word *magnetic*—from the Latin word *magneticus*—which described, at that time, the special forces inherent in naturally occurring iron, iron oxides, and other substances.

The operational definition of diamagnetism, quoted in most electromagnetism and materials sciences textbooks for a century and a half, is still valid. A substance shaped in the form of a bar or needle and suspended freely between the poles of magnet is called paramagnetic or magnetic if it orients itself longitudinally with the magnetic lines of force, i.e., pointing towards the poles; it is called diamagnetic if it orients itself equatorially, i.e., perpendicular to the lines of force. Already in 1848 Mrs. Somerville wrote "Substances affected after the manner of bismuth (when suspended between the poles of an electromagnet) are said to be diamagnetic." To this day the semimetal bismuth is taken as the prototype of a diamagnetic substance.

Although the concept of magnetism of materials predated quantum mechanics by some 80 years, magnetism is a purely quantum-mechanical effect. Strictly classical systems in equilibrium cannot display a magnetic moment, even in the presence of a magnetic field. Understanding diamagnetism, therefore, requires a complete quantum-mechanical treatment.

1-56081-063-7/92/$5.00+.50

1. THERMODYNAMIC ARGUMENTS

In the presence of an applied external magnetic field **B**, the free energy $F(T,\mathbf{B})$ of a given system at a given temperature T can be expanded in a Taylor series of the form

$$F(T,\mathbf{B})=F_0(T)-\mathbf{M}_0(T)\cdot\mathbf{B} -(1/2\mu_0)\sum_{ij} B_i\chi_{ij}(T)B_j. \qquad (1)$$

In Eq. (1) the vector $\mathbf{M}_0(T)$ is called the temperature-dependent intrinsic magnetization, and the second-rank quantity $\chi(T)$ is the magnetic susceptibility tensor. The magnetization $\mathbf{M}(T)$ is defined by

$$\mathbf{M}(T)=\mathbf{M}_0(T)+(1/\mu_0)\chi\cdot\mathbf{B}. \qquad (2)$$

Materials for which $\mathbf{M}_0(T)$ is nonzero at a particular T are called *ferromagnets* or *ferrimagnets* at that temperature, irrespective of the properties of $\chi(T)$. Such is the case, for instance, of the metals Fe, Co, Ni, and Gd and the compounds CrO_2, EuS, and Fe_3O_4, *inter alia*, at low enough temperatures.

If, on the other hand, $\mathbf{M}_0(T)=0$, then the magnetic properties are dominated by $\chi(T)$. In particular, for isotropic systems where χ is a diagonal tensor, $\chi_{ij}(T) = \chi(T)\delta_{ij}$, if $\chi(T)$ is positive, then the substance is said to be *paramagnetic*, whereas if $\chi(T)$ is negative, then it is labeled *diamagnetic*. The definitions can be extended to particular directions of anisotropic systems.

Examples of paramagnets are Fe, Co, Ni, and Gd at high temperatures, most rare-earth compounds, and gaseous molecular oxygen. Examples of diamagnets include the noble gases He, Ne, Ar, Kr, and Xe, most molecular gases, liquids, and solids, and the semimetals As, Sb, Bi, and graphite.

It should be pointed out that the free energy of a diamagnet increases in the presence of a magnetic field, and therefore the system evolves so as to avoid high-field regions of space; there is a net repulsive force, caused by the magnetic field, acting on the diamagnet. Conversely the free energy of paramagnet decreases in the presence of a magnetic field, with an attendant attractive force.

2. DIAMAGNETISM OF ATOMS, MOLECULES, AND MOLECULAR SOLIDS

Lenz's law in electromagnetism (see ELECTRODYNAMICS, CLASSICAL) establishes that when the magnetic flux through an electrical circuit is changed, an induced current appears so as to oppose the flux change. In classical systems that current rapidly decays through dissipation, and a new state of equilibrium is achieved with no current flowing through the circuit, and with the new value of the magnetic flux permanently established. In quantized systems with a discrete, nondegenerate energy spectrum, such as a closed-shell atom or molecule, the induced current persists for as long as the applied field is present, and the net magnetic field—the sum of the applied field and the internal one generated by the electronic current in the atom—is always smaller than the applied field.

In addition, closed-shell systems, because of their nondegenerate ground state, possess no intrinsic magnetic moment and are therefore diamagnetic. (It should be remembered that intrinsic magnetic moments are caused by nonvanishing orbital or spin angular momenta which are able to reorient themselves in the presence of a magnetic field; that capability of reorientation requires degeneracy of the ground-state manifold.)

2.1 Hamiltonian of Electrons in a Magnetic Field

A well-established result, valid in both classical and quantum mechanics, states that a system of electrons in a magnetic field **B**, defined by a vector potential **A**, such that

$$\nabla\times\mathbf{A}=\mathbf{B}, \qquad (3)$$

is described by the following Hamiltonian:

$$H=\frac{1}{2m}\sum_i [\mathbf{p}_i+e\mathbf{A}(\mathbf{r}_i)]^2+V-\sum_i e\phi(\mathbf{r}_i) +\frac{e}{m}\sum_i \boldsymbol{\sigma}_i\cdot\nabla\times\mathbf{A}(\mathbf{r}_i). \qquad (4)$$

In Eq. (4) m is the mass of the electron, e is the (positive) charge of the proton, V is the internal electrostatic energy of the electronic system, ϕ is the electrostatic (scalar) potential, and $\mathbf{r}_i$, $\mathbf{p}_i$, and $(e/m)\,\boldsymbol{\sigma}_i$ are the position, the momentum, and the spin magnetic moment of the ith electron, respectively.

For a single *atom* or *ion* in a uniform magnetic field of magnitude B oriented in the z direction, it is convenient to choose the so-called symmetric gauge

$$\mathbf{A}=\tfrac{1}{2}\mathbf{B}\times\mathbf{r}, \tag{5}$$

where the origin of the gauge is located at the nucleus. Under such conditions the Hamiltonian (4) reduces to

$$H=\sum_i \left[\frac{1}{2m}p_i^2+e\phi(\mathbf{r}_i)+\frac{eB}{2m}(L_{zi}+2\sigma_{zi}) + \frac{e^2B^2}{8m}(x_i^2+y_i^2)\right]+V. \tag{6}$$

The terms in the summation correspond, for each electron, to the kinetic energy, the outside potential energy, the coupling of the magnetic moments arising from the orbital (L_{zi}) and the spin (σ_{zi}) magnetic moments of the electrons to the magnetic field, and the weak quadratic term, proportional to B^2.

2.2 Langevin Theory for Closed-Shell Systems

Diamagnetism arises in systems where the contribution from the third term in Eq. (6)—linear in B—vanishes, i.e., for atoms and molecules with a closed-shell structure. The magnetic-field dependence of the energy originates then exclusively from the term quadratic in B. For a system of N *atoms* or *ions* per unit volume, each consisting of n electrons which have an average square radial distance from the nucleus equal to $\langle r^2\rangle$, the summation

$$\sum_i (x_i^2+y_i^2)$$

in Eq. (6) can be replaced by $\frac{2}{3}Nn\langle r^2\rangle$. This substitution results in an excess magnetic energy per unit volume of

$$(e^2B^2/12m)\ nN\langle r^2\rangle,$$

and, consequently, in a diamagnetic susceptibility (Langevin, 1905)

$$\chi=-(\mu_0e^2/6m)\ nN\langle r^2\rangle. \tag{7}$$

If the system (solid, liquid, or gas) consists of N_α atoms or ions of the α type per unit volume, and all these species are closed electronic shells, then the diamagnetic susceptibility of the system is given by

$$\chi=-\frac{\mu_0e^2}{6m}\sum_\alpha n_\alpha N_\alpha\langle r_\alpha^2\rangle, \tag{8}$$

where n_α and $\langle r_\alpha^2\rangle$ are, respectively, the number of electrons on the α atom or ion and the mean value of r^2 for the electrons from their nucleus, for that particular species.

It should be noted that the diamagnetism of closed-shell systems is essentially a very weak perturbation of the nondegenerate ground state caused by the presence of the applied magnetic field. It is therefore a purely quantum-mechanical phenomenon—since it requires a discrete, nondegenerate lowest energy state in the spectrum, a feature that can only be achieved quantum mechanically—and an effect that depends extremely weakly on the temperature. This virtual temperature independence is, once again, caused by the same features of the spectrum: Only at inaccessible temperatures are the excited states appreciably populated, the only way in which the various $\langle r_\alpha^2\rangle$ in Eq. (8) can be affected by the temperature.

3. ORBITAL DIAMAGNETISM OF METALLIC ELECTRONS

Metallic electrons are, by their very nature, itinerant. Their wave functions extend essentially over the entire crystal and their energy spectrum is in the form of bands separated, in some cases, by forbidden energy gaps. The single-electron wave functions (see ELECTRONIC STRUCTURE OF SOLIDS) are in the form of Bloch states, i.e., a product of

1. a plane wave of vector **k**,
2. a periodic function with the full periodicity of the crystal lattice, and
3. a spinor.

Each Bloch function is thus labeled by a **k** vector, a band index ν (a positive integer), and a spin index σ (↑ or ↓). The vector **k** is restricted to a finite region in reciprocal space (**k** space), a polyhedron known as the Brillouin zone. The energy ϵ of each state is given by the band-structure function $\epsilon_\nu(\mathbf{k},\sigma)$, which is continuous in **k** space. These features arise from the solution of the Schrödinger equation in a periodic potential, and from the properties of periodic functions and Fourier series in three dimensions.

The occupation of the one-electron energy bands follows Fermi–Dirac statistics (see STATISTICAL MECHANICS, CLASSICAL; STATISTICAL MECHANICS, QUANTUM), with no more than one electron occupying each state. At very low temperatures, $T \to 0$, the Fermi–Dirac distribution is a step function, with all electronic states with energy $\epsilon < \epsilon_F$ being occupied by one electron, and all states with $\epsilon > \epsilon_F$ being empty. The energy ϵ_F is called the Fermi level and, in a metal, falls in the middle of one or more bands, i.e., not in the middle of a forbidden energy gap. The equation

$$\epsilon_\nu(\mathbf{k},\sigma) = \epsilon_F \quad (9)$$

defines, for a metal, a surface in **k** space called the Fermi surface (see ELECTRONIC STRUCTURE OF SOLIDS). This surface consists of one or more sheets (one or more for each band ν which crosses the Fermi level), is defined within the Brillouin zone, and has all the symmetry and periodic properties of the Brillouin zone and **k** space.

Metallic electrons do not form closed electronic shells, and therefore do not fall in the category discussed in the previous section. The fact that the Fermi level falls in the middle of a continuous energy band guarantees that the ground state of the many-electron system is not part of the discrete spectrum, i.e., the ground state is the lowest of a set of continuum states. The response of metallic electrons to an applied magnetic field is, in general, very complex, and contains several contributions of various characteristics which must be treated separately:

1. If the bands of different spin are not identical and are occupied differently then the system develops a spontaneous local magnetization which, depending on the way it distributes itself, results in magnetic ordering (see MAGNETIC ORDERING IN SOLIDS); these ordered states include ferromagnets, ferrimagnets, antiferromagnets, spiral spin arrangements, etc. (see FERROMAGNETISM AND FERRIMAGNETISM).
2. If the bands of different spin are identical—and identically occupied—in the absence of an applied magnetic field, then the electron spin and its associated magnetic moment are responsible for a rearrangement of the band occupation when a magnetic field is applied; the imbalance in the spin distribution results in a temperature-independent positive susceptibility known as Pauli paramagnetism (see PARAMAGNETISM). For simple metals of columns I, II, III, and IV of the periodic table that contain only fully occupied and/or fully empty d and f atomic shells, a commonly used approximation is the so-called *free-electron* model. This model replaces the periodic potential of the lattice by a constant, i.e., the periodic oscillations of the potential are completely smoothed out. As a consequence the periodic factor in the Bloch state reduces to a plane wave of vector $\mathbf{G}_\nu$; the set of $\mathbf{G}_\nu$'s is infinite but discrete and constitutes the so-called the reciprocal lattice. The energy spectrum for the Bloch states in the free-electron model is given by

$$\epsilon_\nu(\mathbf{k},\sigma) = \frac{\hbar^2}{2m} (\mathbf{k}+\mathbf{G}_\nu)^2. \quad (10)$$

In this model the Pauli contribution to the (paramagnetic) susceptibility per unit volume is equal to

$$\chi_{\text{Pauli}} = \mu_0 \frac{\mu^2 m}{\pi^2 \hbar^2} (3\pi^2 Nn)^{1/3}. \quad (11)$$

In Eq. (11) μ is the Bohr magneton, N is the number of metal atoms per unit volume, and n is the number of metallic (conduction) electrons per atom.
3. There is, in addition, an orbital diamagnetic contribution to the susceptibility, known as Landau diamagnetism, discussed below. For the free-electron model this contribution is

$$\chi_{\text{Landau}} = -\mu_0 \frac{\mu^2 m}{3\pi^2 \hbar^2} (3\pi^2 Nn)^{1/3}, \quad (12)$$

i.e., $\chi_{\text{Landau}} = -\frac{1}{3}\chi_{\text{Pauli}}$, which yields a total *paramagnetic* susceptibility

$$\chi_{\text{free electrons}} = \frac{2}{3}\chi_{\text{Pauli}} \quad (13)$$

for the free-electron model. The free-electron model, however, has only a very restricted range of applicability. For some substances—the semimetals As, Sb, Bi, and graphite being the most notable examples—real band-structure effects, caused by the periodic potential, produce a large enhancement of the Landau diamagnetism, by factors of up to several hundred. The diamagnetic contribution thus overwhelms the paramagnetic Pauli term; the net result is a negative χ and a diamagnetic solid.

4. The orbital contribution of metallic electrons also includes a diamagnetic term which oscillates with the magnetic field, and is periodic in $1/B$. This oscillatory contribution is called the de Haas–van Alphen effect. Its periods and amplitudes give very precise information on the geometrical and differential properties of the Fermi surface.

3.1 Quantization of Itinerant Electrons in a Magnetic Field

To study diamagnetic effects in metals the effect of the electron-electron interaction can be neglected, and the electrons may be treated as independent particles. Independent electrons moving in a crystal lattice in a magnetic field are described by Eq. (4), with the electron–electron interaction made to vanish, $V=0$, and with the electrostatic term $[-e\phi(\mathbf{r}_i)]$ representing the periodic potential acting on electron i and caused by the lattice of nuclei and ionic cores. The solution of the Schrödinger equation under these conditions is still a very complex and difficult process. The difficulty arises from conflicting period icities imposed separately by the lattice and the magnetic field. The problem is simple only in the free-electron case, $\phi=0$. The solution was obtained by Landau in 1930, the early days of quantum mechanics. Landau chose the field $\mathbf{B}$ along the z axis and an asymmetric gauge:

$$\mathbf{A}=[0,Bx,0]. \tag{14}$$

The resulting spectrum is

$$\epsilon(\nu,k_z)=(\nu+\tfrac{1}{2})\hbar\omega_c+[\hbar^2/2m]k_z^2, \tag{15}$$

where ν is a nonnegative integer, k_z is the wave vector for motion along the direction of the magnetic field, and ω_c, the so-called cyclotron frequency, is given by

$$\omega_c=eB/m. \tag{16}$$

This spectrum for three-dimensional electrons, which depends only on two quantum numbers (a discrete ν and a continuous k_z), is highly degenerate—each energy is shared by a very large number of states. There is a "hidden" quantum number, k_y, necessary to define the wave function completely. The energy ϵ does not depend on k_y, and all wave functions with the same ν and k_z—no matter what the value of k_y—have the same energy. For each fixed value of the set (ν,k_z), there are

$$L_xL_yeB/2\pi\hbar \tag{17}$$

different k_y states, where L_xL_y is the macroscopic area of the crystal—assumed to be prismatic—perpendicular to the direction of the magnetic field.

Each integer ν defines for the perpendicular motion in the (x,y) plane and for fixed k_z a discrete energy level, called a Landau level.

From this spectrum, Eqs. (15)–(17), its degeneracy, and its dependence on the magnetic field intensity B, all thermodynamic quantities can be calculated for the free-electron model. Some of these quantities are discussed explicitly in the next two subsections.

Extension of Eqs. (15)–(17) to the case of a nonvanishing periodic potential was proposed by Onsager (1952), who based his method on a semiclassical approach. This approach consists of determining electron trajectories and applying to them the Bohr–Sommerfeld "old" quantum-mechanical quantization rules, i.e.,

$$\int_{\text{closed orbit}} \mathbf{p}\cdot d\mathbf{r}-(\nu+\gamma)2\pi\hbar, \tag{18}$$

where ν is a nonnegative integer, γ is a (positive) phase correction which takes the value $\frac{1}{2}$ in the free-electron case, the position $\mathbf{r}$ follows the closed semiclassical trajectories of the electron in the (x,y) plane, $\mathbf{p}$ is the *canonical* momentum

$$\mathbf{p}=\hbar\mathbf{k}-e\mathbf{A}, \tag{19}$$

$\hbar\mathbf{k}$ is interpreted as the *kinetic* momentum, and $-e\mathbf{A}$ is the contribution of the field to the canonical momentum.

The semiclassical equations of motion of electrons in a periodic potential and a magnetic field are given by

$$k_{\text{parallel}}=\text{const}, \tag{20}$$

where k_{parallel} is the component of $\mathbf{k}$ parallel to $\mathbf{B}$,

$$\epsilon_\nu(\mathbf{k},\sigma)=\text{const}, \tag{21}$$

and

$$\hbar\dot{\mathbf{k}} = -e\dot{\mathbf{r}} \times \mathbf{B}, \tag{22}$$

which integrated and with proper choice of the origin yields

$$\hbar\mathbf{k} = -e\mathbf{r} \times \mathbf{B} \tag{23}$$

for the motion perpendicular to **B**.

Insertion of Eqs. (23) and (19) into Eq. (18), use of Stokes's theorem, and application of Eq. (3) yields Onsager's famous result for the quantization of itinerant electron orbits in a uniform magnetic field (Onsager, 1952),

$$\int_{\text{closed orbit}} \mathbf{p} \cdot d\mathbf{r} = e\Phi = (\nu+\gamma)2\pi\hbar, \tag{24}$$

for the motion of the electrons perpendicular to **B**. In Eq. (24) Φ is the magnetic flux encircled by the electron in its trajectory in the plane perpendicular to **B**. In other words the only trajectories that are (quantum-mechanically) allowed are those that, in the plane perpendicular to **B**, encircle a magnetic flux Φ which takes the discrete values

$$\Phi_\nu = (\nu+\gamma)\,\frac{2\pi\hbar}{e}. \tag{25}$$

Because the magnetic flux is quantized, and the value of the magnetic field B is externally determined, the areas of the allowed orbits in real space, A_ν, must take well-defined, discrete values. In addition, Eqs. (20), (21), and (23) state that the orbits in real space in the plane perpendicular to **B** are proportional to—and rotated by an angle of $\pi/2$ from—the corresponding trajectories in **k** space. The cross sections (20) at a constant k_{parallel} of the surface of constant energy in **k** space, Eq. (21), are therefore proportional to the areas A_ν of the allowed orbits in real space. The constant of proportionality between areas in **k** space and real space, as given by Eq. (23), is proportional to B^2, i.e., equal to $(eB/\hbar)^2$. Therefore, the allowed cross-sectional areas S_ν of the surfaces of constant energy in **k** space are given by

$$S_\nu = (\nu+\gamma)\,\frac{2\pi e}{\hbar}\,B. \tag{26}$$

Because the cross-sectional areas in **k** space S_ν perpendicular to **B** are quantized—take only the discrete values given by Eq. (26)—the energies for the motion perpendicular to **B** are also quantized. Two successive levels, which differ by $\Delta\nu = \pm 1$, are separated by an energy $\Delta\epsilon$, which defines, for that particular value of the energy and that particular value of k_z, a cyclotron frequency ω_c^* and an effective cyclotron mass m_c^*:

$$\Delta\epsilon = \hbar\omega_c^* = \hbar eB/m_c^*. \tag{27}$$

When the Onsager rules (26) are applied to the free-electron model, the results (15)–(17) are obtained if $\gamma = \frac{1}{2}$. For this particular case $\omega_c^* = \omega_c$, as given by Eq. (16), and $m_c^* = m$, i.e., the cyclotron mass is equal to the electron mass.

It should be mentioned that transition between two adjacent Landau levels, with conservation of k_{parallel}, can be induced, for electrons near the Fermi level, by microwave or far-infrared radiation; when the frequency ω of that radiation is in the vicinity of ω_c, a resonance—the so-called cyclotron resonance or diamagnetic resonance—is observed.

3.2 Landau Diamagnetism

For noninteracting electrons, the total energy of the system—the metal—is obtained by adding the individual energies of each electron. Since electrons satisfy Fermi–Dirac statistics (see STATISTICAL MECHANICS, QUANTUM), in the limit $T \to 0$ that total energy is obtained by summing one-electron energies for all states with energy below the Fermi level ϵ_F.

The term $\frac{1}{2}$ in the factor $\nu+\frac{1}{2}$ in Eq. (15)—or equivalently the term γ in the factor $\nu+\gamma$ in Eqs. (25) and (26)—, which is ordinarily called the zero-point energy, guarantees that although upon the application of a magnetic field some states increase their energy and others decrease it, the energy sum over states of contiguous occupation up to the Fermi level always shows an *increase* by an amount proportional to B^2, thus yielding a diamagnetic contribution to the susceptibility. This is the Landau diamagnetic contribution to the susceptibility of metallic electrons.

The summation is, in general, difficult to perform. It can be done numerically—and laboriously—for some specific cases, where the number of carriers is small, i.e., for the semimetals. For the free-electron model the summation can be accomplished in closed form. The result is that quoted in Eq. (12).

3.3 The de Haas–van Alphen Effect

The de Haas–van Alphen effect is an oscillatory variation in the magnetic susceptibility of metals as a function of the magnitude of the *static* magnetic field intensity B. It is an effect observed in strong magnetic fields, for pure specimens, and at low temperatures. These three conditions arise from the necessity to have the Landau levels widely separated, not blurred by electron collision effects with impurities, and populated with a sharp discontinuity at the Fermi level, i.e., a discontinuity in occupation of essentially 1 between the last "occupied" and the first "empty" levels. These conditions may be summarized by two inequalities:

$$k_B T \ll \hbar\omega_c^*, \quad \omega_c^*\tau \gg 1, \tag{28}$$

where ω_c^* is the cyclotron frequency defined by Eq. (27), k_B is Boltzmann's constant, and τ, the relaxation time, is the median time between electron collisions (see THERMODYNAMICS, EQUILIBRIUM; TRANSPORT PHENOMENA).

The *loci* of allowed quantum states—in **k** space—for itinerant electrons in a periodic lattice and a magnetic field are "tubes," i.e., cylindrical shells of well-determined, discrete cross sections, given by Eq. (26). The cylinders, not necessarily of circular cross section, are inside each other in order of either increasing or decreasing energy. For the free-electron model the tubes are concentric circular cylinders of axis parallel to **B**. These highly degenerate tubes, when the conditions (28) are satisfied, are populated by electrons if the energy of their states is lower than ϵ_F, i.e., states are occupied for that part of those tubes that lies inside the Fermi surface (9). Only *some* parts of *some* tubes are populated by electrons. The rest—the states for which $\epsilon > \epsilon_F$—are allowed but empty states. This distribution is illustrated in Fig. 1.

As the magnitude B of the field increases the tubes get more widely separated and contain more states—as indicated by Eqs. (26) and (17). The tubes thus get repopulated at each increasing value of the field. For each orientation (θ,ϕ) of the magnetic field **B** there are discrete values of the field intensity B_ν for which a given tube gets completely depopulated, i.e., it "leaves" the Fermi surface and lies completely outside. These values of the field B_ν satisfy the condition

$$S_\nu = (\nu+\gamma)\,\frac{2\pi e}{\hbar}\,B_\nu = S_{\text{Fermi surface}}(\theta,\phi), \tag{29}$$

where $S_{\text{Fermi surface}}(\theta,\phi)$ is the *extremal* cross-sectional area of the Fermi surface in the direction perpendicular to **B**. This phenomenon of allowed-energy tubes "peeling off" the Fermi surface results in a contribution to the free energy, the magnetization, and the susceptibility of the sample which is periodic in $1/B$, with period

$$\Delta\left(\frac{1}{B}\right) = \frac{2\pi e}{\hbar S_{\text{Fermi surface}}(\theta,\phi)}, \tag{30}$$

i.e., a $1/B$ period inversely proportional to the extremal cross-sectional area of the Fermi surface in the direction (θ,ϕ) perpendicular to **B**.

If the Fermi surface is complex and exhibits several sheets and/or several extremal cross-sectional areas for a given sheet, the magnetization becomes a very complicated oscillatory function. The analysis of the various periods in $1/B$ permits the determination of the corresponding cross-sectional areas.

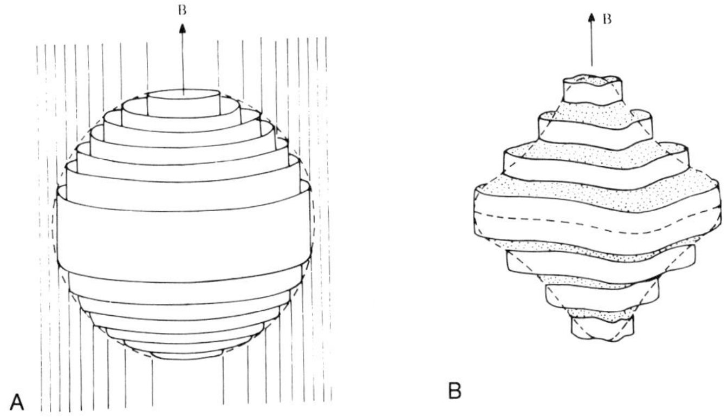

FIG. 1. Schematic representation of the quantization of metallic electrons in a magnetic field. The depicted tubes are the *loci* of allowed states in **k** space. Only the occupied portions of the tubes are drawn. (a) The free-electron model. (b) An arbitrary potential. [From Falicov (1973)].

Studies of the variation of the oscillations with orientation (θ,ϕ) of the magnetic field lead to accurate determination of the geometry and topology of the Fermi surface of the metal under consideration.

The amplitude of the oscillations and its variation with temperature yield information about the separation between the levels—the cyclotron mass m_c^* at the Fermi level for a particular orientation (θ,ϕ), an intrinsic property of the metal—as well as the relaxation time τ, a specific property of the sample under study. The de Haas–van Alphen effect provides the most useful tool to investigate the electronic properties of metals. It yields extremely accurate information on the structural and dynamical properties of itinerant-electron states in solids.

4. DIAMAGNETISM IN SUPERCONDUCTORS

Superconductors are perfect diamagnets for a given range of intensities of applied magnetic fields. They are also perfect conductors, i.e., they have zero resistivity. Although the zero resistivity is a necessary condition for perfect diamagnetism—since it is required for the induced currents in Lenz's law not to decay—by itself it only implies that the magnetic field in the bulk of a superconductor cannot be changed, whatever its value. The expulsion of the field from the bulk of a superconductor is an additional property, known as the Meissner effect (Meissner and Ochsenfeld, 1933). It is illustrated in Fig. 2.

Superconductivity (see SUPERCONDUCTIVITY, LOW TEMPERATURE) is a phase, a state of matter (in the sense that ice and steam are phases of water and diamond and graphite are phases of pure carbon) observed *only* in some solids, mostly metals. The superconducting state has several characteristic properties:

1. When it exists for a given substance, it exists only at temperatures below a so-called transition temperature T_c, and in general down to the absolute zero of the temperature scale 0 K. Historically, substances with the highest known values of T_c have been first pure metals (Hg with $T_c = 4.2$ K, Pb with $T_c = 7.2$ K, and Nb with $T_c = 9.5$ K), later transition-metal compounds (NbN with $T_c = 16.0$ K, V_3Si with T_c=17.1 K, Nb_3Sn with $T_c = 18.05$ K, and Nb_3Ge with $T_c = 23.2$ K), and, since 1986, the so-called ***high-T_c*** materials, all four- or five-element compounds which contain copper and oxygen (La–Ba–Cu–O with T_c of approximately 30 K, rare-earth–Ba–Cu–O with T_c of approximately 92 K, Bi–Sr–Ca–Cu–O with T_c of approximately 105 K, and—the known record at this time—Tl–Ca–Ba–Cu–O with $T_c = 125$ K) (see SUPERCONDUCTIVITY, LOW-TEMPERATURE; HIGH-TEMPERATURE SUPERCONDUCTIVITY).
2. It exhibits zero dc resistivity, i.e., infinite conductivity for zero-frequency measurements (an effect discovered in mercury by Kamerlingh Onnes in 1911),

$$\rho(\omega=0;T<T_c)=0. \tag{31}$$

3. It exhibits, for weak magnetic fields, perfect diamagnetism, i.e., its magnetic susceptibility is given by

$$\chi=-1 \tag{32}$$

which means that magnetic flux lines are completely expelled from the superconductor and that there is a force pushing superconductors away from magnetic fields. This is the Meissner effect discovered by Meissner and Ochsenfeld (1933).
4. In addition to the effect of high tempera-

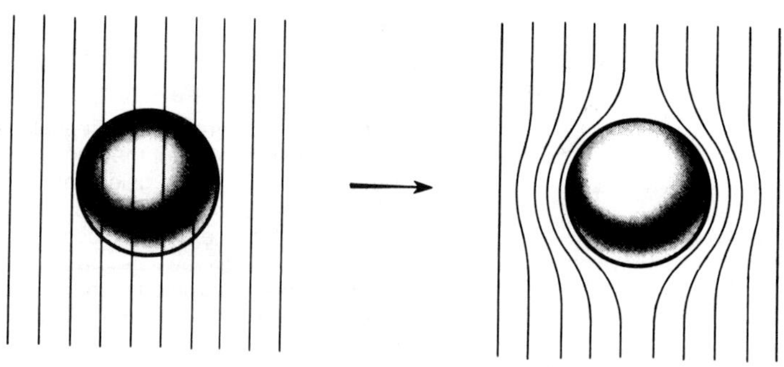

FIG. 2. Schematic representation of the Meissner effect. When a superconducting sphere is cooled in a constant applied magnetic field, below the transition temperature T_c the lines of induction **B** are ejected from the sphere. [From Kittel (1986)].

tures, superconductivity can be destroyed (with a return to the normal metallic state) by either a large enough electric current $I > I_c$ or a large enough applied magnetic field $B > H_c$. The quantities I_c and H_c are called the critical current and the critical magnetic field, respectively.

5. Superconductivity is a macroscopic quantum phenomenon with amplitudes and phases associated with an order parameter ψ. Therefore, interference and diffraction effects can be achieved, in particular the Josephson effect (Josephson, 1962). These effects can be fruitfully employed in processing, storing, and retrieving information, i.e., in computer technology (see SUPERCONDUCTING DEVICES).

4.1 The Meissner Effect

Meissner and Ochsenfeld (1933) discovered that, when a superconductor is cooled below T_c in the presence of a magnetic field, at the transition the lines of induction **B** are completely expelled from the bulk of the sample. In other words, inside the superconductor Eq. (32) is obtained or, equivalently,

$$\mathbf{B} = 0. \tag{33}$$

This result holds up to a given value of the applied field B. In the so-called type-I superconductors there is only one critical field, H_c. For $B > H_c$ superconductivity is destroyed, the flux fully penetrates the sample, and the metal returns to its normal state.

In type-II superconductors there is no field penetration for $B < H_{c1}$ and the superconductor acts as a perfect diamagnet. For intermediate field strengths $H_{c1} < B < H_{c2}$, the magnetic flux lines partially penetrate the superconductor but do not destroy the superconducting state. Since, in general, H_{c2} is very large, this feature of type-II superconductors makes them useful as materials for magnets (see SUPERCONDUCTING MAGNETS). Finally for $B > H_{c2}$ superconductivity is again destroyed with a return of the substance to its normal state. This behavior is illustrated in Fig. 3.

4.2 The London Theory

The London theory (London and London, 1935) is completely classical. It incorporates into Maxwell's equations, phenomenologically, the electrodynamic characteristics of superconductors. The zero-resistivity property (31) requires that, since

$$\rho \mathbf{j} = \mathbf{E},$$

the electric field **E** should vanish everywhere in a superconductor. The Meissner effect, Eqs. (32) and (33), implies vanishing magnetic field **B** in the bulk. These two conditions led London, in 1950, to *postulate* that in a superconductor the current density **j**, instead of being proportional to **E**, as is the case in other media, is proportional to the vector potential **A**,

$$\mathbf{j} = -[1/\mu_0 \lambda_L^2]\mathbf{A}, \tag{34}$$

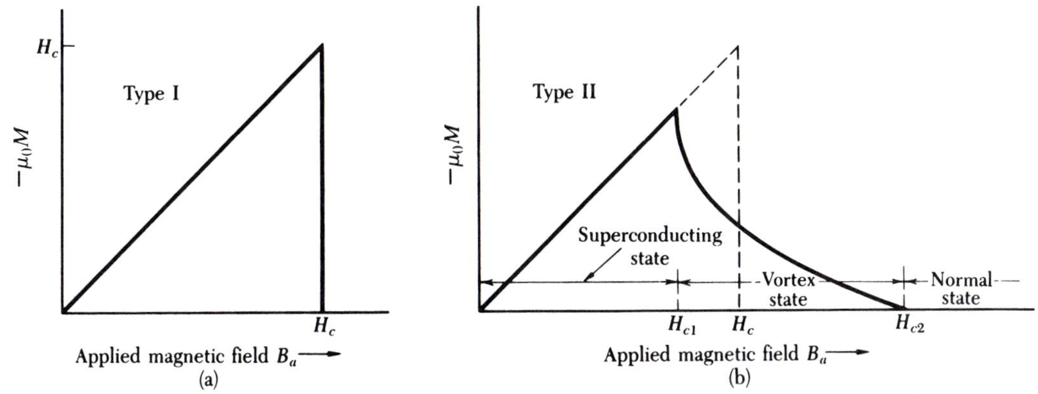

FIG. 3. (a) Magnetization vs applied magnetic field for a type I superconductor exhibiting the Meissner effect (perfect diamagnetism) below H_c. (b) Magnetization vs applied magnetic field for a type II superconductor. [From Kittel (1986)].

where the constant of proportionality defines the length λ_L, the *London penetration depth.* The London equation (34), by taking the curl on both sides, may also be expressed as

$$\nabla\times\mathbf{j}=[1/\mu_0\lambda_L^2]\mathbf{B}. \tag{35}$$

Since the vector potential **A** is *not*, in general, a physically measurable quantity, and the current density **j** *always is*, Eq. (34) in effect selects a particular gauge, the so-called London gauge. Equation (34) must be consistent with Maxwell's equations. Therefore, it implies that in the absence of charges $\nabla\cdot\mathbf{A} = 0$, and that the normal components of **A** are zero on any external surface through which no external current is fed. Additional conditions on the gauge, compatible with the physical boundary conditions, should be imposed on problems with multiply connected geometries.

It is easily seen that the London equation implies the Meissner effect, and that λ_L is the characteristic length describing the penetration and exponential decay of the magnetic field **B** into a superconductor. From Maxwell's equation,

$$\nabla\times\mathbf{B}=\mu_0\mathbf{j}, \tag{36}$$

when the curl is taken on both sides and use is made of Eq. (35) and of Maxwell's other equation, $\nabla\times\mathbf{B} = 0$, one obtains

$$\nabla^2\mathbf{B}=\mathbf{B}/\lambda_L^2. \tag{37}$$

Equation (37) does not allow a constant-**B** solution, unless $\mathbf{B}=0$. In addition Eq. (36) guarantees that $\mathbf{j}=0$ wherever **B** vanishes. In a superconductor extending over the half-space $x \geqslant 0$, if the field at the boundary $x=0$ is parallel to the surface and of magnitude B_0, a solution of Eq. (37) yields

$$B(x)=B_0\exp(-x/\lambda_L), \tag{38}$$

i.e., the field decays exponentially into the superconductor, a perfect diamagnet, with characteristic length given by the *London penetration depth.*

4.3 The Ginzburg–Landau Theory

4.3.1 General Formulation. In 1950 Ginzburg and Landau (GL) proposed a macroscopic, phenomenological theory of superconductivity, which was independent of the microscopic aspects of the phenomenon. The theory is quantum-mechanical, in the sense that it includes coherent, macroscopic quantum effects. It is a pioneering theory which, independent of the mechanisms responsible for superconductivity, is still valid today. It contains such diverse phenomena as magnetic-field penetration depths, coherence lengths, magnetic-field flux quantization, magnetic-field dependence of the superconducting order parameter, and the Josephson effect. It can be applied to all superconductors, as well as to superfluid ^{3}He, and has become the prototype theory to study a whole class of phenomena related to spatial dependence in second-order phase transitions.

The GL theory introduces a complex order parameter ψ which is allowed to vary in space. Originally GL interpreted ψ as an amplitude, and $|\psi|^2$ as the density of the "superconducting" electrons (they envisioned a superconductor as two interpenetrating electron fluids: the nondissipative, nonresistive superconducting electron fluid, and the dissipative, resistive "normal" electron fluid). In 1959, however, Gor'kov proved that for temperatures below and close to T_c, equations identical to those of GL could be obtained from the Bardeen–Cooper–Schrieffer (BCS) microscopic theory of superconductivity (1957), and that the GL parameters ψ could be interpreted (except for a trivial constant of proportionality) as what in the BCS theory is known as the energy-gap parameter.

The starting point of the GL theory is the introduction of a magnetic Helmholtz free energy F_{SH} for the superconductor, derived from plausibility arguments:

$$F_{SH}=\int_{\text{superconductor}} d^3r\left[F_{N0}+\Delta F(|\psi|^2)+\left(\frac{1}{2m}\right)\left|-i\hbar\nabla\psi-e^*\mathbf{A}\psi\right|^2+\left(\frac{1}{2\mu_0}\right)B^2(\mathbf{r})\right]. \tag{39}$$

Here F_{N0} is the free-energy density of the normal state in the absence of a magnetic field; ΔF is the difference of free-energy densities between the superconducting and the normal states (also in the absence of a field) and is a function of $|\psi|^2$. The third term is the gauge-invariant "superconducting kinetic energy," and the last term is the magnetic-field energy in the superconductor. The vector potential is $\mathbf{A}$, $\mathbf{B}$ is the magnetic field, and e^* is an effective charge, known now to be $-2e$, twice the charge of the electron (the charge of a "Cooper pair" in the BCS theory). All terms in Eq. (39) are functions of the position $\mathbf{r}$, and change with the magnitude and direction of the magnetic field.

Because the proper variables of the magnetic Helmholtz free energy are the temperature T and the magnetization $\mathbf{M}$, where

$$\mathbf{M}=(1/\mu_0)\int_{\text{all space}} d^3r[\mathbf{B}(\mathbf{r})-\mathbf{B}_a],$$

$$\mathbf{B}_a=\text{applied magnetic field,} \tag{40}$$

F_{SH} is not continuous at the critical fields. The function which is continuous at H_{c1} and H_{c2}, and whose proper variables are T and $\mathbf{B}_a$, is the Gibbs free energy G_{SH}, given by

$$G_{SH}=F_{SH}-\mathbf{M}\cdot\mathbf{B}_a. \tag{41}$$

Substitution of Eqs. (39) and (40) into Eq. (41) yields

$$G_{SH}=G_{NH}+\int_{\text{superconductor}} d^3r[\Delta F(|\psi|^2) + (1/2m)|-i\hbar\nabla\psi-e^*\mathbf{A}\psi|^2] + \frac{1}{2\mu_0}\int\int_{\text{all space}} d^3r[\mathbf{B}(\mathbf{r})-\mathbf{B}_a]^2, \tag{42}$$

where

$$G_{NH}=\int_{\text{superconductor}} d^3r\left[F_{N0}+\frac{B_u^2}{2\mu_0}\right].$$

It should be noted that the last term in Eq. (42) is to be integrated over *the whole space* (both in the superconductor and outside). Minimization of G with respect to the four functions ψ and $\mathbf{A}$ (or equivalently ψ and $\mathbf{B}$) yields the GL equations:

$$\nabla^2\mathbf{A}=\frac{ie^*\hbar}{m}(\psi^*\nabla\psi-\psi\nabla\psi^*)+\frac{\mu_0e^{*2}}{m}|\psi|^2\mathbf{A}, \tag{43}$$

$$\nabla\times\mathbf{A}=\mathbf{B}_a \text{ on surface,} \tag{44}$$

$$\frac{\partial\Delta F}{\partial\psi^*}+\frac{1}{2m}(-i\hbar\nabla-e^*\mathbf{A})^2\psi=0, \tag{45}$$

$$(i\hbar\nabla\psi+e^*\mathbf{A}\psi)_{\text{perpendicular}}=0, \tag{46}$$

where the London gauge

$$\nabla\cdot\mathbf{A}=0$$

has been chosen.

4.3.2 Penetration Depth and Coherence Length. In singly connected samples with no penetration of the magnetic flux into the bulk superconductor, the phase of ψ can be chosen so that ψ is real throughout the sample. In particular for a one-dimensional, singly connected problem, with quantities varying along the x axis, and with magnetic field and vector potential given by

$$\mathbf{B}=[0,0,B(x)],$$

$$\mathbf{A}=[0,A(x),0],$$

Eqs. (43)-(46) become

$$\frac{d^2A}{dx^2}=(\mu_0e^{*2}/m)\psi^2A, \tag{47}$$

$$\frac{dA}{dx}=H_0 \text{ on surface,} \tag{48}$$

$$\frac{\partial\Delta F}{\partial\psi}+\frac{e^{*2}}{m}A^2\psi=\frac{\hbar^2}{m}\frac{d^2\psi}{dx^2}, \tag{49}$$

$$\frac{d\psi}{dx}=0 \text{ on surface.} \tag{50}$$

For the free-energy difference ΔF, the original GL derivation used a power-series expansion in $|\psi|^2$, and neglected all terms higher than the second. That expansion is still commonly used, and is known to be valid for superconductors at temperatures close to T_c:

$$\Delta F=\frac{H_{cb}^2}{2\mu_0}\left[-2\left(\frac{\psi}{\psi_T}\right)^2+\left(\frac{\psi}{\psi_T}\right)^4\right], \tag{51}$$

where H_{cb} is the thermodynamic bulk critical field, and ψ_T is the equilibrium value of ψ in the bulk, at temperature T, in the absence of a magnetic field.

The problem of the superconducting half-space discussed above—in Eq. (38)—in connection with the London equation can be solved in the GL theory. Integration of Eqs. (47)-(51), under the assumption of small changes in ψ near the surface, yields for $x>0$

$$\frac{\psi(x,B_a)}{\psi_T}=1-\frac{\kappa_0}{(2-\kappa_0^2)\sqrt{8}}\left(\frac{B_a}{H_{cb}}\right)^2[e^{-\sqrt{2}\kappa_0x/\lambda_L} - \tfrac{1}{2}\kappa_0e^{-2x/\lambda_L}] \tag{52}$$

and

$$B(x) \approx B_0 \exp(-x/\lambda_L). \quad (53)$$

Equation (53) reproduces, approximately, Eq. (38). In the GL equations λ_L, the London penetration depth, is given by

$$\lambda_L^2 = m/\mu_0 e^{*2} \psi_T^2. \quad (54)$$

In Eq. (52) κ_0 is a dimensionless constant:

$$\kappa_0 = (\sqrt{2} e^*/\hbar) \lambda_L^2 H_{cb}. \quad (55)$$

Two remarks are necessary at this point. First, there are two length scales in the problem: the decay length for magnetic fields, λ_L, and the decay length $\lambda_L \sqrt{2} \kappa_0$ for the order parameter ψ, given by the first exponent in Eq. (52). Second, Gor'kov (1959) has shown that

$$\kappa_0 \approx 0.96 \lambda_L \xi_0^{-1}, \quad (56)$$

where ξ_0 is the coherence length introduced by Pippard (1953) in connection with the general electromagnetic properties of superconductors (see SUPERCONDUCTIVITY, LOW TEMPERATURE). Values of κ_0 are small (<0.707) for the soft, type-I superconductors (0.01 for A1; 0.3 for Pb), whereas it takes large values (>0.707) for the hard, type-II superconductors [~ 8 for V; extremely large for the new, high-T_c materials (see HIGH-TEMPERATURE SUPERCONDUCTIVITY)].

A type-I superconductor excludes a magnetic field from its bulk completely. If the magnetic field B is increased there is a value H_c for which the superconductivity is suddenly destroyed, the system returns to the normal state, and the magnetic field penetrates the specimen completely. A type-II superconductor excludes the field completely up to a value H_{c1}. Above H_{c1} the field is partially excluded, although the specimen remains superconducting and exhibits zero resistivity. At a higher field, H_{c2}, the flux penetrates completely, superconductivity is destroyed, and the specimen returns to its normal state.

4.3.3 Flux Quantization. In many applications (thin specimens, weak magnetic fields, etc.), the order parameter ψ can be considered to have a constant magnitude $n^{1/2}$, although its phase $\theta(\mathbf{r})$ can vary appreciably in space,

$$\psi = n^{1/2} e^{i\theta(\mathbf{r})}. \quad (57)$$

From standard quantum-mechanical arguments the electrical supercurrent is given in this case by the usual formula

$$\mathbf{j} = \frac{1}{2m} [\psi^* (-i\hbar \nabla - e^* \mathbf{A}) \psi + \psi (i\hbar \nabla - e^* \mathbf{A}) \psi^*] = \frac{ne^*}{m} (\hbar \nabla \theta - e^* \mathbf{A}). \quad (58)$$

Deep inside any superconductor the magnetic field $\mathbf{B}$ and the electric current $\mathbf{j}$ are zero, and therefore from Eq. (58) one obtains

$$e^* \mathbf{A} = \hbar \nabla \theta. \quad (59)$$

In a multiply connected sample one can find a closed path C which encircles a nonsuperconducting region where there may be a magnetic field. Line integration of Eq. (59) over that path, use of Stokes's theorem, and knowledge that ψ must be single valued yields

$$\int_{\text{closed } C} \mathbf{A} \cdot d\mathbf{s} = \int_{\text{area } C} \nabla \times \mathbf{A} \cdot d\boldsymbol{\sigma} = \int_{\text{area } C} \mathbf{B} \cdot d\boldsymbol{\sigma} = \Phi = \frac{\hbar}{e^*} \int_{\text{closed } C} \nabla \theta \cdot d\mathbf{s} = \frac{\hbar}{e^*} \times 2\pi\nu, \quad (60)$$

where Φ is the magnetic-field flux, and ν is an arbitrary integer. In other words Eq. (60), taking $|e^*| = 2e$, can be written

$$\Phi = \nu \Phi_0 = \nu \cdot 2.0678 \times 10^{-15} \text{ T m}^2, \quad (61)$$

i.e., if a closed path without currents can be established deep inside a multiply connected superconductor, then the magnetic-field flux encircled by that path is quantized in units of Φ_0. The unit of flux, Φ_0, is called a *fluxoid*.

4.3.4 Phase-Current Relationship; The Josephson Effect. From the GL equations it can be easily seen that the order parameter ψ has an indeterminate *arbitrary, constant* phase. In a given superconductor (called 1) its phase θ_1 is completely arbitrary. If, however, there is nearby a second superconductor (called 2), which is *weakly* connected to the first one, although both phases, θ_1 and θ_2, are indeterminate by the *same* additive constant, the *phase difference* between the two,

$$\delta=\theta_2-\theta_1,$$

is an observable meaningful quantity. As can be seen from Eq. (58), a variation of θ over space is responsible for the existence of a supercurrent. Similarly a phase difference between two weakly coupled, spatially close superconductors produces a current flow between them given by

$$J=J_0 \sin\delta, \qquad (62)$$

where J_0, a constant, describes the maximum possible current which may flow between the two specimens. Equation (62) is Josephson's dc equation relating current and phase difference (Josephson, 1962). It is implicit in the GL equations and applies to any system with a macroscopic, quantum-mechanical, complex order parameter. It can be interpreted in terms of the standard quantum-mechanical uncertainty relation between particle number and wave-function phase.

4.3.5 Magnetic-Field Dependence of the Order Parameter. Detailed solutions of Eqs. (47)–(51) for thin films clearly exhibit a field dependence of the *amplitude* of the order parameter $|\psi|$ on the applied magnetic field strength B_a. As the field is increased the value of $|\psi|$ decreases, and there is a value H_f for which it goes (either continuously or discontinuously) to zero and the film becomes normal. It is found that H_f—the film critical field—depends on H_{cb}, the film thickness d, and the London penetration depth, and that the $|\psi|$ transition to zero at H_f is discontinuous if

$$d/\lambda_L > \sqrt{5}.$$

These results, and many others obtained from the solution of the GL equations for a variety of geometries and situations, have been confirmed by superconducting tunneling experiments.

4.3.6 Quantum Interference Phenomena. Given the following facts:

1. the order parameter ψ is complex;
2. ψ must be single valued;
3. the magnetic field **B** couples to it in a gauge-invariant form and therefore is directly related to the phase θ of ψ; and
4. the GL equations are nonlinear;

it is possible to obtain a large number of interference and diffraction effects which can be fruitfully used in designing interesting electronic devices (see SUPERCONDUCTING DEVICES). It can be said that, in understanding the origins of the diamagnetism in superconductors, scientists and engineers have promoted quantum mechanics to the macroscopic, everyday-use level.

5. CONCLUSION

All diamagnetic effects are quantum-mechanical. The persistence of the currents induced by Lenz's law is caused exclusively by the phase coherence of the quantum-mechanical wave functions, either at the microscopic level—as evidenced in the diamagnetism of atoms, molecules, and metallic electrons—or in their macroscopic manifestations—as in superconductors. One of the principal effects is the gauge-invariant way in which quantum-mechanical phases couple to magnetic fields. This coupling produces, in addition to steady currents that result in a screening of the magnetic field, i.e., normal diamagnetism, a wealth of refraction, interference, and diffraction effects which are directly observable experimentally. The de Haas–van Alphen effect, discussed in Sec. 3, and the Josephson effect and flux quantization in superconductors, presented briefly in Sec. 4, are all manifestations of the relationship between electrical current and quantum-mechanical phase, coherent coupling of magnetic fields to the phases of quantum-mechanical wave functions or order parameters, and the relationship between electrical currents and magnetic fields as given by Maxwell's equations.

GLOSSARY

Antiferromagnet: A substance in which local magnetic moments are arranged in alternating orientations, resulting in no net macroscopic moment.

Bardeen, Cooper, and Schrieffer Theory: See BCS theory.

BCS Theory: A microscopic theory of superconductivity formulated by Bardeen, Cooper, and Schrieffer in 1957.

Bohr–Sommerfeld Quantization Rules: Rules by which the energy of bound electronic states were quantized in the "old" quantum mechanics.

Brillouin Zone: The polyhedron in reciprocal space over which the wave vector **k** is allowed to vary.

Coherence Length: The distance over which the coherence of the quantum-mechanical phase of the superconducting order parameter is preserved.

Cyclotron Resonance: Resonance absorption of electromagnetic radiation obtained when the electromagnetic frequency is equal to the cyclotron frequency of electrons moving in a magnetic field; it is also sometimes called diamagnetic resonance.

de Haas–van Alphen Effect: Oscillatory contribution to the magnetic susceptibility, periodic in $1/B$, caused by the interrelationship of the quantized energy levels and the Fermi surface.

Diamagnet: A substance or body that repels or expels magnetic flux.

Diamagnetic Resonance: See cyclotron resonance.

Fermi–Dirac Statistics: The thermal-equilibrium statistics that applies to particles which satisfy the Pauli exclusion principle, electrons among them.

Fermi Level: The value of the energy that separates occupied from empty states in the limit $T \to 0$ of Fermi–Dirac statistics.

Fermi Surface: The surface of constant energy in **k** space corresponding to the Fermi level.

Ferrimagnet: A substance in which the arrangement of local magnetic moments exhibits both a uniform value—a macroscopic moment—and a spatially oscillatory component.

Ferromagnet: A substance in which the microscopic magnetic moments are arranged all parallel to one another, with a consequent macroscopic moment.

Fluxoid: The quantum of magnetic flux in multiply connected superconductors (see flux quantization).

Flux Quantization: The property of multiply connected superconductors that restricts the values of the magnetic-field flux inside the sample to integral multiples of a fixed quantity, called the fluxoid.

Ginzburg–Landau Theory (Ginzburg–Landau Equations): A phenomenological quantum theory that describes the electromagnetic properties of superconductors and which allows for spatial variations and coherence effects.

Josephson Effect: Coherent tunneling of electrons between two superconductors separated by a thin nonsuperconducting—normally insulating—layer.

London Theory (London Equation): A macroscopic classical theory which describes the electromagnetic properties of superconductors.

Meissner Effect: The complete exclusion of magnetic flux (perfect diamagnetism) from the body of a superconductor at low enough values of the applied magnetic field.

Paramagnet: A substance that exhibits neither a permanent macroscopic magnetic moment nor long-range order of local moments, and which attracts magnetic flux.

Penetration Depth: The distance over which an outside magnetic field decays into the bulk of a superconductor.

Reciprocal Lattice: The discrete, infinite set of points in reciprocal space which defines the vectors used in the Fourier series for a given periodic structure in three dimensions.

Reciprocal Space (k Space): A mathematical space, with the dimension of inverse length, over which wave vectors and periodicities of Fourier series are defined.

List of Works Cited

Bardeen, J., Cooper, L. N., Schrieffer, J. R. (1957), *Phys. Rev.* **108**, 1175–1204.

Falicov, L. M. (1973), in: *Electrons in Crystalline Solids*, Vienna: International Atomic Energy Agency, pp. 207–279.

Faraday, M. (1846), *Philos. Trans. R. Soc. London* **2**, 2149.

Ginzburg, V. L., Landau, L. D. (1950), *Zh. Eksp. Teor. Fiz.* **20**, 1064–1068.

Gor'kov, L. P. (1959), *Zh. Eksp. Teor. Fiz.* **36**, 1918–1923.

Kittel, C. (1986), *Introduction to Solid State Physics*, 6th ed. New York: Wiley, Chaps. 9, 12, and 14.

Josephson, B. D. (1962), *Phys. Lett.* **1**, 251–253.

Kamerlingh Onnes, H. (1911), *Proc. K. Ned. Akad. Wet.* **14**, 113–114, 818–820.

Landau, L. D. (1930), *Z. Phys.* **64**, 629–637.

Langevin, P. (1905), *J. Phys. (Paris)* **4**, 678–693.

London, F., London, H. (1935), *Proc. R. Soc. London Ser. A* **149**, 71–88.

London, F. (1950), *Superfluids*, New York: Wiley.

Meissner, W., Ochsenfeld, R. (1933), *Naturwissenschaften* **21**, 787–788.

Onsager, L. (1952), *Philos. Mag.* **43**, 1006–1008.

Pippard, A. B. (1953), *Proc. R. Soc. London Ser. A* **216**, 547–568.

Somerville, M. F. G. (1848), *On the Connection of the Physical Sciences*, 7th ed. London: Buckley, Chap. xxiii, p. 369.

Further Reading

General

Kittel, C. (1986), *Introduction to Solid State Physics*, 6th ed. New York: Wiley, Chaps. 9, 12, and 14.

Van Vleck, J. H. (1932), *The Theory of Electric and Magnetic Susceptibilities*, London: Oxford University (reprinted in 1952).

White, R. M. (1970), *Quantum Theory of Magnetism*, New York: McGraw-Hill, Chap. 3.

Diamagnetism of Metals

Cracknell, A. P., Wong, K. C. (1973), *The Fermi Surface*, Oxford: Clarendon.

Falicov, L. M. (1973), in: *Electrons in Crystalline Solids*, Vienna: International Atomic Energy Agency, pp. 207–279.

Pippard, A. B. (1960), *Rep. Prog. Phys.* **23**, 176–266.

Pippard, A. B. (1969), in: Ziman, J. M. (Ed.), *The Physics of Metals. 1 Electrons*, Cambridge: Cambridge University Press, pp. 113–174.

Shoenberg, D. (1957), in: Gorter, C. J. (Ed.), *Progress in Low Temperature Physics*, Vol. 2, New York: Interscience, pp. 226–245.

Wilson, A. H. (1954), *The Theory of Metals*, 2nd ed. London: Cambridge University Press, pp. 150ff.

Ziman, J. M. (1970), *Electrons in Metals: A Short Guide to the Fermi Surface*, London: Taylor and Francis.

Diamagnetism of Superconductors

Chandrasekhar, B. S. (1969), in: Parks, R. D. (Ed.), *Superconductivity*, Vol. 1. New York: Dekker, pp. 1–49.

Tinkham, M. (1975), *Introduction to Superconductivity*, New York: McGraw-Hill.

Werthamer, N. R. (1969), in: Parks, R. D. (Ed.), *Superconductivity*, Vol. 1. New York: Dekker, pp. 321–370.

ERRATUM

CARBON MATERIALS (Vol. 3, pp. 21–40). OTTO J. VOHLER, FERDINAND VON STURM, and ERHARD WEGE.

Page 37. The entries in Table 4 were incorrectly arranged. The correct form follows.

Table 4. Mechanical properties of commercial carbon fibers.

	Density (g/cm^3)	Tensile strength (kN/mm^2)	Young's modulus (kN/mm^2)	Breaking strain (%)
Pan-based				
HT (high strength)	1.75	3	200	1.5
HM (high modulus)	1.95	2	400	0.6
Pitch-based				
Anisotropic MPP (mesophase based)	1.95	2	400	0.8
Isotropic	1.60	1	70	3.5

1-56081-063-7/92/$5.00+.50

LIST OF RECOMMENDED UNITS AND SYMBOLS

RECOMMENDED UNITS AND CONVERSION FACTORS

The SI system provides six basic units: meter m, kilogram kg, second s, ampere A, kelvin K, candela cd, and mole mol.

Some important derived units are also allowed and bear special names, e.g.:

1 N (newton) = 1 kg m s^{-2}
1 J (joule) = 1 N m = 1 kg m^2 s^{-2}
1 W (watt) = 1 J s^{-1} = 1 kg m^2 s^{-3}
1 Pa (pascal) = 1 N m^{-2} = 1 kg m^{-1} s^{-2}

For mass, the gram g, or the metric ton t which equals 1000 kg, may be used instead of kilogram kg.

The so-called "long ton" (UK) and "short ton" (US) have been abandoned and will not be used in the *Encyclopedia of Applied Physics*.

For pressure, the bar (name and symbol alike) may be used, and for temperature, the degree celsius °C.

From all of these, decimal multiples or fractions can be derived:

power of ten	prefix	symbol	power of ten	prefix	symbol
10	deca	da	10^{-1}	deci	d
10^2	hecto	h	10^{-2}	centi	c
10^3	kilo	k	10^{-3}	milli	m
10^6	mega	M	10^{-6}	micro	μ
10^9	giga	G	10^{-9}	nano	n
10^{12}	tera	T	10^{-12}	pico	p
10^{15}	peta	P	10^{-15}	femto	f
10^{18}	exa	E	10^{-18}	atto	a

For mass, multiples or fractions are derived from g, not from kg (since the latter already contains a prefix), e.g., mg.

Units with the prefixes are considered as one entity and can, therefore, be raised to any power, e.g., cm^3.

The liter is now considered as synonymous with dm^3 (which does not hold in the older literature!). Please use the capital letter L as unit symbol (following a recent IUPAC recommendation). The use of the Ångström unit is discouraged; it should be replaced by fractional meters (1 Å = 100 pm = 0.1 nm).

SELECTED QUANTITIES, UNITS, AND SYMBOLS

Name of quantity[a]	Symbol[b]	SI unit[c]	Name of unit	Other units
Space and time				
length*	l	m	meter[d]	
breadth, width	b	m		
height	h	m		
radius	r	m		
thickness	d	m		
area	A,S	m^2		a (are) h (hectare)
volume	V	m^3		L, l(liter)[e]
plane angle	$\alpha,\beta,\gamma,$ ϑ,φ	1, rad	radian	° (degree) ′ (minute) ″ (second)
solid angle	ω,Ω	1, sr	steradian	
wavelength	λ	m		
wave number	σ,ν	m^{-1}		
time*	t	s	second	min (minute) h (hour) d (day)
frequency	ν, f	s^{-1}		Hz (hertz)[f]
relaxation time	τ	s		
velocity	u,v	$m\ s^{-1}$		km/h
acceleration	a	$m\ s^{-2}$		
Mechanics				
mass*	m	kg	kilogram	g (gram) t (tonne)[g]
(mass) density	ρ	$kg\ m^{-3}$		g/cm^3
momentum	$\mathbf{p}$	$kg\ m\ s^{-1}$		
angular momentum	$\mathbf{L}$	$kg\ m^2\ s^{-1}$		
force	$\mathbf{F}$	N	newton[f]	
moment of force	$\mathbf{M}$	N m		
weight	G, W	N		
pressure	p	Pa	pascal	bar (bar)[h]
energy	E, W	J	joule	W h (watt hour)[i] eV (electron volt)[j]
work	W,A	J		
power	P	W	watt	J/s, V A[k]
Molecular Physics and Thermodynamics				
thermodynamic temperature*	T	K	kelvin	°C (degrees Celsius)
Celsius temperature	ϑ,t			°C
number of entities	N			
Avogadro constant[l]	N_A,L	mol^{-1}	(particles) per mole	
Boltzmann constant[m]	k,k_B	$y\ K^{-1}$		
Planck constant[n]	h	y s		
(molar) gas constant[o]	R	$y\ mol^{-1}\ K^{-1}$		
(quantity of) heat	Q	J		
entropy[p]	S	$J\ K^{-1}$		
internal energy[p]	U	J		
Helmholtz function,[p] (Helmholtz) free energy, Helmholtz energy	F,A	J		
enthalpy[p]	H	J		
Gibbs function,[p] (Gibbs) free energy, Gibbs energy	G	J		
heat capacity[p]	C_p,C_V	$J\ K^{-1}$		

Name of quantity[a]	Symbol[b]	SI unit[c]	Name of unit	Other units
Chemical Physics				
amount of substance*	n	mol	mole	
relative atomic mass	A_r	1		
relative molecular mass	M_r	1		
atomic mass constant	m_u	kg		u (atomic mass unit)[q]
mass of a portion (of substance B)	$m_B\, m(B)$	kg		g (gram)
molar mass (of substance B)	$M_B, M(B)$	kg mol^{-1}		
concentration (of substance B)	$c_B, c(B)$	mol m^{-3}		mol/L
mole fraction[r] (of substance B)	$\kappa_B, \kappa(B)$	1		
mass fraction (of substance B)	$\omega_B, \omega(B)$	1		%,%,ppm,ppb
volume fraction (of substance B)	$\varphi_B, \varphi(B)$ $\phi_B, \phi(B)$	1		%,%,ppm,ppb
mass concentration	ρ	kg m^{-3}		g/L
molality		mol kg^{-1}		mmol/kg
volume concentration[s]	σ	1		
molar volume	V_m	m^3 mol^{-1}		L/mol
molar heat capacity	C_m	J mol^{-1} K^{-1}		
molar conductivity	Λ_m	S m^2 mol^{-1}	(S:siemens)	
Faraday constant[t]	F	C mol^{-1}		
Electricity and Magnetism				
quantity of electricity[u]	Q	C	coulomb	
charge density	ρ	C m^{-3}		
electric potential	ϕ, V	V	volt	
electric potential difference, voltage	$U, \Delta\phi, \Delta V$	V		
electric dipole moment	$\mathbf{p}, \mathbf{p}_e$	C m		
electric current*	I	A	ampere	
electric current density	j	A m^{-2}		
electric field strength	$\mathbf{E}$	V m^{-1}		
electric displacement	$\mathbf{D}$	C m^{-2}		
capacitance	C	F	farad	
permittivity	ε	F m^{-1}		
relative permittivity	ε	1		
dielectric polarization	$\mathbf{P}$	C m^{-2}		
electric susceptibility	χ_e	1		
polarization (of a particle)	α	m^2 C V^{-1}		
magnetic flux	Φ	Wb	weber	
magnetic flux density	$\mathbf{B}$	T	tesla	
magnetic field strength	$\mathbf{H}$	A m^{-1}		
permeability	μ	H m^{-1}, N A^{-2}	(H:henry)	
relative permeability	μ_r	1		
magnetization	M	A m^{-1}		
magnetic susceptibility	χ	1		
molar magnetic susceptibility	χ_m	m^3 mol^{-1}		
(electrical) resistance	R	Ω	ohm	
(electrical) conductance	G	S	siemens	
(electrical) resistivity	ρ	Ω m		
(electrical) conductivity	κ	S m^{-1}		
self-inductance	L	H	henry	
Radiation				
radiant energy	Q, W, Q_e	J		
luminous intensity*	I	cd	candela	
radiant intensity	I_e	W sr^{-1}, W		
emissivity, emittance	ε	1		

Name of quantity[a]	Symbol[b]	SI unit[c]	Name of unit	Other units
absorptance	α	1		
reflectance	ρ, R	1		
transmittance	τ	1		
absorption coefficient:				
-linear (decadic)	a	m^{-1}		
-molar (decadic)	ε	$m^2\ mol^{-1}$		
refractive index	n	1		
molar refraction	R_m	$m^3\ mol^{-1}$		
angle of optical rotation	α	1, rad		
Transport Properties				
flux of quantity X	J_X, J	(varies)		
mass flow rate	q_m, $\dot{m}$	$kg\ s^{-1}$		
volume flow rate	q_V, $\dot{V}$	$m^3\ s^{-1}$		
heat flow rate	Φ	W		
thermal conductivity	κ, k, λ	$W\ m^{-1}\ K^{-1}$		
coefficient of heat transfer	h	$W\ m^{-2}\ K^{-1}$		
thermal diffusivity	a	$m^2\ s^{-1}$		
diffusion coefficient	D	$m^2\ s^{-1}$		
thermal diffusion coefficient	D_T	$m^2\ s^{-1}$		
viscosity	η, μ	Pa s		
kinematic viscosity	ν	$m^2\ s^{-1}$		

[a]SI base quantities are marked by asterisks (*)

[b]Recommended by IUPAC

[c]SI base units as well as derived and supplementary units are listed; all are to be used with prefixes as needed

[d]do not use "metre"

[e]do not use "litre"; $1\ L = 10^{-3}\ m^3$

[f]$1\ Hz = 1\ s^{-1}$; $1\ N = 1\ kg\ m\ s^{-2}$

[g]Formerly metric ton; $1\ t = 10^3\ kg$

[h]$1\ bar = 10^5\ Pa$

[i]$1\ W\ h = 3.6 \times 10^3\ J$

[j]$1\ eV = 1.602\ 189 \times 10^{-19}\ J$

[k]$1\ W = 1\ J/s = 1\ VA$

[l]$N_A = 6.022\ 136\ 7 \times 10^{23}\ mol^{-1}$

[m]$k = 1.380\ 658 \times 10^{-23}\ J\ K^{-1}$

[n]$h = 6.626\ 075\ 5 \times 10^{-34}\ J\ s$

[o]$R = 8.314\ 510\ J\ mol^{-1}\ K^{-1}$

[p]Molar quantities can be distinguished from the quantity of a system by adding the subscript m; e.g., molar internal energy U_m, in $J\ mol^{-1}$

[q]$1\ u = 1.660\ 565\ 5 \times 10^{-27}\ kg$

[r]A more accurate, but rather uncommon, name is "amount-of-substance fraction"

[s]σ refers to the total volume of a mixture, whereas the volume fraction φ relates the volume of a substance to the volume of several components before mixing

[t]$F = 9.648\ 530\ 9 \times 10^4\ C\ mol^{-1}$

[u]Also called electric charge